XLVII International Symposium on Multiparticle Dynamics (ISMD 2017)

AA004379

EPJ Web of Conferences Volume 172 (2018)

Tlaxcala City, Mexico
11 - 15 September 2017

Editors:

A. Ayala Mercado
E. Cuautle Flores

ISBN: 978-1-5108-5662-2

Printed from e-media with permission by:

Curran Associates, Inc.
57 Morehouse Lane
Red Hook, NY 12571

proceedings.com

Some format issues inherent in the e-media version may also appear in this print version.

This work is licensed under a Creative Commons Attribution license:
http://creativecommons.org/licenses/by/2.0/
You are free to:
Share – copy and redistribute the material in any medium or format.
Adapt – remix, transform, and build upon the material for any purpose, even commercial.
The licensor cannot revoke these freedoms as long as you follow the license terms.
Under the following terms:
You must give appropriate credit, provide a link to the license, and indicate if changes were made.
You may do so in any reasonable manner, but not in any way that suggests the licensor endorses you or your use. The copyright is retained by the corresponding authors.

Printed by Curran Associates, Inc. (2018)

For additional information, please contact EDP Sciences – Web of Conferences at the address below.

EDP Sciences – Web of Conferences
17, Avenue du Hoggar
Parc d'Activité de Courtabœuf
BP 112
F-91944 Les Ulis Cedex A
France

Phone: +33 (0) 1 69 18 75 75
Fax: +33 (0) 1 69 28 84 91

contact-edps@webofconferences.org

Additional copies of this publication are available from:

Curran Associates, Inc.
57 Morehouse Lane
Red Hook, NY 12571 USA
Phone: 845-758-0400
Fax: 845-758-2633
Email: curran@proceedings.com
Web: www.proceedings.com

TABLE OF CONTENTS

edp sciences

Web of Conferences

Statement of Peer review

In submitting conference proceedings to *Web of Conferences*, the editors of the proceedings certify to the Publisher that

1. They adhere to its **Policy on Publishing Integrity** in order to safeguard good scientific practice in publishing.
2. All articles have been subjected to peer review administered by the proceedings editors.
3. Reviews have been conducted by expert referees, who have been requested to provide unbiased and constructive comments aimed, whenever possible, at improving the work.
4. Proceedings editors have taken all reasonable steps to ensure the quality of the materials they publish and their decision to accept or reject a paper for publication has been based only on the merits of the work and the relevance to the journal.

Title, date and place of the conference

XLVII International Symposium on Multiparticle Dynamics, Tlaxcala City, Mexico, September 10-15, 2017

Proceedings editor(s): Alejandro Ayala

Eleazar Cuautle Flores

Date and editor's signature 12/14/2017

PHENIX Measurements of Heavy Flavor in Small Systems

Alexandre Lebedev[1,⋆]

[1] *Iowa State University, Ames, IA 50011-3160, USA*

Abstract. The study of heavy flavor production in proton-nucleus and nucleus-nucleus collisions is a sensitive probe of the hot and dense matter created in such collisions. Installation of silicon vertex detectors in the PHENIX experiment, and increased performance of the BNL RHIC collider allowed collection of large amount of data on heavy flavor production in small colliding systems.
In this talk we will present recent PHENIX results on open heavy flavor and quarkonia production in p+p, p+A, d+A, and He3+A colliding systems in a broad rapidity range, and discuss how these measurements help us to better understand all stages of nuclear collisions at high energy.

1 J/ψ polarization in $p + p$ collisions at $\sqrt{s} = 510$ GeV

In most genaral case the angular distribution of the positive lepton from the J/ψ decay can be written as

$$\frac{dN}{d\cos\theta d\phi} \propto 1 + \lambda_\theta \cos^2\theta + \lambda_{\theta\phi}\sin 2\theta\cos\phi + \lambda_\phi \sin^2\theta\cos 2\phi \quad (1)$$

where λ_θ is the polar, λ_ϕ is the azimuthal, and $\lambda_{\theta\phi}$ is the "mixed" angular decay coefficients.

The angles ϕ and θ are measured relative to a reference frame defined such that the $\hat{x}$ and $\hat{z}$-axes lie in the production plane, formed by the momenta of the colliding protons and the particle produced. The direction of the $\hat{z}$-axis within the production plane is arbitrary. The simplest frame to study the particle wave function is the one in which the density matrix has only diagonal elements, or the single and double spin-flip terms are zero. This simplest frame is also called the natural frame and is identified when the azimuthal coefficients in (1) are zero. The three most common frames used in particle angular distribution studies are (Fig. 1):

The Helicity frame (HX) [1], traditionally used in collider experiments, takes the $\hat{z}$-axis as the spin-1 particle momentum direction

The Collins-Soper frame (CS) [2], widely used in Drell-Yan measurements, choses the $\hat{z}$-axis as the difference between the momenta of the colliding hadrons boosted into the spin-1 particle rest frame

⋆e-mail: lebedev@iastate.edu

© The Authors, published by EDP Sciences. This is an open access article distributed under the terms of the Creative Commons Attribution License 4.0 (http://creativecommons.org/licenses/by/4.0/).

The Gottfried-Jackson frame (GJ) [3], typically used in fixed target experiments, takes the $\hat{z}$-axis as the beam momentum boosted into the spin-1 particle rest frame. At forward angles in a collider environment, the definition of the GJ frame depends heavily on which beam is used in the definition. If the beam circulating in the same direction as the J/ψ momentum is chosen (GJ forward), the resulting $\hat{z}$-axis is nearly collinear with the $\hat{z}$-axis of the HX and CS frames and points in the same direction. In GJ backward frame (beam circulating in the direction opposite to J/ψ momentum is chosen) the $\hat{z}$-axis points in the opposite direction.

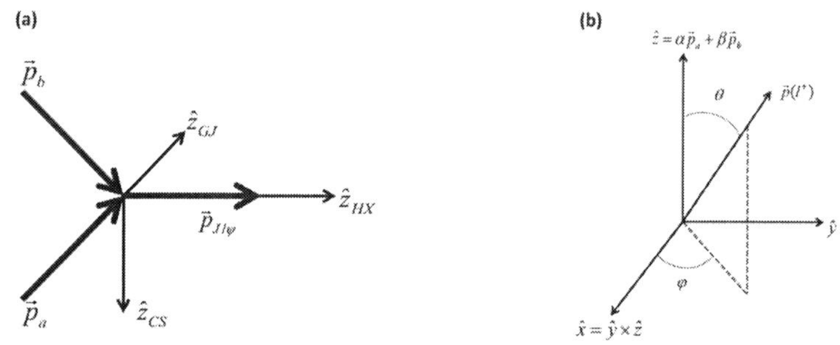

Figure 1. Reference frames and coordinate system used in this analysis. The $\hat{x}$ and $\hat{z}$-axes are chosen to lie in the production plane determined by the colliding hadrons and the particle produced (a J/ψ in this figure) with momenta $\vec{p}_a$, $\vec{p}_b$ and $\vec{p}_{J/\psi}$. The left diagram shows the production plane and the direction convention for the $\hat{z}$ in the Collins-Soper (CS), Helicity (HX) and Gottfried-Jackson (GJ) reference frames. The angles θ and ϕ in the diagram on the right represent the direction of the positive decay lepton in the corresponding reference frame.

J/ψ mesons were reconstructed by calculating the invariant mass of all unlike-sign muon pairs after analysis cuts. Combinatorial random background was estimated by like-sign dimuons calculated as $2\sqrt{N^{++}N^{--}}$, where N^{++} and N^{--} are number of positive and negative same-sign pairs respectively, and subtracted. Mass distributions for each bin in p_T and rapidity are then fit using a double Gaussian as signal and exponential background in order to remove dimuons from Drell-Yan and correlated open-heavy flavor decays. The number of J/ψ's is obtained directly by integrating the dimuon invariant mass distribution in a mass interval from 2.5 to 3.7 GeV/c^2 after background subtraction. Background subtraction was performed for each individual $\cos\theta$-ϕ bin.

The J/ψ angular distributions are divided into 12 bins in $\cos\theta$ and 10 bins in ϕ. Combinatorial and correlated background is subtracted bin-by-bin, and angular distributions are then corrected for acceptance, which is calculated assuming no polarization, that is $\lambda_\theta = \lambda_{\theta\phi} = \lambda_\phi = 0$. This is done for each of the three transverse momentum bins in each polarization frame. The angular decay coefficients obtained in this way were used in the next iteration to calculate acceptance, and the procedure was repeated until it converged.

The resulting angular decay coefficient λ_θ and frame-independent coefficient $\tilde{\lambda}$ are shown in Fig. 2 and Fig. 3 respectively, for four reference frames as a function of transverse momentum. In all frames the polar coefficient λ_θ is strongly negative at low p_T and becomes close to zero at high p_T, while the azimuthal coefficient λ_ϕ is close to zero at low p_T, and becomes slightly negative at high p_T. The frame-independent coefficient $\tilde{\lambda}$ is strongly negative at all p_T in all frames.

The measured polar coefficient λ_θ is compared to theoretical prediction for prompt J/ψ in Helicity frame calculated in the NRQCD factorization by H. .S. Chung et al, [4] and H. Shao et al., [5] in

Fig. 2. At high p_T both predicitions are in good agreement with the data, while at low p_T a strong deviation can be seen. No theoretical calculation is available for the frame-independent coefficient $\tilde{\lambda}$ or for other reference frames. The reported experimental results represent a challenge for the theory and provide a basis for better understanding of quarkonium production in high energy $p+p$ collisions.

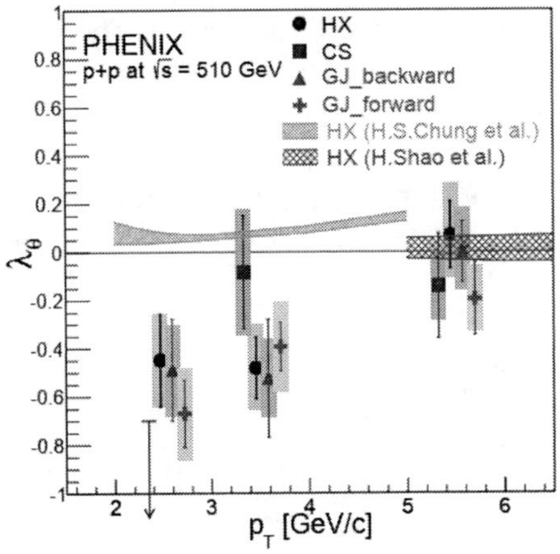

Figure 2. Polar angular decay coefficient λ_θ as a function of transverse mometum for four reference frames and three p_T bins. Black circles: HX frame; blue squares: CS frame; red triangles: GJ Backward; green crosses: GJ Forward frames. Shaded error boxes show systematic uncertainty. Points are shifted in p_T for clarity. Down-pointing arrow indicates 90% confidence limit. The data are compared with NRQCD theorectcal predictions in Helicity frame by H. S. Chung et al. [4] and H. Shao et al. [5].

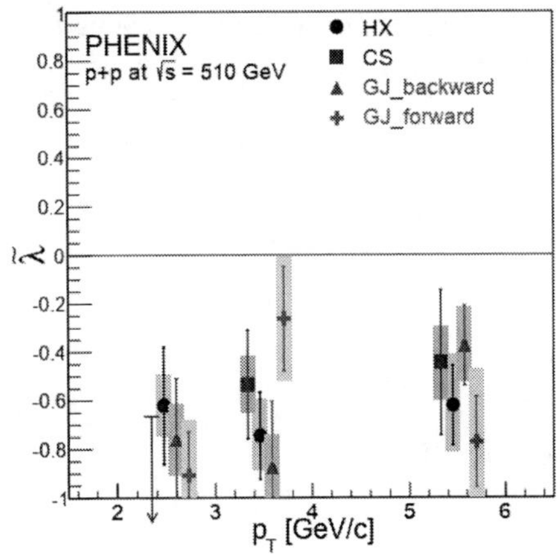

Figure 3. Frame-independent angular decay coefficient $\tilde{\lambda}$ as a function of transverse mometum for the four reference frames and three p_T bins. Black circles: HX frame; blue squares: CS frame; red triangles: GJ Backward; green crosses: GJ Forward frames. Shaded error boxes show systematic uncertainty. Points are shifted in p_T for clarity. Down pointing arrow indicates 90% confidence limit.

2 J/ψ and $\psi(2S)$ in $p/d/He^3 + Al/Au$ collisions at $\sqrt{s} = 200$ GeV

Heavy charm quarks are produced in initial hard scatterings, and their production can be calulated in perturbative QCD approach, so the initial state effects are well under control. $c\bar{c}$ pairs form J/ψ and $\psi(2S)$ at the later stages of the collisions. Binding energies of these two charmonium states are very different (640MeV and 50MeV). Thus, the difference between these two states can help us to understand the final state effiects in nucleus-nucleus collisions.

The PHENIX experiment has measured [7] previously $\psi(2S)$ nuclear modification factor in d+Au collisions at $\sqrt{s} = 200$ GeV/c. That measurement was done at mid-rapidity. The magnitude and the trend with centraility of $\psi(2S)$ meson suppression turned out to be very different than that of J/ψ [6]. In most central d+Au collisions $\psi(2S)$ suppression was measured to be almost 3 times larger than that of J/ψ. Since the initial state factors in a nucleus, such as shadowing and energy loss, should be the same or very similar for $c\bar{c}$ precursor pairs for both $\psi(2S)$ and J/ψ, something different must be happening in the later stages of the collision. This measurement became possible after installation of the Forward Silicon Vertex Detector (FVTX), which allowed better mass resolution by precisely measuring opening angle of an opposite sign muon pair.

The new PHENIX measurement of $\psi(2S)$ production in p+p, p+Au and p+Al collisions at $\sqrt{s}$ = 200 GeV/c at forward/bacward rapidity may shed light on the relatively large suppression of $\psi(2S)$. Fig. 4 shows the fraction of $\psi(2S)/J/\psi$ as a function of transverse momentum (p_T) in proton-proton collisions. Integrated over all p_T, this fraction is around 2%, which is in agreement with the world data at different $\sqrt{s}$. Fig. 5 compares the new PHENIX result at forward rapidity (solid black circles) with the PHENIX result at mid-rapidity (open gray circles) and world data as a function of transverse momentum. The green band in Fig. 5 shows Color Evaporation Model (CEM) prediction. The $p + p$ measurement serves as a baseline for calculating nuclear modification factor in the proton-nucleus collisions.

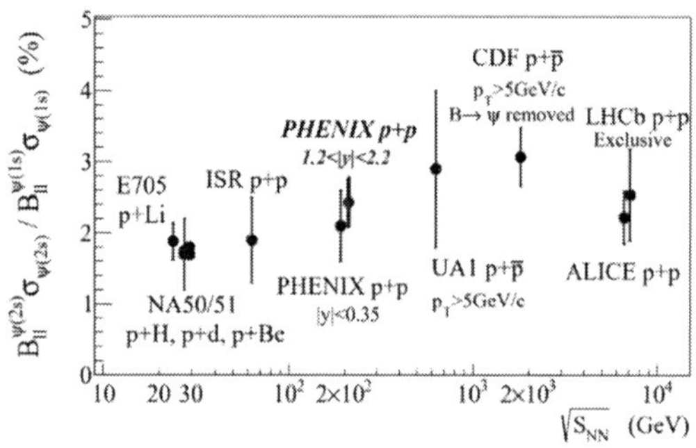

Figure 4. Comparison of world data on the ratio of $\psi(2S)/J/\psi$ mesons in di-lepton decays as a function of $\sqrt{s}$ in proton-proton collisions.

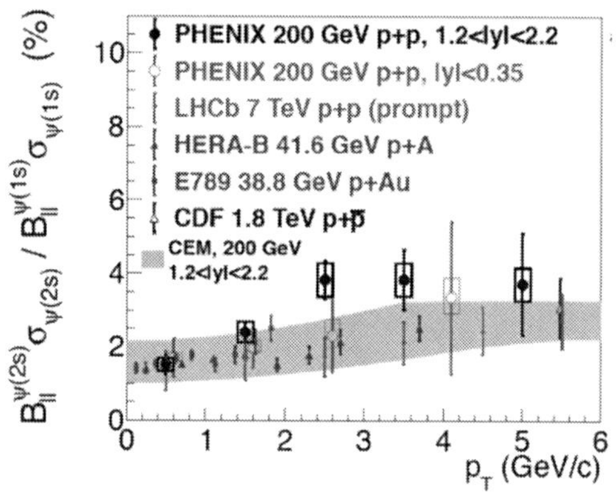

Figure 5. Comparison of world data on the ratio of $\psi(2S)/J/\psi$ mesons as a function of p_T along with calculation from a color evaporation model (green band).

Fig. 6 shows relative suppression of $\psi(2S)/J/\psi$ as a function of rapidity for p+Au and p+Al collisions, as well as the old d+Au result at mid-rapidity. In the proton-going direction (positive rapidity) suppression of $\psi(2S)$ is, essentially, the same as that of J/ψ, while in the nucleus-going direction (negative rapidity) $\psi(2S)$ are much more suppressed both in p+Au and p+Al collisions. In other words, $\psi(2S)$'s are strongly suppressed in both directions, while J/ψ's are strongly suppressed in proton-going direction, and moderately in nucleus-going direction. The same figure also shows a calculation based on breakup by co-moving particles (dashed lines). The calculation qualitatively explains the observed rapidity dependence, but underestimated the magnitude.

Fig. 7 shows relative suppression of $\psi(2S)/J/\psi$ as a function of co-mover density, which is expressed as particle multiplicity $dN/d\eta$ divided by nuclear overlap $< S_T >$, calculated using Glauber model. The results obtained by the ALICE experiment [8], and LHCb experiments also show the increased suppression with increasing co-mover density observed by the PHENIX experiment.

3 Open charm and bottom via di-electrons in $p + p$ and $d + Au$ collisions

Beauty production in p+p collisions at $\sqrt{s}$ = 200 GeV was studied using opposite sign di-electrons using simultaneous fit of invariant mass and transverse momentum distributions [9]. The spectrum of di-electrons produced in p+p collisions is well understood in terms of hadronic cocktail (at low mass) and Drell-Yan, charm, and beauty decays (at high mass). After the yield from vector, pseudo-scalar mesons, and Drell-Yan is subtracted from the measured di-electron spectrum, we are left with

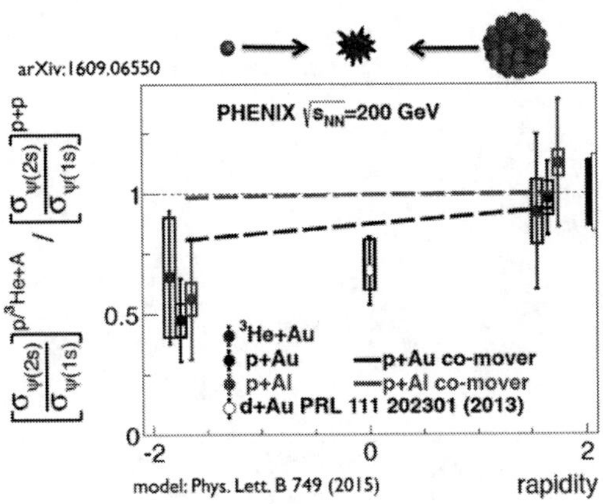

Figure 6. The double ratio of $\psi(2S)/J/\psi$ mesons measured in $p/d/^3He + A$ collisions to the same ratio in $p+p$ collisions as a function of rapidity, along with a calculation based on breakup by co-moving particles.

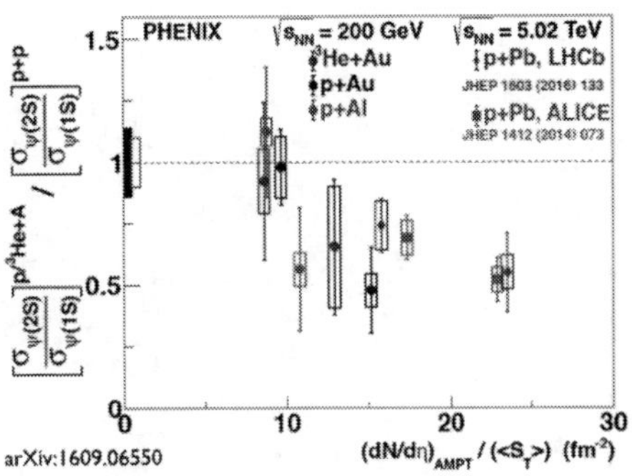

Figure 7. The double ratio of $\psi(2S)/J/\psi$ mesons measured in $p/^3He + A$ collisions to the same ratio in $p + p$ collisions as a function of co-moving particle density.

di-elctrons from semi-leptonic decays of charm and beauty. The main idea behind charm/beauty separation is that di-electrons from charm dominate low invariant mass and low p_T region of this two-dimasional space, while di-electrons from beauty dominate either high mass and low p_T, or low mass and high p_T regions. One-dimensional projections of two-dimensional fits to invariant mass axis are shown in Fig. 8.

This approach is model-dependent, since the decay electron spectra depend on original charm/beauty distributions. However, for heavier quarks (beauty) the effect of the original quark distribution gets smeared by the decay kinematics, and the results do not depend on the chosen model. This is not true in case of charm. We used three models, tuned PYTHIA [10], POWHEG [11], and MC@NLO [12] to check model deendence of the results.

The integrated $c\bar{c}$ and $b\bar{b}$ cross-sections obtained for $p + p$ and $d + Au$ collisions are shown in Fig. 9. The $b\bar{b}$ cross-section shows, as expected, no model dependence, while $c\bar{c}$ cross-section is strongly dependent on rapidity shape of the original quark distribution, which is different in PYTHIA, POWHEG, and MC@NLO. The same model dependence is seen for both $p+p$ and $d+Au$ collisions.

We also looked at the ratio (or nuclear modification factor) of cross-sections of $c\bar{c}$ and $b\bar{b}$ in $d+Au$ and $p + p$ collisions, and this is shown in Fig. 10. This ratio is similar for all the event generators and no deviation from unity is observed.

4 Conclusions

We have reviewed recent PHENIX measurements of heavy flavor in small systems.

First full measurement of J/ψ polarization at RHIC energies represents a challenge to theory and could provide a basis for better understanding of quarkonia production mechanisms.

A wealth of data on J/ψ and $\psi(2S)$ production show enhanced relative suppression of $\psi(2S)$ in nucleus-going direction. Qualitatively it can explained by co-mover dissociation.

Charm and bottom cross-section measurement in $p+p$ and $d+Au$ shows no significant modification in $d + Au$ within experimental uncertainties.

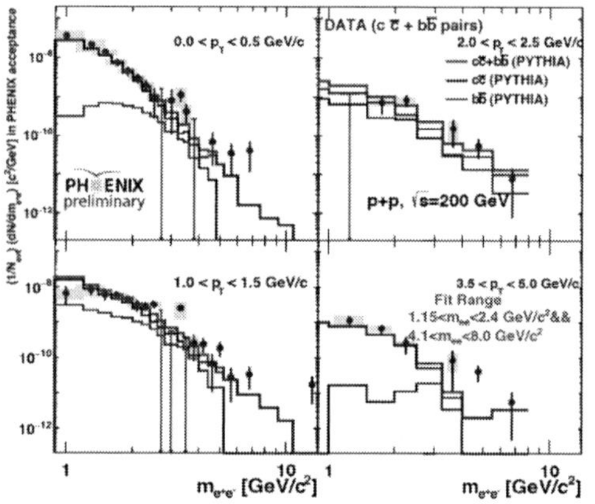

Figure 8. Double differential e^+e^- pair yields from heavy flavor decays fitted to simulated distributions from MC@NLO in four p_T bins. Black: $c\bar{c}$ yield, blue: $b\bar{b}$ yield, red: the sum.

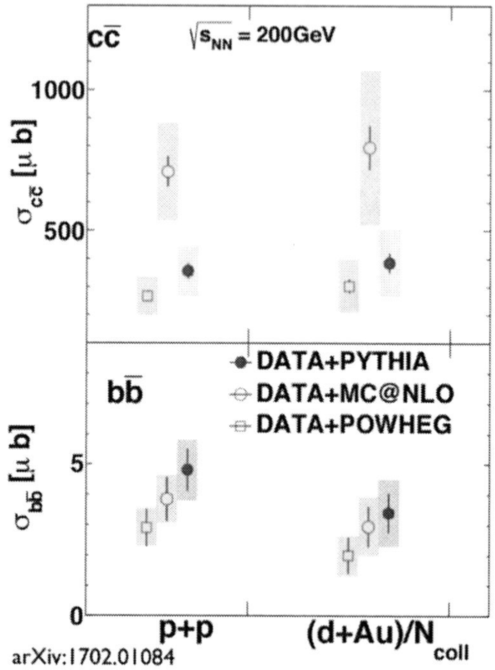

Figure 9. The extracted cross-sections of $c\bar{c}$ (top) and $b\bar{b}$ (bottom) in $p + p$ (left) and $d + Au$ (right) collisions. The $d + Au$ cross-section has been scaled down by number of binary collisions N_{coll} to represent the equivalent nucleon-nucleon cross-section. cross-section obtained using PYTHIA, POWHEG, and MC@NLO models.

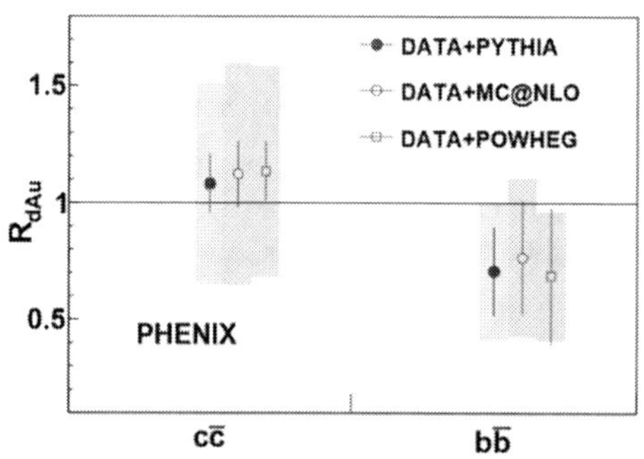

Figure 10. The nuclear modification factor R_{dAu} of $c\bar{c}$ and $b\bar{b}$ pairs constructed using cross-sections. The $d + Au$ cross-sections are scled by number of binary collisions $N_{coll} = 7.6 \pm 0.4$.

References

[1] M. Jacob and G. C. Wick, Ann. Phys. **7**, 404 (1959)
[2] J. C. Collins and D. E Soper, Phys. Rev. **D16**, 2219 (1977)
[3] K. Gottfried and J. D. Jackson, Nuovo Cimento, **33**, 309 (1964)
[4] H. S.Chung, S. Kim, J. Lee, and C. Yu, Phys. Rev. D **83**, 037501 (2011)
[5] H.-S. Shao, H. Han, Y.-Q.'Ma, C. Meng, Y.-J. Zhang, and K.-T. Chao, J. High Energy Phys. **05**, 103, (2015)
[6] A. Adare et al., (PHENIX Collaboration), Phys. Rev. Lett. **1**07, 142301 (2013)
[7] A. Adare et al., (PHENIX Collaboration), Phys. Rev. Lett. **1**11, 202301 (2013)
[8] B. Abelev et al., (ALICE Collaboration), J. High Energy Phys. **1**412, 73 (2014)
[9] A. Adare et al., (PHENIX Collaboration) Phys. Rev. C **9**1, 014907 (2015)
[10] T. Sjostrand, S. Mrenna, and P. Z. Skands, J. High Energy Phys. **5**, 026 (2006)
[11] S. Frixione, P. Nason, and G. Ridolfi, J. High Energy Phys. **0**9, 0126 (2007)
[12] S. Frixione and B. Webber, J. High Energy Phys. **6**, 029 (2002)

Sound waves in hadronic matter

Grzegorz Wilk[1,*] and Zbigniew Włodarczyk[2,**]

[1] *National Centre for Nuclear Research, Hoża 69, 00-681 Warsaw, Poland*
[2] *Institute of Physics, Jan Kochanowski University, Świętokrzyska 15, 25-406 Kielce, Poland*

Abstract. We argue that recent high energy CERN LHC experiments on transverse momenta distributions of produced particles provide us new, so far unnoticed and not fully appreciated, information on the underlying production processes. To this end we concentrate on the small (but persistent) log-periodic oscillations decorating the observed p_T spectra and visible in the measured ratios $R = \sigma_{data}(p_T)/\sigma_{fit}(p_T)$. Because such spectra are described by quasi-power-like formulas characterised by two parameters: the power index n and scale parameter T (usually identified with temperature T), the observed log-periodic behaviour of the ratios R can originate either from suitable modifications of n or T (or both, but such a possibility is not discussed). In the first case n becomes a complex number and this can be related to scale invariance in the system, in the second the scale parameter T exhibits itself log-periodic oscillations which can be interpreted as the presence of some kind of sound waves forming in the collision system during the collision process, the wave number of which has a so-called self similar solution of the second kind. Because the first case was already widely discussed we concentrate on the second one and on its possible experimental consequences.

1 Introduction

The high energy CERN LHC experiments measure, among other things, transverse momenta distributions of produced particles [1–5]. We would like to draw attention the fact that, apparently, some features of the results obtained have so far remained unnoticed or not fully appreciated. They are presented in Fig. 1. In panel (*a*) one can see the typical pp large p_T cross section which is best described by the Tsallis quasi power-like distribution where q is the so-called nonextensivity parameter and T_0 is a scale factor sometimes identified with the temperature (we use units with $c = k_B = h = 1$).

$$f(p_T) = \frac{n-1}{nT_0}\left(1 + \frac{p_T}{nT_0}\right)^{-n}, \quad n = \frac{1}{q-1}, \tag{1}$$

However, if the ratios $R = \sigma_{data}(p_T)/\sigma_{fit}(p_T)$ are plotted log-periodic oscillations are seen in the observed p_T spectra, see panels (*b*) – (*d*). They are rather small but they show up in the results from different experiments (panel (*b*)), at different energies (panels (*c*) – (*d*) and in different systems (panels (*e*) – (*f*)). Panel (*e*) presents a comparison of p_T distributions for $p+p$ and $Pb+Pb$ collisions whereas

* e-mail: grzegorz.wilk@ncbj.gov.pl
** e-mail: zbigniew.wlodarczyk@ujk.kielce.pl

© The Authors, published by EDP Sciences. This is an open access article distributed under the terms of the Creative Commons Attribution License 4.0 (http://creativecommons.org/licenses/by/4.0/).

panel (*f*) compares the respective factors R showing that the effect increases with the centrality of the nuclear collision [6, 7]. In panel (*f*) we also present a comparison with the proposed alternative to our description (using a two component, *soft+hard*, picture of the production process, each in the form of Eq. 1) [8, 9]). As can be seen it is not able to describe the observed effect in the whole region of pT. Because it turns out that, additionally, these oscillations cannot be erased by any reasonable change of fitting parameters, we shall assume that this is a real effect which should be investigated in detail. Note that log periodic oscillations can originate either from suitable modifications of the power index n (or nonextensivity parameter q in our case) or from some suitable modification of the scale parameter T (or from both cases, but such possibility will not be discussed).

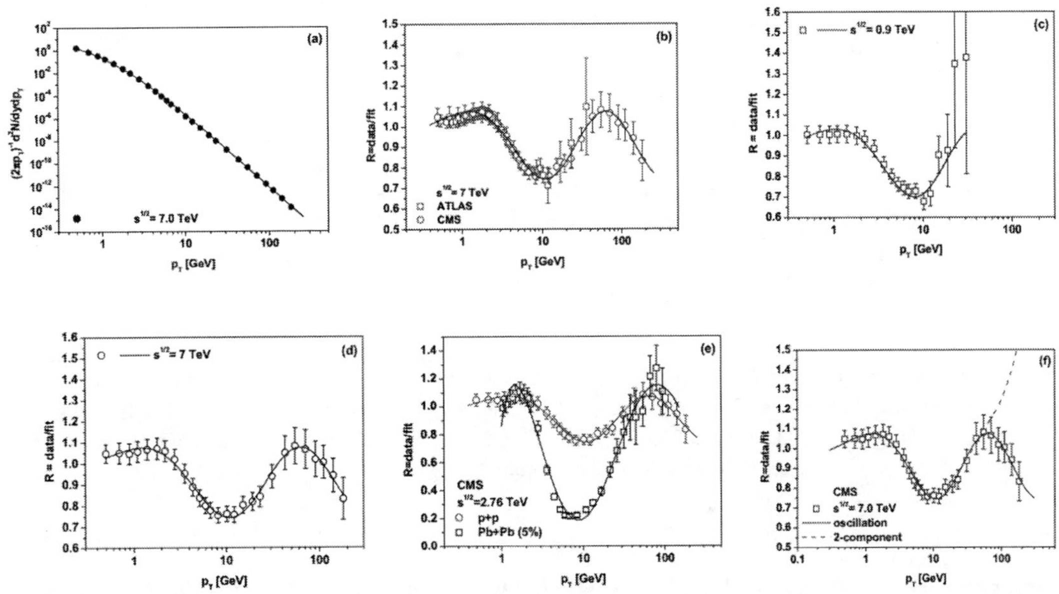

Figure 1. (Color online) (*a*) - Example of a typical $p + p$ large p_T distribution observed in LHC experiments. (*b*) - Log periodically oscillating factor R for different LHC experiments. (*c*) and (*d*) - The same factor R at two different energies. (*e*) - Example of a $Pb + Pb$ large p_T cross section. (*f*) - The R factor for $Pb + Pb$ collisions; the prediction of the two-component approach proposed in [8, 9] is shown as the red line. For more details see text and [12, 13].

The first possibility may be related to scale invariance in the system [10] and was already widely discussed by us in [11–13]. Such log-periodic oscillations are an immanent feature of any power-like distribution [10]. In [6] we have shown that they also appear in quasi power-like distributions of the Tsallis type. In general they are attributed to a discrete scale invariance (connected with a possible fractal structure of the process under consideration) and are described by introducing a complex power index n, or q in our case [10]. Such a case was discussed by us in [6, 7, 11–13] and has a number of observed consequences, like a complex heat capacity of the system, complex probability and complex multiplicative noise, all of them discussed briefly in [11].

The other possibility is that the scale parameter T has some specific log-periodic oscillations. To the best of our knowledge such a possibility has not so far been addressed. We discussed it in [12, 13] where it was shown that the observed transverse momenta distributions presented in Fig. 1 can be

 https://doi.org/10.1051/epjconf/201817201002

described using the following log-oscillating (as a function of transverse momentum p_T) temperature T and keeping the power index n constant and real:

$$T(p_T) = a + b \sin\left[c \ln(p_T + d) + e\right]. \tag{2}$$

It was shown there that the origin of such logarithmic oscillations (and, respectively, the meaning

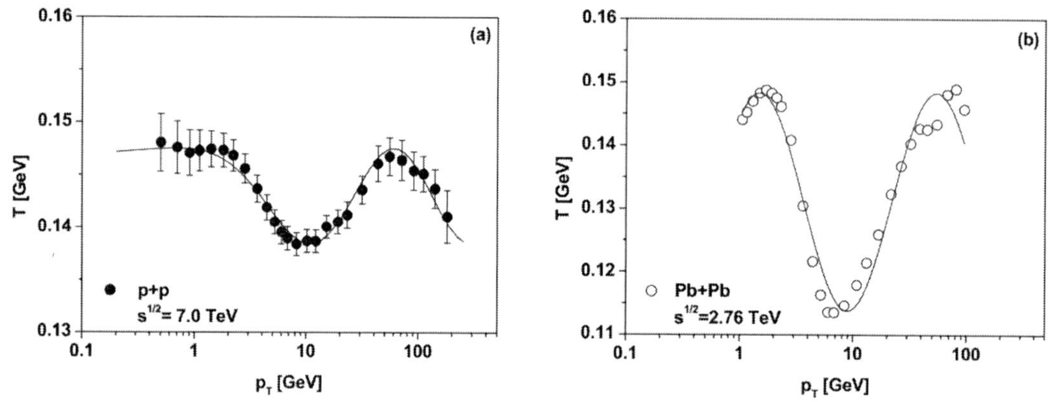

Figure 2. (Color online) Example of the observed log-periodicity of the scale parameter T for $p + p$ (panel (a)) and $Pb + Pb$ (panel b)) collisions (deduced from ALICE data [4, 5], see [12, 13] for details). The fits follow Eq. (2) with parameters $a = 0.143$, $b = 0.0045$, $c = 2.0$, $d = 2.0$, $e = -0.4$ for $p + p$ collisions and $a = 0.131$, $b = 0.019$, $c = 1.7$, $d = 0.05$, $e = 0.98$ for $Pb + Pb$ collisions.

of the parameters a, b, c, d and e in Eq. (2)) comes from the possible energy dependence of the noise defining the corresponding stochastic processes in the well known stochastic equation for the evolution of temperature T written in the Langevin formulation with p_T-dependent noise, $\xi(t, p_T)$, and allowing for time-dependent $p_T = p_T(t)$ [14]:

$$\frac{dT}{dt} + \frac{1}{\tau} T + \xi(t, p_T) = \Phi. \tag{3}$$

Here τ is the relaxation time, $\xi(t, p_T)$ is the p_T-dependent noise. Assuming additionally that we have time-dependent observed transverse momentum, $p_T = p_T(t)$, and that it increases following the scenario of the preferential growth of networks [15] we get

$$\frac{dp_T}{dt} = \frac{1}{\tau_0}\left(\frac{p_T}{n} \pm T\right) \tag{4}$$

(where n coincides with the power index in Eq. (1); τ_0 is some characteristic time step). The stationary dependence of $T(p_T)$ represented by Eq. (2) is obtained in two cases (cf. Section 3.2 in [13] or Section 5.2 in [12]): either the noise term increases logarithmically with transverse momentum while the relaxation time τ remains constant:

$$\xi(t, p_T) = \xi_0(t) + \frac{\omega^2}{n} \ln(p_T), \tag{5}$$

or, equivalently, one keeps the p_T-independent white noise, $\xi(t, p_T) = \xi_0(t)$, constant but allows for the relaxation time becoming p_T-dependent, for example assuming that

$$\tau = \tau(p_T) = \frac{n\tau_0}{n + \omega^2 \ln(p_T)} \tag{6}$$

(in both cases ω is some new parameter, cf. [12, 13] for details). It turns out that to fit data one needs only a rather small admixture of the stochastic processes with noise depending on the transverse momentum (defined by the ratio $b/a \sim 3\%$). The main contribution comes from the usual energy-independent Gaussian white noise. It should be stressed that whereas each of the proposed approaches is based on a different dynamical picture, nevertheless they are numerically equivalent.

2 Oscillating T as a signal of sound waves in hadronic matter

Whereas the occurrence of a complex power index n is a rather ubiquitous phenomenon in situations where one encounters scaling phenomena, a varying scale parameter (temperature) T is not so widely known or used. However, the possible oscillatory character of T is not so exotic because it can be found in hydrodynamical models widely used to describe strongly interacting systems and also in astrophysics in the description of dense stars. Such models have recently become increasingly sophisticated and popular in response to the continuous support received from experiment. For example, a specially strong case is the observation of the so called elliptic flow of secondaries produced in multiparticle production experiments. This phenomenon is not easy to explain in other approaches, which are more successful in the description of the distributions of the measured transverse momenta but use a fixed scale parameter T. Also, a number of observations strongly indicate that hadronic matter produced in heavy ion collisions at RHIC and the LHC behaves as a kind of perfect fluid. The observation of sound waves would therefore provide yet further support for a hydrodynamical description of the multiparticle production process. Such waves could arise because in the initial phase of the collision process a number of highly energetic partons are created which subsequently lose their energy. This can proceed in two ways: either by exciting modes of the medium in *collisional energy loss*, or by radiating gluons in *radiative energy loss*. The further dissipation of the released energy resulting in thermalization depends on the character of the medium: in a *weakly coupled medium* it proceeds through a cascade of collisions among quarks and gluons in the quark-gluon plasma, whereas in a *strongly coupled medium* the released energy is dissipated directly into thermal excitations and sound waves[1]. Actually, in relativistic heavy ion collisions we may also have hard parton-parton collisions in which the outgoing partons have to traverse the surrounding fluid before escaping and forming jets which subsequently hadronize. Such partons may therefore form Mach shock waves during their passages, and this, in turn, will affect the transverse momentum distribution of the observed final particles.

In this work we present yet another possible consequence of the hydrodynamical picture of the production process, namely the possible formation and propagation of sound waves in hadronic matter. So far we have only shown that to fit data the temperature T has to show log-periodic oscillations in transverse momentum, i.e., $T = T(p_T)$. To gain further insight into the collision process one should therefore form the Fourier transformation of such dependencies (as, for example, was done recently in [17]). This will allow us to reconstruct the space-time structure of the collision process[2].

[1] Such phenomena were, in fact, already studied in [16].

[2] Some explanatory remarks are in order here. In [17] investigations were concentrated only on the possible fluctuations of T whereas we also have at our disposal their additional log-periodic oscillations. The intrinsic fluctuations of T investigated in [17] are in our case accounted for by using the Tsallis distribution which emerges from the usual Boltzmann distribution

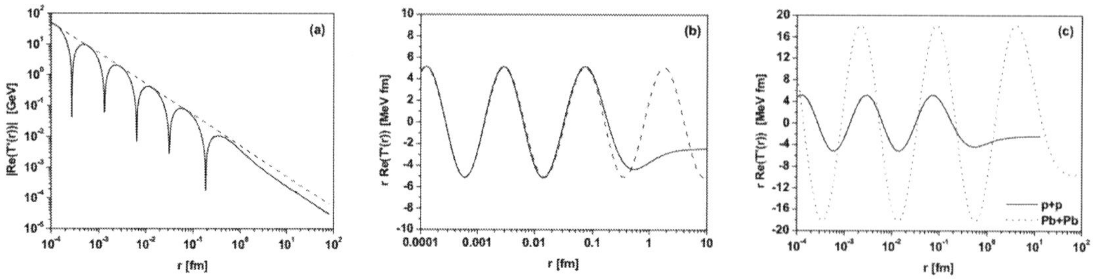

Figure 3. (Color online) Panels (*a*) and (*b*): The results of the Fourier transform of $T(p_T)$ from Eq. (2) describing the results inferred from the CMS data [1, 2]. Panel (*c*): Oscillating behaviour of $rT'(r)$ for $p + p$ collisions at 7 TeV presented in previous panels compared with similar results for the most central $Pb + Pb$ collisions at 2.76 TeV (inferred from the ALICE data [4, 5]). See text for details.

To this end we investigate the use of the Fourier transform of the log-normal oscillations of $T(p_T)$ in the form presented in Eq. (2):

$$T(r) = \sqrt{\frac{2}{\pi}} \int_0^{\infty} T(p_T)\, e^{ip_T r} dp_T. \tag{7}$$

Because the oscillations are in transverse momentum p_T, r is defined in the plane perpendicular to the collision axis and located at the collision point and denotes the distance from the collision axis. The results are shown in Fig. 3. In panel (*a*) we have $T'(r)$ (continuous line) as a function of r confronted with $T'(r) = 0.0051/r$ dependence (dashed line). In panel (*b*) the continuous line represents $rT'(r)$ versus r and is confronted with the dashed line denoting the function $rT'(r) = 5.1\sin[(2\pi)/3.2\ln(1.24r)]$ fitting it for small values of r. Panel (*c*) presents a comparison of the oscillations of $T(r)$ for $p+p$ and the most central $Pb+Pb$ collisions deduced from the ALICE data [4, 5] at 2.76 TeV [7] (the parameters of $T(p_T)$ used in both cases are the same as used in panel (*f*) of Fig. 1. For the most central $Pb + Pb$ collisions the amplitude is ~ 3.6 bigger and the period of the oscillations is ~ 1.15 longer than in pp collisions. With decreasing centrality the amplitude in central $Pb + Pb$ collisions decreases smoothly reaching practically the same value as for $p + p$ collisions [7]. In the case considered here, the region of regular oscillations can be fitted by

$$(p+p): \qquad rT'(r) = 5.1\sin\left[\frac{2\pi}{3.2}\ln(r) + 0.42\right]; \tag{8}$$

$$(Pb+Pb): \qquad rT'(r) = 18.53\sin\left[\frac{2\pi}{3.68}\ln(r) - 0.51\right]. \tag{9}$$

To see how such a log-periodically oscillating structure of $T(r)$ occurs let us study the flow of a compressible fluid in a cylindrical source assuming small oscillations and oscillatory motion with small amplitude in a compressible fluid (i.e., a *sound wave*, at each point in the fluid it causes alternate compression and rarefaction of the matter). Since the oscillations are small, their velocity υ is also

when one allows the scale parameter (temperature) to fluctuate according to a gamma distribution [18]. Because in [17] non-equilibrated systems were described using the Boltzmann transport equation and their space variations were investigated by means of the Fourier transformations, the log-periodic oscillations were out of reach there.

small and the term $(\mathbf{v}\cdot\mathbf{grad})\mathbf{v}$ in Euler's equation may be neglected. For the same reason, the relative changes in the fluid density and pressure are small and we may write [19]

$$P = P_0 + P', \qquad \rho = \rho_0 + \rho', \tag{10}$$

where P_0 and ρ_0 are the constant equilibrium density and pressure, and P' and ρ' are their variations in the sound wave ($\rho' << \rho,\ \ P' << P$). Neglecting small quantities of the second order (P', ρ' and v are regarded to be of the first order) the equation of continuity, $\partial\rho/\partial t + \mathbf{div}\cdot(\rho\mathbf{v}) = 0$ becomes

$$\frac{\partial\rho'}{\partial t} + \rho_0\mathbf{div}(\mathbf{v}) = 0 \tag{11}$$

and Euler's equation reduces in this approximation to

$$\frac{\partial\mathbf{v}}{\partial t} + \frac{1}{\rho_0}\mathbf{grad}P' = 0. \tag{12}$$

The linearized equations (11) and (12) are applicable to the propagation of sound waves if $v << c$ (where c is the velocity of sound), which means that $P' << P_0$. Note that a sound wave in an ideal fluid is adiabatic, therefore the small change P' in the pressure is related to the small change ρ' in the density by (c is the velocity of sound)

$$P' = \left(\frac{\partial P}{\partial\rho_0}\right)_S \rho' = c^2\rho' \qquad \text{where} \qquad c = \sqrt{\left(\frac{\partial P}{\partial\rho}\right)_S}. \tag{13}$$

Substituting ρ' from Eq. (13) into Eq. (11) one gets

$$\frac{\partial P'}{\partial t} + \rho_0\left(\frac{\partial P}{\partial\rho_0}\right)_S \mathbf{div}\cdot\mathbf{v} = 0 \tag{14}$$

which, together with Eq. (12), using the unknowns $\mathbf{v}$ and P', provides a complete description of the sound wave we are looking for. To express all the unknowns in terms of one of them, it is convenient to introduce the velocity potential f by putting $\mathbf{v} = \mathbf{grad}f$. From Eq. (12) we have the relation between P' and f:

$$P' = -\rho\frac{\partial f}{\partial t}. \tag{15}$$

which, used together with Eq. (14), results in the following cylindrical wave equation which the potential f must satisfy,

$$\frac{1}{r}\frac{\partial}{\partial r}\left(r\frac{\partial f}{\partial r}\right) - \frac{1}{c^2}\frac{\partial^2 f}{\partial t^2} = 0. \tag{16}$$

It can be shown that this represents a travelling longitudinal plane sound wave with velocity $\mathbf{v}$ in the direction of propagation. It is related to the pressure P' and the density ρ' in a simple manner, namely

$$v = \frac{P'}{\rho c} = c\frac{\rho'}{\rho}. \tag{17}$$

To relate the above results with the temperature note that we can write $T(r)$ as consisting of a constant term, T_0, and some oscillating addition, $T'(r)$:

$$T(r) = T_0 + T'(r) \qquad \text{where} \qquad T' = \left(\frac{\partial T}{\partial P}\right)_S P'. \tag{18}$$

Using the well known thermodynamic formula $(\partial T/\partial P)_S = (T/c_P)(\partial V/\partial T)_P$ and Eq. (17) one obtains that

$$T' = \frac{c\kappa T}{c_P}v \qquad \text{where} \qquad \kappa = \frac{1}{V}\left(\frac{\partial V}{\partial T}\right)_P \tag{19}$$

where κ is the coefficient of thermal expansion and c_P denotes the specific heat at constant presure [19]. In the case of a monochromatic wave, when $f(r,t) = f(r)\exp(-i\omega t)$, we have that

$$\frac{\partial^2 f(r)}{\partial r^2} + \frac{1}{r}\frac{\partial f(r)}{\partial r} + K^2 f(r) = 0, \qquad K = K(r) = \frac{\omega}{c(r)} \tag{20}$$

where K is the wave number which in inhomogeneous media depends on r. For the wave number given by

$$K(r) = \frac{\alpha}{r} \tag{21}$$

the solution of Eq. (20) is given by a log-periodic oscillation in the form

$$f(r) \propto \sin[\alpha \ln(r)]. \tag{22}$$

Because $\mathbf{v} = \mathbf{grad} f$ we have $f(r) \propto vr$. Therefore, using Eq. (19), we can write that

$$rT'(r) \propto \frac{c\kappa T_0}{c_P} f(r) = \frac{c\kappa T_0}{c_P}\sin[\alpha \ln(r)]. \tag{23}$$

This is the solution we have used in describing the $T'(r)$ deduced from data and presented in Fig. 3.

It should be mentioned at this point that the above problem can be considered from yet another point of view. Namely, it turns out that, according to [20–23], Eq. (20) with the wave number (21) has a so-called *self similar solution of the second kind*. Such a solution is known from other branches of physics and is connected with the so called *intermediate asymptotic* encountered whenever dependence on the initial conditions disappears (because sufficient time has already passed from the beginning of the process considered), but our system has not yet reached the state of equilibrium [20–23]. Introducing the variable

$$\xi = \ln r \tag{24}$$

we find that Eq. (20) for the wave number (21) represents a travelling wave equation,

$$\frac{\partial^2 F(\xi)}{\partial \xi^2} + \alpha^2 F(\xi) = 0, \tag{25}$$

the solution of which is

$$F(\xi) \propto \sin(\alpha\xi). \tag{26}$$

Interestingly enough, this self similarity of Eq. (20), which can be written as $F(\xi + \ln\lambda) = F(\xi)$, can be confronted with a kind of scale invariance of this equation, namely with the fact that $f(\lambda \cdot r) = f(r)$.

To summarize this part: The space picture of the collision (in the plane perpendicular to the collision axis and located at the collision point) presented in Fig. 3 (panels (*b*) and (*c*)) shows us the existence of some regular (on the logarithmic scale) structure for small distances. For $p + p$ collisions it starts to weaken quite early (at $r \sim 0.1$ fm) and essentially disappears when r reaches the dimension of the nucleon, i.e., for $r \sim 1$ fm. For $Pb + Pb$ collisions it seems to last longer, to around $r \sim 10$ fm (i.e., to a typical dimension of the nucleus).

3 Possible experimental consequences

What can be deduced from our results? Note that the longer period of the oscillations in the $Pb + Pb$ collisions means that the values of the parameter α in Eq. (22) in nuclear collisions are smaller than those for $p + p$ collisions. Furthermore, considering the form of K from Eq. (20) or Eq. (21), and remembering that $\omega/c(r) = \alpha/r$, one may deduce that the velocity of sound, $c(r) = (\omega/\alpha)r$, is greater in the nuclear environment (for $Pb + Pb$) than in the $p + p$ case. This, in turn, means that the refractive index $n(r) = c_0/c(r)$ at position r in nuclear collisions is smaller than in proton collisions. To summarize: In both cases we encounter an inhomogeneous medium with r-dependent velocity of sound, $c(r)$, and refractive index, $n(r)$.

These findings seem to agree with the fact that in nuclear collisions one really observes a higher speed of sound as demonstrated by the NA61/SHINE collaboration at SPS energies [24, 25] (note, however, that what is measured is a parameter in the equation of state of hadronic matter described by a hydrodynamical model, c_s^2). This is not so unexpected because, considering the connection of the isothermal compressibility of nuclear matter, $\kappa_T = -(1/V)(\partial V/\partial P)_T$, and fluctuations of the multiplicity of produced secondaries represented by the relative variance, ϖ, of multiplicity fluctuations, one finds that [26, 27]

$$T\kappa_T\rho_0 = \frac{\langle N^2\rangle - \langle N\rangle^2}{\langle N\rangle} = \varpi \tag{27}$$

(where $\rho_0 = \langle N\rangle/V$ denotes the equilibrium density for N particles with mass m located in volume V). This allows the velocity of sound to be expressed by fluctuations of multiplicity:

$$c = \sqrt{\frac{\gamma}{\kappa_T\rho_0 m}} = \sqrt{\frac{\gamma T}{\varpi m}} \quad \text{or} \quad \varpi = \frac{\gamma T}{m}\cdot\frac{1}{c^2} \quad \text{where} \quad \gamma = \frac{c_P}{c_V}. \tag{28}$$

Note that a higher velocity of sound c corresponds to lower fluctuations of multiplicity ϖ. From the experimental data shown in Fig. 4 [24, 25, 28] one has that

$$\frac{c_{Pb+Pb}}{c_{p+p}} \simeq 1.04 \quad \text{and} \quad \frac{\varpi_{p+p}}{\varpi_{Pb+Pb}} \simeq 1.29 \pm 0.04 \tag{29}$$

From them, using Eq. (28), one can deduce the expected ratio of the factors γT for $p+p$ and $Pb+Pb$ collisions at a beam energy of 158 GeV:

$$\frac{(\gamma T)_{p+p}}{(\gamma T)_{Pb+Pb}} \sim 1.2. \tag{30}$$

This could be the subject of further experimental investigations.

We close with the remark that these results may be connected with the pair correlation function, $g^{(2)}$, because the scaled variance can be written as [29]

$$\rho_0\kappa_T T = 1 + \rho_0\int d\vec{r}\left[g^{(2)}(r) - 1\right]. \tag{31}$$

As shown in [30], for central nuclear collisions the number of binary collisions exceeds that of wounded nucleons (each nucleon participates in a number of collisions with other nucleons). This results in the correlation function becoming negative which, in turn, leads to a diminishing of fluctuations of multiplicity (because the variance of the total multiplicity from a number of particular collisions is smaller that the sum of the variances of independent nucleon-nucleon collisions).

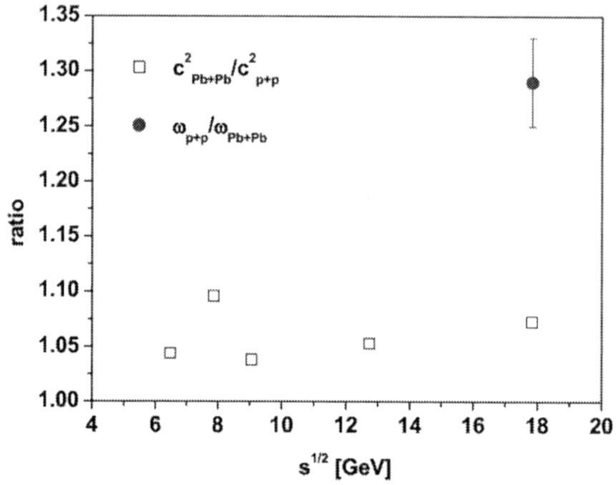

Figure 4. (Color online) The ratio c^2_{Pb+Pb}/c^2_{p+p} presented in [24, 25] compared with the ratio $\omega_{p+p}/\omega_{Pb+Pb}$ taken from [28].

4 Summary

Data from all LHC experiments strongly suggest that transverse momentum distributions are not only described by a quasi-power law (Tsallis distribution with power index n and scale parameter T usually identified with temperature) but also that these distributions are additionally decorated with some characteristic log-periodic oscillations clearly visible for large p_T events. Whereas this is a rather subtle effect, it is very persistent. It shows itself in all experiments, at all energies (provided that the range of transverse momenta observed is large enough) and also in reactions with nuclei where they grow with increasing centrality of the collision, and finally, they cannot be erased by any reasonable change of fitting parameters. All these points strongly suggest that these oscillations deserve to be taken seriously as a real effect which should be studied in detail.

Such studies reveal that either the system and/or the underlying physical mechanisms have characteristic *scale invariance behavior* (resulting in the power index n becoming complex) or that we observe a *sound wave in hadronic matter* (resulting in the temperature oscillations) which has a self similar solution (in log-periodic form). In the former case the discrete scale invariance and its associated complex exponents n can appear spontaneously, without a pre-existing hierarchical structure [10]. In the latter case the corresponding wave equation has *self-similar solutions of the second kind* connected with the so called *intermediate asymptotic* (observed in phenomena which do not depend on the initial conditions because sufficient time has already passed, although the system considered is still out of equilibrium) [20–23]. This suggests that both in $p + p$ and $Pb + Pb$ collisions one deals with an inhomogeneous medium with the density and the velocity of sound both depending on the position and this can have some interesting experimental consequences.

This research was supported in part (GW) by the National Science Center (NCN) under contract 2016/22/M/ST2/00176. We would like to thank warmly Dr Nicholas Keeley for reading the manuscript.

References

[1] V. Khachatryan, et al., (CMS Collaboration), J. High Energy Phys. **02** (2010) 041.
[2] V. Khachatryan, et al., (CMS Collaboration), J. High Energy Phys. **08** (2011) 086.
[3] G. Aad, et al., (ATLAS Collaboration), New J. Phys. **3** (2011) 053033.
[4] B. Abelev et al., (ALICE Collaboration), Eur. Phys. J. C **73** (2013) 2662.
[5] B. Abelev et al., (ALICE Collaboration), Phys. Lett. B **720** (2013) 52.
[6] G. Wilk and Z. Włodarczyk, Physica A **413** (2014) 53..
[7] M. Rybczyński, G. Wilk and Z. Włodarczyk, Eur. Phys. J. Web Conf. **90** (2015) 1002.
[8] G. G. Barnaföldi, K. Ürmössy, G. Biró, J. Phys. Conf. Ser. **612** (2015) 012048.
[9] K. Ürmössy, G. G. Barnaföldi, Sz. Harangozó, T. S. Biró, Z. Xu, J. Phys. Conf. Ser. **805** (2017) 012010.
[10] D. Sornette, Phys. Rep. **239** (1998) 239.
[11] G. Wilk and Z. Włodarczyk, Entropy **17** (2015) 384.
[12] G. Wilk and Z. Włodarczyk, Acta Phys. Pol. B **46** (2015) 1103.
[13] G. Wilk and Z. Włodarczyk, Chaos Solit. Frac. **81** (2015) 487.
[14] N.G. van Kampen, Stochastic Processes in Physics and Chemistry, Elsevier Science Publishers B.V., North-Holland, Amsterdam 1987, Chap. VIII.
[15] G. Wilk and Z. Włodarczyk, Acta Phys. Pol. B **35** (2004) 871.
[16] D.A. Fogaça, L.G. Ferreira Filho, F.S. Navarra, Nucl. Phys. A **819** (2009) 150.
[17] T. Bhattacharyya, P. Garg, R. Sahoo and P. Samantray, Eur. Phys. J. A **52** (2016) 283.
[18] G. Wilk and Z. Włodarczyk, Phys. Rev. Lett. **84** (2000) 2770.
[19] L. D. Landau and E. M. Lifshitz, *Fluid mechanics, Pergamon Press*, Oxford (1987).
[20] G.I. Barenblatt and Ya.B. Zeldovich, Usp. Mat. Nauk **26** (2) (158) 115 (1971) (in Russian); Russian Math. Surv. 26 (2) (1971) 45 (English Translation).
[21] G.I. Barenblatt and Ya.B. Zeldovich, Ann. Rev. Fluid Mech. **4** (1972) 285.
[22] G.I. Barenblatt, *Scaling, self-similarity, and intermediate asymptotics*, Cambridge University Press 1996.
[23] G.I. Barenblatt, *Scaling*, Cambridge University Press 2013.
[24] K. Grebieszkow, PoS DIS2014 (2014) 018; arXiv:1407.3690 [hep-ex].
[25] M. Gaździcki, M. Gorenstein and P. Seyboth, Acta Phys. Polon. B **42** (2011) 307.
[26] T.L. Hill, *Statistical Mechanics*, (McGraw-Hill, New York, 1956) (also reprinted (Dover, New York, 1987)).
[27] R. Balescu, *Equilibrium and Nonequilibrium Statistical Mechanics*, (New York: Wiley Interscience, 1975).
[28] A. Aduszkiewicz et al. (NA61/SHINE Collaboration), Eur. Phys. J. C **76** (2016) 635.
[29] F.H. Stillinger, J. Chem. Phys. **109** (1998) 3983.
[30] M. Rybczyński and Z. Włodarczyk, J. Phys. Conf. Ser. **5** (2005) 238.

Phenomenology of anomalous chiral transports in heavy-ion collisions

Xu-Guang Huang[1,2]

[1] *Physics Department and Center for Particle Physics and Field Theory, Fudan University, Shanghai 200433, China*
[2] *Key Laboratory of Nuclear Physics and Ion-beam Application (MOE), Fudan University, Shanghai 200433, China*

Abstract.
High-energy Heavy-ion collisions can generate extremely hot quark-gluon matter and also extremely strong magnetic fields and fluid vorticity. Once coupled to chiral anomaly, the magnetic fields and fluid vorticity can induce a variety of novel transport phenomena, including the chiral magnetic effect, chiral vortical effect, etc. Some of them require the environmental violation of parity and thus provide a means to test the possible parity violation in hot strongly interacting matter. We will discuss the underlying mechanism and implications of these anomalous chiral transports in heavy-ion collisions.

1 Introduction

High-energy heavy-ion collisions can generate deconfined quark-gluon plasma (QGP). In addition, strong magnetic fields can also be created in, e.g., non-central heavy-ion collisions [1, 2]. This is because that the two colliding nuclei generate two electric currents in opposite directions and thus produce a magnetic field perpendicular to the reaction plane which is the plane defined by the impact parameter and the beam direction. Strong electric fields can also be generated owing to event-by-event fluctuations or in asymmetric collisions like Cu + Au collision [3, 4]. Similarly, in a non-central heavy-ion collisions, a finite angular momentum can be retained in the collision region which can induce a finite vorticity. Recent numerical simulations and experimental measurement of the Λ hyperon spin polarization showed that such generated vorticity can be very large [5–7]. Thus the heavy-ion collisions provide a unique terrestrial environment to study QCD matter in strong electromagnetic field and under fast (local) rotation. We will in the next section discuss the generation of the magnetic field and flow vortcity in heavy-ion collisions.

Once coupled to the chiral anomaly, the electromagnetic field and flow vorticity can induce a number of novel transport phenomena — the anomalous chiral transports (ACTs). One famous example is the chiral magnetic effect (CME) which is the generation of an electric current along the direction of the external magnetic field in a matter in which the right-handed fermions and left-handed fermions have unequal chemical potentials [8, 9]. Other ACTs include the chiral separation effect (CSE) [10], chiral vortical effects (CVEs) [11–13], chiral electric separation effects (CESE) [14, 15], etc. New collective modes can arise through the interplay among these ACTs, e.g., the chiral magnetic wave

© The Authors, published by EDP Sciences. This is an open access article distributed under the terms of the Creative Commons Attribution License 4.0 (http://creativecommons.org/licenses/by/4.0/).

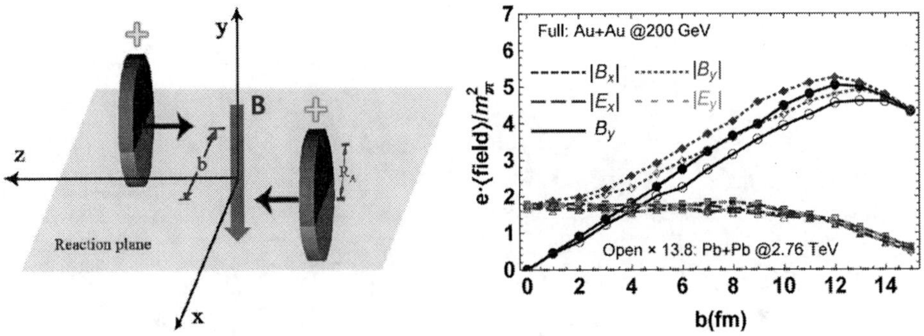

Figure 1. (Left) The geometrical illustration of a typical non-central heavy-ion collision. The magnetic field $\vec{B}$ is expected to be perpendicular to the reaction plane. (Right) The electromagnetic fields at $t = 0$ and $\vec{r} = \vec{0}$ versus the impact parameter b. Figures are modified from Ref. [2, 20].

(CMW) [16] and chial vortical wave (CVW) [17]. Recently, the experiments performed at RHIC and LHC generated data with features consistent with the expectation of the CME and CVE as well as the CMW. However, there exist elliptic-flow induced background effects which strongly mask the signals and make the interpretation of the experimental results ambiguous. We will give a pedagogical introduction to the underlying mechanism of the ACTs and then discuss their experimental searches.

2 Magnetic fields and vorticity in heavy-ion collisions

2.1 Magnetic fields

In Fig. 1 (left), we illustrate a typical non-central heavy-ion collision event. Two nuclei are accelerated to very high speed and then collide. During the collision, the two nuclei produce two counter-propagating currents which in turn induce a magnetic field whose direction is perpendicular to the reaction plane. (Strictly speaking, only after the event average the direction of the magnetic field is perpendicular to the reaction plane. On the event-by-event basis, the direction of the magnetic field fluctuates as a consequence of the fluctuation of the proton position in the nucleus [18, 19].) The question is: how strong the magnetic field can be? We can make a rough estimate by using the Biot-Savart formula, $eB \sim \gamma\alpha_{\rm EM} Z/R_A^2$, where γ is the Lorentz gamma factor associated with the moving nuclei, $\alpha_{\rm EM}$ is the fine structure constant, Z is the charge number of the nucleus, and R_A is the radius of the nucleus. For RHIC Au + Au collisions at $\sqrt{s} = 200$ GeV, the magnetic field is very large, $\sim 10^{19}$ Gauss, and for LHC Pb + Pb collisions at $\sqrt{s} = 2.76$ TeV the magnetic field can reach the order of 10^{20} Gauss.

More realistic numerical simulations indeed confirm that the high-energy heavy-ion collisions can generate extremely strong magnetic fields as well electric fields. In Fig. 1 (right), we present the numerical result for the magnetic fields as well as the electric fields at $t = 0$ (defined as the time when the two nuclei overlap maximally) and at $\vec{r} = \vec{0}$ (the center of the overlapping zone) [2, 20]. The simulations show some interesting features of the electromagnetic fields in heavy-ion collisions. (1) When $b < 2R_A$, $\langle e\vec{B}\rangle$ ($\langle\cdots\rangle$ means event average) is perpendicular to the reaction plane and its strength is roughly linearly proportional to b. (2) The strength $e|\vec{B}|$ is roughly proportional to the beam energy $\sqrt{s}$. (3) The event-by-event fluctuation of the proton position induces a finite electric field at the center $\vec{r} = \vec{0}$ which after the event average goes to zero. But in some asymmetric collisions, e.g., the Cu + Au collision, there is a persistent electric field whose strength is of the same order as the magnetic

 https://doi.org/10.1051/epjconf/201817201003

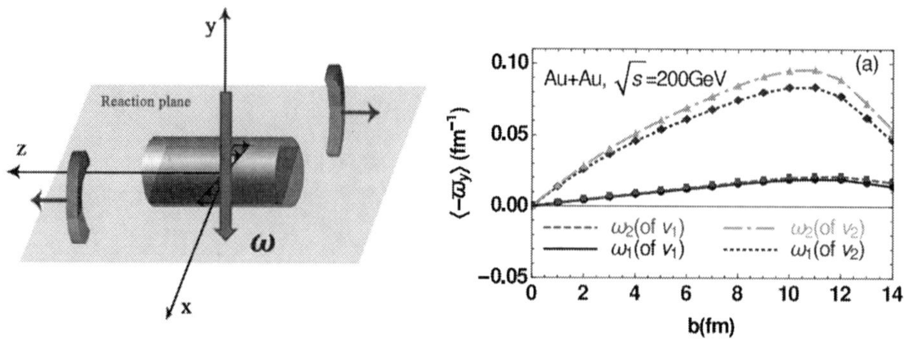

Figure 2. (Left) The illustration of the appearance of the flow vorticity in a non-central heavy-ion collision. (Right) The vorticity at $\tau = \tau_0$ (the initial proper time) and $\eta = 0$ averaged over the overlapping region and then averaged over events for RHIC Au + Au collisions at $\sqrt{s} = 200$ GeV. Only the y-component of the vorticity is sizable, other components are negligibly small. Different curves correspond to different definitions of vorticity and velocity, see Ref. [5] for details.

field [3, 4]. (4) On event-by-event basis, the direction of the magnetic field fluctuates which leads to a suppression in the correlation between the azimuthal angle Ψ_B and the second harmonic plane angle like the participant plane angle Ψ_2. (5) The initial magnetic field is dominated by the spectator protons, the participants and the hot quarks do not contribute much. This causes a very fast decay of the magnetic field as a consequence of the fast fly-away of the spectators from the collision zone. However, if the produced matter is a conductor, the Faraday effect may retain the magnetic field in the medium for a much longer while [21].

The main message here is that there exist strong transient electromagnetic fields in high-energy heavy-ion collisions which are actually the strongest fields that we have ever known in our current universe. The strength of such magnetic fields is much larger than the current masses squared of light quarks, m_u^2, m_d^2, and thus is capable of inducing significant quantum effects in the QGP.

2.2 Vorticity

The existence of strong magnetic fields in heavy-ion collisions suggests the existence of fast rotation and/or large flow vorticity in the QGP. In classical physics, the Larmor's theorem relates the motion of a charged particle in a magnetic field $\vec{B}$ to the motion in a rotating frame with angular velocity $\vec{\Omega} = -q\vec{B}/(2m)$ with m and q the mass and charge of the particle. In quantum physics, the Einstein-de-Haas effect is the generation of a rotation by magnetizing a ferromagnetic material which is a consequence of the spin polarization by the magnetic field. Besides, it is also natural to expect that heavy-ion collisions could create flow vorticity. Let us consider a non-central heavy-ion collision with impact parameter b as shown in Fig. 2 (left). Before the collision, the total angular momentum J_0 with respect to the collision center is roughly given by $Ab\sqrt{s}/2$ with A the total nucleon number in one nucleus. After the collision, a fraction of J_0 will be kept in the produced quark-gluon matter and manifests itself as a shear of the longitudinal flow because the equation of state of the quark-gluon matter is too soft to support a global rigid rotation. Thus the nonzero local vorticity arises as a consequence of this shear flow and the direction of the vorticity would be merely along the direction of angular momentum which is perpendicular to the reaction plane.

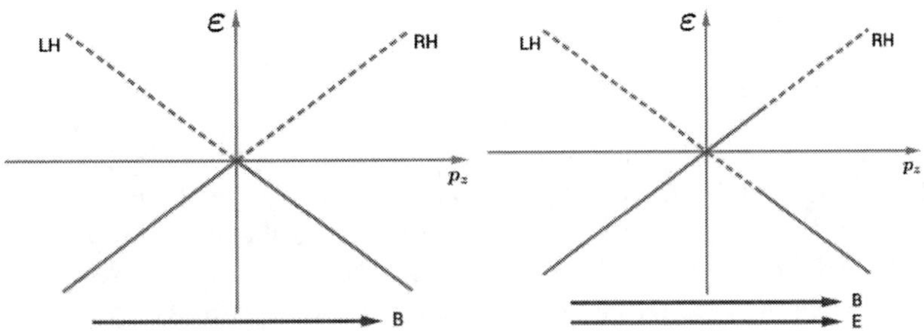

Figure 3. (Left) Lowest Landau level in a strong magnetic field. (Right) The electric field pumps LH fermions into RH fermions near the zero node which leads to the chiral anomaly.

We have performed more realistic numerical simulations recently [5, 22]; see also [6]. In Fig. 2 (right), we show the vorticity as a function of the impact parameter for RHIC Au + Au collisions at $\sqrt{s} = 200$ GeV. The vorticity is defined as $\vec{\omega}_1 = \vec{\nabla} \times \vec{v}$ and $\vec{\omega}_2 = \gamma^2 \vec{\nabla} \times \vec{v}$ where $\vec{v}_1$ represents the velocity of particle flow while $\vec{v}_2$ represents the velocity of energy flow. According to our simulation, the vorticity generated in the heavy-ion collisions is indeed very large and it shows up some interesting features. (1) After suitably averaged over the collision region and then over many events, the vorticity is perpendicular to the reaction plane. But both the magnitude and the azimuthal direction of the vorticity suffer from the event-by-event fluctuation, and as a consequence, the vorticity direction is blurred from being perfectly perpendicular to the reaction plane. (2) The vorticity is larger for larger impact parameter b when $b \lesssim 2R_A$ and when $b > 2R_A$ it drops. (3) Although the total angular momentum J_0 increases with $\sqrt{s}$, the event-averaged vorticity at mid rapidity decreases with increasing $\sqrt{s}$. This counterintuitive behavior is mainly because that the quark-gluon medium at mid rapidity behaves closer to the Bjorken boost invariant picture at higher $\sqrt{s}$ and thus supports lower vorticity.

The vorticity can be measured through measuring the spin polarization of hadrons, e.g., the Λ hyperon [23–26]. The underlying mechanism is the spin-orbit coupling. The vorticity induces orbital motion of the quarks which then polarizes the spin of quarks through spin-orbit coupling. The spin polarization of quarks can be inherited by hadrons of finite spin, e.g., the Λ hyperon, after hadronization, which can in turn be measured. Recently, the STAR collaboration at RHIC measured the Λ and $\bar{\Lambda}$ spin polarization with respect to the reaction plane from which they deduced an averaged vorticity over the beam energy region 7.7 GeV< $\sqrt{s}$ <200 GeV: $\omega \sim (9 \pm 1) \times 10^{21}/s$ [7]. This is the highest vorticity comparing to the vorticity of all other known fluids.

3 Anomalous chiral transports

Perhaps the most intriguing effects that the strong electromagnetic fields and the strong vorticity can induce are the anomalous chiral transports (ACTs). The famous examples of ACTs are the chiral magnetic effect, chiral vortical effects, chiral separation effects, etc. We in this section will give a pedagogical discussion about the underlying mechanisms of the ACTs; more discussions can be found in Refs. [20, 27, 28]

Let us start with the $U_A(1)$ chiral anomaly which is the basis of the ACTs. Let us consider a massless Dirac fermion of charge $e > 0$ in a strong constant magnetic field along the z direction. This is the usual Landau problem in quantum mechanics. The energy spectrum consists of different Landau levels,

$$\mathcal{E}_n^2 = p_z^2 + 2neB, \quad n = 0, 1, 2, \cdots, \tag{1}$$

where n labels the Landau levels. The lowest Landau level (LLL), $n = 0$, is special, see Fig. 3 (left). First, the LLL is gapless while all the higher Landau levels are gapped with a magnitude of $\sqrt{2neB}$. Thus when eB is large, we need to consider only the LLL. Second, the spin of LLL is fully polarized, namely, the LLL is non-degenerate in spin. All the states of the LLL are of spin up. In a many-body picture, this means that the LLL fermions are all have spin up. Third, the dynamics of LLL is 1+1 dimensional because the transverse motion of the LLL fermion is frozen. We define the chirality for each LLL fermion according to its momentum direction with respect to its spin direction. If p_z is parallel to the spin, we call it a right-handed (RH) fermion; if p_z is opposite to its spin, we call it a left-handed (LH) fermion.

In this situation, the numbers of RH and LH fermions are separately conserved, namely, $\partial_\mu J^\mu_{R/L} = 0$ with $J^\mu_{R/L} = (1/2)\bar{\psi}\gamma^\mu(1 \pm \gamma_5)\psi$, or equivalently, $\partial_\mu J^\mu_V = 0$ and $\partial_\mu J^\mu_A = 0$, where the vector and axial currents are defined as $J^\mu_V = J^\mu_R + J^\mu_L$ and $J^\mu_A = J^\mu_R - J^\mu_L$.

Now we apply an electric field in the same direction as the magnetic field, see Fig. 3 (right). Near the node $p_z = 0$, the electric field can drive the downward moving particles to upward moving and thus tune some LH fermions to RH fermions. In such case, although the sum of the numbers of RH and LH fermions, $N_V = N_R + N_L$, is still conserved, the difference $N_A = N_R - N_L$ is not conserved. We can calculate the changing rate of N_A in the following way. Suppose the RH and LH fermions have Fermi surfaces $p_F^{R/L}$. Then

$$N_{R/L} = V\frac{p_F^{R/L}}{2\pi}\frac{eB}{2\pi}, \tag{2}$$

where $eB/(2\pi)$ is the transverse density of state and V is the spatial volume. Thus,

$$\frac{dN_{R/L}}{dt} = V\frac{\dot{p}_F^{R/L}}{2\pi}\frac{eB}{2\pi} = \pm V\frac{eE}{2\pi}\frac{eB}{2\pi}, \tag{3}$$

which implies $dN_V/dt = 0$ and $dN_A/dt = Ve^2EB/(2\pi^2)$ which, written in a differential form, is essentially

$$\partial_\mu J^\mu_A = \frac{e^2}{2\pi^2}\vec{E}\cdot\vec{B}. \tag{4}$$

This result is know as chiral or axial anomaly. Although we obtain Eq. (4) by considering strong magnetic field so that only the LLL is occupied, the result is actually true for weak magnetic field too, because the higher Landau levels are degenerate in chirality so that they do not contribute to Eq. (4).

Now we remove the electric field and calculate the RH and LH currents along the magnetic field, see Fig. 4 (left). A current is equal to the density that the current carries times the velocity of the constitute particles. For massless particle, the velocity is one. Thus we have

$$J_{R/L} = \pm n_{R/L} = \pm\frac{p_F^{R/L}}{2\pi}\frac{eB}{2\pi}, \tag{5}$$

where the minus sign for LH current is because LH fermions move opposite to the direction of the magnetic field. This gives

$$J_V = \frac{p_F^R - p_F^L}{2\pi}\frac{eB}{2\pi} = \frac{\mu_A}{2\pi^2}eB, \tag{6}$$

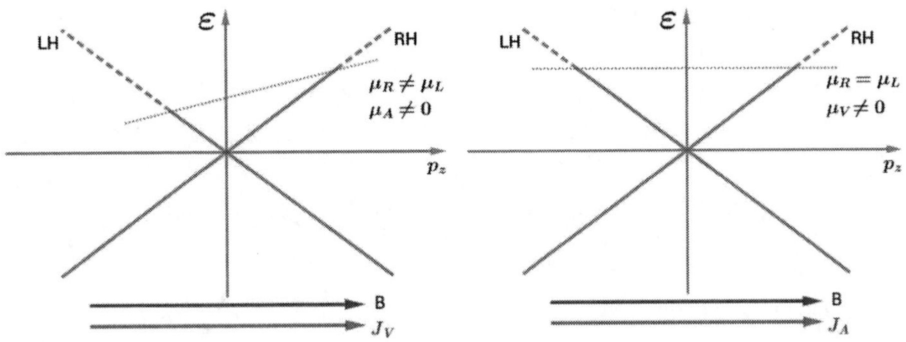

Figure 4. (Left) The arising of the chiral magnetic effect. (Right) The arising of the chiral separation effect.

where we have defined the axial chemical potential $\mu_A = (\mu_R - \mu_L)/2$. This current is the chiral magnetic current [8, 9]. Similarly,

$$J_A = \frac{p_F^R + p_F^L}{2\pi}\frac{eB}{2\pi} = \frac{\mu_V}{2\pi^2}eB, \tag{7}$$

where $\mu_V = (\mu_R + \mu_L)/2$. This is the chiral separation effect (CSE) [10]; it appear even when $p_F^R = P_F^L$, see Fig. 4 (right). We note that the CME has very intriguing properties. First, it is a macroscopic quantum effect. Second, its occurrence requires P and CP violation in the medium. Third, the generation of the CME current is time reversal even, namely, there is no associated entropy production. Thus the CME current is a kind of superconducting current. We also emphasize that the CME conductivity is fixed by the chiral anomaly and thus is free of renormalization.

In classical physics, a charged particle of mass m in a magnetic field is equivalent to being in a rotating frame with frequency $eB/(2m)$. This suggests that there may exist effects analogous to CME and CSE but induced by rotation or vorticity. Let us consider a massless particle in a rotating frame. The particle feels a Coriolis force, $\vec{F} = 2p\dot{\vec{x}} \times \vec{\omega} + O(\omega^2)$, where $\vec{\omega}$ is the rotating frequency. We have assumed that ω is so small that we neglect the centrifugal force which is $O(\omega^2)$. As the Coriolis force is very similar with the Lorentz force (replacing eB by $2p\omega$), we can consider the "Landau level problem" in rotating frame. Let us again consider only the LLL. Comparing to the magnetic case, the only difference is that the expression for the density is modified: $n_{R/L} = (2\pi)^{-2}\int_0^\infty dp_z 2p_z\omega\theta(p_F^{R/L} - p_z) = (p_F^{R/L})^2\omega/(2\pi)^2$. Thus the currents:

$$J_V = n_R - n_L = \frac{\mu_V\mu_A}{\pi^2}\omega, \tag{8}$$

$$J_A = n_R + n_L = \frac{\mu_V^2 + \mu_A^2}{2\pi^2}\omega. \tag{9}$$

This is the chiral vortical effects [11–13]. More rigorous derivation shows that there is an additional term, $T^2\omega/6$, in J_A which may be related to the gravitational anomaly [29].

The electric field can also lead to anomalous transport, that is the chiral electric separation effect (CESE) [14]. It is not directly related to the chiral anomaly and its appearance requires both P and C violation. The CESE represents an axial current along the direction of the electric field. Its expression

Table 1. Table of anomalous chiral transports

	$\vec{E}$	$\vec{B}$	$\vec{\omega}$
$\vec{J}_V$	σ	$\frac{e\mu_A}{2\pi^2}$	$\frac{\mu_V\mu_A}{\pi^2}$
$\vec{J}_A$	$\propto \frac{\mu_V\mu_A}{T^2}\sigma$	$\frac{e\mu_V}{2\pi^2}$	$\frac{T^2}{6}+\frac{\mu_V^2+\mu_A^2}{2\pi^2}$
Collective modes:	chiral electric wave	chiral magnetic wave	chiral vortical wave

for two flavor QCD up to leading-log accuracy is given by [15]

$$\vec{J}_A \approx 14.5163\mathrm{Tr}(Q_e Q_A)\frac{\mu_V\mu_A}{T^2}\frac{eT}{g^4\ln(1/g)}\vec{E}, \tag{10}$$

where Q_e and Q_A are charge matrix and axial matrix in flavor space, g is the strong coupling constant.

There can emerge collective modes from the coupled evolution of the axial and vector charges via CME and CSE, or vector CVE and axial CVE, or CESE and the usual Ohm's law. For example, the continuation equations for vector and axial charges can be written in terms of RH and LH charges

$$\partial_t J^0_{R/L} + \vec{\nabla}\cdot\vec{J}_{R/L} = 0. \tag{11}$$

Now we substitute the CME and CSE expressions, and consider small departures from equilibrium for $J^0_{R,L}$ and $\mu_{R,L}$ and keep linear terms in the small departures:

$$\partial_t\delta J^0_R + \frac{e^2}{4\pi^2\chi_R}\vec{B}\cdot\vec{\nabla}\delta J^0_R = 0, \tag{12}$$

$$\partial_t\delta J^0_L - \frac{e^2}{4\pi^2\chi_L}\vec{B}\cdot\vec{\nabla}\delta J^0_L = 0, \tag{13}$$

where $\chi_R = \partial J^0_R/\partial\mu_R$ and $\chi_L = \partial J^0_L/\partial\mu_L$ are susceptibilities for RH and LH chiralities. These two equations express two collective, gapless, wave modes which are called the chiral magnetic waves (CMWs) [16]. Similarly, if we consider the CESE and the Ohm's law, we can find new collective modes, the chiral electric waves and axial or vector density waves [14]; if we consider the two CVEs, we can find chiral vortical waves [17]. Further more, if one consider, e.g. both magnetic field and vorticity, there emerge more collective modes. Finally, we summarize the anomalous chiral transports in Table (1).

4 Anomalous chiral transports in heavy-ion collisions

If the ACTs occur in the heavy-ion collisions, what will be their experimental signals? As the magnetic fields are, on average over events, perpendicular to the reaction plane, the CME will drive a charge dipole in the fireball which will be converted to a charge dipole in momentum space and can be tested via detecting the final charged hadron distribution. Similarly, the CVE will induce a baryonic dipole with respect to the reaction plane and can be tested by detecting the final baryon number distribution. The CESE, together with the CME, may drive a charge quadrupole in, e.g. Cu + Au collisions and thus may be tested. The CMW may induce a splitting between the elliptic flows of positively and negatively charged particles and the CVW may induce a splitting between the elliptic flows of baryons and antibaryons. We will focus on the CME.

Because μ_A is fluctuating over events, the CME-induced charged dipole is also fluctuating. To measure this dipolar fluctuation, one can use the following charge-dependent two-particle azimuthal correlation [30]:

$$\gamma_{\alpha\beta} \equiv \langle\cos(\phi_\alpha + \phi_\beta - 2\Psi_{\rm RP})\rangle, \tag{14}$$

where α and β denote the charges, ϕ_α and ϕ_β are the corresponding azimuthal angles, $\Psi_{\rm RP}$ is the reaction plane angle, and $\langle\cdots\rangle$ is event average. It is easy to see that a charge separation with respect to the reaction plane will give a positive γ_{+-} and γ_{-+} (will be referred to as $\gamma_{\rm OS}$) while a negative γ_{++} or γ_{--} (will be referred to as $\gamma_{\rm SS}$). In real experiments, the measurements are done with three-particle correlations where the third hadron (of arbitrary charge) is used to reconstruct the reaction plane.

The correlation $\gamma_{\alpha\beta}$ was measured by STAR Collaboration at RHIC for Au + Au and Cu + Cu collisions at $\sqrt{s}$ = 200 GeV [31, 32], by PHENIX Collaboration at RHIC for Au + Au collisions at $\sqrt{s}$ = 200 GeV [33], by ALICE Collaboration at LHC for for Pb + Pb collisions at $\sqrt{s}$ = 2.76 TeV [34], and by CMS Collaboration at LHC for Pb+Pb collisions at $\sqrt{s}$ = 5.02 TeV [35]. At mid-central collisions, these measurements show clear positive $\gamma_{\rm OS}$ and negative $\gamma_{\rm SS}$ with features in consistence with the expectation of the CME. The STAR Collaboration also measured $\gamma_{\alpha\beta}$ at different beam energies [36] and found that $\gamma_{\alpha\beta}$ persists as long as the beam energy is larger than 19.6 GeV.

However, the interpretation of the data is ambiguous owing to possible background effects that are not correlated to CME. For example, the transverse momentum conservation (TMC) can induce a back-to-back correlation which contributes to $\gamma_{\alpha\beta}$ [37–39], which can be subtracted by making a difference $\Delta\gamma \equiv \gamma_{\rm OS} - \gamma_{\rm SS}$ as the TMC is charge blind. There are also charge dependent background effects, like the local charge conservation (LCC) or neutral cluster decay [40, 41] which gives a finite contribution to $\Delta\gamma$: $\Delta\gamma^{\rm LCC} \approx Mv_2/N$ where M is the number of hadrons in a local neutral cell, N is the multiplicity, and v_2 is the elliptic flow. To disentangle the possible CME signal which is proportional to the magnetic field squared and the flow-related backgrounds which is proportional to v_2, one can either vary the backgrounds with the signal fixed or vary the signal with the backgrounds fixed. The former approach was carried out by using central U + U collisions [42] where sizable v_2 can appear due to the prolate shape of uranium nucleus but the magnetic field should be small [19]. The experimental result from the STAR Collaboration shows that in the 0 – 1% most central events of U + U collisions at $\sqrt{s}$ = 193 GeV there is indeed sizable v_2 and very small $\Delta\gamma$ [43]. However, it was found later that the total multiplicity is far less correlated to the number of binary collisions than expected [44], thus the shape selection is hard to control. The latter approach can be achieved by using the isobaric collisions, such as $^{96}_{44}$Ru and $^{96}_{40}$Zr [42, 45–47]. It is expected that Ru + Ru and Zr + Zr collisions at the same beam energy and same centrality will produce similar elliptic flow but a 10% difference in the magnetic fields. Thus if the γ correlation contains contribution from CME, the isobaric collisions could have a chance to see it by comparing the same γ correlation in Ru + Ru and Zr + Zr collisions. The detailed simulation in Ref [45] indeed found that the two collisions at $\sqrt{s}$ = 200 GeV have more than 10% difference in the CME signal and less than 2% difference in the elliptic-flow-driven backgrounds for the centrality range of 20 – 60% by assuming that the CME contribution to the γ correlation is 1/3. Such a difference is feasible in current of heavy-ion collisions and RHIC have planed the isobaric run in 2018, so let us look forward to their result which will be a very important step towards the detection of the CME in heavy-ion collisions.

Recently, the CMS Collaboration reported the result of $\gamma_{\alpha\beta}$ in high multiplicity events in p + Pb and Pb + Pb collisions at $\sqrt{s}$ = 5.02 TeV [35], in which they found similar $\gamma_{\alpha\beta}$'s in p + Pb and in Pb + Pb as functions of multiplicity. These results suggest that the gamma correlation at 5.02 TeV or above in both p + Pb and Pb + Pb collisions are dominated by background effects. This is somehow compatible with the measurement of STAR collaboration in which the so-called H correlation in measured at various energies through the Beam Energy Scan (BES) program [36]. If one extrapolates the H correlation from lower energies to $\sqrt{s} = 5.02$ TeV, one can find that H correlation is actually small there. See relevant articles in this volume. More studies are definitely needed.

To end this section, we discuss the possible observables for other ACTs. The CESE may lead to observable effect in, e.g., Cu + Au collisions where persist electric fields from Au to Cu nuclei exist [48]. The CVE can induce baryon number separation with respect to the reaction plane [49] and CVW may cause a v_2 splitting between Λ and $\bar{\Lambda}$ [17]. Recently, the STAR Collaboration has reported the first measurement of CVE in Au + Au collisions at $\sqrt{s} = 200$ GeV [50] and the data is consistent with the expectation of CVE. But, similar with the case of CME and CMW detections, the data contains contributions from background effects like the transverse momentum conservation and the local baryon number conservation, and we need more efforts to understand the data.

Acknowledgments: This work is supported by NSFC with Grants No. 11535012 and No. 11675041 and the One Thousand Young Talents Program of China.

References

[1] V. Skokov, A. Y. Illarionov and V. Toneev, Int. J. Mod. Phys. A **24**, 5925 (2009).
[2] W. T. Deng and X. G. Huang, Phys. Rev. C **85**, 044907 (2012).
[3] Y. Hirono, M. Hongo and T. Hirano, Phys. Rev. C **90**, 021903 (2014).
[4] W. T. Deng and X. G. Huang, Phys. Lett. B **742**, 296 (2015).
[5] W. T. Deng and X. G. Huang, Phys. Rev. C **93**, 064907 (2016).
[6] Y. Jiang, Z. W. Lin and J. Liao, Phys. Rev. C **94**, 044910 (2016) Erratum: [Phys. Rev. C **95**, 049904 (2017)].
[7] L. Adamczyk *et al.* [STAR Collaboration], Nature **548**, 62 (2017).
[8] D. E. Kharzeev, L. D. McLerran and H. J. Warringa, Nucl. Phys. A **803**, 227 (2008).
[9] K. Fukushima, D. E. Kharzeev and H. J. Warringa, Phys. Rev. D **78**, 074033 (2008).
[10] D. T. Son and A. R. Zhitnitsky, Phys. Rev. D **70**, 074018 (2004).
[11] J. Erdmenger, M. Haack, M. Kaminski and A. Yarom, JHEP **0901**, 055 (2009).
[12] N. Banerjee, J. Bhattacharya, S. Bhattacharyya, S. Dutta, R. Loganayagam and P. Surowka, JHEP **1101**, 094 (2011).
[13] D. T. Son and P. Surowka, Phys. Rev. Lett. **103**, 191601 (2009).
[14] X. G. Huang and J. Liao, Phys. Rev. Lett. **110**, 232302 (2013).
[15] Y. Jiang, X. G. Huang and J. Liao, Phys. Rev. D **91**, 045001 (2015).
[16] D. E. Kharzeev and H. U. Yee, Phys. Rev. D **83**, 085007 (2011).
[17] Y. Jiang, X. G. Huang and J. Liao, Phys. Rev. D **92**, 071501 (2015).
[18] J. Bloczynski, X. G. Huang, X. Zhang and J. Liao, Phys. Lett. B **718**, 1529 (2013).
[19] J. Bloczynski, X. G. Huang, X. Zhang and J. Liao, Nucl. Phys. A **939**, 85 (2015).
[20] X. G. Huang, Rept. Prog. Phys. **79**, 076302 (2016).
[21] K. Tuchin, Phys. Rev. C **82**, 034904 (2010) Erratum: [Phys. Rev. C **83**, 039903 (2011)].
[22] W. T. Deng and X. G. Huang, J. Phys. Conf. Ser. **779**, 012070 (2017).
[23] Z. T. Liang and X. N. Wang, Phys. Rev. Lett. **94**, 102301 (2005) Erratum: [Phys. Rev. Lett. **96**, 039901 (2006)].
[24] F. Becattini, F. Piccinini and J. Rizzo, Phys. Rev. C **77**, 024906 (2008).
[25] X. G. Huang, P. Huovinen and X. N. Wang, Phys. Rev. C **84**, 054910 (2011).
[26] L. G. Pang, H. Petersen, Q. Wang and X. N. Wang, Phys. Rev. Lett. **117**, 192301 (2016).
[27] D. E. Kharzeev, J. Liao, S. A. Voloshin and G. Wang, Prog. Part. Nucl. Phys. **88**, 1 (2016).

[28] K. Hattori and X. G. Huang, Nucl. Sci. Tech. **28**, 26 (2017).
[29] K. Landsteiner, E. Megias and F. Pena-Benitez, Phys. Rev. Lett. **107**, 021601 (2011).
[30] S. A. Voloshin, Phys. Rev. C **70**, 057901 (2004).
[31] B. I. Abelev *et al.* [STAR Collaboration], Phys. Rev. Lett. **103**, 251601 (2009).
[32] B. I. Abelev *et al.* [STAR Collaboration], Phys. Rev. C **81**, 054908 (2010).
[33] N. N. Ajitanand, R. A. Lacey, A. Taranenko and J. M. Alexander, Phys. Rev. C **83**, 011901 (2011).
[34] B. Abelev *et al.* [ALICE Collaboration], Phys. Rev. Lett. **110**, 012301 (2013).
[35] V. Khachatryan *et al.* [CMS Collaboration], Phys. Rev. Lett. **118**, 122301 (2017).
[36] L. Adamczyk *et al.* [STAR Collaboration], Phys. Rev. Lett. **113**, 052302 (2014).
[37] S. Pratt, arXiv:1002.1758 [nucl-th].
[38] S. Pratt, S. Schlichting and S. Gavin, Phys. Rev. C **84**, 024909 (2011).
[39] A. Bzdak, V. Koch and J. Liao, Phys. Rev. C **83**, 014905 (2011).
[40] S. Schlichting and S. Pratt, Phys. Rev. C **83**, 014913 (2011).
[41] F. Wang, Phys. Rev. C **81**, 064902 (2010).
[42] S. A. Voloshin, Phys. Rev. Lett. **105**, 172301 (2010).
[43] G. Wang [STAR Collaboration], Nucl. Phys. A **904-905**, 248c (2013).
[44] L. Adamczyk *et al.* [STAR Collaboration], Phys. Rev. Lett. **115**, 222301 (2015).
[45] W. T. Deng, X. G. Huang, G. L. Ma and G. Wang, Phys. Rev. C **94**, 041901 (2016).
[46] V. Koch, S. Schlichting, V. Skokov, P. Sorensen, J. Thomas, S. Voloshin, G. Wang and H. U. Yee, Chin. Phys. C **41**, 072001 (2017).
[47] X. G. Huang, W. T. Deng, G. L. Ma and G. Wang, Nucl. Phys. A **967**, 736 (2017).
[48] G. L. Ma and X. G. Huang, Phys. Rev. C **91**, 054901 (2015).
[49] D. E. Kharzeev and D. T. Son, Phys. Rev. Lett. **106**, 062301 (2011).
[50] F. Zhao [STAR Collaboration], Nucl. Phys. A **931**, 746 (2014).

New experimental constrains on chiral magnetic effect using charge-dependent azimuthal correlation in pPb and PbPb collisions at the LHC

On behalf of the CMS Collaboration

Zhoudunming Tu*

[1]*6100 Main Street*
[2]*HBH 112*
[3]*Rice University, Houston, TX 77002*

Abstract. Studies of charge-dependent azimuthal correlations for the same- and opposite-sign particle pairs are presented in PbPb collisions at 5 TeV and pPb collisions at 5 and 8.16 TeV, with the CMS experiment at the LHC. The azimuthal correlations are evaluated with respect to the second- and also higher-order event planes, as a function of particle pseudorapidity and transverse momentum, and event multiplicity. By employing an event-shape engineering technique, the dependence of correlations on azimuthal anisotropy flow is investigated. Results presented provide new insights to the origin of observed charge-dependent azimuthal correlations, and have important implications to the search for the chiral magnetic effect in heavy ion collisions.

1 Introduction

In relativistic heavy ion collisions, the metastable gluon field with nontrivial topological configurations may form, which can cause parity and charge conjugation parity violating effects [1–4]. The interactions between chiral quarks and these gluon field can lead to a chirality imbalance, where the number of left- and right-handed particles are not the same. Given the extremely strong magnetic field that can be produced by noncentral heavy ion collisions, charged-particles would tend to move along or opposite to the direction of the magnetic field. This phenomenon is known as the "chiral magnetic effect" (CME).

The CME has been extensively studied both experimentally and theoretically. The measurement was first done by the STAR Collaboration and later ALICE Collaboration [5–9] using a charge-dependent azimuthal correlation with respect to the reaction plane. The results are found to be consistent with the CME expectations. However, not long after this observation, it has been realized that there is non negligible flow-related background contaminated in this correlator. One of the proposed background mechanisms is related to the local charge conservation coupled with an anisotropic flow at the freezeout surface, where short-range correlation caused by jets or resonance decays that coupled with a strong elliptic flow can play an important role in the charge-dependent three-particle correlator, γ [10–12]. However, this speculation has never been directly confirmed using the experimental data.

*e-mail: kongkong@rice.edu

© The Authors, published by EDP Sciences. This is an open access article distributed under the terms of the Creative Commons Attribution License 4.0 (http://creativecommons.org/licenses/by/4.0/).

Recently, the CMS experiment reported a new measurement of studying the CME with γ-correlator in high-multiplicity pPb and peripheral PbPb collisions, where similar signal has been observed between pPb and PbPb systems [13, 14]. Because of the small magnetic field and its decorrelation to the second-order event plane in pA collisions [13], the CME is not expected to be observed in such small systems. Therefore, this data poses a challenge to the interpretation of the CME in AA collisions. It has been suggested that the lifetime of the strong magnetic field will not survive long enough at LHC energies in order to have CME to be present even in AA collisions. However, assuming there is no CME present at LHC energies, the CMS data still cannot be explained from a pure background model if the background is dominated by a flow only scenario, $\sim v_2/N$, where N is charged-particle multiplicity. Therefore, understanding the background related to this γ-correlator is also extremely important for searching the CME.

The very recent analysis done by the CMS Collaboration has developed a new experimental approach to study the background correlation, together with an upper limit for the CME signal using an event shape engineering (ESE) technique at LHC energies. This analysis introduces a new correlator, $\gamma_{123} \equiv \left\langle \cos(\phi_\alpha + 2\phi_\beta - 3\Psi_3) \right\rangle$, which is expected to be a CME free correlator because of the decorrelation between the event planes Ψ_2 and Ψ_3 [15]. However, this can be extremely useful for testing whether the dominated source of background is related to the local charge conservation and anisotropic flow, because one would expect, $\Delta\gamma_{112}/v_2\Delta\delta \approx \Delta\gamma_{123}/v_3\Delta\delta$, where Δ represents the difference between same and opposite sign pairs, $\delta \equiv \left\langle \cos(\phi_\alpha - \phi_\beta) \right\rangle$ is a two-particle correlation between the α and β particles and γ_{112} is the same as the conventional γ-correlator (hereafter γ_{112}- and γ-correlator are used intermittently). The advantage of using this correlator is that it provides an independent constrain on the background mechanism without involving any theoretical models. Furthermore, not only the background mechanism has been explored, an upper limit on the v_2-independent component fraction with respect to the γ_{112} correlator that is directly related to the CME, has been set for both pPb and PbPb collisions using an ESE method. Instead of changing the centrality or multiplicity, the ESE method provides an independent handle on the initial geometry of the collisions, which is proportional to the elliptic flow, so that the γ_{112} correlator can be explicitly studied as a function of v_2 without changing the magnetic field [9, 14, 16]. This talk focused mainly on the new results that published from CMS Collaboration in Ref. [14], while details of the CMS detector can be found in Ref. [17].

2 Results

2.1 Higher-harmonic correlator

In Fig. 1, the same sign (SS) and opposite sign (OS) three-particle correlators γ_{112} (upper) and γ_{123} (middle), and two-particle correlator, δ (lower) are shown as a function of $|\Delta\eta|$ for multiplicity range $185 \leq N_{\text{trk}}^{\text{offline}} < 250$ in pPb collisions at $\sqrt{s_{_{\text{NN}}}} = 8.16$ TeV (left) and PbPb collisions at 5.02 TeV (right). For pPb collisions, the solid markers denote the results that are obtained with particle c from the Pb-going direction, and the open markers are for the p-going direction. It is shown that the γ_{123} and δ correlator are both charge dependent at similar $|\Delta\eta|$ range as it is found in γ_{112}, where no charge separation is expected for the γ_{123} correlator. Furthermore, both the OS and SS two-particle correlator, δ, in pPb collisions are found to have a larger magnitude than that in PbPb collisions.

In order to explore the multiplicity dependence of the charge-dependent correlators, the SS and OS three-particle correlators, γ_{112} (upper) and γ_{123} (middle), and two-particle correlator, δ (lower), averaged over $|\Delta\eta| < 1.6$ as a function of $N_{\text{trk}}^{\text{offline}}$ in pPb collisions at $\sqrt{s_{_{\text{NN}}}} = 8.16$ TeV and PbPb collisions at 5.02 TeV, are shown in Fig. 2. The published results from pPb collisions at 5.02 TeV are also shown [13]. In pPb collisions, the particle c for three-particle correlators is taken from the

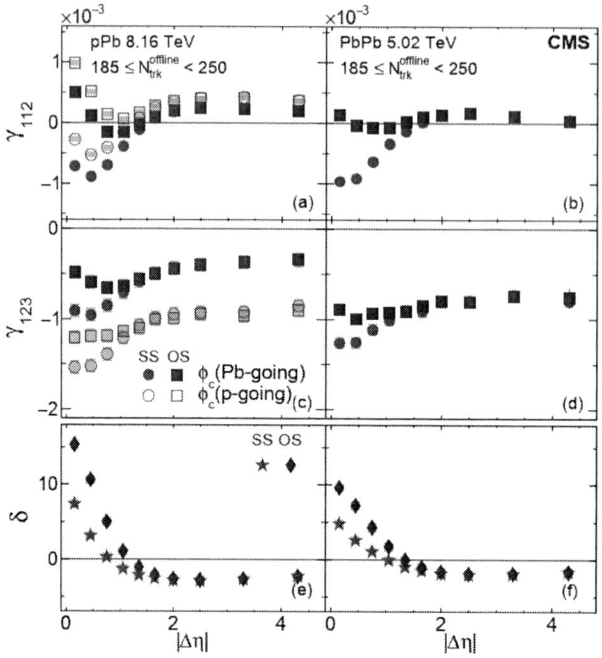

Figure 1. The same sign (SS) and opposite sign (OS) three-particle correlators, γ_{112} (upper) and γ_{123} (middle), and two-particle correlator, δ (lower), as a function of $|\Delta\eta|$ for $185 \leq N_{\text{trk}}^{\text{offline}} < 250$ in pPb collisions at $\sqrt{s_{\text{NN}}}$ = 8.16 TeV (left) and PbPb collisions at 5.02 TeV (right) [14]. The pPb results obtained with particle c in Pb-going (solid markers) and p-going (open markers) sides are shown separately. The SS and OS two-particle correlators are denoted by different markers for both pPb and PbPb collisions. Statistical and systematic uncertainties are indicated by the error bars and shaded regions, respectively.

Pb-going direction of the HF detectors. The average over $|\Delta\eta| < 1.6$ is weighted by the density of particle pairs in $|\Delta\eta|$. Hereafter, all figures that averaged over $|\Delta\eta| < 1.6$ are weighted similarly.

From 5.02 TeV to 8.16 TeV, the SS and OS three-particle correlator, γ_{112}, in pPb collisions have been found to be the same within uncertainties and both agree with results from PbPb collisions, which indicates the γ_{112} correlator is mostly dominated by background correlation at both collision energies and systems. On the other hand, both OS and SS of γ_{123} in pPb collisions are different than that in PbPb collisions. For two-particle correlator, δ, the SS shows similar magnitude between pPb and PbPb collisions, while OS shows different values.

However, the individual SS and OS correlators also have charge-independent correlations (e.g., momentum conservation) and the difference in OS and SS should largely cancel out those effects. In Fig. 3, the difference between OS and SS three-particle correlators, γ_{112} (upper) and γ_{123} (middle), and two-particle correlators, δ (lower), averaged over $|\Delta\eta| < 1.6$ as a function of $N_{\text{trk}}^{\text{offline}}$ in pPb collisions at $\sqrt{s_{\text{NN}}}$ = 8.16 TeV and PbPb collisions at 5.02 TeV, are shown. The results in terms of γ_{112} from 5.02 TeV pPb collisions are also shown for comparison [13]. Note that the $\Delta\delta$ is shown very differently between pPb and PbPb collisions, with pPb collisions a larger magnitude. Given the v_2, however, is larger in PbPb than in pPb collisions at the same multiplicity, a similar value between pPb and PbPb collisions in terms of the $\Delta\gamma_{112}$ can be explained by the difference in v_2 and $\Delta\delta$ correlator, which is expected from the background mechanism of local charge conservation coupled with the anisotropy flow.

 https://doi.org/10.1051/epjconf/201817201004

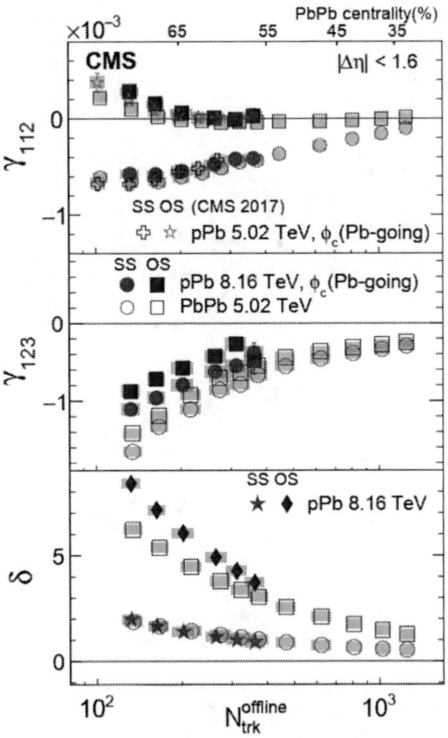

Figure 2. The SS and OS three-particle correlators, γ_{112} (upper) and γ_{123} (middle), and two-particle correlator, δ (lower), averaged over $|\Delta\eta| < 1.6$ as a function of $N_{\rm trk}^{\rm offline}$ in pPb collisions at $\sqrt{s_{\rm NN}}$ = 8.16 TeV and PbPb collisions at 5.02 TeV [14]. The SS and OS two-particle correlators are denoted by different markers for pPb collisions. The results of γ_{112} for pPb collisions at 5.02 TeV from CMS Collaboration (CMS 2017 [13]), are also shown for comparison. Statistical and systematic uncertainties are indicated by the error bars and shaded regions, respectively.

To test if the γ_{112} is entirely from background, the ratios of $\Delta\gamma_{112}/v_2\Delta\delta$ and $\Delta\gamma_{123}/v_3\Delta\delta$ are compared in Fig. 4 as a function of multiplicity, where these ratios are expected to be harmonic independent to the first order as they are related to the particle productions and detector acceptance [11, 14]. These ratios are found to be the same for both harmonic order n=2 and 3, and for both collision systems, which indicates again a pure background scenario for the γ_{112} correlator. In addition, these ratios are almost invariant for different multiplicity ranges, which would have been different if the correlation is dominated by the event plane correlation between the second- and the third-order [15]. The measurement of this ratio, usually regarded as κ, has been constrained for the background only scenario in a data-driven way. For future CME searches, this has provided a new experimental approach to constrain the κ parameter, which can be used to estimate the background contribution together with a measurement of v_2 and δ correlator.

2.2 Event shape engineering (ESE)

It has been shown in the previous section that the γ_{112} is consistent with a pure background scenario, where the background has been found to be an interplay between local charge conservation and the

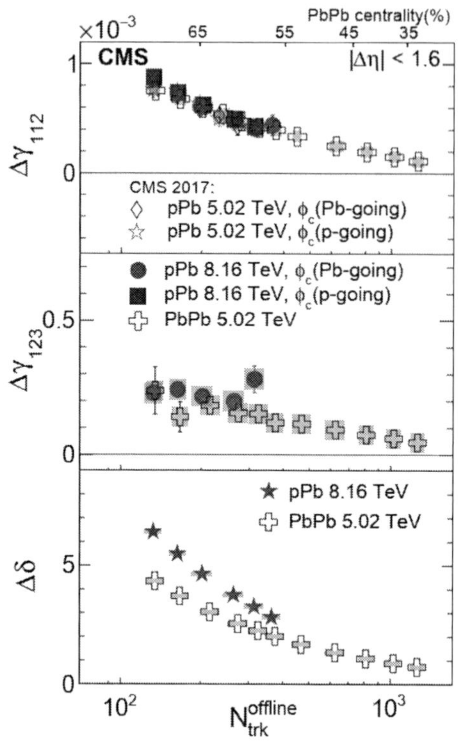

Figure 3. The difference of the OS and SS three-particle correlators, γ_{112} (upper) and γ_{123} (middle), and two-particle correlator, δ (lower), averaged over $|\Delta\eta| < 1.6$ as a function of $N_{\rm trk}^{\rm offline}$ in pPb collisions at $\sqrt{s_{\rm NN}} = 8.16$ TeV and PbPb collisions at 5.02 TeV [14]. The pPb results are obtained with particle c from Pb- and p-going sides separately. The $\Delta\delta$ correlator is denoted by a different marker for pPb collisions. The results of γ_{112} for pPb collisions at 5.02 TeV from CMS Collaboration (CMS 2017 [13]), are also shown for comparison. Statistical and systematic uncertainties are indicated by the error bars and shaded regions, respectively.

anisotropy flow. However, the v_2-independent component, which is believed to be related to the CME, still remains unknown. Using an ESE technique, events in the same multiplicity or centrality range with different average v_2 values can be selected without changing the magnetic field, a consequence of the initial-state fluctuation. As a result, the charge-dependent correlator $\Delta\gamma_{112}$ can be studied explicitly as a function of v_2, and its intercept when $v_2 = 0$.

In Fig. 6, the ratios between $\Delta\gamma_{112}$ and $\Delta\delta$ correlators, averaged over $|\Delta\eta| < 1.6$ as a function of v_2 (evaluated as the average v_2 in each q_2 event class), for different centrality classes in PbPb collisions at $\sqrt{s_{\rm NN}}$=5.02 TeV, are shown. The $\Delta\gamma_{112}$ and $\Delta\delta$ as a function of v_2 separately can be found in Ref. [14]. After taking the ratio, the v_2 dependence of the $\Delta\delta$ correlator is expected to be removed, and the intercepts can be interpreted as the v_2-independent component that is related to the CME (scaled by $1/\Delta\delta$). As one can see from Fig. 6, this ratio is linear as a function of v_2 and the intercepts are found to be consistent with zero, which is expected from a pure background scenario without any CME signal. Similar results have been found in pPb collisions [14].

In order to quantify the intercepts and the possible remaining CME fraction in terms of the $\Delta\gamma_{112}$ correlator, the intercepts, $b_{\rm norm}$, from Fig. 6 (left) and the upper limits at 95% confidence level (CL) on the v_2-independent fraction, $f_{\rm norm}$ (right), are shown as a function of multiplicity in pPb and PbPb

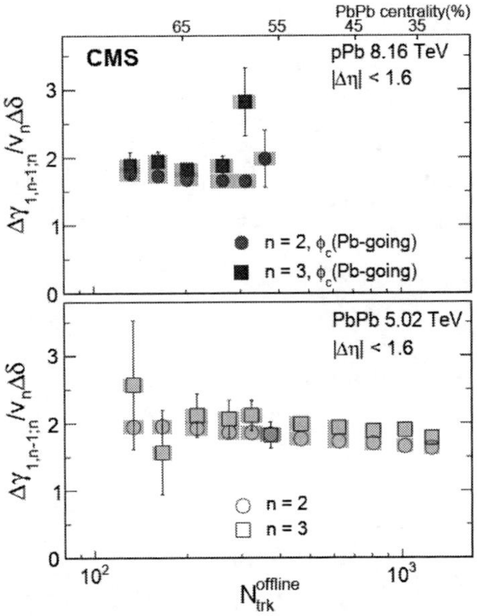

Figure 4. The ratio of $\Delta\gamma_{112}$ and $\Delta\gamma_{123}$ to the product of v_n and δ, averaged over $|\Delta\eta| < 1.6$, in pPb collisions for the Pb-going direction at $\sqrt{s_{NN}}$ = 8.16 TeV (upper) and PbPb collisions at 5.02 TeV (lower) [14]. Statistical and systematic uncertainties are indicated by the error bars and shaded regions, respectively.

collisions. After combining all the measured multiplicities and centralities, the upper limit at 95% CL is found to be 6.6% and 3.8% for pPb and PbPb collisions, respectively. Note that the dominant uncertainty comes from the systematics instead of statistics, where more data is not going to improve the precision of this measurement.

A similar study has been recently done by the ALICE Collaboration [9], where details of extracting the v_2-independent component are slightly different. Because of the finite resolution of the v_2 measurement, the expected CME signal that can be experimentally measured also depends on the value of v_2. However, this correlation is only accessible via model-dependent simulations, and the upper limits that extracted from CMS Collaboration [14] may increase by ~20% depending on specific models with different initial-state fluctuations, which would still be at a few % level. Therefore, the precision measurement of the charge-dependent three-particle azimuthal correlations in pPb and PbPb collisions are consistent with a pure background scenario, posing a constrain to the possible CME signal at LHC energies.

3 Summary

Charge-dependent azimuthal correlations of same- and opposite-sign (SS and OS) pairs with respect to the second- and third-order event planes have been studied in pPb collisions at $\sqrt{s_{NN}}$= 8.16 TeV and PbPb collisions at 5.02 TeV by the CMS experiment at the LHC. With the independent constrain from the charge-dependent three-particle correlator with respect to the third-order event plane, it has been found that the background mechanism is consistent with local charge conservation coupled with the anisotropy flow. In addition, the ratio, divided the OS and SS difference of the three-particle correlator by the product of v_n harmonic of the corresponding order and the difference of the two-particle correlator, is experimentally constrained at the value around 2 at LHC energies, which provides a

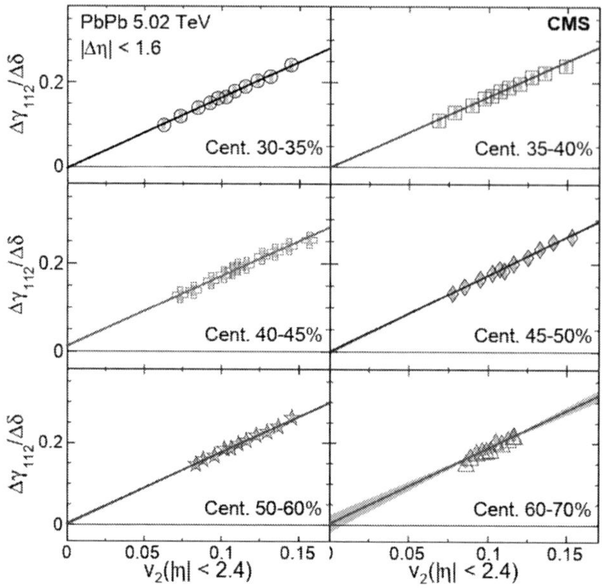

Figure 5. The ratio between the difference of the OS and SS three-particle correlators and the difference of OS and SS in δ correlators, $\Delta\gamma_{112}/\Delta\delta$, averaged over $|\Delta\eta| < 1.6$ as a function of v_2 evaluated in each q_2 class, for different centrality classes in PbPb collisions at $\sqrt{s_{NN}}$=5.02 TeV [14]. Statistical and systematic uncertainties are indicated by the error bars and shaded regions, respectively. A one standard deviation uncertainty from the fit is also shown.

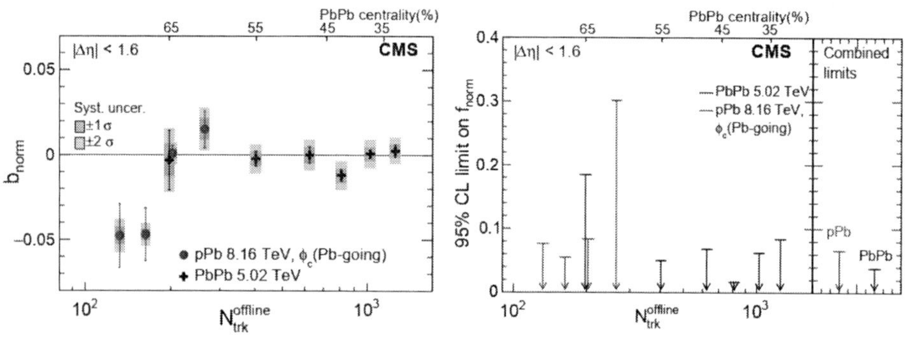

Figure 6. Extracted intercept parameter b_{norm} (left) and corresponding upper limit of the fraction of v_2-independent γ_{112} correlator component (right), averaged over $|\Delta\eta| < 1.6$, as a function of $N_{\mathrm{trk}}^{\mathrm{offline}}$ in pPb collisions at $\sqrt{s_{NN}}$ = 8.16 TeV and PbPb collisions at 5.02 TeV [14]. Statistical and systematic uncertainties are indicated by the error bars and shaded regions in the top panel, respectively.

new experimental approach to the understanding of background contribution in terms of the charge-dependent correlators. Moreover, using an event shape engineering technique, the upper limit on the v_2-independent fraction that is related to the CME, has been found to be 6.6% and 3.8% for pPb and PbPb collisions, respectively, at 95% confidence level after combining all the measured multiplicity and centrality ranges. Therefore, this measurement not only provides a constrain on the magnitude of the possible CME signal at LHC energies, but also a new baseline and experimental approach of searching the CME signal for future attempts.

References

[1] T.D. Lee, Phys. Rev. D **8**, 1226 (1973)
[2] T.D. Lee, G.C. Wick, Phys. Rev. D **9**, 2291 (1974)
[3] P.D. Morley, I.A. Schmidt, Z. Phys. C **26**, 627 (1985)
[4] D. Kharzeev, R.D. Pisarski, M.H.G. Tytgat, Phys. Rev. Lett. **81**, 512 (1998), hep-ph/9804221
[5] B.I. Abelev et al. (STAR), Phys. Rev. Lett. **103**, 251601 (2009), 0909.1739
[6] L. Adamczyk et al. (STAR), Phys. Rev. Lett. **113**, 052302 (2014), 1404.1433
[7] L. Adamczyk et al. (STAR), Phys. Rev. C **88**, 064911 (2013), 1302.3802
[8] B. Abelev et al. (ALICE), Phys. Rev. Lett. **110**, 012301 (2013), 1207.0900
[9] S. Acharya et al. (ALICE) (2017), 1709.04723
[10] S. Schlichting, S. Pratt, Phys. Rev. C **83**, 014913 (2011), 1009.4283
[11] A. Bzdak, V. Koch, J. Liao, Lect. Notes Phys. **871**, 503 (2013), 1207.7327
[12] A. Bzdak, V. Koch, J. Liao, Phys. Rev. C **83**, 014905 (2011), 1008.4919
[13] V. Khachatryan et al. (CMS), Phys. Rev. Lett. **118**, 122301 (2017), 1610.00263
[14] A.M. Sirunyan et al. (CMS) (2017), 1708.01602
[15] G. Aad et al. (ATLAS), Phys. Rev. C **90**, 024905 (2014), 1403.0489
[16] F. Wen, L. Wen, G. Wang (2016), 1608.03205
[17] S. Chatrchyan et al. (CMS), JINST **3**, S08004 (2008)

EPJ Web of Conferences **172**, 01005 (2018) https://doi.org/10.1051/epjconf/201817201005
ISMD 2017

Chiral magnetic effect search in p+Au, d+Au and Au+Au collisions at RHIC

Jie Zhao[1,⋆] (for the STAR collaboration)

[1] *Department of Physics and Astronomy, Purdue University*

Abstract. Metastable domains of fluctuating topological charges can change the chirality of quarks and induce local parity violation in quantum chromodynamics. This can lead to observable charge separation along the direction of the strong magnetic field produced by spectator protons in relativistic heavy-ion collisions, a phenomenon called the chiral magnetic effect (CME). A major background source for CME measurements using the charge-dependent azimuthal correlator ($\Delta\gamma$) is the intrinsic particle correlations (such as resonance decays) coupled with the azimuthal elliptical anisotropy (v_2). In heavy-ion collisions, the magnetic field direction and event plane angle are correlated, thus the CME and the v_2-induced background are entangled. In this report, we present two studies from STAR to shed further lights on the background issue. (1) The $\Delta\gamma$ should be all background in small system p+Au and d+Au collisions, because the event plane angles are dominated by geometry fluctuations uncorrelated to the magnetic field direction. However, significant $\Delta\gamma$ is observed, comparable to the peripheral Au+Au data, suggesting a background dominance in the latter, and likely also in the mid-central Au+Au collisions where the multiplicity and v_2 scaled correlator is similar. (2) A new approach is devised to study $\Delta\gamma$ as a function of the particle pair invariant mass (m_{inv}) to identify the resonance backgrounds and hence to extract the possible CME signal. Signal is consistent with zero within uncertainties at high m_{inv}. Signal at low m_{inv}, extracted from a two-component model assuming smooth mass dependence, is consistent with zero within uncertainties.

1 Introduction

Quark interactions with topological gluon configurations can induce chirality imbalance and local parity violation in quantum chromodynamics (QCD) [1–3] . In relativistic heavy-ion collisions, this can lead to observable electric charge separation along the direction of the strong magnetic field produced by spectator protons [4–6]. This is called the chiral magnetic effect (CME). An observation of the CME-induced charge separation would confirm a fundamental property of QCD. The measurements of the charge separation can provide a means to studying the non-trivial QCD topological structures. Extensive theoretical and experimental efforts have been devoted to the search for CME [6] .

The commonly used observable to measure charge separation is the three-point correlator difference [7], $\Delta\gamma \equiv \gamma_{OS} - \gamma_{SS}$. Here $\gamma = \langle\cos(\alpha+\beta-2\psi_2)\rangle$, α and β are the azimuthal angles of two charged particles and ψ_2 is that of the second-order harmonic plane; γ_{OS} stands for the γ of opposite electric charge sign (OS) and γ_{SS} for that of same-sign pairs (SS). Significant $\Delta\gamma$ has indeed been observed

⋆e-mail: zhao656@purdue.edu

© The Authors, published by EDP Sciences. This is an open access article distributed under the terms of the Creative Commons Attribution License 4.0 (http://creativecommons.org/licenses/by/4.0/).

in heavy-ion collisions [8–13]. One of the difficulties in its interpretation as from the CME is the major charge-dependent background contributions to the $\Delta\gamma$ observable [14–16], such as those from resonance decays. The $\Delta\gamma$ variable is ambiguous between an OS pair from the CME back-to-back perpendicular to ψ_2 and an OS pair from a resonance decay along ψ_2. More resonances are produced along the ψ_2 than perpendicular to it, the relative difference of which is quantified by the elliptical anisotropy parameter v_2 of the resonances. The CME background arises from the coupling of this elliptical anisotropy and the intrinsic decay correlation [17].

The CME and the v_2-related background are driven by different physics: the CME is sensitive to the magnetic field, $\vec{B}$, while the v_2-related background is connected to the participant plane, ψ_{PP}. In non-central heavy-ion collisions, the ψ_{PP}, although fluctuating [18], is generally aligned with the reaction plane (span by the impact parameter direction and the beam), thus generally perpendicular to $\vec{B}$. The $\Delta\gamma$ measurement is thus *entangled* by the two contributions: the possible CME and the v_2-induced background. In small-system collisions, however, the ψ_{PP} is determined purely by geometry fluctuations, uncorrelated to the impact parameter or the $\vec{B}$ direction [19]. As a result any CME would average to zero $\Delta\gamma$ in small-system collisions. Background sources, on the other hand, contribute to small-system collisions similarly as to heavy-ion collisions when measured with respect to the event plane reconstructed from mid-rapidity particle momenta (as a proxy to the ψ_{PP}). Due to the fluctuating nature of the small systems, the event planes reconstructed over a large pseudorapidity gap can be uncorrelated, and therefore measurements of mid-rapidity particle correlations with respect to a forward event plane are sensitive neither to CME nor to the background.

Recent CMS data show that the correlator signal from p+Pb is comparable to the signal from Pb+Pb collisions at similar multiplicities [19, 20]. This indicates significant background contributions in Pb+Pb collisions at LHC energy. It is predicted that the CME would decrease with the collision energy due to the more rapidly decaying $\vec{B}$ at higher energies [6]. Hence, the similarity between small-system and heavy-ion collisions at the LHC may be expected, and the situation at RHIC could be different [6]. In this report, we present results from similar control experiments using p+Au and d+Au collisions at RHIC. These results are analyzed as a function of the final-state particle multiplicity to shed light on the background to the CME measurements in heavy-ion collisions at RHIC.

Because the main background contributions come from resonance decays, we devise a new analysis approach exploiting the particle pair invariant mass, m_{inv}, to identify the backgrounds and hence to extract the possible CME signal. The $\Delta\gamma$ signal is reported at large m_{inv} where resonance background contributions are small. A two-component model fit to the low mass region is also reported along with the fitted possible CME signals.

2 Experiment setup and data analysis

The data reported here were collected by the STAR experiment [21] at Brookhaven National Laboratory from 2003 (d+Au), 2011 (Au+Au) and 2015 (p+Au). The main subsystems used for the data analysis are the Time Projection Chamber (TPC) [22] and the Time-Of-Flight detector (TOF) [23], both with 2π azimuthal coverage at mid-rapidity.

Charged particle identification is performed by ionization energy loss, dE/dx, measured by TPC and/or particle velocity measured by TOF. The correlations are reported for charged particles with pseudorapidity $|\eta| < 1$ and transverse momentum $0.2 < p_T < 2.0$ GeV/c , and for identified $\pi^{\pm}$ with $0.2 < p_T < 1.8$ GeV/c .

We use the three-particle correlator:

$$\gamma \equiv \langle\cos(\phi_\alpha + \phi_\beta - 2\psi_2)\rangle = \langle\cos(\phi_\alpha + \phi_\beta - 2\phi_c)\rangle / v_2. \quad (1)$$

Here ϕ_α and ϕ_β are the azimuthal angle of the two particles, ϕ_c is that of the third particle which serves as a measure of the ψ_{PP}. The imprecision in determining the ψ_{PP} by a single particle is corrected by the resolution factor, equal to the particle's elliptic flow anisotropy $v_{2,c}$. The $\Delta\gamma$ from the difference of γ_{OS} and γ_{SS} correlator is used to quantify the charge dependent signal. We use particles in the full TPC acceptance ($|\eta| < 1$) for all three particles. No η gap ($\Delta\eta$) is applied between any pair among the three particles. The $v_{2,c}$ used for resolution correction should, ideally, be free of non-flow. We obtain $v_{2,c}$ by the two-particle cumulant, applying a cut of $\Delta\eta > 1.0$ to reduce non-flow contaminations.

3 Results from small systems

The three-particle correlator from small systems are analyzed as functions of rapidity-gap and multiplicity to shed light on the background contaminations to the CME measurements in heavy-ion collisions. Figure 1 shows $\Delta\gamma$ as a function of the $\Delta\eta$ between particle α and β in p+Au and d+Au collisions, and in peripheral Au+Au collisions for comparison. The distributions from p+Au and d+Au collisions are similar to those in peripheral Au+Au collisions.

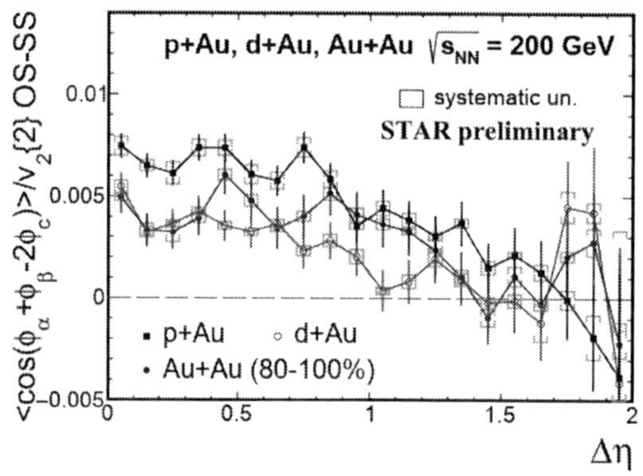

Figure 1: (Color online) The $\Delta\gamma$ in p+Au and d+Au collisions as a function of $\Delta\eta$ between particle α and β, compared to that in peripheral Au+Au collisions.

Figure 2 (Left) shows the γ_{SS} and γ_{OS} results as functions of particle multiplicity, N, in p+Au and d+Au collisions at $\sqrt{s_{NN}} = 200$ GeV. Here N is taken as the geometric mean of the multiplicities of particle α and β. The systematic uncertainties on γ are estimated by varying the track quality cuts, the correction method used for the detector non-uniform azimuthal acceptance, and the p_T range of the reference particle c. For comparison the corresponding Au+Au results are also shown. The trends of the correlator magnitudes are similar, decreasing with increasing N. The γ_{SS} results seem to follow a smooth trend in N over all systems. The γ_{OS} results are less so; the small system data appear to differ somewhat from the heavy-ion data over the range in which they overlap in N. Since the particle pseudorapidity density, $dN/d\eta$, is asymmetric in small systems, there are relatively more small $\Delta\eta$ pairs than in Au+Au collisions, hence larger correlations. We have checked this effect by weighting particles in computing the correlator such that the weighted $dN/d\eta$ is symmetric, and found it insufficient to account for the observed difference.

Similar to LHC, the small system data at RHIC are found to be comparable to Au+Au results at similar multiplicities. Since the p+Au and d+Au data are all backgrounds, the $\Delta\gamma$ should be approximately proportional to the averaged v_2 of the background sources, in turn the v_2 of final-state particles.

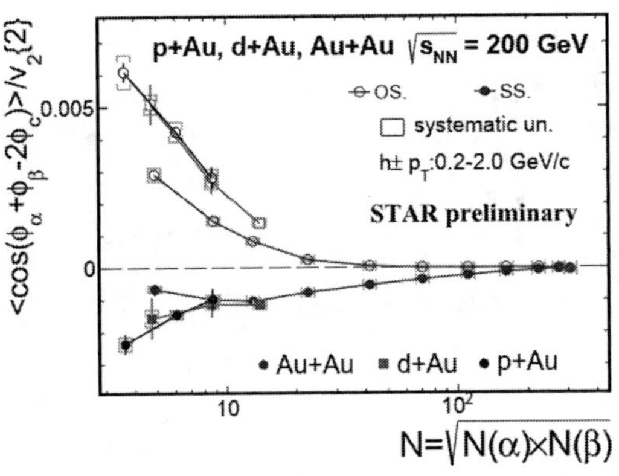

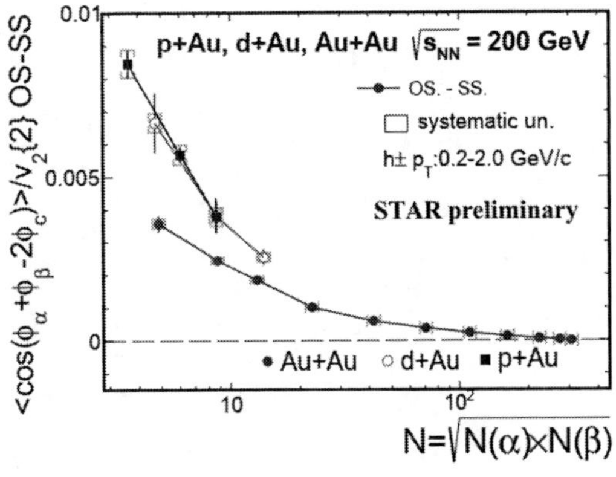

Figure 2: (Color online) The γ_{SS}, γ_{OS} (Left panel) and $\Delta\gamma$ (Right panel) correlators in p+Au and d+Au collisions as a function of multiplicity, compared to those in Au+Au collisions. Particles α, β and c are from the TPC pseudorapidity coverage of $|\eta| < 1$ with no η gap applied. The $v_{2,c}\{2\}$ is obtained by two-particle cumulant with η gap of $\Delta\eta > 1.0$. Statistical uncertainties are shown by the vertical bars and systematic uncertainties are shown by the caps.

It should also be proportional to the number of background sources, and, because $\Delta\gamma$ is a pair-wise average, inversely proportional to the total number of pairs. As the number of background sources likely scales with multiplicity, we have $\Delta\gamma \propto v_2/N$. Therefore, to gain more insight, we scale the $\Delta\gamma$ by N/v_2:

$$\Delta\gamma_{\rm scaled} = \Delta\gamma \times N/v_2 \,. \tag{2}$$

Since there is no distinction between the two particles (α,β) and particle c, the v_2 in Eq. (2) is the same as $v_{2,c}\{2\}$. The $v_2\{2\}$ with a η gap of 1.0 is shown as a function of N in p+Au, d+Au, and Au+Au collisions in Fig. 3 (Left). Figure 3 (Right) shows the scaled $\Delta\gamma_{\rm scaled}$ as a function of N in p+Au and d+Au collisions, and compares to that in Au+Au collisions. AMPT simulation results for d+Au and Au+Au are also plotted for comparison. The AMPT simulations can account for about 2/3 of the STAR data, and are approximately constant over N. The $\Delta\gamma_{\rm scaled}$ in p+Au and d+Au collisions are compatible or even larger than that in Au+Au collisions. Since in p+Au and d+Au collisions only the background is present, the data suggest that the peripheral Au+Au measurement may be largely, if not all, background. For both small-system and heavy-ion collisions, the $\Delta\gamma_{\rm scaled}$ is approximately constant over N. It may not be strictly constant because the correlations caused by decays ($\Delta\gamma_{\rm bkgd} \propto \langle\cos(\alpha+\beta-2\phi_{\rm res})\rangle \times v_{2,\rm res}$), depends on the $\langle\cos(\alpha+\beta-2\phi_{\rm res})\rangle$ which is determined by the parent kinematics and can be somewhat N-dependent. Given that the background is large suggested by the p+Au and d+Au data, the approximate N-independent $\Delta\gamma_{\rm scaled}$ in Au+Au collisions is consistent with the background scenario.

In terms of CME, on the other hand, the number of local topological domains, $N_{\rm domain}$, is likely proportional to the collision volume, so $N_{\rm domain} \propto N$. Since the topological charge randomly fluctuates, the charge asymmetry $a_1 \propto \sqrt{N_{\rm domain}}/N$. As a result, the $\Delta\gamma \propto a_1^2$ would be inversely proportional to N, similar to the background scenario. The CME-induced charge separation magnitude is expected to be proportional to $\langle B^2\cos 2(\psi_B - \psi_2)\rangle$ [6], where ψ_B is the azimuthal angle of $\vec{B}$. The quantity, $\langle B^2\cos 2(\psi_B - \psi_2)\rangle$, may have similar N-dependence as the v_2 but the present theoretical uncertainties are large [24, 25]. Thus, the N-dependence of the CME and the background may, unfortunately, be similar. Given the present uncertainties, a small finite CME signal is not disallowed

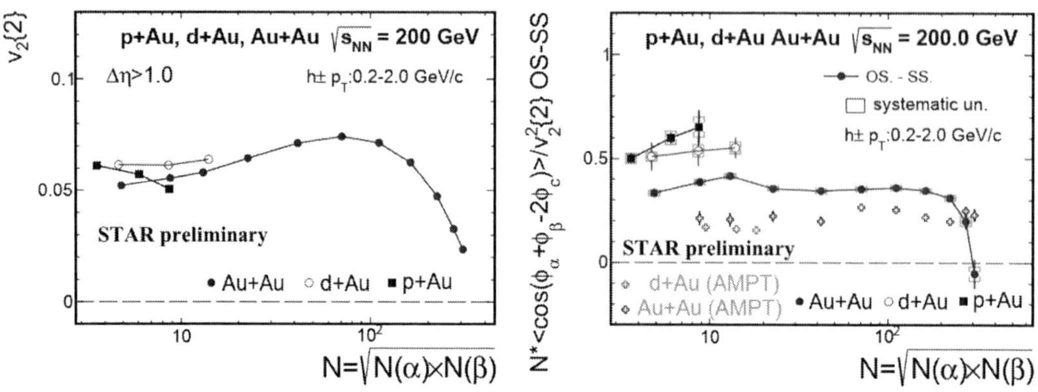

Figure 3: (Color online) (Left panel) The measured two-particle cumulant $v_{2,c}\{2\}$ of charged particles ($0.2 < p_T < 2.0$ GeV/c) with η gap of 1.0, and (Right panel) the scaled three-particle correlator difference in p+Au and d+Au collisions as a function of N, compared to those in Au+Au collisions. AMPT simulation results for d+Au and Au+Au are also plotted for comparison.

in the measured $\Delta\gamma$ in heavy-ion collisions. The present analysis does not currently allow conclusive statements to be made regarding the presence of CME.

4 Results from the invariant mass method

As discussed in the introduction, the major background arises from resonance decay correlations coupled with the elliptical anisotropy. Figure 4 (Left) shows the OS and SS $\pi\pi$ pair difference as a function of m_{inv} from minimum bias p+p and peripheral Au+Au collisions at 200 GeV [26]. Many resonances have broad mass distributions [27]. They are often identified statistically in relativistic heavy ion collisions. However statistical identification of resonances does not help eliminate their contribution to the $\Delta\gamma$ variable. Most of the $\pi\pi$ resonances are located in the low m_{inv} region [26]. With increasing mass, the data show that resonance contributions to the difference between OS and SS pairs decrease. It is possible to exclude them "entirely" by applying a lower cut on m_{inv}. In the low mass region we analyze the $\Delta\gamma$ correlator as a function of m_{inv}, which may help to understand the background sources and isolate the possible CME [28].

Figure 4 (Right) shows the $\Delta\gamma$ results with and without applying an invariant mass cut of $m_{inv} >$ 1.5 GeV/c^2 . The results are summarized in Table 1. With the m_{inv} cut, the $\Delta\gamma$ is significantly reduced from the inclusive measurement. The $\Delta\gamma$ with the large m_{inv} cut is consistent with zero within the current uncertainty.

Table 1: The inclusive $\Delta\gamma$ over all mass and $\Delta\gamma$ at $m_{inv} > 1.5$ GeV/c^2 for different centralities in Au+Au collisions at 200 GeV.

Centrality	$\Delta\gamma$ in all mass (A)	$\Delta\gamma$ at $m_{inv} > 1.5$ GeV/c^2 (B)	B/A
50-80%	$(7.45 \pm 0.21) \times 10^{-4}$	$(1.3 \pm 5.7) \times 10^{-5}$	$(1.8 \pm 7.6)\%$
20-50%	$(1.82 \pm 0.03) \times 10^{-4}$	$(7.7 \pm 9.0) \times 10^{-6}$	$(4.3 \pm 4.9)\%$
0-20%	$(3.70 \pm 0.67) \times 10^{-5}$	$(-0.1 \pm 1.8) \times 10^{-5}$	$(-3.8 \pm 49)\%$

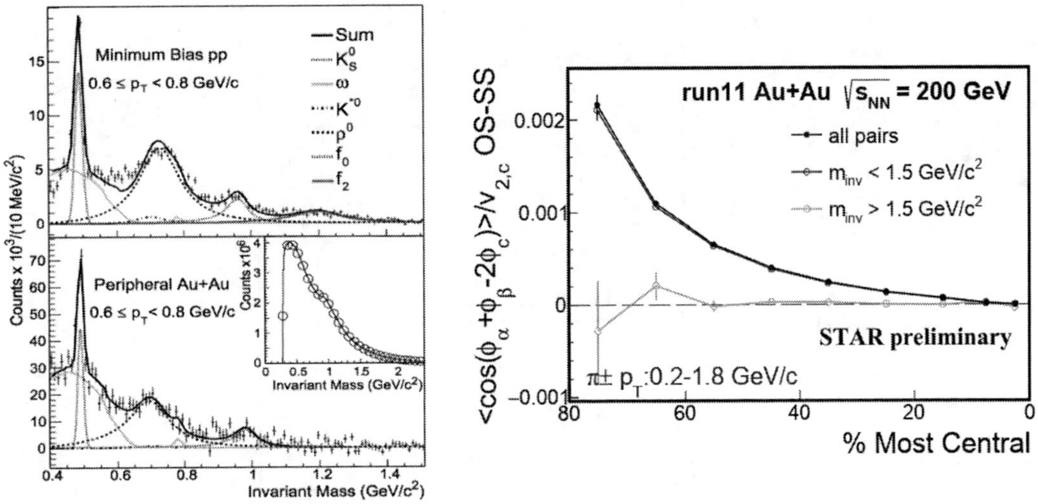

Figure 4: (Left panel) The raw π-π invariant mass (m_{inv}) distributions after subtraction of the like-sign reference distribution for minimum bias p+p (top) and peripheral Au+Au (bottom) collisions. The insert plot corresponds to the raw π^+-π^- m_{inv} (solid line) and the like-sign reference distributions (open circles) for peripheral Au+Au collisions [26]. (Right panel) The inclusive $\Delta\gamma$ over all mass (black) and at $m_{inv} > 1.5$ GeV/c^2 (green) as a function of centrality in Au+Au collisions at 200 GeV.

CME is expected to be a low p_T phenomenon [8, 29]; its contribution to high mass may be small. In order to extract CME at low mass, resonance contributions need to be subtracted. To this end, we show in Fig. 5 (a) the relative OS and SS pair abundance difference, $r = (N_{OS} - N_{SS})/N_{OS}$, and in Fig. 5 (b) the $\Delta\gamma$ correlator as a function of m_{inv} from mid-central (20-50%) Au+Au collisions at 200 GeV. The data show resonance structures in r and $\Delta\gamma$ as functions of m_{inv}: a clear resonance peak from K_s^0 decay is observed, and possible ρ and f^0 peaks are also visible. The $\Delta\gamma$ correlator traces the distributions of those resonances.

The $\Delta\gamma$ in Fig. 5 (b) may be composed of two components, a resonance decay background and a CME signal:

$$\Delta\gamma = r(m_{inv}) \cdot R(m_{inv}) + \mathrm{CME}(m_{inv}). \tag{3}$$

Here CME(m_{inv}) represents the CME contribution. The background depends on $r(m_{inv})$, with a smooth response function $R(m_{inv})$, so the background component is peaky in m_{inv}. We assume that the CME component is smooth in m_{inv}. If the CME contribution was appreciable, then the ratio of $\Delta\gamma/r$ shown in Fig. 5 (c) would reveal a structure resembling the inverse shape of r [28]. However, a more or less smooth dependence is observed; no evidence of an inverse shape of the resonance mass distribution is observed in the ratio of $\Delta\gamma/r$, suggesting insignificant CME signal contributions.

In order to isolate the possible CME from the resonance contributions, the two-component model is used to fit the $\Delta\gamma$ as a function of m_{inv}. We use the first-order polynomial function for $R(m_{inv})$,

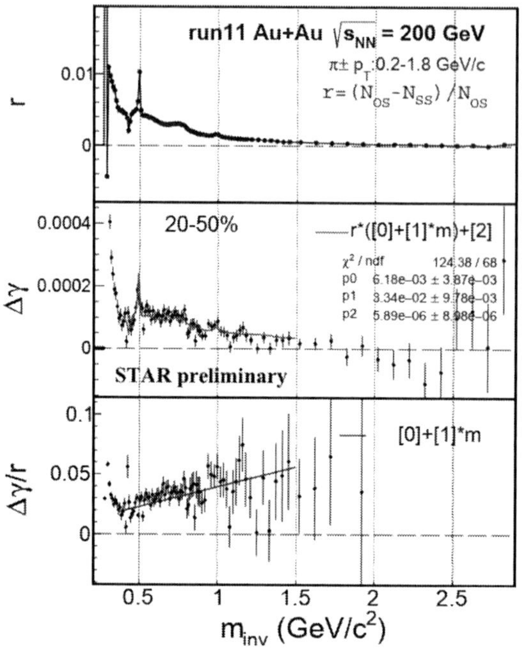

Figure 5: (Color online) Pair invariant mass (m_{inv}) dependence of the relative excess of OS over SS charged π pair multiplicity, $r = (N_{OS} - N_{SS})/N_{OS}$ (top panel), event-plane dependent azimuthal correlator difference, $\Delta\gamma = \gamma_{OS} - \gamma_{SS}$ (middle panel), and the ratio of $\Delta\gamma/r$ (bottom panel) in 20-50% Au+Au collisions at 200 GeV. Errors shown are statistical. The red curve in the middle panel shows the two-component model fit assuming a constant CME contribution independent of m_{inv}; The blue curve in the bottom panel shows the corresponding resonance response function.

motivated by the data in Fig. 5 (c) and MC simulation [28]. At present, no theoretical calculation is available on the m_{inv} dependence of the CME, therefore we consider two cases: (i) a constant CME distribution independent of m_{inv}, and (ii) an exponential CME distribution in m_{inv}. The curve superimposed in Fig. 5 (b) shows the fit of case (i). The extracted average $\Delta\gamma$ from CME contributions, and their strengths relative to the inclusive $\Delta\gamma$ measurement are tabulated in Table 2. The extracted CME signals from the two-component model fits are at present consistent with zero within uncertainties. Future theoretical calculations of the CME mass dependence would be valuable.

Table 2: Average $\Delta\gamma$ signal (corresponding to the CME contribution in the two-component fit model) extracted from the model fit at $m_{inv} < 1.5$ GeV/c^2 in mid-central ($20-50\%$) Au+Au collisions at 200 GeV, with two assumptions for the CME m_{inv} dependence: a constant independent of m_{inv} and an exponential in m_{inv}.

$\Delta\gamma$ (inclusive)	$(1.82 \pm 0.03) \times 10^{-4}$	
	constant CME	exponential CME in m_{inv}
average signal $\Delta\gamma$ (fit)	$(5.9 \pm 9.0) \times 10^{-6}$	$(3.0 \pm 2.0) \times 10^{-5}$
fit/inclusive	$(3.2 \pm 4.9)\%$	$(16 \pm 11)\%$

5 Summary

The chiral magnetic effect (CME) can produce charge separation perpendicular to the reaction plane. Charge separation measurements by the three-particle correlators ($\Delta\gamma$) are contaminated by a major background arising from particle correlations coupled with elliptical anisotropy (v_2). To shed more light on the background and to reduce/eliminate background contamination in charge separation measurements, we have studied the small-system p+Au and d+Au collisions in comparison to Au+Au collisions, and the $\Delta\gamma$ correlator as a function of the particle pair invariant mass (m_{inv}).

With respect to the second-order harmonic plane (ψ_2), the p+Au and d+Au charge dependent correlations are backgrounds. Peripheral Au+Au data are similar to those of p+Au and d+Au. The scaled correlator ($\Delta\gamma \times N/v_2$) from peripheral to mid-central Au+Au collisions is approximately constant over multiplicity. Similar dependence is found in AMPT. These data indicate a dominant contribution from background to the peripheral and mid-central heavy-ion collisions, and do not show clear evidence for the presence of the CME in those collisions.

A new method exploiting the particle pair invariant mass is used to identify the resonance background and the possible CME. In order to exclude the resonance contributions, we apply a lower cut on the m_{inv}. At high mass ($m_{inv} > 1.5$ GeV/c^2), $\Delta\gamma$ is consistent with zero within uncertainty. In the low mass region ($m_{inv} < 1.5$ GeV/c^2), data show resonance structures in $\Delta\gamma$ as function of m_{inv}. A two-component fit is devised where the background component is peaky following the resonance contributions and we assume that the CME signal is smooth in m_{inv}. Two functional forms are assumed in the present study for the CME as function of m_{inv}, a constant and an exponential. The extracted CME signals from the two-component model fit are consistent with zero within the current uncertainties. Theoretical guidance on the mass dependence of CME would be valuable to further our understanding.

6 Acknowledgments

This work was partly supported by the U.S. Department of Energy (Grant No. de-sc0012910).

References

[1] T.D. Lee, G.C. Wick, Phys. Rev. **D9**, 2291 (1974)
[2] D. Kharzeev, R.D. Pisarski, M.H.G. Tytgat, Phys. Rev. Lett. **81**, 512 (1998), hep-ph/9804221
[3] D. Kharzeev, R.D. Pisarski, Phys. Rev. **D61**, 111901 (2000), hep-ph/9906401
[4] K. Fukushima, D.E. Kharzeev, H.J. Warringa, Phys. Rev. **D78**, 074033 (2008), 0808.3382
[5] B. Muller, A. Schafer, Phys. Rev. **C82**, 057902 (2010), 1009.1053
[6] D.E. Kharzeev, J. Liao, S.A. Voloshin, G. Wang, Prog. Part. Nucl. Phys. **88**, 1 (2016), 1511.04050
[7] S.A. Voloshin, Phys. Rev. **C70**, 057901 (2004), hep-ph/0406311
[8] B.I. Abelev et al. (STAR), Phys. Rev. **C81**, 054908 (2010), 0909.1717
[9] B.I. Abelev et al. (STAR), Phys. Rev. Lett. **103**, 251601 (2009), 0909.1739
[10] L. Adamczyk et al. (STAR), Phys. Rev. **C89**, 044908 (2014), 1303.0901
[11] L. Adamczyk et al. (STAR), Phys. Rev. Lett. **113**, 052302 (2014), 1404.1433
[12] B. Abelev et al. (ALICE), Phys. Rev. Lett. **110**, 012301 (2013), 1207.0900
[13] R.A.L. N. N. Ajitanand, S. Esumi (PHENIX), Proc. of the RBRCWorkshops **96**, 230 (2010)
[14] F. Wang, Phys.Rev. **C81**, 064902 (2010), 0911.1482

[15] A. Bzdak, V. Koch, J. Liao, Phys.Rev. **C81**, 031901 (2010), 0912.5050
[16] S. Schlichting, S. Pratt, Phys.Rev. **C83**, 014913 (2011), 1009.4283
[17] F. Wang, J. Zhao, Phys. Rev. **C95**, 051901 (R) (2017), 1608.06610
[18] B. Alver et al. (PHOBOS), Phys. Rev. Lett. **98**, 242302 (2007), nucl-ex/0610037
[19] V. Khachatryan et al. (CMS), Phys. Rev. Lett. **118**, 122301 (2017), 1610.00263
[20] A.M. Sirunyan et al. (CMS) (2017), 1708.01602
[21] K.H. Ackermann et al. (STAR), Nucl. Instrum. Meth. **A499**, 624 (2003)
[22] M. Anderson et al., Nucl. Instrum. Meth. **A499**, 659 (2003), nucl-ex/0301015
[23] W.J. Llope et al., Nucl. Instrum. Meth. **A522**, 252 (2004), nucl-ex/0308022
[24] W.T. Deng, X.G. Huang, Phys. Rev. **C85**, 044907 (2012), 1201.5108
[25] J. Bloczynski, X.G. Huang, X. Zhang, J. Liao, Phys. Lett. **B718**, 1529 (2013), 1209.6594
[26] J. Adams et al. (STAR), Phys. Rev. Lett. **92**, 092301 (2004), nucl-ex/0307023
[27] K.A. Olive et al. (Particle Data Group), Chin. Phys. **C38**, 090001 (2014)
[28] J. Zhao, H. Li, F. Wang (2017), 1705.05410
[29] D.E. Kharzeev, L.D. McLerran, H.J. Warringa, Nucl. Phys. **A803**, 227 (2008), 0711.0950

Asymptotic Analysis of the parton branching equation at LHC Energies

W. Y. Wang[1,⋆], *H. P.* Lau[1], *Q.* Leong[1], *A. H.* Chan[1], and *C. H.* Oh[1]

[1] *Physics Department, National University of Singapore, Singapore*

Abstract. An asymptotic solution to the QCD parton branching equation is derived using the method of Laplace transformation and saddle point approximation. The distribution is applied to charged particle multiplicity distributions in proton-proton collisions at $\sqrt{s} = 0.9, 2.36$, and 7 TeV for $|\eta| < 0.5, 1.0, 1.5, 2.0, 2.4$, and 8 TeV for $|\eta| < 0.5, 1.0, 1.5$, as well as 13 TeV data for $|\eta| < 0.8$ and 2.5.

1 Introduction

The charged particle multiplicity distribution is a global observable of the final state of high energy collisions. The shape of the distribution offers information on the dynamics of particle production processes as well as any correlation in particle production. Photon counting distributions in optical experiment and theory has shown that the independent production of photons lead to a Poisson distributed counting distribution. In the context of high energy proton-proton (pp) and electron-positron (e^+e^-) collisions, the charged-particle multiplicity distributions are most often described using the Negative Binomial Distribution (NBD) [1] and even the double NBD [2]. The NBD describes cascades (or showers) of particles originating from an initial number of ancestor particles produced after the collision [3].

The total multiplicity distribution of partons within a jet [4] can be expressed as Markov branching processes of quarks (q) and gluons (g) [5]. Considering the processes $g \to g+g$, $q \to q+g$, $g \to q+q$, and $g \to g+g+g$ with average branching probabilities of A, $\tilde{A}$, B, and C respectively, the QCD parton branching differential equation for producing m quarks and n gluons after branching [6–8] is given as:

$$\begin{aligned}\frac{\partial P_{m,n}(t)}{\partial t} = &- \tilde{A}mP_{m,n}(t) + \tilde{A}mP_{m,n-1}(t) \\ &- AnP_{m,n}(t) + A(n-1)P_{m,n-1}(t) \\ &- BnP_{m,n}(t) + B(n+1)P_{m,n+1}(t) \\ &- CnP_{m,n}(t) + C(n-2)P_{m,n-2}(t),\end{aligned} \tag{1}$$

where

$$t = \frac{6}{11N_c - 2N_f} \ln\left[\frac{\ln\left(Q^2/\mu^2\right)}{\ln\left(Q_0^2/\mu^2\right)}\right] \tag{2}$$

⋆e-mail: w.y.wang@nus.edu.sg

© The Authors, published by EDP Sciences. This is an open access article distributed under the terms of the Creative Commons Attribution License 4.0 (http://creativecommons.org/licenses/by/4.0/).

and Q is the initial parton invariant mass, Q_0 is the hadronisation mass, μ is a QCD mass scale of a few GeV, N_c is the number of colours, and N_f is the number of flavours.

The above equation yields the Generalised Multiplicity Distribution (GMD) [6–8] as a possible exact solution in the case where $B = C = 0$. Given an initial m and k' number of quarks (q) and gluons (g), the probability of obtaining n final state particles is given by the GMD:

$$P_{GMD}(n; p, k, k') = \frac{\Gamma(n+k)}{\Gamma(n-k'+1)\Gamma(k'+k)}(1-p)^{n-k'}(p)^{k'+k}, \tag{3}$$

where $k = m\tilde{A}/A$, $p = \exp(-At) = (k'+k)/(\overline{n}+k)$, and $\overline{n}$ is the average number of observed final state particles. Special cases of the GMD are the NBD when $k' = 0$, and Fury-Yule Distribution [9] when $k = 0$. Furthermore, the NBD converges to the PD for $k \to \infty$.

However, an exact solution including all the processes in equation 1 has never been derived. In this paper, an asymptotic solution is obtained using the method of Laplace transformation and the saddle point approximation. The result corrects an error by A. H. Chan and C. K. Chew [10] in the use of the Stirling's formula. The resulting asymptotic multiplicity distribution is then applied to measurements of charged particle multiplicity distributions in pp collisions by the ALICE [11], ATLAS [12], and CMS [13] collaborations at the LHC.

2 Asymptotic solution of the QCD parton branching equation

Using a continuous-variable approximation, where n is replaced by $x \in \mathbb{R}$ and performing a Laplace transform with initial condition , equation 1 becomes

$$\begin{aligned}-P(x, t=0) + sP^*(x,s) = &- A^\dagger P^*(x,s) + A^\dagger P^*(x-1,s) \\ &- AxP^*(x,s) + A(x-1)P^*(x-1,s) \\ &- BxP^*(x,s) + B(x+1)P^*(x+1,s) \\ &- CxP^*(x,s) + C(x-2)P^*(x-2,s),\end{aligned} \tag{4}$$

with $A^\dagger = \tilde{A}m$.

When $x > k'$, the initial condition ensures that $P(x > k', t = 0) = 0$. In the case where $x = k'$,

$$-P(x=k', t=0) + sP^*(k',s) = -\left(Ak' + A^\dagger + Bk' + Ck'\right)P^*(k',s), \tag{5}$$

$$P^*(k',s) = \frac{1}{s + A^\dagger + (A+B+C)k'}. \tag{6}$$

To further simplify equation 4 when $x > k'$, let $\ln(P^*(x,s)) = L(x)$ and approximate $L(x+1) \sim L(x)$ and $L(x-2) \sim L(x-1)$ as we expect large number of gluons to be created at high energies. Equation 4 then becomes

$$\begin{aligned}se^{L(x)} = &- (Ax + A^\dagger + Bx + Cx)e^{L(x)} \\ &+ A(x-1)e^{L(x)-L'(x)} + A^\dagger e^{L(x)-L'(x)} \\ &+ B(x+1)e^{L(x)} + C(x-2)e^{L(x)-L'(x)},\end{aligned} \tag{7}$$

where a Taylor expansion has been made for all the $L(x-1)$ terms about $L(x)$ up to the first differential coefficient.

Rearranging, the equation

$$-\frac{\partial L(x)}{\partial x} = \ln\left[\frac{s + A^{\dagger} - B + (A + C)x}{(A + C)x - A + A^{\dagger} - 2C}\right] \tag{8}$$

can be obtained.

To simplify the analysis, let

$$\begin{aligned} \alpha &= A^{\dagger} - B, \\ \beta &= A + C, \\ \gamma &= A^{\dagger} - A - 2C, \\ D &= A + B + C. \end{aligned} \tag{9}$$

Integrating equation 8, and letting $L_1(k', s) = \ln P^*(x = k', s)$ (Eq. 6),

$$L_1(x, s) = -\int_{k'}^{x} \ln\left(\frac{s + \alpha + \beta\omega}{\gamma + \beta\omega}\right) d\omega - \ln\left(s + A^{\dagger} + Dk'\right), \tag{10}$$

A better approximation can then be obtained by expanding the Taylor series up to the second term,

$$-\frac{\partial L(x)}{\partial x} + \frac{1}{2}\frac{\partial^2 L(x)}{\partial x^2} = \ln\left(\frac{s + \alpha + \beta x}{\gamma + \beta x}\right). \tag{11}$$

Differentiating equation 8, and substituting into equation 11,

$$-\frac{\partial L(x)}{\partial x} = \ln\left(\frac{s + \alpha + \beta x}{\gamma + \beta x}\right) - \frac{\beta}{2}\left(\frac{1}{\gamma + \beta x} - \frac{1}{s + \alpha + \beta x}\right), \tag{12}$$

which on integrating gives

$$L_2(x, s) = -\int_{k'}^{x} \ln\left(\frac{s + \alpha + \beta\omega}{\gamma + \beta\omega}\right) d\omega + \ln\left[\frac{(\gamma + \beta x)^{1/2}(s + \alpha + \beta k')^{1/2}}{(\gamma + \beta k')^{1/2}(s + \alpha + \beta x)^{1/2}(s + A^{\dagger} + Dk')}\right]. \tag{13}$$

The required asymptotic solution is obtained by the inverse Laplace transform

$$P_2(x, t) = \frac{1}{2\pi i}\int_{C-i\infty}^{C+i\infty} g(s)e^{st-f(s)} ds \tag{14}$$

where

$$g(s) = \frac{(\gamma + \beta x)^{1/2}(s + \alpha + \beta k')^{1/2}}{(\gamma + \beta k')^{1/2}(s + \alpha + \beta x)^{1/2}(s + A^{\dagger} + Dk')} \tag{15}$$

and

$$f(s) = \int_{k'}^{x} \ln\left(\frac{s + \alpha + \beta\omega}{\gamma + \beta\omega}\right) d\omega. \tag{16}$$

The integral of equation 14 is expected to be dominated by the highest saddle point at $s = s_0$ for large t. Using the saddle point approximation, the asymptotic solution is given by

$$P_2(x, t) = \frac{g(s_0)e^{s_0 t - f(s_0)}}{\sqrt{-2\pi f''(s_0)}}. \tag{17}$$

Substituting the corresponding expressions for $g(s_0)$, $f(s_0)$, and $f''(s_0)$, equation 17 becomes

$$P_2(x,t) = \frac{\left[\frac{(\gamma+\beta x)(s_0+\alpha+\beta k')}{(\gamma+\beta k')(s_0+\alpha+\beta x)}\right]^{1/2} \frac{1}{s_0+A^\dagger+Dk'} e^{s_0 t - \int_{k'}^{x} \ln\left(\frac{s_0+\alpha+\beta\omega}{\gamma+\beta k'}\right) d\omega}}{\sqrt{2\pi \int_{k'}^{x} \frac{1}{(s_0+\alpha+\beta\omega)^2} d\omega}}, \tag{18}$$

which can be further evaluated by looking for the stationary function of the exponential term in the above equation

$$\frac{\partial}{\partial s_0}\left[e^{s_0 t - \int_{k'}^{x} \ln\left(\frac{s_0+\alpha+\beta\omega}{\gamma+\beta k'}\right) d\omega}\right] = 0. \tag{19}$$

This gives the expression

$$s_0 = \frac{(x-k')\beta}{1-e^{-\beta t}} - \alpha - \beta x, \tag{20}$$

which by letting $k = \frac{\gamma}{\beta+1}$, we obtain

$$P_2(x,t) = \frac{e^{-t(\alpha+\beta k'+\beta)}\left(1-e^{-\beta t}\right)^{x-k'}}{\frac{B}{\beta}(1+k')(1-e^{-\beta t}) + (x-k')e^{-\beta t}} \times (x-k') \\ \times \frac{(k+x-1)^{k+x-1/2}}{(2\pi)^{1/2}(x-k')^{x-k'+1/2}(k+k'-1)^{k+k'-1/2}}. \tag{21}$$

For large values of x, the Stirling's approximation is given by $x! = \frac{x^{x+1/2}}{e^x}(2\pi)^{1/2}$. Taking note that $x! = \Gamma(x+1)$ and replacing x by n, the final form of the asymptotic solution to the QCD parton branching equation is obtained as

$$P(n,t) = \frac{e^{-t(\alpha+\beta k'+\beta)}(1-e^{-\beta t})^{n-k'}}{\frac{B}{\beta}(1+k')(1-e^{-\beta t}) + (n-k')e^{-\beta t}} \times (n-k') \times \frac{\Gamma(n+k)}{\Gamma(n-k'+1)\Gamma(k'+k)}. \tag{22}$$

Replacing α, β, γ, and D with their corresponding expressions in equation 9 and letting $A = At$, $A^\dagger = A^\dagger t$, $B = Bt$, and $C = Ct$,

$$P(n) = \frac{e^{-(A^\dagger+A-B+C)}e^{-(A+C)k'}\left(1-e^{-(A+C)}\right)^{n-k'}}{\frac{B}{A+C}(1+k')\left(1-e^{-(A+C)}\right) + (n-k')e^{-(A+C)}} \\ \times (n-k') \times \frac{\Gamma\left(n+\frac{A^\dagger-C}{A+C}\right)}{\Gamma(n-k'+1)\Gamma\left(k'+\frac{A^\dagger-C}{A+C}\right)}. \tag{23}$$

With the last substitution, the branching probabilities $A, A^\dagger, B, C$ are now reinterpreted as the branching probabilities scaled by the QCD evolution parameter. The ratios between the branching probabilities (e.g. $A^\dagger/A$) remain fixed regardless of the scale.

Applied to charged particle multiplicity data, equation 23 represents the probability of obtaining n charged particles from the hadronisation of gluons in QCD jets, via the branching processes associated with the parameters A, $A^\dagger$, B, and C.

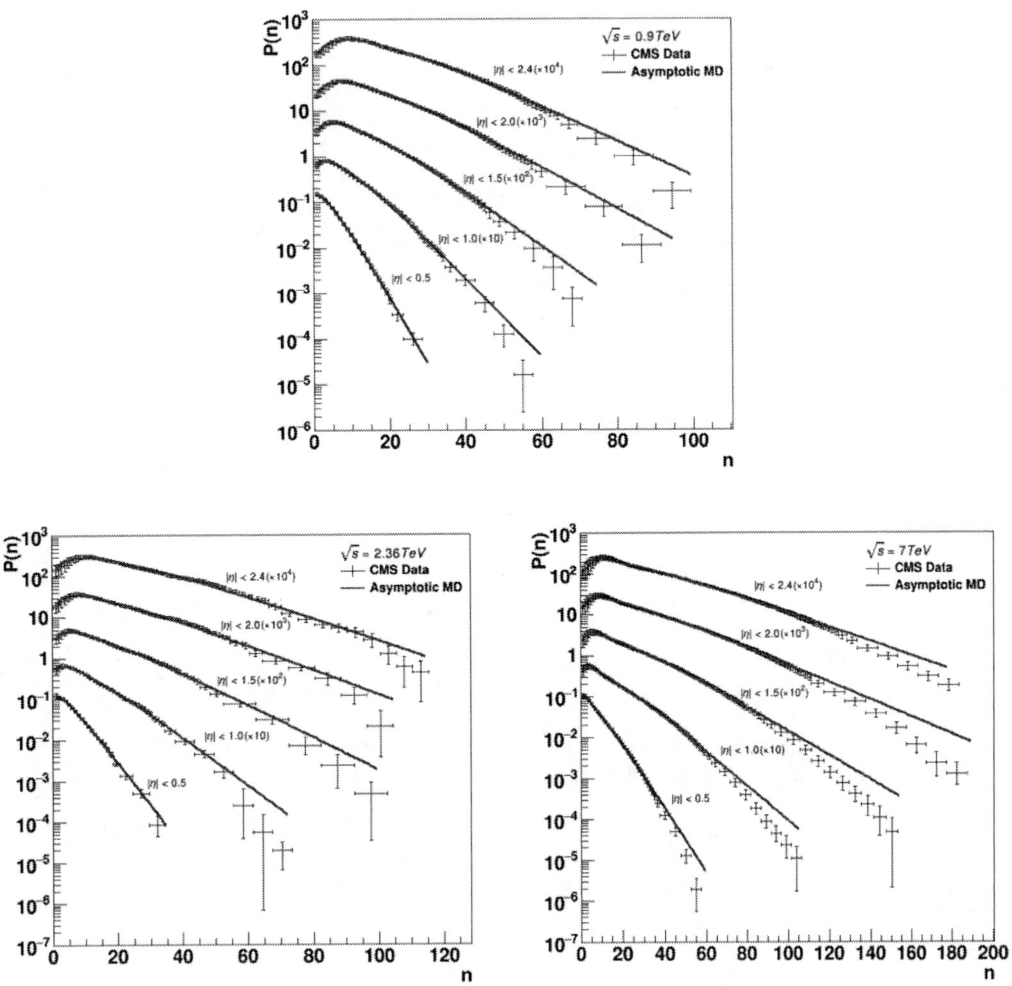

Figure 1. The asymptotic multiplicity distribution compared to CMS charged particle multiplicity data at (top) $\sqrt{s} = 0.9$, (bottom left) 2.36, and (bottom right) 7 TeV for all pseudorapidity windows $|\eta| < 0.5, 1.0, 1.5, 2.0, 2.4$. The values for successively larger pseudorapidities are scaled by factors of 10 for clarity in presentation.

3 Results

Figure 1 shows charged particle multiplicity distributions in *pp* collisions for five pseudorapidity windows $|\eta| < 0.5, 1.0, 1.5, 2.0$, and 2.4 as measured by the CMS collboration at $\sqrt{s} = 0.9, 2.36$, and 7 TeV. Comparison is made with the best fit asymptotic multiplicity distribution. The corresponding parameter values are given in table 1.

Similarly, figure 2 (top) shows charged particle multiplicity distributions for $|\eta| < 0.5, 1.0$, and 1.5 at $\sqrt{s} = 8$ TeV measured by the ALICE collaboration. Figure 2 (bottom) shows the same distributions for $|\eta| < 0.8$ and 2.5 at $\sqrt{s} = 13$ TeV measured by the ATLAS collaboration.

The asymptotic multiplicity distribution is found to describe data from CMS collaboration well. However it is noted that there is increasing difficulty in describing the tail of the distribution as the

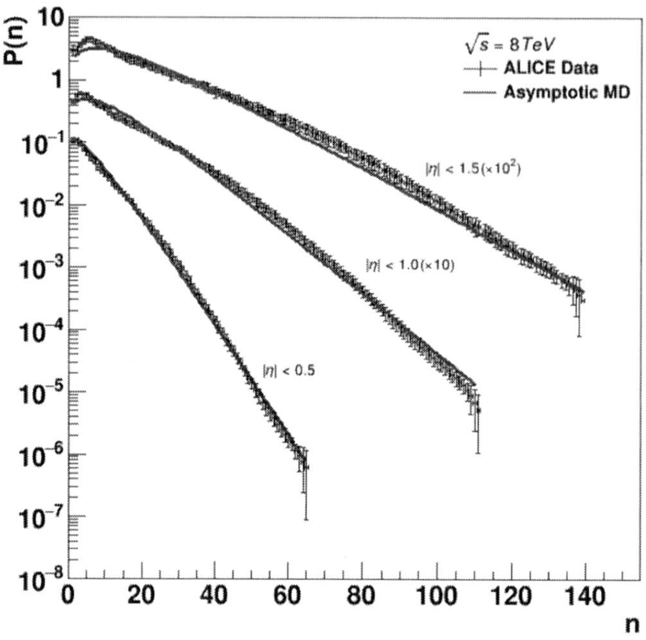

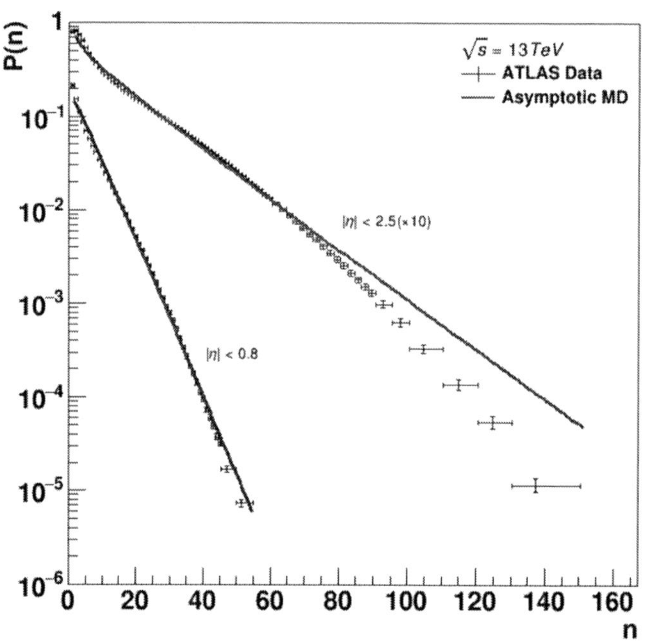

Figure 2. The asymptotic multiplicity distribution compared to charged particle multiplicity distributions measured by (top) ALICE at $\sqrt{s}$ = 8 TeV for $|\eta| < 0.5, 1.0, 1.5$ and by (bottom) ATLAS at $\sqrt{s}$ = 13 TeV for $|\eta| < 0.8, 2.5$. The values for successively larger pseudorapidities are scaled by factors of 10 for clarity in presentation.

pseudorapidity range increases. This suggests that the asymptotic multiplicity distribution is insufficient to describe all the processes involved in pp collisions and the subsequent fragmentation. At $\sqrt{s}$ = 8 TeV the asymptotic multiplicity distribution appears to describe the tail of the distribution well. However, it poorly models the data at mid-multiplicity ($n \sim 70$) and low-multiplicity ($n \sim 10$) regimes. At $\sqrt{s}$ = 13 TeV, the data is well modelled for $|\eta| < 0.8$ but described poorly for $|\eta| < 2.5$, where the tail of the distribution is overestimated from $n \sim 70$.

Referring to table 1, the values of k', B and C generally tend to 0 or some small value for centre-of-mass energies of up to 8 TeV. This implies that quark pair creation and the four-gluon vertex occur at a relatively insignificant rate during the branching process. Quark bremsstrahlung and gluon fission dominate the branching process, and the entire process occurs with no initial gluons. Hence at the above centre-of-mass energies, charged particle multiplicities are still better described by the NBD ($k' = 0$). However the non-zero k' values at $\sqrt{s}$ = 13 TeV seem to suggest that more initial gluons are produced at higher centre-of-mass energies.

4 Conclusion

From the QCD parton branching equation, the asymptotic multiplicity distribution is derived using the method of Laplace transform and saddle point approximation. Revision of the derivation from previous work is done with the correct Stirling's formula. Starting with an initial number of gluons k', the distribution describes the probability of producing n gluons via four processes that contribute to the overall quark and gluon distribution inside QCD jets, with each process occuring with a different probability.

Table 1. Parameter values for the asymptotic multiplicity distribution.

$\sqrt{s}$ (TeV)	$\|\eta\|$	k'	$A^{\dagger}$	A	B	C	$\chi2/dof$
0.9 (CMS)	0.5	0	2.1	1.13	0.0	0.06	0.42/17
	1.0	0	3.1	1.59	0.0	0.06	10.43/34
	1.5	0	3.8	1.92	0.0	0.05	16.40/46
	2.0	0	4.3	2.15	0.1	0.06	17.21/56
	2.4	0	4.6	2.29	0.1	0.06	24.91/62
2.36 (CMS)	0.5	0	2.2	1.44	0.0	0.04	1.92/17
	1.0	0	3.2	1.95	0.0	0.04	15.32/34
	1.5	0	3.6	2.32	0.1	0.05	14.20/44
	2.0	0	3.9	2.54	0.2	0.1	23.68/54
	2.4	0	4.6	2.65	0.1	0.05	35.94/62
7 (CMS)	0.5	0	2.4	1.73	0.00	0.04	20.60/35
	1.0	0	3.4	2.27	0.00	0.05	77.55/64
	1.5	0	4.0	2.61	0.00	0.06	140.55/89
	2.0	0	4.2	2.97	0.00	0.02	112.71/109
	2.4	0	4.3	3.16	0.04	0.02	97.66/121
8 (ALICE)	0.5	0	2.47	1.65	0.00	0.00	55.79/60
	1.0	0	3.60	2.15	0.00	0.05	206.26/106
	1.5	0	4.3	2.44	0.00	0.10	286.34/134
13 (ATLAS)	0.8	0.997	0.17	1.70	0.00	0.02	45.63/42
	2.5	0.859	0.00	2.85	0.00	0.00	336.50/76

The asymptotic multiplicity distribution is applied to charged particle multiplicity distribution data at $\sqrt{s} = 0.9, 2.36, 7, 8$, and 13 TeV from the ALICE, ATLAS, and CMS collaborations. The parameters k', $A^\dagger$, A, B, and C are estimated and the result is found to describe the observed multiplicity well at $\sqrt{s} = 0.9, 2.36$, and 7 TeV well except for the tail of the distribution. The parameter k' reduces to 0 (yielding NBD), suggesting that branching processes up to 7 TeV are dominated by gluon fission and quark bremsstrahlung with negligible initial gluon densities.

The asymptotic multiplicity distribution is found to be inadequate in modelling observed multiplicities at $\sqrt{s} = 8$ TeV and more so at 13 TeV, where it fails to describe both the mid- and high-multiplicity regions at larger pseudorapidity intervals. This hints at the presence of additional processes which should be accounted for.

5 Acknowledgements

The authors would like to thank the National University of Singapore, and the support and helpful discussions with colleagues. We would also like to congratulate the ALICE, ATLAS, and CMS collaboration for their analysis on the multiplicity distributions and thank them for the well organised release of data points to public.

References

[1] P. Ghosh, Phys. Rev. D **85**, 054017 (2012)
[2] ALICE Collaboration (2017), `arXiv:1708.01435v1`
[3] A. Giovannini, L. Van Hove, A. Phys. Pol. B **19**, 495 (1988)
[4] K. Konishi, A. Ukawa, G. Veneziano, Nucl. Phys. B **157**, 45 (1979)
[5] A. Giovannini, Nucl. Phys. B **161**, 429 (1979)
[6] C.K. Chew, D. Kiang, H. Zhou, Phys. Lett. B **186**, 411 (1987)
[7] A.H. Chan, C.K. Chew, Phys. Rev. D **41**, 851 (1990)
[8] A.H. Chan, C.K. Chew, Z. Phys. C **55**, 503 (1992)
[9] R.C. Hwa, C.S. Lam, Phys. Lett. B **173**, 346 (1986)
[10] A.H. Chan, C.K. Chew, J. Math. Phys. **33**, 9 (1992)
[11] ALICE Collaboration, Eur. Phys. J. C **77**, 33 (2017)
[12] ATLAS Collaboration, Phys. Lett. B **758**, 67 (2016)
[13] CMS Collaboration, JHEP **01**, 79 (2011)

What can we learn from femtoscopic and angular correlations of identified particles in ALICE?

Łukasz Kamil Graczykowski[1,⋆]
for the ALICE Collaboration

[1] *Faculty of Physics, Warsaw University of Technology*
Koszykowa 75, 00-662 Warszawa, Poland

Abstract. Two-particle correlations have proven to be a robust tool which allow the exploration of many physics phenomena present in high-energy particle collisions. In this work, two techniques, namely femtoscopic and angular correlations of identified particles, are discussed the most recent results of both of them obtained from pp and Pb–Pb collisions recorded by the ALICE detector are presented.

1 Introduction

ALICE (A Large Ion Collider Experiment) [1] is one of the four large detectors located at the LHC (Large Hadron Collider) facility [2] at CERN – the European Organization for Nuclear Research. The detector was specifically designed to investigate ultra-relativistic Pb–Pb collisions, where the very high temperatures and energy densities allow for the production of the Quark–Gluon Plasma (QGP) [3–7]. In addition, the ALICE detector has an extensive physics program analyzing proton-proton and proton-lead collisions which are also delivered by the LHC.

The ALICE detector has some unique features. Among the most important is excellent tracking and particle identification (PID) over a broad momentum range which makes a variety of different measurements possible, not only connected to the studies of the QGP. Two-particle correlation measurements, namely femtoscopy and angular correlations, have been performed for different particle pairs and allowed for calculations of previously poorly known pair cross sections and shed new light on particle fragmentation models. Both techniques reported in this paper use the same definition of correlation function, expressed as a ratio of correlated pairs from the same event to the reference distribution containing pairs of particles coming from different events. The correlation functions are normalized in the way that the reported distributions represent the ratio of probabilities.

2 Femtoscopy

Femtoscopy is the term describing two-particle correlations measured as a function of relative momentum $q = 2 \cdot k^* = p_1 - p_2$ which are used to gather information about the spatio-temporal properties

⋆e-mail: lukasz.graczykowski@pw.edu.pl

© The Authors, published by EDP Sciences. This is an open access article distributed under the terms of the Creative Commons Attribution License 4.0 (http://creativecommons.org/licenses/by/4.0/).

of the collision. The femtoscopic correlation is related to the source and the final-state interaction via the Koonin-Pratt equation [8, 9]

$$C(k^*) = \int S(\vec{r}^*) \left|\Psi_{-\vec{k}^*}(\vec{r}^*)\right| d^3\vec{r}^*, \tag{1}$$

where $S(\vec{r}^*)$ is the source function, typically a one-dimensional Gaussian $S(\vec{r}^*) \sim e^{-|\vec{r}^*|^2/(4R^2)}$ (where R is the width corresponding to the source size, sometimes referred to as the "HBT radius"), $|\vec{r}^*|$ is the relative distance in the pair-rest frame (PRF) and $\Psi_{-\vec{k}^*}(\vec{r}^*)$ is the two-particle interaction kernel (solution to the scattering problem). Generally, $\Psi_{-\vec{k}^*}(\vec{r}^*)$ may include effects arising due to the effects of quantum statistics and final-state interactions, depending on the studied pair.

Traditionally, in femtoscopy the form of $\Psi_{-\vec{k}^*}(\vec{r}^*)$ is known and the parameters of $S(\vec{r}^*)$, which describe the shape and evolution of the source, are extracted. In fact, Eq. (1) allows us to utilize femtoscopic technique in a novel way, constraining $S(\vec{r}^*)$ from other measurements and extracting information on $\Psi_{-\vec{k}^*}(\vec{r}^*)$. In particular, if only the strong interaction is present, $\Psi_{-\vec{k}^*}(\vec{r}^*)$ has the following form of a superposition of a plane wave and a spherical wave

$$\Psi_{-\vec{k}^*}(\vec{r}^*) = e^{-i\vec{k}^*\cdot\vec{r}^*} + f(k^*)\frac{e^{ik^*\cdot r^*}}{r^*}, \tag{2}$$

where $f(k^*)$ is the s-wave scattering amplitude which has a different form depending on the particle pair under study. Analytical solutions of Eq. (1) with $\Psi_{-\vec{k}^*}(\vec{r}^*)$ accounting for the strong interaction, in the effective range approximation, were introduced for different pair systems by R. Lednicky [10, 11].

This paper focuses on two such measurements are presented. Firstly, correlations of baryons with anti-baryons are discussed in Sec. 2.1 and secondly, correlations between neutral and charged kaons are presented in Sec. 2.2.

2.1 Baryon–anti-baryon pairs

One of the most important problems in QCD is to understand the interaction between two baryons. While significant amount of experimental data exists for baryon–baryon pairs [12], no measurements exist for baryon–anti-baryon systems other than $\mathrm{p\bar{p}}$, $\mathrm{p\bar{d}}$, and $\mathrm{p\bar{n}}$. UrQMD, a current re-scattering model widely used in the field of collisions of heavy ions, assumes baryon–anti-baryon interaction from scaling of $\mathrm{p\bar{p}}$ interaction parameters on the basis of the Additive Quark Model [13].

In general, femtoscopy can be used to extract the interaction parameters of such systems. The scattering amplitude $f(k^*)$ in Eq. (2) in this scenario becomes (in the effective range approximation)

$$f(k^*) = \left[\frac{1}{f_0} + \frac{1}{2}d_0 k^{*2} - ik^*\right]^{-1}, \tag{3}$$

where f_0 is the scattering length and d_0 is the effective range of the interaction. An attempt to experimentally constrain the parameters of $\mathrm{p\bar{\Lambda}}$ pairs using femtoscopy was made by the STAR experiment at RHIC [14]. In particular, both real and imaginary parts of f_0 were extracted manifesting the presence of a significant non-elastic channel, annihilation, while the extracted source size ("HBT radius") was significantly lower than the results from baryon–baryon pairs at similar transverse mass m_T. These results seemed to be in contradiction to hydrodynamic model calculations which predict $1/m_\mathrm{T}$ scaling which has been confirmed by RHIC and LHC results of meson and baryon correlations. This inconsistency was studied later in Ref. [15] which concludes that the so-called residual correlations[1] were

[1]Correlations arising from admixture in the sample of particles of the same type being products of heavier baryon decays.

not properly included in the original study. The results were re-analyzed with the proper accounting for those correlations.

In this paper, the recent ALICE measurements extending the femtoscopic studies of baryon–anti-baryon interactions on $\Lambda\overline{\Lambda}$ pairs are discussed. In addition, a procedure to fit simultaneously all measured baryon–anti-baryon systems is introduced. The analysis was carried out for $p\overline{p}$, $p\overline{\Lambda}$ ($\overline{p}\Lambda$), and $\Lambda\overline{\Lambda}$ pairs on Pb–Pb collisions at the LHC recorded at the center-of-mass energies of $\sqrt{s_{NN}}$ = 2.76 TeV and $\sqrt{s_{NN}}$ = 5.02 TeV, respectively. In addition, six centrality ranges were introduced, giving in total 36 measured baryon–anti-baryon correlation functions. The χ^2 was calculated globally, representing the minimum of the fit to all functions together. For each pair a proper form of correlation function was chosen, taking into account the strong interaction as well as Coulomb repulsion in case of the $p\overline{p}$ system. Residual correlations were also taken into account with the fractions of residual pairs estimated from the AMPT model [16] and decay kinematics from the THERMINATOR 2 model [17]. The parameters of the interaction of residual pairs are the same as those for experimentally measured pairs, and are free parameters of the fit. In addition, the scaling of "HBT radii" with m_T and the cube root of multiplicity (relating to different centralities) was assumed in order to constrain the radius. Finally, three sets of interaction parameters were introduced, namely for $p\overline{\Lambda}$ ($\overline{p}\Lambda$) and $\Lambda\overline{\Lambda}$ pairs as well as parameters of an effective potential for other baryon–anti-baryon pairs not measured directly but present as residual correlation contributions.

Figure 1 presents baryon–anti-baryon femtoscopic correlation functions measured in Pb–Pb collisions at both studied collision energies. The Lednicky fit shows a good description of experimental data. The scattering parameters – the results of the fit, compared to measurements from other experiments and model predictions, are shown in Fig. 2.

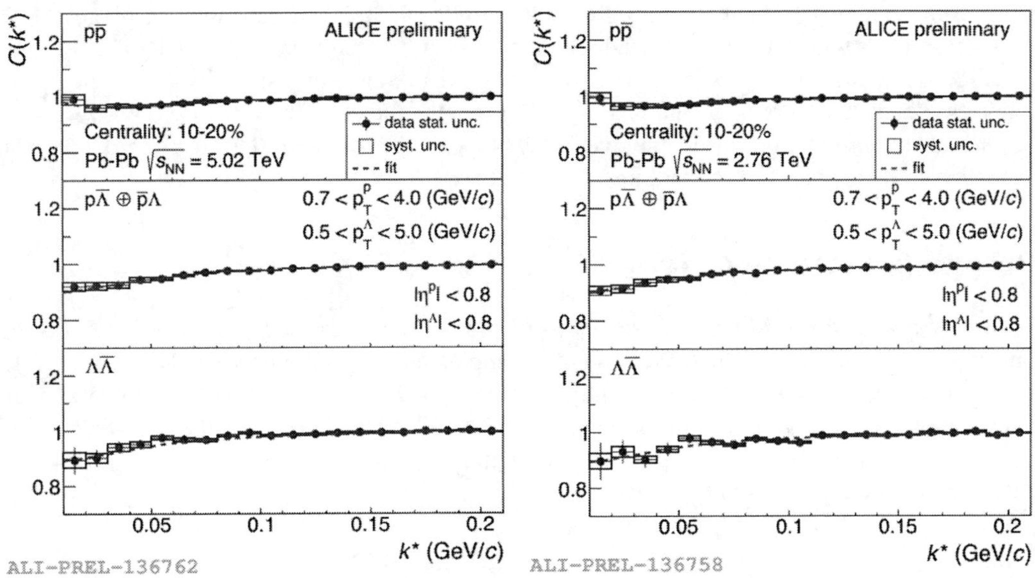

Figure 1. Femtoscopic correlation functions for $p\overline{p}$, $p\overline{\Lambda} + \overline{p}\Lambda$, and $\Lambda\overline{\Lambda}$ pairs from Pb–Pb collisions at $\sqrt{s_{NN}}$ = 5.02 TeV (left) and $\sqrt{s_{NN}}$ = 2.76 TeV (right). Red lines represent Lednicky fits.

There are a few conclusions that can be drawn from these results: (I) All the extracted scattering parameters are similar to each other meaning that the interaction for different baryon–anti-baryon

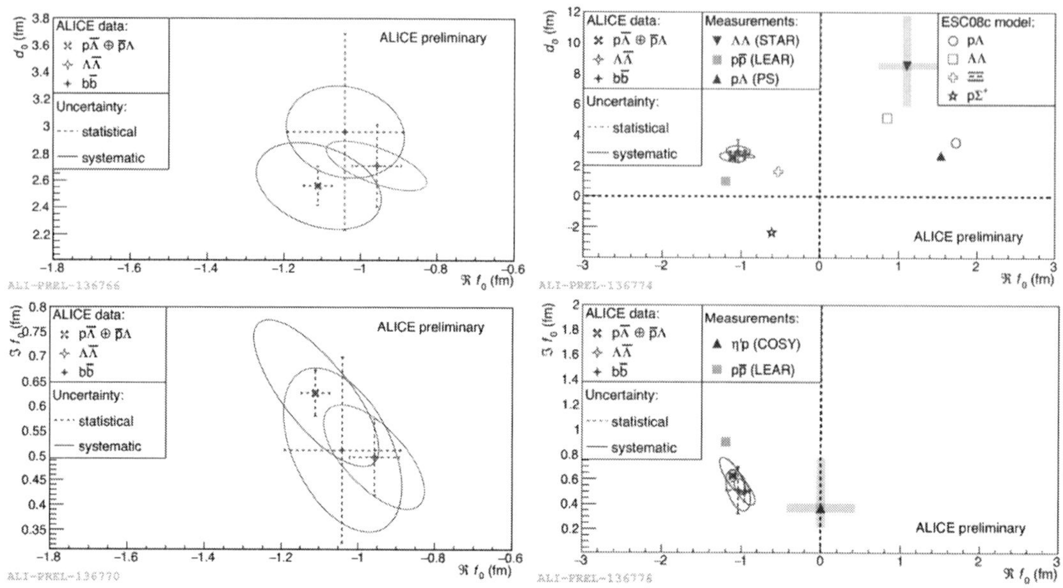

Figure 2. (Left) Scattering parameters f_0 and d_0 for $\mathrm{p}\overline{\Lambda} + \overline{\mathrm{p}}\Lambda$, $\Lambda\overline{\Lambda}$ pairs and effective scattering parameters describing other baryon–anti-baryon pairs. (Right) Comparison to other measurements and model predictions.

pairs is comparable. (II) The real part of the scattering length $\Re f_0$ is negative which suggests either the existence of baryon–anti-baryon bound states or a repulsive strong interaction. (III) The imaginary part of the scattering length $\Im f_0$ is positive, reflecting the presence of a significant inelastic channel (in this case baryon–anti-baryon annihilation). (IV) Finally, the effective range d_0 is not zero which suggests that the assumptions made in previous analyses, where it was fixed to zero in order to reduce the number of free parameters, are not justified [14, 15].

2.2 Neutral kaon-charged kaon pairs

The first measurements of $\mathrm{K}^0_\mathrm{S}\mathrm{K}^\pm$ femtoscopic correlations were performed by ALICE in Pb–Pb collisions at $\sqrt{s_{\mathrm{NN}}} = 5.02$ TeV [18]. In addition, the complementary preliminary results in $\sqrt{s} = 7$ TeV pp collisions from ALICE have also recently become available. The only allowed process (source of correlation) which contributes to the femtoscopic correlation function for these particle pairs is the strong final-state interaction. It proceeds through the $\mathrm{a}_0(980)$ near-threshold resonance with the scattering amplitude $f(k^*)$ from Eq. (2) of the form:

$$f(k^*) = \frac{\gamma_{\mathrm{a_0}\to\mathrm{K}\overline{\mathrm{K}}}}{m^2_{\mathrm{a_0}} - s - i(\gamma_{\mathrm{a_0}\to\mathrm{K}\overline{\mathrm{K}}}k^* + \gamma_{\mathrm{a_0}\to\pi\eta}k_{\pi\eta})}, \tag{4}$$

where $m_{\mathrm{a_0}}$ is the mass of the $\mathrm{a_0}$ resonance, $\gamma_{\mathrm{a_0}\to\mathrm{K}\overline{\mathrm{K}}}$ and $\gamma_{\mathrm{a_0}\to\pi\eta}$ are the $\mathrm{a_0}$ coupling parameters of the $\mathrm{a_0}$ resonance to the $\mathrm{K}^0_\mathrm{S}\mathrm{K}^-$ (or $\overline{\mathrm{K}}^0\mathrm{K}^+$) and $\pi\eta$ channels, respectively. The $k_{\pi\eta}$ is the momentum in the $\mathrm{a_0} \to \pi\eta$ channel while s is defined as $s = 4(m^2_{\mathrm{K}^0} + k^{*2})$. In this analysis several sets of coupling parameters of $\mathrm{a_0}$ (in the order of lower to higher values) taken from [19] ("Martin"), [20] ("Antonelli"), [21] ("Achasov1"), and [22] ("Achasov2") were studied.

The main motivation for studying the $K^0_S K^{\pm}$ final-state interaction is the possibility of constraining the coupling parameters of a_0 which in fact is considered in the literature as a potential tetraquark state, a hadron composed of 4 constituent quarks, or a "$\overline{K} - K$ molecule" [19–24].

The advantage of ALICE is the possibility of performing the same analysis on both collision systems, where different underlying physics phenomena are present. This allows to cross-check the two results and better constrain the extracted a_0 parameters. The tetraquark hypothesis could also be supported if the measurements lead to the same results.

Figures 3 and 4 show femtoscopic correlation functions (corrected for min-jet underlying correlations) of $K^0_S K^{\pm}$ pairs measured in Pb–Pb and pp collision systems, respectively, in different pair transverse momentum $k_T = |p_{T,1} + p_{T,2}|/2$ ranges. Blue points represent the experimental data while the red lines correspond to the fit function. One can clearly see from both figures that the a_0 parameterization of the femtoscopic signal describes very precisely the experimental result. The extracted source sizes R are compared to the results of identical kaon femtoscopy from ALICE [25, 26], which is shown in Fig. 5. The "Martin" parameterization [19], which contains the lowest values of a_0 coupling parameters, gives radii which lie significantly below those of identical kaon femtoscopy. Therefore, higher values of a_0 resonance are favored by the experimental data.

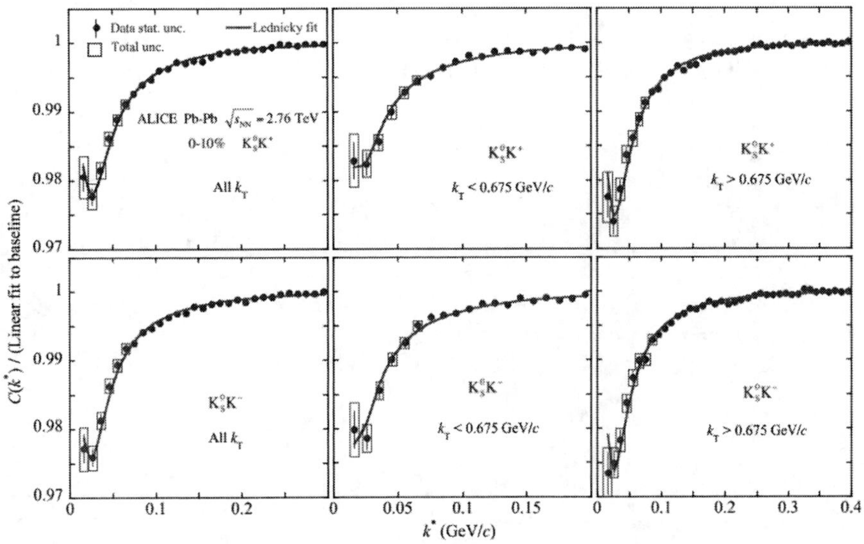

Figure 3. Correlation functions for $K^0_S K^{\pm}$ pairs from Pb–Pb collisions at $\sqrt{s_{NN}} = 2.76$ TeV divided by the linear baseline functions with fits using a_0 parameters from [22] ("Achasov2"). Figure from [18].

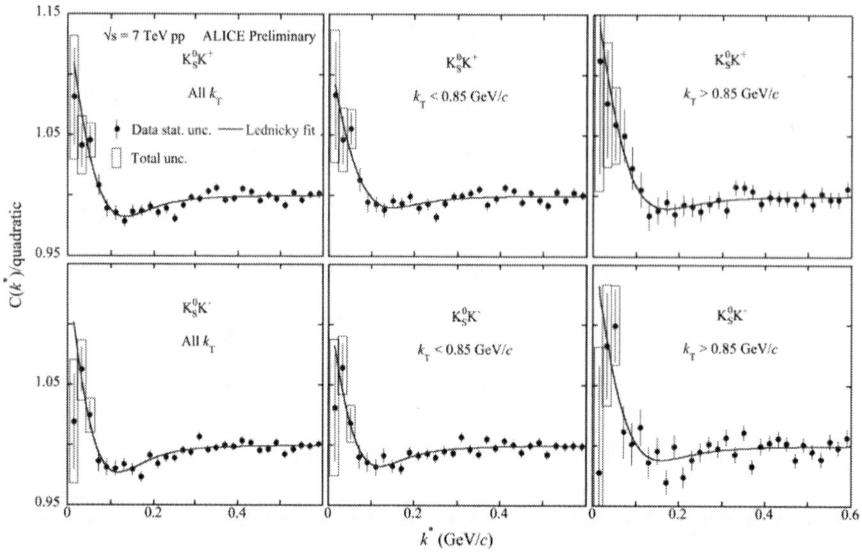

Figure 4. Correlation functions for $K_S^0K^{\pm}$ pairs from pp collisions at $\sqrt{s}$ = 7 TeV divided by the quadratic baseline functions with fits using a_0 parameters from [22] ("Achasov2").

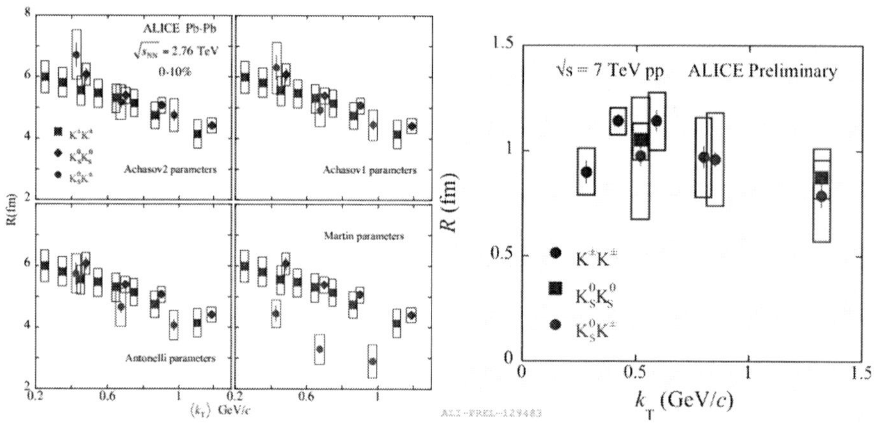

Figure 5. Radius R from $K_S^0K^{\pm}$ femtoscopy in Pb–Pb collisions at $\sqrt{s_{NN}} = 2.76TeV$ (left, plots from [18]) and pp collisions at $\sqrt{s}$ = 7 TeV (right) averaged over the results from fits to $K_S^0K^+$ and $K_S^0K^+$ correlation functions in both collision systems. The a_0 parameters used in the fit come from [19–22] for Pb–Pb data and from [22] ("Achasov2") for pp data.

3 Angular correlations

Two-particle correlations measured as a function of relative pseudorapidity $\Delta\eta = \eta_1 - \eta_2$ and azimuthal angle $\Delta\varphi = \varphi_1 - \varphi_2$, also referred to as $\Delta\eta\Delta\varphi$ correlations, allow the exploration of a wide spectrum of physics phenomena occurring during high-energy particle collisions. In this paper, the most important findings of the angular correlations analysis in pp collisions at $\sqrt{s} = 7$ TeV [27] are presented.

The correlation analysis was performed separately for particle–particle and particle–anti-particle cases. Four particle species, namely pions, charged kaons, protons, and lambdas, were measured at relatively low transverse momenta (up to $p_{\mathrm{T}} = 2.5$ GeV/c) in the central rapidity region $|\eta| < 0.8$. Particle purities exceed 99% for pions and protons, and are around 96% for kaons and 95% for lambdas.

The full two-dimensional $\Delta\eta\Delta\varphi$ distributions can be found elsewhere [27]. In this paper, we focus on one-dimensional $\Delta\eta$-integrated distributions which are directly compared to contemporary Monte Carlo models: PYTHIA6.4 Perugia-0 and PYTHIA6.4 Perugia-2011 [28, 29], PYTHIA8 Monash [30, 31], and PHOJET 1.12 [32]. The experimental results and Monte Carlo calculations for identical particle pairs are shown in Fig. 6, for particle–anti-particle pairs in Fig. 7, while the comparison among different baryon pairs is shown in Fig. 8. One can clearly notice from those plots that all correlations functions except the ones for baryon–baryon (anti-baryon–anti-baryon) case have a typical shape with a near-side peak around $\Delta\varphi = 0$ and an away-side correlation around $\Delta\varphi \approx \pi$. This difference is especially pronounced when the measured correlation functions to Monte Carlo models are compared – none of them is able to reproduce the experimental result even qualitatively. Studies performed with EPOS-LHC [33] and HERWIG [34] also do not agree with the measurements.

One has to note that studies similar to those with similar findings were performed in e^+e^- collisions at the SLAC-PEP accelerator at $\sqrt{s} = 29$ GeV/c collision energy [35]. The reported results were explained on the basis of local conservation of baryon number, i.e. in a single parton fragmentation. In this scenario, the production of two baryons in a single fragmentation requires the production of two anti–baryons, which in total results in four heavy particles. This is a very unlikely scenario at this collision energy, where the initial parton energy is small with respect to energies at the LHC. This hypothesis was supported by the models which seemed to describe those data at the time. Current results from the LHC however suggest that there should be an additional, yet unknown, mechanism which suppresses the probability of producing two baryons close in phase-space. The ruled-out scenarios include: (I) Fermi-Dirac quantum statistics – effect present for proton–lambda pairs, (II) Coulomb repulsion – lambda hyperons are neutral particles, (III) strong final-state interaction – produces a positive correlation, and (IV) local baryon number conservation – all studied models have this mechanism implemented.

4 Summary

The analysis of two-particle correlations, both in relative momentum (femtoscopy) and $\Delta\eta\Delta\varphi$ space (angular correlations), is an interesting topic in high-energy collisions of protons and heavy ions. Reported results show that both techniques are powerful tools which can be used to study a variety of different phenomena. Applications of femtoscopy beyond traditional measurements of the system size allows for the extraction of previously unknown parameters of the strong interaction for certain particle pairs. In this paper, we show the first measurements of $\Lambda\overline{\Lambda}$ interaction parameters obtained together with those for the $p\overline{\Lambda}$ system as well as the effective baryon–anti-baryon interaction for those pairs which cannot be measured directly. The conclusion which can be drawn from this study is that the interaction between those systems is similar, it includes a significant annihilation channel, and the strong force itself is either repulsive or bound states are created. The femtoscopic analysis of neutral kaons paired with charged kaons allows for a constraint of the parameters of the a_0 resonance and are

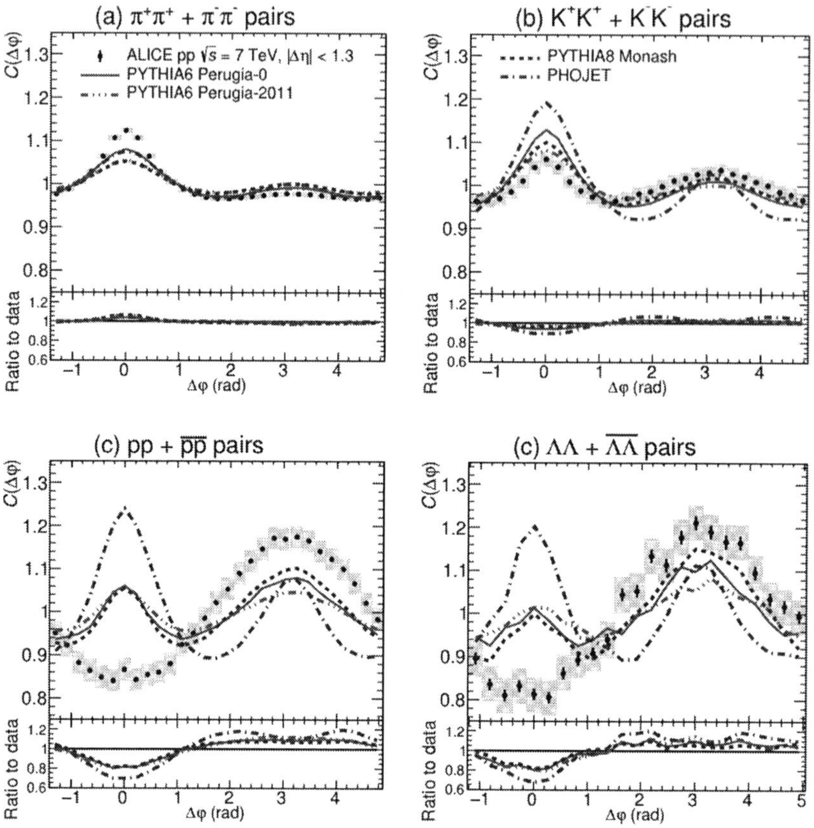

Figure 6. Correlation functions integrated over $\Delta\eta$ for pairs of identical (a) pions, (b) charged kaons, (c) protons, and (d) lambdas from pp collisions at $\sqrt{s}$ = 7 TeV compared to Monte Carlo models. Figure from [27].

compatible with the hypothesis that this particle is indeed a tetraquark state. The angular correlations of baryon–baryon (anti-baryon–anti–baryon) pairs reveal a surprising anti-correlation effect, not explainable by any of currently used Monte Carlo models and the understanding of this result remains an open question.

Acknowledgements

The author acknowledges support from the Polish National Science Centre grants nos. UMO-2015/19/D/ST2/01600, UMO-2014/13/B/ST2/04054, and UMO-2016/22/M/ST2/00176, as well as from the Polish Ministry of Science and Higher Education.

References

[1] K. Aamodt et al. (ALICE), JINST **3**, S08002 (2008)
[2] L. Evans, P. Bryant, JINST **3**, S08001 (2008)
[3] E.V. Shuryak, Phys. Lett. **B78**, 150 (1978), [Yad. Fiz.28,796(1978)]

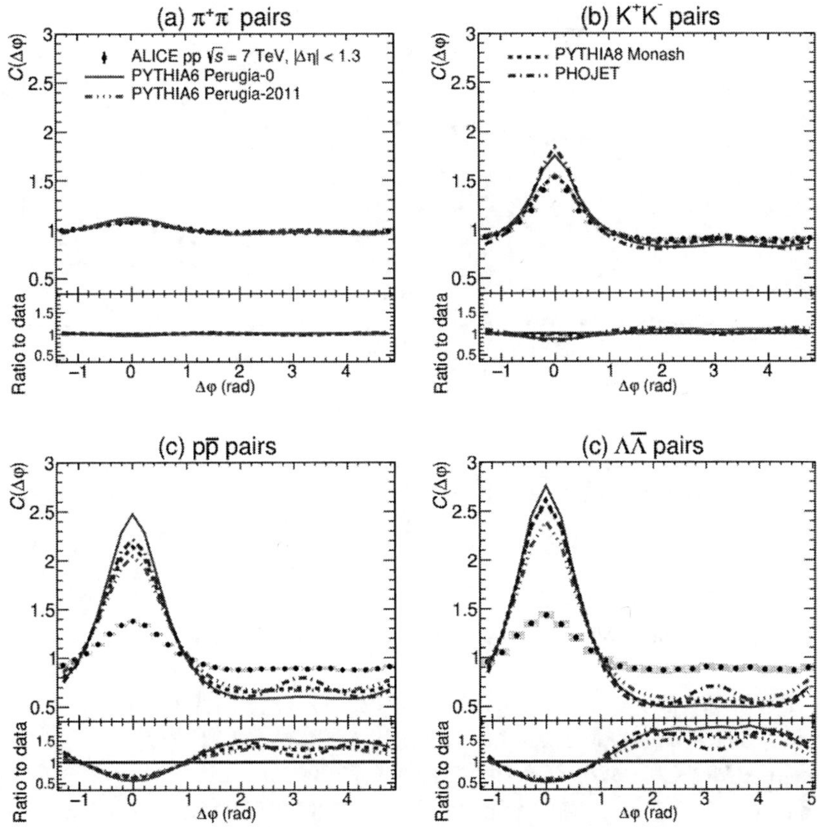

Figure 7. Correlation functions integrated over $\Delta\eta$ for particle–anti-particle pairs of (a) pions, (b) charged kaons, (c) protons, and (d) lambdas from pp collisions at $\sqrt{s}$ = 7 TeV compared to Monte Carlo models. Figure from [27].

[4] J. Adams et al. (STAR), Nucl. Phys. **A757**, 102 (2005)
[5] K. Adcox et al. (PHENIX), Nucl. Phys. **A757**, 184 (2005)
[6] B.B. Back et al., Nucl. Phys. **A757**, 28 (2005)
[7] I. Arsene et al. (BRAHMS), Nucl. Phys. **A757**, 1 (2005)
[8] S.E. Koonin, Phys. Lett. **B70**, 43 (1977)
[9] S. Pratt, T. Csorgo, J. Zimanyi, Phys. Rev. **C42**, 2646 (1990)
[10] R. Lednicky, V.L. Lyuboshits, Sov. J. Nucl. Phys. **35**, 770 (1982), [Yad. Fiz.35,1316(1981)]
[11] R. Lednicky, Nucl. Phys. **A774**, 189 (2006)
[12] C. Patrignani et al. (Particle Data Group), Chin. Phys. **C40**, 100001 (2016)
[13] M. Bleicher et al., J. Phys. **G25**, 1859 (1999)
[14] J. Adams et al. (STAR), Phys. Rev. **C74**, 064906 (2006)
[15] A. Kisiel, H. Zbroszczyk, M. Szymański, Phys. Rev. **C89**, 054916 (2014)
[16] Z.W. Lin, C.M. Ko, B.A. Li, B. Zhang, S. Pal, Phys. Rev. **C72**, 064901 (2005)

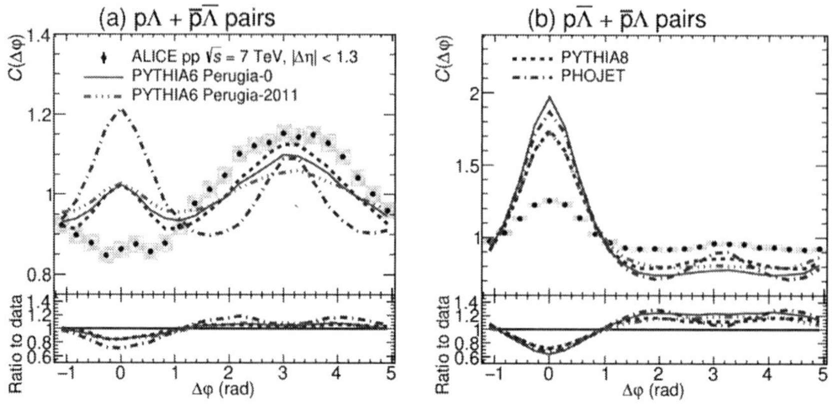

Figure 8. Correlation functions integrated over $\Delta\eta$ for (left) particle–particle and (right) particle–anti-particle correlations of protons with lambdas from pp collisions at $\sqrt{s}$ = 7 TeV compared to Monte Carlo models. Figure from [27].

[17] M. Chojnacki, A. Kisiel, W. Florkowski, W. Broniowski, Comput. Phys. Commun. **183**, 746 (2012)

[18] S. Acharya et al. (ALICE), Phys. Lett. **B774**, 64 (2017), 1705.04929

[19] A.D. Martin, E.N. Ozmutlu, E.J. Squires, Nucl. Phys. **B121**, 514 (1977)

[20] A. Antonelli, eConf **C020620**, THAT06 (2002), [,65(2002)]

[21] N.N. Achasov, V.V. Gubin, Phys. Atom. Nucl. **65**, 1528 (2002), [Phys. Rev.D63,094007(2001)]

[22] N.N. Achasov, A.V. Kiselev, Phys. Rev. **D68**, 014006 (2003)

[23] E. Santopinto, G. Galata, Phys. Rev. **C75**, 045206 (2007)

[24] R.L. Jaffe, Phys. Rev. **D15**, 267 (1977)

[25] B. Abelev et al. (ALICE), Phys. Lett. **B717**, 151 (2012)

[26] B. Abelev et al. (ALICE), Phys. Rev. **D87**, 052016 (2013)

[27] J. Adam et al. (ALICE), Eur. Phys. J. **C77**, 569 (2017)

[28] T. Sjostrand, S. Mrenna, P.Z. Skands, JHEP **05**, 026 (2006)

[29] P.Z. Skands, Phys. Rev. **D82**, 074018 (2010)

[30] T. Sjostrand, S. Mrenna, P.Z. Skands, Comput. Phys. Commun. **178**, 852 (2008)

[31] P. Skands, S. Carrazza, J. Rojo, Eur. Phys. J. **C74**, 3024 (2014)

[32] R. Engel, Z. Phys. **C66**, 203 (1995)

[33] T. Pierog, I. Karpenko, J.M. Katzy, E. Yatsenko, K. Werner, Phys. Rev. **C92**, 034906 (2015)

[34] G. Corcella, I.G. Knowles, G. Marchesini, S. Moretti, K. Odagiri, P. Richardson, M.H. Seymour, B.R. Webber, JHEP **01**, 010 (2001)

[35] H. Aihara et al. (TPC/Two Gamma), Phys. Rev. Lett. **57**, 3140 (1986)

Measurements of event properties and multi-differential jet cross sections and impact of CMS measurements on Proton Structure and QCD parameters

Anterpreet Kaur⋆ for the CMS Collaboration
Panjab University, Chandigarh, India

Abstract. We present results on the measurements of characteristics of events with jets including jet-charge, investigations of shapes and jet mass distributions. The measurements are compared to theoretical predictions including those matched to parton shower and hadronization. Multi-differential jet cross sections are also presented over a wide range in transverse momenta from inclusive jets to multi-jet final states. These measurements have an impact on the determination of the strong coupling constant as well as on parton distribution functions (PDFs) and are helpful in the treatment of heavy flavours in QCD analyses. We also show angular correlations in multi-jet events at highest center-of-mass energies and compare the measurements to theoretical predictions including higher order parton radiation and coherence effects. Measurements of cross sections of jet and top-quark pair production are in particular sensitive to the gluon distribution in the proton, while the electroweak boson production - inclusive or associated with charm or beauty quarks - gives insight into the flavour separation of the proton sea and to the treatment of heavy quarks in PDF-related studies.

1 Introduction

Jets are narrow collimated cones of stable particles (mainly hadrons) produced by the hadronization of partons (quarks or gluons) in high-energy particle interactions. The strong interactions between the partons are described by the theory called Quantum Chromodynamics (QCD). In a hard scattering process, partons are produced with high-transverse momentum (p_T) and give rise to a parton shower (PS) by emission of additional gluons and production of quark-antiquark pairs. This phase can be described by perturbative QCD (pQCD) approximations. In the non-perturbative phase, the colored partons hadronize into colorless objects in the form of collimated sprays known as jets. Jets are measured in particle detectors and studied in order to determine the properties of the original quarks. The distribution of particles and energy within a jet can also reveal information about both the production mechanism and underlying physics.

CMS [1] is one of the two multipurpose experiments at the Large Hadron Collider (LHC) at CERN, designed to see a wide range of particles and phenomena produced in high-energy proton-proton (pp) collisions in the LHC. This paper presents several studies of jet production performed by CMS with data from proton-proton collisions taken at centre-of-mass energies ($\sqrt{s}$) of 8 and 13 TeV during Run 1 and 2 of the LHC respectively, from 2012 until 2016. The CMS detector covers a solid

⋆e-mail: anterpreet.kaur@cern.ch

© The Authors, published by EDP Sciences. This is an open access article distributed under the terms of the Creative Commons Attribution License 4.0 (http://creativecommons.org/licenses/by/4.0/).

angle of almost 4π and consists of precise tracking devices around the interaction point, surrounded by calorimeters and muon chambers. In the CMS experiment, all particles are reconstructed and identified using a particle-flow (PF) algorithm, which combines the information from the individual subdetectors [2]. The four-vectors of particle candidates, reconstructed by the above technique, are used as input to the anti-k_t jet clustering algorithm [3]. The clustering is performed within the FastJet package [4] using four-momentum summation. A factorized jet calibration procedure with corrections for pile-up and jet energy scale are applied. The data is also corrected for detector effects using an unfolding procedure which allows direct comparisons with theory. Theoretical predictions are obtained from various Monte Carlo (MC) event generators. To compare with measurements, the parton-level calculations must be complemented with corrections for nonperturbative (NP) effects that involve the modeling of hadronization (HAD) and multiparton interactions (MPI).

2 QCD multijet production

Inclusive jet production (p + p $\rightarrow$ jet + X) is a key process to test predictions of perturbative QCD (pQCD) over a wide region in phase space. It probes the parton-parton interaction and is sensitive to the value of the strong coupling constant, α_S and also provides the important constraints on the description of the proton structure, expressed by the parton distribution functions (PDFs).

2.1 Inclusive jet production

The double-differential inclusive jet cross sections [5] has been measured using data collected at $\sqrt{s}$ = 8 TeV corresponding to an integrated luminosity of 19.7 fb^{-1}. The measurement is presented as a function of jet transverse momentum p_T and absolute rapidity $|y|$ and covers a large range in jet p_T from 74 GeV up to 2.5 TeV, in six rapidity bins up to $|y|$ = 3.0. The region of low jet p_T, in particular the range from 21 to 74 GeV, has also been studied up to $|y|$ = 4.7, using a dedicated low-pileup 5.6 pb^{-1} data sample. Perturbative QCD, supplemented by a small nonperturbative and electroweak corrections, describes the data over a wide range of jet p_T and y as shown in Figure 1 (left). Using the probed p_T range and six different rapidity bins, the high-p_T jet cross section measurements are used to extract the value of the strong coupling constant. The best fitted value is $\alpha_s(M_Z) = 0.1164^{+0.0060}_{-0.0043}$ using the CT10 NLO PDF set. The running of the strong coupling constant as a function of the energy scale Q, $\alpha_S(Q)$, measured for nine different values of energy scale between 86 GeV and 1.5 TeV, is in good agreement with previous experiments and extends the measurement to the highest values of the energy scale as shown in Fig. 3. This measurement also probes hadronic parton-parton interaction over a wide range of x and Q. The QCD analysis of these data together with HERA DIS measurements illustrates the potential of the high-p_T jet cross sections to provide important constraints on the gluon PDF in a new kinematic regime.

Similarly a measurement of the double-differential cross section as a function of jet p_T and absolute rapidity $|y|$ has been performed for two jet sizes R = 0.4 and 0.7 using data collected at $\sqrt{s}$ = 13 TeV [6]. Data samples corresponding to integrated luminosities of 71 pb^{-1} and 44 pb^{-1} are used for absolute rapidities $|y| < 3$ and for the forward region $3.2 < |y| < 4.7$, respectively. The comparison between data and predictions at NLO in perturbative QCD including corrections for nonperturbative and electroweak effects are shown in Fig. 1 (right) and it is observed that jet cross sections for the larger jet size of R = 0.7 are accurately described, while for R = 0.4 theory overestimates the cross section by 5–10% almost globally. This measurement is a first indication that jet physics is as well understood at $\sqrt{s}$ = 13 TeV as at smaller centre-of-mass energies in the phase space accessible with the new data.

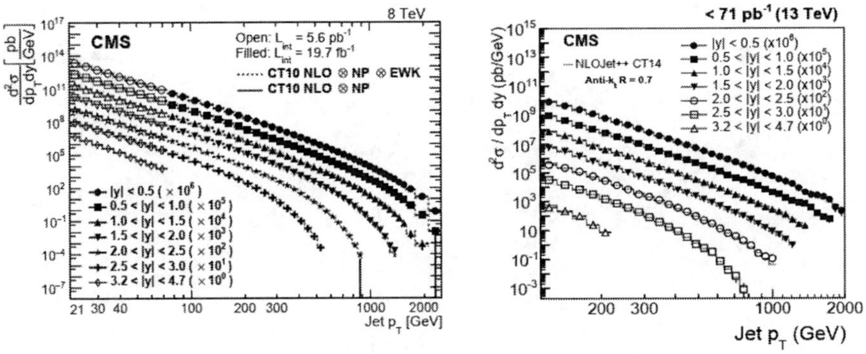

Figure 1. Double-differential inclusive jet cross sections as function of jet p_T for six different $|y|$ bins at an interval of $\Delta\,|y| = 0.5$. Left [5] : Data (open points representing the low-p_T analysis, filled points representing the high-p_T one) and NLO predictions based on the CT10 PDF set corrected for the nonperturbative factor for the low-p_T data (solid line) and the NP and electroweak correction factors for the high-p_T data (dashed line), at $\sqrt{s} = 8$ TeV are shown. Right [6] : Data (points) and predictions from NLOJET++ based on the CT14 PDF set corrected for the NP and electroweak effects (line), at $\sqrt{s} = 13$ TeV are shown.

2.2 Triple differential dijets

Measurements of dijet cross sections can be used to thoroughly test predictions of perturbative QCD (pQCD) at high energies and to constrain parton distribution functions (PDFs). A measurement of the triple-differential dijet cross section has been performed using CMS data collected at $\sqrt{s} = 8$ TeV, as a function of the average transverse momentum $p_{T,avg} = \frac{1}{2}(p_{T,1} + p_{T,2})$ of the two leading jets, half of their rapidity separation $y^{\star} = \frac{1}{2}|y_1 - y_2|$, and the boost of the dijet system $y_b = \frac{1}{2}|y_1 + y_2|$ [7]. The data are found to be well described by NLO predictions corrected for nonperturbative and electroweak effects, except for highly boosted event topologies that suffer from large uncertainties in parton distribution functions (PDFs) as shown in Fig. 2 (left). For large values of y_b, the momentum fractions carried by the incoming partons must correspond to one large and one small value, while for small y_b the momentum fractions must be approximately equal. In addition, for high transverse momenta of the jets, x values are probed above 0.1, where the proton PDFs are less precisely known. From Fig. 2 (right), it is observed that the precise data constrain the PDFs, especially in the highly boosted regime that probes the highest fractions x of the proton momentum carried by a parton. Also in a simultaneous fit, the strong coupling constant $\alpha_s(M_Z)$ is extracted together with the PDFs and the value obtained at the mass of the Z boson is $\alpha_s(M_Z) = 0.1199 \pm 0.0015\ (\text{exp})^{+0.0031}_{-0.0020}$ (theo).

2.3 Inclusive multijets

The investigation of inclusive multijet event cross sections ($\sigma_{jet} \propto \alpha_S$) as a function of jet p_T and rapidity y provides essential information about the PDFs and the strong coupling constant. Moreover, the ratios of such cross sections, $R_{mn} = \frac{\sigma_{m-jet}}{\sigma_{n-jet}} \propto \alpha^{m-n}$, with m > n, provide an ideal tool to determine the strong coupling constant $\alpha_s(M_Z)$ as numerous theoretical and experimental uncertainties cancel. The inclusive 2- and 3-jet event cross sections as well as cross section ratio (R_{32}) have been measured using data corresponding to an integrated luminosity of 19.7 fb^{-1}at $\sqrt{s} = 8$ TeV, as a function of the average transverse momentum, $H_{T,2}/2 = \frac{1}{2}(p_{T,1} + p_{T,2})$ of the two leading jets [9]. R_{32} as a function of $H_{T,2}/2$ is extracted from the data by dividing the differential cross sections for each bin in $H_{T,2}/2$.

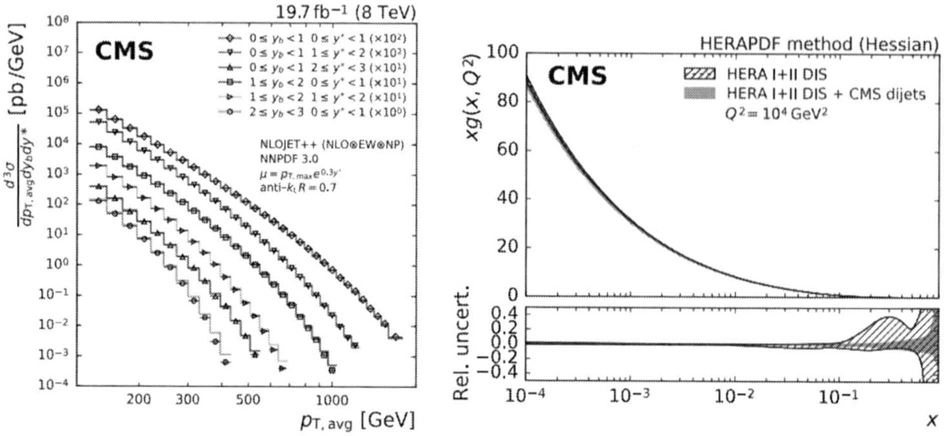

Figure 2. [7] Left : The triple-differential dijet cross section in six bins of $y^\star$ and y_b. The data are indicated by different markers for each bin. The theoretical predictions, obtained with NLOJet++ and NNPDF 3.0, and complemented with EW and NP corrections, are depicted by solid lines. Apart from the boosted region, the data are well described by the predictions at NLO accuracy over many orders of magnitude. Right : The gluon PDFs as a function of x as derived from HERA inclusive DIS data alone (hatched band) [8] and in combination with CMS dijet data (solid band). The PDFs are shown at the scale $Q^2 = 10^4$ GeV^2 with their total uncertainties.

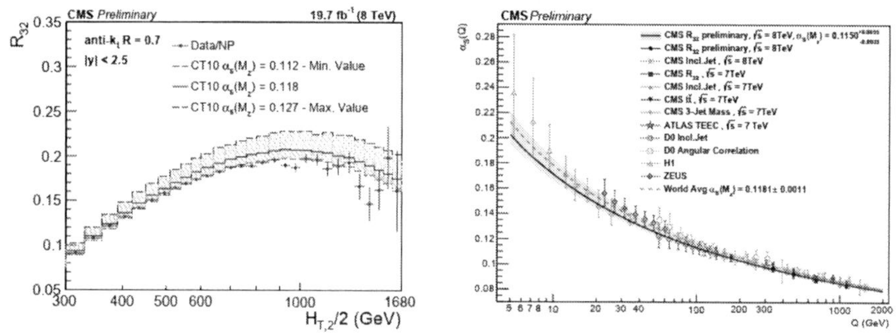

Figure 3. [9] Left : Cross section ratio R_{32} as a function of $H_{T,2}/2$ calculated from data (solid circles) in comparison to that from NLO pQCD (lines). The NLO predictions using the CT10 NLO PDF set corrected with NP corrections are shown for a series of values assumed for $\alpha_s(M_Z)$ (dashed lines) together with the central prediction (solid line). Right : The running $\alpha_S(Q)$ as a function of the scale Q is shown. The dashed line represents the evolution of the world average [10] along with the results from other measurements of CMS [5, 11–14], ATLAS [15], D0 [16, 17], H1 [18, 19], and ZEUS [20].

Fig. 3 (left) presents this ratio as obtained from unfolded data in comparison to that from NLO pQCD using CT10 NLO PDF set corrected with NP corrections. From a fit of the ratio of the 3-jet over 2-jet event cross section using the MSTW2008 PDF set, the extracted value of strong coupling constant at the scale of the Z boson mass is $\alpha_s(M_Z) = 0.1150 \pm 0.0010\,(\text{exp}) \pm 0.0013\,(\text{PDF}) \pm 0.0015\,(\text{NP})\,^{+0.0050}_{-0.0000}\,(\text{scale})$. As shown in Fig. 3 (right), this value is in agreement with the world average value of $\alpha_s(M_Z) = 0.1181 \pm 0.0011$ derived in Ref. [10] as well as previous determinations obtained by the ATLAS and CMS collaborations [5, 11–15].

2.4 Azimuthal correlations

The two-final state partons, at leading order (LO) in pQCD, are produced back-to-back in the transverse plane and thus the azimuthal angular separation between the two highest p_T jets in the transverse plane, $\Delta\phi_{1,2} = |\phi_{jet1} - \phi_{jet2}|$, equals π. The production of a third or more high-p_T jets leads to a deviation from π in the azimuthal angle. The measurement of the azimuthal angular correlation (or decorrelation from π) in multijet topologies is proven to be an interesting tool to test theoretical predictions of multijet production processes. So measurements of the normalized inclusive 2-jet, 3-jet, and 4-jet cross sections differential in $\Delta\phi_{1,2}$ and of the normalized inclusive 3-jet, and 4-jet cross sections differential in $\Delta\phi_{2j}^{min}$ (minimum azimuthal angular separation between any two of the three or four leading p_T jets,) have been performed for several regions in the leading-jet transverse momentum p_T^{max} [21]. The measurements are performed using data collected during 2016 corresponding to an integrated luminosity of 35.9 fb^{-1}of proton-proton collisions at $\sqrt{s} = 13$ TeV. The unfolded, normalized, inclusive 2-jet (left), 3-jet (middle), and 4-jet (right) cross section differential in $\Delta\phi_{1,2}$ is shown in Fig. 4 for the various p_T^{max} regions considered in the analysis. In the 2-jet case the $\Delta\phi_{1,2}$ distributions are strongly peaked at π and become steeper with increasing p_T^{max}. The $\Delta\phi_{1,2}$ distributions, in the 3-jet case become flatter at π since the dijet events are missing, and in the 4-jet case they become even more flat. The extension of $\Delta\phi_{1,2}$ correlations in inclusive 3-jet and 4-jet topologies are new results.

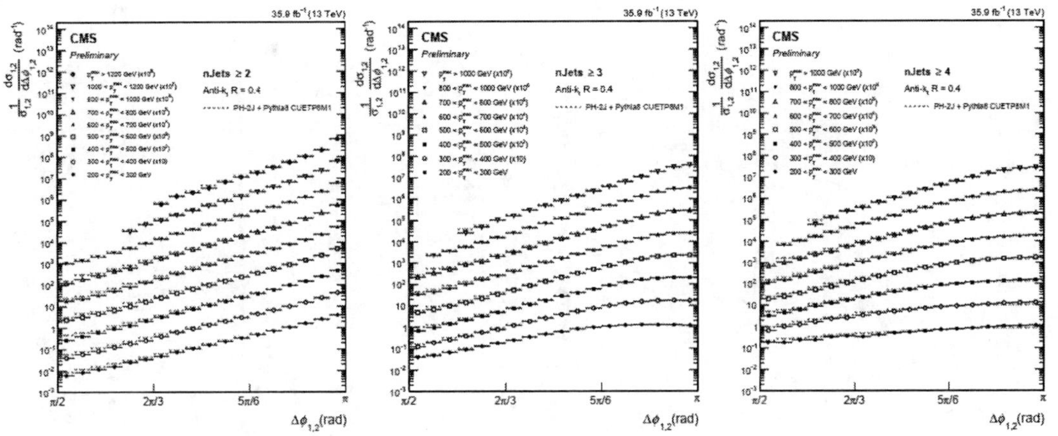

Figure 4. [21] Normalized inclusive 2-jet (left), 3-jet (middle) and 4-jet (right) cross section differential in $\Delta\phi_{1,2}$ for nine p_T^{max} regions, scaled by multiplicative factors for presentation purposes. Overlaid on the data are predictions from the PH-2J + PYTHIA8 event generator.

3 Jet charge and mass

At leading order (LO) in QCD, the type of partons that initiate jets can be distinguished and can be referred as quark jets, antiquark jets, or gluon jets. To distinguish signal from background, or to characterize a new particle, it is often important to identify the object initiating a jet by means of the properties of the reconstructed particles that define the jet. In particular, the electric charge quantum number of the original parton from which a jet is initiated can be estimated from a momentum-weighted sum of the charges of the particles in the jet. This observable is sensitive to the charge of the initiating quark or gluon and can provide constraints on models of jet formation. A study of the jet charge distribution, unfolded for detector effects, with dijet events in pp collisions at $\sqrt{s} = 8$ TeV has been performed in various ranges ofp_T [22]. This measurement is carried out for different definitions of jet charge as defined in equation 1, to gain a better understanding of the underlying models that can be used to improve the predictions of MC event generators. Figure 5 (left) compares data with the normalized charge distribution of the leading jet with $\kappa = 0.6$, initiated by either an up quark (u), down quark (d), or a gluon (g) in PYTHIA6. The charge distribution for jets initiated by quarks with positive electric charge peaks at positive values, with a mean of 0.166e, as opposed to that for jets initiated by negatively charged quarks, with a mean of -0.088e and gluons, with a mean of 0.013e, where e is the proton charge. This suggests that the jet charge can be used to differentiate statistically jets from quarks of different electric charge, or to distinguish jets initiated by a gluon or a quark.

$$
\begin{aligned}
Q^{\kappa} &= \frac{1}{(p_T^{jet})^{\kappa}} \sum_i Q_i (p_T^i)^{\kappa}, \\
Q_L^{\kappa} &= \sum_i Q_i \left(p_{\parallel}^i\right)^{\kappa} / \sum_i \left(p_{\parallel}^i\right)^{\kappa}, \\
Q_T^{\kappa} &= \sum_i Q_i \left(p_{\perp}^i\right)^{\kappa} / \sum_i \left(p_{\perp}^i\right)^{\kappa} \qquad (1)
\end{aligned}
$$

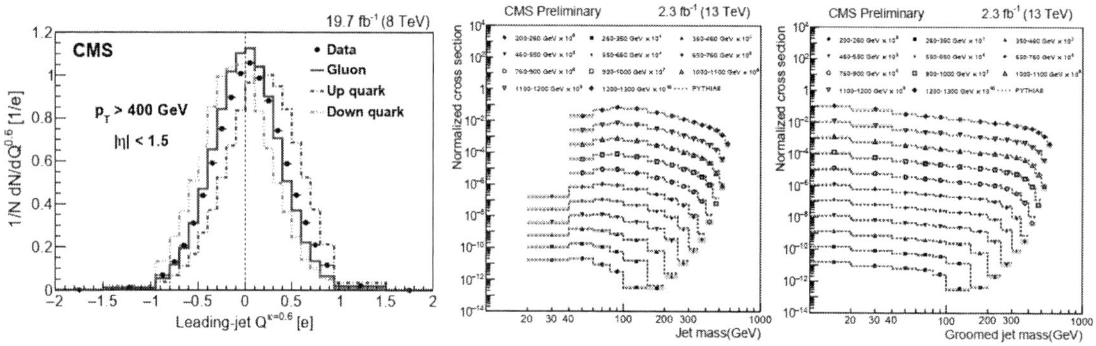

Figure 5. Comparison of jet charge distribution, Q^{κ} for leading jets and $\kappa = 0.6$, between data (points) and u, d, and g distributions from PYTHIA6 (left [22]). Results [23] of unfolding ungroomed (middle) and groomed (right) jets for all p_T bins. Bins with total uncertainty larger than 60% are not shown. The data are shown in markers for each p_T bin, scaled by a factor for better visibility. The predictions from PYTHIA8 are shown as a dashed red line.

Jet mass is more sensitive to the internal structure of jets, which is theoretically described by QCD parton showering. The recent techniques make use of the mass of the jets to investigate signals of physics beyond the standard model (BSM). The jet mass generated by QCD radiation is typically lower than the jet mass generated by hadronically-decaying heavy SM particles that have 2- and 3-prong decays such as W, Z, and H bosons, or top quarks. This can be exploited to identify BSM particles that decay into highly Lorentz-boosted SM particles ("boosted object"). CMS has performed a study [23] of double-differential jet cross section in bins of the ungroomed jet p_T in conjunction with the ungroomed and groomed jet mass using the 'soft drop" grooming algorithm [24], which is the same as the "mass drop" grooming algorithm for our choice of parameters. This observable is sensitive to the physics modeling, and could be used in future global fits for parameter tuning.

This analysis improves over previous results by using a more theoretically controlled jet grooming algorithm, as well as by unfolding in both p_T and mass. For ungroomed jets shown in Fig. 5 (middle), measurements from PYTHIA8 are found to predict the jet mass spectrum within uncertainties in the data for intermediate masses ($0.1 < m/p_T < 0.3$). For groomed jets as in Fig. 5 (right), the Sudakov peak is suppressed and the precision in the intermediate mass region ($0.1 < m/p_T < 0.3$) improves, since the grooming algorithm removes the portions of the jet arising from soft radiation.

4 Differential Z and W cross sections

Electroweak gauge boson production in pp collisions is an important probe of SM gauge boson interactions. The final states of W and Z decays have unique experimental signatures and the production cross sections for these processes are theoretically well understood and sensitive to deviations from the SM. The study of the cross sections and differential cross sections for a Z boson, decaying to dilepton channel, produced in association with jets in pp collisions at $\sqrt{s}$ = 8 TeV [25] as well as at $\sqrt{s}$ = 13 TeV [26] has been done. The cross sections are determined as functions of various relevant kinematic variables of jets. The predictions of a number of multileg generators with leading or next-to-leading order accuracy are compared with the measurements. The comparison shows the importance of including multiparton contributions in the matrix elements and the improvement in the predictions when next-to-leading order terms are included.

Similarly the differential cross sections have been measured for a W boson produced in association with jets in pp collisions at $\sqrt{s}$ = 8 TeV [27] and at $\sqrt{s}$ = 13 TeV [28]. The measured differential cross sections are compared with tree-level and higher-order recent event generators, as well as next-to-leading-order and next-to-next-to-leading-order theoretical predictions. The agreement of the generators with the measurements builds confidence in their use for the simulation of W+jets background processes in searches for new physics at the LHC.

5 Heavy flavour jet physics

The associated production of vector bosons, W or Z, and jets originating from heavy-flavour quarks is a large background source in measurements of several SM processes, Higgs boson studies, and many searches for physics beyond the SM. The study of events with one or two well-identified and isolated leptons accompanied by heavy-flavour jets is crucial to refine the theoretical calculations in perturbative QCD, as well as to validate associated Monte Carlo techniques.

The cross section of the production of a Z boson associated with at least one jet originated by a c-quark in pp collisions at $\sqrt{s}$ = 8 TeV corresponding to an integrated luminosity of 19.7 ± 0.5 fb^{-1}, is measured [29]. The cross sections ratio of the production of a Z boson and at least one c- or b-quark jet is also determined. The measured Z+c production cross section is σ (pp $\to$ Z + c + X) = 8.6 ± 0.5 (stat.) ± 0.7 (syst.) pb, and the cross sections ratio is σ (pp $\to$ Z + c + X)/σ (pp $\to$ Z + b + X) = 2.0 ± 0.2 (stat.) ± 0.2 (syst.). The measurements of Z+c production cross section and the cross sections ratio, inclusively and differentially as a function of p_T^{jet} are in agreement with the LO predictions from MadGraphand NLO predictions from MG5_AMC. Predictions from the MCFM program are lower than the measured Z+c cross section, both inclusive and differentially, as shown in 6 (left). A better description is reached in terms of the Z+c/Z+b cross sections ratio shown in 6 (right).

6 Top physics

The study of top production has a great discovery potential for physics beyond the standard model. It helps to test the production mechanisms, to check the validity of QCD, and to provide an important

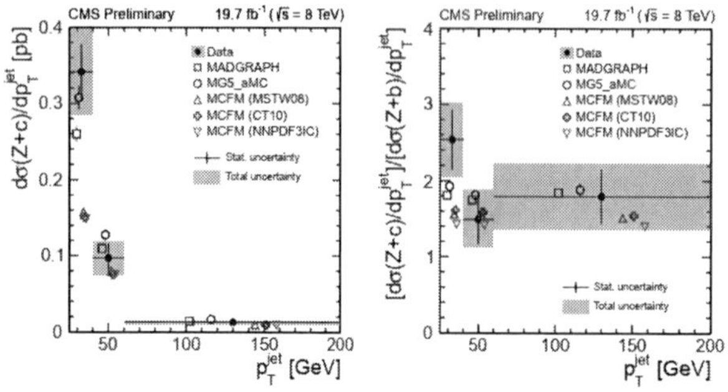

Figure 6. [29] Differential Z+c cross section (left) and Z+c/Z+b cross sections ratio (right) as a function of the transverse momentum of the jet. The solid rectangles indicate the total (statistical plus systematic) experimental uncertainty. Statistical and systematic uncertainties in the theoretical predictions are shown added in quadrature.

source of background in searches for physics beyond the standard model. Top quarks can decay leptonically t → Wb → (lν) (b-jet), and hadronically t → hadrons (three jets). In case of top pair production, di-leptonic decays of both quarks can be selected by requiring one electron and one muon of opposite charge, and at least two jets in the final state. The case when one decay is leptonic and the other one is hadronic is selected by asking for one electron or one muon, and at least four jets in the final state.

6.1 Differential $t\bar{t}$ production

Normalized double-differential cross sections for top quark pair ($t\bar{t}$) production are measured in pp collisions at $\sqrt{s}$ = 8 TeV corresponding to an integrated luminosity of 19.7 fb^{-1}[30]. The measurement is performed in the dilepton $e^{\pm}\mu^{\mp}$ final state. The $t\bar{t}$cross section is determined as a function of various pairs of observables characterizing the kinematics of the top quark and $t\bar{t}$ system. The data is in agreement with calculations using pQCD at next-to-leading and approximate next-to-next-to-leading orders as well as with the fixed-order computations using MC Carlo event generators. The measured double-differential cross sections have been incorporated into a PDF fit, together with other data from HERA and the LHC. Including the $t\bar{t}$ data, there is a significant reduction in the uncertainties in the gluon distribution at large values of x and can be seen in Fig. 7 (left). The constraints provided by these data are competitive with those from inclusive jet data. This improvement exceeds that from using single-differential $t\bar{t}$ cross section data, thus strongly suggesting the use of the double-differential $t\bar{t}$ measurements in PDF fits.

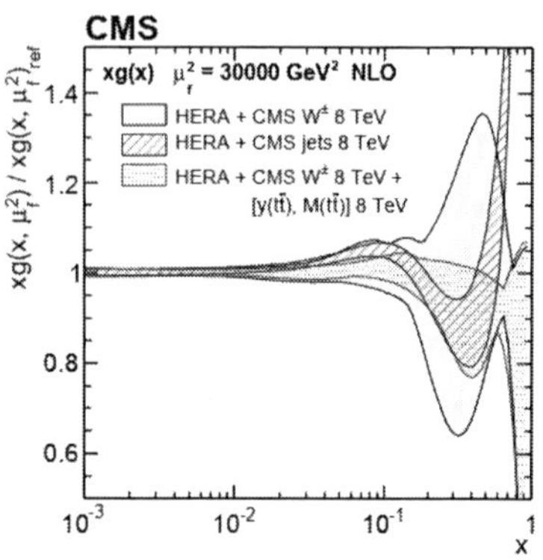

Figure 7. [30] The gluon distribution at μ^{2f} = 30000 GeV^2 , as obtained from the PDF fit to the HERA DIS data and CMS W$^{\pm}$ boson charge asymmetry measurements (shaded area), the CMS inclusive jet production cross sections (hatched area), and the W$^{\pm}$ boson charge asymmetry plus the double-differential $t\bar{t}$cross section (dotted area).

6.2 t-channel single top quark production

The cross section of the t-channel single top quark production is measured using events with one muon and jets in the final state [31]. The cross section for the production of single top quarks and the ratio of the top quark to top antiquark production are measured together in a simultaneous fit where the results are used to evaluate the production cross section of single top antiquarks. Several kinematic variables are then combined into a multivariate discriminator to distinguish signal from background events. A fit to the distribution of the discriminating variable yields a total cross section of 238 ± 13 (stat) ± 29 (syst) pb and a ratio of top quark and top antiquark production of $R_{t-ch.}$ = 1.81 ± 0.18 (stat) ± 0.15 (syst).From the total cross section the absolute value of the CKM matrix element V_{tb} is calculated to be 1.05 ± 0.07 (exp) ± 0.02 (theo). All results are in agreement with the standard model predictions and previous measurements.

6.3 Top mass

The precise measurement of top quark mass (m_t), a fundamental parameter of the standard model (SM), can benefit the calculation of radiative corrections to several observables. The current m_t measurement methods exploit either the full kinematic reconstruction of top quark pair ($t\bar{t}$) events or alternative event topologies or observables that are sensitive to m_t variations. The latter can avoid intrinsic hadronization-related uncertainties and provide further understanding of the relation between the experimental results obtained for m_t and the mass parameters employed in theoretical calculations. The first measurement of the differential $t\bar{t}$ cross section has been performed in the ℓ + jets channel as a function of the leading-jet mass m_{jet} in the highly boosted top quark regime [32]. The measurement is carried out in a fiducial region with fully merged top quark decays in hadronic final states, corrected to the particle level. The predictions exceed the measurements of differential cross section as a function of the leading-jet mas as shown in Fig. 8 (left) whereas the normalized differential cross section shown in Fig. 8 (right) agrees with predictions from simulations, indicating the good quality of modelling the jet mass in highly boosted top quark decays. The normalized measurement is used to extract a value of m_t, in order to estimate the current sensitivity of the data. The value obtained is, m_t = 170.8 ± 9.0 GeV which is consistent with the current LHC and Tevatron average of 173.34 ± 0.27 (stat) ± 0.71 (syst) GeV [33].

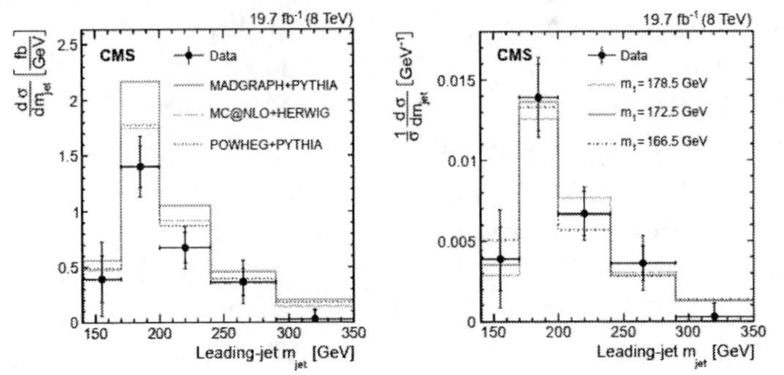

Figure 8. [32] Left : Measured particle-level differential $t\bar{t}$ cross sections as a function of the leading-jet mass, compared to predictions from MadGraph+pythia, powheg+pythia, and MC@NLO+Herwig generators (lines). Right : The normalized particle-level $t\bar{t}$ differential cross section compared to predictions from MadGraph+pythia for three values of m_t.

7 Summary

Jet production in proton-proton collisions is one of the main phenomenological predictions of perturbative Quantum Chromodynamics (pQCD). This paper summarizes the contribution of jets in the measurements of characteristics of events, ranging from jet-charge to shapes as well as jet mass distributions. The measurements are compared to theoretical predictions including those matched to parton shower and hadronization. Along with this multi-differential jet cross-sections are presented over a wide range in transverse momenta from inclusive jets to multi-jet final states. These measurements have an impact on the determination of the strong coupling constant α_S as well as on parton density functions (PDFs). The electroweak boson production, whether inclusive or associated with charm or beauty quarks, gives insight into the flavour separation of the proton sea. Also the measurements of cross sections of jet and top-quark pair production are in particular sensitive to the gluon distribution in the proton and α_S.

References

[1] CMS Collaboration, JINST **3**, S08004 (2008).
[2] CMS Collaboration, JINST **12**, no. 10, P10003 (2017) arXiv:1706.04965 [physics.ins-det].
[3] M. Cacciari, G. P. Salam and G. Soyez, JHEP **0804**, 063 (2008) arXiv:0802.1189 [hep-ph].
[4] M. Cacciari, G. P. Salam and G. Soyez, Eur. Phys. J. C **72**, 1896 (2012) arXiv:1111.6097 [hep-ph].
[5] CMS Collaboration, JHEP **1703**, 156 (2017) arXiv:1609.05331 [hep-ex].
[6] CMS Collaboration, Eur. Phys. J. C **76**, no. 8, 451 (2016) arXiv:1605.04436 [hep-ex].
[7] CMS Collaboration, arXiv:1705.02628 [hep-ex].
[8] H1 and ZEUS Collaborations, Eur. Phys. J. C **75**, no. 12, 580 (2015) arXiv:1506.06042 [hep-ex].
[9] CMS Collaboration, CMS-PAS-SMP-16-008, http://cds.cern.ch/record/2253091
[10] Particle Data Group, Chin. Phys. C **40**, no. 10, 100001 (2016).
[11] CMS Collaboration, Eur. Phys. J. C **73**, no. 10, 2604 (2013) arXiv:1304.7498 [hep-ex].
[12] CMS Collaboration, Phys. Lett. B **728**, 496 (2014) arXiv:1307.1907 [hep-ex].
[13] CMS Collaboration, Eur. Phys. J. C **75**, no. 6, 288 (2015) arXiv:1410.6765 [hep-ex].
[14] CMS Collaboration, Eur. Phys. J. C **75**, no. 5, 186 (2015) arXiv:1412.1633 [hep-ex].
[15] ATLAS Collaboration, Phys. Lett. B **750**, 427 (2015) arXiv:1508.01579 [hep-ex].
[16] D0 Collaboration, Phys. Rev. D **80**, 111107 (2009) arXiv:0911.2710 [hep-ex].
[17] D0 Collaboration, Phys. Lett. B **718**, 56 (2012) arXiv:1207.4957 [hep-ex].
[18] H1 Collaboration, Eur. Phys. J. C **75**, no. 2, 65 (2015) arXiv:1406.4709 [hep-ex].
[19] H1 Collaboration, Eur. Phys. J. C **77**, no. 4, 215 (2017) arXiv:1611.03421 [hep-ex].
[20] ZEUS Collaboration, Nucl. Phys. B **864**, 1 (2012) arXiv:1205.6153 [hep-ex].
[21] CMS Collaboration, CMS-PAS-SMP-16-014, http://cds.cern.ch/record/2257685
[22] CMS Collaboration, JHEP **1710**, 131 (2017) arXiv:1706.05868 [hep-ex].
[23] CMS Collaboration, CMS-PAS-SMP-16-010, http://cds.cern.ch/record/2273393
[24] A. J. Larkoski *et al.*, JHEP **1405**, 146 (2014) arXiv:1402.2657 [hep-ph].
[25] CMS Collaboration, JHEP **1704**, 022 (2017) arXiv:1611.03844 [hep-ex].
[26] CMS Collaboration, CMS-PAS-SMP-15-010, http://cds.cern.ch/record/2114819
[27] CMS Collaboration, Phys. Rev. D **95**, 052002 (2017) arXiv:1610.04222 [hep-ex].
[28] CMS Collaboration, CMS-PAS-SMP-16-005, http://cds.cern.ch/record/2204927
[29] CMS Collaboration, CMS-PAS-SMP-15-009, http://cds.cern.ch/record/2202823
[30] CMS Collaboration, Eur. Phys. J. C **77**, no. 7, 459 (2017) arXiv:1703.01630 [hep-ex].
[31] CMS Collaboration, Phys. Lett. B **772**, 752 (2017) arXiv:1610.00678 [hep-ex].
[32] CMS Collaboration, Eur. Phys. J. C **77**, no. 7, 467 (2017) arXiv:1703.06330 [hep-ex].
[33] ATLAS and CDF and CMS and D0 Collaborations, arXiv:1403.4427 [hep-ex].

EPJ Web of Conferences **172**, 02002 (2018) https://doi.org/10.1051/epjconf/201817202002
ISMD 2017

Cross-section measurements of final states with photons and jets with the ATLAS experiment

Miguel Villaplana Pérez[1,⋆], on behalf of the ATLAS Collaboration

[1] *INFN Milano*

Abstract.
The ATLAS Collaboration has performed precise measurements of the cross-section of final states with photons and/or jets at centre-of-mass energies of 8 and 13 TeV. The results are compared with state-of-the-art theory predictions and with predictions of several Monte Carlo generators. We also present new measurements of transverse energy-energy correlations and their associated asymmetries in multi-jet events at 8 TeV. Both measurements are used to extract the strong coupling constant and test the renormalization group equations.

1 Introduction

The modeling of a proton-proton collision at the Large Hadron Collider (LHC) can be divided in four main steps. The hard-scatter process is described by perturbative quantum chromodynamics (pQCD) and it depends on the parton distribution function (PDF) of the protons. The partons resulting from the hard-scatter process are taken to the non-perturvatibe regime through parton emission and splitting in the fragmentation process. Hadronisation evolves the confined quarks to the state we observe in the ATLAS detector [1]. Color-connected beam–beam remnants and multiple-parton interactions within the colliding protons, the "underlying event", are also described by non-perturbative models. Photons are a hard colourless probe of the hard-scatter process, which provides a good test of pQCD calculations and helps constrain the proton PDF. The analyses described in this paper include both photons coming from the matrix element (direct production) and from the fragmentation process. An isolation requirement is applied in order to reduce the background resulting from photons produced during hadronisation.

In the ATLAS experiment, detector-level jets are formed by clustering topological energy clusters from both the hadronic and electromagnetic calorimeters. On the simulations side, the particles resulting from the fragmentation and the hadronisation are clustered into particle-level jets. When only parton-level predictions are available, these are first clustered into jets, which are then corrected to account for the effect of fragmentation and hadronisation. Monte Carlo (MC) simulations are used to correct the effects of detector efficiency and resolution in both photon and jet measurements in order to compare them to theory predictions at particle level. This allows ATLAS data not only to be confronted to PDF models and perturbative calculations but also to test the modeling of factorisation and hadronisation.

⋆e-mail: Miguel.Villaplana.Perez@cern.ch

© The Authors, published by EDP Sciences. This is an open access article distributed under the terms of the Creative Commons Attribution License 4.0 (http://creativecommons.org/licenses/by/4.0/).

The production of jets and prompt isolated photons at hadron colliders provides a stringent test of pQCD and can be used to probe the proton structure. The production of prompt photons in association with jets provides an additional testing ground for pQCD with a hard colourless probe less affected by hadronisation effects than jet production. Jet production can also be used to probe the gluon density function of the proton. Specific topologies can be used to extract the strong coupling constant.

This paper presents recent measurements of the cross-section of final states involving photons or jets at center-of-mass energies of 8 and 13 TeV performed by the ATLAS Collaboration. We also present new measurements of transverse energy-energy correlations (TEEC) and their associated asymmetries (ATEEC) in multi-jet events at 8 TeV and the extraction of the strong coupling constant from them.

2 Photon production

2.1 Inclusive isolated photon production

Inclusive isolated photon production at 13 TeV has been studied using a data set with an integrated luminosity of 3.2 fb^{-1} [2]. Cross-sections as a function of E_T^γ are measured in four different regions of η^γ for photons with $E_T^\gamma > 125$ GeV and $|\eta^\gamma| < 2.37$. Values of E_T^γ up to 1.5 TeV are measured. The predictions of the Pythia [3] and Sherpa [4] Monte Carlo models give a good description of the shape of the measured cross-section distributions except for $E_T^\gamma \gtrsim 500$ GeV in the regions $|\eta^\gamma| < 0.6$ and $0.6 < |\eta^\gamma| < 1.37$.

Figure 1 shows the individual components of the systematic uncertainties added in quadrature in representative phase-space regions. Photon energy scale is the dominant uncertainty at high E_T^γ, while at low E_T^γ it is the uncertainty related to the background subtraction method that dominates. Uncertainties are larger in the forward regions.

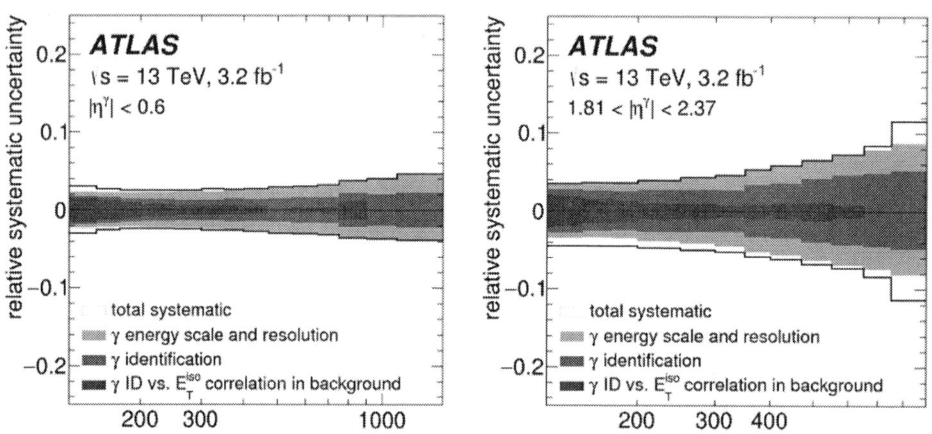

Figure 1. Total relative systematic uncertainty, excluding that in the luminosity measurement, as a function of E_T^γ (black histogram) as well as the contributions from the photon energy scale and resolution (yellow histogram), the photon identification (green histogram) and the photon identification vs. E_T^{iso} correlation in the background (red histogram) in $|\eta^\gamma| < 0.6$ (right) and $1.81 < |\eta^\gamma| < 2.37$ (left). The histograms show the stacked contributions [2].

The NLO pQCD predictions, using Jetphox [5] and based on different sets of proton PDFs, provide an adequate description of the data within the experimental and theoretical uncertainties. For most of the phase space the theoretical uncertainties are larger than those of experimental nature and dominated by the terms beyond NLO, from which it is concluded that NNLO pQCD corrections are needed to make an even more stringent test of the theory. The ratios of the theoretical predictions based on different PDF sets to the measured cross-sections are found to be very similar, the differences being much smaller than the theoretical scale uncertainties.

NNLO predictions have been also compared to ATLAS inclusive photon measurements at 8 TeV [7]. These new calculations, displayed in figure 2, show a trend to be above the data at high E_T. Accounting for both NNLO QCD and electroweak effects provides an improved prediction. The theoretical uncertainty is reduced by a factor close to 3 with respect to NLO predictions.

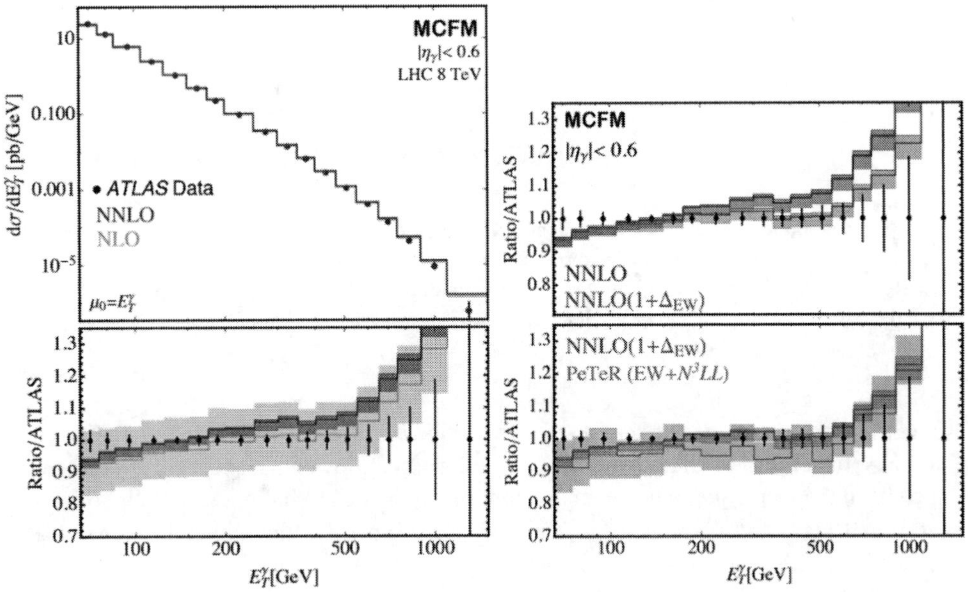

Figure 2. (Left) A comparison of the MCFM predictions [7] for the transverse momentum of the photon to ATLAS 8 TeV data. (Right) Upper: the effect of including electroweak corrections in addition to the NNLO predictions provided by MCFM. Lower: a comparison of the NNLO+EW prediction of MCFM with the N3LL+EW prediction of PeTeR [8].

2.2 Photon pair production

Measurements of the production cross-section of two isolated photons at a center-of-mass energy of 8 TeV have been published [9]. The uncertainties are dominated by systematic effects that have been reduced, compared to the previous ATLAS measurement at 7 TeV, due to an improved method to estimate the background and improved corrections to the modeling of the calorimeter isolation variable in simulated samples.

Predicted cross-sections from fixed-order QCD calculations implemented in DIPHOX [10] and RESBOS [11] at NLO, and in 2γNNLO [12] at NNLO, shown in figure 3, are lower than the data. The relative errors associated to the predictions from DIPHOX (2γNNLO) are 10%–15% (5%–10%).

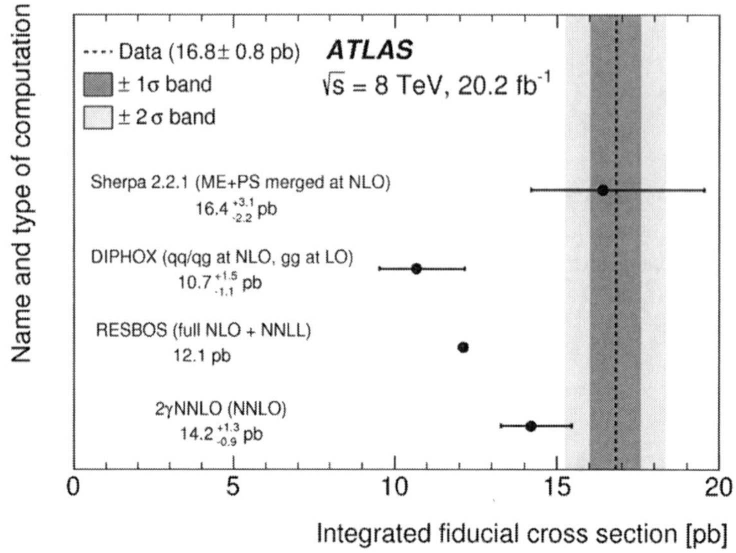

Figure 3. Measured fiducial cross-section compared to the predictions from Sherpa 2.2.1, Diphox, Resbos and 2γNNLO. The uncertainties in the theoretical predictions are shown as horizontal bars. Only the central value is shown for Resbos. The green (yellow) band represents the one- (two-)standard deviation uncertainty, including both the statistical and systematic uncertainties in the measurement added in quadrature [9].

Differential cross-sections are measured as functions of six observables: the diphoton invariant mass, the absolute value of the cosine of the scattering angle with respect to the direction of the proton beams, the opening angle between the photons in the azimuthal plane, the diphoton transverse momentum and two related variables (a_T and ϕ^{η}), with uncertainties typically below 5% per bin, reaching as high as 25% in a few bins with low numbers of data events.

The effects of infrared emissions, probed precisely by measuring the cross-section as functions of a_T (see figure 4) and ϕ^{η}, are well reproduced by the inclusion of soft-gluon resummation at the NNLL accuracy. However, in most parts of the phase space, the predictions above are unable to reproduce the data. The discrepancies can reach a factor of 2 in many regions, beyond the theoretical uncertainties, which are typically below 20%.

The predictions of a parton-level calculation of varying jet multiplicity up to NLO matched to a parton-shower algorithm in SHERPA 2.2.1 provide an improved description of the data compared to all the other computations considered in the study and are in good agreement with the measurements, for both the integrated and differential cross-sections.

2.3 Photon plus jet production

Measurements of the cross-sections for the production of an isolated photon in association with one, two or three jets have been studied with the ATLAS detector at 8 TeV [16]. The photon is required to have $E_T^{\gamma} > 130$ GeV and $|\eta^{\gamma}| < 2.37$. The jets are reconstructed using the anti-kt algorithm with radius parameter R = 0.6.

The cross-sections for photon plus one jet are measured as functions of E_T^{γ} and p_T^{jet1} with $p_T^{jet1} > 100$ GeV; the measurements extend up to values of $E_T^{\gamma}(p_T^{jet1})$ of 1.1 TeV (1.2 TeV). The

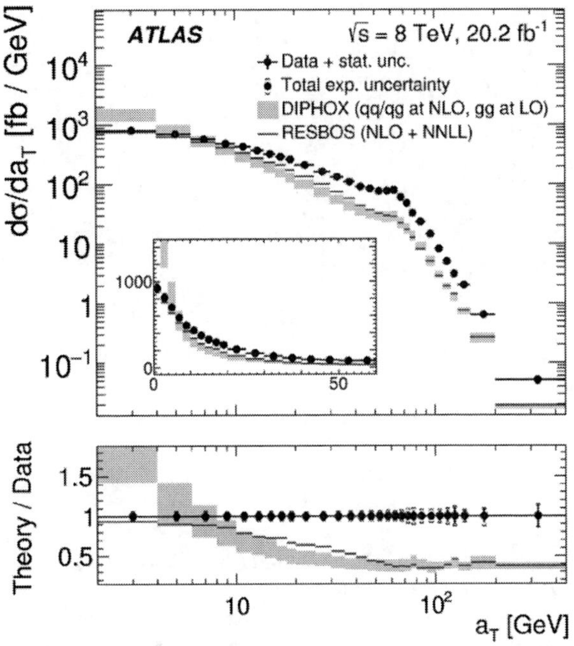

Figure 4. Differential cross-sections as a function of a_T compared to the predictions from Diphox and Resbos. At the bottom, the ratio of the prediction to the data is shown. The bars and bands around the data and theoretical predictions represent the statistical and systematic uncertainties. Only the central values are shown for Resbos. Negative cross-section values are obtained with Diphox in the first bin of a_T and therefore are not shown [9].

dependence on $m_{\gamma-jet1}$ and $|cos\,\theta^*|$ is also measured for $m_{\gamma-jet1} > 467$ GeV and extends up to $m_{\gamma-jet1}$ of 2.45 TeV. The NLO QCD predictions from Jetphox, corrected for hadronisation and underlying-event effects, give a good description of the measured cross-section distributions in both shape and normalisation. In particular, the measured dependence on $|cos\,\theta^*|$ and its scale dependence is consistent with the dominance of processes in which a quark is being exchanged; the experimental (theoretical) uncertainty in $d\sigma/d|cos\theta^*|$ amounts to $\approx 3\%$ (10%).

Photon plus two-jet production is investigated by measuring cross-sections as functions of E_T^{γ}, p_T^{jet2} and angular correlations between the final-state objects for $p_T^{jet1} > 100$ GeV and $p_T^{jet2} > 65$ GeV. The NLO QCD predictions from Blackhat [17] provide a good description of the measurements except for $E_T^{\gamma} > 750$ GeV, as shown on the left panel of figure 5. The predictions from Sherpa, which include higher-order tree-level matrix elements, are found to be superior to those from Pythia, based on $2 \to 2$ processes, in describing the distributions in p_T^{jet2} and the angular correlations. The patterns of QCD radiation around the photon and the leading jet are compared by measuring the production of the subleading jet in an annular region centred on the given final-state object, β_{object}. The cross-sections as functions of β_{γ} and β_{jet1} are observed to be different. The ratio of the cross-sections (see figure 5) shows enhancements in the directions towards the beams ($\beta = 0$ and π).

Photon plus three-jet production is characterised by measurements of cross-sections as functions of p_T^{jet3} and angular correlations for $p_T^{jet1} > 100$ GeV, $p_T^{jet2} > 65$ GeV and $p_T^{jet3} > 50$ GeV. The NLO QCD predictions from Blackhat provide an adequate description of the measurements. Whereas the prediction from Sherpa for p_T^{jet3} is superior to that from Pythia, both give adequate descriptions of the angular correlations.

The dynamics of isolated-photon production in association with a jet have also been studied using 3.2 fb^{-1} of ATLAS data at 13 TeV [18]. Photons are required to have transverse energies above 125 GeV. Jets are identified using the anti-kt algorithm with radius parameter R=0.4 and required to have transverse momenta above 100 GeV. Measurements of isolated-photon plus jet cross-sections are

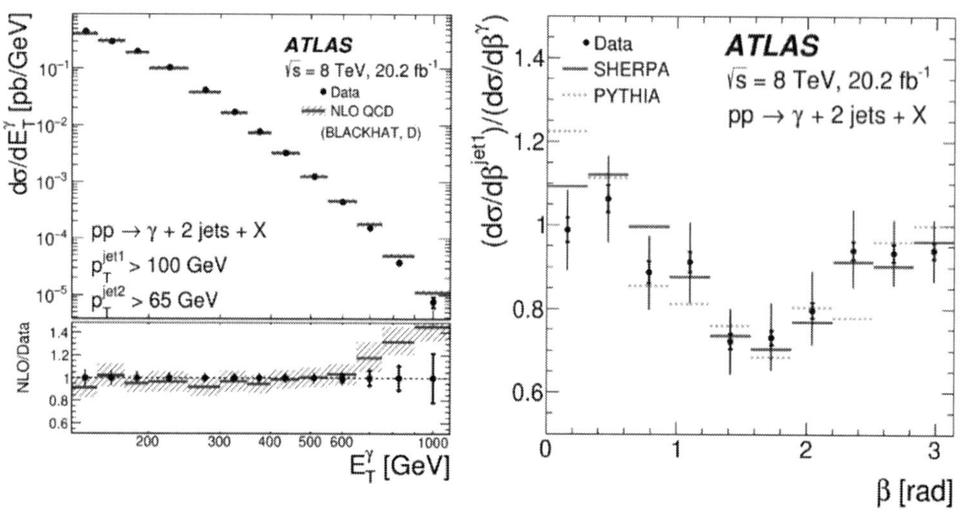

Figure 5. (Left) Measured cross-section for isolated-photon plus two-jet production (dots) as functions of E_T^γ. The NLO QCD predictions from Blackhat corrected for hadronisation and underlying-event effects and using the CT10 PDF set (solid lines) are also shown. These predictions include only the direct contribution (D). The bottom part shows the ratio of the NLO QCD prediction to the measured cross-section. (Right) Ratio of the measured cross-section $d\sigma/d\beta^{jet1}$ and $d\sigma/d\beta^\gamma$ (dots); the ratios for the Sherpa and Pythia predictions are shown as solid and dashed lines, respectively. In both figures, the inner (outer) error bars represent the statistical uncertainties (the statistical and systematic uncertainties added in quadrature) and the shaded band represents the theoretical uncertainty. For most of the points, the inner error bars are smaller than the marker size and, thus, not visible [16].

presented as functions of the leading photon transverse energy, the leading jet transverse momentum, the angular separation in azimuth between the photon and the jet, the photon-jet invariant mass and the scattering angle in the photon-jet centre-of-mass system. The measurements extend up to values of 1.5 TeV in E_T^γ and p_T^{jet}, and the dependence on $m_{\gamma-jet}$ and $|cos\ \theta^*|$ is measured for $m_{\gamma-jet} > 450$ GeV.

The predictions of the tree-level plus parton-shower Monte Carlo models Pythia and LO Sherpa give a reasonable description of the shape of the data, except for p_T^{jet} in the case of Pythia. The fixed-order NLO QCD calculations of Jetphox, corrected for hadronisation and underlying-event effects, and the multi-leg NLO QCD plus parton-shower calculations of Sherpa describe the measured cross-sections within the experimental and theoretical uncertainties. The comparison of predictions based on different parameterisations of the proton PDFs shows that the description of the data achieved does not depend significantly on the specific PDF set used. The only meaningful prediction for $d\sigma/d\Delta\phi^{\gamma-jet}$ is that of NLO Sherpa, which is able to reproduce the data down to $\Delta\phi^{\gamma-jet} = \pi/2$ due to the inclusion of the matrix elements for $2 \to n$ processes with n = 4 and 5. The measured dependence on $|cos\ \theta^*|$, shown in figure 6, is consistent with the dominance of processes in which a quark is being exchanged; the experimental (theoretical) uncertainty on $d\sigma/d|cos\ \theta^*|$ amounts to 3-4% (10% for Jetphox and 15-25% for NLO Sherpa).

All these studies provide stringent tests of pQCD and validate the description of the dynamics of isolated photon in association with jets production in pp collisions at 8 and 13 TeV.

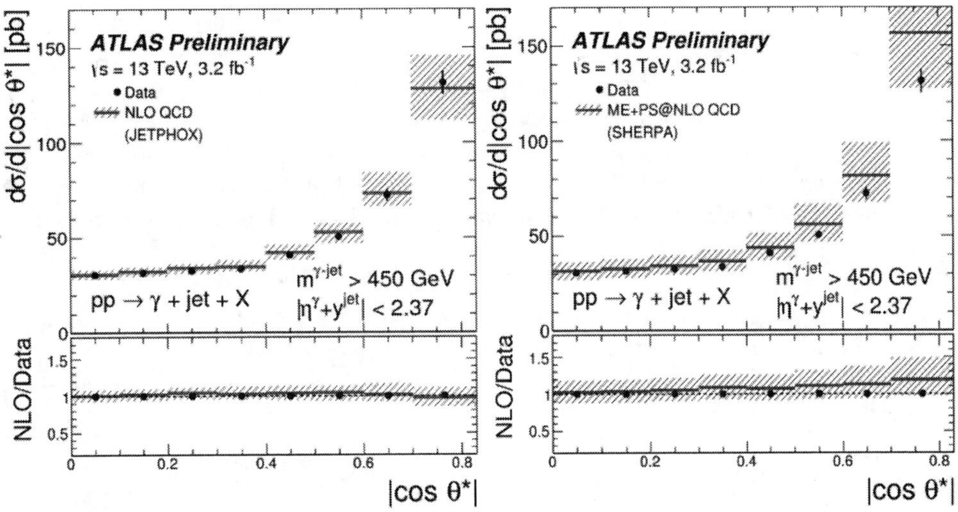

Figure 6. Measured cross-section for isolated-photon plus one-jet production (dots) as a function of $|cos\ \theta^*|$. For comparison, the NLO QCD predictions from JETPHOX corrected for hadronisation and underlying-event effects (left) and the multi-leg NLO QCD plus parton shower predictions from SHERPA (right) are shown with solid lines. The bottom part of each figure shows the ratios of the predictions to the measured cross-section. The inner (outer) error bars represent the statistical uncertainties (the statistical and systematic uncertainties added in quadrature) and the hatched band displays the theoretical uncertainty. For most of the points, the inner error bars are smaller than the marker size and, thus, not visible [18].

3 Jet production

The double-differential inclusive jet cross-sections have been measured using the ATLAS 8 TeV data set. Jets are reconstructed with the anti-kt algorithm with jet radius parameter values of R = 0.4 and R = 0.6, in the kinematic region of the jet transverse momentum from p_T = 70 GeV to about 2.5 TeV and jet rapidities |y| < 3 [13]. The cross-sections are measured double-differentially in the jet transverse momentum and rapidity. The dominant systematic uncertainty arises from the jet energy calibration. Compared to previous jet cross-section measurements a significant reduction of the uncertainties is achieved.

A quantitative comparison of the measurements to fixed-order NLO QCD calculations, corrected for non-perturbative and electroweak effects, shows overall fair agreement (with p-values in the percent range) when considering jet cross-sections in individual jet rapidity bins treated independently. Some tension between data and theory is observed in the central rapidity region for anti-kt jets with R = 0.6. Strong tension between data and theory is observed when considering data points from all jet transverse momentum and rapidity regions, with a full treatment of the correlations. This tension can be reduced, but not completely resolved, using alternative correlation scenarios for the experimental and theoretical two-point systematic uncertainties. The remaining tension could be due either to the breakdown of the assumptions that need to be made in the treatment of two-point systematic uncertainty components, or to an incomplete theoretical description, such as missing higher-order corrections.

The inclusive jet and dijet cross-sections have also been measured using 3.2 fb^{-1} of ATLAS data at 13 TeV [14]. Jets are identified with the anti-kt algorithm with a jet radius parameter value of

R = 0.4. The inclusive jet cross-sections are measured double-differentially as a function of the jet transverse momentum, covering the range from 100 GeV to 3.5 TeV, and the absolute jet rapidity up to |y|<3. In addition, the double-differential dijet production cross-sections are presented as a function of the dijet mass, covering the dijet mass from 300 GeV to 9 TeV and the half absolute rapidity separation between the two leading jets, y^*, up to $y^* < 3$. Figure 7 shows the individual components of the systematic uncertainties added in quadrature for the inclusive jet cross-section measurements in representative phase-space regions. The dominant systematic uncertainty arises from the jet energy scale.

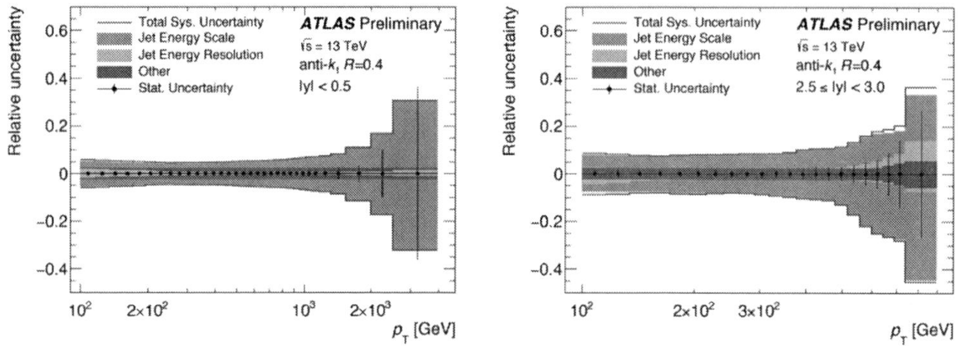

Figure 7. Relative systematic uncertainty for the inclusive jet cross-section as a function of the jet p_T for $|y| < 0.5$ (left) and $2.5 < |y| < 3$ (right). The individual uncertainties are shown in different colors: the jet energy scale, jet energy resolution and the other uncertainties (jet cleaning, luminosity and unfolding bias). The total systematic uncertainty, calculated by adding the individual uncertainties in quadrature, is shown as a green line. The statistical uncertainty is shown as vertical black bars [14].

NLO and NNLO pQCD calculations for the inclusive jet measurement, corrected for non-perturbative and electroweak effects, are compared to the measured cross-sections in figure 8. Similarly to the study at 8 TeV, a quantitative comparison of the measurements to fixed-order NLO QCD calculations shows overall fair agreement when considering jet cross-sections in individual jet rapidity bins independently. In the inclusive jet measurement, a strong tension (with p-values $\ll 10^{-3}$) between data and theory is observed when considering data points from all jet transverse momentum and rapidity regions. No significant deviations between the inclusive jet cross-sections and the fixed-order NNLO QCD calculations corrected for non-perturbative and electroweak effects are observed when using p_T^{jet} as QCD scale.

4 Transverse energy-energy correlations and the extraction of α_s

TEEC and ATEEC in multi-jet events have been measured using the ATLAS 8 TeV data set [15]. The data, binned in six intervals of the sum of transverse momenta of the two leading jets, $H_{T2} = p_{T1} + p_{T2}$, are corrected for detector effects and compared to the predictions of pQCD, corrected for hadronisation and multi-parton interaction effects. The results show that the data are compatible with the theoretical predictions, within the uncertainties.

The data are used to determine the strong coupling constant α_s and its evolution with the interaction scale $Q = (p_{T1} + p_{T2})/2$ by means of a χ^2 fit to the theoretical predictions for both TEEC and ATEEC in each energy bin. Additionally, global fits to the TEEC and ATEEC data are performed, as it is shown in figure 9, leading to

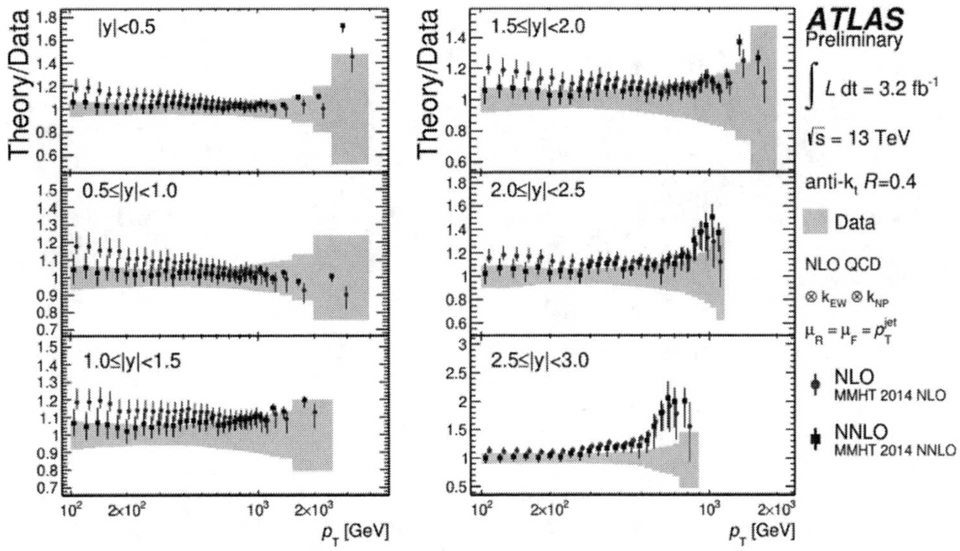

Figure 8. Ratios of the NLO and NNLO pQCD predictions to the measured inclusive jet cross-sections are shown as a function of the jet p_T in six |y| bins for anti-kt jets with R = 0.4 [14]. The NLO predictions are calculated using NLOJET++ with MMHT 2014 NLO PDF set. The NNLO predictions are provided by the authors of ref. [19, 20] using NNLOJET with p_T^{jet} as the QCD scale and the MMHT 2014 NNLO PDF set. Non-perturbative and electroweak corrections were applied to the predictions. The NLO and NNLO uncertainties are shown by the color lines. The grey band shows the uncertainty including both, systematic (JES, JER, unfolding, jet cleaning, luminosity) and statistical uncertainties.

$$\alpha_s(m_Z) = 0.1162 \pm 0.0011(exp.)^{+0.0076}_{-0.0061}(scale) \pm 0.0018(PDF) \pm 0.0003(NP),$$
$$\alpha_s(m_Z) = 0.1196 \pm 0.0013(exp.)^{+0.0061}_{-0.0013}(scale) \pm 0.0017(PDF) \pm 0.0004(NP),$$

respectively. Conservatively, the values obtained using the NNPDF 3.0 PDF set are chosen, as they provide the largest PDF uncertainty among the different PDF sets investigated. These two values are in good agreement with the determinations in previous experiments and with the current world average $s(m_Z) = 0.1181 \pm 0.0011$. The correlation coefficient between the two determinations is equal to 0.60. The present results are limited by the theoretical scale uncertainties, which amount to 6% of the value of $s(m_Z)$ in the case of the TEEC determination and to 4% in the case of the ATEEC. This uncertainty is expected to decrease as higher orders are calculated for the perturbative expansion.

5 Summary

High-precision measurements involving photons and jets on 8 TeV data are being complemented by the first results at 13 TeV. The ATLAS Collaboration has achieved a strong performance in jet and photon reconstruction and many of the results are now dominated by the uncertainties on the predictions and the PDFs. This highlights the importance of the NNLO predictions, which will reduce the renormalisation and factorisation scale uncertainty. Complex measurements of final states involving photons and jets explore regions of phase-space where the current theory struggles to match the data, providing valuable physics input to PDF and α_s fits, and to the description of various aspects of QCD radiation. Diphoton production processes are particularly sensitive to higher-order QCD phenomenology. For the future, the systematic exploration of more complex variables and final states,

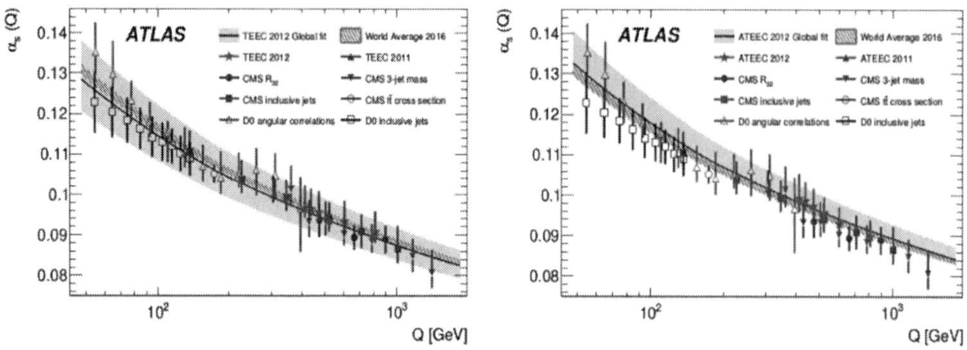

Figure 9. Comparison of the values of $\alpha_s(Q)$ obtained from fits to the TEEC (left) and the ATEEC (right) functions at the energy scales given by $< H_{T2} > /2$ (red star points) with the uncertainty band from the global fit (orange full band) and the 2016 world average (green hatched band) [15]. Determinations from other experiments are also shown as data points. The error bars, as well as the orange full band, include all experimental and theoretical sources of uncertainty. The strong coupling constant is assumed to run according to the two-loop solution of the RGE.

combining several beam energies, will be crucial for fine-tuning descriptions of the Standard Model, and providing precise background estimates for new physics searches.

References

[1] The ATLAS Collaboration, JINST *3* (2008)
[2] The ATLAS Collaboration, Physics Letters B *770* 473–493 (2017)
[3] Torbjörn Sjöstrand et al, Computer Physics Communications *Volume 178 Issue 11* 852-867 (2008)
[4] T. Gleisberg et al, JHEP 02 (2009) 007
[5] Stefano Catani et al, JHEP05(2002)028
[6] L.A. Harland-Lang et al, Eur. Phys. J. C 75 (2015) 204
[7] John M. Campbellet al, Phys. Rev. Lett. 118, 222001
[8] Schwartz, M.D. J. High Energ. Phys. (2016) 2016: 5
[9] The ATLAS Collaboration, PHYSICAL REVIEW D 95, 112005 (2017)
[10] T.Binoth et al, Eur. Phys. J. C 16, 311 (2000).
[11] C.Balaza et al, Phys. Lett. B 637, 235 (2006)
[12] S.Catani et al, Phys. Rev. Lett.108, 072001 (2012)
[13] The ATLAS Collaboration, J. High Energ. Phys. (2017) 2017: 20
[14] The ATLAS Collaboration, ATLAS-CONF-2017-048
[15] The ATLAS Collabotarion, arXiv:1707.02562 [hep-ex]
[16] The ATLAS Collaboration, Nucl.Phys. B918 (2017) 257-316
[17] C.F. Berger et al, Phys. Rev. D 78 (2008) 036003,
[18] The ATLAS Collaboration, ATLAS-CONF-2017-059
[19] J. Currie et al, Phys. Rev. Lett. 118 (2017)
[20] J. Currie et al, arXiv: 1704.00923 [hep-ex].

Locating the critical end point using the linear sigma model coupled to quarks.

Alejandro Ayala[1,3], *J. A.* Flores[2], *L. A.* Hernández[1], and *S.* Hernández-Ortiz[1,*]

[1] *Instituto de Ciencias Nucleares, Universidad Nacional Autónoma de México, Apartado Postal 70-543, México Distrito Federal 04510, México.*
[2] *Departamento de Física, Facultad de Ciencias, Universidad Nacional Autónoma de México, Avenida Universidad 3000, Ciudad Universitaria, 04510 Distrito Federal, México.*
[3] *Centre for Theoretical and Mathematical Physics, and Department of Physics,University of Cape Town, Rondebosch 7700, South Africa.*

Abstract. We use the linear sigma model coupled to quarks to compute the effective potential beyond the mean field approximation, including the contribution of the ring diagrams at finite temperature and baryon density. We determine the model couplings and use them to study the phase diagram in the baryon chemical potential-temperature plane and to locate the Critical End Point.

1 Introduction

The study of strongly interacting matter under extreme conditions, such as high temperature (T) and baryon chemical potential (μ_B), is of great importance in today's physics. One of the principal goals is to gather accurate knowledge of the phase diagram in the $\mu - T$ plane. In this work we use the linear sigma model coupled to quarks, including the plasma screening effects, to explore the effective QCD phase diagram, focusing on the chiral symmetry restoration. To do so, we fix the coupling constants using the physical values of the model parameters, such as the vacuum pion and sigma masses, the critical temperature T_c at $\mu_B = 0$ and the conjectured maximum value of μ_B of the transition line at $T = 0$.

2 Linear Sigma Model coupled to quarks

In order to study the spontaneous breaking of chiral symmetry with the intention of sketching the QCD phase diagram at finite temperature and quark chemical potential. We use the *Linear Sigma Model* coupled to quarks, that is an effective model that accounts for the physics of spontaneous symmetry breaking. The Lagrangian for the linear sigma model when the two lightest quarks are included is given by

$$\mathcal{L} = \frac{1}{2}(\partial_\mu\sigma)^2 + \frac{1}{2}(\partial_\mu\vec{\pi})^2 + \frac{a^2}{2}(\sigma^2 + \vec{\pi}^2) - \frac{\lambda}{4}(\sigma^2 + \vec{\pi}^2)^2 + i\bar{\psi}\gamma^\mu\partial_\mu\psi - g\bar{\psi}(\sigma + i\gamma_5\vec{\tau}\cdot\vec{\pi})\psi, \quad (1)$$

[*] e-mail: saul.hernandezr@correo.nucleares.unam.mx

© The Authors, published by EDP Sciences. This is an open access article distributed under the terms of the Creative Commons Attribution License 4.0 (http://creativecommons.org/licenses/by/4.0/).

where ψ is an SU(2) isospin doublet, $\vec{\pi} = (\pi_1, \pi_2, \pi_3)$ is an isospin triplet and σ is an isospin singlet. λ is the boson's self-coupling and g is the fermion-boson coupling. $a^2 > 0$ is the mass parameter. To allow for an spontaneous breaking of symmetry, we let the σ field to develop a vacuum expectation value v

$$\sigma \rightarrow \sigma + v, \tag{2}$$

which can later be taken as the order parameter of the theory. After this shift, the Lagrangian can be rewritten as

$$\begin{aligned}\mathcal{L} &= -\frac{1}{2}[\sigma\partial_\mu^2\sigma] - \frac{1}{2}\left(3\lambda v^2 - a^2\right)\sigma^2 \\ &- \frac{1}{2}[\vec{\pi}\partial_\mu^2\vec{\pi}] - \frac{1}{2}\left(\lambda v^2 - a^2\right)\vec{\pi}^2 + \frac{a^2}{2}v^2 \\ &- \frac{\lambda}{4}v^4 + i\bar{\psi}\gamma^\mu\partial_\mu\psi - gv\bar{\psi}\psi + \mathcal{L}_I^b + \mathcal{L}_I^f,\end{aligned} \tag{3}$$

where $\mathcal{L}_I^b$ and $\mathcal{L}_I^f$ are given by

$$\begin{aligned}\mathcal{L}_I^b &= -\frac{\lambda}{4}\Big[(\sigma^2 + (\pi^0)^2)^2 + 4\pi^+\pi^-(\sigma^2 + (\pi^0)^2 + \pi^+\pi^-)\Big], \\ \mathcal{L}_I^f &= -g\bar{\psi}(\sigma + i\gamma_5\vec{\tau}\cdot\vec{\pi})\psi.\end{aligned} \tag{4}$$

Equation (4) describes the interactions among the σ, $\vec{\pi}$ and ψ fields after symmetry breaking. From Eq. (3) one can see that the sigma, the three pions and the quarks have masses given by

$$m_\sigma^2 = 3\lambda v^2 - a^2, \quad m_\pi^2 = \lambda v^2 - a^2, \quad m_f = gv, \tag{5}$$

respectively. We study the behavior of the effective potential, which we deduce in the next section in detail, in order to analyze the chiral symmetry restoration conditions in terms of temperature and quark chemical potential.

3 Effective potential

In this section, we compute the $T-$ and $\mu-$dependent effective potential up to ring diagrams in order to account for the plasma screening effects. The tree level potential is given by

$$V^{\text{tree}}(v) = -\frac{a^2}{2}v^2 + \frac{\lambda}{4}v^4, \tag{6}$$

whose minimum is given by

$$v_0 = \sqrt{\frac{a^2}{\lambda}}, \tag{7}$$

since $v_0 \neq 0$, we notice that the symmetry is spontaneously broken. To include quantum corrections at finite temperature and density, we work within the imaginary-time formalism of thermal field theory. The general expression for the one-loop boson contribution can be written as

$$V^{(1)\text{b}}(v, T) = T\sum_n \int \frac{d^3k}{(2\pi)^3} \ln D(\omega_n, \vec{k})^{1/2}, \tag{8}$$

where

$$D(\omega_n, \vec{k}) = \frac{1}{\omega_n^2 + k^2 + m_b^2}, \tag{9}$$

is the free boson propagator with m_b^2 being the square of the boson's mass and $\omega_n = 2n\pi T$ the Matsubara frequencies for boson fields.

For a fermion field with mass m_f, the general expression for the one-loop correction at finite temperature and quark chemical potential μ_q is

$$V^{(1)\mathrm{f}}(v, T, \mu_q) = -T \sum_n \int \frac{d^3k}{(2\pi)^3} \mathrm{Tr}[\ln S(\tilde{\omega}_n - i\mu_q, \vec{k})^{-1}], \tag{10}$$

where

$$S(\tilde{\omega}_n - i\mu_q, \vec{k}) = \frac{1}{\gamma_0\left(\tilde{\omega}_n - i\mu_q\right) + \not{k} + m_f}, \tag{11}$$

is the free fermion propagator and $\tilde{\omega}_n = (2n+1)\pi T$ are the Matsubara frequencies for fermion fields. The ring diagrams term is given by

$$V^{\mathrm{Ring}}(v, T, \mu_q) = \frac{T}{2} \sum_n \int \frac{d^3k}{(2\pi)^3} \ln(1 + \Pi(m_b, T, \mu_q) D(\omega_n, \vec{k})), \tag{12}$$

where $\Pi(m_b, T, \mu_q)$ is the boson's self-energy. In order to compute the self-energy for one boson field, we include all the contribution from the Feynman rules. Therefore, the self-energy is written as

$$\Pi(T, \mu_q) = \sum_{i=\sigma,\pi^0,\pi^\pm} \Pi_i(T) + \sum_{j=u,d} \Pi_j(T, \mu_q), \tag{13}$$

where

$$\begin{aligned}
\Pi_\sigma(T) &= \frac{\lambda}{4}\left[12I\left(\sqrt{m_\sigma^2 + \Pi_\sigma}\right) + 4I\left(\sqrt{m_{\pi^0}^2 + \Pi_{\pi^0}}\right) + 8I\left(\sqrt{m_{\pi^\pm}^2 + \Pi_{\pi^\pm}}\right)\right], \\
\Pi_{\pi^\pm}(T) &= \frac{\lambda}{4}\left[4I\left(\sqrt{m_\sigma^2 + \Pi_\sigma}\right) + 4I\left(\sqrt{m_{\pi^0}^2 + \Pi_{\pi^0}}\right) + 16I\left(\sqrt{m_{\pi^\pm}^2 + \Pi_{\pi^\pm}}\right)\right], \\
\Pi_{\pi^0}(T) &= \frac{\lambda}{4}\left[4I\left(\sqrt{m_\sigma^2 + \Pi_\sigma}\right) + 12I\left(\sqrt{m_{\pi^0}^2 + \Pi_{\pi^0}}\right) + 8I\left(\sqrt{m_{\pi^\pm}^2 + \Pi_{\pi^\pm}}\right)\right],
\end{aligned} \tag{14}$$

with

$$I(x) = \frac{1}{2\pi^2} \int dk \frac{k^2}{\sqrt{k^2 + x}} n\left(\sqrt{k^2 + x}\right), \tag{15}$$

and $n(x)$ being the Bose-Einstein distribution. On the other hand, the fermion contribution is given by

$$\begin{aligned}
\Pi_j(T, \mu_q) = -g^2 T \sum_n \int \frac{d^3k}{(2\pi)^3} &\mathrm{Tr}[S(\tilde{\omega}_n - i\mu_q, \vec{k}, m_f) \\
&\times S(\tilde{\omega}_n - i\mu_q - \tilde{\omega}_m, \vec{k} - \vec{p}, m_f)].
\end{aligned} \tag{16}$$

As we work close to the phase transition, a good approximation is to take the fermion's mass ($m_f = 0$) and the boson's mass including thermal correction ($m_i^2 + \Pi_i = 0$) to be small. Therefore, the total self-energy for one boson is

$$\Pi(T,\mu_q) = -N_f N_c g^2 \frac{T^2}{\pi^2}[\mathrm{Li}_2(-e^{\mu_q/T}) + \mathrm{Li}_2(-e^{-\mu_q/T})] + \frac{\lambda T^2}{2}. \tag{17}$$

With the boson self-energy at hand we can study the properties of the effective potential. In order to work with analytical expressions we turn to study two cases: first the high temperature approximation ($T \gg m_b,\ \mu_q$) and then the low temperature approximation ($T \ll m_b,\ \mu_q$). In the following we compute explicitly both cases.

3.1 High temperature approximation

As long as T is the largest of the energy scales, a high temperature approximation is suited to study chiral symmetry restoration. Let's start from Eq. (8), the one-loop correction for boson fields. First we compute the sum over Matsubara frequencies

$$V^{(1)\mathrm{b}}(v,T) = \frac{1}{2\pi^2}\int dk\, k^2\Big\{\frac{\sqrt{k^2+m_b^2}}{2} + T\ln\Big(1-e^{-\sqrt{k^2+m_b^2}/T}\Big)\Big\}. \tag{18}$$

Notice that Eq. (18) has two pieces, the first one is the *vacuum* contribution and the second one is the *matter* contribution, namely, the T-dependent correction. In order to compute the vacuum term, we regularize and renormalize the former employing dimensional regularization and the Minimal Subtraction scheme (MS), with the renormalization scale $\tilde{\mu} = e^{-1/2}a$. For the matter term, we take the approximation $m_b/T \ll 1$ and we include only the most dominant terms [1, 2]. Taking all of this into account, the one-loop contribution to the effective potential from fermion fields is given by

$$\begin{aligned} V^{(1)\mathrm{b}}_{\mathrm{HT}}(v,T) = &-\frac{m_b^4}{64\pi^2}\Big[\ln\Big(\frac{4\pi a^2}{m_b^2}\Big) - \gamma_E + \frac{1}{2}\Big] \\ &-\frac{m_b^4}{64\pi^2}\ln\Big(\frac{m_b^2}{(4\pi T)^2}\Big) - \frac{\pi^2 T^4}{90} \\ &+\frac{m_b^2 T^2}{24} - \frac{m_b^3 T}{12\pi}. \end{aligned} \tag{19}$$

For the case of the fermion one-loop contribution, we follow the procedure outlined for the boson case. For the matter term, we compute the integral in momentum taking into account the approximation where $m_f/T \ll 1$ and $\mu_q/T < 1$, and we consider only the dominant terms. After computing the momentum integral we get

$$\begin{aligned} V^{(1)\mathrm{f}}_{\mathrm{HT}}(v,T) = &\frac{m_f^4}{16\pi^2}\Big[\ln\Big(\frac{4\pi a^2}{m_f^2}\Big) - \gamma_E + \frac{1}{2}\Big] \\ &+\frac{m_f^4}{16\pi^2}\Big[\ln\Big(\frac{m_f^2}{(4\pi T)^2}\Big) - \psi^0\Big(\frac{1}{2}+\frac{i\mu}{2\pi T}\Big) - \psi^0\Big(\frac{1}{2}-\frac{i\mu}{2\pi T}\Big)\Big] \\ &- 8m_f^2T^2\Big[\mathrm{Li}_2(-e^{\mu_q/T}) + \mathrm{Li}_2(-e^{-\mu_q/T})\Big] \\ &+ 32T^4\Big[\mathrm{Li}_4(-e^{\mu_q/T}) + \mathrm{Li}_4(-e^{-\mu_q/T})\Big]. \end{aligned} \tag{20}$$

For the purposes of considering the plasma screening effects, we go beyond the mean field approximation. These can be accounted for by means of the ring diagrams. Since we are working in the high temperature approximation, we notice that the lowest Matsubara mode is the most dominant term [3]. Therefore, we do not need to compute the other modes and Eq. (12) becomes

$$\begin{aligned} V^{\text{Ring}}(v,T,\mu_q) &= \frac{T}{2}\int \frac{d^3k}{(2\pi)^3}\ln(1+\Pi(T,\mu_q)D(\vec{k})) \\ &= \frac{T}{4\pi^2}\int dk\, k^2\Big\{\ln(k^2+m_b^2+\Pi(T,\mu_q)) - \ln(k^2+m_b^2)\Big\}. \end{aligned} \tag{21}$$

From Eq. (21), we see that both integrands are almost the same except that one is modified by the self-energy and the other one is not. Thus, after integration, we obtain that the ring diagrams contribution is

$$V^{\text{Ring}}(v,T,\mu_q) = \frac{T}{12\pi}(m_b^3 - (m_b^2+\Pi(T,\mu_q))^{3/2}). \tag{22}$$

With these pieces at hand, we can write the effective potential up to the ring diagrams contribution in the high temperature approximation. The effective potential in the high temperature approximation is given by

$$\begin{aligned} V^{\text{eff}}_{\text{HT}}(v,T,\mu_q) &= -\frac{(a^2+\delta a^2)}{2}v^2 + \frac{(\lambda+\delta\lambda)}{4}v^4 \\ &+ \sum_{b=\sigma,\tilde{\pi}}\Big\{-\frac{m_b^4}{64\pi^2}\Big[\ln\Big(\frac{a^2}{4\pi T^2}\Big) - \gamma_E + \frac{1}{2}\Big] \\ &- \frac{\pi^2 T^4}{90} + \frac{m_b^2T^2}{24} - \frac{(m_b^2+\Pi(T,\mu_q))^{3/2}T}{12\pi}\Big\} \\ &+ \sum_{f=u,d}\Big\{\frac{m_f^4}{16\pi^2}\Big[\ln\Big(\frac{a^2}{4\pi T^2}\Big) - \gamma_E + \frac{1}{2} \\ &- \psi^0\Big(\frac{1}{2}+\frac{\mathrm{i}\mu_q}{2\pi T}\Big) - \psi^0\Big(\frac{1}{2}-\frac{\mathrm{i}\mu_q}{2\pi T}\Big)\Big] \\ &- 8m_f^2T^2\Big[\mathrm{Li}_2(-e^{\mu_q/T}) + \mathrm{Li}_2(-e^{-\mu_q/T})\Big] \\ &+ 32T^4\Big[\mathrm{Li}_4(-e^{\mu_q/T}) + \mathrm{Li}_4(-e^{-\mu_q/T})\Big]\Big\}. \end{aligned} \tag{23}$$

Notice that the potentially dangerous pieces coming from linear or cubic powers of the boson mass, that could become imaginary for certain values of v, are removed or replaced by the contribution of the ring diagrams [4].

3.2 Low temperature approximation

To have access to the region in the QCD phase diagram where μ_B is large and T is small, we compute the effective potential in the approximation where T is the soft scale in the system. We call this the low temperature approximation. For boson fields case, we include a boson chemical potential. We relate this to the conservation of an average number of particles and not to a conserved charge. Therefore, the one-loop contribution for boson fields after the sum over Matsubara frequencies is

$$V_{\rm LT}^{(1){\rm b}}(v,T,\mu_b) = \frac{1}{2\pi^2}\int dk\, k^2\Big\{\sqrt{k^2+m_b^2} + T\ln\Big(1-e^{-(\sqrt{k^2+m_b^2}-\mu_b)/T}\Big) + T\ln\Big(1-e^{-(\sqrt{k^2+m_b^2}+\mu_b)/T}\Big)\Big\}. \tag{24}$$

In Eq. (24), the matter contribution has two terms corresponding to particles and anti-particles. In this work, we follow the procedure used in Ref. [5]. The general idea consists on developing a Taylor series around $T = 0$ of the following expression

$$V_{\rm LT}^{(1){\rm b}}(v,T,\mu_b) = \int_{\frac{\mu_b-m_b}{T}}^{\infty} V_0^{\rm b}(v,\mu_b+xT)h_B(x)dx, \tag{25}$$

where $h_B(x)$ is the first derivative of the Bose-Einstein distribution and $V_0^{\rm b}(v,\mu_b+xT)$ is the one-loop boson contribution evaluated at $T = 0$, which is

$$V_0^{(1){\rm b}}(v,\mu_b) = -\frac{m_b^4}{64\pi^2}\Big[\ln\Big(\frac{4\pi a^2}{(\mu_b+\sqrt{\mu_b^2-m_b^2})^2}\Big) - \gamma_E + \frac{1}{2}\Big] + \frac{\mu_b\sqrt{\mu_b^2-m_b^2}}{96\pi^2}(2\mu_b^2-5m_b^2). \tag{26}$$

From here, the Taylor series give us an expression of one-loop matter contribution from one boson field in the low temperature approximation

$$V_{\rm LT}^{(1){\rm b}}(v,T,\mu_b) = V_0^{\rm b}(v,\mu_b) + \frac{\pi^2T^2}{12}\frac{\partial^2}{\partial T^2}V_0^{\rm b}(v,\mu_b) + \frac{7\pi^4T^4}{1260}\frac{\partial^4}{\partial T^4}V_0^{\rm b}(v,\mu_b). \tag{27}$$

For fermion fields, we implement the low temperature approximation in the same way as we did for boson fields. We now develop a Taylor series around $T = 0$ and we get the one-loop contribution for one fermion field in the low temperature approximation

$$V_{\rm LT}^{(1){\rm f}}(v,T,\mu_q) = V_0^{\rm f}(v,\mu_q) + \frac{\pi^2T^2}{6}\frac{\partial^2}{\partial T^2}V_0^{\rm f}(v,\mu_q) + \frac{\pi^4T^4}{360}\frac{\partial^4}{\partial T^4}V_0^{\rm f}(v,\mu_q). \tag{28}$$

Equations (6), (27) and (28) provide the full expression for the effective potential in the low temperature approximation, which is given by

$$
\begin{aligned}
V_{\rm LT}^{\rm eff}(v,T,\mu_q,\mu_b) &= -\frac{(a^2+\delta a^2)}{2}v^2 + \frac{(\lambda+\delta\lambda)}{4}v^4 \\
&- \sum_{i=\sigma,\bar{\pi}} \Big\{ \frac{m_i^4}{64\pi^2}\Big[\ln\Big(\frac{4\pi^2 a^2}{(\mu_b+\sqrt{\mu_b^2-m_i^2})^2}\Big) - \gamma_E + \frac{1}{2}\Big] \\
&- \frac{\mu_b\sqrt{\mu_b^2-m_i^2}}{24\pi^2}(2\mu_b^2-5m_i^2) - \frac{T^2\mu_b}{12}\sqrt{2\mu_b^2-5m_i^2} \\
&- \frac{\pi^2 T^4\mu_b}{180}\frac{(2\mu_b^2-3m_i^2)}{(\mu_b^2-m_i^2)^{3/2}}\Big\} + N_c\sum_{f=u,d}\Big\{\frac{m_f^4}{16\pi^2}\Big[\ln\Big(\frac{4\pi^2 a^2}{(\mu_q+\sqrt{\mu_q^2-m_f^2})^2}\Big) \\
&- \gamma_E + \frac{1}{2}\Big] - \frac{\mu_q\sqrt{\mu_q^2-m_f^2}}{24\pi^2}(2\mu_q^2-5m_f^2) - \frac{T^2\mu_q}{6}\sqrt{\mu_q^2-m_f^2} \\
&- \frac{7\pi^2T^4\mu_q}{360}\frac{(2\mu_q^2-3m_f^2)}{(\mu_q^2-m_f^2)^{3/2}}\Big\}
\end{aligned} \tag{29}
$$

We are now able to analyze the QCD phase transition in the regions of the QCD phase diagram where the temperature is larger than the quark chemical potential and where the temperature is smaller than the quark chemical potential. To do so, first we need to determine the value of all the parameters involved in the linear sigma model under appropriate conditions. In the following section we proceed in this direction to determine the values of those parameters and in particular of the couplings λ and g.

4 Coupling Constants

The effective potential has tree free parameters which should be fixed: the two coupling constants λ and g and the square mass parameter a^2. In order to determine the square mass parameter, we use that the vacuum boson masses in Eq. (5) are related by the expression

$$
a = \sqrt{\frac{m_\sigma^2 - 3m_\pi^2}{2}}. \tag{30}
$$

Therefore, we can fix a by using the vacuum sigma and pion masses. To fix the coupling constants we use physical inputs from QCD matter around the phase transition in the high and low temperature domains.

First from LQCD computations [6], we know that at $\mu_q \equiv \mu_B/3 = 0$, the QCD phase transition is a crossover (in our case a second order transition), and happens for the case of two light flavors at $T_0^c \simeq 170$ MeV. On the other hand, at very low values of T and high values of μ_q the transition is first order. In addition, from the analysis based on the Hagedorn's limiting temperature [7] at finite μ_B, we know that the critical value for the transition curve to intersect the horizontal axis in the QCD diagram is $\mu_B \simeq m_B$, where $m_B \simeq 1$ GeV is the typical value of the baryon mass. In one or the other case, since the pion field is a Goldstone mode, the thermal pion mass evaluated at the minima of the potential always vanishes. Then, to fix the coupling constants we use as inputs the values of temperature and quark chemical potential in two extreme points along the transition curve, namely, when the restoration of chiral symmetry is at $\mu_q = 0$ (refer as point A) and when it is at $T = 0$ (refer as point B).

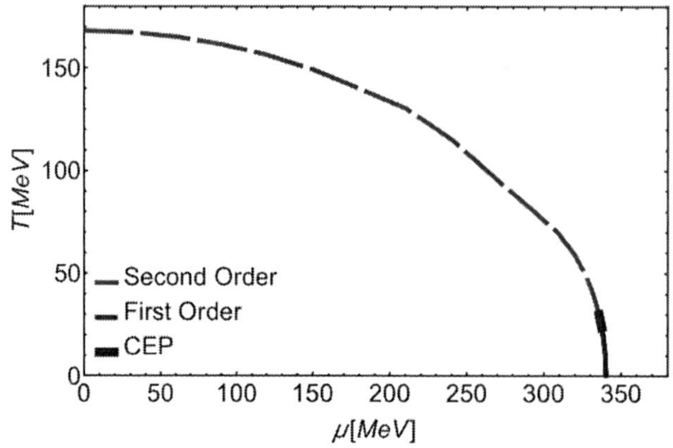

Figure 1. QCD phase diagram obtained from the solutions to the equations that determine the coupling constants. This are computed with $T_0^c(\mu_q = 0) = 170$ MeV and $\mu_q^c(T = 0) = 340$ MeV. The second order transitions are indicated by red line and the first order transitions by the blue line. These areas represent the results directly obtained from our analysis.

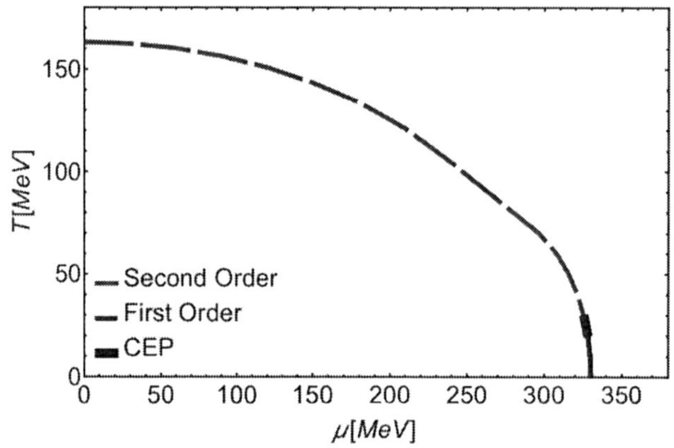

Figure 2. QCD phase diagram obtained from the solutions to the equations that determine the coupling constants. This are computed with $T_0^c(\mu_q = 0) = 165$ MeV and $\mu_q^c(T = 0) = 330$ MeV. The second order transitions are indicated by red line and the first order transitions by the blue line. These areas represent the results directly obtained from our analysis.

At point (A), the phase transition is second order, hence the square of the pion thermal mass, evaluated at $v = 0$ and $T = T_c^0$, is given by

$$m_\pi^2(0, T_0^c, \mu_q = 0) = -a^2 + \Pi(T_0^c, \mu_q = 0) = 0. \tag{31}$$

At point (B), the phase transition is first order (degenerates minimal), therefore the minimum we are considering is the one with a vacuum expectation value different from zero, which we call v_1. This last condition can be written as

Table 1. Physical inputs from points A and B and for vaccum bosons masses

Input	Fig. 2	Fig. 2
Vaccum m_π	139 MeV	139 MeV
Vaccum m_σ	475 MeV	475 MeV
$T_c^0(\mu_q = 0)$	170 MeV	165 MeV
$\mu_c^0(T = 0)$	340 MeV	330 MeV
(λ, g)	(0.87,1.57)	(0.91,1.61)

$$m_\pi^2(v_1, 0, \mu_q^c) = \lambda v_1 - a^2 + \Pi(0, \mu_q^c) = 0, \tag{32}$$

In Eq. (32), we notice that a new unknown appears: v_1, that is, the value of the non-vanishing minimum. In order to guarantee that the order of the transition in points A and B is consistent with the physical input, we need to add counter-terms δa^2 and $\delta\lambda$ to the bare constants a^2 and λ, respectively, in the tree level potential

$$\begin{aligned} V^{\rm tree} &= -\frac{a^2}{2}v^2 + \frac{\lambda}{4}v^4 \\ &\to -\frac{(a^2 + \delta a^2)}{2}v^2 + \frac{(\lambda + \delta\lambda)}{4}v^4. \end{aligned} \tag{33}$$

Therefore the set of conditions necessary to determine v_1 and the counter-terms is

$$\frac{\partial^2 V^{\rm eff}}{\partial v^2}(v = 0, T = T_c, \mu_q = 0) = 0, \tag{34}$$

$$\begin{aligned} \frac{\partial V^{\rm eff}}{\partial v}(v = 0, T = 0, \mu_q = \mu_q^c) = 0, \quad \frac{\partial V^{\rm eff}}{\partial v}(v = v_1, T = 0, \mu_q = \mu_q^c) = 0, \\ V^{\rm eff}(v = 0, T = 0, \mu_q = \mu_q^c) = V^{\rm eff}(v = v_1, T = 0, \mu_q = \mu_q^c). \end{aligned} \tag{35}$$

The expressions in Eq. (34) indicate that the effective potential is second order when $\mu = 0$ and $T = T_c$ and the three expressions in Eq. (35) indicate that the effective potential has two degenerated minima at the phase transition and thus that the transition is first order when $T = 0$ and $\mu_q = \mu_q^c$. The above set of conditions, Eqs. (31), (32) and (35), represent the five algebraic equations that determine the values of λ and g. Finally, we can explore the QCD phase diagram.

5 Results

Figures 1–2 show the phase diagrams thus obtained. These are computed using the inputs in Table 1. We find that at high (low) temperature and low (high) quark chemical potential the phase transitions are second (first) order. The second order transitions are indicated by the red line and the first order transitions by the blue line. These areas represent the results directly obtained from our analysis. In all cases, we locate the CEP's region at low temperatures and high quark chemical potential.

6 Summary

In this work we have used the linear sigma model with quarks to explore the QCD phase diagram from the point of view of chiral symmetry restoration. We have computed the finite temperature effective potential up to the contribution of the ring diagrams to account for the plasma screening effects. For high quark chemical potential we introduced a boson chemical potentials linked to the high baryon abundanceand.

Our approach was to determine the model's couplings using physical inputs such as the vacuum pion and sigma masses, the LQCD value for the critical temperature at $\mu_q = 0$ and the conjectured end point value of μ_B of the transition line at $T = 0$. The set of conditions that determine the couplings enforce the requirement that at high temperature the transition is second order whereas at low temperature is first order.

In order to provide a more robust CEP's location, we need to include the temperature and density modifications to the couplings which has been shown useful to describe the inverse magnetic catalysis phenomenon [8]. For more explicit details, the reader us referred to Ref [9].

Acknowledgments

Support for this work has been received in part by UNAM-DGAPA-PAPIIT grant number IN101515 and by Consejo Nacional de Ciencia y Tecnología grant number 256494.

References

[1] A. Ayala, J. D. Castaño-Yepes, J. J. Cobos-Martínez, S. Hernández-Ortiz, A. J. Mizher, A. Raya, Int. J. Mod. Phys. A **31**, 1650199 (2016).

[2] A. Ayala, A. Bashir, J.J. Cobos-Martínez, S. Hernández-Ortiz, A. Raya, Nucl. Phys. B **897**, 77-86 (2015).

[3] M. Le Bellac, *Thermal Field Theory* (Cambridge Monographs on Mathematical Physics), Cambridge University Press, 1996.

[4] L. Dolan and R. Jackiw, Phys. Rev. D **9**, 3320-3341 (1974).

[5] C. O. Dib, O. R. Espinosa, Nucl. Phys. B **612**, 492-518 (2001).

[6] T. Bhattacharya *et al.*, Phys. Rev. Lett. **113**, 082001 (2014).

[7] R. Hagedorn, Nuovo Cim. Suppl. **3** (1965) 147; Nuovo Cim. A**56** (1968) 1027; See also K, Fukushima, T. Hatsuda, Rept. Prog. Phys. **74**, 014001 (2011).

[8] A. Ayala, L. A. Hernández, M. Loewe, A. Raya, J. C. Rojas, R. Zamora, Phys. Rev. D **96**, 034007 (2017); A. Ayala, C. A. Dominguez, L. A. Hernández, M. Loewe, A. Raya, J. C. Rojas, C. Villavicencio, Phys. Rev. D **94**, 054019 (2016); A. Ayala, M. Loewe, R. Zamora, Phys. Rev. D **91**, 016002 (2015); A. Ayala, M. Loewe, A. J. Mizher, R. Zamora, Phys. Rev. D **90**, 036001 (2014).

[9] A. Ayala, S. Hernández-Ortiz and L. A. Hernández, arXiv:1710.09007 (2017).

EPJ Web of Conferences **172**, 02004 (2018) https://doi.org/10.1051/epjconf/201817202004
ISMD 2017

Future Measurements of the Nucleon Elastic Electromagnetic Form Factors at Jefferson Lab

Gerard Gilfoyle[1,⋆]

[1] *Physics Department, University of Richmond, Richmond, VA, USA*

Abstract. The elastic, electromagnetic form factors are fundamental observables that describe the internal structure of protons, neutrons, and atomic nuclei. Jefferson Lab in the United States has completed the 12 GeV Upgrade that will open new opportunities to study the form factors. A campaign to measure all four nucleon form factors (electric and magnetic ones for both proton and neutron) has been approved consisting of seven experiments in Halls A, B, and C. The increased energy of the electron beam will extend the range of precision measurements to higher Q^2 for all four form factors together. This combination of measurements will allow for the decomposition of the results into their quark components and guide the development of a QCD-based understanding of nuclei in the non-perturbative regime. I will present more details on the 12 GeV Upgrade, the methods used to measure the form factors, and what we may learn.

1 Introduction

The elastic electromagnetic form factors are fundamental observables in electron scattering that encode information about the distribution of electric charge and current in the nucleus and its constituents. The study of nuclei with electromagnetic probes has a long, distinguished history in nuclear physics including the Nobel Prize awarded to Robert Hofstadter in 1961 [1]. With the completion of the 12 GeV Upgrade at the Thomas Jefferson National Accelerator Facility (JLab) in Newport News, VA, USA we now stand on the doorstep of a new chapter in our understanding of atomic nuclei [2]. The JLab Program Advisory Committee (PAC) has approved a campaign of seven experiments to run in three different experimental halls to measure the elastic, electric and magnetic form factors for both the neutron and proton. The focus of the campaign will be mapping out the quark substructure of the nucleon far beyond our current range and to test the fundamental theory of the strong force, Quantum Chromodynamics (QCD), in the non-perturbative region.

In this paper we present a status report on our current understanding of the elastic, electromagnetic form factors (EEFFs) of the nucleon and what we may learn with the extended reach of this new generation of experiments at JLab. We then describe how that reach will be extended in the JLab campaign. We end with a summary and conclusions.

⋆e-mail: ggilfoyl@richmond.edu

© The Authors, published by EDP Sciences. This is an open access article distributed under the terms of the Creative Commons Attribution License 4.0 (http://creativecommons.org/licenses/by/4.0/).

2 Where We Are Now.

In this section we present some of the necessary background to understand the EEFFs and then discuss selected topics in our current experimental and theoretical understanding. The electromagnetic form factors are the essential components of the cross section in the region where the one-photon exchange is applicable. That cross section can be expressed in terms of the Dirac (F_1) and Pauli (F_2) form factors

$$\frac{d\sigma}{d\Omega} = \sigma_{Mott}\left[(F_1^2 + \kappa^2\tau F_2^2) + 2\tau(F_1 + \kappa F_2)^2 \tan^2\left(\frac{\theta_e}{2}\right)\right] \tag{1}$$

where

$$\sigma_{Mott} = \frac{\alpha^2 E' \cos^2(\frac{\theta_e}{2})}{4E^3 \sin^4(\frac{\theta_e}{2})} \tag{2}$$

and κ is the anomalous magnetic moment, E (E') is the incoming (outgoing) electron energy, θ_e is the scattered electron angle and $\tau = Q^2/4M^2$. The Dirac and Pauli form factors can be recast into electric and magnetic parts - the Sachs form factors .

$$\frac{d\sigma}{d\Omega} = \frac{\sigma_{Mott}}{\epsilon(1+\tau)}\left(\epsilon G_E^2 + \tau G_M^2\right) \tag{3}$$

where

$$G_E = F_1 - \kappa\tau F_2 \quad \text{and} \quad G_M = F_1 + \kappa F_2 \quad \text{and} \quad \epsilon = \left[1 + 2(1+\tau)\tan^2\frac{\theta_e}{2}\right]^{-1} \tag{4}$$

There is an extensive data set for the EEFFs at lower 4-momentum transfer squared or Q^2, but the data become sparse at increasing Q^2. The proton magnetic form factor G_M^p has the best data set at this point. Figure 1a shows the world's data for G_M^p relative to the dipole form factor. The dipole form factor is $G_D = (1 + Q^2/\Lambda)^{-2}$ where $\Lambda = 0.71$ GeV^2. The measurements extend out to $Q^2 \approx 30$ GeV^2

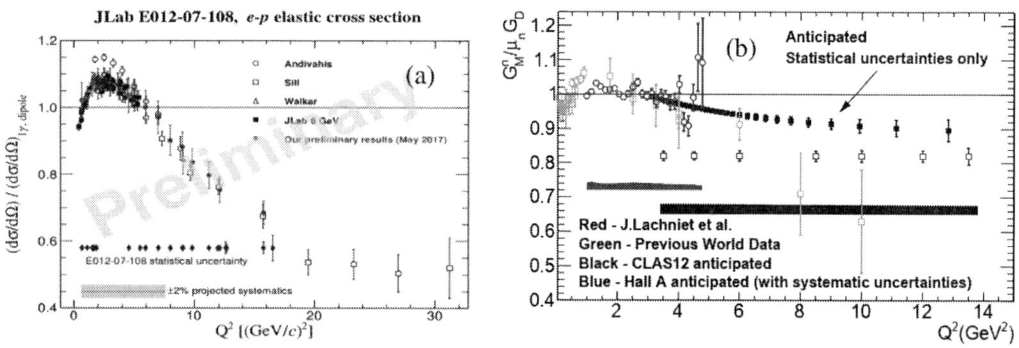

Figure 1. The world's data for $G_M^p/\mu_p G_D$ data are shown in panel (a) [3]. The red points are preliminary results from Experiment E12-07-108. The world's data for $G_M^n/\mu_n G_D$ are shown in panel (b) [4]. The measured points are the green, open squares and the red, open circles. The uncertainties on the red points are statistical ones. The red bar chart represents the systematic uncertainties on the red data points. The other data points are anticipated ones for future experiments and are discussed below.

and are particularly dense for $Q^2 < 8$ GeV^2. Above 8 GeV^2 they are sparse and the uncertainties grow. This is where the JLab program will have the greatest impact. The red points in Figure 1a are preliminary results from JLab experiment E12-07-108 which completed data collection in 2017. The uncertainties on the red points are much larger than what is expected for the final results ($\approx 2\%$). The more-difficult-to-measure neutron magnetic form factor G_M^n measurements are more limited in Q^2 coverage and are shown in Figure 1b. The measured results are the green, open squares and the red, open circles. The data are precise out to $Q^2 \approx 5$ GeV^2 with a few more points with large uncertainties extending out to $Q^2 = 10$ GeV^2.

The Sachs nucleon electric form factors G_E^p and G_E^n are now routinely shown as the ratio relative to the magnetic form factors $\mu_p G_E/G_M$. The world's data for both proton and neutron ratios are shown in Figure 2. The proton ratio is extracted with the recoil polarization method (which will be discussed

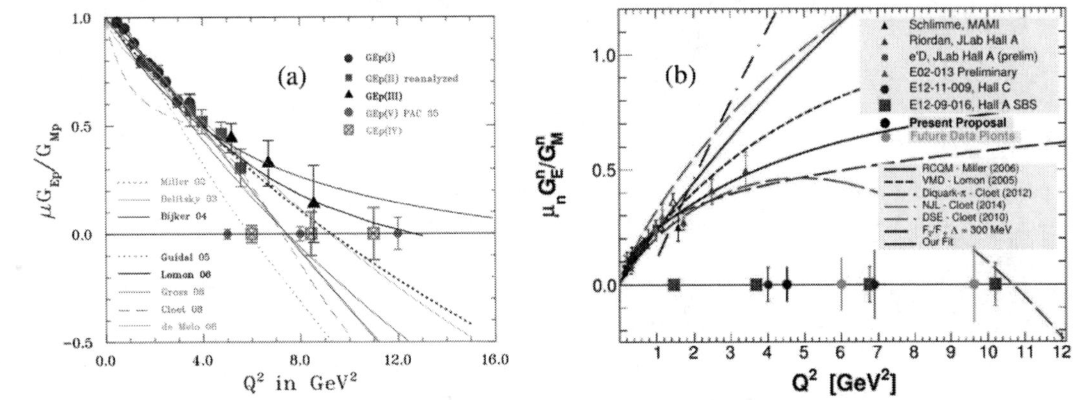

Figure 2. World's data for the proton form factor ratio $\mu_p G_E^p/G_M^p$ data using the recoil polarization method are shown in panel (a) [5, 6]. World's data for $\mu_n G_E^n/G_M^n$ are shown in panel (b) [7]. In both cases the points plotted along the axis represent the anticipated Q^2 and uncertainty on future measurements.

below) and drops linearly from $Q^2 = 0$. The data extend out to $Q^2 \approx 8$ GeV^2. This behavior is in sharp contrast to earlier measurements using the Rosenbluth separation methods which showed a Q^2 dependence largely constant at unity although with large uncertainties [8]. This radical difference is now mostly attributed to two-pion exchange [9]. There also appears to be a zero-crossing around the highest Q^2 point. The significance of this feature is discussed below. The neutron electric form factor data is extracted with both the double-spin asymmetry and the recoil polarization methods (more discussion below). The data here are even more limited. They extend only out to $Q^2 = 3.5$ GeV^2 and are steadily rising. It is worth noting here that many of the measurements discussed above have been made over the last twenty years and have generated renewed interest in the EEFFs especially at large Q^2. We next explore what these new data have revealed and where they may lead.

By measuring all four nucleon EEFF's and invoking charge symmetry the quark Dirac and Pauli form factors can be extracted [10].

$$F^u_{1(2)} = 2F^p_{1(2)} + F^n_{1(2)} \qquad \text{and} \qquad F^d_{1(2)} = 2F^n_{1(2)} + F^p_{1(2)} \tag{5}$$

One can now make a flavor decomposition and determine the Q^2-dependence of the individual valence quarks [11]. The result of this process is shown in Figure 3 [10]. There are large differences between the u- and d-quark distributions both in size and shape. The u-quark form factors in Figure 3 are scaled by Q^4 and still rise steadily across the full range while the d-quark form factors saturate at around $Q^2 = 1.4$ GeV^2. To normalize the two quark distributions to each other at low-Q^2 requires applying factors (0.75 and 2.5 in Figure 3) that differ by more than a factor of three. In addition, the Q^2 dependence of the form factor ratio F_2/F_1 is predicted by perturbative QCD to follow a $1/Q^2$ form (see Figure 1 in Reference [10]). This feature differs sharply from the results in Figure 3 and also

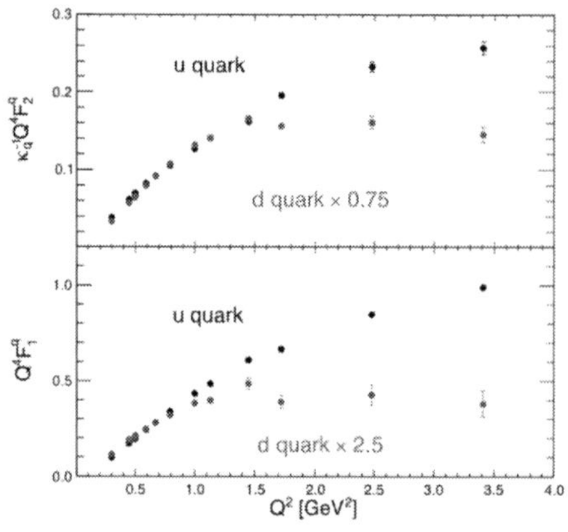

Figure 3. Flavor decomposition of the nucleon elastic form factors showing the Q^2 behavior of the Dirac and Pauli form factors for the $u-$ and $d-$quarks [10]. The form factors are multiplied by Q^4 to show their properties more clearly.

for the behavior of the proton and neutron form factors. A calculation of the flavor form factors has found encouraging agreement with the data using the Nambu-Jona-Lasinio model with the empirical nucleon form factors and treating the pion degrees of freedom as a perturbation on the quark core [12]. The quark form factors are extracted assuming charge symmetry. No model parameters are adjusted to fit the data. Diquark correlations arise naturally in this model and may provide an explanation of the $u-$ and d-quark Q^2 dependence in Figure 3.

Another theoretical approach to describing the EEFFs is shown in Figure 4 which also shows the proton form factor ratio data. The Dyson-Schwinger equations (DSEs) are the equations of motion of quantum field theory and form an infinite set of coupled, integral equations. They are inherently

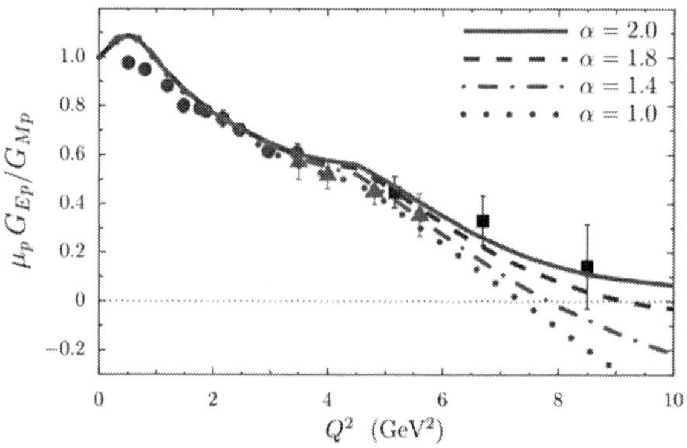

Figure 4. A calculation of the proton form factor ratio $\mu_p G_E^p/G_M^p$ using the Dyson-Schwinger approach[13]. The different curves are linked to different shapes of the mass function $M(p)$ which describes how the proton acquires mass through the dynamical chiral symmetry breaking mechanism.

relativistic, non-perturbative, and deeply connected to QCD. Choosing the proper subset of equations is an essential step in connecting the DSEs to measurement. In Reference [13], Cloët *et al.* model the nucleon dressed quark operator as a quark-diquark combination and include a damping factor to alter the shape of the momentum dependence of the dressed-quark mass function $M(p)$. This mass function describes how the nucleons acquire mass from dynamical chiral symmetry breaking. The authors calculate the ratio of the proton form factors $\mu_p G_E^p / G_M^p$ and observe a zero crossing in this ratio near the Q^2 limit of the existing data. The different curves shown in Figure 4 are for different values of the damping factor and show how changing the shape of $M(p)$ changes the Q^2 value of the crossover point. In other words, measuring this crossover is revealing how the proton acquires mass. A precise measurement of the the proton form factor ratio at higher Q^2 is needed to determine this crossover.

In the discussion of Figure 3 above it was noted that the data are not well described by perturbative QCD. Another calculation of the quark form factors is described in Reference [14] within the framework of the light-front holographic QCD (LFHQCD). This approach takes advantage of a striking relationship between light-front dynamics, its holographic mapping to gravity in a higher-dimensional anti-de Sitter space, and conformal quantum mechanics [14]. To obtain good agreement with the data an admixture of a five quark state is included in the proton (30%) and the neutron (40%). A third parameter r is introduced to account for $SU(6)$ spin-flavor symmetry breaking and adjusted to fit the data. The results for the neutron form factor ratio $\mu_n G_E^n / G_M^n$ are shown in Panel (a) in Figure 5. The fitted value of $r = 2.08$ is required to produce the green curve which agrees well with the data. The orange curve shows the same calculation with no $SU(6)$ spin-flavor symmetry breaking (*i.e.* $r = 1.0$). Note the calculation (green curve) approaches an asymptotic value at large Q^2. In Panel (b) a

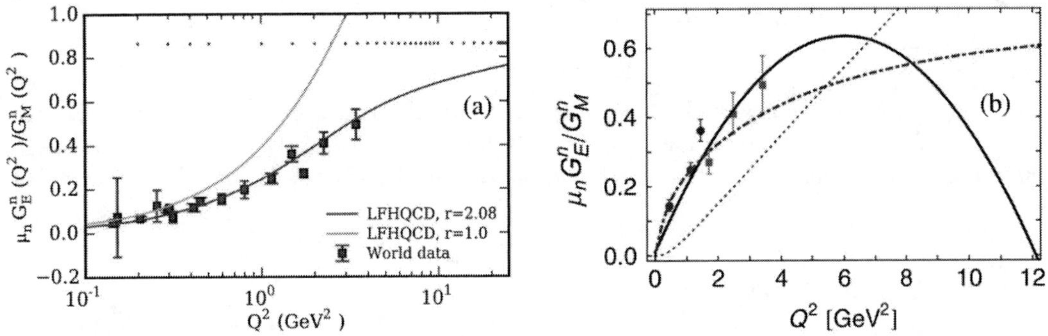

Figure 5. Panel (a) shows the neutron form factor ratio $\mu_n G_E^n / G_M^n$ calculated from the LFHQCD model (green curve) [14]. The model approaches an asymptotic value as $Q^2 \to \infty$ of 0.86 ± 0.11 (green points). Panel (b) shows the result of a calculation using the DSE approach (black curve) [15]. The red, dot-dashed curve is a parameterization of the data [16]. The blue, dashed curve is a contact interaction calculation [17].

prediction of the same ratio using the Dyson-Schwinger approach discussed above is shown [15, 18]. Like a similar calculation of the proton ratio shown in Figure 4 above, it has a zero crossing at high Q^2 where the LFHQCD calculation is approaching its asymptotic value of 0.86 ± 0.11. This zero crossing is a striking difference with the LFHQCD calculation and presents an opportunity to probe fundamental hadronic physics and confinement. Data at higher Q^2 are needed to distinguish between these two calculations.

Lattice QCD calculations are now becoming feasible in the few-GeV^2 range, and over the next decade these calculations will become increasingly precise [19]. The elastic form factors for both the proton and neutron are an important, early test case of the accuracy of the lattice calculations. With all four of them, one can extract the isovector combination of the form factors [20] which are easier to calculate on the lattice because they lack disconnected contributions [21]. Figure 6 shows a recent calculation. The blue points are the results of the calculation and the red curve is a parameterization of the data. There is excellent agreement between the data parameterization and the calculations, but note the limited Q^2 range - only out to $Q^2 = 1.3$ GeV^2. Higher Q^2 data will continue to challenge these calculations. Finally, Generalized Parton Distributions combine information from the EEFFs

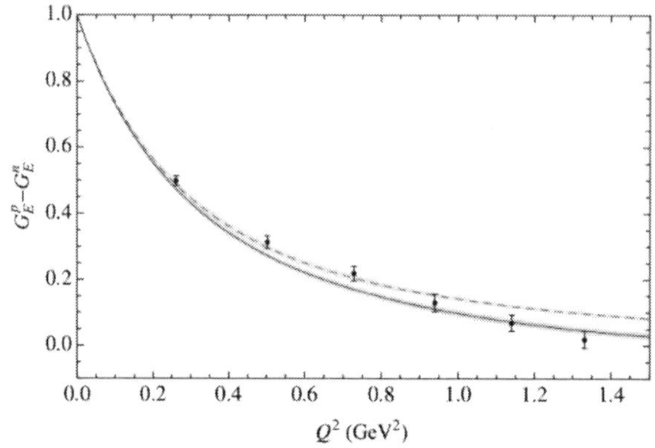

Figure 6. The results of a lattice QCD calculation of the isovector combination of the form factors ($G_E^p - G_E^n$) in shown [22]. The blue points are the calculation and the red curve is a parameterization of the data. The green dashed curve is a fit to the calculated points (blue).

and parton distribution functions (PDFs) and provide a deeper picture of nucleon structure including the role of orbital angular momentum [23]. The elastic form factors are a limiting case of the GPDs and provide an essential constraint on GPD models. We now discuss how those measurements will be made.

3 Where We Are Going

The campaign to measure the EEFFs at JLab begins with CEBAF - a superconducting, continuous-wave, electron accelerator with beams up to 12 GeV, an energy resolution of 2×10^{-4}, and polarization greater than 80%. The 12 GeV Upgrade has recently been completed and the accelerator is now in production running. There are four end stations for experiments (Halls, A, B, C, and D). The EEFF campaign will be carried out in three of those halls (A, B, C). The lineup of approved experiments is shown in Table 1. All of the experiments in the table are based on measurements performed during the earlier 6-GeV era at JLab. More than 200 days of beamtime have been approved to measure the elastic electromagnetic form factors. Details of the experimental program will be given below.

The proton magnetic form factor G_M^p is the best-known of the EEFFs and that understanding will be strengthened by JLab Experiment E12-07-108 in Hall A. A precise measurement of the ep elastic cross will be made using the HRS spectrometers [3]. The range of very precise cross sections will be extended up to $Q^2 = 15$ GeV^2 with expected systematic uncertainties of 2%. Data collection was completed in 2017 and the analysis is underway. Preliminary results (with increased uncertainties) are shown in Figure 1a. This first measurement is crucial because it sets the scale for the remaining EEFF measurements.

The measurement of the proton form factor ratio $\mu_p G_E^p / G_M^p$ (E12-07-109) will also be performed in Hall A using the recoil polarization method on a proton cryo-target in the reaction $^1\mathrm{H}(\vec{e}, e'\vec{p})$ [6].

Quantity	Method	Target	Q^2(GeV^2)	Hall	Beam Days
G_M^p *	Elastic scattering	LH_2	$7-15.5$	A	24
G_E^p/G_M^p	Recoil Polarization	LH_2	$5-12$	A	45
G_M^n	$E-p/e-n$ ratio	LD_2-LH_2	$3.5-13.0$	B	30
G_M^n	$E-p/e-n$ ratio	LD_2, LH_2	$3.5-13.5$	A	25
G_E^n/G_M^n	Double polarization asymmetry	polarized ^{3}He	$5-8$	A	50
G_E^n/G_M^n	Recoil Polarization	LD_2	$4-7$	C	50
G_E^n/G_M^n	Recoil Polarization	LD_2	4.5	A	5

Table 1. Listing of approved experiments for measuring the elastic electromagnetic form factors.

This experiment takes advantage of the relationship

$$\frac{G_E^p}{G_M^p} = -\frac{P_t}{P_l}\frac{E+E'}{2M}\tan\left(\frac{\theta_e}{2}\right) \tag{6}$$

where P_t and P_l are the transverse and longitudinal polarizations (measured with a proton polarimeter), E and E' are the electron incident and scattered energies, M is the proton mass, and θ_e is the electron scattering angle. The experiment will rely on the Big Cal lead-glass spectrometer for the electron arm and the new Super BigBite magnetic spectrometer for the proton polarimeter. This experiment has been given a high-impact rating by the JLab PAC and is expected to run in the next 3-4 years. The light blue, filled circles in Figure 2a located on the axis represent the Q^2 values where the measurements will be made and the expected uncertainties. The uncertainties are low and extend out to $Q^2 = 12$ GeV^2. The results could precisely map the location of the zero crossing in the proton form factor ratio where the DSE calculation (recall Figure 4) showed the position of this crossover is sensitive to the shape of the proton mass function $M(p)$. This measurement has the potential to probe a core component of nucleon structure and QCD.

The neutron magnetic form factor G_M^n will be measured in two experiments in Halls A and B. Precise neutron measurements are notoriously difficult so measuring G_M^n with two separate experimental setups provides better control of the systematic uncertainties. The measurement in Hall B (E12-07-104) will use the new CLAS12 detector to extract the ratio R of quasi-elastic (QE) $e-n$ to $e-p$ scattering on deuterium

$$R = \frac{\frac{d\sigma}{d\Omega}[^2\mathrm{H}(e,e'n)_{QE}]}{\frac{d\sigma}{d\Omega}[^2\mathrm{H}(e,e'p)_{QE}]} = a \times \frac{\sigma_{Mott}\left(\frac{(G_E^n)^2+\tau(G_M^n)^2}{1+\tau} + 2\tau\tan^2\frac{\theta_e}{2}(G_M^n)^2\right)}{\frac{d\sigma}{d\Omega}[^1\mathrm{H}(e,e')p]} \tag{7}$$

where a is a nuclear correction [4]. The denominator in Equation 7 is the well-known $e-p$ cross section (whose precision will improve with the G_M^p measurement described above). The numerator contains the desired quantity G_M^n and the neutron electric form factor G_M^n which is small over the range of the G_M^n measurement. Our understanding of G_E^n will be folded into the systematic uncertainty on G_M^n. Using a ratio measurement here reduces the vulnerability to a number of sources of uncertainty (radiative corrections, Fermi motion correction, *etc*). A precise measurement of the neutron detection efficiency (NDE) is required so a dual $LD_2 - LH_2$ target has been proposed to use the $p(e,e'\pi^+n)$ reaction as a source of tagged neutrons to measure NDE at the same time the production data are collected. The systematic uncertainties are expected to be less than 2.5% across the full range out to $Q^2 = 13$ GeV^2. These data will more than double the existing range of precise measurements of G_M^n

as shown by the black points in Figure 1b. The black bar in the figure represents the maximum value of the expected systematic uncertainty.

The second measurement of G^n_M will be performed in Hall A (E12-09-019) [24]. This experiment will also exploit the ratio method on deuterium as in Hall B using the Super BigBite spectrometer for the electron arm and the HCal hadron calorimeter paired with a large bending magnet (BigBen) to separate neutral and charged particles. A separate calibration measurement of the NDE will be made using photons produced by a radiator inserted into the beam and incident on a hydrogen target. The ^{1}H(γ, π^+)n reaction will provide a source of tagged photons and a cut on the pion momentum will select two-body reactions (the end-point method). The expected systematic uncertainties are below 2.1%. The planned Q^2 coverage and uncertainties are represented by the blue, open squares in Figure 1b. The experiment is expected to run in Hall A in the next 2-3 years.

There are also two experiments approved to measure the neutron form factor ratio $\mu_n G^n_E/G^n_M$ in Halls A and C. The Hall A measurement (E12-09-016) will measure a double polarization asymmetry A^V_{en} from $^3\vec{\mathrm{He}}(\vec{e}, e'n)pp$ using a polarized beam and polarized ^{3}He to provide the neutron target [25]. This asymmetry depends on the neutron form factor ratio and a set of known kinematic factors. It is the following

$$A^V_{en} = -\frac{2\sqrt{\tau(\tau+1)}\tan(\theta_e/2)\cos\phi^*\sin\theta^* G^n_E/G^n_M + 2\tau\sqrt{1+\tau+(\tau+1)^2\tan^2(\theta_e/2)}\tan(\theta_e/2)\cos\theta^*}{(G^n_E/G^n_M)^2 + \tau/\epsilon} \tag{8}$$

where θ^* is the polar angle and ϕ^* is the azimuthal angle of the target polarization in the laboratory frame with respect to the axis of the momentum transfer (the other quantities are defined above).

The electron arm will use the Super BigBite spectrometer (same as used in the proton form factor ratio and G^n_M measurements in Hall A) and the HCal hadron calorimeter will detect the scattered neutron (same as used in G^n_M measurement in Hall A). The planned Q^2 coverage and uncertainties are shown as the red, filled squares in Figure 2b. The Q^2 range of the data for this ratio will be extended by nearly a factor of three expanding the opportunity to do a flavor decomposition out to $Q^2 = 10$ GeV^2. These data will challenge the DSE and LFHQCD calculations of Figure 5. The experiment is expected to run in the next 2-3 years.

The second measurement of the neutron form factor ratio is in Hall C (E12-11-009) and is based on the recoil polarization mechanism used for the proton form factor ratio measurement in Hall A (see Equation 6) [26]. In this experiment a deuterium nucleus serves as the neutron target for the ^{2}H$(\vec{e}, e'\vec{n})p$ reaction. The electron is detected in the Super High Momentum Spectrometer and a neutron polarimeter measures the polarization of the scattered neutron. As with the G^n_M measurements, two experiments have been approved for the neutron form factor ratio measurements to provide better understanding of the systematic errors. The Q^2 points and expected uncertainties for this measurement are shown as the blue, filled circles in Figure 2b. At the highest Q^2 point the statistical uncertainties exceed 10% and this method is near the limit of it's applicability. This limit has motivated the last experiment in this report. A short (five days) run has been approved to test the feasibility of using a charge exchange reaction in the neutron polarimeter to expand the reach of the recoil polarization method to higher Q^2 (E12-17-004) [7]. A single point at $Q^2 = 4.5$ GeV^2 will be measured and is shown as the black, filled circle in Figure 2b. This short run is expected in the next 2-3 years while the Hall C neutron ratio method is anticipated after 2020.

4 Summary and Conclusions

In this paper we have discussed the recent gains in our understanding of the fundamental observables the elastic electromagnetic form factors. These studies already have a long history, but new features have been revealed over the last fifteen years. This new knowledge is built on new technologies in superconducting electron accelerators, high luminosity detectors, and cryogenic and polarized targets. Their application has lead to new discoveries that overturned our previous understanding. Jefferson Lab has a program to measure all of the EEFFs precisely and over a broad kinematic range. This opens the door to extending our understanding of nuclear structure at the quark level (*e.g.* flavor decomposition to extract individual quark form factors) and to stringently challenge our theoretical understanding of QCD with data (*e.g.* distinguish between the DSE approach and the LFHQCD model in Figure 5). This campaign has just begun and we expect much of it to play out in the next few years.

References

[1] R. Hofstadter, H. Fechter, J. McIntyre, Phys. Rev. **91**, 422 (1953)
[2] Nuclear Science Advisory Committee, *The 2015 Long Range Plan for Nuclear Science* (US Department of Energy, 2015)
[3] S. Gilad et al., Experiment E12-07-108, Jefferson Lab, Newport News, VA (2007)
[4] G. Gilfoyle et al., Experiment E12-07-104, Jefferson Lab, Newport News, VA (2007)
[5] V. Punjabi, C.F. Perdrisat, M.K. Jones, E.J. Brash, C.E. Carlson, Eur. Phys. J. **A51**, 79 (2015), 1503.01452
[6] P. Perdrisat et al., Experiment E12-07-109, Jefferson Lab, Newport News, VA (2007)
[7] J. Annand et al., Experiment E12-17-004, Jefferson Lab, Newport News, VA (2017)
[8] A. Puckett et al., Phys. Rev. Lett. **104**, 242301 (2010)
[9] D. Adikaram et al., Phys. Rev. Lett. **114**, 062003 (2015)
[10] G.D. Cates, C.W. de Jager, S. Riordan, B. Wojtsekhowski, Phys. Rev. Lett. **106**, 252003 (2011)
[11] G. Miller, Phys. Rep. **194**, 1 (1990)
[12] I. Cloët et al., Phys. Rev. **C90**, 045202 (2014)
[13] I. Cloët et al., Phys. Rev. Lett. **111**, 101803 (2013)
[14] R. Sufian et al., Phys. Rev. **D95**, 014011 (2017)
[15] J. Segovia, I. Cloët, C. Roberts, S. Schmidt, Few-Body Syst. **55**, 1185 (2014)
[16] J.J. Kelly, Phys. Rev. C **70**, 068202 (2004)
[17] D.J. Wilson, I.C. Cloet, L. Chang, C.D. Roberts, Phys. Rev. **C85**, 025205 (2012), 1112.2212
[18] C. Roberts, **arXiv:1509.02925 [nucl-th]** (2015), 1509.02925
[19] J.J. Dudek, R.G. Edwards, K. Orginos, D.G. Richards, J.Phys.Conf.Ser. **299**, 012007 (2011)
[20] A.W. Thomas, W. Weise, *The Structure of the Nucleon* (Wiley-VCH, 2001)
[21] J.P.A. Committee, Tech. rep., Jefferson Laboratory (2007)
[22] P. Shanahan et al., Phys. Rev. **D90**, 034502 (2014)
[23] M. Diehl, P. Kroll, Eur. Phys. J. C **73**, 2397 (2013)
[24] B. Quinn et al., Experiment E12-09-019, Jefferson Lab, Newport News, VA (2009)
[25] B. Wojtsekhowski et al., Experiment E12-09-016, Jefferson Lab, Newport News, VA (2009)
[26] B. Anderson et al., Experiment E12-11-009, Jefferson Lab, Newport News, VA (2011)

The Charm and Beauty of Strong Interactions

Bruno El-Bennich[1,⋆]

[1]*Laboratório de Física Teórica e Computacional, Universidade Cruzeiro do Sul, Rua Galvão Bueno 868, 01506-000 São Paulo, SP, Brazil*

Abstract. We briefly review common features and overlapping issues in hadron and flavor physics focussing on continuum QCD approaches to heavy bound states, their mass spectrum and weak decay constants in different strong interaction models.

1 Hadron and Flavor Physics: a twofold motivation, a unified approach?

Hadron and flavor physics are often thought of as different research fields though they have in common the underlying theory which describes strong interactions in flavored and non-flavored hadrons, namely Quantum Chromodynamics (QCD). In particular *confinement*, one of QCD's paradigmatic hallmarks, may not be appreciated enough in flavor physics. Undoubtedly, in nonrelativistic QCD and effective field theory approaches to quarkonia and the fauna of new XYZ states the main focus is on spectroscopy and production mechanisms, where most efforts concentrate on nonrelativistic perturbative aspects, effective theories or potentials models [1–4].

While certain quarkonia provide a relatively clean environment to probe strong dynamics, and the mass spectrum of their confined quarks is reasonably well described in nonrelativistic perturbative QCD (pQCD), many of the heavier XYZ states close to threshold may not be pure $\bar{q}q$ states but admixtures of meson molecules and a core of quarks. Moreover, it is desirable to unify the meson *and* baryon spectrum with a unique confining framework that correctly describes the dynamical chiral symmetry breaking (DCSB) pattern which occurs to varying degrees in hadrons. In particular, DCSB may still play an important role in charmonia where nonperturbative $O(\Lambda_{\mathrm{QCD}}/m_c)$ contributions can be non negligible. On the other hand, the study of weak B-meson decays, once driven by factorization theorems [5] and soft collinear effective field theory [6], appears to be nowadays dominated by the quest for new physics largely motivated by recurrent announcements of the LHCb collaboration [7–9]. Still, it is exactly in the case of heavy-light mesons where our understanding of their QCD dynamics and emerging properties are the least well understood within continuum QCD approaches [10–19]. The reason for this is obviously the important difference in quark masses and the energetic light mesons produced in their decays, both of which are at the origin of a wide array of energy scales [20, 21] which effective theories can merely separate.

More generally, the main object of study in flavor physics is the origin of CP-violating phases in weak decays of flavored quarks *confined* in hadrons, where the Cabibbo-Kobayashi-Maskawa (CKM) mechanism has been established as the dominant Standard Model source of CP violation in heavy-meson decays. On the other hand, the overwhelming bulk of visible matter is made of baryons containing light quarks. DCSB due to gluon self-interactions has by now been established to be the most important mass generating mechanism for visible matter in the Universe, which implies that for most observed matter the Higgs mechanism is almost irrelevant. Thus, both field's major preoccupation is the origin and generation of matter and pose the fundamental questions: *What is the origin of matter and its preponderance over antimatter? What is the origin of the mass of baryonic matter?*

⋆e-mail: bruno.bennich@cruzeirodosul.edu.br

© The Authors, published by EDP Sciences. This is an open access article distributed under the terms of the Creative Commons Attribution License 4.0 (http://creativecommons.org/licenses/by/4.0/).

While the first question may not be completely answered within the Standard Model, as the CKM mechanism in the quark sector is not sufficient to explain our matter-dominated Universe [22], the origin of mass due to DCSB has by now been firmly established [23–25]. Moreover, there are good reasons to believe DCSB and confinement are intimately related [26–28]. Experimental data ought to teach us something about the interaction that holds quark-antiquark bound states together in different constellations, namely in ground states and in radially excited or larger angular-momentum states. Quarkonia and heavy-flavored mesons consist of strongly interacting $\bar{q}q$ pairs that offer different perspectives on confinement: in the former case one tests QCD's running coupling at much shorter distances than for lighter mesons whose mass is chiefly due to DCSB, whereas in the latter case the same interaction is scrutinized in an asymmetric system giving rise to disparate energy scales where truncation effects are more evident [29].

At any rate, in calculations of hadronic observables a reliable description of DCSB and confinement cannot be realized with perturbative methods. Focusing on continuum QCD, so far the most successful approaches are based on QCD's equations of motion in Euclidean space, given by the infinite tower of coupled Dyson-Schwinger integral equations [26, 30, 31], and on the functional renormalization group [32]. We here review recent results on the quarkonia and flavored meson mass spectrum and corresponding weak decay constants obtained in the combined framework of the Dyson-Schwinger equation (DSE) and Bethe-Salpeter equation (BSE) and stress the overlap of contemporary hadron physics studies, such as the experimental program of the 12 GeV upgrade at Jefferson Lab, with that of flavor physics facilities.

2 Spectroscopy from continuum QCD

Charmonia and Bottomonia have first been explored within a rainbow-ladder truncation of the gap and Bethe-Salpeter equations in the work by Jain and Munczek [33]. Driven by the ample growth of experimental data on quarkonia and in particular the discovery of many XYZ states by B-factories, CLEO-c, BES II and BES III, more sophisticated BSE computations of the quarkonia spectra followed the initial steps and at least the masses of the lower J^{PC} states compare very well with experimental data [16, 17, 19, 34–38].

Recently, the gap equation and BSE for heavier quarks have been revisited in the context of a given vector × vector contact interaction (CI), which introduces a mass scale and has zero range [39–42]. As for related Nambu-Jona-Lasinio models [43], these CI models are not confining yet the implementation of a proper-time regularization scheme [44] prevents the quarks from going on their mass shell and hence emulates confinement. The spectrum produced with this interaction model compares quantitatively well with that obtained using more sophisticated finite-range QCD interaction models mentioned above.

In a functional approach to QCD, the quark propagator for a given flavor f is computed from the DSE,

$$S_f^{-1}(p) = Z_2^f\,(i\gamma\cdot p + m_f^{\rm bm}) + Z_1^f\, g^2 \int_k^\Lambda D^{\mu\nu}(k-p)\frac{\lambda^a}{2}\gamma_\mu\, S(k)\,\Gamma_\nu^f(k,p)\frac{\lambda^a}{2}, \qquad (1)$$

where $\int_k^\Lambda$ represents a Poincaré invariant regularization of the integral and $Z_{1,2}^f(\mu^2,\Lambda^2)$ are, respectively, the vertex and quark wave-function renormalization constants at the renormalization scale μ. Note that the regularization mass scale is much larger than the renormalization scale: $\Lambda \gg \mu$. The current-quark mass, $Z_2^f(\mu,\Lambda) m_{\rm bm}^f(\Lambda)$, receives self-energy corrections where the integral in the second term of Eq. (1) is over the dressed gluon propagator, $D_{\mu\nu}(k)$, the quark-gluon vertex, $\Gamma_\nu^f(k,p)$, and λ^a are the SU(3) color matrices for quarks in the fundamental representation. The solution to the quarks's

DSE (1) contains a scalar and vector term,

$$S(p) = -i\gamma \cdot p\, \sigma_V(p^2) + \sigma_S(p^2) \equiv \left[i\gamma \cdot p\, A(p^2) + B(p^2)\right]^{-1} . \tag{2}$$

The full nonperturbative quark-gluon vertex, $\Gamma^a_\nu(k,p)$, plays a crucial role for DCSB [45–53] and when only the leading Dirac covariant, γ_μ, is included, the lack of strength to produce sufficient DCSB in the quark mass must be compensated by an enhanced gluon-dressing function. This is the premise of the leading *rainbow-ladder* truncation scheme which satisfies the axialvector Ward-Green-Takahashi identity and thus correctly describes chiral symmetry and its breaking pattern [54].

Typically, one neglects three-gluon interactions in this "Abelian" truncation and assuming the corresponding Ward-Green-Takahashi identity holds, i.e. $Z_1^f = Z_2^f$, one re-expresses the kernel of Eq. (1) as [55, 56],

$$Z_1^f g^2 D_{\mu\nu}(q)\, \Gamma^f_\mu(k,p) \;=\; \left(Z_2^f\right)^2 \mathcal{G}(q^2)\, D^{\text{free}}_{\mu\nu}(q)\, \gamma_\mu\,, \tag{3}$$

where $D^{\text{free}}_{\mu\nu}(q) := (g_{\mu\nu} - q_\mu q_\nu/q^2)/q^2$, $q = k - p$, is the perturbative gluon propagator in Landau gauge. An effective model coupling, whose momentum dependence mirrors that of DSE- and lattice-QCD results and yields successful explanations of numerous hadron observables, is given by the sum of two scale-distinct contributions [57]:

$$\frac{\mathcal{G}(q^2)}{q^2} = \frac{8\pi^2}{\omega^4} D e^{-q^2/\omega^2} + \frac{8\pi^2 \gamma_m \mathcal{F}(q^2)}{\ln\left[\tau + (1 + q^2/\Lambda^2_{\text{QCD}})^2\right]} . \tag{4}$$

The first term is an infrared-massive and -finite *ansatz* for the interaction, where $\gamma_m = 12/(33 - 2N_f)$, $N_f = 4$, $\Lambda_{\text{QCD}} = 0.234$ GeV; $\tau = e^2 - 1$; and $\mathcal{F}(q^2) = [1 - \exp(-q^2/4m_t^2)]/q^2$, $m_t = 0.5$ GeV. The parameters ω and D control the width and strength of the interaction, respectively, yet they should not be thought of as independent [56].

An even more dramatic simplification is the CI ansatz in the DSE, namely the following substitution,

$$g^2 D_{\mu\nu}(q)\, \Gamma^f_\nu(k,p) \quad\Longrightarrow\quad \left(\frac{4\pi\alpha_{\text{IR}}}{m_g^2}\right)_f \gamma_\mu\,, \tag{5}$$

where $m_g \simeq 800$ MeV is a gluon mass scale generated dynamically in QCD[1], and α_{IR} is a parameter that determines the interaction strength [42]. With this simplification, the solution to the gap equation becomes,

$$S_f^{-1}(p) = i\gamma \cdot p + M_f\,. \tag{6}$$

The constituent quark mass in Eq. (6) is momentum independent, a feature known from nonrelativistic and relativistic constituent quark models [59–66]. As we shall see, in conjunction with the BSE (7), this approximation still produces a realistic description of static observables and in particular of the mass spectrum of non-exotic ground-state mesons[2]. We stress this is not only true for heavy quarkonia but also for the Goldstone bosons, since the CI-BSE in rainbow-ladder truncation satisfies the axialvector Ward-Green-Takahashi identity when treated in a symmetry-preserving manner [42, 70]. For a given quantum state, the homogeneous BSE is generally given by,

$$[\Gamma_M(k,P)]^{fg}_{AB} = \int_q^\Lambda [K(k,q,P)]^{fg}_{AC,DB} \left[S_f(q_+)\, \Gamma^{fg}_M(q,P)\, S_g(q_-)\right]_{CD}, \tag{7}$$

[1] See, e.g., discussion in Ref. [58].

[2] This is also true for baryons; see Refs. [67–69].

$m_{0^{-(+)}}$	CI [41]	CI [42]	DSE [16, 17]	DSE [35]	DSE [36]	DSE [19]	PDG [72]
D	—	1.869	2.115	—	—	1.868	1.864
D_s	—	1.977	2.130	—	—	1.872	1.968
$\eta_c(1S)$	2.950	—	3.065	2.925	2.980	2.672	2.983
$\eta_c(2S)$	—	—	3.784	3.684	—	3.256	3.639
$\eta_b(1S)$	9.345	—	—	9.414	9.390	9.424	9.399
$\eta_b(2S)$	—	—	—	9.987	—	9.820	9.999
$m_{1^{-(-)}}$	CI [41]	CI [42]	DSE [16, 17]	DSE [35]	DSE [36]	DSE [19]	PDG [72]
D^*	—	2.011	—	—	—	—	2.010
D_s^*	—	2.098	—	—	—	—	2.112
J/ψ	3.129	—	3.114	3.113	3.070	2.840	3.097
$\psi(2S)$	—	—	3.760	3.676	—	3.294	3.686
$\Upsilon(1S)$	9.460	—	9.634	9.490	9.460	9.463	9.460
$\Upsilon(2S)$	—	—	10.140	10.089	—	9.838	10.023

Table 1: Comparative juxtaposition of the charmonium and bottomonium mass spectrum. The J^{PC} meson masses are obtained with different interaction models in a rainbow-ladder truncation of Eq. (7). Details about the exact form of these interaction ansätze are found in the references of each column. All masses are in [GeV] and C-parity is only a good quantum number for neutral $\bar{q}q$ eigenstates. Averaged experimental values by the Particle Data Group [72] are listed in the last column.

where $\mathcal{M}$ represents the Dirac spinor structure of the Bethe-Salpeter amplitude $\Gamma_{\mathcal{M}}(k;P)$, $K(q,k;P)$ is the fully amputated quark-antiquark scattering kernel, $[A,B,C,D,E,F]$ collectively denote color and spinor indices and f,g are flavor indices. Moreover, q is the relative quark-antiquark momentum and $q_\pm = q \pm \eta_\pm P$ with the constraint $\eta_+ + \eta_- = 1$. In a Poincaré covariant form of this BSE the solutions do not depend on the choice for $\eta_\pm$.

To obtain the spectrum, one introduces a fictitious eigenvalue, $\lambda_n(P^2)$, into the bound-state equation, such that,

$$\lambda_n(P^2)\,\Gamma^{fg}_{\mathcal{M}}(k,P) = \int_q^\Lambda K^{fg}(k,q,P)\,S_f(q_+)\,\Gamma^{fg}_{\mathcal{M}}(q,P)\,S_g(q_-)\,, \tag{8}$$

where color and spin indices have been suppressed. The solution in a particular J^{PC} channel is found for $\lambda_n(M_n^2) - 1 = 0$ which determines the mass, M_n, of the nth bound state with physical solutions for $\lambda_0(M_0^2) = \lambda_1(M_1^2) = \lambda_2(M_2^2) = \cdots = \lambda_i(M_i^2) = 1$. Here, the solution for $P^2 = -M_0^2$ is associated with the hadron's ground-state mass and M_i are the masses of radial excitations ordered as, $M_i < M_{i+1}$, and $\lambda_0(M_i^2) > \lambda_1(M_i^2) > \cdots > \lambda_i(M_i^2)$ [29, 71].

In Table 1 we compare a selection of 0^{-+} and 1^{--} meson masses obtained with the CI and finite-range interaction models in the DSE and BSE kernels of Eqs. (1) and (7). Note that the renormalized running quark mass, $m(\mu) = Z_2^f(\mu,\Lambda)/Z_4^f(\mu,\Lambda)m_f^{\rm bm}$ where Z_4^f is the renormalization constant in the QCD Lagrangian, can be set to reproduce the experimental η_c or J/ψ masses or to minimize the variation of calculated masses within a range of experimental values. Therefore, different calculations of, e.g., the $\eta_c(1S)$ lead to variations of the order of 3–5% except for the value of Ref. [19] which is due to their charm current-quark mass, $m_c(19\text{ GeV}) = 975$ MeV, using the Alkofer-Watson-Weigel interaction. This value is chosen much smaller, $m_c(19\text{ GeV}) = 695$ MeV, when the Maris-Tandy model is employed, yet while the D mass is in very good agreement with experiment, other charmonia masses are underestimated. A very similar picture emerges upon examination of the vector-meson states and radial excitations with small variations throughout the models. The agreement with data is obvious for the bottomonia with a maximal variation of $\approx$ 2–3%.

Remarkably, the CI model yields mass values that are consistent with the more sophisticated interaction ansätze and in very good agreement with measured values. This is a feature of the CI — it is a faithful approximation of the QCD dynamics for static and dynamic observables for $Q^2 \lesssim M_n^2$ [40, 67, 73]. Notable deviations of predictions of the CI and constituent quark models from measurements can be tested in elastic and transition form factors at large Q^2 [74, 75]. Finally, we observe that, so far, none of the BSE truncations provides a solution for the B mesons.

3 Weak decay constants of ground states and radial excitations

An important observable in the context of flavor physics and CP violation is the weak decay constant of mesons. The axial-vector Ward-Green-Takahashi identity implies a defining relation between the kernels in Eqs. (1) and (7) for pseudoscalar mesons. The identity can be used to prove a series of Goldberger-Treiman type of relations [76], the most important of which relates the two-body problem to the one-body scalar amplitude in the chiral limit ($\mathcal{M} := 0^-$):

$$f^0_{0^-} E_{0^-}(k;0) \;=\; B^0(k^2) \tag{9}$$

Here, $E_{0^-}(k;0)$ is the leading covariant of the pseudoscalar meson's Bethe-Salpeter amplitude and the superscript denotes the chiral limit of the quark propagator's scalar piece. The weak decay constant, f_{0^-}, is given by,

$$f_{0^-} P_\mu = \langle 0 \,|\, \bar{q}_f \gamma_5 \gamma_\mu q_g \,|\, 0^- \rangle \;=\; \left(Z_2^f Z_2^g\right)^{\frac{1}{2}} \mathrm{tr}_{\mathrm{CD}} \int_q^\Lambda i\gamma_5\gamma_\mu\, S_f(q_+)\, \Gamma_{0^-}(q;P)\, S_g(q_-) \tag{10}$$

where the trace is over Dirac and color indices.

While the chiral relation (9) does not hold anymore for heavy-light mesons, it nevertheless is crucial to preserve the chiral properties of the light quark in calculating their masses and weak decay constants. Typically, studies that fail to implement the DCSB properties of light quarks in these systems use quark propagators with momentum-independent constituent-quark masses, $M_{u,d} \simeq 250 - 350$ MeV [61–66]. This leads to considerable model dependence of heavy-light form factors and couplings [20].

An interesting feature is that at first the weak decay constants of light 0^- and 1^- mesons increase with the light current-quark mass [77], yet tend to level off between the strange- and charm-quark mass. For heavy-heavy and heavy-light mesons the decay constants fall off with increasing meson mass as $f_{\mathcal{M}} \propto 1/\sqrt{m_{\mathcal{M}}}$ which is consistent with heavy-quark effective field theory [78].

This behavior seems to occur as low as $q_f = c$ for $q_f \bar{u}$ mesons and the inflection point, somewhere above the strange mass, depends on the interaction ansatz one employs. Since the decay constants depend on the norm of the Bethe-Salpeter amplitudes and are thus proportional to the derivative of the quark propagators, it is not surprising that they are more sensitive to the model details than the masses. The transition region between the strange and charm flavors is interesting in many aspects, as it also describes the mass region where the effects of DCSB and explicit chiral symmetry breaking due to the Higgs mechanism are of the same order. A convenient parameter to study the effect of DCSB

$f_{0^{-(+)}}$	CI [41]	CI [42]	DSE [16, 17]	DSE [36]	DSE [19]	Reference
D	—	0.207	0.204	—	0.323	0.204
D_s	—	0.240	0.249	—	0.269	0.258
$\eta_c(1S)$	0.360	—	0.389	0.371	0.322	0.361
$\eta_c(2S)$	—	—	0.105	—	0.137	—
$\eta_b(1S)$	0.781	—	—	0.768	0.378	—
$\eta_b(2S)$	—	—	—	—	0.263	—
$f_{1^{-(-)}}$	CI [41]	CI [42]	DSE [16, 17]	DSE [36]	DSE [19]	Reference
D^*	—	0.281	—	—	—	0.278
D_s^*	—	0.276	—	—	—	0.322
J/ψ	0.291	—	0.433	0.361	0.383	0.416
$\psi(2S)$	—	—	0.176	—	0.031	0.295
$\Upsilon(1S)$	0.310	—	—	0.666	0.393	0.715
$\Upsilon(2S)$	—	—	0.564	—	0.283	0.497

Table 2: Comparison of weak decay constants of 0^- and 1^- $D_{(s)}$ mesons and quarkonia in CI and DSE-BSE approaches. All decay constant values are in [GeV] and the original values of Refs. [36, 41] were rescaled by $\sqrt{2}$. The reference values for quarkonia in the last column are inferred from data [72] on ${}^1S_0 \to \gamma\gamma$ and ${}^3S_1 \to e^+e^-$ decays; for the D and D_s mesons they are extracted from the measurement of $f_{D^+}|V_{cd}|$ and $f_{D_s^+}|V_{cs}|$, respectively [72]; the D^* and D_s^* decay constants are those of recent lattice-QCD simulations [81].

is the renormalization-point invariant ratio $\zeta := \sigma_f / M_f^E$ where σ_f is defined as a constituent-quark σ term via the Hellmann-Feynman theorem [21, 79],

$$\sigma_f := m_f(\mu) \frac{\partial M_f^E}{\partial m_f(\mu)} , \tag{11}$$

and M_f^E is the definition of the Euclidean constituent-quark mass: $(M^E)^2 := \{p^2 | p^2 = M^2(p^2)\}$. For light quarks the ratio ζ must vanish as their constituent-quark mass owes predominantly to DCSB, whereas for very heavy quarks, $\zeta \to 1$. It turns out that this ratio is 0.5 for $m_q(19\text{ GeV}) \approx 0.5$ GeV, which is about halfway between the strange- and charm-quark mass at this renormalization scale [79]. Obviously, where $\zeta \simeq 0.5$ depends on the truncation and interaction model employed and should be investigated further.

The chiral properties of pseudoscalar mesons have consequences for their radially excited states, as the following relation for their weak decay constants holds [80],

$$f_{0_n^-} m_{0_n^-}^2 = 2\, m(\mu)\, \rho_{0_n^-}(\mu) , \tag{12}$$

with the pseudoscalar projection of the pseudoscalar meson's Bethe-Salpeter wave function onto the origin in configuration space:

$$\rho_{0_n^-}(\mu) = -i \left(Z_4^f Z_4^g \right)^{\frac{1}{2}} \mathrm{Tr}_{\mathrm{CD}} \int_q^{\Lambda} \gamma_5\, S_f(q_+)\, \Gamma_{0_n^-}^{fg}(q,P)\, S_g(q_-) . \tag{13}$$

In QCD, the ultraviolet behaviour of the quark-antiquark scattering kernel and the fact that the Bethe-Salpter wave function $\chi_{0_n^-}(k,P) = S_f(k_+)\Gamma^{fg}_{0_n^-}(k,P)S_g(k_-)$ is a finite matrix-valued function ensures that $\rho_{0_n^-}(\mu)$ is finite and does not vanish in the chiral limit [76]. Since in this limit the ground-state pseudoscalar decay constant is also characterized by $0 < f^0_{0^-} < \infty$, it follows that $m_{0^-} = 0$ as expected for a Goldstone boson. Now, we turn our attention to the radially excited states. By assumption, $m_{0_n^-} > m_{0^-}$ and thus $m_{0_n^-} \neq 0$ for $n > 0$ in the chiral limit, and one deduces from Eq. (12) that $f_{0_n^-} \equiv 0$, $\forall n > 0$. Thus, in the chiral limit the probability for a radially excited state to decay via an electroweak current vanishes. Obviously, the decay constants of the first radial excitation with heavier quarks are not zero; it was noted that [in GeV]: $f_{\pi(1300)} = -8.3 \times 10^{-4}$; $f_{K(1460)} = -0.017$; $f_{(\bar{s}s)_1} = -0.0216$; and $f_{\eta_c(2S)} = 0.105$. Here again, the weak decay constants are slightly negative near and beyond the chiral limit but reach a turning point in between the s- and c-quark mass.

We conclude this section with a selective comparison of weak decay constants for charmed mesons, charmonia and their radially excited states obtained in different approaches in Table 2. Comparing decay constants of radially excited states with those of their ground states shows that the former remain suppressed even at the charm-mass scale. We remark that theoretical values for f_{η_c} and $f_{J/\psi}$ vary considerably, up to $\approx 30\%$ in the case of the J/ψ. This reflects again the dependence of the decay constants on the details of the interaction. As a generalization one may add that reproducing the mass spectrum of the lower-angular momentum quarkonia is not difficult — capturing their internal dynamics and electroweak properties is a more challenging task [82], which is exacerbated in unequal mass systems. This mirrors our experience with light hadrons whose mass spectrum is easier to understand than their space- and time-like electromagnetic form factors [23, 24, 83–85].

4 Final remarks

Flavor physics has entered a new era with the plethora of high-precision luminosity measurements at LHC*b* and provides a fertile playground for *indirect* tests of the Standard Model *as well as* an opportune field for complementary nonperturbative QCD studies in charm and beauty mesons. Much of our current knowledge about the flavor sector and the weak mixing angles is an important guideline to interpret weak interactions of any new possible particle, yet nonperturbative effects are still the major source of uncertainty in extracting electroweak phases. Thus, what are seemingly two unrelated fields, flavor and hadron physics, share in common the difficulties of what is frequently called *soft physics*.

Quarkonia, on the other hand, offer additional insights into bound-state dynamics, notably in the domain of two heavy nonrelativistic quarks almost adequately described by a perturbative interaction. Yet, their rich spectrum defies the simpler picture of quark models with such forms of matter as quark-gluon hybrids, mesonic molecules and tetraquarks.

In concluding, let us consider the example of a weak B-meson decay into two daughter mesons, e.g. $B(p_1) \to \pi(p_2)\pi(q)$, $q = p_2 - p_1$. In QCD factorization [5], the weak decay amplitude, up to corrections of order $\Lambda_{\rm QCD}/m_b$, is expressed as,

$$\langle \pi(p_2)\,\pi(q)|Q_i(\mu)|B(p_1)\rangle = F_0^{B\to\pi}\int_0^1 T_i^I(x)\,\phi_\pi(x)\,dx \;+\; \int_0^1 T_i^{II}(\xi,x,y)\,\phi_B(\xi)\phi_\pi(x)\phi_\pi(y)\,d\xi dx dy\,, \quad (14)$$

wherein $Q_i(\mu)$ is a dimension-six four-quark operator of the effective weak Hamiltonian[3] taken at the renormalization scale μ, $T_i^{I,II}$ are perturbative hard-scattering kernels, $F_0^{B\to\pi}$ is the scalar heavy-to-light transition form factor and ϕ_B and ϕ_π are leading-twist light-front distribution amplitudes. It

[3] The corresponding Wilson coefficient, $C_i(\mu)$, in the operator-product expansion has been omitted here.

plainly follows from Eq. (14) that the nonperturbative contributions to these amplitudes are encoded in form factors and partonic distribution amplitudes. For the latter, it has often been assumed that for any scale μ the asymptotic limit, $\phi^{\rm asy}(x) = 6x(1-x)$, is a good approximation. At best, the first few terms of their Gegenbauer expansion, when known, are included. However, as demonstrated in Ref. [86], $\phi^{\rm asy}(x)$ is a poor approximation of the complete distribution amplitude even at the scales currently accessible to experiments. It follows that in weak decays of heavy mesons and quarkonia it is advisable to use the broader $\phi(x)$ amplitudes computed from a large numbers of moments in Refs. [36, 86–88].

This is just one example of how recent developments in continuum QCD approaches within the DSE-BSE framework can be very profitable to precision calculation of weak heavy-meson decays. Likewise, the time-like transition form factors, $F^{D(B)\to\pi}(q^2)$, are extracted from complicated hadronic matrix elements for which progress is underway [89], yet more effort is essential; see the discussion in Ref. [20]. These examples, amongst many others, and the discussions in the preceding sections serve to highlight common difficulties and issues faced in hadron and flavor physics. Many theoretical and computational problems of flavored observables will only be overcome with more sophisticated DSE and BSE kernels. This, on the other hand, provides hadron physicists with challenges and opportunities in the years to come.

Acknowledgements

This overview is based on the presentation at the "XLVII International Symposium on Multiparticle Dynamics" (ISMD), Tlaxcala, Mexico and on related talks at the ECT* workshop "The Charm and Beauty of Strong Interactions" in Trento, Italy, and during the "Hadrons and Their Properties as a Problem in Strong QCD" workshop at Peking University. The hospitality of the ISMD organizers and friendly atmosphere in Tlaxcala were strongly appreciated. Work supported by the Brazilian research agencies FAPESP and CNPq under grant nos. 2016/03154-7 and 458371/2014-9, respectively.

References

[1] G. T. Bodwin, E. Braaten and G. P. Lepage, Phys. Rev. D **51**, 1125 (1995) Erratum: [Phys. Rev. D **55**, 5853 (1997)].
[2] N. Brambilla *et al.*, Eur. Phys. J. C **71**, 1534 (2011).
[3] N. Brambilla, G. Krein, J. Tarrús Castellà and A. Vairo, Phys. Rev. D **93** (2016) no.5, 054002.
[4] N. Brambilla, G. Krein, J. Tarrús Castellà and A. Vairo, arXiv:1707.09647 [hep-ph].
[5] M. Beneke, G. Buchalla, M. Neubert and C. T. Sachrajda, Phys. Rev. Lett. **83**, 1914 (1999).
[6] C. W. Bauer, S. Fleming, D. Pirjol and I. W. Stewart, Phys. Rev. D **63**, 114020 (2001).
[7] R. Aaij *et al.* [LHCb Collaboration], JHEP **1602**, 104 (2016).
[8] A. Abdesselam *et al.* [Belle Collaboration], arXiv:1604.04042 [hep-ex].
[9] R. Aaij *et al.* [LHCb Collaboration], JHEP **1708**, 055 (2017).
[10] B. El-Bennich, A. Furman, R. Kamiński, L. Leśniak and B. Loiseau, Phys. Rev. D **74**, 114009 (2006).
[11] O. Leitner, J.-P. Dedonder, B. Loiseau and B. El-Bennich, Phys. Rev. D **82**, 076006 (2010).
[12] B. El-Bennich, M. A. Ivanov and C. D. Roberts, Phys. Rev. C **83**, 025205 (2011).
[13] B. El-Bennich, G. Krein, L. Chang, C. D. Roberts and D. J. Wilson, Phys. Rev. D **85**, 031502 (2012).
[14] B. El-Bennich, E. Rojas, M. A. Paracha and J. P. B. C. de Melo, AIP Conf. Proc. **1625**, 80 (2014).

[15] B. El-Bennich, M. A. Paracha, C. D. Roberts and E. Rojas, Phys. Rev. D **95**, no. 3, 034037 (2017).
[16] E. Rojas, B. El-Bennich and J. P. B. C. de Melo, Phys. Rev. D **90**, 074025 (2014).
[17] F. F. Mojica, C. E. Vera, E. Rojas and B. El-Bennich, Phys. Rev. D **96**, no. 1, 014012 (2017).
[18] M. Gómez-Rocha, T. Hilger and A. Krassnigg, Phys. Rev. D **92**, no. 5, 054030 (2015).
[19] T. Hilger, M. Gómez-Rocha, A. Krassnigg and W. Lucha, Eur. Phys. J. A **53**, no. 10, 213 (2017).
[20] B. El-Bennich, M. A. Ivanov and C. D. Roberts, Nucl. Phys. Proc. Suppl. **199**, 184 (2010).
[21] B. El-Bennich, C. D. Roberts and M. A. Ivanov, PoS QCD-**TNT-II**, 018 (2012).
[22] H. R. Quinn, Phys. Today **56N2**, 30 (2003).
[23] I. C. Cloët and C. D. Roberts, Prog. Part. Nucl. Phys. **77**, 1 (2014).
[24] G. Eichmann, H. Sanchis-Alepuz, R. Williams, R. Alkofer and C. S. Fischer, Prog. Part. Nucl. Phys. **91**, 1 (2016).
[25] G. S. Bali *et al.* [RQCD Collaboration], Phys. Rev. D **93**, no. 9, 094504 (2016).
[26] A. Bashir, L. Chang, I. C. Cloët, B. El-Bennich, Y. X. Liu, C. D. Roberts and P. C. Tandy, Commun. Theor. Phys. **58**, 79 (2012).
[27] E. A. O'Malley, W. Kamleh, D. Leinweber and P. Moran, Phys. Rev. D **86**, 054503 (2012).
[28] N. Brambilla *et al.*, Eur. Phys. J. C **74**, no. 10, 2981 (2014).
[29] B. El-Bennich, G. Krein, E. Rojas and F. E. Serna, Few Body Syst. **57**, no. 10, 955 (2016).
[30] G. Krein, C. D. Roberts and A. G. Williams, Int. J. Mod. Phys. A **7**, 5607 (1992).
[31] C. S. Fischer and R. Alkofer, Phys. Rev. D **67**, 094020 (2003).
[32] J. M. Pawlowski, Annals Phys. **322**, 2831 (2007).
[33] P. Jain and H. J. Munczek, Phys. Rev. D **48**, 5403 (1993).
[34] M. Blank and A. Krassnigg, Phys. Rev. D **84**, 096014 (2011).
[35] C. S. Fischer, S. Kubrak and R. Williams, Eur. Phys. J. A **51**, 10 (2015).
[36] M. Ding, F. Gao, L. Chang, Y. X. Liu and C. D. Roberts, Phys. Lett. B **753**, 330 (2016).
[37] T. Hilger, C. Popovici, M. Gómez-Rocha and A. Krassnigg, Phys. Rev. D **91**, no. 3, 034013 (2015).
[38] T. Hilger, M. Gómez-Rocha and A. Krassnigg, Phys. Rev. D **91**, no. 11, 114004 (2015).
[39] Marco A. Bedolla, J. J. Cobos-Martínez, and Adnan Bashir, Phys. Rev. D **92**, no. 5, 054031 (2015).
[40] M. A. Bedolla, K. Raya, J. J. Cobos-Martínez and A. Bashir, Phys. Rev. D **93**, no. 9, 094025 (2016).
[41] K. Raya, M. A. Bedolla, J. J. Cobos-Martínez and A. Bashir, arXiv:1711.00383 [nucl-th].
[42] F. E. Serna, B. El-Bennich and G. Krein, Phys. Rev. D **96**, no. 1, 014013 (2017).
[43] Y. Ninomiya, W. Bentz and I. C. Cloët, Phys. Rev. C **96**, no. 4, 045206 (2017).
[44] D. Ebert, T. Feldmann and H. Reinhardt, Phys. Lett. B **388**, 154 (1996).
[45] R. Alkofer, C. S. Fischer, F. J. Llanes-Estrada and K. Schwenzer, Annals Phys. **324**, 106 (2009).
[46] A. C. Aguilar and J. Papavassiliou, Phys. Rev. D **83**, 014013 (2011).
[47] A. C. Aguilar, J. C. Cardona, M. N. Ferreira and J. Papavassiliou, Phys. Rev. D **96**, no. 1, 014029 (2017).
[48] E. Rojas, J. P. B. C. de Melo, B. El-Bennich, O. Oliveira and T. Frederico, JHEP **1310**, 193 (2013).
[49] E. Rojas, B. El-Bennich, J. P. B. C. de Melo and M. A. Paracha, Few Body Syst. **56**, no. 6-9, 639 (2015).
[50] A. Bashir, R. Bermudez, L. Chang and C. D. Roberts, Phys. Rev. C **85**, 045205 (2012).

[51] R. Bermudez, L. Albino, L. X. Gutiérrez-Guerrero, M. E. Tejeda-Yeomans and A. Bashir, Phys. Rev. D **95**, no. 3, 034041 (2017).
[52] S. X. Qin, L. Chang, Y. X. Liu, C. D. Roberts and S. M. Schmidt, Phys. Lett. B **722**, 384 (2013).
[53] A. K. Cyrol, M. Mitter, J. M. Pawlowski and N. Strodthoff, arXiv:1706.06326 [hep-ph].
[54] A. Bender, C. D. Roberts and L. Von Smekal, Phys. Lett. B **380**, 7 (1996).
[55] J. C. R. Bloch, Phys. Rev. D **66**, 034032 (2002).
[56] D. Binosi, L. Chang, J. Papavassiliou and C. D. Roberts, Phys. Lett. B **742**, 183 (2015).
[57] S. x. Qin, L. Chang, Y. x. Liu, C. D. Roberts and D. J. Wilson, Phys. Rev. C **85**, 035202 (2012).
[58] P. Boucaud, J. P. Leroy, A. L. Yaouanc, J. Micheli, O. Pène and J. Rodríguez-Quintero, Few Body Syst. **53**, 387 (2012).
[59] P. G. Ortega, J. Segovia, D. R. Entem and F. Fernández, Phys. Rev. D **95**, no. 3, 034010 (2017).
[60] P. G. Ortega, J. Segovia, D. R. Entem and F. Fernández, Phys. Rev. D **94**, no. 11, 114018 (2016).
[61] B. El-Bennich, O. Leitner, J.-P. Dedonder and B. Loiseau, Phys. Rev. D **79**, 076004 (2009).
[62] B. El-Bennich, J. P. B. C. de Melo, B. Loiseau, J.-P. Dedonder and T. Frederico, Braz. J. Phys. **38**, 465 (2008).
[63] B. El-Bennich, J. P. B. C. de Melo and T. Frederico, Few Body Syst. **54**, 1851 (2013).
[64] E. O. da Silva, J. P. B. C. de Melo, B. El-Bennich and V. S. Filho, Phys. Rev. C **86**, 038202 (2012).
[65] J. P. B. C. de Melo, K. Tsushima, B. El-Bennich, E. Rojas and T. Frederico, Phys. Rev. C **90**, no. 3, 035201 (2014).
[66] G. H. S. Yabusaki, I. Ahmed, M. A. Paracha, J. P. B. C. de Melo and B. El-Bennich, Phys. Rev. D **92**, no. 3, 034017 (2015).
[67] C. Chen, L. Chang, C. D. Roberts, S. Wan and D. J. Wilson, Few Body Syst. **53**, 293 (2012).
[68] S. S. Xu, C. Chen, I. C. Cloët, C. D. Roberts, J. Segovia and H. S. Zong, Phys. Rev. D **92**, no. 11, 114034 (2015).
[69] Y. Lu, C. Chen, C. D. Roberts, J. Segovia, S. S. Xu and H. S. Zong, Phys. Rev. C **96**, no. 1, 015208 (2017).
[70] F. E. Serna, M. A. Brito and G. Krein, AIP Conf. Proc. **1701**, 100018 (2016).
[71] B. El-Bennich and E. Rojas, EPJ Web Conf. **113**, 05003 (2016).
[72] C. Patrignani *et al.* [Particle Data Group], Chin. Phys. C **40** (2016) no. 10, 100001.
[73] J. Segovia, C. Chen, C. D. Roberts and S. Wan, Phys. Rev. C **88**, no. 3, 032201 (2013).
[74] I. C. Cloët, G. Eichmann, B. El-Bennich, T. Klähn and C. D. Roberts, Few Body Syst. **46**, 1 (2009).
[75] L. Chang, I. C. Cloët, B. El-Bennich, T. Klähn and C. D. Roberts, Chin. Phys. C **33**, 1189 (2009).
[76] P. Maris, C. D. Roberts and P. C. Tandy, Phys. Lett. B **420**, 267 (1998).
[77] P. Maris and P. C. Tandy, Nucl. Phys. Proc. Suppl. **161**, 136 (2006).
[78] A. V. Manohar and M. B. Wise, Camb. Monogr. Part. Phys. Nucl. Phys. Cosmol. **10**, 1 (2000).
[79] A. Höll, P. Maris, C. D. Roberts and S. V. Wright, Nucl. Phys. Proc. Suppl. **161**, 87 (2006).
[80] A. Höll, A. Krassnigg and C. D. Roberts, Phys. Rev. C **70**, 042203 (2004).
[81] D. Bečirević, V. Lubicz, F. Sanfilippo, S. Simula and C. Tarantino, JHEP **1202**, 042 (2012).
[82] K. Raya, M. Ding, A. Bashir, L. Chang and C. D. Roberts, Phys. Rev. D **95**, no. 7, 074014 (2017).
[83] T. Horn and C. D. Roberts, J. Phys. G **43**, no. 7, 073001 (2016).
[84] I. G. Aznauryan *et al.*, Int. J. Mod. Phys. E **22**, 1330015 (2013).

[85] C. Chen, B. El-Bennich, C. D. Roberts, S. M. Schmidt, J. Segovia and S. Wan, arXiv:1711.03142 [nucl-th].

[86] L. Chang, I. C. Cloët, J. J. Cobos-Martínez, C. D. Roberts, S. M. Schmidt and P. C. Tandy, Phys. Rev. Lett. **110**, no. 13, 132001 (2013).

[87] J. Segovia, L. Chang, I. C. Cloët, C. D. Roberts, S. M. Schmidt and H. s. Zong, Phys. Lett. B **731**, 13 (2014).

[88] B. L. Li, L. Chang, M. Ding, C. D. Roberts and H. S. Zong, Phys. Rev. D **94**, no. 9, 094014 (2016).

[89] V. Lubicz *et al.* [ETM Collaboration], Phys. Rev. D **96**, no. 5, 054514 (2017).

Gluon condensate and the Polyakov loop

Juan Pablo Carlomagno[1,2,3,*] and *Juan Cristóbal* Rojas[4,**]

[1] *IFLP, CONICET – Dpto. de Física, Universidad Nacional de La Plata, C.C. 67, 1900 La Plata, Argentina*
[2] *CONICET, Rivadavia 1917, 1033 Buenos Aires, Argentina*
[3] *CONICET, Rivadavia 1917, 1033 Buenos Aires, Argentina*
[4] *Departamento de Física, Universidad Católica del Norte, Angamos 0610, Antofagasta, Chile*

Abstract. The temperature dependence of the gluon condensate is deduced from the Polyakov loop effective potential. It is shown that this approach provides a simple picture for the electric gluon condensate around the deconfinement temperature, showing that it drops to zero in a temperature range which is in good agreement with different pure gauge lattice results.

1 Introduction

One of the main aspects of quantum Chromodynamics (QCD) is the appearance of nontrivial gluon and chiral condensates.

A fundamental feature of QCD, regarded as a classical theory, is the fact that the trace of the energy-momentum tensor is zero, but, when it is considered as a full quantum theory, we have that such trace becomes non-zero, since there is an anomaly, implying a nonzero gluon condensate [1, 2].

It is well established from Quantum Field Theory at Finite Temperature that, in presence of a thermal bath, the energy-momentum tensor can be split in a zero temperature part and a finite temperature contribution. The zero temperature contains the ultraviolet terms which give the anomalous term. On the other side, the finite temperature part is connected to the thermodynamic contribution to the energy density.

In this work, we compute the temperature dependence of the gluon condensate from the effective potential of a pure gauge theory. Our hypothesis is that the effective potential of self-interacting gluons comes from the Polyakov Loop potential.

2 Polyakov Loop

The order parameter which describes the deconfinement phase transition is the Polyakov loop [3, 4]

$$\Phi(x) = \frac{1}{N_c}\mathrm{Tr}\,\mathcal{P}\exp\left(i\int_0^{1/T} dx_4\,A_4\right). \tag{1}$$

*e-mail: carlomagno@fisica.unlp.edu.ar
**e-mail: jurojas@ucn.cl

© The Authors, published by EDP Sciences. This is an open access article distributed under the terms of the Creative Commons Attribution License 4.0 (http://creativecommons.org/licenses/by/4.0/).

The center symmetry plays a key role since the Polyakov loop is not gauge invariant under transformations which belong to the gauge center. So, if the have $\Phi = 0$, it means that there is a manifest $Z(N)$ symmetry, meaning that the system is confined. On the other side, if $\Phi \neq 0$, it means that the symmetry is broken and the system is deconfined.

Gauge freedom allow us to rotate the gauge field A_4 in order that it becomes static and diagonal. In that gauge fixing, the parameter Φ, becomes the diagonal matrix $\exp(iA_4/T)$, with $A_4 = iA_0 = ig\,\delta_{\mu 0}\,A^{\mu}_{a}\lambda^{a}/2$, where A^{μ}_{a} are the $SU(3)$ color gauge fields. The traced Polyakov loop (PL) is given by $\Phi = \frac{1}{3}\mathrm{Tr}\ \exp(iA_4/T)$.

We work in the Polyakov gauge, in which the matrix $\Phi = \phi_3\lambda_3 + \phi_8\lambda_8$ is diagonal [5]. The charge conjugation properties of the QCD Lagrangian imply that the mean field traced Polyakov loop field must be a real quantity [6]. Assuming that ϕ_3 and ϕ_8 are real-valued, ϕ_8 has to be zero [7], and therefore

$$\Phi = \frac{1}{3}[1 + 2\cos(\phi_3/T)]\,. \tag{2}$$

There is no unique choice for the the effective potential that describes the effective gauge field self-interactions. The Ansatz used come from the properties of pure gauge QCD. For example, it can be inspired by the logarithm of the Haar measure, the Jacobian which comes from the $SU(3)$ integration [7, 8]

$$\frac{\mathcal{U}_{\log}(\Phi,T)}{T^4} = -\frac{1}{2}\,a(T)\,\Phi^2 + b(T)\,\log\left(1 - 6\,\Phi^2 + 8\,\Phi^3 - 3\,\Phi^4\right)\,. \tag{3}$$

Another possibility is basically a Ginzburg-Landau Ansatz [9, 10]

$$\frac{\mathcal{U}_{\mathrm{poly}}(\Phi,T)}{T^4} = -\frac{b_2(T)}{2}\,\Phi^2 - \frac{b_3}{3}\,\Phi^3 + \frac{b_4}{4}\,\Phi^4\,. \tag{4}$$

There is also a mixed model proposed by Fukushima [11, 12]

$$\mathcal{U}_{\mathrm{Fuku}}(\Phi,T) = -b\,T\left[54\,\exp(-a/T)\,\Phi^2 + \log\left(1 - 6\,\Phi^2 + 8\,\Phi^3 - 3\,\Phi^4\right)\right]\,. \tag{5}$$

The coefficients $a_i(T), b_i(t)$ are obtained by fitting to the pure lattice QCD data and imposing physical conditions. For details see references given above.

3 Thermal Gluon Condensate

The pure gauge QCD Lagrangian is given by

$$\mathcal{L} = -\frac{1}{4g_s^2}G^a_{\mu\nu}G^{a\,\mu\nu}\,, \tag{6}$$

where

$$G^a_{\mu\nu} = \partial_\mu A^a_\nu - \partial_\nu A^a_\mu - f^a_{bc}A^b_\mu A^c_\nu. \tag{7}$$

We follow [13], where is proposed the following decomposition

$$\mathcal{Z} = \int\left[d\tilde{A}\right]\exp\left\{\sum\!\!\!\!\!\!\int \mathcal{L}_{eff}\right\} \approx \int\left[d\tilde{A}\right]\exp\left\{\frac{-1}{16\pi\alpha_s(T)}\ \sum\!\!\!\!\!\!\int\ G^a_{\mu\nu}G^{a\,\mu\nu}\right\}\,. \tag{8}$$

On the other side, we have the effective lagrangian

$$\mathcal{U} = -T\ln\mathcal{Z}\,, \tag{9}$$

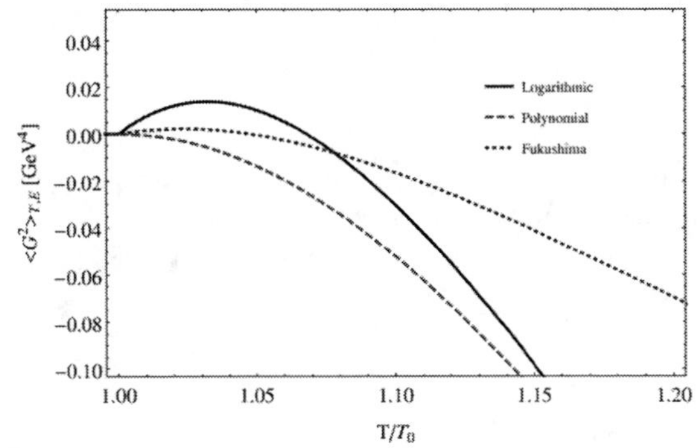

Figure 1. Electric part of the gluon condensate for the PL potentials.

A direct comparison lead us to

$$\langle G^a_{\mu\nu} G^{a\,\mu\nu} \rangle_T = 16\pi \frac{\partial\,\mathcal{U}}{\partial \alpha_s^{-1}} \,. \tag{10}$$

The gluon tensor can be decomposed in an "electric" and "magnetic" terms:

$$E_i = G_{i4},\; B_k = \frac{1}{2}\varepsilon_{ijk} G_{ij} \quad \rightarrow \quad \langle \frac{\alpha_s}{\pi} G^a_{\mu\nu} G^{a\,\mu\nu} \rangle = \langle \frac{\alpha_s}{\pi} G^a_{\mu\nu} G^{a\,\mu\nu} \rangle_E + \langle \frac{\alpha_s}{\pi} G^a_{\mu\nu} G^{a\,\mu\nu} \rangle_B . \tag{11}$$

From lattice results [14], it can be realized that the magnetic part is not affected by temperature, then we can decompose the gluon condensate as

$$\langle \frac{\alpha_s}{\pi} G^a_{\mu\nu} G^{a\,\mu\nu} \rangle \equiv \langle G^2 \rangle = \langle G^2 \rangle_0 + \langle G^2 \rangle_{T,E} \,, \tag{12}$$

from (10) we have

$$\langle G^2 \rangle_{T,E} = -\frac{4}{\pi} \alpha_s(T)^2 \left(\frac{\partial \alpha_s(T)}{\partial T} \right)^{-1} \frac{\partial\,\mathcal{U}(\Phi, T)}{\partial T} \,. \tag{13}$$

It remains to determine the value of $\alpha(T)$. If we assume that the temperature corresponds to the energy scale of the system, which is the natural criteria near T_0, the critical temperature. We choose two parametrizations for α which come from Schwinger-Dyson Analysis, given in [15],

$$\alpha_s^{(1)}(T) \quad = \quad \frac{1}{1.16 + T^4/\Lambda_1^4} \left[3.49 \left(1.16 - 0.07 \left(\frac{T^2}{\Lambda_1^2} \right)^{\frac{2}{3}} \right) + \left(\frac{T^2}{\Lambda_1^2} + 2 \right) \frac{T^4}{\Lambda_1^4} \alpha_s^{\beta_1} \right] , \tag{14}$$

where

$$\alpha_s^{\beta_1} = \frac{4\pi}{11 \ln(T^2/\Lambda_1^2)} \left[1 - \frac{102}{11^2} \frac{\ln(T^2/\Lambda_1^2)}{T^2/\Lambda_1^2} \right] .$$

And

$$\alpha_s^{(2)}(T) \quad = \quad \frac{1}{15 + T^2/\Lambda_2^2} \left[15 \times 2.6 + \frac{4\pi}{11} \left(\frac{1}{\ln(T^2/\Lambda_2^2)} - \frac{1}{T^2/\Lambda_2^2 - 1} \right) \frac{T^2}{\Lambda_2^2} \right] , \tag{15}$$

with $\Lambda_2 = 0.33$ GeV.

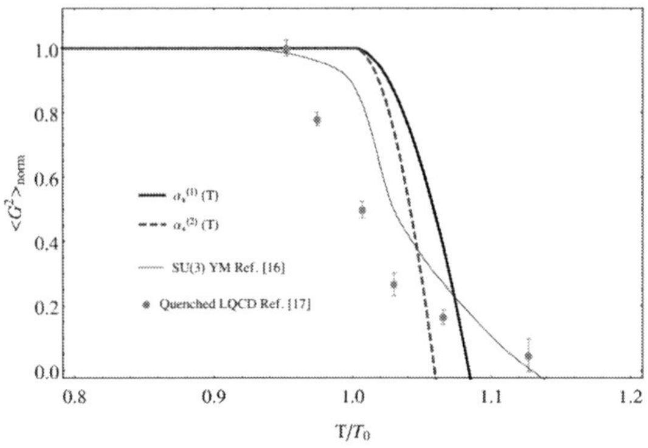

Figure 2. Normalized electric gluon condensate as a function of the reduced temperature for the polynomial potential and both strong coupling constants.

4 Results and Conclusions

Now is possible to plot the behavior of $\langle G^2\rangle_{T,E}$ using the different potentials (3), (4) and (5), those are depicted in Figure (1). First, it must be noticed that deconfinement occurs above the critical temperature T_0, implying that the effective potential must be zero below T_0. There we find that the best fit is given by the polynomial effective potential, since the Fukushima potential has an unstable behavior.

We have analyzed the temperature dependence of the gluon condensate for the three most used PL potentials in the literature, and proposed two parametrizations for the temperature dependence of the strong coupling constant. We showed that the polynomial effective potential has the best thermal behavior, while the qualitative dependence for both of the strong coupling parametrizations is equivalent. Finally, we conclude saying that the predictions obtained within this approach are in better agreement with estimates from lattice QCD than other more complex effective theories, and provides an accurate simple picture for the description of the electric gluon condensate around the deconfinement critical temperature.

References

[1] C. G. Callan, Jr., S. R. Coleman and R. Jackiw, Annals Phys. **59**, 42 (1970)
[2] H. Leutwyler, "Deconfinement and Chiral Symmetry in QCD 20 Years Later", Vol. 2, P. M. Zerwas and H. A. Kastrup (Eds.), World Scientific, Singapore, 1993, pp. 693-716
[3] G. 't Hooft, Nucl. Phys. B **138**, 1 (1978)
[4] A. M. Polyakov, Phys. Lett. **72B**, 477 (1978)
[5] D. Diakonov and M. Oswald, Phys. Rev. D **70**, 105016 (2004)
[6] A. Dumitru, R. D. Pisarski and D. Zschiesche, Phys. Rev. D **72**, 065008 (2005)
[7] S. Roessner, C. Ratti and W. Weise, Phys. Rev. D **75**, 034007 (2007)
[8] B. J. Schaefer, J. M. Pawlowski and J. Wambach, Phys. Rev. D **76**, 074023 (2007)
[9] O. Scavenius, A. Dumitru and J. T. Lenaghan, Phys. Rev. C **66**, 034903 (2002)
[10] C. Ratti, M. A. Thaler and W. Weise, Phys. Rev. D **73**, 014019 (2006)
[11] K. Fukushima, Phys. Lett. B **591**, 277 (2004)
[12] K. Fukushima, Phys. Rev. D **77**, 114028 (2008) Erratum: [Phys. Rev. D **78**, 039902 (2008)]
[13] N. O. Agasian, JETP Lett. **104** no.2, 71 (2016)
[14] M. D'Elia, A. Di Giacomo and E. Meggiolaro, Phys. Rev. D **67**, 114504 (2003)
[15] A. Deur, S. J. Brodsky and G. F. de Teramond, Prog. Part. Nucl. Phys. **90**, 1 (2016)

Strong magnetic fields in a nonlocal Polyakov chiral quark model

D. Gómez Dumm[1,2,⋆], *M.F.* Izzo Villafañe[1,2], *S.* Noguera[3], *V.P.* Pagura[3], and *N.N.* Scoccola[2,4,5].

[1] *IFLP, CONICET – Departamento de Física, Fac. de Cs. Exactas, Universidad Nacional de La Plata, C.C. 67, (1900) La Plata, Argentina*
[2] *CONICET, Rivadavia 1917, (1033) Buenos Aires, Argentina*
[3] *Departamento de Física Teórica and IFIC, Centro Mixto Universidad de Valencia-CSIC, E-46100 Burjassot (Valencia), Spain*
[4] *Physics Department, Comisión Nacional de Energía Atómica, Av. Libertador 8250, (1429) Buenos Aires, Argentina*
[5] *Universidad Favaloro, Solís 453, (1078) Buenos Aires, Argentina*

Abstract. We study the behavior of strongly interacting matter under an external constant magnetic field in the context of nonlocal chiral quark models that incorporate a coupling to the Polyakov loop. We find that at zero temperature the behavior of the quark condensates shows the expected magnetic catalysis effect, our predictions being in good quantitative agreement with lattice QCD results. On the other hand when the analysis is extended to the case of finite temperature our results show that nonlocal models naturally lead to the Inverse Magnetic Catalysis effect for both the chiral restoration and deconfinement transition temperatures.

1 Introduction

The study of the behavior of strongly interacting matter under intense external magnetic fields has gained increasing interest in the last few years. In fact, this topic has important applications e.g. in the description of compact objects like magnetars [1], the analysis of heavy ion collisions at very high energies [2] and the exploration of phases at the early Universe [3]. Since these studies require to deal with quantum chromodynamics (QCD) in nonperturbative regimes, present theoretical analyses are based either in the predictions of effective models or in the results obtained through lattice QCD (LQCD) calculations. In particular, the features of QCD phase transitions under external magnetic fields deserve significant interest. Recent reviews on this subject can be found in Refs. [4–6]. In view of the difficulty of theoretical calculations, most works concentrate on the case in which one has a uniform and static external magnetic field $\vec{B}$. At zero temperature and chemical potential, both the results of low-energy effective models of QCD and LQCD calculations indicate that the chiral quark condensates should behave as increasing functions of B, which is usually known as "magnetic catalysis". On the contrary, close to the chiral restoration temperature, LQCD calculations carried out with realistic quark masses [7, 8] show that light quark-antiquark condensates behave as nonmonotonic functions of the external magnetic field, and this leads to a decrease of the transition temperature when the magnetic field is increased. This effect is known as "inverse magnetic catalysis" (IMC). In addition,

⋆e-mail: dumm@fisica.unlp.edu.ar

© The Authors, published by EDP Sciences. This is an open access article distributed under the terms of the Creative Commons Attribution License 4.0 (http://creativecommons.org/licenses/by/4.0/).

LQCD calculations predict an entanglement between the chiral restoration and deconfinement critical temperatures [7]. These findings become a challenge to model calculations. Indeed, most naive effective approaches to low energy QCD (Nambu–Jona-Lasinio model, chiral perturbation theory, MIT bag model, quark-meson models) predict that the chiral transition temperature should grow with B, i.e., they do not find IMC. Although many scenarios have been considered in the last few years to account for the IMC [5], the mechanism behind this effect is not yet fully understood.

The aim of this contribution is to present the results of some recent analyses [9, 10] of the behavior of strongly interacting matter under a uniform, static magnetic field in the framework of nonlocal chiral quark models. We will show that nonlocal models are able to describe, at the mean field level, not only the IMC effect but also the entanglement between chiral restoration and deconfinement transition temperatures when a coupling to the Polyakov loop is incorporated. The "nonlocal Polyakov-Nambu-Jona-Lasinio" (nlPNJL) models considered here are a sort of nonlocal extensions of the PNJL model that intend to provide a more realistic effective approach to QCD. In fact, nonlocality arises naturally in the context of successful descriptions of low-energy quark dynamics [11, 12], and it has been shown [13] that nonlocal models can lead to a momentum dependence in quark propagators that is consistent with LQCD results. It is also found that in this framework one obtains an adequate description of the properties of light mesons at both zero and finite temperature/density (see e.g. [14] and references therein).

The article is organized as follows. In Sect. II we introduce the formalism to deal with a nonlocal NJL-like model in the presence of the magnetic field at zero temperature. Afterwards we extend this formalism to a finite temperature system, taking into account the coupling to the Polyakov loop. In Sect. III we quote our numerical results, discussing the behavior of the different relevant quantities as functions of the magnetic field and/or temperature. In Sect. IV we present our conclusions.

2 Theoretical formalism

Let us start by stating the Euclidean action for our nonlocal NJL-like two-flavor quark model,

$$S_E = \int d^4x \left\{ \bar{\psi}(x)\left(-i\partial\!\!\!/ + m_c\right)\psi(x) - \frac{G}{2} j_a(x) j_a(x) \right\} . \tag{1}$$

Here m_c is the current quark mass, which is assumed to be equal for u and d quarks. The currents $j_a(x)$ are given by

$$j_a(x) = \int d^4z\, \mathcal{G}(z)\, \bar{\psi}(x + \frac{z}{2})\, \Gamma_a\, \psi(x - \frac{z}{2}) , \tag{2}$$

where $\Gamma_a = (\mathbf{1}, i\gamma_5\vec{\tau})$, and the function $\mathcal{G}(z)$ is a nonlocal form factor that characterizes the effective interaction. We introduce now in the effective action Eq. (1) a coupling to an external electromagnetic gauge field $\mathcal{A}_\mu$. For a local theory this can be done by performing the replacement

$$\partial_\mu \;\to\; D_\mu \equiv \partial_\mu - i\,\hat{Q}\mathcal{A}_\mu(x) , \tag{3}$$

where $\hat{Q} = \mathrm{diag}(q_u, q_d)$, with $q_u = 2e/3$, $q_d = -e/3$, is the electromagnetic quark charge operator. In the case of the nonlocal model under consideration, the inclusion of gauge interactions implies a change not only in the kinetic terms of the Lagrangian but also in the nonlocal currents in Eq. (2). One has

$$\psi(x - z/2) \;\to\; \mathcal{W}(x, x - z/2)\; \psi(x - z/2) , \tag{4}$$

and a related change holds for $\bar{\psi}(x + z/2)$ [13]. Here the function $\mathcal{W}(s,t)$ is defined by

$$\mathcal{W}(s,t) \;=\; \mathrm{P}\,\exp\left[-i \int_s^t dr_\mu\, \hat{Q}\,\mathcal{A}_\mu(r)\right] , \tag{5}$$

where r runs over an arbitrary path connecting s with t. As it is usually done, we take it to be a straight line path.

To proceed it is convenient to bosonize the fermionic theory, introducing scalar and pseudoscalar fields $\sigma(x)$ and $\vec{\pi}(x)$ and integrating out the fermion fields. The bosonized action can be written as

$$S_{\rm bos} = -\ln\det\mathcal{D}_{x,x'} + \frac{1}{2G}\int d^4x\left[\sigma(x)\sigma(x) + \vec{\pi}(x)\cdot\vec{\pi}(x)\right], \tag{6}$$

where

$$\mathcal{D}_{x,x'} \;=\; \delta^{(4)}(x-x')\left(-i\not{D} + m_c\right)\mathcal{G}(x-x')\gamma_0\,\mathcal{W}(x,\bar{x})\,\gamma_0[\sigma(\bar{x}) + i\gamma_5\vec{\tau}\cdot\vec{\pi}(\bar{x})]\,\mathcal{W}(\bar{x},x'), \tag{7}$$

with $\bar{x} = (x+x')/2$. We will consider the case of a constant and homogenous magnetic field orientated along the 3-axis, choosing the Landau gauge, in which one has $\mathcal{A}_\mu = B\,x_1\,\delta_{\mu 2}$. We assume that the field σ has a nontrivial translational invariant mean field value $\bar{\sigma}$, while the mean field values of pseudoscalar fields π_i are zero. The assumption that $\bar{\sigma}$ is independent of x does not imply that the resulting quark propagator will be translational invariant. It just states that the deviations from translational invariance driven by the magnetic field are not affected by the dynamics of the theory. In this way, within the mean field approximation (MFA) we get

$$\mathcal{D}^{\rm MFA}_{x,x'} \;=\; {\rm diag}(\mathcal{D}^{{\rm MFA},u}_{x,x'}\,\mathcal{D}^{{\rm MFA},d}_{x,x'}), \tag{8}$$

where

$$\mathcal{D}^{{\rm MFA},f}_{x,x'} = \delta^{(4)}(x-x')\left(-i\not{\partial} - q_f B\,x_1\gamma_2 + m_c\right) + \bar{\sigma}\,\mathcal{G}(x-x')\,\exp\left[i(q_f B/2)\,(x_1+x_1')\,(x_2-x_2')\right]. \tag{9}$$

To deal with this operator it is convenient to introduce its Ritus transform $\mathcal{D}^{{\rm MFA},f}_{\bar{p},\bar{p}'}$, defined by

$$\mathcal{D}^{{\rm MFA},f}_{\bar{p},\bar{p}'} = \int d^4x\,d^4x'\;\bar{\mathbb{E}}_{\bar{p}}(x)\,\mathcal{D}^{{\rm MFA},f}_{x,x'}\,\mathbb{E}_{\bar{p}'}(x'), \tag{10}$$

where $\mathbb{E}_{\bar{p}}(x)$ and $\bar{\mathbb{E}}_{\bar{p}}(x)$, with $\bar{p} = (k, p_2, p_3, p_4)$, are Ritus functions [15]. The index k is an integer that will label the Landau energy levels. Using the properties of Ritus functions, and after some calculation, we obtain

$$\mathcal{D}^{{\rm MFA},f}_{\bar{p},\bar{p}'} = (2\pi)^4\delta_{kk'}\,\delta(p_2-p_2')\,\delta(p_3-p_3')\,\delta(p_4-p_4')\,\mathcal{D}^f_{k,p_\parallel}, \tag{11}$$

where

$$\mathcal{D}^f_{k,p_\parallel} \;=\; P_{k,s_f}\left(-s_f\sqrt{2k|q_fB|}\,\gamma_2 + p_\parallel\cdot\gamma_\parallel\right) + \sum_{\lambda=\pm} M^{\lambda,f}_{k,p_\parallel}\,\Delta^\lambda. \tag{12}$$

Here we have introduced the definitions $s_f = {\rm sign}(q_fB)$, $p_\parallel = (p_3, p_4)$, $\gamma_\parallel = (\gamma_3,\gamma_4)$, $\Delta^+ = {\rm diag}(1,0,1,0)$, $\Delta^- = {\rm diag}(0,1,0,1)$ and $P_{k,\pm 1} = (1-\delta_{k0})\,\mathcal{I} + \delta_{k0}\,\Delta^\pm$. In addition, we denote

$$M^{\lambda,f}_{k,p_\parallel} = \frac{4\pi}{|q_fB|}(-1)^{k_\lambda}\int\frac{d^2p_\perp}{(2\pi)^2}\left(m_c + \sigma\,g(p_\perp^2 + p_\parallel^2)\right)\exp(-p_\perp^2/|q_fB|)\,L_{k_\lambda}(2p_\perp^2/|q_fB|). \tag{13}$$

where we have used the definitions $k_\pm = k - 1/2 \pm s_f/2$ and $p_\perp = (p_1, p_2)$, while $g(p^2)$ is the Fourier transform of $\mathcal{G}(x)$ and $L_m(x)$ are Laguerre polynomials, with the usual convention $L_{-1}(x) = 0$.

Using the fact that $\mathcal{D}^{{\rm MFA},f}$ is diagonal in Ritus space the corresponding contribution to the MFA action can be readily calculated. We obtain

$$\frac{S^{\rm MFA}_{\rm bos}}{V^{(4)}} \;=\; \frac{\bar{\sigma}^2}{2G} - N_c\sum_{f=u,d}\frac{|q_fB|}{2\pi}\int\frac{d^2p_\parallel}{(2\pi)^2}\left[\ln\left(p_\parallel^2 + M^{\lambda_f,f\,2}_{0,p_\parallel}\right) + \sum_{k=1}^{\infty}\ln\Delta^f_{k,p_\parallel}\right], \tag{14}$$

where $\lambda_f = +(-)$ for $s_f = +1\,(-1)$, and $\Delta^f_{k,p_\parallel}$ is defined by

$$\Delta^f_{k,p_\parallel} = \left(2k|q_f B| + p_\parallel^2 + M^{+,f}_{k,p_\parallel} M^{-,f}_{k,p_\parallel}\right)^2 + p_\parallel^2 \left(M^{+,f}_{k,p_\parallel} - M^{-,f}_{k,p_\parallel}\right)^2 . \tag{15}$$

Here it is seen that the functions $M^{\pm,f}_{k,p_\parallel}$ play the role of constituent quark masses in the presence of the external magnetic field.

We extend now the analysis to a system at finite temperature. This is done by using the standard Matsubara formalism. To account for confinement effects, we also include the coupling of fermions to the Polyakov loop (PL), assuming that quarks move on a constant color background field $\phi = ig\,\delta_{\mu 0}\, G^\mu_a \lambda^a/2$, where G^μ_a are the SU(3) color gauge fields. We work in the so-called Polyakov gauge, in which the matrix ϕ is given a diagonal representation $\phi = \phi_3\lambda_3 + \phi_8\lambda_8$, taking the traced Polyakov loop $\Phi = \frac{1}{3}\mathrm{Tr}\,\exp(i\phi/T)$ as an order parameter of the confinement/deconfinement transition. Since — owing to the charge conjugation properties of the QCD Lagrangian— the mean field traced Polyakov loop is expected to be a real quantity, and ϕ_3 and ϕ_8 are assumed to be real valued [16], one has $\phi_8 = 0$, $\Phi = [1 + 2\cos(\phi_3/T)]/3$. Finally, we include in the Lagrangian a Polyakov-loop potential $\mathcal{U}(\Phi, T)$, which accounts for effective gauge field self-interactions. The resulting scheme is usually denoted as nonlocal Polyakov-Nambu-Jona-Lasinio (nlPNJL) model [17, 18].

Concerning the PL potential, its functional form is usually based on properties of pure gauge QCD. In this work we will mostly focus on a potential given by a polynomial function based on a Ginzburg-Landau ansatz [19], namely

$$\frac{\mathcal{U}_{\rm poly}(\Phi, T)}{T^4} = -\frac{b_2(T)}{2}\,\Phi^2 - \frac{b_3}{3}\,\Phi^3 + \frac{b_4}{4}\,\Phi^4 , \qquad b_2(T) = \sum_{n=0}^{3} a_n \left(\frac{T_0}{T}\right)^n . \tag{16}$$

The parameters a_i and b_i can be fitted to pure gauge lattice QCD results imposing the presence of a first-order phase transition at T_0, which is a further parameter of the model. In the absence of dynamical quarks, from lattice calculations one expects a deconfinement temperature $T_0 = 270$ MeV. However, it has been argued that in the presence of light dynamical quarks this temperature scale should be considerably reduced. Following Ref. [20] we use here $T_0 = 210$ MeV, taking the values of a_i and b_i from Ref. [19].

In this way, the grand canonical thermodynamic potential of the system under the external magnetic field is found to be given by

$$\Omega^{\rm MFA}_{B,T} = \frac{\bar\sigma^2}{2G} - T\sum_{n=-\infty}^{\infty}\sum_{c,f}\frac{|q_f B|}{2\pi}\int\frac{dp_3}{2\pi}\left[\ln\left(p_{\parallel_{nc}}^2 + M^{\lambda_f,f\;2}_{0,p_{\parallel_{nc}}}\right) + \sum_{k=1}^{\infty}\ln\left(\Delta^f_{k,p_{\parallel_{nc}}}\right)\right] + \mathcal{U}(\Phi, T) , \tag{17}$$

where we have defined $p_{\parallel_{nc}} = (p_3\,,\,(2n+1)\pi T + \phi_c)$. The sums over color and flavor indices run over $c = r, g, b$ and $f = u, d$, respectively, while the color background fields are $\phi_r = -\phi_g = \phi_3$, $\phi_b = 0$. As usual in nonlocal models, it is seen that $\Omega^{\rm MFA}$ turns out to be divergent, thus it has to be regularized. We take here the usual subtraction prescription

$$\Omega^{\rm MFA,reg}_{B,T} = \Omega^{\rm MFA}_{B,T} - \Omega^{\rm free}_{B,T} + \Omega^{\rm free,reg}_{B,T} , \tag{18}$$

where the "free" contribution stands for the potential obtained in absence of the strong current-current coupling (i.e. setting $\bar\sigma = 0$), but keeping the interaction with the magnetic field and the PL. For this "free" piece the Matsubara sum can be performed analytically, leading to

$$\Omega^{\rm free,reg}_{B,T} = -\frac{N_c}{2\pi^2}\sum_f (q_fB)^2\left[\zeta'(-1,x_f)+\frac{x_f^2}{4}-\frac{1}{2}(x_f^2-x_f)\ln x_f\right] - T\sum_{f,c}\frac{|q_fB|}{\pi}\sum_{k=0}^{\infty}\alpha_k\int\frac{dp}{2\pi}\,\ln\left\{1+\exp\left[-(\epsilon^f_{kp}+i\phi_c)/T\right]\right\}. \tag{19}$$

Here we have defined $x_f = m_c^2/(2|q_fB|)$, $\alpha_k = 2-\delta_{k0}$, $\epsilon^f_{kp} = (2k|q_fB|+p^2+m_c^2)^{1/2}$, while $\zeta'(-1,x_f) = d\zeta(z,x_f)/dz|_{z=-1}$, where $\zeta(z,x_f)$ is the Hurwitz zeta function. Owing to the presence of the background field, one has now a set of two coupled "gap equations"

$$\frac{\partial\Omega^{\rm MFA,reg}_{B,T}}{\partial\bar\sigma} = 0\,, \qquad \frac{\partial\Omega^{\rm MFA,reg}_{B,T}}{\partial\Phi} = 0\,. \tag{20}$$

Given $\Omega^{\rm MFA,reg}_{B,T}$, the magnetic field dependent quark condensates $\langle\bar q_f q_f\rangle$ can be calculated by taking the derivatives with respect to the corresponding current quark masses. To make contact with the LQCD results quoted in Ref. [8] we define the quantities

$$\Sigma^f_{B,T} = -\frac{2\,m_c}{S^4}\left[\langle\bar q_f q_f\rangle^{\rm reg}_{B,T} - \langle\bar q q\rangle^{\rm reg}_{0,0}\right] + 1\,, \tag{21}$$

where $S = (135\times 86)^{1/2}$ MeV. The subindex f can be omitted for $B=0$, owing to isospin symmetry. We also introduce the definitions $\Delta\Sigma^f_{B,T} = \Sigma^f_{B,T} - \Sigma^f_{0,T}$, $\bar\Sigma_{B,T} = (\Sigma^u_{B,T}+\Sigma^d_{B,T})/2$ and $\Delta\bar\Sigma_{B,T} = (\Delta\Sigma^u_{B,T} + \Delta\Sigma^d_{B,T})/2$, which correspond to the subtracted normalized flavor condensate, the normalized flavor average condensate and the subtracted normalized flavor average condensate, respectively.

3 Numerical results

To obtain numerical predictions for the behavior of the above defined quantities as functions of the temperature and the external magnetic field, it is necessary to specify the particular shape of the nonlocal form factor $g(p^2)$. We consider here two often-used forms, namely a Gaussian function and a "5-Lorentzian" function, defined by

$$g(p^2) = \exp(-p^2/\Lambda^2)\,, \qquad g(p^2) = \frac{1}{1+(p^2/\Lambda^2)^5}\,, \tag{22}$$

respectively. Notice that in these form factors we introduce an energy scale Λ, which acts as an effective momentum cut-off. This has to be taken as an additional parameter of the model. Given the nonlocal form factor, one has to determine the values of the parameters m_c, G and Λ. We consider different parameter sets, obtained by requiring that the model leads to the empirical values of the pion mass and decay constant, as well as some phenomenologically acceptable value of the quark condensate at $B=0$ and $T=0$. We take in particular $(-\langle\bar q q\rangle^{\rm reg}_{0,0})^{1/3} = 220$, 230 and 240 MeV. The corresponding parameter sets for the Gaussian and 5-Lorentzian form factors can be found e.g. in Ref. [10].

Let us start by discussing our results for zero temperature. In Fig. 1 we show the predictions of our model for $\Delta\bar\Sigma_{B,0}$ (left panel) and $\Sigma^u_{B,0} - \Sigma^d_{B,0}$ (right panel) as functions of eB. LQCD data from Ref. [8] are also displayed in both cases for comparison. Solid, dashed and dotted curves correspond to $(-\langle\bar q_f q_f\rangle^{\rm reg}_{0,0})^{1/3} = 220$, 230 and 240 MeV, respectively, for the case of a Gaussian form factor. It can be seen that the predictions for $\Delta\bar{\Sigma}_{B,0}$ are very similar for all parametrizations considered, and show a very good agreement with LQCD results. In the case of $\Sigma^u_{B,0} - \Sigma^d_{B,0}$, although the overall agreement with LQCD calculations is still good, we find some dependence on the parameterization. For the case of the 5-Lorentzian form factor our results turn out to be qualitatively similar to those in Fig. 1.

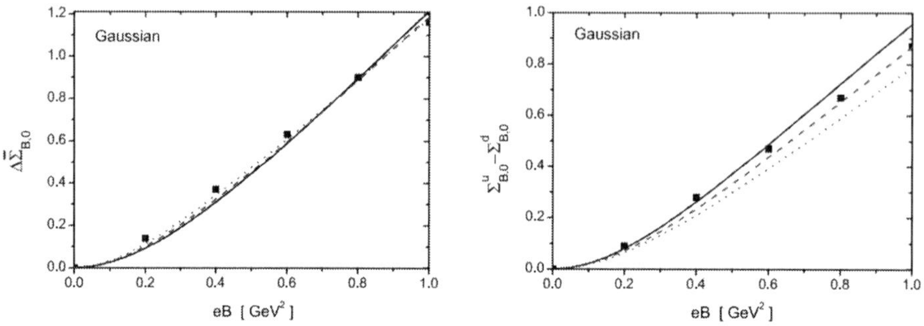

Figure 1. Normalized condensates as functions of the magnetic field at $T = 0$, for the case of a Gaussian form factor. Left panel: subtracted flavor average; right panel: flavor difference [see Eq. (21) and the text below]. Solid (black), dashed (red) and dotted (blue) curves correspond to parameterizations leading to $(-\langle\bar{q}q\rangle^{\rm reg}_{0,0})^{1/3} = 220$, 230 and 240 MeV, respectively. Full square symbols indicate LQCD results taken from Ref. [8].

We turn now to our numerical results for a system at finite temperature. In the upper panels of Fig. 2 we show the behavior of the averaged chiral condensate $\bar{\Sigma}_{B,T}$ and the traced Polyakov loop Φ as functions of the temperature, for three representative values of the external magnetic field B, namely $B = 0$, 0.6 and 1 GeV^2. The curves correspond to parameter sets leading to $(-\langle\bar{q}q\rangle^{\rm reg}_{0,0})^{1/3} = 230$ MeV. Given a value of B, it is seen from the figure that for the cases of both Gaussian and 5-Lorentzian form factors the chiral restoration and deconfinement transitions proceed as smooth crossovers, at approximately the same critical temperatures. For definiteness we take these temperatures from the maxima of the chiral and PL susceptibilities, which we define as the derivatives $\chi_{\rm ch} = -\partial[(\langle\bar{u}u\rangle^{\rm reg}_{B,T} + \langle\bar{d}d\rangle^{\rm reg}_{B,T})/2]/\partial T$ and $\chi_\Phi = \partial\Phi/\partial T$, respectively. Our results for the behavior of the susceptibilities as functions of the temperature, for $B = 0$, 0.6 and 1 GeV^2, are shown in the lower panels of Fig. 2.

Table 1. Critical temperatures for $B = 0$ and various parametrizations.

	Gaussian			5-Lorentzian		
$(-\langle q\bar{q}\rangle^{\rm reg}_{0,0})^{1/3}$ (MeV)	220	230	240	220	230	240
Chiral T_c (MeV)	182.1	179.1	177.4	177.0	177.0	177.8
Deconfinement T_c (MeV)	182.1	178.0	175.8	174.8	174.7	175.5

The chiral restoration and deconfinement critical temperatures obtained in absence of external magnetic field for different parametrizations are quoted in Table 1. It is seen that in all cases the splitting between both critical temperatures is below 5 MeV, which is consistent with the results obtained in lattice QCD. From Table 1 it is also seen that the values of critical temperatures do not vary significantly with the parametrization (recalling that in all cases the parameters have been fixed to reproduce the empirical values of the pion mass and decay constant). On the other hand, the critical temperatures in Table 1 are found to be somewhat higher than those obtained from LQCD, which lie

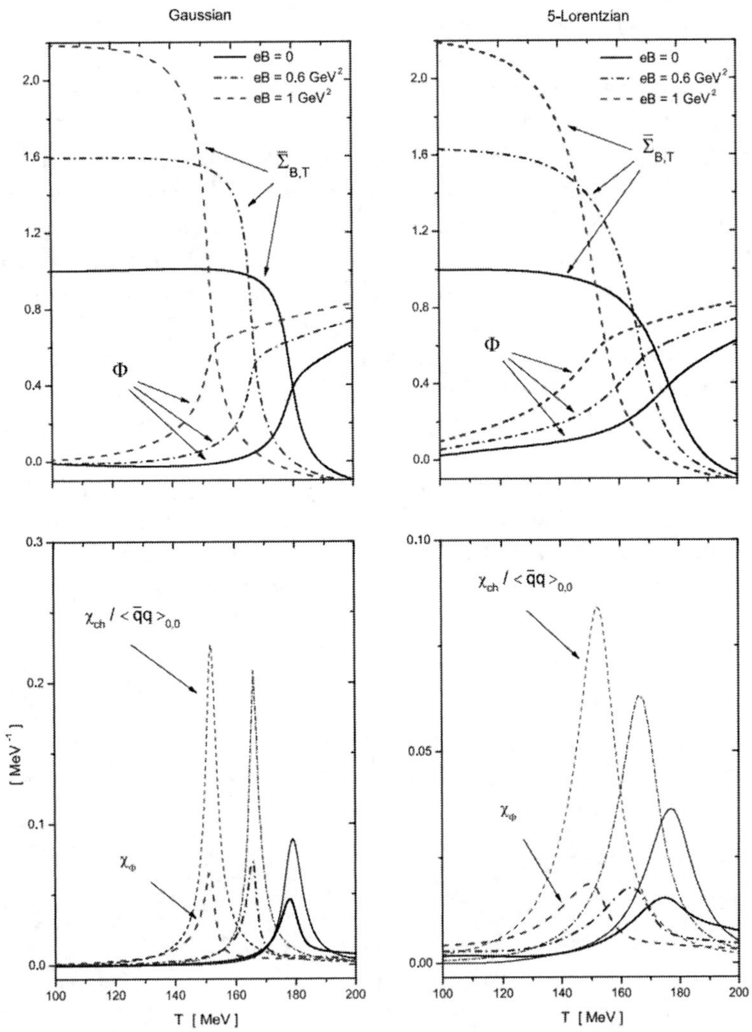

Figure 2. Upper panels: normalized flavor average condensate and traced Polyakov loop as functions of the temperature, for three representative values of eB. Lower panels: behavior of the corresponding chiral and PL susceptibilities as functions of the temperature.

around 160 MeV [21, 22]. It is worth noticing that in absence of the interaction with the Polyakov loop the values of T_c drop down to about 130 MeV [9].

Let us discuss the effect of the magnetic field on the phase transition features. From Fig. 2 it is seen that the splitting between the chiral restoration and deconfinement critical temperatures remains very small in the presence of the external field (in fact, a detailed analysis shows that the splitting gets reduced for larger values of eB). In addition, it is seen that the nonlocal NJL models show inverse magnetic catalysis. Indeed, contrary to what happens e.g. in the standard local NJL model [4–6], in our models the chiral restoration critical temperature becomes lower as the external magnetic field is increased. This is related with the fact that the condensates do not show in general a monotonic increase with B for a fixed value of the temperature. The situation is illustrated in Fig. 3, where we show the behavior of the averaged difference $\Delta\bar{\Sigma}_{B,T}$ as a function of eB, for $T = 0$ and for values of the temperature in the critical region. The curves correspond to models with Gaussian (left) and Lorentzian (right) form factors, with $(-\langle\bar{q}q\rangle^{\rm reg}_{0,0})^{1/3} = 230$ MeV. For these parametrizations the critical temperatures for $B = 0$ are slightly below 180 MeV (see Table 1). While for $T = 0$ the value of $\Delta\bar{\Sigma}_{B,0}$ shows a monotonic growth with the external magnetic field, it is seen that when the temperatures get closer to the critical values the curves have a maximum and then start to decrease for increasing B. This is the typical behavior associated to IMC and observed from lattice QCD results, see e.g. Fig. 2 of Ref. [8]. Qualitatively similar results are found for the other parametrizations in Table I. Finally, in Fig. 4 we plot our results for the chiral restoration critical temperatures $T_c(B)$, normalized to the corresponding values at vanishing external magnetic field. The figure includes the curves for nonlocal NJL models with Gaussian (left) and 5-Lorentzian (right) form factors and different parameter sets (see caption), and the gray bands in both panels show the results obtained in LQCD, taken from Ref. [8]. From the figure it is clearly seen that the inverse magnetic catalysis effect is observed for all considered parametrizations. In addition, for a given form factor the effect is found to be stronger for parameter sets leading to a lower absolute value of the chiral quark condensates. As a general conclusion, it can be stated that the behavior of the critical temperatures with the external magnetic field is compatible with LQCD results, for phenomenologically adequate values of the chiral condensate.

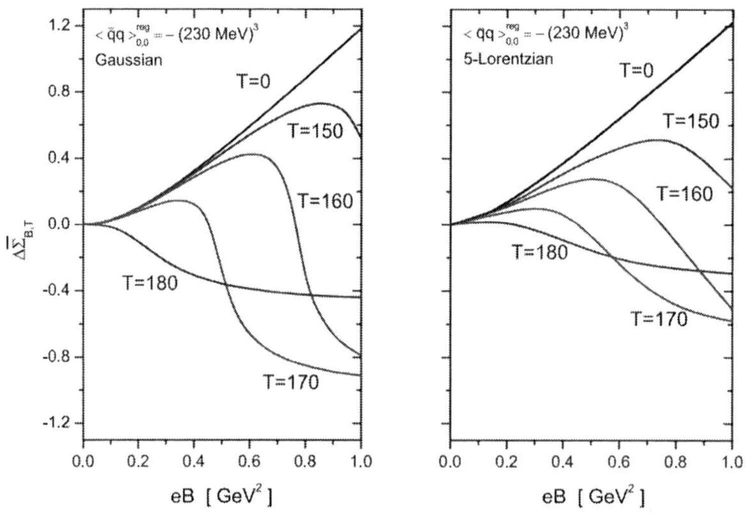

Figure 3. Subtracted normalized flavor average condensate as a function of eB for different representative temperatures. Left and right panels correspond to Gaussian and 5-Lorentzian form factors, respectively, with $(-\langle\bar{q}q\rangle^{\rm reg}_{0,0})^{1/3} = 230$ MeV. Temperature values are given in MeV.

4 Summary & conclusions

We have studied the behavior of strongly interacting matter under a uniform static external magnetic field in the context of a nonlocal chiral quark model. In this approach, which can be viewed as

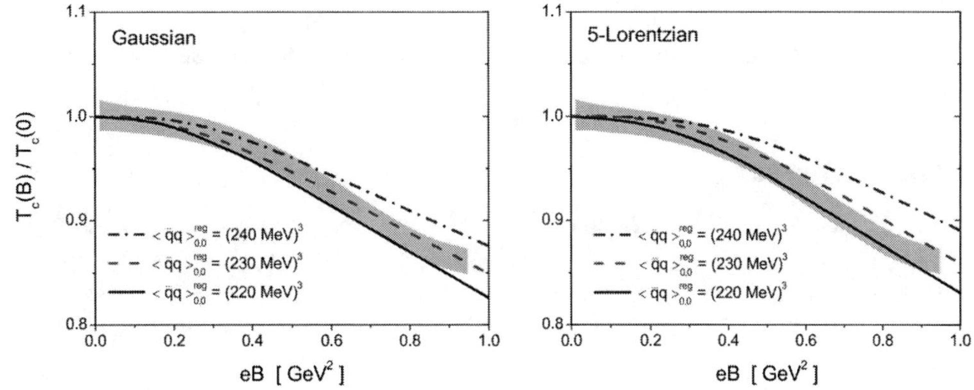

Figure 4. Normalized critical temperatures as functions of eB for various model parametrizations. For comparison, LQCD results of Ref. [8] are indicated by the gray band. Left and right panels correspond to Gaussian and 5-Lorentzian form factors, respectively.

an extension of the Polyakov-Nambu-Jona-Lasinio model, the effective couplings between quark-antiquark currents include nonlocal form factors that regularize ultraviolet divergences in quark loop integrals and lead to a momentum-dependent effective mass in quark propagators. We have worked out the formalism introducing Ritus transforms of Dirac fields, which allow to obtain closed analytical expressions for the gap equations, the chiral quark condensate and the quark propagator.

We have considered the case of Gaussian and Lorentzian form factors, choosing some sets of model parameters that allow to reproduce the empirical values of the pion mass and decay constants. At zero temperature, with these parameterizations we have calculated the behavior of the subtracted flavor average condensate $\Delta\bar{\Sigma}_{B,0}$ and the normalized condensate difference $\Sigma^u_{B,0} - \Sigma^d_{B,0}$ as functions of the external magnetic field B. Our results show the expected effect of magnetic catalysis (condensates behave as growing functions of B), the curves being in quantitative agreement with lattice QCD calculations with slight dependence on the parametrization.

Finally we have extended the calculations to finite temperature systems, including the couplings of fermions to the Polyakov loop. We have defined chiral and PL susceptibilities in order to study the chiral restoration and deconfinement transitions, which turn out to proceed as smooth crossovers for the considered polynomial PL potential. From our numerical calculations, on one hand it is seen that, for all considered values of B, both transitions take place at approximately the same temperature, in agreement with LQCD predictions. On the other hand, it is found that for temperatures close to the transition region $\Delta\bar{\Sigma}_{B,T}$ becomes a nonmonotonic funtion of B, which eventually leads to the phenomenon of inverse magnetic catalysis, i.e., a decrease of the critical temperature when the magnetic field gets increased. This feature is also in qualitative agreement with LQCD expectations. Moreover, for some parameterizations we find a remarkably good quantitative agreement with the results from LQCD calculations for the behavior of the normalized critical temperatures with B (see Fig. 4). The values of the critical temperature at $T = 0$, which show some dependence on the parameterization, lie also within the range estimated by LQCD results.

It is interesting to compare the nonlocal models with approaches in which IMC is obtained by considering some dependence of the effective couplings on B and/or T [23, 24]. The naturalness of the IMC behavior in our framework can be understood by noticing that for a given Landau level the associated nonlocal form factor turns out to be a function of the external magnetic field, according to the convolution in Eq. (13). Since the form factors can be identified with some gluon-mediated effective interaction, the dependence on the magnetic field can be seen as originated by the backreaction of the quarks on the gluon fields.

Acknowledgements

This work has been supported in part by CONICET and ANPCyT (Argentina), grants PIP14-492, PIP12-449, and PICT14-03-0492, by UNLP (Argentina), Project X718, by the Mineco (Spain), contract FPA2013-47443-C2-1-P, FPA2016-77177-C2-1-P, by Centro de Excelencia Severo Ochoa Programme, grant SEV-2014-0398, and by Generalitat Valenciana (Spain), grant PrometeoII/2014/066.

References

[1] R. C. Duncan and C. Thompson, Astrophys. J. 392, L9 (1992); C. Kouveliotou et al., Nature **393**, 235 (1998).
[2] D. E. Kharzeev, L. D. McLerran and H. J. Warringa, Nucl. Phys. A **803**, 227 (2008); V. Skokov, A. Y. Illarionov, and V. Toneev, Int. J. Mod. Phys. A **24**, 5925 (2009); V. Voronyuk *et al.*, Phys. Rev. C **83**, 054911 (2011).
[3] T. Vachaspati, Phys. Lett. **B265**, 258 (1991); K. Enqvist and P. Olesen, Phys. Lett. **B319**, 178 (1993).
[4] D. E. Kharzeev, K. Landsteiner, A. Schmitt and H. U. Yee, Lect. Notes Phys. **871**, 1 (2013).
[5] J. O. Andersen, W. R. Naylor and A. Tranberg, Rev. Mod. Phys. **88**, 025001 (2016).
[6] V. A. Miransky and I. A. Shovkovy, Phys. Rept. **576**, 1 (2015).
[7] G. S. Bali *et al.*, JHEP **1202**, 044 (2012).
[8] G. S. Bali *et al.*, Phys. Rev. D **86**, 071502 (2012).
[9] V. P. Pagura, D. Gomez Dumm, S. Noguera and N. N. Scoccola, Phys. Rev. D **95**, 034013 (2017).
[10] D. Gomez Dumm, M. F. I. Villafañe, S. Noguera, V. P. Pagura and N. N. Scoccola, arXiv:1709.04742 [hep-ph].
[11] T. Schafer and E. V. Shuryak, Rev. Mod. Phys. **70**, 323 (1998).
[12] C. D. Roberts and A. G. Williams, Prog. Part. Nucl. Phys. **33**, 477 (1994); C. D. Roberts and S. M. Schmidt, Prog. Part. Nucl. Phys. **45**, S1 (2000).
[13] S. Noguera and N. N. Scoccola, Phys. Rev. D **78**, 114002 (2008).
[14] G. A. Contrera, D. G. Dumm and N. N. Scoccola, Phys. Rev. D **81**, 054005 (2010).
[15] V. I. Ritus, Sov. Phys. JETP **48**, 788 (1978).
[16] S. Roessner, C. Ratti and W. Weise, Phys. Rev. D **75**, 034007 (2007).
[17] G. A. Contrera, D. Gomez Dumm and N. N. Scoccola, Phys. Lett. B **661**, 113 (2008).
[18] T. Hell, S. Roessner, M. Cristoforetti and W. Weise, Phys. Rev. D **79**, 014022 (2009).
[19] C. Ratti, M. A. Thaler and W. Weise, Phys. Rev. D **73**, 014019 (2006).
[20] B. -J. Schaefer, J. M. Pawlowski and J. Wambach, Phys. Rev. D **76**, 074023 (2007); B. -J. Schaefer, M. Wagner and J. Wambach, Phys. Rev. D **81**, 074013 (2010).
[21] Y. Aoki *et al.*, JHEP **0906**, 088 (2009); S. Borsanyi *et al.*, JHEP **1009**, 073 (2010).
[22] A. Bazavov *et al.*, Phys. Rev. D **85**, 054503 (2012).
[23] A. Ayala, M. Loewe, A. J. Mizher and R. Zamora, Phys. Rev. D **90**, 036001 (2014).
[24] R. L. S. Farias, K. P. Gomes, G. I. Krein and M. B. Pinto, Phys. Rev. C **90**, 025203 (2014).

EPJ Web of Conferences **172**, 03001 (2018) https://doi.org/10.1051/epjconf/201817203001
ISMD 2017

Measurements of the vector boson production with the ATLAS detector

A. Lapertosa[1,2,⋆], on behalf of the ATLAS Collaboration

[1] *Università degli Studi di Genova*
[2] *INFN Sezione di Genova*

Abstract. Measurements of the Drell-Yan production of W and Z bosons at the LHC provide a benchmark of our understanding of perturbative QCD and probe the proton structure in a unique way. The ATLAS collaboration has performed new high precision measurements at a center-of-mass energy of 7 TeV. The measurements are performed for W^+, W^- and Z bosons integrated and as a function of the boson or lepton rapidity and the Z mass. Unprecedented precision is reached and strong constraints on Parton Distribution Functions, in particular the strange density are found. Z boson cross sections are also measured at center-of-mass energies of 8 TeV and 13 TeV, and cross-section ratios to the top-quark pair production have been derived. This ratio measurement leads to a cancellation of systematic effects and allows for a high precision comparison to the theory predictions. The production of jets in association with vector bosons is a further important process to study perturbative QCD in a multi-scale environment. The ATLAS collaboration has performed new measurements of Z boson plus jets cross sections, differential in several kinematic variables, in proton-proton collision data taken at a center-of-mass energy of 13 TeV. The measurements are compared to state-of-the art theory predictions. They are sensitive to higher-order pQCD effects, probe flavour and mass schemes and can be used to constrain the proton structure. In addition, a new measurement of the splitting scales of the k_t jet-clustering algorithm for final states containing a Z boson candidate at a center-of-mass energy of 8 TeV is presented.

1 Introduction

The production of W and Z bosons via the Drell-Yan process is widely studied at the LHC experiments with proton-proton collision, including the ATLAS experiment [1]. Those processes have a clear signature (especially in the lepton decay channels: $Z \to ll$ and $W \to l\nu$) and are characterized by small background contamination. Due to the high production cross section, it is possible to collect large statistics samples of events containing vector bosons.

W and Z boson events, in turn, are common backgrounds in Standard Model measurements, including Higgs boson measurements, and in new physics searches. As a consequence, an improved description of the vector boson production has a positive impact in many ATLAS physics analyses.

Nevertheless, studying in details the production of vector bosons is useful to test the Standard Model accuracy and extract its parameters. It is possible to probe the proton structure, costraining

⋆e-mail: alessandro.lapertosa@cern.ch

© The Authors, published by EDP Sciences. This is an open access article distributed under the terms of the Creative Commons Attribution License 4.0 (http://creativecommons.org/licenses/by/4.0/).

Parton Distribution Functions (PDFs), and to test the perturbative QCD theory, comparing experimental results to state-of-the-art predictions.

In this review, a selection of seven of the most recent ATLAS results on vector boson physics will be presented: W and Z boson production cross section at 7 TeV [2] and 13 TeV [3]; ratio of $t\bar{t}$ over Z boson production cross section at 7, 8 and 13 TeV [4]; Z+jets production cross section at 13 TeV [5]; triple differential Z boson cross section at 8 TeV [6]; angular coefficients of Z boson leptons [7] and splitting scales of anti-k_t algorithm clustered jets [8], both measured with 8 TeV data.

2 W and Z boson cross section at 7 TeV

Several high precision vector boson cross section measurements have been performed on 4.6 fb^{-1} 7 TeV data, both inclusive and differential. Very good agreement is found between the ATLAS results and the Standard Model in terms of lepton universality, with the measurement of the $W \to e\nu$/W$\to \mu\nu$ ratio and the $Z \to ee$/Z$\to \mu\mu$ ratio, both found to be compatible with unity (see Fig.1).

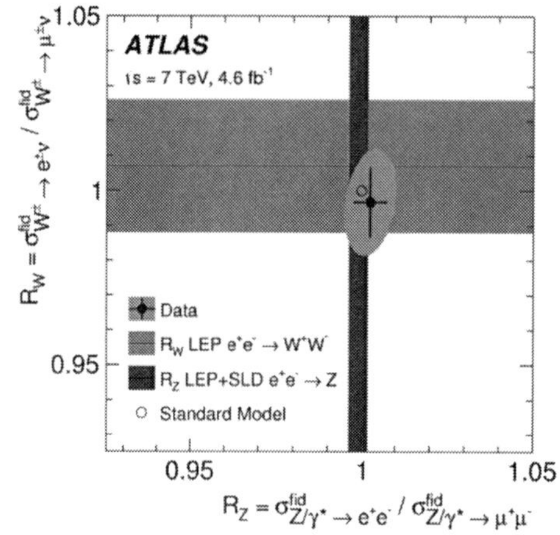

Figure 1. Measurement of the electron-to-muon cross-section ratios for the W and Z bosons. The green shaded ellipse represents the 68% CL for the correlated measurement of W and Z ratios, while the black error bars give the one-dimensional standard deviation. The orange and blue shaded bands represent the combination of the ratios as measured at the LEP and SLAC e^+e^- colliders. The SM expectation is indicated with an open circle. Taken from Ref. [2].

The measurement of the $W \to l\nu$/Z$\to ll$ ratio is sensitive to the strangeness content of the proton. The experimental precision reached is comparable and sometimes better than the one on theoretical predictions. The ATLAS result is compared with predictions obtained adopting different PDFs; in all cases, the predicted cross section ratio values are higher than those measured in data (see Fig. 2), clearly pointing out the need for improvements in the description of the proton structure.

The W boson charge asymmetry A_l, measured as a function of the pseudorapidity of the lepton $|\eta_l|$, defined as:

$$A_l = \frac{d\sigma_{W^+}/d|\eta_l| - d\sigma_{W^-}/d|\eta_l|}{d\sigma_{W^+}/d|\eta_l| + d\sigma_{W^-}/d|\eta_l|}, \tag{1}$$

is found to be in good agreement with the predictions (see Fig. 3), with an accurancy better than 1% across the whole range considered ($|\eta| < 2.5$).

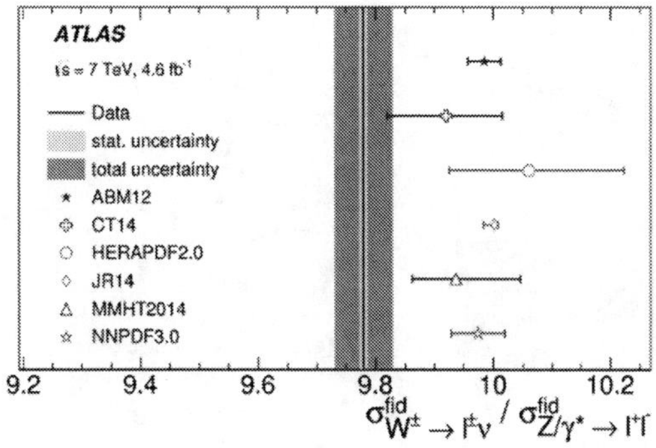

Figure 2. Fiducial cross sections times leptonic branching ratios of the $W \to l\nu/Z \to ll$ ratio. The data (solid blue line) are shown with the statistical (yellow band) and the total uncertainties (green band). The uncertainties of the theoretical calculations correspond to the PDF uncertainties only. Taken from Ref. [2].

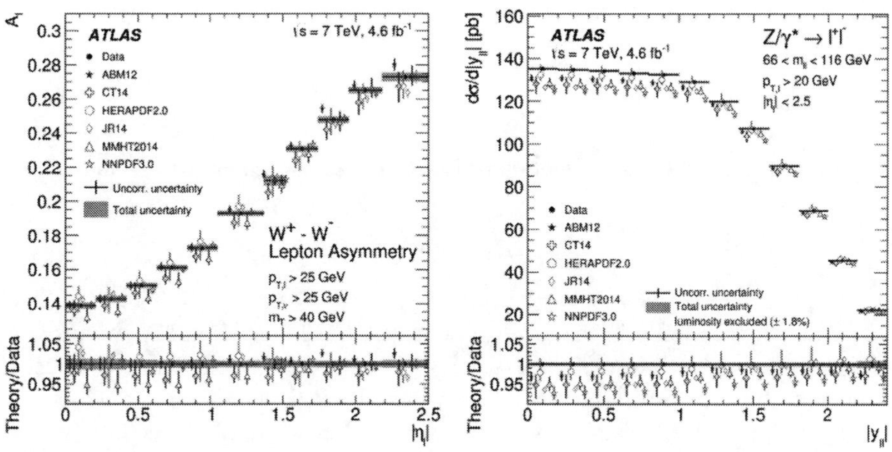

Figure 3. Left: lepton charge asymmetry A_l production as a function of the lepton pseudorapidity. Right: Z boson differential cross-section measurement as a function of the leptons rapidity $|y_{ll}|$. Predictions computed at NNLO QCD with NLO EW corrections using various PDF sets (open symbols) are compared to the data (full points). The ratio of theoretical predictions to the data is also shown. The predictions are displaced within each bin for better visibility. The theory uncertainty corresponds to the quadratic sum of the PDF uncertainty and the statistical uncertainty of the calculation. Taken from Ref. [2].

The Z boson production cross section measured as a function of the leptons rapidity $|y_{ll}|$ is found to be in agreement with predictions for values of $|y_{ll}| > 1$, while for $|y_{ll}| < 1$ the predictions are underestimating the data results (see Fig. 3). This discrepancy can be interpreted as enhanced PDF strangeness content.

Therefore, the results have been used to further constrain the proton PDF: with a profile fit, the new version of the ATLAS based PDF set has been derived, namely *ATLAS-epWZ16*. The new PDF

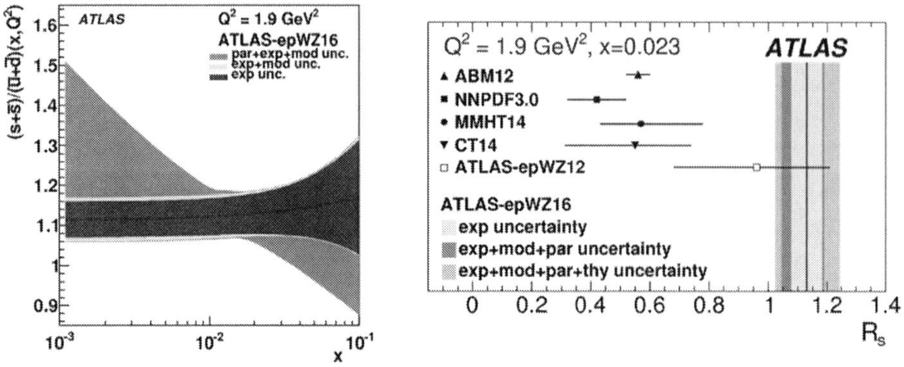

Figure 4. Left: R_s ratio of the PDFs from the present *ATLAS-epWZ16* determination shown in the region of maximum sensitivity of the ATLAS data, $10^{-3} < x < 10^{-1}$. Uncertainty bands represent the experimental (exp), model (mod) and parameterization (par) components in red, yellow and green, respectively. Right: R_s in the region of ATLAS best sensitivity, $Q^2 = 1.9$ GeV^2 and $x = 0.023$. Bands: present result and its uncertainty contributions from experimental data, QCD fit, and theoretical uncertainties; Closed symbols with horizontal error bars: predictions from different NNLO PDF sets; Open square: previous ATLAS resuls. Taken from Ref. [2].

shows unsuppressed strangeness as a function of *Bjorken-x*; in particular, the variable R_s, built as:

$$R_s = \frac{s + \bar{s}}{\bar{u} + \bar{d}}, \tag{2}$$

is found to be higher than unity in the x range considered ($10^{-3} < x < 10^{-1}$), with larger model uncertainties in the high and low x regions (see Fig. 4). The ATLAS data have the best sensitivity in the region with $Q^2 = 1.9$ GeV^2 and $x = 0.023$, where the R_s is measured to be 1.13 ± 0.05 (exp) $\pm\, 0.02$ (mod) $^{+0.01}_{-0.06}$ (par). With this high precision measurements, ATLAS is definitely accessing the proton structure.

3 W and Z boson cross section at 13 TeV

The W^+, W^-, $W^\pm$ and Z boson production cross sections have been measured separately in the lepton decay channels with 81 pb^{-1} 13 TeV data, within the fiducial region defined by the detector acceptance. The ATLAS results are compared to theoretical predictions obtained with different PDFs, showing good agreement with many of them. The precision obtained is of the order of 2% on W boson and 1% on Z boson cross section, excluding the 2.1% uncertainty due to the luminosity determination (see Fig. 5).

The cross section ratios benefit from significant cancellation of experimental uncertainties, resulting in higher constraining power. The W^+/W^- ratio measured by ATLAS is found to be sistematically lower than the predictions, showing better agreement with the CT14 and MMHT14 PDFs. The W/Z ratio is found to be compatible with the different predictions, within the uncertainty. Both resuts are shown in Fig. 6.

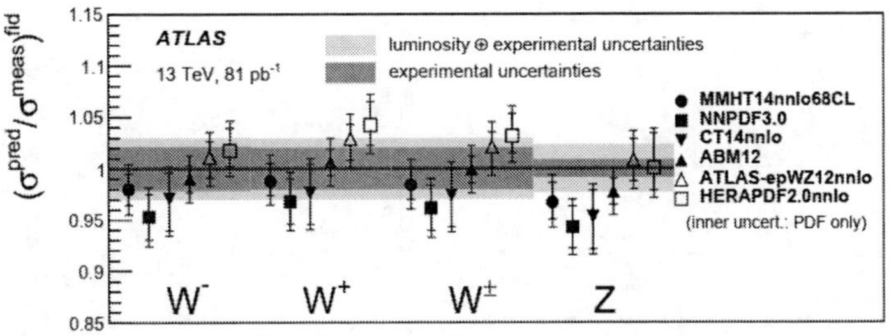

Figure 5. Ratio of the predicted to measured fiducial cross section for the combined electron and muon channels using various PDFs. The inner (outer) band corresponds to the experimental uncertainty without (with) the luminosity uncertainty. The inner error bar of the predictions represents the PDF uncertainty while the outer error bar includes the sum in quadrature of all other systematic uncertainties. Taken from Ref. [3].

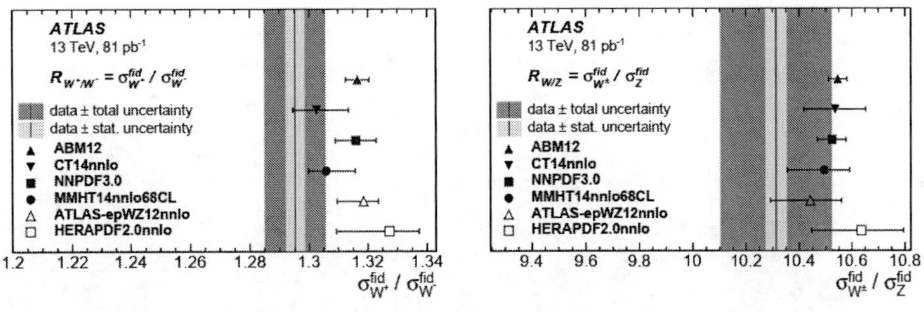

Figure 6. Ratios (red line) of W^+ to W^- boson (left) and W to Z boson (right) combined production cross sections in the fiducial region compared to predictions based on different PDF sets. The inner (yellow) shaded band corresponds to the statistical uncertainty while the outer (green) band shows statistical and systematic uncertainties added in quadrature. The theory predictions are given with only the corresponding PDF uncertainties shown as error bars. Taken from Ref. [3].

4 $t\bar{t}$ over Z boson cross section at 7, 8 and 13 TeV

The $t\bar{t}/Z$ cross section ratio has been measured at 7, 8 and 13 TeV: results are given directly for each considered $\sqrt{s}$ and also as double ratios of the two processes at different $\sqrt{s}$. The results are corrected to a common phase space and, furthermore, the systematic uncertainties are properly correlated among measurements. The cross section ratio $R_{t\bar{t}/Z}$ is defined as:

$$R_{t\bar{t}/Z} = \frac{\sigma_{t\bar{t}}}{0.5(\sigma_{Z\to ee} + \sigma_{Z\to\mu\mu})}, \quad (3)$$

where $\sigma_{Z\to ee}$ and $\sigma_{Z\to\mu\mu}$ are the Z boson cross section measured in the electron and the muon channel, respectively.

The results are compared to calculations performed at next-to-next-to-leading-order accuracy using recent PDF sets. The final results uncertainty is further reduced thanks to the cancellation of systematic uncertainties in the ratios, with the data results being more precise than predictions, especially in the 7 and 8 TeV cases (see Fig. 7).

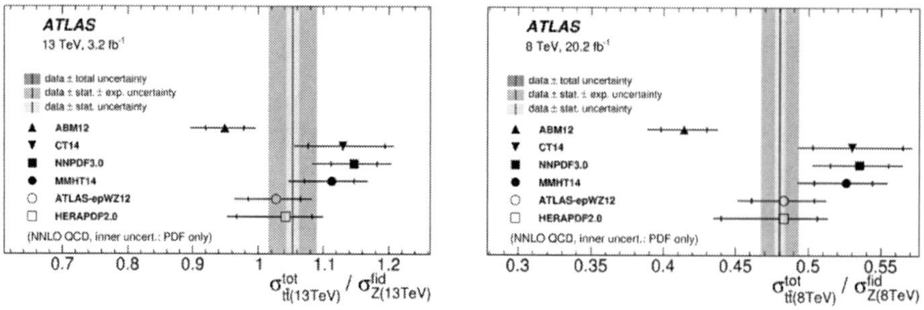

Figure 7. The $t\bar{t}/Z$ cross section ratio for $\sqrt{s} = 7$ TeV (left) and 8 TeV (right): results are compared to predictions based on different PDF sets. The inner shaded band corresponds to the statistical uncertainty, the middle band to the statistical and experimental systematic uncertainties added in quadrature, while the outer band shows the total uncertainty, including the luminosity uncertainty which almost entirely cancel in these ratios. The theory predictions are given with the corresponding PDF uncertainties shown as inner bars while the outer bars include all other uncertainties added in quadrature. Taken from Ref. [4].

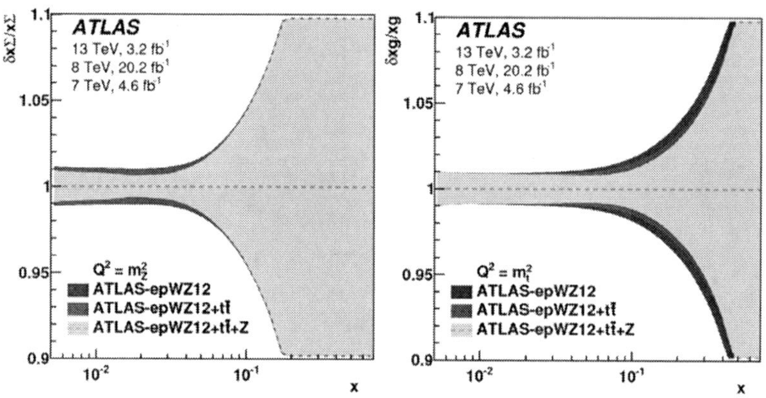

Figure 8. Impact of the ATLAS Z boson and $t\bar{t}$ cross-section data on the determination of PDFs. The bands represent the uncertainty for the *ATLAS-epWZ12* PDF set and the uncertainty of the profiled *ATLAS-epWZ12* PDF set using Z boson and $t\bar{t}$ data as a function of x for the total light-quark-sea distribution (left) for the gluon density (right). The plots also show the impact of only including the ATLAS $t\bar{t}$ data set. Taken from Ref. [4].

The six measured cross sections ($t\bar{t}$ and Z cross sections at 7, 8 and 13 TeV) are used together within a profile fit to further constrain the *ATLAS-epWZ12* PDF: significant reductions of the uncertainties are obtained for the gluon distribution function for x values near 0.1 and for the light-quark sea for $x < 0.02$ (see Fig. 8).

5 Z+jets production cross section at 13 TeV

The measurements of the production cross section of a Z boson in association with jets is performed on 3.16 fb^{-1} 13 TeV data. Inclusive and differential cross sections are measured for events containing a Z boson decaying to electrons or muons and produced in association with up to seven jets with $p_T > 30$ GeV and $|y| < 2.5$. The data results are compared to different order predictions: in general,

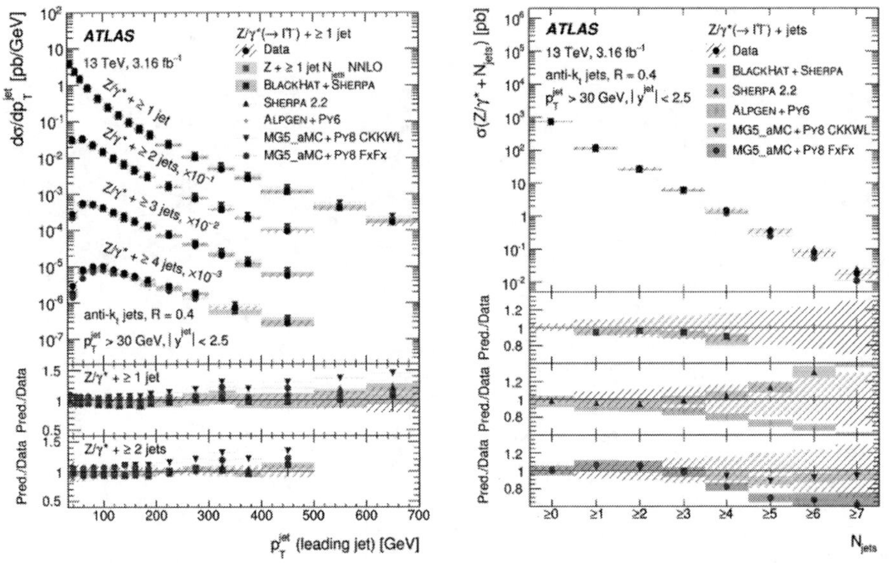

Figure 9. Left: measured cross section as a function of the leading jet p_T for inclusive Z+ $\geq$ 1,2,3,4 jet events. Right: measured cross section as a function of the inclusive jet multiplicity for Z+jets events. The data are compared to the predictions from Z + $\geq$ 1 jet Njetti NNLO (p_T spectra only), BLACKHAT+SHERPA, SHERPA 2.2, ALPGEN+PY6, MG5_aMC+PY8 CKKWL, and MG5_aMC+PY8 FxFx. The error bars correspond to the statistical uncertainty, and the hatched bands to the data statistical and systematic uncertainties (including luminosity) added in quadrature. Taken from Ref. [5].

good agreement is found within the uncertainties with the generators which use next-to-leading-order matrix elements and the more recent next-to-next-to-leading-order fixed-order predictions.

In particular, leading-order predictions are observed to have a too hard jet p_T spectrum compared to data, for values of jet p_T > 200 GeV. Regarding the jet multiplicity observable, the models considered are sensibly underestimating or overestimating the data results for high jet multiplicity (number of jets higher than 4). Both results are shown in Fig. 9.

6 Triple differential Z boson cross section at 8 TeV

The Z boson differential cross section in the lepton decay channel has been measured with 20.2 fb^{-1} 8 TeV data. The results are presented as a function of the invariant mass m_{ll}, the dilepton rapidity $|y_{ll}|$, and the angular variable $cos\theta^*$, between the outgoing lepton and the incoming quark in the Collins-Soper frame.

The cross sections are used to determine the Z boson forward-backward asymmetry A_{FB}, defined as:

$$A_{FB} = \frac{d^3\sigma(cos\theta^* > 0) - d^3\sigma(cos\theta^* < 0)}{d^3\sigma(cos\theta^* > 0) + d^3\sigma(cos\theta^* < 0)}, \quad (4)$$

as a function of m_{ll} and $|y_{ll}|$.

The forward-backward asymmetry is found to change sign at the Z boson mass and to increase with $|y_{ll}|$, as expected. A_{FB} values are compatible with 0 for m_{ll} close to Z boson mass, while they are positive (negative) for events with m_{ll} lower (higher) than the Z boson mass.

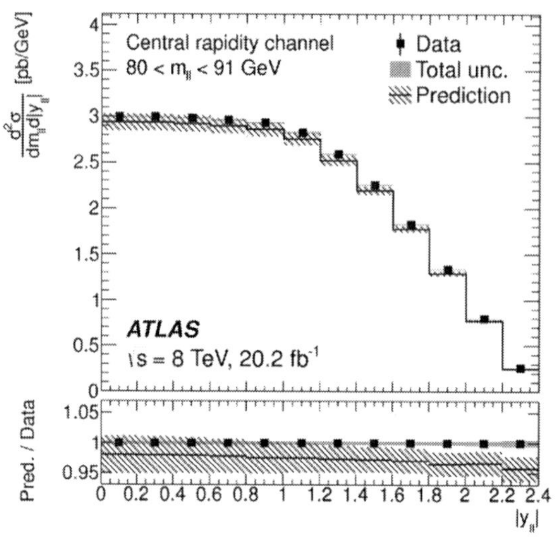

Figure 10. Fiducial cross section as a function of m_{ll}, the dilepton rapidity $|y_{ll}|$ in the invariant mass range close to the Z boson mass. The data are shown as solid markers and the prediction from Powheg is shown as the solid line. The inner error bars represent the data statistical uncertainty and the solid band shows the total experimental uncertainty. The contribution to the uncertainty from the luminosity measurement is excluded. The hatched band represents the statistical and PDF uncertainties in the prediction. Taken from Ref. [6].

The measurements achieve high-precision (1% level, excluding the uncertainty in the integrated luminosity) and are in agreement with predictions, with a slight tendency of the prediction to underestimate the data results, as can be seen in Fig. 10.

7 Drell-Yan lepton angular correlations

The angular correlations between lepton pairs has been measured with 20.3 fb^{-1} 8 TeV data, in the invariant mass region close to Z boson mass. The complete set of angular coefficients A_{0-7} is extracted by polinomial decomposition of the lepton polar θ and azimuthal ϕ angular variables, measured in the Z boson Collins-Soper frame.

Results are compared to several available next-to-next-to-leading-order predictions. Evidence is found for nonzero $A_{5,6,7}$ observables, consistent with expectations. The observable $A_0 - A_2$, built as the difference between A_0 and A_2, is found to be higher than zero, as expected, when introducing higher order corrections. All the predictions are observed to be a factor of 2 lower than the data results (see Fig. 11).

8 Anti-k_t jet splitting scales

Splitting scales of the jets produced in association with a Z boson, clusterized with the anti-k_t algorithm [9], are measured with 20.2 fb^{-1} 8 TeV data. Starting from charged particle tracks as input, jets are clusterized according to the minimum distance criteria: minimum distance between the tracks or between the tracks and the beam axes. After the jet clusterization, the splitting scales $\sqrt{d_k}$ are defined as the square root of d_k, where:

$$d_k = min(d_{ij}, d_{ib}), \quad (5)$$

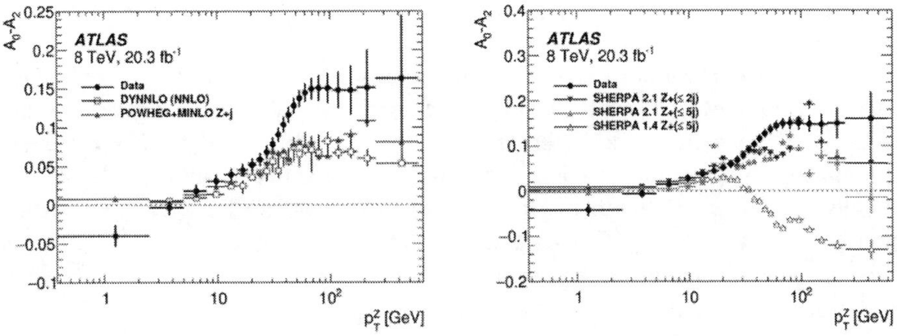

Figure 11. Distributions of the angular coefficients $A_0 - A_2$ as a function of p_T^Z. The results are compared to the DYNNLO and POWHEG MINLO predictions (left) and various SHERPA predictions (right). The error bars for the calculations show the total uncertainty for DYNNLO, but only the statistical uncertainties for POWHEG MINLO and SHERPA. Taken from Ref. [7].

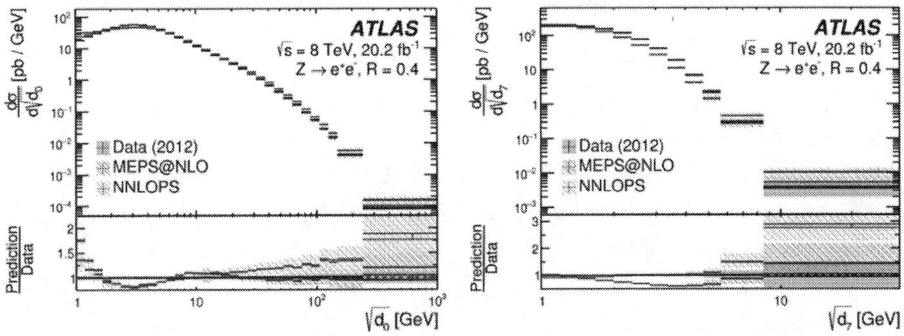

Figure 12. Charged-only distributions for the 0-th (left) and the 7-th (right) splitting scales in the electron channel using the jet-radius parameter R=0.4. The size of the error bars reflects the statistical uncertainty, while the combined statistical and systematic uncertainty is indicated by the grey band. Theoretical predictions from SHERPA with NLO multijet merging (MEPS@NLO) and from POWHEG+PYTHIA8 with NNLO matching (NNLOPS) are displayed including error bands for the generator uncertainties. Taken from Ref. [8].

with d_{ij} defined as:

$$d_{ij} = min(p_{T,i}^2, p_{T,j}^2) \times \frac{\Delta R_{i,j}^2}{R^2}, \tag{6}$$

and d_{ib} as:

$$d_{ib} = p_{T,i}^2. \tag{7}$$

With this definition, the 0-th order splitting scale corresponds to the p_T of the leading jet, while the N-th order of the splitting scale can be interpreted as the distance at which a N jet event can be resolved as a $N + 1$ jet event.

Data results are compared to next- and next-to-next-to-leading-order predictions: significant deviations are found between data and state-of-the-art predictions in various regions of the observables (see Fig. 12). Compatible results are found between the electron and muon channel Z boson decay, and between clusterization performed with the anti-k_T radius parameter equal to 0.4 and to 1.0.

9 Conclusion

High precision ATLAS measurements of vector boson observables have been performed with 7, 8 and 13 TeV data. In several measurements, the experimental precision has been higher than the prediction one: in particular, the uncertainty had been further reduced in the W^+/W^-, W/Z and $t\bar{t}/Z$ cross section ratio measurements, due to the factorization of some systematic uncertainties.

The results have be exploited to improve the *ATLAS-epWZ12* PDF precision and to derive the new version of the ATLAS based PDF, *ATLAS-epWZ16*, having reduced uncertainty compared to the previous version and showing unsuppressed strangeness content.

Several results are found to confirm the needs for higher order corrections. The remaining discrepancies will motivate further work to improve the precision and the modelling of the predictions. ATLAS precision measurements are therefore a powerful mean to improve our understanding of perturbative QCD.

References

[1] ATLAS Collaboration, JINST **3**, S08003 (2008)
[2] ATLAS Collaboration, Eur. Phys. J. C **77**, 367 (2017), 1612.03016
[3] ATLAS Collaboration, Phys. Lett. B **759**, 601 (2016), 1603.09222
[4] ATLAS Collaboration, JHEP **02**, 117 (2017), 1612.03636
[5] ATLAS Collaboration, Eur. Phys. J. C **77**, 361 (2017), 1702.05725
[6] ATLAS Collaboration, Submitted to JHEP (2017), 1710.05167
[7] ATLAS Collaboration, JHEP **08**, 159 (2016), 1606.00689
[8] ATLAS Collaboration, JHEP **08**, 026 (2017), 1704.01530
[9] M. Cacciari, G. P. Salam, and G. Soyez, JHEP **0805**, 063 (2008), 0802.1189v2

Measurement of hadronic form factors at BESIII

Haiming HU[1,*] on behalf of the BESIII Collaboration

[1] *Institute of High Energy Physics, CAS, Beijing, China.*

Abstract. The measurements of hadronic form factors of three modes using the data samples collected with the BESIII detector at BEPCII collider are presented. The cross section of $e^+e^- \to p\bar{p}$ at 12 energies from 2232.4 to 3671.0 MeV are measured, the electromagnetic form factor is deduced, and the ratio $|G_E/G_M|$ is extracted by fitting the polar angle distribution. The preliminary results about the form factors of $e^+e^- \to \Lambda_c^+\bar{\Lambda}_c^-$ will also be described. The cross section of $e^+e^- \to \pi^+\pi^-$ between effective center-of-mass energy 600 and 900 MeV is measured by the ISR return method using the data set with the integrated luminosity of 2.93 fb^{-1} taken at $\psi(3773)$ peak, the pion form factor is extracted.

1 Introduction

The standard model (SM) is the basic theory of elementary particles, its one of the main subjects is to explore the inner structure of hadrons (mesons and baryons). Two or three valence quarks (also sea quarks and gluons) are confined within hadrons, and the interaction among the quarks and gluons are nonperturbative, which leads to the complex multi-parton structures. At present, many properties of hadrons could not be accurately calculated and explained from the first principle of the quantum chromodynamics (QCD). In the study of the strong interaction and hadronic structures, experiments always take crucial role.

BEPCII is a double-ring e^+e^- collider running at center-of-mass energies between 2.0 and 4.6 GeV and reached a peak luminosity of $10^{33}\text{cm}^{-2}\text{s}^{-1}$ at center-of-mass energy of 3770 MeV. The BESIII detector is located at the BEPCII [1], which is designed to fulfill the requirement of the tau-charm physics [2].

The exclusive hadronic production cross sections in the e^+e^- annihilation are usually expressed as the product of the phase-space factor and form factors. For almost processes, the phase-space factors are ordinary, and the special dynamics reflect in the form factors. From this point of view, if the form factors are derived from theory and confirmed by experiments, it means the laws and properties of strong interactions are understood. One of the quantities can be directly measured in experiments is the (differential) cross section.

Proton and pion are two simplest hadrons composed of the light quarks u and d, the measurements of form factors to the processes $e^+e^- \to p\bar{p}$ and $\pi^+\pi^-$ have special significance. Many experiments did the comprehensive studies to measure the form factors of proton and pion [3, 4]. BESIII also measured the form factor of strange baryon pair $\Lambda\bar{\Lambda}$, and observed the non-zero cross section near the

*e-mail: huhm@ihep.ac.cn

© The Authors, published by EDP Sciences. This is an open access article distributed under the terms of the Creative Commons Attribution License 4.0 (http://creativecommons.org/licenses/by/4.0/).

threshold [5]. Λ_c^+ and $\bar{\Lambda}_c^-$ are the grounds of the lightest singly charmed baryons, the properties of the cross section of the process $e^+e^- \to \Lambda_c^+\bar{\Lambda}_c^-$ and $\Lambda_c^\pm$ decay branch fractions were known less, BESIII did some preliminary measurements using the newly collected data.

This paper will briefly introduce the measurement results of form factors for the following three reaction channels: (1) $e^+e^- \to p\bar{p}$ at twelve center-of-mass energies between 2232.4 and 3671.0 MeV [6]; (2) $e^+e^- \to \Lambda_c^+\bar{\Lambda}_c^-$ at energies 4574.5, 4580.0, 4590.0 and 4599.5 MeV [7]; (3) $e^+e^- \to \pi^+\pi^-$ between 600 and 900 MeV by the initial state radiation (ISR) return method using the data collected at the peak of $\psi(3773)$ [8].

2 Measurement of $e^+e^- \to p\bar{p}$

The differential and total Born cross sections of the non-pointlike fermion pair can be expressed as the function of the electromagnetic form factors $|G_E|$ and $|G_M|$ [9],

$$\frac{d\sigma_{\text{Born}}(s)}{d\Omega} = \frac{\alpha^2\beta C}{4s}[|G_M(s)|^2(1+\cos^2\theta_p) + \frac{4m_p^2}{s}|G_E(s)|^2\sin^2\theta_p], \tag{1}$$

$$\sigma_{\text{Born}}(s) = \frac{4\pi\alpha^2\beta}{3s}C[|G_M|^2 + \frac{2m_p^2}{s}|G_E|^2]. \tag{2}$$

The meaning of the symbols in above equations are explained in Ref. [6]. The form factors $|G_E|$ and $|G_M|$ can be obtained respectively by fit the measured polar angle θ_p distribution. The earlier experiments measured the $\sigma_{\text{Born}}(s)$ under the simple assumption $|G_E| = |G_M| \equiv |G|$ due to the limited statistics of the data samples. BESIII did the same measurement for a comparison.

Experimentally, the Born cross section of $e^+e^- \to p\bar{p}$ is measured by following expression

$$\sigma_{\text{Born}} = \frac{N_{\text{obs}} - N_{\text{bkg}}}{\mathcal{L}\cdot\varepsilon\cdot(1+\delta)}, \tag{3}$$

where N_{obs} is the observed number of candidate $p\bar{p}$ pair, N_{bkg} the number of remnant background events, $\mathcal{L}$ the integrated luminosity, ε the detection efficiency determined with MC generator, and $(1+\delta)$ the initial state radiation correction factor. The $|G|$ can be deduced with the measured cross section by following relation

$$|G|^2 = \frac{3s}{4\pi\alpha^2\beta C(1+2m_p^2/s)}\sigma_{\text{Born}}. \tag{4}$$

The measured results by BESIII and other experiments for the invariant $p\bar{p}$ masses from 2.20 to 3.70 GeV/c^2 are shown in Figure 1.

Some times, the ratio $R = |G_E/G_M|$ is more interested in understanding the electric and magnetic structures, the Eq.(1) can be rewritten as the form:

$$F(\cos\theta_p) = N_{\text{norm}}[1+\cos^2\theta_p + \frac{4m_p^2}{s}R^2\sin^2\theta_p], \tag{5}$$

where N_{norm} is the overall normalization factor. The efficiency corrected distributions of $\cos\theta_p$ and fit result for data at c.m. energy 3080.0 MeV of the invariant masse of $p\bar{p}$ is shown in left of Figure 2. The comparisons of form factor and ratio $|G_E/G_M|$ measured with BESIII data to the previous experimental measurements are shown in right of Figure 2. Compared to the BaBar results [10], the precision of the Born cross section measured with BESIII is improved by 30% for data sets with $\sqrt{s} \leq 3080.0$ MeV, and the precision of the form factor is improved.

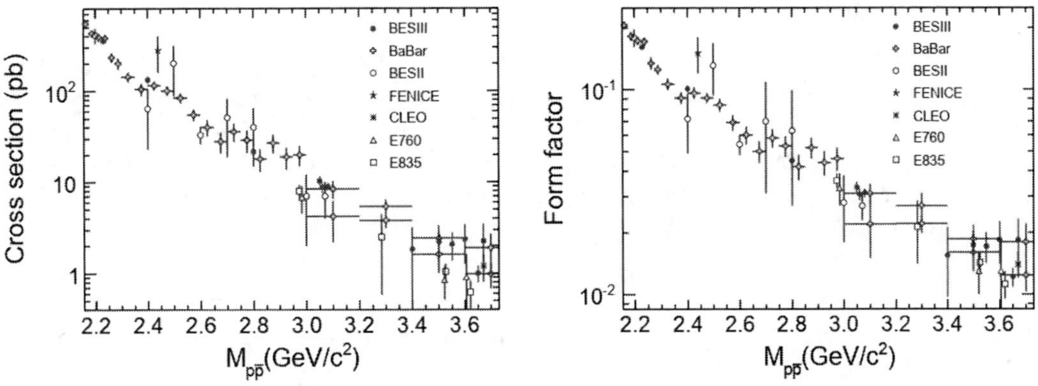

Figure 1. Comparison of the Born cross section (left) and the effective form factor $|G|$ (right) between BESIII measurement and other experiments for the invariant $p\bar{p}$ masses from 2.20 to 3.70 GeV/c^2.

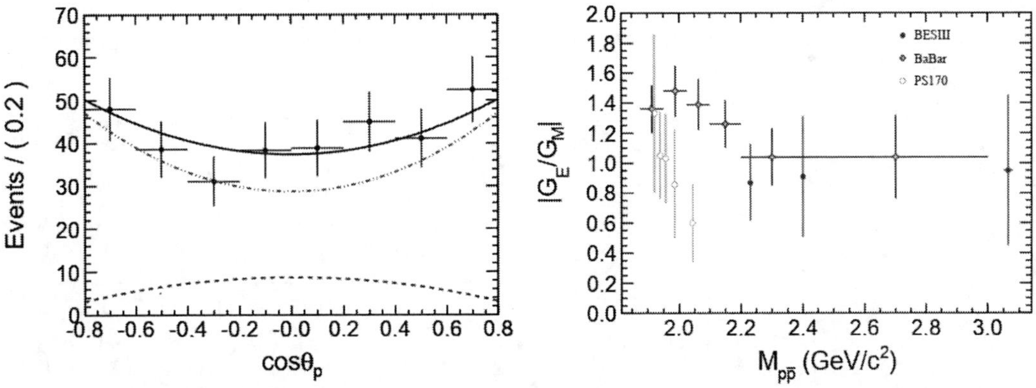

Figure 2. Left: The efficiency corrected distributions of $\cos\theta_p$ and fit result for data at c.m. energy 3080.0 MeV of the invariant masse of $p\bar{p}$; Right: the ratio $|G_E/G_M|$ at different energies measured by BESIII, BaBar and PS170.

3 Measurement of $e^+e^- \to \Lambda_c^+\Lambda_c^-$

Λ_c is the lightest single charmed baryon. It was the first hint of charmed baryon in the decay process $\Sigma_c^{++} \to \Lambda_c^+\pi^+$ observed at BNL in 1975 [11], and was firstly evidenced at Fermi Laboratory in 1976 [12], then was firstly established at MarkII in 1980 [13]. Most of the heavier charmed baryon will eventually decay to Λ_c, including the doubly charmed baryon Σ_{cc}^{++}, the error of $B(\Lambda_c^+ \to pK^-\pi^+)$ dominance the uncertainty of V_{ub} via Λ_b decay.

The form factor of charmed baryon in the process $e^+e^- \to \Lambda_c^+\Lambda_c^-$ near its production threshold is of interested in theory, which leads to various models [14]. Belle Collaboration ever measured the cross section of this channel using initial state radiation (ISR) technique [15]. BESIII also measured the form factor as well as the branching fractions of 12 hadronic decay modes and the semi-leptonic decay $\Lambda_c^+ \to \Lambda e^+\nu_e$ and $\Lambda_c^+ \to \Lambda\mu^+\nu_\mu$ at center-of-mass energy 4574.5, 4580.0, 4590.0 and 4599.5 MeV, the detailed description can be found in references [7].

3.1 The form factor

The ten Cabibbo-favored hadronic decay modes $\Lambda_c^+ \to pK^-\pi^+$, pK_S^0, $\Lambda\pi^+$, $pK^-\pi^+\pi^0$, $pK_S^0\pi^0$, $\Lambda\pi^+\pi^0$, $pK_S^0\pi^+\pi^-$, $\Lambda\pi^+\pi^+\pi^-$, $\Sigma^0\pi^+$ and $\Sigma^+\pi^+\pi^-$, as well as the corresponding charge-conjugate modes are analyzed. The measurement of each mode produce the corresponding cross section, and the total cross section is obtained from the weighted average over the 20 individual measurements [16]. The measured cross section are shown in the left of Figure 3 together with the Belle data [15] for comparisons. The polar angle θ distribution of the produced fermion pair $\Lambda_c^\pm$ has the form

$$f(\theta) \propto (1 + \alpha_{\Lambda_c} \cos^2\theta). \tag{6}$$

Fitting the angular distribution $f(\theta)$ can be obtained the parameter α_{Λ_c}, and it connects to the ratio $|G_E/G_M|$ in following expression:

$$|G_E/G_M|^2(1-\beta^2) = (1-\alpha_{\Lambda_c})/(1+\alpha_{\Lambda_c}). \tag{7}$$

The measured angular distribution and the corresponding data fit result at 4599.5 MeV are shown in the right of Figure 3, and the values of fit results of α_{Λ_c} and $|G_E/G_M|$ are listed in Table 1.

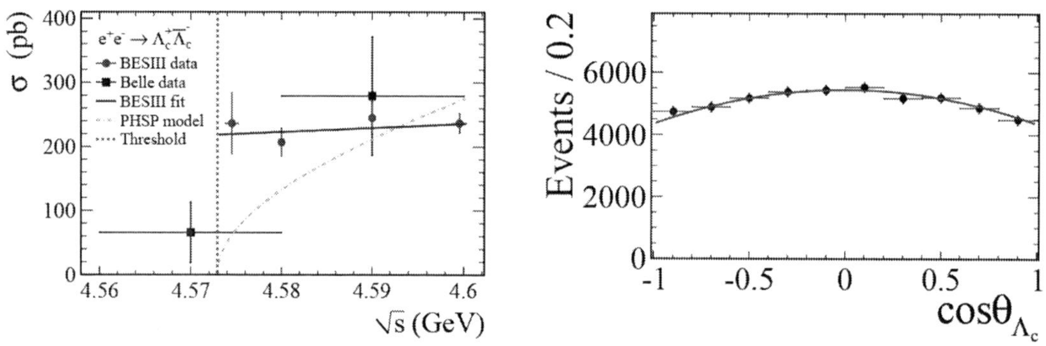

Figure 3. Left: the cross section of $e^+e^- \to \Lambda_c^+\Lambda_c^-$ measured by BESSIII and Belle. The blue solid curve represents the input line-shape for KKMC when determine the f_{ISR}. The dash-dot cyan curve denotes the prediction of the phase space model. Right: the angular distribution and the corresponding data fit result at 4599.5 MeV.

Table 1. Shape parameters of the angular distribution α_{Λ_c} and $|G_E/G_M|$ for the $e^+e^- \to \Lambda_c^+\Lambda_c^-$

$\sqrt{s}$ *(MeV)*	α_{Λ_c}	$\lvert G_E/G_M \rvert$
4574.5	$-0.13 \pm 0.12 \pm 0.08$	$1.14 \pm 0.14 \pm 0.07$
4599.5	$-0.20 \pm 0.04 \pm 0.02$	$1.23 \pm 0.05 \pm 0.03$

3.2 The hadronic decay branch ratio

The 12 Cabibbo-favored hadronic decay modes listed in Table 2 are measured using the data sample collected at 4599.5 MeV with an integrated luminosity of 566.9 pb^{-1}. To identify the $\Lambda_c^+\bar{\Lambda}_c^-$ signal candidates, the tagged techniques [17, 18] are adopted, and a least-squares fitter, which considers the statistic and systematic correlations among the different hadronic decay modes, are used to obtain the branch ratio of the 12 decay modes globally. The goodness-of-fit is evaluated as $\chi^2/ndf = 9.9/(24 - 13) = 0.9$, the extracted values are shown in Table 2.

Table 2. The preliminary results of the 12 hadronic decay branch ratio of Λ_c^+

Channel	BESIII preliminary results (%)	PDG (%)
pK_S^0	$1.52 \pm 0.08 \pm 0.03$	1.15 ± 0.30
$pK^-\pi^+$	$5.84 \pm 0.27 \pm 0.23$	5.0 ± 1.3
$pK_S^0\pi^0$	$1.87 \pm 0.13 \pm 0.05$	1.65 ± 0.50
$pK_S^0\pi^+\pi^-$	$1.53 \pm 0.11 \pm 0.09$	1.30 ± 0.53
$pK^-\pi^+\pi^0$	$4.53 \pm 0.23 \pm 0.30$	3.4 ± 1.0
$\Lambda\pi^+$	$1.24 \pm 0.07 \pm 0.03$	1.07 ± 0.28
$\Lambda\pi^+\pi^0$	$7.01 \pm 0.37 \pm 0.19$	3.6 ± 1.3
$\Lambda\pi^+\pi^-\pi^+$	$3.81 \pm 0.24 \pm 0.18$	2.6 ± 0.7
$\Sigma^0\pi^+$	$1.27 \pm 0.08 \pm 0.03$	1.05 ± 0.28
$\Sigma^+\pi^0$	$1.18 \pm 0.10 \pm 0.03$	1.00 ± 0.34
$\Sigma^+\pi^+\pi^-$	$4.25 \pm 0.24 \pm 0.20$	3.6 ± 1.0
$\Sigma^+\omega$	$1.56 \pm 0.20 \pm 0.07$	2.7 ± 1.0

3.3 The semi-leptonic decay branch ratio

The decay mode to be analyzed is $\Lambda_c^+ \to \Lambda l^+ \nu_l$, where $l = e, \nu$. The candidate of $\bar{\Lambda}_c^-$ is reconstructed with the single tag (ST) by the 11 hadronic decay modes: $\bar{\Lambda}_c^- \to \bar{p}K_S^0$, $\bar{p}K^+\pi^-$, $\bar{p}K_S^0\pi^0$, $\bar{p}K^+\pi^-\pi^0$, $\bar{p}K_S^0\pi^+\pi^-$, $\bar{\Lambda}K^-$, $\bar{\Lambda}K^-\pi^0$, $\bar{\Lambda}K^-\pi^+\pi^-$, $\bar{\Sigma}^0\pi^-$, $\bar{\Sigma}^-\pi^0$ and $\bar{\Sigma}\pi^+\pi^-$, where the intermediate particles K_S^0, $\bar{\Lambda}$, $\bar{\Sigma}^0$, $\bar{\Sigma}^-$ and π^0 are reconstructed by their following decays: $K_S^0 \to \pi^+\pi^-$, $\bar{\Lambda} \to \bar{p}\pi^+$, $\bar{\Sigma}^0 \to \gamma\bar{\Lambda}$ and $\bar{\Lambda} \to \bar{p}\pi^+$, $\bar{\Sigma}^- \to \bar{p}\pi^0$ and $\pi^0 \to \gamma\gamma$. The signal candidates for Λ_c^+ are selected from the remaining tracks recoiling against the ST $\bar{\Lambda}_c^-$ candidates. The measured results of the branch ratio of the Λ_c^+ semi-leptonic decay are

$$B_r(\Lambda_c^+ \to \Lambda e^+ \nu_e) = (3.63 \pm 0.38 \pm 0.20), \tag{8}$$

$$B_r(\Lambda_c^+ \to \Lambda \mu^+ \nu_\mu) = (3.49 \pm 0.48 \pm 0.27), \tag{9}$$

which verify the lepton universality in baryon decays within errors:

$$B_r(\Lambda_c^+ \to \Lambda \mu^+ \nu_\mu)/B_r(\Lambda_c^+ \to \Lambda e^+ \nu_e) = (0.96 \pm 0.16 \pm 0.04) \tag{10}$$

4 Measurement of $e^+e^- \to \pi^+\pi^-$

4.1 The theoretical formula

The cross section of $e^+e^- \to \pi^+\pi^-$ in Born (bare) level at the effective center-of-mass energy $\sqrt{s'}$ can be expressed in term of the form factor $F_\pi(s')$ [19, 20],

$$\sigma^0_{\pi\pi}(s') = \frac{\pi\alpha^2\beta_\pi^3}{3s'}|F_\pi(s')|^2. \tag{11}$$

In fact, the measured cross section using the experimental data contains the vacuum polarization and the final state radiation (FSR) effects, the corresponding term in the theory is the so called dressed one $\sigma_{\pi\pi}^{dressed}$. The bare and dressed cross sections are related by following relation [21]:

$$\sigma^{\text{bare}} = \frac{\sigma^{\text{dressed}}}{|1 - \Pi(s)|^2} = \sigma_{\pi\pi}^{dressed}\left[\frac{\alpha(0)}{\alpha(s)}\right]^2, \tag{12}$$

where $\Pi(s)$ is the vacuum polarization factor [22, 23]. The form factor $|F_\pi|^2$ can be extracted using the measured $\sigma_{\pi\pi}^{dressed}$.

4.2 The experimental methods

The Born cross section of $e^+e^- \to \pi^+\pi^-$ was measured by two independent normalization schemes with BESIII data [8].

4.2.1 The method I

The following formula is used to measure the Born cross section:

$$\sigma^{\text{bare}}_{\pi\pi(\gamma_{\text{FSR}})} = \frac{N_{\pi\pi\gamma} \cdot (1 + \delta^{\pi\pi}_{FSR})}{\mathcal{L} \cdot \epsilon^{\pi\pi\gamma}_{\text{global}} \cdot H(s) \cdot \delta_{vac}}, \tag{13}$$

where $N_{\pi\pi\gamma}$ is the number of signal events, and H the radiator function, $\epsilon^{\pi\pi\gamma}_{\text{global}}$ the global efficiency of the signal, δ_{vac} the vacuum polarization correction, $\delta^{\pi\pi}_{\text{FSR}}$ the final state radiation correction factor. The tree-level diagrams for ISR and FSR contribution to $e^+e^- \to \pi^+\pi^-\gamma$ can be found in Ref. [24].

4.2.2 The method II

The Born cross section and can also be measured by

$$\sigma^{\text{bare}}_{\pi\pi(\gamma_{\text{FSR}})} = \frac{N_{\pi\pi\gamma}}{N_{\mu\mu\gamma}} \frac{\epsilon^{\mu\mu\gamma}_{\text{global}}}{\epsilon^{\pi\pi\gamma}_{\text{global}}} \frac{1 + \delta^{\mu\mu}_{\text{FSR}}}{1 + \delta^{\pi\pi}_{\text{FSR}}} \sigma^{\text{bare}}_{\mu\mu}. \tag{14}$$

In this way, the common factors $\mathcal{L}$, H and δ_{vac} for $e^+e^- \to \pi^+\pi^-\gamma_{ISR}$ and $e^+e^- \to \mu^+\mu^-\gamma_{ISR}$ can be canceled. Where, $\epsilon^{\mu\mu\gamma}_{\text{global}}$ is the global efficiency, $\delta^{\mu\mu}_{\text{FSR}}$ the FSR correction factor, $\sigma^{\text{bare}}_{\mu\mu}$ the bare cross section of $e^+e^- \to \mu^+\mu^-$, which corresponds to a QED process and it can be easily measured in experiment.

4.3 Data analysis

The events type $e^+e^- \to \pi^+\pi^-\gamma_{ISR}$ in the invariant mass range $600 < m_{\pi\pi} < 900$ MeV are selected from the data sample taken at center-of-mass energy $\sqrt{s} = 3773$ MeV[8]. The integrated luminosity of the data set $\mathcal{L} = (2931.8 \pm 0.2_{\text{stat}} \pm 13.8_{\text{sys}})\text{pb}^{-1}$. The track-based muon-pion separation based on the Artificial Neural Network (ANN), which is trained using $\pi^+\pi^-\gamma$ and $\mu^+\mu^-\gamma$ MC samples, is used to suppress the dominant background $e^+e^- \to \mu^+\mu^-\gamma_{ISR}$. The possible remaining backgrounds are are subtracted using Mont Carlo method.

4.4 Results

The results for $\sigma^{bare}(e^+e^- \to \pi^+\pi^-(\gamma_{\text{FSR}}))$ measured with method I and II are consistent within errors. The cross section and the form factor as the functions of $\sqrt{s'} = m_{\pi\pi}$ is illustrated in Fig. 4. The cross section is corrected for vacuum polarization effect, and the final state correction is considered. The result for $|F_\pi|^2$ which includes vacuum polarizations, but final state radiation effects are excluded, and the exactly the same fit formula and fit procedure are applied as described in Ref. [4]. Fig. 5 shows the difference between fit and data. The comparisons between the BESIII fit and the previous measurements, BaBar [4], KLOE [25–27].

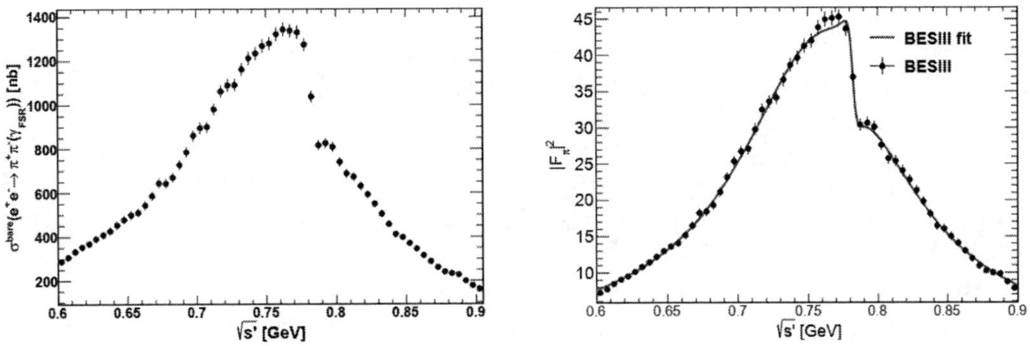

Figure 4. The bare $e^+e^- \to \pi^+\pi^-(\gamma_{\rm FSR})$ cross section (left), and the form factor and the fitted line (right).

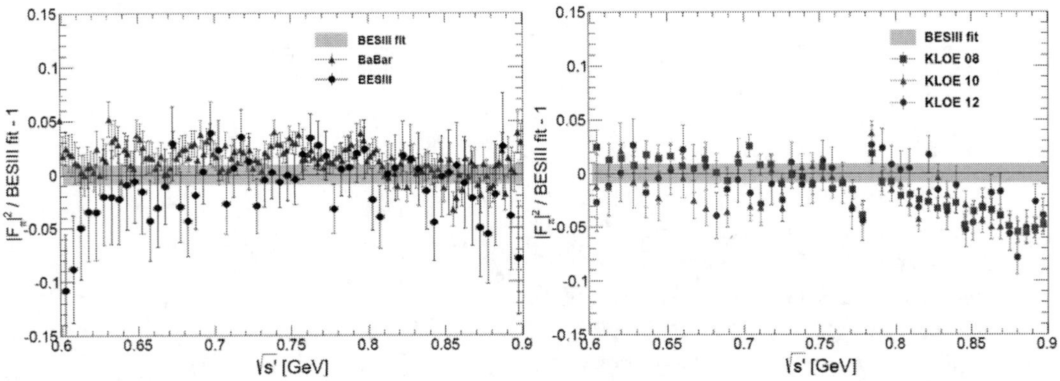

Figure 5. Relative difference of the form factor squared from BaBar (left), KLOE (right) and the BESIII fit. Statistic and systematic uncertainties are included in the data points.

5 Summary

The Born cross section of $e^+e^- \to p\bar{p}$ is measured, and the effective form factor $|G|$ is extracted under the assumption $|G_E| = |G_M|$ between 2232.4 MeV and 3671.0 MeV. The precision of the Born cross section with $\sqrt{s} \leq 3.08$ GeV is between 6.0% and 18.9%. The $|G_E/G_M|$ ratios and $|G_M|$ are extracted at the energies $\sqrt{s}$ = 2232.4 and 2400.0 MeV, and a combined data sample with energies of 3050.0, 3060.0 and 3080.0 MeV. The measured $|G_E/G_M|$ ratios are close to unity which are consistent with the BaBar experiment in the same q^2 region.

Using the method of ISR return, a measurement of the cross section $\sigma^{\rm bare}(e^+e^- \to \pi^+\pi^-(\gamma_{\rm FSR}))$ is performed with an accuracy of 0.9% in the dominant $\rho(770)$ mass region between 600 and 900 MeV. The two-pion contribution to $(g-2)_\mu$ is determined to be $a_\mu^{\pi\pi,LO}(600-900\ {\rm MeV}) = (370.0 \pm 2.5_{\rm stat} \pm 3.3_{\rm sys}) \cdot 10^{-10}$. It is found that a deviation of more than 3σ between the SM prediction of $(g-2)_\mu$ and its direct measurement is confirmed.

Based on the data sets collected near the production threshold, the Born cross section and electromagnetic form factor ratios of $e^+e^- \to \Lambda_c^+\bar{\Lambda}_c^-$ are presented. Using the data at 4599.5 MeV, the absolute hadronic branching fractions of twelve Cabibbo-favored decays of Λ_c^+ baryon and the branching fraction of the semi-leptonic decay $\Lambda_c^+ \to \Lambda l^+ \nu_l$ were measured.

References

[1] M. Ablikim et al. [BESIII Colaboration], Nucl. Instrum. Meth. A **614**, 345 (2010).
[2] K. Chao and Y. Wang, Physics at BESIII, Modn. Phys. A Supplement, **V24**, (2009).
[3] B. Aubert *et al.* Phys. Rev. Lett. **103**, 231801 (2009),
J. P. Lees *et al.* Phys. Rev. D **86**, 032013 (2012).
[4] F. Ambrosino *et al.* Phys. lett. B **670**, 285 (2009),
F. Ambrosino *et al.* Phys. lett. B **700**, 102 (2011),
D. Babusci *et al.* Phys. lett. B **720**, 336 (2013).
[5] M. Ablikim *et al.* [BESIII Colaboration], arXiv:1709.10236 [hep-ex].
[6] M. Ablikim *et al.* [BESIII Colaboration], Phys. Rev. D **91**, 112004 (2015).
[7] M. Ablikim *et al.* (BESIII Collaboration) Phys. Rev. Lett. **115**, 221805 (2015),
M. Ablikim *et al.* (BESIII Collaboration) Phys. Rev. B **767**, 42 (2017).
[8] M. Ablikim *et al.* [BESIII Colaboration], Phys. Lett. **753**, 629 (2016).
[9] N. Cabibbo and R. Gatto, Phys. Rev. **124** 1577 (1961).
[10] J. P. Lees *et al.* (BaBar Collaboration), Phys. Rev. D **87**, 092005 (2013)
[11] Phys. Rev. Lett. **34**, 1125 (1975)
[12] Phys. Rev. Lett. **37**, 882 (1975)
[13] Phys. Rev. Lett. **44**, 10 (1980)
[14] O. D. Dalkarov *et al.* Nucl. Phys. A**833**, 104 (2010).
B. El-Bennich *et al.* Phys. Rev. C**79**, 054001 (2009).
J. Haidenbauer *et al.* Phys. Lett. B **643**, 29 (2006).
R. Baldini *et al.* Eur. Phys. J. A **39**, 315 (2009).
R. Baldini *et al.* Eur. Phys. J. A **48**, 33 (2012).
[15] G. Pakhlova *et al.* (Belle Collaboration), Phys. Rev. Lett. **101** 172001 (2008)
[16] M. Schmelling, Phys. Scripta **51**, 676 (1995)
[17] R. M. Baltrusaiti *et. al.* (MARK-III Collaboration), Phys. Rev. Lett. **56**, 2140 (1986)
[18] J. Adler *et. al.* (MARK-III Collaboration), Phys. Rev. Lett. **60**, 89 (1988)
[19] B. L. Ioffe, V.A. Khooze, and L.N. Lipatov, Hard processes, Volume 1, Phenomenology Quark-Parton Model, (1984), North Holland.
[20] F. E. Close, An Introduction to Quarks and Partons, (1979), Academic Press.
[21] S. Actis et al. arXiv:0912.0749 [hep-ph].
[22] M. E. Peskin and D. V. Schroeder, An Introduction to Quantum field Theory, (1995), Westview Press.
[23] A. Höfer, et al. Eur. Phys. J. C **24**, 51, (2002)
[24] S. Pacetti, Nucl. Phys. B (Proc. Suppl.)**162**, 172(2006).
[25] F. Ambrosino et al. [KLEO Collaboration], Phys. Lett. B **670**, 285 (2009).
[26] F. Ambrosino et al. [KLEO Collaboration], Phys. Lett. B **700**, 102 (2011).
[27] D. Babusci et al. [KLEO Collaboration], Phys. Lett. B **720**, 336 (2013).

The Fragmentation Region of High Energy Nucleus-Nucleus and Hadron-Nucleus Collisions

Larry McLerran[1,⋆]

[1] *Institute for Nuclear Theory, University of Washington, Physics/Astronomy Bldg. Box 351550, Seattle WA 98195-1550*

Abstract. The fragmentation region of particles produced in high energy nuclear collisions provides a laboratory for studying high baryon number density systems. This talk outlines work in progress that attempts to compute properties of the matter produced in these collisions at the highest energies.

1 Introduction

The earliest studies of the production of the Quark Gluon Plasma in ultra-relativistic heavy ions collisions concerned the fragmentation region[1] . The energy considered was high enough so that the fragmentation regions of the two nuclei would be separated. This requirement is easiest to understand in the rest frame of one of the target nucleus. A produced fragment of the projectile with a gamma factor γ and a formation time t_{for} must be produced outside the target,

$$\gamma t_{for} >> R \tag{1}$$

Using a formation time of order the inverse transverse mass, this becomes

$$E/M_T^2 >> R \tag{2}$$

For collisions with energies of the order of 10's to 100's of GeV, a typical transverse mass is a few hundred MeV, and we see that for a large nucleus such as gold this is satisfied if the projectile has an energy larger than about 15 GeV, corresponding to a center of mass energy of the order of 30 GeV per nuclei. This is the energy range where a central region of low baryon density begins to emerge.

In these early estimates, the energy densities were estimated in the fragmentation region to be a few GeV/Fm^3, baryon densities several times that of nuclear matter and lifetimes of order the produced matter of order 10 Fm/c. It was argued that similar energy densities and lifetimes might be made in the central region[1]. Bjorken then developed an elegant and simple picture for the central region[2], and formulated hydrodynamic equations for the description of thermalized matter produced in ultra-relativistic heavy ion collisions. These developments led to a standard picture of heavy ion collisions which is illustrated ibn Fig. 1.

There is now a consistent space-time picture that descirbes the thermalized matter formed in the central region of heavy ion collisions. This picture uses the color charges associated with the Color

⋆e-mail: lmcler@uw.edu

© The Authors, published by EDP Sciences. This is an open access article distributed under the terms of the Creative Commons Attribution License 4.0 (http://creativecommons.org/licenses/by/4.0/).

Figure 1. A conceptual picture for the production of matter in ultra-relativistic heavy ion collisions.

Glass Condensate[3]-[4] of the heavy nuclei to produce highly coherent color electric and magnetic fields, the Glasma[5]-[6], that ultimately evolves into a thermalized Quark Gluon Plasma[7]-[8].

What I will present in the talk is an outline of work which I am doing with Srimoyee Sen and Soeren Schlicting, to generalize the computation of the initial distribution of matter in the central region to the baryon rich fragmentation region[9]. This has an extra complication relative to the Glasma computation, since one must determine the evolution of the distribution of baryon matter, in addition to computing the formed color fields and their gluon radiation.

What I will consider in this talk is the fragmentation region produced in collisions at asymptotically high energies. In principle, these considerations might be extended to lower energies where the fragmentation regions do not fully separate but where the high energy description of the collision is valid. This problem is more complicated and we do not consider it here.

In Fig. 2, I show the distribution of baryons and of mesons as seen in a high energy collision. The fragmentation region is the region of a few units of rapidity around that of the prjectile and target where the baryon number is concentrated.

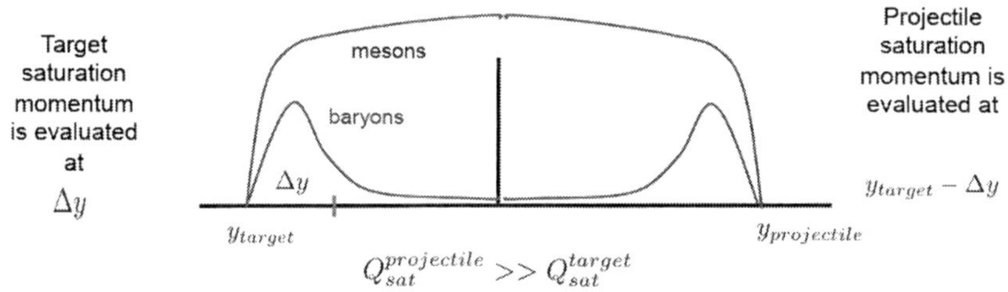

Figure 2. The rapidity distribution of particles and the fragmentation regions for ultra-relativistic nuclear collisions.

2 Asymmetric Saturation Momenta of the Target and Projectile in the Fragmentation Region.

Consider the kinematic region of the target. We expect the saturation momentum of the target to be of order

$$Q_{target}^{sat2} \sim A^{1/3} \Lambda_{QCD}^2 e^{\kappa \Delta y} \tag{3}$$

where Δy is the width of the fragmentation region and κ is a constant of order $\kappa \sim .2 - .3$. On the other hand, the saturation momentum of the projectile is much larger since it is evolved in rapidity from the rapidity of the projectile to that of the target

$$Q_{proj}^{sat2} \sim A^{1/3}\Lambda_{QCD}^{2} e^{\kappa(y_{prj} - y_{target})} \tag{4}$$

At LHC energies, the target saturation momenta is of the order of a GeV but the projectile is of order $5 - 10\ GeV$. This means the projectile is black to the partons in the target up to a scale of momentum which is the projectile saturation momentum. The dominant particle production occurs in the region of momentum between these two saturation momenta. This is a region where the color sources produce a weak field A « 1/g and there is not much interaction of produced particles in this kinematic region, at least when the degrees of freedom correspond to classical fields. At momenta scales less than the saturation momenta of the target, there are strong fields and classical time evolution of classical fields. This latter region is that of the Glasma.

We can estimate the transverse momentum distribution of particles in various transverse momentum intervals[10]-[11]. For transverse momenta scales less than that of the target saturation momenta, the distribution will go to a constant,

$$\frac{dN}{dyd^2p_T} \sim cons, \quad p_T < Q_{target}^{sat} \tag{5}$$

In the region of transverse momentum between the two scales, the produced gluons reflects the bremstrahlung distribution of the target gluons

$$\frac{dN}{dyd^2p_T} \sim \frac{Q_{target}^{sat2}}{p_T^2}, \quad Q_{target}^{sat} < p_T < Q_{proj}^{sat} \tag{6}$$

For large transverse momenta, the distribution ahs the well known $1/p_T^4$ particle production cross section

$$\frac{dN}{dyd^2p_T} \sim \frac{Q_{target}^2 Q_{proj}^2}{p_T^4} \quad Q_{proj}^{sat} < p_T \tag{7}$$

These transverse momentum distributions have consequences for the average transverse momentum and the multiplicity. Up to slowly growing terms, the total multiplicity is not affected by the increase of the projectile saturation momentum,

$$\frac{dN}{dy} \sim Q_{target}^2 \tag{8}$$

On the other hand the transverse momentum per particle is greatly enhanced and is proportional to the saturation momentum of the projectile

$$< p_T^2 > \sim Q_{proj}^2 \tag{9}$$

The average transverse momentum per particle in the fragmentation region should increase by a factor of 100 between RHIC and LHC energies. Of course this is tempered somewhat by the fragmentation of the initially produced gluons.

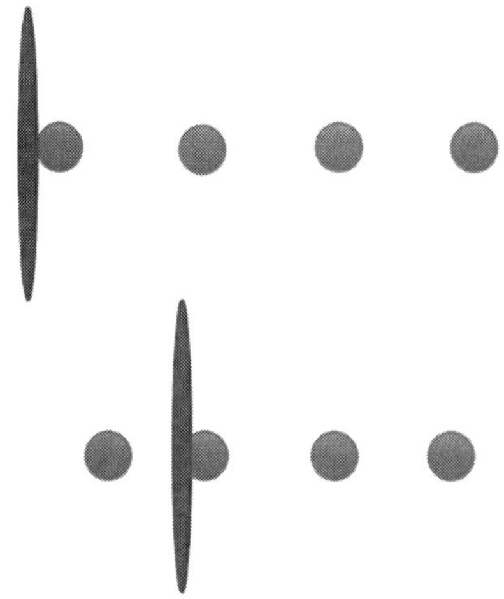

Figure 3. The sequential collision of a projectile nucleus with target baryons.

3 Compression of Baryon Number

Due to the fact that the target nucleons are hit sequentially by the Lorentz contracted projectile, there is compression of the target baryon number. This sequential collision is shown in Fig. 3. Due to the sequential nature of the collision, the nuclear separation is reduced by $\Delta z/z = 1 - v$ where v is the velocity of the struck projectile baryon[1]. Some of this compression is artificial since it is associated with a moving distribution of baryons, and some of the apparent compression is simply Lorentz contraction. This reduces the compression by a gamma factors of the struck baryon, so that the physically relevant compression is

$$\Delta z/Z \sim 1/\gamma_{baryon} \tag{10}$$

We can estimate the typical gamma factor. In the collision, as on mass shell target baryon passes through the projectile sheet. It acquires a transverse momentum of order Q^{sat}_{proj}. The longitudinal p^- of the baryon is unchanged since the projectile field is x^+ independent. Therefore, the typical change in p^+ is $p^+ \sim Q^{sat2}_{proj}/p^-$. This means there is a typical induced rapidity of order

$$\Delta y \sim \frac{1}{2} ln(p^+/p^-) \sim ln(Q^{proj}_{sat}/m) \tag{11}$$

corresponding to a Lorentz gamma factor fo Q^{proj}_{sat}/M. We will take M to be a typical momentum scale for the quark constituents of the nucleus when the nucleus is at rest,

The typical baryon number density after the collision is therefore of order

$$N_B/V \sim Q^{proj}_{sat} \rho_{nucleus}/M \tag{12}$$

For gluons when they are initially produced, their typical longitudinal size scale is also of order the saturation momentum. Therefore the gluon density and the compressed baryon density have the same dependence upon Q^{proj}_{sat} Therefore

$$N_B/N_{gluon} \sim cons \tag{13}$$

that is indepdnent of beam energy.

Now of course the expansion dynamics might change this result. Naively, this may not be such a large effect, because the dominant gluon production is in the region of weak field strength. To a first approximation, the multiplicity of these gluons might not change so much. Nevertheless, there may be a some partial dilution of this factor due to final state interaction. A measure of this quantity as a function of energy is therefore a probe of the importance of final state interactions and thermalization in the fragmentation region.

4 The Ba-Glasma

We shall refer to this baryon rich matter produced in the fragmentation region of the collision of two ultra-relativistic nuclei as the Ba-Glasma to differentiate it form the Glasma of the low baryon number density region. To set up the problem of computation for the baryon density and the gluon density in the collision from first principles in QCD is not so difficult. For the baryons, we can take a distribution of quarks in the initial state. We let them collide with a sheet of projectile nucleus color charge. This is a simple external field problem, and can be solved for the resulting baryon density.

Once the baryon density of the target is specified and the color field of the projectile assumed to be known, one can compute the distribution for produced gluons in the region that dominates the multiplicity, $Q^{sat}_{targ} < p_T < Q^{sat}_{proj}$. In this region the field is weak and can be computed to lowest order in the strength of the target color field but all order in the projectile source strength. This is presumably possible because the projectile source strength is a delta function and should generate a field that is a gauge rotation of a small fluctuation field induced by the target source strength.

5 Acknowledgements

I thank my colleagues Soeren Schlichting and Srimoyee Sen with whom work is progressing along the lines outlined in this talk L. McLerran is supported under Department of Energy under grant number DOE grant No. DE-FG02-00ER41132.

References

[1] R. Anishetty, P. Koehler and L. D. McLerran, Phys. Rev. D **22**, 2793 (1980). doi:10.1103/PhysRevD.22.2793

[2] J. D. Bjorken, Phys. Rev. D **27**, 140 (1983). doi:10.1103/PhysRevD.27.140

[3] L. D. McLerran and R. Venugopalan, Phys. Rev. D **49** (1994) 2233 doi:10.1103/PhysRevD.49.2233 [hep-ph/9309289].

[4] L. D. McLerran and R. Venugopalan, Phys. Rev. D **49** (1994) 3352 doi:10.1103/PhysRevD.49.3352 [hep-ph/9311205].

[5] A. Kovner, L. D. McLerran and H. Weigert, Phys. Rev. D **52** (1995) 6231 doi:10.1103/PhysRevD.52.6231 [hep-ph/9502289].

[6] A. Kovner, L. D. McLerran and H. Weigert, Phys. Rev. D **52** (1995) 3809 doi:10.1103/PhysRevD.52.3809 [hep-ph/9505320].

[7] T. Epelbaum and F. Gelis, Phys. Rev. Lett. **111** (2013) 232301 doi:10.1103/PhysRevLett.111.232301 [arXiv:1307.2214 [hep-ph]].

[8] J. Berges, B. Schenke, S. Schlichting and R. Venugopalan, Nucl. Phys. A **931** (2014) 348 doi:10.1016/j.nuclphysa.2014.08.103 [arXiv:1409.1638 [hep-ph]].

[9] L. McLerran, S. Sen and S. Schlichting, work in progress.

[10] Y. V. Kovchegov and A. H. Mueller, Nucl. Phys. B **529** (1998) 451 doi:10.1016/S0550-3213(98)00384-8 [hep-ph/9802440].
[11] A. Dumitru and L. D. McLerran, Nucl. Phys. A **700** (2002) 492 doi:10.1016/S0375-9474(01)01301-X [hep-ph/0105268].

Nucleon Spectroscopy with CLAS

Steffen Strauch[1,*] for the CLAS Collaboration

[1]*University of South Carolina, Columbia, SC 29208*

Abstract. Meson photoproduction is an important tool in the study of nucleon resonances. The spectrum of broad and overlapping nucleon excitations can be greatly clarified by use of polarization observables. The N^* program at Jefferson Lab with the CEBAF Large Acceptance Spectrometer (CLAS) includes experimental studies with linearly- and circularly-polarized tagged-photon beams, longitudinally- and transversely-polarized nucleon targets, and recoil polarizations. Selected results from these experimental studies will be presented.

1 Introduction

The nature of QCD confinement continues to provide an ongoing challenge to our understanding of non-perturbative QCD. Studying the nucleon spectrum provides one approach to learn more about this unique feature since the location and properties of excited states reflect the dynamics and relevant degrees-of-freedom within hadrons. Symmetric quark models with three independent quark degrees of freedom predict an overabundance of higher-lying excited states compared to what has been observed until now [1]. Lattice QCD calculations predict states with the same quantum numbers as constituent quark models with an underlying $SU(6) \times O(3)$ symmetry and find a counting of levels that is consistent with those models [2]. Quark-diquark models with a 'frozen' diquark predict fewer states. To clarify the nucleon resonance spectrum with the help of new experimental data is essential to inform studies of QCD.

Nucleon resonances have hadronic decay channels and are typically short-lived. They have large widths and overlap in the resonance spectrum. Identifying resonances and extracting resonance information in partial-wave analyses (PWA) of experimental data is challenging and requires not only unpolarized cross-section data, but also phase information from polarization observables. The photoproduction of a single pion, for example, is fully described by four complex amplitudes. It requires a minimum of eight carefully chosen observables at each energy and angle to determine the amplitudes of the reaction completely [3]. The photoproduction of two mesons requires even more observables [4]. Most present resonance information originate from πN reactions. Some resonances, however, couple weakly to the πN channel and multiple final states need to be investigated to fully explore the resonance spectrum.

The CLAS Collaboration has performed a series of photoproduction experiments off nucleons as part of the N^* program at the Thomas Jefferson National Accelerator Facility to provide the needed

*e-mail: strauch@sc.edu

© The Authors, published by EDP Sciences. This is an open access article distributed under the terms of the Creative Commons Attribution License 4.0 (http://creativecommons.org/licenses/by/4.0/).

data. These measurements cover a variety of final states including single pseudoscalar-mesons, vector-mesons, double-pions, and hyperons. The observables include cross sections, single- and double-polarization observables. The incident bremsstrahlung photon beams were energy-tagged [5] and either unpolarized, circularly, or linearly polarized. A central component of this program are experiments off polarized protons [6–10] and neutrons [11]. These experiments were made possible by the construction of the frozen-spin polarized butanol (FROST) [12] and hydrogen-deuteride (HDice) [13] targets. Final-state particles were detected in the CEBAF Large Acceptance Spectrometer (CLAS) [14] over a wide angular range, excluding the very beam-forward and backward directions. A toroidal magnetic field and packages of drift chambers allowed for the determination of the particle momentum. Time-of-flight detectors helped in the particle identification. The CLAS detector has in the meantime been replaced by CLAS12. Few selected results from these CLAS photoproduction experiments off polarized nucleons are discussed in the following.

2 CLAS Results

2.1 Pion Photoproduction

The double-polarization observable E in the $\gamma p \rightarrow \pi^+ n$ reaction can be determined in an experiment that measures the helicity asymmetry of circularly-polarized photons incident on longitudinally-polarized protons. The CLAS collaboration measured this process in the center-of-mass energy range from W = 1.25 GeV to W = 2.23 GeV [15]. Figure 1 shows angular distributions of the data for selected W bins. The upper panels show that previous PWA results describe the new data where the PWA have been well constrained by previous data at low energies, however they fail to describe the data at higher energies. The data have subsequently been included in new analyses of the SAID, Jülich-Bonn, and Bonn-Gatchina groups. The groups achieved an overall satisfying description of the data, lower panels of Fig. 1, that was obtained by a series of smaller updates of nucleon resonance parameters. The process led to an improved consistency of helicity amplitudes among the various analyses. One particularly interesting result is strengthened evidence for the poorly known $\Delta(2200)\frac{7}{2}^-$ resonance in improving the Bonn-Gatchina fit at the highest energies [16]. The mass of the $\Delta(2200)\frac{7}{2}^-$ resonance is significantly higher than the mass of its parity partner $\Delta(1950)\frac{7}{2}^+$, which is the lowest-mass Δ^* resonance with spin-parity $J^P = \frac{7}{2}^+$. This finding has relevance for the hypothesis of chiral-symmetry restoration at high mass.

Other single-pion photoproduction channels are also being studied. For example, results from $\gamma p \rightarrow \pi^0 p$ off transversally polarized protons have been obtained in an analysis of the target T and beam-target F polarization observables [20]. The preliminary results agree with world data in the range of overlap at lower energies, but provide new information up to center-of-mass energies of $W \approx 2.5$ GeV.

The lack of free-neutron targets makes experiments off the neutron difficult and results for observables off the proton dominate the data bases. However, information from both isospin channels are necessary to completely specify the amplitudes of the reaction. Novel data have been obtained in quasielastic photoproduction reactions off polarized neutrons from the CLAS g14 experiments. Figure 2 shows CLAS results for the polarization observable E in the $\gamma n(p) \rightarrow \pi^- p(p)$ reaction [21] as blue data points. The data are compared to updated fits of the SAID and Bonn Gatchina analyses. Three selected W bins show also previous fits and demonstrate how much the new data helped constraining those new analyses. The inclusion of these results in the new analyses has resulted in revised $\gamma n N^*$ couplings and, in the case of the $N(2190)\frac{7}{2}^-$, to convergence among different partial-wave analyses. The $\gamma n N^*$ couplings are sensitive to the dynamical process of N^* excitation and provide important guides to nucleon structure models.

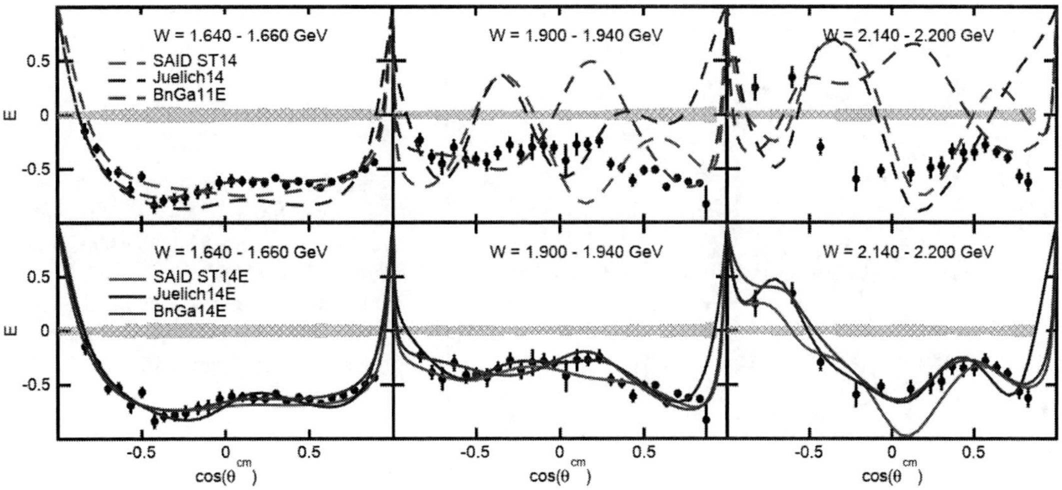

Figure 1. CLAS data for the beam-target polarization observable E in the $\vec{\gamma}\vec{p} \to \pi^+ n$ reaction for selected W bins. The shaded bands indicate systematic uncertainties. The curves are results of partial-wave analyses before (upper panels) and after (lower panels) including the new CLAS data. The previous analyses are from SAID ST14 [17], Jülich14 [18], and BnGa11E [19]. Figure from Ref. [15].

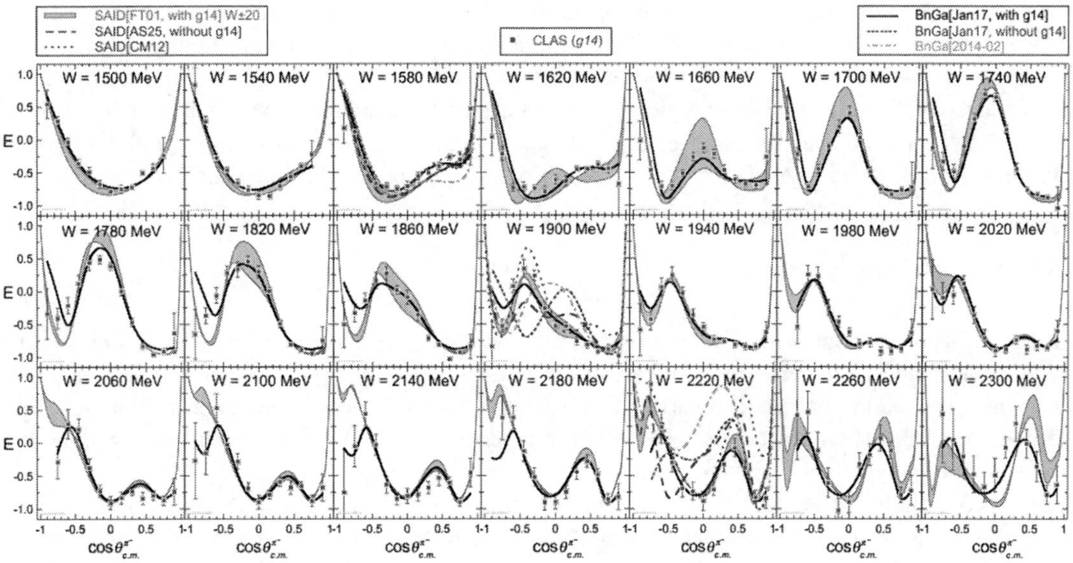

Figure 2. CLAS data for the beam-target polarization observable E in the $\vec{\gamma}\vec{n} \to \pi^- p$ reaction (blue data points). SAID and Bonn-Gatchina analyses that include these data are shown as red bands and solid black lines, respectively. Figure from Ref. [21].

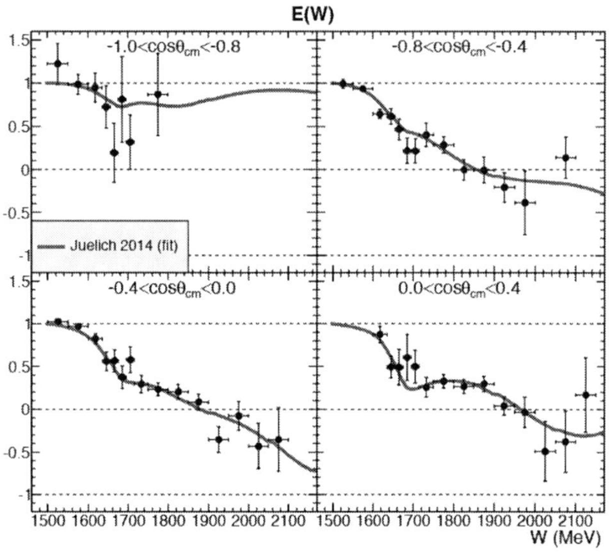

Figure 3. Energy dependence of the helicity asymmetry E in the $\vec{\gamma}\vec{p} \rightarrow \eta p$ reaction for various angular bins. The red curves show calculations of the Jülich-Bonn dynamical coupled-channel model that were fit to the CLAS data (black points). Figure from Ref. [22].

2.2 Eta Photoproduction

The η meson is an isospin-zero particle and η photoproduction selectively excites isospin-$\frac{1}{2}$ resonances of the nucleon resonance spectrum. The CLAS collaboration has measured the polarization observable E in the $\gamma p \rightarrow \eta p$ reaction from threshold to $W = 2.15$ GeV [22]. Figure 3 shows the energy dependence of E. In the threshold region E approaches unity. The data were fit with the Jülich-Bonn dynamical coupled-channel model (red curves). The fit describes the data quite well without the need for an additional narrow resonance near 1.68 GeV, which was previously suggested [22].

2.3 Hyperon Photoproduction

The analysis of the angular distribution of the parity-violating weak decay of hyperons allows for the reconstruction of the hyperon recoil polarization in the reactions $\gamma p \rightarrow K^+\Lambda$ and $\gamma p \rightarrow K^+\Sigma^0$. This gives access to additional recoil-polarization observables. CLAS measurements of the Λ induced polarization P [23, 24] as well as the polarization-transfers C_x and C_z from circularly-polarized photons to the recoiling hyperons [25] were critical input to a coupled-channel analysis by the Bonn-Gatchina group [26, 27]. In particular, the analysis found further evidence for the, at the time, poorly known $N(1900)\frac{3}{2}^+$ resonance. This resonance is predicted by symmetric three-quark models, but is not expected to exist in earlier quark-diquark models. New hyperon-photoproduction data with linearly-polarized photons off an unpolarized proton target have been obtained up to $W \approx 2.2$ GeV. Together with the recoil polarization of the hyperon, this gives access to five polarization observables: Σ, P, T, O_x, and O_z. Figure 4 shows as an example the beam-recoil polarization observable O_z for the $K^+\Lambda$ channel as a function of center-of-mass energy W for eight angular bins of $\cos\theta_K$ of the kaon's polar angle. The data are compared with predictions of the ANL-Osaka (red bands) coupled-channel model and the 2014 Bonn-Gatchina partial-wave analysis (green bands). None of those describe the new data well. A refit of the BnGa analysis to the data gives an excellent description. This refit includes two new nucleon resonances with spin and parity of $J^P = \frac{3}{2}^+$ and $\frac{5}{2}^+$. However, no strong evidence for their existence can be derived from the data.

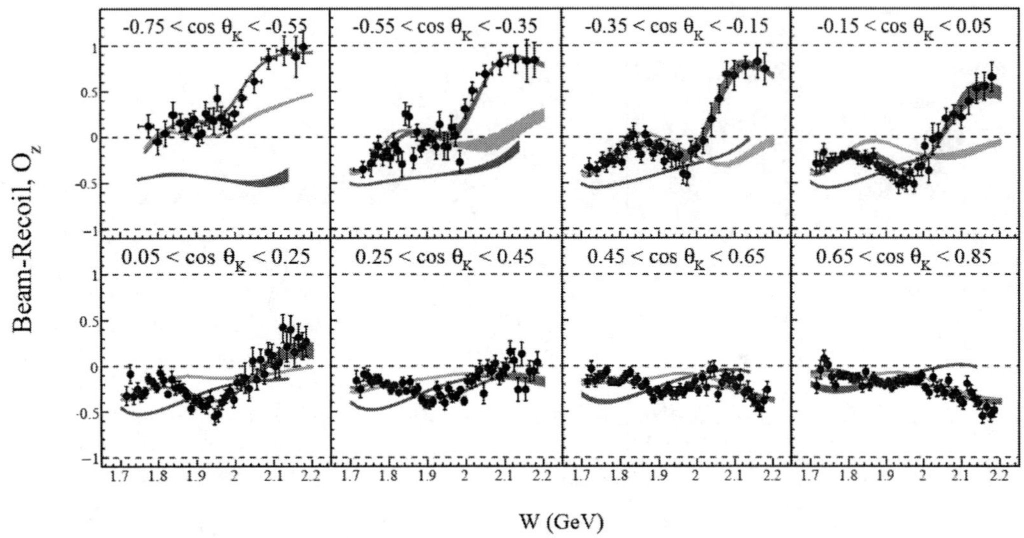

Figure 4. CLAS data of the beam-recoil double-polarization observable O_z in the $\vec{\gamma}p \to K\vec{\Lambda}$ reaction. The red and green curves show predictions from the ANL-Osaka coupled-channel and Bonn-Gatchina partical-wave analyses, respectively. The blue curves show a Bonn-Gatchina refit to the new data. Figure from Ref. [28].

2.4 Double-Pion Photoproduction

Higher-lying nucleon resonances above about 1.6 GeV have significant branching ratios for $N^* \to \pi\Delta$ or $N^* \to \rho N$ that lead to $\pi\pi N$ final states. This reaction channel and the study of the sequential decays contain potentially rich information about the nucleon-resonance spectrum. Many polarization observables are accessible in two-pion photoproduction off the nucleon [4]. The CLAS Collaboration was first to study the beam-helicity asymmetry $I^{\odot}$ for the $\gamma p \to \pi^+\pi^- p$ reaction. The measurements covered energies between $W = 1.35$ and 2.30 GeV [29]. Many more polarization observables were measured with the FROST experiment: the target observables P_i, beam-target observables $P_i^{\odot}$ for circularly polarized photons, and beam-target observables $P_i^{s,c}$ for linearly polarized photons; $i = x, y$ are for reactions off transversely- and $i = z$ off longitudinally-polarized nucleons. Figure 5 shows as one example preliminary results for observable P_x as a function of the azimuthal angle ϕ^* in seven $\cos\theta_{CM}$ bins for $W \approx 1600$ MeV. See Fig. 6 for the definition of the angles. This observable is accessible in measurements with unpolarized photons off transversally-polarized protons. The observable is an odd function of the azimuthal angle ϕ^*. This is highlighted by a fit of a sum of $\sin n\phi^*$ terms to the data (red curves). The data are compared to bin-averaged results of a calculation by Fix and Arenhövel [30] based on an effective Lagrangian approach (green curves). The model results resemble key features of the preliminary data, but seem to be off in phase. Once finalized, the data will strongly constrain models.

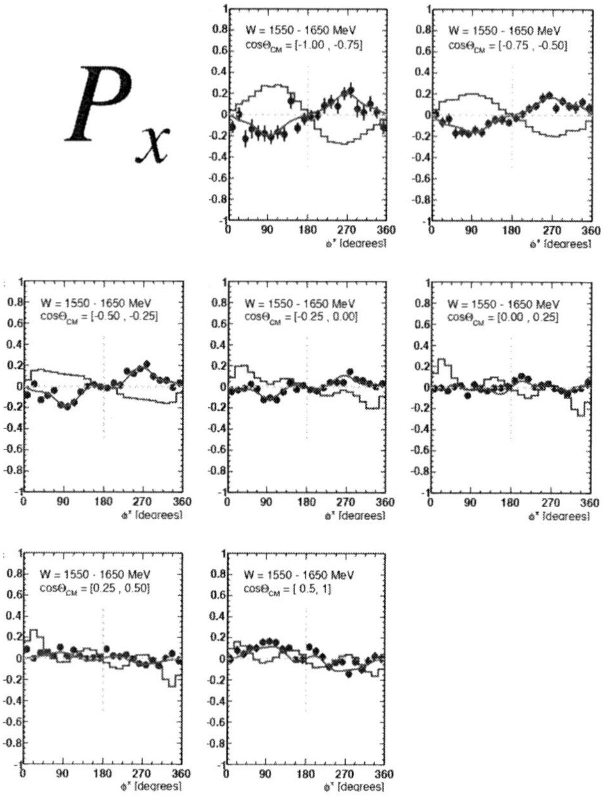

Figure 5. Preliminary ϕ^* angular distributions for the target-polarization observable P_x in the $\gamma p \to \pi^+\pi^- p$ reaction for a center-of-mass energy of $W = 1550 - 1650$ MeV and seven bins in $\cos\theta_{CM}$ from CLAS. Angles are defined in Fig. 6. The green cuves show bin-averaged results of a model calculation from A. Fix [30]. The red curves are fits of a sum of $\sin n\phi^*$ terms to the data. Figure from [31].

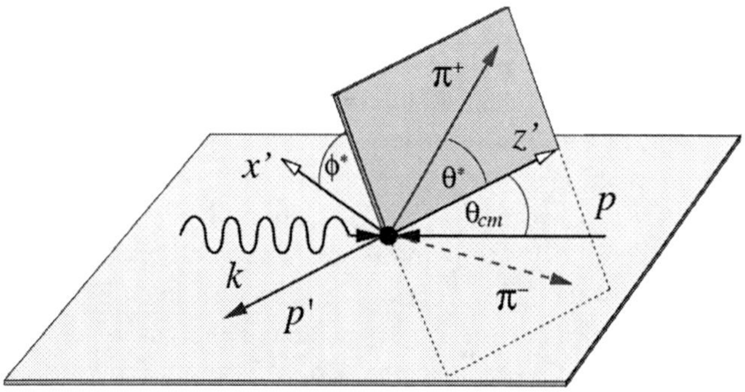

Figure 6. Cartoon of the $\gamma p \to \pi^+\pi^- p$ reaction with definitions of kinematic variables.

3 Conclusion

The Particle Data Group routinely updates its compilation of N- and Δ-resonance information. Major input came from analyses of CLAS data. CLAS data continue to make an impact as new results for polarization observables in photoproduction reactions off protons and neutrons become available. The new data will contribute to complete or nearly-complete experiments, and will challenge previously poorly constrained models. It is very likely that these data will have a tremendous impact on the understanding of baryon resonances and may provide evidence for new states found in coupled-channel analyses.

This work was supported in parts by the U.S. National Science Foundation: NSF PHY-1505615.

References

[1] V. Credé, W. Roberts, Rept. Prog. Phys. **76**, 076301 (2013)
[2] R.G. Edwards, J.J. Dudek, D.G. Richards, S.J. Wallace, Phys. Rev. **D84**, 074508 (2011), 1104.5152
[3] W.T. Chiang, F. Tabakin, Phys. Rev. **C55**, 2054 (1997)
[4] W. Roberts, T. Oed, Phys.Rev. **C71**, 055201 (2005), nucl-th/0410012
[5] D. Sober et al., Nucl. Instrum. Meth. A **440**, 263 (2000)
[6] Jefferson Lab Experiment E02-112, "Search for Missing Nucleon Resonances in Hyperon Photoproduction", P. Eugenio, F. Klein, and L. Todor, spokespersons
[7] Jefferson Lab Experiment E03-105, "Pion Photoproduction from a Polarized Target", N. Benmouna, W. Briscoe, G. O'Rielly, I. Strakovsky, S. Strauch, spokespersons
[8] Jefferson Lab experiment E04-102, "Helicity Structure of Pion Photoproduction", D. Crabb, M. Khandaker, and D. Sober, spokespersons
[9] Jefferson Lab experiment E05-012, "Measurement of polarization observables in η-photoproduction with CLAS", M. Dugger and E. Pasyuk, spokespersons
[10] Jefferson Lab Experiment E06-013, "Measurement of $\pi^+\pi^-$ Photoproduction in Double-Polarization Experiments using CLAS", M. Bellis, V. Credé, S. Strauch, spokespersons
[11] Jefferson Lab Experiment E06-101, "N* Resonances in Pseudoscalar-meson photo-production from Polarized Neutrons in $\vec{H} \cdot \vec{D}$ and a complete determination of the $\gamma n \rightarrow K^0\Lambda$ amplitude", F. Klein and A.M. Sandorfi, spkespersons
[12] C. Keith, J. Brock, C. Carlin, S. Comer, D. Kashy, J. McAndrew, D. Meekins, E. Pasyuk, J. Pierce, M. Seely, Nucl. Instrum. Meth. A **684**, 27 (2012)
[13] M.M. Lowry et al., Nucl. Instrum. Meth. A **815**, 31 (2016)
[14] B.A. Mecking et al., Nucl. Instrum. Meth. A **503**, 513 (2003)
[15] S. Strauch et al. (CLAS), Phys. Lett. **B750**, 53 (2015), 1503.05163
[16] A.V. Anisovich, V. Burkert, J. Hartmann, E. Klempt, V.A. Nikonov, E. Pasyuk, A.V. Sarantsev, S. Strauch, U. Thoma, Phys. Lett. **B766**, 357 (2017), 1503.05774
[17] R.L. Workman, M.W. Paris, W.J. Briscoe, I.I. Strakovsky, Phys. Rev. **C86**, 015202 (2012)
[18] D. Rönchen, M. Döring, F. Huang, H. Haberzettl, J. Haidenbauer, C. Hanhart, S. Krewald, U.G. Meißner, K. Nakayama, Eur. Phys. J. **A50**, 101 (2014), 1401.0634
[19] A.V. Anisovich, R. Beck, E. Klempt, V.A. Nikonov, A.V. Sarantsev, U. Thoma, Eur. Phys. J. **A48**, 15 (2012), 1112.4937
[20] H. Jiang, Ph.D. thesis, University of South Carolina (2017)
[21] D. Ho et al. (CLAS), Phys. Rev. Lett. **118**, 242002 (2017), 1705.04713
[22] I. Senderovich et al. (CLAS), Phys. Lett. **B755**, 64 (2016), 1507.00325
[23] J.W.C. McNabb et al. (CLAS), Phys. Rev. **C69**, 042201 (2004), nucl-ex/0305028
[24] M.E. McCracken et al. (CLAS), Phys. Rev. **C81**, 025201 (2010), 0912.4274
[25] R.K. Bradford et al. (CLAS), Phys. Rev. **C75**, 035205 (2007), nucl-ex/0611034
[26] A.V. Sarantsev, V.A. Nikonov, A.V. Anisovich, E. Klempt, U. Thoma, Eur. Phys. J. **A25**, 441 (2005), hep-ex/0506011

[27] V.A. Nikonov, A.V. Anisovich, E. Klempt, A.V. Sarantsev, U. Thoma, Phys. Lett. **B662**, 245 (2008), 0707.3600

[28] C.A. Paterson et al. (CLAS), Phys. Rev. **C93**, 065201 (2016), 1603.06492

[29] S. Strauch et al. (CLAS Collaboration), Phys. Rev. Lett. **95**, 162003 (2005), hep-ex/0508002

[30] A. Fix, H. Arenhövel, Eur.Phys.J. **A25**, 115 (2005), nucl-th/0503042

[31] L.A. Net, Ph.D. thesis, University of South Carolina (2017)

Dynamics of inhomogeneous chiral condensates

Juan Pablo Carlomagno[1,2,3,⋆], *Gastão* Krein[4,⋆⋆], *Daniel* Kroff[4,5,⋆⋆⋆], and *Thiago* Peixoto[4,6,⋆⋆⋆⋆]

[1] *IFLP, CONICET Departamento e Física, Universidad Nacional de La Plata C.C. 67, 1900 La Plata, Argentina*
[2] *CONICET, Rivadavia 1917, 1033 Buenos Aires, Argentina*
[3] *ICTP South American Institute for Fundamental Research Rua Dr. Bento Teobaldo Ferraz, 271 - Bloco II, 01140-070 São Paulo, SP, Brazil*
[4] *Instituto de Física Teórica, Universidade Estadual Paulista Rua Dr. Bento Teobaldo Ferraz, 271 - Bloco II, 01140-070 São Paulo, SP, Brazil*
[5] *Centre de Physique Théorique, Ecole Polytechnique, CNRS, Université Paris-Saclay F-91128 Palaiseau, France*
[6] *Instituto Federal do Norte de Minas Gerais, Campus Almenara Rodovia BR 367 Km 07 S/N, 39900-000 Almenara, MG, Brazil*

Abstract. We study the dynamics of the formation of inhomogeneous chirally broken phases in the final stages of a heavy-ion collision, with particular interest on the time scales involved in the formation process. The study is conducted within the framework of a Ginzburg-Landau time evolution, driven by a free energy functional motivated by the Nambu–Jona-Lasinio model. Expansion of the medium is modeled by one-dimensional Bjorken flow and its effect on the formation of inhomogeneous condensates is investigated. We also use a free energy functional from a nonlocal Nambu–Jona-Lasinio model which predicts metastable phases that lead to long-lived inhomogeneous condensates before reaching an equilibrium phase with homogeneous condensates.

1 Introduction

The study of the phase diagram of strongly interacting matter as a function of the temperature and baryon chemical potential has been an active field of research since the early days of nuclear physics. In the early days, the field was driven by studies of cold nuclear matter and aimed at understanding the structure of atomic nuclei and the central regions of neutron stars. With the discovery of the quark-gluon substructure of nucleons and the establishment of quantum chromodynamics (QCD) as the theory of the strong interactions in the 1970s, the field entered into a new era. The asymptotic freedom property of QCD predicts the possibility of creating through high-energy heavy-ion collisions a state of matter resembling the one that presumably existed in the early universe, in that quarks and gluons are no longer confined in the protons and neutrons as in cold nuclear matter [1].

⋆e-mail: juanpablocarlomagno@gmail.com
⋆⋆e-mail: gkrein@ift.unesp.br
⋆⋆⋆e-mail: daniel.kroff-fogaca@polytechnique.edu
⋆⋆⋆⋆e-mail: thiagop@ift.unesp.br

© The Authors, published by EDP Sciences. This is an open access article distributed under the terms of the Creative Commons Attribution License 4.0 (http://creativecommons.org/licenses/by/4.0/).

In the last few decades, great experimental effort has been devoted to the extraction from heavy-ion collisions observables relevant for the phase diagram of QCD. For the high temperature region of the phase diagram, results coming from RHIC-BNL and LHC-CERN suggest that a new state of matter, the quark-gluon plasma (QGP), is indeed formed after the collision event [2]. Yet, it is not clear what is the nature of the phase transition between ordinary nuclear matter and the QGP. The beam energy scan program at RHIC, and programs at the new accelerator facilities FAIR and NICA that are expected to become operational in a few years, aim at getting clues on the nature of QCD phase transition, in particular on the chiral critical point and on its vicinity. One exciting possibility is the search of signals of the formation of spatially varying chiral condensates which break translational invariance, a subject that has raised a great deal of interest in the last years — for a review and references, see Ref. [3]. In addition to the expected tricritical endpoint of the first order chiral phase transition, model calculations suggest the existence of a so-called Lifshitz point, where two homogeneous phases and one inhomogeneous phase meet [4, 5]. Interestingly, even for ordinary cold nuclear matter there seems to be a possibility that an inhomogeneous phase can be realized at densities a few times larger than the ground-state density [6].

Possible observable consequences of inhomogeneous phases have been suggested in the context of neutron stars. Ref. [7] proposes that the existence of an inhomogeneous chiral phase leads to a novel cooling scenario for neutron stars, finding a neutrino emissivity with efficiency comparable with the usual quark cooling or pion cooling. Consequences of an inhomogeneous chiral phase for the equation of state of matter in neutron stars have been investigated in Refs. [8, 9]. On the other hand, to the best of our knowledge, there has been no investigation on the possibility of formation of an inhomogeneous chiral phase in the matter produced in the late stage of a heavy-ion collision. Motivated by this fact, in the present communication we present results of a study of the dynamics of the formation of inhomogeneous chirally broken phases, with the aim of getting insight on the possibility of formation of such phases at the final stages of a heavy-ion collision. We are particularly interested in the time scales involved in the formation process. Our study is conducted within the framework of the phenomenological time-dependent Ginzburg-Landau equation.

2 Dynamics of order parameters

It is well known that static phase transitions in many-particle systems with short-range interactions depend on the spatial dimensionality and symmetry properties of the order parameters characterizing the different phase transitions in the system. On the other hand, the dynamics of phase transitions depends, in addition, on whether there are conservation laws associated with the order parameters and also on the couplings among the order parameters and conserved densities. Like in the case of static phase transitions, dynamical phase transitions can be grouped into universality classes, or models, in the nomenclature of Hohemberg and Halpering [10]. We are interested in the dynamics of the chiral order parameters. When neglecting the coupling of the order parameters to conserved densities, like the baryon density and the energy-momentum tensor, one is lead to assume that the dynamics belongs to the model A dynamical universality class [10–14].

Denoting by $\phi(\mathbf{x}, t)$ the order parameter field, a protype equation of motion for $\phi(\mathbf{x}, t)$ is the phenomenological time-dependent Ginzburg-Landau (GL) equation

$$\frac{\partial \phi(\mathbf{x}, t)}{\partial t} = -\Gamma \frac{\delta H[\phi]}{\delta \phi(\mathbf{x}, t)} + \zeta(\mathbf{x}, t), \tag{1}$$

where $H[\phi]$ is the Ginzburg-Landau-Wilson (GLW) Hamiltonian, which is typically of the form

$$H[\phi] = \int d\mathbf{x}\, \mathcal{H}(\phi), \qquad \mathcal{H}(\phi) = \frac{1}{2}\kappa(\nabla\phi)^2 + U_0(\phi), \tag{2}$$

with κ a positive constant and $U_0(\phi)$ contains nonderivative polinomial terms, the lowest order being

$$U_0(\phi) = \frac{1}{2} r_0 \phi^2 + \frac{1}{4} u_0 \phi^4, \tag{3}$$

where r_0 and u_0 are constants that can depend on the temperature, the latter being necessarily positive at this level of truncation. For $r_0 < 0$, $U_0(\phi)$ has a double-well shape. In addition, $\zeta(x, t)$ is the noise field describing thermal fluctuations and Γ is an Onsager (dissipation) coefficient. For simplicity, we consider a simple white-noise correlation function for the noise field:

$$\langle \zeta(\mathbf{x}, t) \rangle_\zeta = 0, \qquad \langle \zeta(\mathbf{x}, t) \zeta(\mathbf{x}', t') \rangle_\zeta = 2\Gamma T \, \delta(t - t') \, \delta(\mathbf{x} - \mathbf{x}'), \tag{4}$$

where $\langle \cdots \rangle_\zeta$ means average over noise realizations and T is the external temperature.

From the Fokker-Planck equation associated with the GL equation of Eq. (1), it is not difficult to show that the equilibrium probability distribution of field configurations $\phi(\mathbf{x})$ is the Boltzmann factor

$$P_{\rm eq}[\phi] = e^{-H[\phi]/T}. \tag{5}$$

The equilibrium partition function is then given by the functional integral

$$Z[\phi] = \int \mathcal{D}\phi \, e^{-H[\phi]/T}. \tag{6}$$

The meaning of this is that a correlation function, e.g. a two-point correlation function $\langle \phi(\mathbf{x}_1) \phi(\mathbf{x}') \rangle$, defined as

$$\langle \phi(\mathbf{x}) \phi(\mathbf{x}') \rangle = \frac{1}{Z} \int \mathcal{D}\phi \, \phi(\mathbf{x}) \phi(\mathbf{x}') \, e^{-H[\phi]/T}, \tag{7}$$

can also be calculated via the long-time solutions of the GL equation as

$$\langle \phi(\mathbf{x}) \phi(\mathbf{x}') \rangle = \lim_{t \to \infty} \frac{1}{N_r} \sum_{r=1}^{N_r} \phi_r(\mathbf{x}, t) \phi_r(\mathbf{x}', t), \tag{8}$$

where $\phi_r(\mathbf{x}, t)$ is a solution of the GL equation of Eq. (1), with the index r indicating r-th noise realization, and N_r is the total number of realizations (N_r is supposed to be large). Despite straightforward, the numerical solution of Eq. (1) requires some care with respect to the lattice spacing and time steps used in the discretization of the derivatives. One way to minimize the dependence on the lattice spacing is to use a renormalization procedure, as explained e.g. in Refs. [15–17]. When considering the dynamics of a conserved density, e.g. the baryon number, one would be lead to consider a model B dynamics [18, 19].

Equations of the Ginzburg-Landau type can be derived from a microscopic model via the closed-time-path (CTP) effective action formalism [20]. In general, the equations obtained with the CTP formalism contain nonlocal (non-Markovian) dissipation kernels and colored noise fields, as well as the possibility of field-dependent noise terms (multiplicative noise) [15, 21–23]. In a phenomenological approach, which we employ in the present study, one takes the equilibrium Hamiltonian $H[\phi]$ and so the description is valid for situations not too far from equilibrium. In addition, the dissipation coefficient Γ that enters Eq. (1) must be obtained from independent estimates.

To illustrate a typical solution $\phi_r(\mathbf{x}, t)$ of Eq. (1) for a given noise realization, let us consider a double-well potential $U_0(\phi)$. We take as an initial configuration $\phi(\mathbf{x}, 0)$ that assumes a random value at each point $\mathbf{x} = (x, y, x)$, with amplitude $\phi_0(\mathbf{x}, 0)/\phi_0 = 0.1$ and zero average, where $\phi_0 = (|r_0|/u_0)^{1/2}$ is the positive minimum of $U_0(\phi)$. Figure 1 shows density plots of a typical field configuration

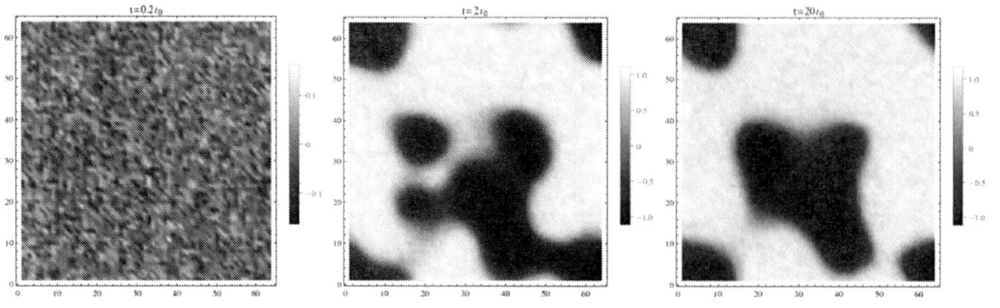

Figure 1. Typical spatial profiles of the order parameter — see text for explanations. Figure taken from Ref. [24].

$\phi_r(x, y, z = 0, t)/\phi_0$ at different time instants t/t_0, where t_0 is a typical time scale. At $t = 0$, left panel of the figure, one has the initial random configuration with zero noise average. As t/t_0 increases, central e right panels, one starts seeing very clearly the inhomogeneous nature of the solutions. For t/t_0 very large, the system will eventually evolve to a nearly homogeneous equilibrium configuration for which $\phi_r(\mathbf{x}, t)/\phi_0$ is positive or negative, with $|\phi_r(\mathbf{x}, t)| \simeq \phi_0$. The equilibrium value is not precisely the "bare vacuum" ϕ_0 because of the thermal fluctuations.

3 Chiral order parameters – statics

We consider NJL-type models with vanishing quark masses. The Lagrangian density defining the models is written generically as

$$\mathcal{L} = \bar{\psi}(x)\, i\gamma^\mu \partial_\mu\, \psi(x) + \frac{G}{2} J_a(x) J_a(x), \tag{9}$$

where ψ stands for the $N_f = 2$ fermion doublet, and $J_a(x)$ are currents of the form

$$J_a(x) = \int d^4z\, \mathcal{G}(z)\, \bar{\psi}\left(x + \frac{z}{2}\right) \Gamma_a\, \psi\left(x - \frac{z}{2}\right), \tag{10}$$

where $\Gamma_a = (\mathbf{1}, i\gamma_5\boldsymbol{\tau})$, and $\mathcal{G}(z)$ is a form factor that characterizes the effective interaction. For the traditional, local NJL model, teh form factor is given by $\mathcal{G}(z) = \delta(z)$. One should be aware of the fact that NJL-type models are meant to model the chiral transition of QCD, but they miss two prominent features of QCD, asymptotic freedom and quark-gluon confinement, that certainly are very important at high densities. Models developed in the past that incorporate those features, as those of Refs. [25–27], have achieved relative success in describing vacuum hadron properties, and provide a good alternative for studying the confined phase. Still another alternative, although technically more elaborate, is the use of Dyson-Schwinger equations of QCD at finite density [28], like the recent study in Ref. [29].

To obtain an equilibrium GLW Hamiltonian as a functional of the order parameters to be used in Eq. (1), we perform a standard bosonization of the theory by integrating out the quark fields in favor of scalar and pseudoscalar fields. We employ the mean field approximation, in that the bosonic fields are replaced by their vacuum expectation values $\phi(\mathbf{x}) = (\sigma(\mathbf{x}), \pi(\mathbf{x}))$. Following Refs. [4, 30], the mean field thermodynamic potential is expanded around the symmetric ground state in powers of the order parameters and their spatial gradients and define it as being the GLW Hamiltonian density

$$\mathcal{H}_{\rm NJL}(\phi) = \frac{\alpha_2}{2}\phi^2 + \frac{\alpha_4}{4}(\phi^2)^2 + \frac{\alpha_{4b}}{4}(\nabla\phi)^2 + \frac{\alpha_6}{6}(\phi^2)^3 + \frac{\alpha_{6b}}{6}(\phi,\nabla\phi)^2 + \frac{\alpha_{6c}}{6}\left[\phi^2(\nabla\phi)^2 - (\phi,\nabla\phi)^2\right] + \frac{\alpha_{6d}}{6}(\Delta\phi)^2, \tag{11}$$

where the α_n, $n = 2, 4, 4b, 6c, 6d$ depend upon the temperature T and baryon chemical potential μ; their expressions are given in Ref. [30] and their explicit behavior as a function of T and μ will be presented elsewhere [31].

For the local NJL model, one has [30] $\alpha_{4b} = \alpha_4$, and $\alpha_{6b}/5 = \alpha_{6c}/3 = 2\alpha_{6d} \equiv \alpha_6$, which is the case investigated in Ref. [4]. In this case, Eq (11) has a critical point in the $T \times \mu$ phase diagram where $\alpha_2 = \alpha_4 = 0, \alpha_6 > 0$. By exploring the phase diagram in the vicinity of this critical point, the author of Ref. [4] found that, for one dimensional modulations of the order parameter $\phi(z)$ (neglecting the coupling to the π order parameter), that the $\alpha_4/\alpha_6 \times \mathrm{sgn}(\alpha_2)(|\alpha_2|/\alpha_6)^{1/2}$ plane is divided into three regions, as shown in Fig. 2: one in which chiral symmetry is broken and $\phi \neq 0$ is spatially homogeneous, a second contiguous region in which the symmetry is still broken but $\phi = \phi(z)$ is an inhomogeneous configuration, and a third, contiguous to the second, where chiral symmetry is restored, $\phi = 0$. The profile of the inhomogeneous configurations can be written in terms of a Jacobi Elliptic function [4, 32–34]

$$\phi(z) = \sqrt{\nu}q\,\mathrm{sn}(qz;\nu), \tag{12}$$

where the parameters ν and q depend upon the α_n. On the line $\alpha_4 = -\sqrt{36/5\alpha_2\alpha_6}$, where the transition between the homogeneous and inhomogeneous phases takes place, $\nu = 1$ [4, 35], and $\phi(z) = q\tanh(qz)$. On this line, the homogeneous and inhomogeneous solutions coexist.

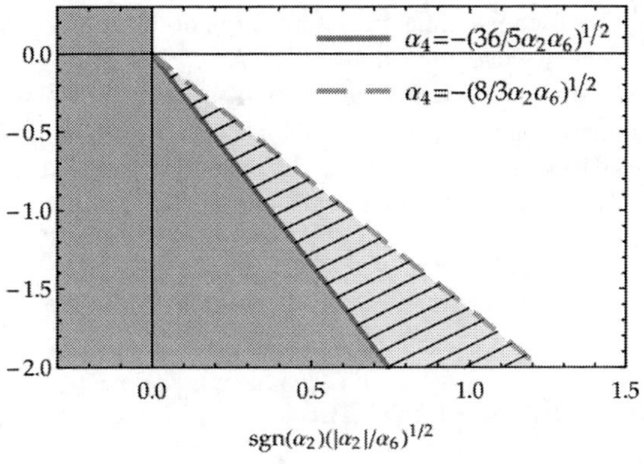

Figure 2. Phase diagram in parameter space of the local NJL model adapted from Ref. [4]. Grey region: the order parameter is spatially homogeneous. Crosshatched region: the order parameter assumes the configuration given in Eq. (12). White region: chiral symmetry is restored, $\phi = 0$.

Next, we consider the phase diagram from a nonlocal NJL model for a particular set of parameters [36]. The model is completely determined by the form factor $\mathcal{G}(z)$ and the coupling constant G. The form factor is taken to be a Gaussian; its Fourier transform is $g(p) = \exp(-p^2/\Lambda^2)$, where the parameter Λ is related to the range of the interaction. Those parameters are fixed by fitting the pion decay constant and the quark condensate in the chiral limit: $f_\pi = 86$ MeV and $\langle\bar{q}q\rangle = -(270\,\mathrm{MeV})^3$. The numerical values of G and Λ are $G = 14.668\,\mathrm{GeV}^{-2}$ and $\Lambda = 1.046$ GeV. Fig. 3 shows the corresponding phase diagram for a spatial modulation in the form of a dual chiral density wave, i.e. a plane wave for the scalar and pseudoscalar fields — see Ref. [36] for details. Although a stable inhomogeneous phase exists for very high densities only, there exists, however, a local stable minimum of the GLW Hamiltonian of Eq. (11) with an inhomogeneous condensate, but it is located within the extremely narrow band enclosed by the dotted and dash-dotted lines. This latter situation is very interesting, in that for a T and μ near the tricritical and Lifshitz points — e.g. the point $(\mu, T) = (190\,\text{MeV}, 50\,\text{MeV})$ identified in the figure — there appear inhomogeneous order parameters as solutions of the corresponding Ginzburg-Landau equation that are long-lived, with a life time much larger than Γ, as we discuss in the next section. It is important to note that the shape of the phase diagram and the location of the inhomogeneous minimum depend heavily on the value of the chiral condensate.

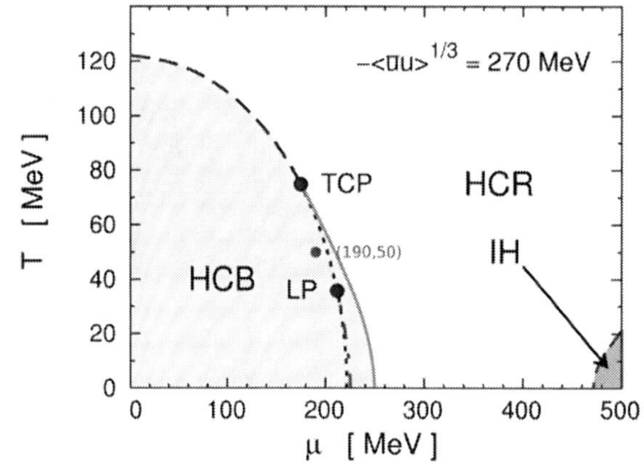

Figure 3. $T \times \mu$ phase diagram for a particular parametrization of the nonlocal NJL model [36]. Solid (dashed) lines indicate first (second) order phase transitions. The dotted line corresponds to the homogeneous chiral restoration transition. The dash-dotted line is a boundary of a region in which there exists a local inhomogeneous minimum of the thermodynamical potential. HCB (HCR) and IH stand for homogeneous chirally broken (restored) and inhomogeneous phases. TCP and LP stand for tricritical and Lifshitz points. The values $(\mu = 190\,\text{MeV}, T = 50\,\text{MeV})$ are those used in the evaluation of the α_n in Eq. (11) for the dynamical simulation. Figure adapted from Ref. [36].

4 Chiral order parameters – dynamics

We have two order parameters, σ and π. The corresponding time-dependent GL equations are coupled due to the the nonlinear terms in $H_{\text{NJL}}[\sigma, \pi]$; they are given by

$$\frac{\partial \sigma(\mathbf{x}, t)}{\partial t} = -\Gamma_\sigma \frac{\delta H_{\text{NJL}}[\sigma, \pi]}{\delta \sigma(\mathbf{x}, t)} + \zeta_\sigma(\mathbf{x}, t), \qquad \frac{\partial \pi^a(\mathbf{x}, t)}{\partial t} = -\Gamma_\pi \frac{\delta H_{\text{NJL}}[\sigma, \pi]}{\delta \pi^a(\mathbf{x}, t)} + \zeta_\pi(\mathbf{x}, t), \tag{13}$$

where ζ_σ and ζ_π are white-noise fields. We start presenting solutions for the local NJL model and a static medium — expansion of the medium is considered in the following section. We take $\pi = 0$ and, to avoid proliferation of different choices of parameters, we study the line $\alpha_4 = -\sqrt{36/5\,\alpha_2\alpha_6}$; as discussed above, on this line the solution given in Eq. (12) becomes a hyperbolic tangent. Dimensionful quantities are rescaled with with the appropriate powers of Γ. We choose $\alpha_2 = \alpha_6^{-1} = \Gamma^{-2}$, then $\alpha_4 = -\sqrt{36/5}$. The initial profile $\phi(z, 0)$ is chosen to be a random configuration generated with Gaussian noise, with the same parity of the inhomogeneous equilibrium solution — recall that on the line

$\alpha_4 = -\sqrt{36/5}\,\alpha_2\alpha_6$, the homogeneous and inhomogeneous solutions coexist. As the inhomogeneous solutions occur at relatively low temperatures, the noise field ζ can be neglected.

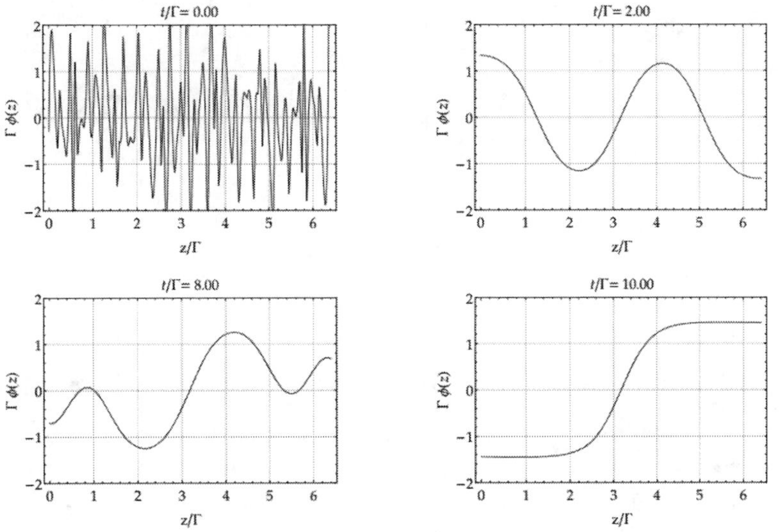

Figure 4. Results for the simulation on the first critical line, $\alpha_4 = -\sqrt{36/5}\,\alpha_2\alpha_6$. The initial condition is generated with a Gaussian noise and it has the parity of the equilibrium solution.

In Fig. 4 we plot the solutions $\phi(z, t)$ at different times t. Clearly, the long-time solution resembles the equilibrium hyperbolic tangent solution, achieved after a time $t \sim 10\,\Gamma$. Since we do not have a microscopic derivation of Eq. (1), the value of Γ must be taken from elsewhere. A good estimate for Γ, calculated from the decay $\sigma \to \pi\pi$, is $\Gamma = 1/3\,\text{fm}$ [22, 37] at $T = 0$, which is the relevant temperature in the present case. For this value of Γ, equilibrium is achieved after a time interval of $t = 30$ fm. In the next section we compare this time scale with the expansion rate. It is important to mention that in principle there is a minimum size of the system for the decay $\sigma \to \pi\pi$ to happen; if the size is too small, the decay cannot happen, there is no dissipation and fluctuation and, therefore, Γ would be zero.

We consider now the coupled system of GL equations, Eq. (13), for the non local NJL model. We evaluate the α_n coefficients of the GLW Hamiltonian in Eq. (11) for ($\mu = 190$ MeV, $T = 50$ MeV). This (μ, T) point, identified on the phase diagram of Fig. (3), is located within the chirally broken homogeneous equilibrium phase, but it lies very close to the tricritical and Lifshitz points. We take as initial profiles $\sigma(z, 0)$ and $\pi(z, 0)$ dual chiral density waves superimposed by Gaussian noise. Fig. 5 shows the solutions for $\sigma(z, t)$ and $\pi(z, t) = \pi^3(z, t)$. Clearly, the long-time solutions are the expected homogeneous solutions $\pi = 0$ and $\sigma \neq 0$, the latter indicating that chiral symmetry is dynamically broken. The interesting feature here is that although the equilibrium order parameters are homogeneous, homogeneity is reached only after a very long time, $t > 10^4\Gamma$. The investigation of the dependence of the lifetime of the inhomogeneous solutions on the size of the system is reserved for a future publication.

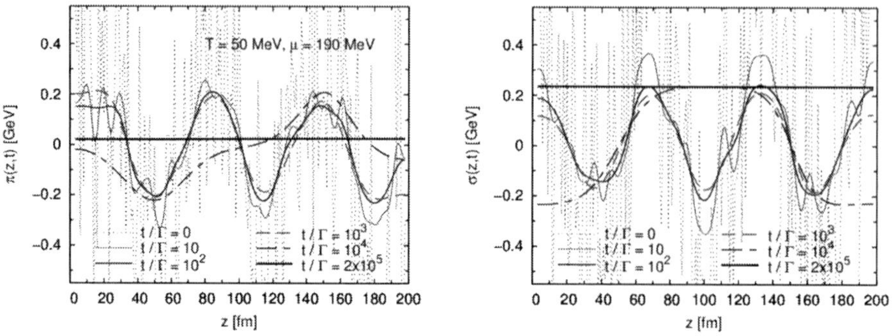

Figure 5. Time dependence of spatial profiles of the pseudoscalar (left panel) and scalar (right panel) order parameters. The initial profiles are shown by the light dotted lines.

5 Chiral order parameters – dynamics with expansion of the medium

Here we address the evolution of the order parameter in an expanding medium using Bjorken flow [38]. A more realistic simulation of the expansion would demand a full hydrodynamical simulation. The appropriate coordinates to describe longitudinal expansion are the so called Milne coordinates [39], the longitudinal proper time $\tau = (t^2 - z^2)^{1/2}$ and space-time rapidity $\eta = 1/2 \log{(t+z)/(t-z)}$, where t and z denote the time and longitudinal coordinates in the laboratory frame. We must transform the derivatives in the GL equation accordingly. The initial conditions are set at a time-like surface $\tau = \tau_0$. The value of τ_0 sets how late the GL evolution starts after the beginning of the expansion.

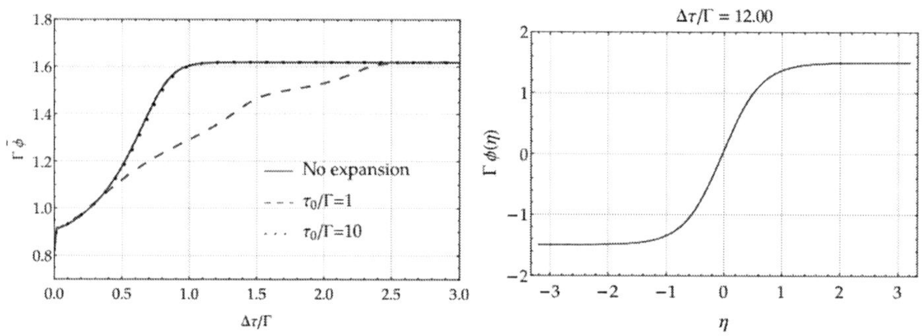

Figure 6. GL evolution of the order parameter for an expanding medium in the local NJL model. The simulation is performed on the first critical line, $\alpha_4 = -\sqrt{36/5}\,\alpha_2\alpha_6$. Left panel: spatial average $\bar{\phi}$, defined in Eq. (14), as a function of elapsed proper time $\Delta\tau = \tau - \tau_0$, for two values of τ_0. Right panel: the order parameter ϕ as a function of the space-time rapidity η for a delayed dynamics, $\tau_0 = 10^4\,\Gamma$, and elapsed proper time $\Delta\tau = 12\,\Gamma$.

We consider the local NJL model only. Initially, we consider the homogeneous equilibrium phase and calculate the time evolution of the spatial average of the order parameter

$$\bar{\phi} = \frac{1}{L}\int_L dz\,\phi(z,\tau), \tag{14}$$

where L is the spatial extension of system. Again we simulate on the first critical line, $\alpha_4 = -\sqrt{36/5}\,\alpha_2\alpha_6$. The result for $\bar{\phi}$ as a function of $\Delta\tau \equiv \tau - \tau_0$ is shown in the left panel of Fig. 6

for $\tau_0 = 10\,\Gamma$ and $\tau_0 = \Gamma$, as well as for the nonexpanding case. As expected, the sooner the GL dynamics starts with respect to the start of the expansion, the more delayed is the approach to the equilibrium configuration compared the to static case. In the right panel of the figure, we present the result for a delayed dynamics, i.e. for a case with $\tau_0 = 10^4\,\Gamma$. The equilibrium configuration is obtained for an elapsed time $\Delta\tau = 12\,\Gamma$.

6 Conclusions and Perspectives

We have presented results of a study of the dynamics of the formation of inhomogeneous chirally broken phases. With the aim of getting insight on the possibility of formation of such phases at the final stages of a heavy-ion collision, we have investigated the time scales involved in the formation process. Our study was based on the framework of the phenomenological time-dependent Ginzburg-Landau equation. The Ginzburg-Landau dynamics depends on two fundamental quantities: the energy functional that gives the equilibrium phase diagram and the Onsager coefficient Γ, a time scale related to fluctuation and dissipation processes. We employed NJL-type models to obtain the energy functional; we considered a local NJL model with point-like interactions [4] and a nonlocal NJL model with finite-range interactions [30]. The results are qualitatively similar for both models, but there are interesting differences. In particular, we have shown that even in the case that the equilibrium phase is a homogeneous chirally broken phase, long-lived inhomogeneous phases can be formed when the model energy functional predicts metastable phases, like in the case of the nonlocal NJL model. For $\Gamma = 1/3$ fm, that is the value found in Ref. [37] from an estimate of the decay $\sigma \to 2\pi$ at zero temperature, we found that, for appropriate initial configurations, well-defined inhomogeneous domains are formed for a time interval of the order of $\Delta t = 10$ fm, in both models. We have also considered expansion of the medium, a crucial feature of heavy-ion collisions that must be taken into account for a realistic description of the dynamics. We modeled the expansion by one-dimensional Bjorken flow and found, as expected, that the expansion delays the formation of inhomogeneous phases.

Our work must be improved on several aspects. First of all, one needs to consider dynamics in three spatial dimensions. One needs also to investigate the role of confinement and asymptotic freedom in the phase diagram. For that, one needs models that incorporate such features and go beyond NJL-type models. In addition, possible observable signals linked to the decay of the inhomogeneous condensates must be found [40]. Finally, one needs to go beyond the phenomenological Ginzburg-Landau approach and derive dynamical equations from microscopic models using, for example, the closed-time-path effective action formalism [20].

Acknowledgments: This work has been partially funded by CONICET under Grant No. PIP 449 (JPC), and the National University of La Plata, Project No. X718 (JPC), and Conselho Nacional de Desenvolvimento Científico e Tecnológico - CNPq, 305894/2009-9 (G.K.), 501881/2014-0 (DK) and Fundação de Amparo à Pesquisa do Estado de São Paulo - FAPESP, 2013/01907-0 (G.K.), 2016/01343- 7 (JPC).

References

[1] E. Shuryak, Rev. Mod. Phys. **89**, 035001 (2017).
[2] B. Muller, J. Schukraft and B. Wyslouch, Ann. Rev. Nucl. Part. Sci. **62**, 361 (2012).
[3] M. Buballa and S. Carignano, Prog. Part. Nucl. Phys. **81**, 39 (2015).
[4] D. Nickel, Phys. Rev. Lett. **103**, 072301 (2009).

[5] D. Nickel, Phys. Rev. D **80**, 074025 (2009).
[6] A. Heinz, F. Giacosa and D. H. Rischke, Nucl. Phys. A **933**, 34 (2015).
[7] T. Tatsumi and T. Muto, Phys. Rev. D **89**, 103005 (2014).
[8] S. Carignano, E. J. Ferrer, V. de la Incera and L. Paulucci, Phys. Rev. D **92**, 105018 (2015).
[9] M. Buballa and S. Carignano, Eur. Phys. J. A **52**, 57 (2016).
[10] P. C. Hohenberg and B. I. Halperin, Rev. Mod. Phys. **49**, 435 (1977).
[11] D. T. Son and M. A. Stephanov, Phys. Rev. D **70**, 056001 (2004).
[12] E. S. Fraga and G. Krein, Phys. Lett. B **614**, 181 (2005).
[13] E. S. Fraga, T. Kodama, G. Krein, A. J. Mizher and L. F. Palhares, Nucl. Phys. A **785**, 138 (2007).
[14] E. S. Fraga, G. Krein and A. J. Mizher, Phys. Rev. D **76**, 034501 (2007).
[15] N. C. Cassol-Seewald, R. L. S. Farias, E. S. Fraga, G. Krein and R. O. Ramos, Physica A **391**, 4088 (2012).
[16] G. Krein, J. Phys. Conf. Ser. **378**, 012032 (2012).
[17] N. C. Cassol-Seewald, R. L. S. Farias, R. Marques de Carvalho and G. Krein, Int. J. Mod. Phys. C **23**, 1240016 (2012).
[18] D. Bower and S. Gavin, Phys. Rev. C **64**, 051902 (2001).
[19] T. Koide, G. Krein and R. O. Ramos, Phys. Lett. B **636**, 96 (2006).
[20] E. A. Calzetta and B. L. B. Hu, *Nonequilibrium Quantum Field Theory* (Cambridge University Press, Cambridge, 2008).
[21] M. Gleiser and R. O. Ramos, Phys. Rev. D **50**, 2441 (1994).
[22] M. Nahrgang, S. Leupold and M. Bleicher, Phys. Lett. B **711**, 109 (2012).
[23] M. Nahrgang, S. Leupold, C. Herold and M. Bleicher, Phys. Rev. C **84**, 024912 (2011).
[24] Thiago Peixoto, *Dynamics of the chiral and deconfinement transitions of quantum chromodynamics with the PNJL model* (IFT-D.006/2014, MSc dissertation, in protuguese).
[25] G. Krein, P. Tang and A. G. Williams, Phys. Lett. B **215**, 145 (1988).
[26] G. Krein, P. Tang, L. Wilets and A. G. Williams, Phys. Lett. B **212**, 362 (1988).
[27] G. Krein, P. Tang, L. Wilets and A. G. Williams, Nucl. Phys. A **523**, 548 (1991).
[28] C. D. Roberts and S. M. Schmidt, Prog. Part. Nucl. Phys. **45**, S1 (2000).
[29] D. Muller, M. Buballa and J. Wambach, Phys. Lett. B **727**, 240 (2013).
[30] J. P. Carlomagno, D. Gomez Dumm and N. N. Scoccola, Phys. Lett. B **745**, 1 (2015).
[31] J. P. Carlomagno and G. Krein, in preparation.
[32] A. I. Buzdin and H. Kachkachi, Phys. Lett. A **225**, 341 (1997).
[33] G. Basar and G. V. Dunne, Phys. Rev. Lett. **100**, 200404 (2008).
[34] G. Basar and G. V. Dunner, Phys. Rev. D **78**, 065022 (2008).
[35] O. Schnetz, M. Thies and K. Urlichs, Ann. Phys. **314**, 425 (2004).
[36] J. P. Carlomagno, D. Gómez Dumm and N. N. Scoccola, Phys. Rev. D **92**, 056007 (2015).
[37] D. H. Rischke, Phys. Rev. C **58**, 2331 (1998).
[38] J. D. Bjorken, Phys. Rev. D **27**, 140 (1983).
[39] V. F. Mukhanov, *Physical Foundations of Cosmology* (Cambridge University Press, Cambridge, 2005).
[40] G. Krein, D. Kroff and T. Peixoto, in preparation.

Small x Asymptotics of the Quark and Gluon Helicity Distributions

Yuri V. Kovchegov[1,⋆], *Daniel* Pitonyak[2,⋆⋆], and *Matthew D.* Sievert[3,⋆⋆⋆]

[1]*Department of Physics, The Ohio State University, Columbus, OH 43210, USA*

[2]*Division of Science, Penn State University-Berks, Reading, PA 19610, USA*

[3]*Theoretical Division, Los Alamos National Laboratory, Los Alamos, NM 87545, USA*

Abstract. We determine the small-x asymptotics of the gluon helicity distribution in a proton at leading order in perturbative QCD at large N_c. To achieve this, we begin by evaluating the dipole gluon helicity TMD at small x. We then construct and solve novel small-x large-N_c evolution equations for the operator related to the dipole gluon helicity TMD. Our main result is the small-x asymptotics for the quark helicity distribution

$$\Delta q \sim \left(\frac{1}{x}\right)^{\alpha_h^q} \quad \text{with} \quad \alpha_h^q = \frac{4}{\sqrt{3}}\sqrt{\frac{\alpha_s N_c}{2\pi}} \approx 2.31\sqrt{\frac{\alpha_s N_c}{2\pi}} \tag{1}$$

and the small-x asymptotics of the gluon helicity distribution

$$\Delta G \sim \left(\frac{1}{x}\right)^{\alpha_h^G} \quad \text{with} \quad \alpha_h^G = \frac{13}{4\sqrt{3}}\sqrt{\frac{\alpha_s N_c}{2\pi}} \approx 1.88\sqrt{\frac{\alpha_s N_c}{2\pi}}. \tag{2}$$

1 Introduction

The main goal of this work is to determine the small-x asymptotics of the quark and gluon helicity PDFs and TMDs. Below we will review our determination of the small-x asymptotics of the quark helicity, and then present the determination of the small-x asymptotics for the gluon helicity distributions. This talk and these proceedings are based on [1–5].

2 Quark Helicity Distribution

In [1], we derived the polarized small-x evolution equations for the TMD quark helicity distribution [6],

$$g_{1L}^q(x, k_T^2) = \frac{1}{(2\pi)^3}\frac{1}{2}\sum_{S_L} S_L \int d^2r\, dr^-\, e^{ixP^+r^-}\, e^{-i\underline{k}\cdot\underline{r}}\, \langle P, S_L|\bar{\psi}(0)\, \mathcal{U}[0,r]\, \frac{\gamma^+\gamma^5}{2}\, \psi(r)\,|P, S_L\rangle_{r^+=0}, \tag{3}$$

⋆e-mail: kovchegov.1@osu.edu
⋆⋆e-mail: dap67@psu.edu
⋆⋆⋆e-mail: sievertmd@lanl.gov

© The Authors, published by EDP Sciences. This is an open access article distributed under the terms of the Creative Commons Attribution License 4.0 (http://creativecommons.org/licenses/by/4.0/).

by relating it to a "polarized dipole amplitude" $G(x_{10}^2, zs)$, giving

$$g_{1L}^{q,S}(x, k_T^2) = \frac{8N_c}{(2\pi)^6} \sum_f \int_{\Lambda^2/s}^{1} \frac{dz}{z} \int d^2x_{01}\, d^2x_{0'1}\, e^{-i\underline{k}\cdot(\underline{x}_{01}-\underline{x}_{0'1})}\, \frac{\underline{x}_{01}\cdot\underline{x}_{0'1}}{x_{01}^2\, x_{0'1}^2}\, G(x_{10}^2, zs = \frac{z}{x}Q^2) \tag{4}$$

in the flavor-singlet case [3]. In the above and throughout this paper, we use light-front coordinates $x^\pm \equiv \frac{1}{\sqrt{2}}(x^0 \pm x^3)$, denote transverse vectors $(x_\perp^1, x_\perp^2)$ by $\underline{x}$ and their magnitudes by $x_T \equiv |\underline{x}|$, and indicate differences in transverse coordinates by the abbreviated notation $\underline{x}_{10} \equiv \underline{x}_1 - \underline{x}_0$. The center-of-mass energy squared for the scattering process is s, the infrared (IR) transverse momentum cutoff is Λ, and z is the fraction of the light-cone momentum of the dipole carried by the polarized (anti-)quark. As is well-known, the TMD (3) contains a process-dependent gauge link $\mathcal{U}[0, r]$. For specificity, in [1] we considered semi-inclusive deep inelastic scattering (SIDIS), although the resulting small-x evolution equations also apply to the collinear quark helicity distribution, which is process independent.

The impact-parameter integrated polarized dipole amplitude is

$$G(x_{10}^2, zs) = \int d^2b_{10}\, G_{10}(zs) \tag{5}$$

with $\underline{b}_{10} = (\underline{x}_1 + \underline{x}_0)/2$. The polarized dipole scattering amplitude $G_{10}(zs)$ was defined as the polarized generalization of the forward dipole S-matrix in terms of Wilson lines [1]:

$$\begin{aligned} G_{10}(zs) &\equiv \frac{1}{2N_c} \Big\langle \mathrm{tr}\left[V_{\underline{0}} V_{\underline{1}}^{pol\,\dagger}\right] + \mathrm{tr}\left[V_{\underline{1}}^{pol} V_{\underline{0}}^{\dagger}\right] \Big\rangle(zs) \\ &\equiv \frac{zs}{2N_c} \Big\langle\!\Big\langle \mathrm{tr}\left[V_{\underline{0}} V_{\underline{1}}^{pol\,\dagger}\right] + \mathrm{tr}\left[V_{\underline{1}}^{pol} V_{\underline{0}}^{\dagger}\right] \Big\rangle\!\Big\rangle(zs), \end{aligned} \tag{6}$$

where the double-angle brackets are defined to scale out the center-of-mass energy zs between the polarized (anti)quark and the target. While the unpolarized Wilson lines in Eq. (6) are the standard eikonal gauge links (in the fundamental representation),

$$V_{\underline{0}} \equiv V_{\underline{x}_0}[+\infty, -\infty] \equiv \mathcal{P}\exp\left[ig \int_{-\infty}^{+\infty} dx^- A^+(x^+ = 0, x^-, \underline{x}_0)\right], \tag{7}$$

the polarized Wilson lines $V_{\underline{1}}^{pol}$ are more complex operators [5]

$$(V_{\underline{x}}^{pol})^g = \frac{ig\, p_1^+}{s} \int_{-\infty}^{+\infty} dx^-\, V_{\underline{x}}[+\infty, x^-]\, F^{12}(x^+ = 0, x^-, \underline{x})\, V_{\underline{x}}[x^-, -\infty]. \tag{8}$$

The expression for the polarized Wilson line in Eq. (8) is only due to the gluon interaction with the target (with the quark exchanges with the target to be included in the future work).

Employing Eq. (8) in Eq. (6) we can finally write down an explicit operator expression for the polarized dipole scattering amplitude (in $A^- = 0$ gauge) illustrated in Fig. 1:

$$G_{10}(zs) \equiv \frac{p_1^+}{2N_c} \int_{-\infty}^{\infty} dx_1^- \Big\langle \mathrm{tr}\left[V_{\underline{0}}\; V_{\underline{1}}[-\infty, x_1^-] \left(ig\, \epsilon_T^{ij} \frac{\partial}{\partial(x_1)_\perp^i} A_\perp^j(x_1^-, \underline{x}_1)\right) V_{\underline{1}}[x_1^-, \infty]\right] + \mathrm{c.c.} \Big\rangle(zs). \tag{9}$$

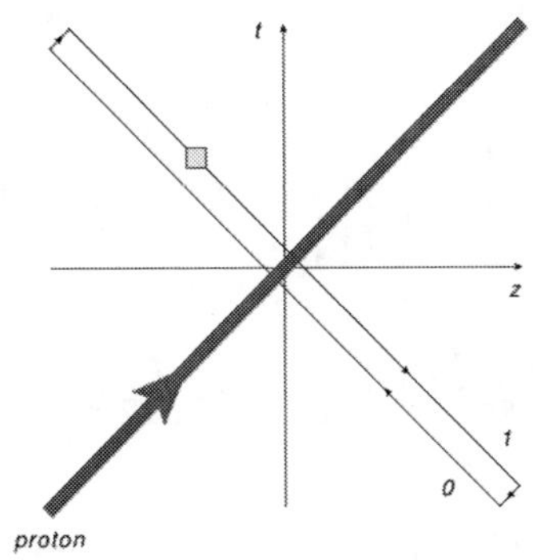

Figure 1. Pictorial representation of the polarized dipole operator in Eq. (9), with the square representing an insertion of the sub-eikonal operator $ig\epsilon_T^{ij}\partial_i A^j$.

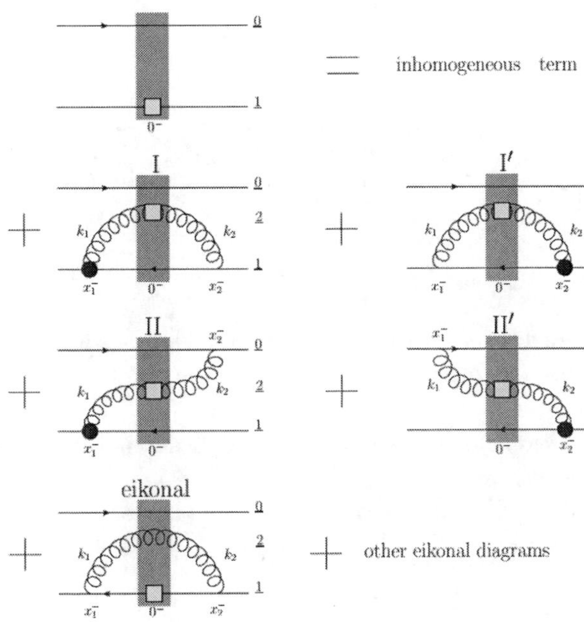

Figure 2. Diagrams illustrating the evolution of the polarized dipole amplitude $G_{10}(zs)$. The blue band represents the classical fields (shock wave), the black vertex represents the sub-eikonal operator insertion from the center of (9), and the gray box represents the polarized Wilson line.

To determine the small-x asymptotics of the quark helicity distribution we need to find the high-energy limit of $G_{10}(zs)$. The latter is determined by deriving and solving the high-energy evolution equation for $G_{10}(zs)$. This equation is

$$G_{10}(zs) = G_{10}^{(0)}(zs) + \frac{\alpha_s N_c}{2\pi^2} \int\limits_{\frac{\Lambda^2}{s}}^{z} \frac{dz'}{z'} \int d^2 x_2 \left\{ \left[\frac{1}{x_{21}^2} - \frac{\underline{x}_{21} \cdot \underline{x}_{20}}{x_{21}^2 \, x_{20}^2} \right] \left\langle \frac{1}{N_c^2} \mathrm{tr} \left[V_{\underline{0}} t^a V_{\underline{1}}^\dagger t^b \right] (U_{\underline{2}}^{pol})^{ba} + \mathrm{c.c.} \right\rangle (z's) \right.$$
$$\left. + \frac{x_{10}^2}{x_{21}^2 \, x_{20}^2} \left\langle \frac{1}{N_c^2} \mathrm{tr} \left[V_{\underline{0}} t^a V_{\underline{1}}^{pol\,\dagger} t^b \right] (U_{\underline{2}})^{ba} - \frac{C_F}{N_c^2} \mathrm{tr} \left[V_{\underline{0}} V_{\underline{1}}^{pol\,\dagger} \right] + \mathrm{c.c.} \right\rangle (z's) \right\}, \tag{10}$$

and it is diagrammatically represented in Fig. 2. The equation resums powers of $\alpha_s \ln^2(1/x)$ in the quark helicity distribution: this is the double logarithmic approximation (DLA).

Equation (10) is not closed: the operator on the right-hand side is not the same as the operator on the left. To obtain a closed system of equations one has to invoke either the large-N_c or the large-N_c&N_f limits (the latter would also require the quark exchange contribution added to the polarized Wilson lines). In the large-N_c limit, with DLA accuracy, the polarized evolution equations are [1, 3]

$$G(x_{10}^2, zs) = G^{(0)}(x_{10}^2, zs) + \frac{\alpha_s N_c}{2\pi} \int\limits_{\frac{1}{x_{10}^2 s}}^{z} \frac{dz'}{z'} \int\limits_{\frac{1}{z's}}^{x_{10}^2} \frac{dx_{21}^2}{x_{21}^2} \left[\Gamma(x_{10}^2, x_{21}^2, z's) + 3G(x_{21}^2, z's) \right], \tag{11a}$$

$$\Gamma(x_{10}^2, x_{21}^2, z's) = G^{(0)}(x_{10}^2, z's) + \frac{\alpha_s N_c}{2\pi} \int\limits_{\frac{1}{x_{10}^2 s}}^{z'} \frac{dz''}{z''} \int\limits_{\frac{1}{z''s}}^{\min\left[x_{10}^2, x_{21}^2 \frac{z'}{z''}\right]} \frac{dx_{32}^2}{x_{32}^2} \left[\Gamma(x_{10}^2, x_{32}^2, z''s) + 3G(x_{32}^2, z''s) \right], \tag{11b}$$

where $G^{(0)}$ are the initial conditions. The polarized dipole G depends upon an auxiliary function Γ, termed the "neighbor dipole amplitude", in which further evolution is constrained by the lifetime of an adjacent dipole.

Equations (11) were solved numerically in [2] and analytically in [4] for the high-energy asymptotics yielding

$$G(x_{10}^2, zs) = \frac{1}{3} G_0 \, (zs \, x_{10}^2)^{\alpha_h^q} \tag{12a}$$

$$\Gamma(x_{10}^2, x_{21}^2, zs) = \frac{1}{3} G_0 \, (zs \, x_{21}^2)^{\alpha_h^q} \left[4 \left(\frac{x_{10}^2}{x_{21}^2} \right)^{\frac{\alpha_h^q}{4}} - 3 \right], \tag{12b}$$

where the exponent of the energy, known as the "quark helicity intercept" in analogy to the Pomeron intercept, is given by

$$\alpha_h^q = \frac{4}{\sqrt{3}} \sqrt{\frac{\alpha_s N_c}{2\pi}} \approx 2.31 \sqrt{\frac{\alpha_s N_c}{2\pi}}. \tag{13}$$

This leads to Eq. (1) above.

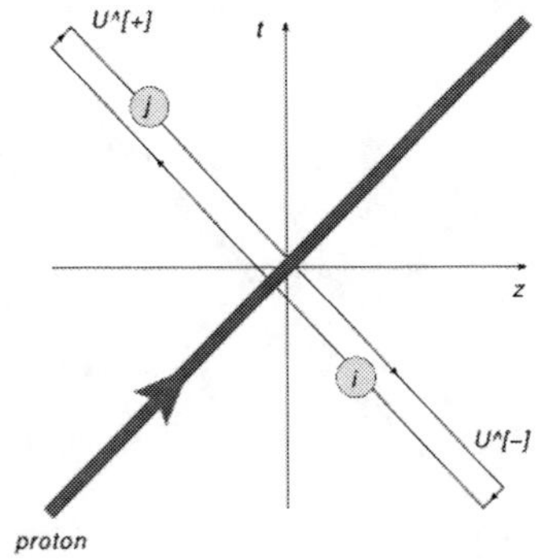

Figure 3. Pictorial representation of the gluon dipole helicity TMD in Eq. (14), with the circles representing insertions of F^{+i} and F^{+j}.

3 Gluon Helicity Distribution

The small-x asymptotics of the gluon helicity can be determined in a similar way [5]. Start with the operator definition of the gluon dipole helicity TMD [7]

$$g_{1L}^{G}(x,k_T^2) = \frac{-2i}{x\,P^+}\,\frac{1}{2}\sum_{S_L} S_L \int \frac{d\xi^-\,d^2\xi}{(2\pi)^3}\, e^{ixP^+\,\xi^- - i\underline{k}\cdot\underline{\xi}} \times \langle P, S_L|\,\epsilon_T^{ij}\,\mathrm{tr}\left[F^{+i}(0)\,\mathcal{U}^{[+]}[0,\xi]\,F^{+j}(\xi)\,\mathcal{U}^{[-]}[\xi,0]\right]|P,S_L\rangle_{\xi^+=0} \tag{14}$$

illustrated in Fig. 3.

It can be shown [4] that at small x this TMD becomes

$$g_{1L}^{G\,dip}(x,k_T^2) = \frac{-8i\,N_c}{g^2(2\pi)^3}\int d^2x_{10}\,e^{i\underline{k}\cdot\underline{x}_{10}}\,k_\perp^i\,\epsilon_T^{ij}\left[\int d^2b_{10}\,G_{10}^j(zs=\tfrac{Q^2}{x})\right] \tag{15}$$

where we have defined another dipole-like polarized operator

$$G_{10}^i(zs) \equiv \frac{1}{2N_c}\left\langle \mathrm{tr}\left[V_{\underline{0}}(V_{\underline{1}}^{pol\,\dagger})_\perp^i\right] + \mathrm{c.c.}\right\rangle(zs) \tag{16}$$

with

$$(V_{\underline{x}}^{pol})_\perp^i \equiv \int\limits_{-\infty}^{+\infty} dx^-\,V_{\underline{x}}[+\infty,x^-]\left(ig\,P^+\,A_\perp^i(x)\right)V_{\underline{x}}[x^-,-\infty]. \tag{17}$$

The operator from Eq. (16) is shown in Fig. 4.

After the integration over all impact parameters, the new polarized dipole amplitude is a vector-valued function of $\underline{x}_{10}$ alone, allowing us to write the decomposition

$$\int d^2b_{10}\,G_{10}^i(zs) = (x_{10})_\perp^i\,G_1(x_{10}^2,zs) + \epsilon_T^{ij}\,(x_{10})_\perp^j\,G_2(x_{10}^2,zs). \tag{18}$$

The gluon dipole helicity TMD then becomes

$$g_{1L}^{G\,dip}(x,k_T^2) = \frac{-N_c}{\alpha_s\,2\pi^4}\int d^2x_{10}\,e^{i\underline{k}\cdot\underline{x}_{10}}\left[1 + x_{10}^2\frac{\partial}{\partial x_{10}^2}\right]G_2(x_{10}^2,zs=\tfrac{Q^2}{x}). \tag{19}$$

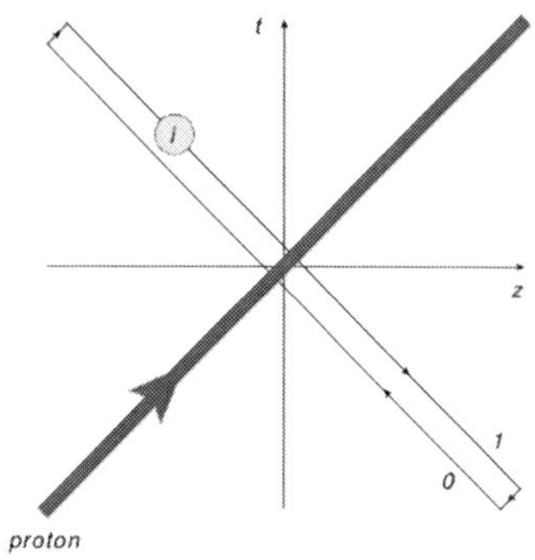

Figure 4. Pictorial representation of the new polarized dipole operator from Eq. (16), with the circle representing an insertion of the sub-eikonal transverse gluon field from Eq. (17).

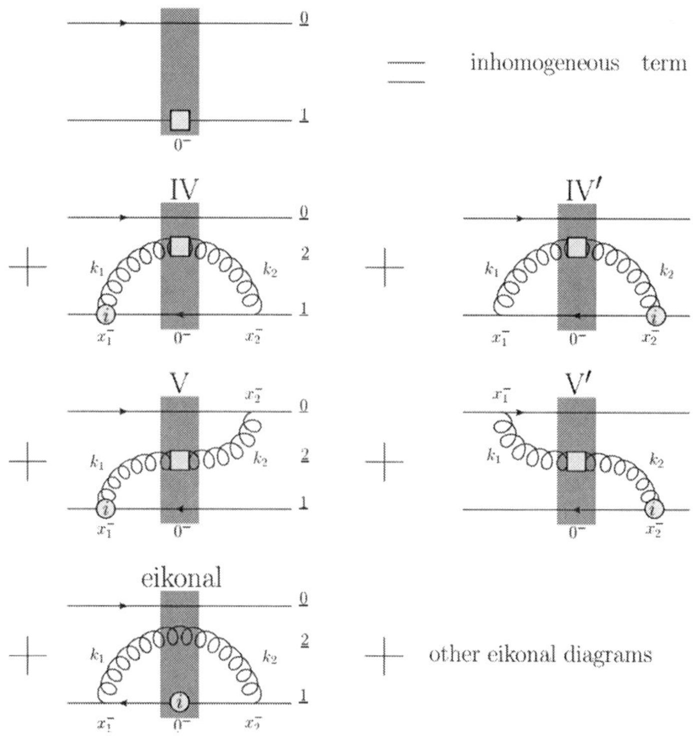

Figure 5. Diagrams illustrating the evolution of the polarized dipole amplitude for the dipole gluon helicity distribution. The blue band represents the classical gluon fields (shock wave), the vertex (*i*) denotes the sub-eikonal operator insertion, and the gray box represents the polarized Wilson line.

The DLA evolution equation for $G^i_{10}(zs)$ is illustrated diagrammatically in Fig. 5. The equation is

$$G^i_{10}(zs) = G^{i\,(0)}_{10}(zs) + \frac{\alpha_s N_c}{2\pi^2} \int\limits_{\frac{\Lambda^2}{s}}^{z} \frac{dz'}{z'} \int d^2 x_2 \tag{20}$$
$$\times \left\{ \ln \frac{1}{x_{21}\Lambda}\, \epsilon_T^{ij} \left[\frac{(x_{21})^j_\perp}{x_{21}^2} - \frac{(x_{20})^j_\perp}{x_{20}^2} \right] \left\langle \frac{1}{2N_c} \mathrm{tr}\left[V_{\underline{0}} V_{\underline{2}}^{pol\,\dagger} \right] + \frac{1}{2N_c} \mathrm{tr}\left[V_{\underline{2}}^{pol} V_{\underline{1}}^{\dagger} \right] + \mathrm{c.c.} \right\rangle (z's) \right\}$$
$$+ \frac{\alpha_s N_c}{2\pi^2} \int\limits_{\frac{\Lambda^2}{s}}^{z} \frac{dz'}{z'} \int \frac{d^2 x_2}{x_{21}^2}\, \theta\!\left(x_{10}^2 - x_{21}^2\right) \theta\!\left(x_{21}^2 - \frac{1}{z's}\right) \left[G^i_{12}(z's) - \Gamma^i_{10,21}(z's) \right],$$

where the have introduced the "neighbor" dipole amplitude Γ^i by analogy with the quark helicity evolution case.

Equation (20) again does not close. To obtain a closed system of equations we take the large-N_c limit. Remembering that we are interested in G_2 to find the gluon helicity TMD in Eq. (19), we project out the G_2 component to write

$$G_2(x_{10}^2, zs) = G_2^{(0)}(x_{10}^2, zs) + \frac{\alpha_s N_c}{2\pi^2} \int\limits_{\frac{\Lambda^2}{s}}^{z} \frac{dz'}{z'} \int d^2 x_2 \ln \frac{1}{x_{21}\Lambda} \frac{\underline{x}_{10} \cdot \underline{x}_{21}}{x_{10}^2\, x_{21}^2} \left[\Gamma_{gen}(x_{20}^2, x_{21}^2, z's) + G(x_{21}^2, z's) \right]$$
$$- \frac{\alpha_s N_c}{2\pi^2} \int\limits_{\frac{\Lambda^2}{s}}^{z} \frac{dz'}{z'} \int d^2 x_2 \ln \frac{1}{x_{21}\Lambda} \frac{\underline{x}_{10} \cdot \underline{x}_{20}}{x_{10}^2\, x_{20}^2} \left[\Gamma_{gen}(x_{20}^2, x_{21}^2, z's) + \Gamma_{gen}(x_{21}^2, x_{20}^2, z's) \right]$$
$$- \frac{\alpha_s N_c}{2\pi} \int\limits_{\frac{1}{x_{10}^2 s}}^{z} \frac{dz'}{z'} \int\limits_{\frac{1}{z's}}^{x_{10}^2} \frac{dx_{21}^2}{x_{21}^2}\, \Gamma_2(x_{10}^2, x_{21}^2, z's), \tag{21a}$$

$$\Gamma_2(x_{10}^2, x_{21}^2, z's) = G_2^{(0)}(x_{10}^2, z's) + \frac{\alpha_s N_c}{2\pi^2} \int\limits_{\frac{\Lambda^2}{s}}^{z'} \frac{dz''}{z''} \int d^2 x_3 \ln \frac{1}{x_{31}\Lambda} \frac{\underline{x}_{10} \cdot \underline{x}_{31}}{x_{10}^2\, x_{31}^2} \left[\Gamma_{gen}(x_{30}^2, x_{31}^2, z''s) + G(x_{31}^2, z's) \right]$$
$$- \frac{\alpha_s N_c}{2\pi^2} \int\limits_{\frac{\Lambda^2}{s}}^{z'} \frac{dz''}{z''} \int d^2 x_3 \ln \frac{1}{x_{31}\Lambda} \frac{\underline{x}_{10} \cdot \underline{x}_{30}}{x_{10}^2\, x_{30}^2} \left[\Gamma_{gen}(x_{30}^2, x_{31}^2, z''s) + \Gamma_{gen}(x_{31}^2, x_{30}^2, z''s) \right]$$
$$- \frac{\alpha_s N_c}{2\pi} \int\limits_{\frac{1}{x_{10}^2 s}}^{z'} \frac{dz''}{z''} \int\limits_{\frac{1}{z''s}}^{\min\left[x_{10}^2,\, x_{21}^2 \frac{z'}{z''}\right]} \frac{dx_{31}^2}{x_{31}^2}\, \Gamma_2(x_{10}^2, x_{31}^2, z''s). \tag{21b}$$

We have defined an impact-parameter integrated amplitude Γ_{gen} by

$$\Gamma_{gen}(x_{20}, x_{21}, z's) = \theta(x_{20} - x_{21})\, \Gamma(x_{20}, x_{21}, z's) + \theta(x_{21} - x_{20})\, G(x_{20}, z's). \tag{22}$$

This function can be easily found using the analytic solution (12) for the asymptotics of G and Γ at high energies.

Equations (21) can be solved analytically [5]. The result it

$$G_2(x_{10}^2, zs) \approx -\frac{19}{32\sqrt{3}} \sqrt{\frac{\alpha_s N_c}{2\pi}} G_0 \ln \frac{1}{x_{10}\Lambda} (zs\, x_{10}^2)^{\frac{13}{4\sqrt{3}} \sqrt{\frac{\alpha_s N_c}{2\pi}}}, \tag{23a}$$

$$\Gamma_2(x_{10}^2, x_{21}^2, z's) \approx -\frac{1}{2\sqrt{3}} \sqrt{\frac{\alpha_s N_c}{2\pi}} G_0 \ln \frac{1}{x_{10}\Lambda} \left[(z's\, x_{10}^2)^{\frac{4}{\sqrt{3}} \sqrt{\frac{\alpha_s N_c}{2\pi}}} (z's\, x_{21}^2)^{-\frac{\sqrt{3}}{4} \sqrt{\frac{\alpha_s N_c}{2\pi}}} + \frac{3}{16} (z's\, x_{21}^2)^{\frac{4}{\sqrt{3}} \sqrt{\frac{\alpha_s N_c}{2\pi}}} (z's\, x_{10}^2)^{-\frac{\sqrt{3}}{4} \sqrt{\frac{\alpha_s N_c}{2\pi}}} \right]. \tag{23b}$$

Substituting the solution Eq. (23a) into Eq. (19) yields the small-x asymptotics of the dipole gluon helicity distribution:

$$g_{1L}^{G\,dip}(x, k_T^2) \sim G_2(x_{10}^2, zs = \tfrac{Q^2}{x}) \sim \left(\frac{1}{x}\right)^{\alpha_h^G} \tag{24}$$

with the gluon helicity intercept

$$\alpha_h^G = \frac{13}{4\sqrt{3}} \sqrt{\frac{\alpha_s N_c}{2\pi}} \approx 1.88 \sqrt{\frac{\alpha_s N_c}{2\pi}}. \tag{25}$$

We conclude that (cf. Eq. (2))

$$\Delta G(x, Q^2) \sim \left(\frac{1}{x}\right)^{\alpha_h^G} \sim \left(\frac{1}{x}\right)^{\frac{13}{4\sqrt{3}} \sqrt{\frac{\alpha_s N_c}{2\pi}}} \sim \left(\frac{1}{x}\right)^{1.88 \sqrt{\frac{\alpha_s N_c}{2\pi}}}. \tag{26}$$

4 Phenomenology of the Quark and Gluon Spin at Small x

In this section we give an estimate for the quark and gluon spin

$$S_q(Q^2) = \frac{1}{2} \int_0^1 dx\, \Delta\Sigma(x, Q^2) \quad \text{and} \quad S_G(Q^2) = \int_0^1 dx\, \Delta G(x, Q^2), \tag{27}$$

based on our quark (13) and gluon (25) helicity intercepts.

In order to calculate S_G, we need input for the gluon helicity PDF $\Delta G(x, Q^2)$, and we focus here on the fit from DSSV14 [8]. We proceed through a simple approach, which we also employed in Ref. [2] for an estimate of the quark spin, and leave a more rigorous phenomenological study for future work. First, we attach a curve $\Delta\tilde{G}(x, Q^2) = N\, x^{-\alpha_h^G}$ (with α_h^G given in (25)) to the DSSV14 result for $\Delta G(x, Q^2)$ at a particular small-x point x_0. We fix the normalization N by requiring $\Delta\tilde{G}(x_0, Q^2) = \Delta G(x_0, Q^2)$. Then we calculate the truncated integral

$$S_G^{[x_{min}]}(Q^2) \equiv \int_{x_{min}}^1 dx\, \Delta G(x, Q^2) \tag{28}$$

of the modified gluon helicity PDF

$$\Delta G_{mod}(x, Q^2) \equiv \theta(x - x_0)\, \Delta G(x, Q^2) + \theta(x_0 - x)\, \Delta\tilde{G}(x, Q^2) \tag{29}$$

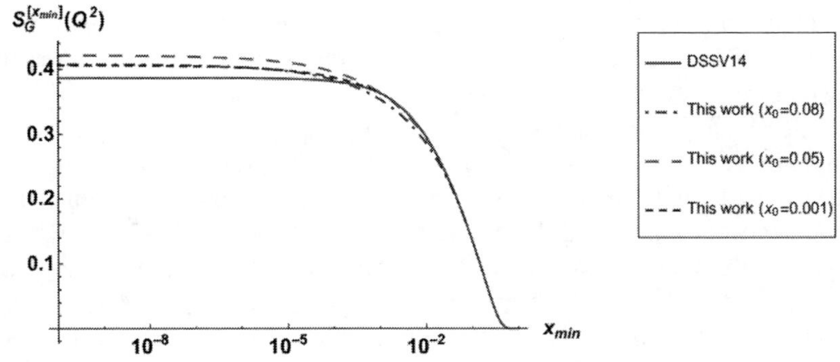

Figure 6. Plot of $S_G^{[x_{min}]}(Q^2)$ vs. x_{min} at $Q^2 = 10\,\mathrm{GeV}^2$. The solid curve is from DSSV14 [8]. The dot-dashed, long-dashed, and short-dashed curves are from various small-x modifications of $\Delta G(x, Q^2)$ at $x_0 = 0.08, 0.05, 0.001$, respectively, using our gluon helicity intercept (see the text for details).

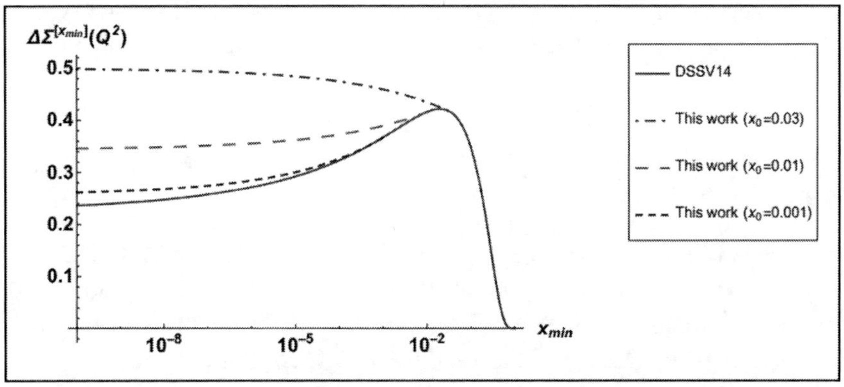

Figure 7. Plot of $\Delta\Sigma^{[x_{min}]}(Q^2)$ vs. x_{min} at $Q^2 = 10\,\mathrm{GeV}^2$. The solid curve is from DSSV14 [8]. The dot-dashed, long-dashed, and short-dashed curves are from various small-x modifications of $\Delta\Sigma(x, Q^2)$ at $x_0 = 0.03, 0.01, 0.001$, respectively, using our helicity intercept (see the text for details).

for different x_0 values. The results are shown in Fig. 6 for $Q^2 = 10\,\mathrm{GeV}^2$ and $\alpha_s \approx 0.25$, in which case $\alpha_h^G \approx 0.65$. We see that the small-x evolution of $\Delta G(x, Q^2)$ gives about a $5 \div 10\%$ increase to the gluon spin, depending on where in x the effects set in and on the parameterization of the gluon helicity PDF at higher x.

The same procedure can be applied to quark spin. The result is shown in Fig. 7 [4], where we plot the truncated integral

$$\Delta\Sigma^{[x_{min}]}(Q^2) \equiv \int_{x_{min}}^{1} dx\, \Delta\Sigma(x, Q^2) \tag{30}$$

with $\Delta\Sigma(x, Q^2) = \Delta u + \Delta\bar{u} + \Delta d + \Delta\bar{d} + \ldots$. We see that the small-$x$ evolution of $\Delta\Sigma(x, Q^2)$ could offer a moderate to significant enhancement to the quark spin, depending on where in x the effects set in and on the parameterization of the helicity PDFs at higher x.

Again we emphasize that the first principles results of this work can be included in future extractions of helicity PDFs, especially once the present large-N_c approximation is relaxed, which will provide strong constraints on the small-x behavior of the quark and gluon spin.

Saturation effects may also impact the amount of spin carried by small-x quarks and gluons. The small-x asymptotics of Δq and ΔG found here are such that $x\Delta G \to 0$ and $x\Delta q \to 0$ as $x \to 0$. Hence the helicity PDFs will not violate unitarity at small x. However, as one can see from the helicity evolution equations including (LLA) saturation effects, as derived in [1, 3], saturation would completely suppress the small-x evolution of helicity PDFs, making the effective α_h^q and α_h^G zero in the saturation region (cf. [9] for the flavor non-singlet unpolarized quark distribution). Therefore, a very small amount of the proton spin should reside in the saturation region. This observation can become an important component of the future small-x helicity PDF phenomenology.

Acknowledgments

This material is based upon work supported by the U.S. Department of Energy, Office of Science, Office of Nuclear Physics under Award Number DE-SC0004286 (YK), within the framework of the TMD Topical Collaboration (DP), and DOE Contract No. DE-AC52-06NA25396 (MS). MS received additional support from the U.S. Department of Energy, Office of Science under the DOE Early Career Program.

References

[1] Y.V. Kovchegov, D. Pitonyak, M.D. Sievert, JHEP **01**, 072 (2016), 1511.06737
[2] Y.V. Kovchegov, D. Pitonyak, M.D. Sievert, Phys. Rev. Lett. **118**, 052001 (2017), 1610.06188
[3] Y.V. Kovchegov, D. Pitonyak, M.D. Sievert, Phys. Rev. **D95**, 014033 (2017), 1610.06197
[4] Y.V. Kovchegov, D. Pitonyak, M.D. Sievert (2017), 1703.05809
[5] Y.V. Kovchegov, D. Pitonyak, M.D. Sievert (2017), 1706.04236
[6] P.J. Mulders, R.D. Tangerman, Nucl. Phys. **B461**, 197 (1996), [Erratum: Nucl. Phys.B484,538(1997)], hep-ph/9510301
[7] C.J. Bomhof, P.J. Mulders, F. Pijlman, Eur. Phys. J. **C47**, 147 (2006), hep-ph/0601171
[8] D. de Florian, R. Sassot, M. Stratmann, W. Vogelsang, Phys. Rev. Lett. **113**, 012001 (2014), 1404.4293
[9] K. Itakura, Y.V. Kovchegov, L. McLerran, D. Teaney, Nucl. Phys. **A730**, 160 (2004), hep-ph/0305332

The Tensor and the Scalar Charges of the Nucleon from Hadron Phenomenology

A. Courtoy[1,⋆]

[1] *Instituto de Física, Universidad Nacional Autónoma de México*
Apartado Postal 20-364, Ciudad de México 01000, México

Abstract. We discuss the impact of the determination of the nucleon tensor charge on searches for physics Beyond the Standard Model. We also comment on the future extraction of the subleading-twist PDF $e(x)$ from Jefferson Lab soon-to-be-released Beam Spin Asymmetry data as well as from the expected data of CLAS12 and SoLID, as the latter is related to the scalar charge. These analyses are possible through the phenomenology of Dihadron Fragmentation Functions related processes, which we report on here as well.

1 Introduction

In these proceedings, we discuss some properties of the nucleons that are of interest for both hadronic physics and physics Beyond the Standard Model (BSM): their tensor and the scalar charges. Both can, in principle, be accessed through the phenomeology of Parton Distribution Functions. The latter are related to the charges via specific sum rules. Due to their chiral-odd nature, the respective parton distributions are involved in processes together with another non-perturbative chiral-odd partner. In semi-inclusive DIS, that partner is a fragmentation function. So to examine collinear/1-dimensional parton distributions, we will consider Dihadron Fragmentation Functions (DiFF), which encode information about the fragmentation process of a quark into a hadron pair (plus something else, undetected). Fragmentation has been hardly studied through models for pion pairs, the main knowledge about those DiFFs being for now fits from data.

The rich phenomenology associated to dihadron fragmentation compares, though with less statistics, to the well-known Semi-Inclusive Deep Inelastic Scattering (SIDIS) in which a single hadron is produced in the current fragmentation region. Once the distribution functions are extracted through SIDIS asymmetries, their x-dependence can be integrated to obtain the charges, as dictated by the sum rules. We further discuss the future of dihadron data and their impact on physics BSM.

2 The Dihadron Way to Distribution Functions

The phenomenological knowledge on DiFFs is based on $e^+ - e^-$ annihilation at Belle. The unpolarized DiFF has been extracted by fitting the pion pair distribution simulated by a Monte Carlo event generator [1, 2] to compensate for the absence of data on multiplicities at that time. The resulting error on

⋆e-mail: aurore@fisica.unam.mx

© The Authors, published by EDP Sciences. This is an open access article distributed under the terms of the Creative Commons Attribution License 4.0 (http://creativecommons.org/licenses/by/4.0/).

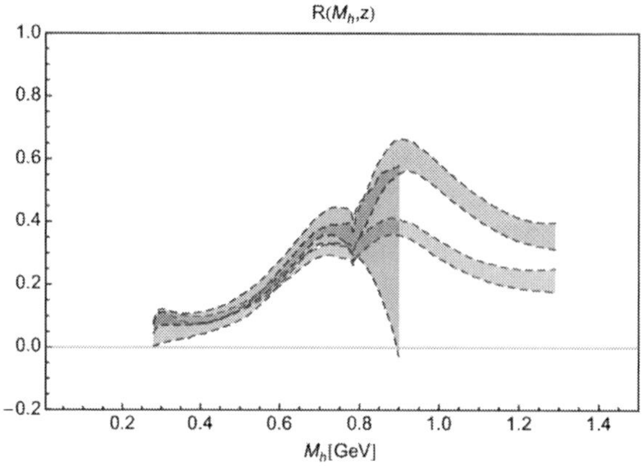

Figure 1. The ratio $R(z, M_h)$ as a function of M_h at $Q_0^2 = 1$ GeV^2 for three different $z = 0.25$ (shortest band), $z = 0.45$ (lower band at $M_h \sim 1.2$ GeV), and $z = 0.65$ (upper band at $M_h \sim 1.2$ GeV), with the value $\alpha_s(M_Z^2) = 0.139$ used in the QCD evolution equations.

the parametrization of the unpolarized DiFF is obviously small. On the other hand, once the unpolarized DiFF parameterized, the chiral-odd DiFF has been extracted from the Artru–Collins asymmetry at Belle. The factorization of the process $e^+e^- \to (\pi^+\pi^-)_{\rm jet}(\pi^+\pi^-)_{\overline{\rm jet}}X$ is valid in the kinematical regime $P_h^2 = M_h^2 \ll Q^2 = -q^2 \geq 0$ and $q = k - k'$ is the space-like momentum transferred, with the total momentum $P_h = P_1 + P_2$.

Belle data for the Artru–Collins asymmetry led to the extraction —and, hence, parameterization— of the leading-twist DiFFs [1, 2], illustrated in Fig. 1 through their ratio

$$R(z, M_h) = \frac{|\mathbf{R}|}{M_h} \frac{H_1^{\sphericalangle u}(z, M_h; Q_0^2)}{D_1^u(z, M_h; Q_0^2)} , \tag{1}$$

the relevant variables being the pion pair invariant mass, M_h, and the sum of the fractional energies carried by the two final hadrons, z.

While the first fitting approach was based on the usual Hessian statistics with a variation of the chisquare equal to unity [1], the newest version of the fitting procedure incorporates a Neural Network preparation, *i.e.* the normal generation of data replicas within the —1σ— experimental errors [2]. This second technique insures the proper treatment of errors outside the data range, thus avoiding the vanishing of the errors at the endpoints, where they should be maximal. We have noticed that both fitting techniques, still based on adhoc functional forms, agree with each other as far as the chiral-odd DiFF is concerned.

Combining the parametrization obtained for DiFFs to HERMES and COMPASS data on dihadron SIDIS off transversly polarized target allowed for the extraction of the last leading-twist PDF, *i.e. the transversity PDF* [2–4]. The structure functions can be written as products of distribution functions and DiFFs:

$$F_{UU} = x \sum_q e_q^2 f_1^q(x; Q^2)\, D_1^q(z, \cos\theta, M_h; Q^2) , \tag{2}$$

$$F_{UT}^{\sin(\phi_R+\phi_S)} = \frac{|\mathbf{R}| \sin\theta}{M_h} x \sum_q e_q^2 h_1^q(x; Q^2)\, H_1^{\sphericalangle q}(z, \cos\theta, M_h; Q^2) , \tag{3}$$

where e_q is the fractional charge of a parton with flavor q and the distribution functions $f_1^q(x)$ and $h_1^q(x)$ are, respectively, the unpolarized and the transversity PDFs.

While the fit of the transversity is an achievement by itself, though affected of a large uncertainty outside the data kinematical range, its first Mellin moment is also of great interest. The tensor charge is obtained by integrating the transversity PDF over the physical support in x,

$$\delta q_v(Q^2) \quad = \quad \int_0^1 dx\, h_1^{q_v}(x, Q^2) \,. \tag{4}$$

The error we mentioned above, consequence of the extrapolation of the PDF outside the data range, is the main source of uncertainty on the determination of the tensor charge. For example, going from a truncated integration over x to a full integration increases the central value up to a factor of 2 and the error up to a factor of 3 for the up quark and factors of 4 and 7 for the down quark tensor charge respectively, as can be seen in Table 3 of Ref. [2]. Another source of error consists in the choice of the functional form for the transversity PDF. This problem has been taken care of, in a first step, considering three different functional forms, each with a growing number of free parameters. The value for the isovector tensor charge $g_T = \delta u_v - \delta d_v$ for the functional form related to the so-called *flexible scenario* with $\alpha_s(M_Z^2) = 0.125$ is, at 1σ,

$$g_T = 0.81 \pm 0.44 \quad \text{at} \quad Q^2 = 4\,\text{GeV}^2 \quad . \tag{5}$$

It is in agreement with lattice determinations as well as with the other extractions from hadronic phenomenology [5–8], though the absolute value is slightly smaller than the lattice's. Due to its non-perturbative nature, the nucleon structure can only be unveiled using complementary methods such as effecive field theories, lattice calculations, models for the nucleon structure, Schwinger-Dyson based techniques and phenomenological extraction of observables from data. As such, the lattice field theory and the data based determinations play a similar, yet uncorrelated, role.

With the future JLab data, we will analyze the 4-dimensional binning spanning a wide kinematical range in (z, M_h, x, Q^2), for both proton and deuteron targets, allowing for a flavor separation of the valence distributions. The development expected from such a measurement and statistical analysis will qualitatively improve the present estimates for a number of processes, *e.g.* we expect about 10% improvement in the errorbars with CLAS12 and SoLID data [9]. We would like to restate that DiFFs are universal and can, as such, be transported to, *e.g.*, SIDIS or proton-proton asymmetries. This characteristic has recently been studied and reinforced through a confirmation on the role of transversity in proton-proton collisions at RHIC [10].

Though in a more elusive way, the scalar charge is related to a twist-3 PDF, $e(x)$. QCD equations of motion allow to decompose the chiral-odd twist-3 distributions into three terms

$$e^q(x) \quad = \quad e^q_{\text{loc}}(x) + e^q_{\text{tw-3}}(x) + e^q_{\text{mass}}(x) \quad . \tag{6}$$

The first term comes from the local operator and is exactly the contribution related to scalar charge,

$$\begin{aligned} e^q_{\text{loc}}(x) &= \frac{1}{2M} \int \frac{d\lambda}{2\pi}\, e^{i\lambda x} \langle P|\bar{\psi}_q(0)\psi_q(0)|P\rangle \quad , \\ &= \frac{\delta(x)}{2M} \langle P|\bar{\psi}_q(0)\psi_q(0)|P\rangle \quad ; \end{aligned} \tag{7}$$

the second term is a genuine twist-3 contribution, *i.e.* pure quark-gluon interaction term ; while the last term is related to the current quark mass.

The sum rule related the chiral-odd twist-3 PDF to the scalar charge is similar to Eq. (4). Only the *local* term contributes to the first Mellin moment, *i.e.* the scalar charge is nothing else than $e(x = 0)$.

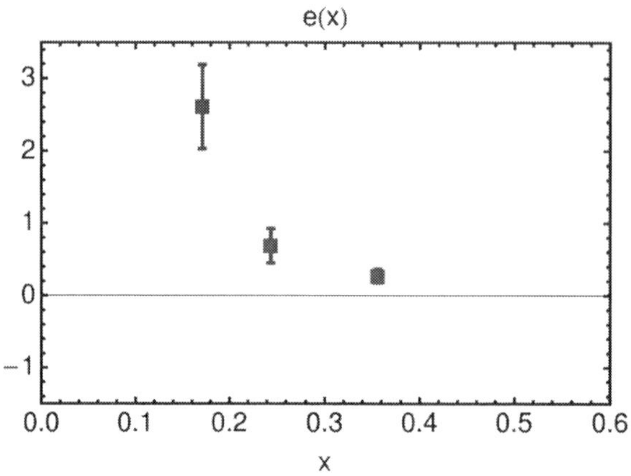

Figure 2. The valence combination $e(x) \equiv 4e^{u_V}(x_i, Q_i^2)/9 - e^{d_V}(x_i, Q_i^2)/9$ (see text). The error bars correspond to the propagation of the experimental and DiFF errors.

That delta-function singularity follows from chiral symmetry and the existence of non-vanishing quark condensate [11].

The chiral-odd twist-3 PDF is accessible in single- [12] and di-hadron SIDIS. In the latter case, the chiral-odd partner of $e(x)$ is the chiral-odd DiFF:

$$F_{LU}^{\sin\phi_R} = -\frac{|\mathbf{R}|\sin\theta}{Q}\, x \sum_q e_q^2 \left[\frac{M}{m_{hh}}\, x\, e^q(x;Q^2)\, H_1^{\sphericalangle q}(z,\cos\theta,M_h;Q^2) + \frac{f_1^q(x;Q^2)}{z}\, \widetilde{G}^{\sphericalangle q}(z,\cos\theta,M_h;Q^2)\right], \tag{8}$$

where the second term of the structure function contains the direct product of the unpolarized PDF and an –unknown– twist-3 DiFF.

An extraction of the twist-3 PDF, $e(x)$, through the analysis of the preliminary data [13] for the $\sin\phi_R$-moment of the beam-spin asymmetry for dihadron Semi-Inclusive DIS at CLAS at 6 GeV was proposed in Ref. [14]. It is illustrated in Fig. 2 in a scenario in which the asymmetry is dominated by the term containing the twist-3 PDF. We find a non-zero $e(x)$ even relaxing that hypothesis. The statistics do not allow for a proper fit for now, what restricts the possibility of extracting the scalar charge accurately.

Soon the CLAS12 data [15] will be analyzed and allow for more statistics and information on that particular function. The determination of the scalar charge will be a delicate task as the value of the function at $x = 0$ is experimentally unaccessible. The solution will be to appeal to other properties of the subleading-twist function together with advanced fitting techniques.

The evaluation of the scalar charge represents a challenge for the hadronic physics community but an opportunity to experimentally explore the consequence of chiral symmetry on the one hand.

3 Hadron phenomenology and BSM

The role of the scalar and tensor aspects of the nucleon becomes clear in beta decay where the hadronic characterization of the process is encoded through, among others, the scalar and tensor form factor — which value at zero momentum transfer is nothing else than the corresponding charge. While there is no leptonic counterpart for both structures in the Standard Model, all quark bilinear Lorentz structures can be introduced through an effective Lagrangian which is relevant for beta decay observables. The beta decay observables, *e.g.* b_0^+ and the Fierz term b, involve products of the Beyond the Standard Model (BSM) couplings, ϵ_i, and the corresponding hadronic charges, g_i. The scalar (S) and tensor (T) operators, in particular, contribute linearly to the beta decay parameters through their interference with the SM amplitude, and they are, therefore, more easily detectable. We focus here on the following low-energy effective interaction (see Ref. [16] for a review)

$$\Delta\mathcal{L}_{\rm eff} \;=\; -C_S\,\bar{p}n\,\bar{e}(\mathbb{I}-\gamma_5)\nu_e - C_T\,\bar{p}\sigma_{\mu\nu}n\,\bar{e}\sigma^{\mu\nu}(\mathbb{I}-\gamma_5)\nu_e \quad , \tag{9}$$

where $C_S = G_F V_{ud}\sqrt{2}\epsilon_S g_S$, and $C_T = 4G_F V_{ud}\sqrt{2}\epsilon_T g_T$ and $G_F \equiv \sqrt{2}g^2/(8M_W^2)$ is the tree-level definition of the Fermi constant. In the SM, the ϵ_S and ϵ_T coefficients vanish leaving the well-known $(V-A)\times(V-A)$ structure generated by the exchange of a W boson.

Precise measurements in beta decay set bounds on the scalar coupling and on the plane $\epsilon_S - \epsilon_T$ through, *i.e.*, b_0^+ from super-allowed Fermi transition as well as the Fierz term b. The latter can be expressed in terms of the factorized coupling [17]

$$b \;=\; \frac{2}{1+3\lambda^2}\left[g_S\epsilon_S - 12 g_T\epsilon_T\lambda\right] < 10^{-3} \quad , \tag{10}$$

with $\lambda = g_A/g_V$. The precision limit on the Fierz term corresponds to future experiments expectations, so does the bound on b_0^+ given at 68% CL in Ref. [18]:

$$b_0^+ \;=\; -2g_S\epsilon_S < 0.0028(26) \quad . \tag{11}$$

To isolate the tensor coupling from Eq. (10), in Refs. [19, 20], a global fit has been performed, establishing strong bounds on the combination $g_T\epsilon_T$:

$$|g_T\epsilon_T| < 6\cdot 10^{-4} \qquad (90\%\ \mathrm{CL})\,, \tag{12}$$

which is expected to be improved by the next generation of experiments.

In order to extract a bound on the Wilson coefficients $\epsilon_{S,T}$ from Eqs. (11, 12) it is necessary to know the value of the hadronic charge $g_{S,T}$. Hence, the sensitivity of beta decay measurements to an exotic scalar or tensor interaction depends on our knowledge of the scalar or tensor charge, respectively. The impact of data based determinations of the tensor charge has been studied in Ref. [9]. Given the sensitivity bounds on the Fierz term b, an error of about 10 – 15% on the tensor charge is estimated to be sufficient. The values for g_T mentioned above are still far from that precision.

Originally, in Ref. [9], an alternative to the standard Hessian evaluation of the errors was considered. The argument that both the lattice QCD and experimental extractions of the couplings from β-decay are affected by systematic or theoretical uncertainty invalidates the assumption of a gaussian distribution of the error around the central value. In order to deal with this situation we followed Ref. [17] and we calculated the confidence interval on ϵ_T using the so-called R-Fit method [21]. In this scheme the theoretical likelihoods do not contribute to the χ^2 of the fit and the corresponding QCD parameters take values within certain "allowed ranges". In our case, this means that g_T is restricted to remain inside a given interval, e.g. $0.37 \leq g_T \leq 1.25$ for the current determination from di-hadron SIDIS. The chisquare function is then given by

$$\chi^2(\epsilon_T) = \min_{g_T}\left(\frac{\left[g_T\epsilon_T\right]^{\rm exp} - g_T\epsilon_T}{\delta\left[g_T\epsilon_T\right]^{\rm exp}}\right)^2 \quad , \tag{13}$$

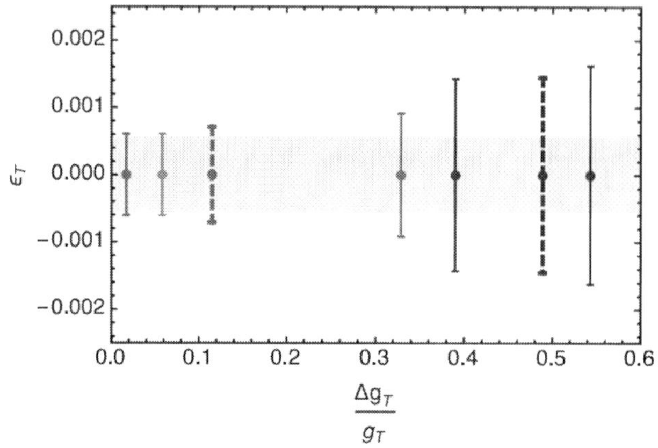

Figure 3. Bounds on ϵ_T obtained from precision measurements of beta decay using all current extractions and lattice QCD evaluations of the tensor charge g_T, plotted vs. the relative error, $\Delta g_T/g_T$: (turqoise) Lattice QCD [23, 24]; (yellow) Lattice QCD [25]; (green) Deeply virtual π^o and η production [6]; (blue) single pion SIDIS [5]; (red) dihadron SIDIS [2]. The dashed lines are future projections based on JLab's upgrade. All results were obtained using in the definition of $\Delta g_T/g_T$, each individual evaluation's g_T. The grey band gives the error assuming $\Delta g_T = 0$, and the average g_T of the three hadronic analyses considered in Ref. [9]. The lattice evaluations from Refs. [23, 24] are indistinguishable.

where the minimization is performed varying g_T within its allowed range. In this approach, the bound on ϵ_T depends only on the lower limit of the tensor charge, as long as the experimental determination of $g_T\epsilon_T$ is compatible with zero at 1σ. In particular, the tensor charge given by Eq. (5) leads to a bound $|\epsilon_T| < 0.00162$. This bound is the largest found through hadron phenomenology (see Fig. 3) but the error on g_T is expected to decrease of about 10% with the new JLab data. For comparison, the bound obtained from the analysis of LHC data carried out in Ref. [22] is $|\epsilon_T| < 0.0013$.

This error analysis presents shortcomings, *e.g.* the fact that Eq. (13) only depends on the lower bound on the tensor charge. The principal criticism would be that the R-fit method is a global method for data fitting, used here for the determination of a single parameter based on a fit output, Eq. (12). Alternative error treatments will be consider in an extensive study of the hadronic structure impact on physics BSM.

On the other hand, a fully 2-dimensional analysis of the $\epsilon_S - \epsilon_T$ plane will be possible with more data for the scalar charge related observables.

4 Conclusions

The phenomenology of Dihadron Fragmentation Functions integrates the idea that more structures and information become accessible when incorporating a dependence on the transverse momentum. The elegance of DiFF results in less intricated and more easily taken care of observables, namely collinear objects and simple products, as opposed to convolutions. This particular feature allows for more flexibility in their statistical analysis. The parameterization of the third leading-twist PDF, *transversity*, marked the first stage of an alternative approach to transverse momentum dependent functions. Then followed the tensor charge determination, made possible thanks to the flavor structure of the analyzed observables. Soon, data for subleading processes involving DiFF will be released, for which we will be able to consider the study of twist-3 PDFs.

The latter encode invaluable information on the low energy dynamics of quarks and gluons, so far hardly explored. The nonperturbative nature of nucleons influences numerous observables that involve them. The possibility of obtaining the scalar and tensor charges directly from experiment with sufficient precision, not having to deal with purely theoretical uncertainties of these quantities, gives a different weight to searches for BSM involving nucleons.

The results presented here concerning the determination of the scalar and tensor charges through Dihadron Fragmentation Functions can be complemented, in the tensor case, by the Deeply Virtual Meson Production analysis [6] and the Single-hadron SIDIS approach [8].

References

[1] A. Courtoy, A. Bacchetta, M. Radici and A. Bianconi, Phys. Rev. D **85** (2012) 114023

[2] M. Radici, A. Courtoy, A. Bacchetta and M. Guagnelli, JHEP **1505** (2015) 123

[3] A. Bacchetta, A. Courtoy and M. Radici, Phys. Rev. Lett. **107** (2011) 012001 doi:10.1103/PhysRevLett.107.012001 [arXiv:1104.3855 [hep-ph]].

[4] A. Bacchetta, A. Courtoy and M. Radici, JHEP **1303** (2013) 119

[5] M. Anselmino, M. Boglione, U. D'Alesio, S. Melis, F. Murgia and A. Prokudin, Phys. Rev. D **87** (2013) 094019

[6] G. R. Goldstein, J. O. G. Hernandez and S. Liuti, arXiv:1401.0438 [hep-ph].

[7] Z. B. Kang, A. Prokudin, P. Sun and F. Yuan, Phys. Rev. D **93** (2016) no.1, 014009

[8] Z. Ye *et al.*, Phys. Lett. B **767** (2017) 91 doi:10.1016/j.physletb.2017.01.046 [arXiv:1609.02449 [hep-ph]].

[9] A. Courtoy, S. Baessler, M. González-Alonso and S. Liuti, Phys. Rev. Lett. **115** (2015) 162001

[10] M. Radici, A. M. Ricci, A. Bacchetta and A. Mukherjee, Phys. Rev. D **94** (2016) no.3, 034012

[11] M. Wakamatsu and Y. Ohnishi, Phys. Rev. D **67** (2003) 114011 doi:10.1103/PhysRevD.67.114011 [hep-ph/0303007].

[12] A. V. Efremov, K. Goeke and P. Schweitzer, Phys. Rev. D **67** (2003) 114014

[13] S. Pisano [CLAS Collaboration], EPJ Web Conf. **73** (2014) 02008.

[14] A. Courtoy, arXiv:1405.7659 [hep-ph].

[15] S. Pisano, A. Courtoy et al.. JLab. Exp. E12-06-112B/E12-09-008B. http://www.jlab.org/exp_prog/proposals/14/E12-06-112B_E12-09-008B.pdf

[16] V. Cirigliano, S. Gardner and B. Holstein, Prog. Part. Nucl. Phys. **71** (2013) 93 doi:10.1016/j.ppnp.2013.03.005 [arXiv:1303.6953 [hep-ph]].

[17] T. Bhattacharya, V. Cirigliano, S. D. Cohen, A. Filipuzzi, M. Gonzalez-Alonso, M. L. Graesser, R. Gupta and H. W. Lin, Phys. Rev. D **85** (2012) 054512

[18] J. C. Hardy and I. S. Towner, Phys. Rev. C **79** (2009) 055502

[19] F. Wauters, A. García and R. Hong, Phys. Rev. C **89** (2014) no.2, 025501 Erratum: [Phys. Rev. C **91** (2015) no.4, 049904]

[20] R. W. Pattie, K. P. Hickerson and A. R. Young, Phys. Rev. C **88** (2013) 048501 Erratum: [Phys. Rev. C **92** (2015) no.6, 069902]

[21] A. Hocker, H. Lacker, S. Laplace and F. Le Diberder, Eur. Phys. J. C **21** (2001) 225

[22] O. Naviliat-Cuncic and M. González-Alonso, Annalen Phys. **525** (2013) 600

[23] J. R. Green, J. W. Negele, A. V. Pochinsky, S. N. Syritsyn, M. Engelhardt and S. Krieg, Phys. Rev. D **86** (2012) 114509 doi:10.1103/PhysRevD.86.114509 [arXiv:1206.4527 [hep-lat]].

[24] T. Bhattacharya, S. D. Cohen, R. Gupta, A. Joseph, H. W. Lin and B. Yoon, Phys. Rev. D **89** (2014) no.9, 094502 doi:10.1103/PhysRevD.89.094502 [arXiv:1306.5435 [hep-lat]].

[25] G. S. Bali *et al.*, Phys. Rev. D **91** (2015) no.5, 054501 doi:10.1103/PhysRevD.91.054501 [arXiv:1412.7336 [hep-lat]].

Are PDFs still consistent with Tevatron data?

Zack Sullivan[1,⋆]

[1] *Department of Physics, Illinois Institute of Technology, Chicago, Illinois 60616-3793, USA*

Abstract. As active data taking has moved to the LHC at CERN, more and more LHC data have been included into fits of parton distribution functions. An anomaly has arisen where formerly excellent agreement between theoretical predictions and experiment in single-top-quark production at the Tevatron is no longer quite as good. Is this indicative of a deeper issue?

1 Introduction

In the years leading up to its discovery at the Fermilab Tevatron [1, 2] t-channel single-top-quark production played a pivotal role in the development and understanding of improved perturbation theory and heavy quark parton distribution functions (PDFs). The analytic connection between this process and deeply inelastic scattering (DIS) couples single-top to choices made in the extraction of PDFs that imply significant constraints on higher order corrections and scale choices. Furthermore, large logarithms that appear in intermediate steps of next-to-leading order (NLO) and next-to-next-to-leading order (NNLO) calculations undergo delicate cancellations that magnify any errors in the PDFs. Hence, t-channel single-top-quark production is one of the most constraining processes on the consistency PDF fits at different orders, and it directly tests the analytic framework of improved perturbation theory.

This paper introduces how modern PDF sets are failing some of the stringent analytic tests set by t-channel single-top-quark production, and speculates on some possible reasons for the failures. In Sec. 2 I explain the constraints imposed by this process and why it is such a stringent test. In Sec. 3 I demonstrate how past PDF sets were successful (or failed in calculable ways), but current PDFs from three collaborations CTEQ, NNPDF, and HERAPDF fail by upwards of 5σ. In Sec. 4 I explore possible explanations, and suggest that we may be seeing first hints that our framework needs to be improved. I conclude with recommendations for where to proceed to solve these issues.

2 t-channel single top, scales, and large logarithms

Data from the CERN Large Hadron Collider (LHC) has become precise enough to require significantly improved theoretical calculations. While the state-of-the-art calculations have shifted to fully differential NNLO, few checks have been performed to confirm that the framework and all of its pieces are both self-consistent and consistent with all previous data. The NNLO calculation of

⋆e-mail: Zack.Sullivan@IIT.edu

© The Authors, published by EDP Sciences. This is an open access article distributed under the terms of the Creative Commons Attribution License 4.0 (http://creativecommons.org/licenses/by/4.0/).

t-channel single-top-quark production with a stable top quark [3] (see Fig. 1 for a leading-order figure) was recently updated to include NNLO decays of the top quark [4]. When coupled to NNLO PDFs, this enables a comparison to experimental data at the Tevatron and LHC with small theoretical uncertainties.

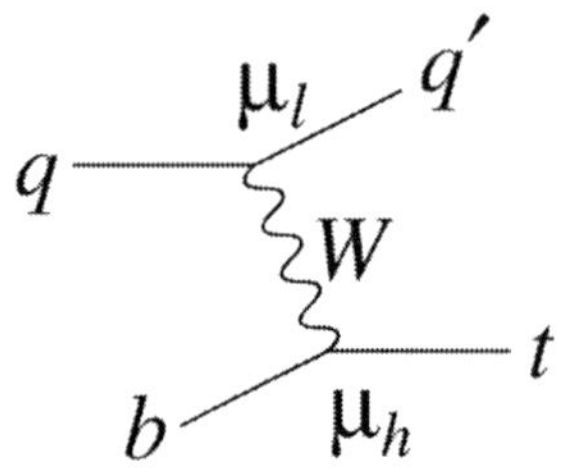

Figure 1. Leading order Feynman diagram for t-channel single-top-quark production. The light-quark and heavy-quark sides of the diagram factorize with independent scale μ_l and μ_h, respectively.

There is only one problem with the calculation in Refs. [3, 4]: both were performed at fixed factorization μ_F and renormalization μ_R scales set to the top quark mass m_t. Why is this a problem? Isn't it true that theoretical calculations should be independent of scale choice (or at least push the effect to higher orders)? Not in this case.

t-channel single-top-quark production is a special process in that it is analytically identical to double deeply-inelastic-scattering (DDIS) through NLO [5, 6]. The reason is simply that you cannot exchange a single gluon between the light-quark line (q–q' in Fig. 1) and the heavy-quark line (b–t in Fig. 1) and conserve color. In principle you could exchange two gluons at NNLO, but Refs. [3, 5, 7] argue that this is numerically a tiny effect, and so preserves an effective separation between sides of the diagram. Hence, you have DDIS with two sets of independent scales: μ_l and μ_h for the light- and heavy-quark lines, respectively.

This critical feature of this factorization is that when you calculate DIS you must choose the same scale to evaluate your calculation as was chosen to extract the PDFs from DIS data. Otherwise, you do not get back to the data. This means there is a unique choice for t-channel single-top: $\mu_l = Q^2$ and $\mu_h = Q^2 + m_t^2$, where Q^2 is the virtuality of the W boson, and the heavy-quark line sees the heavy-quark version of DIS [5]. A further observation is since DIS data is used to fit the PDF, the inclusive cross sections calculated at LO, NLO, or NNLO must all be identical — again you are just undoing the fit at each order. This is a very powerful constraint on the PDF fits.

A second important aspect of t-channel single-top-quark production is related to its name. If one tries to work in a 4-flavor scheme where there is no b quark in the initial state, a large logarithmic divergence of the form

$$\alpha_s \ln\left(\frac{Q^2 + m_t^2}{m_b^2}\right) \tag{1}$$

appears at every order in the perturbative series [5]. Numerically this term is close to 1. Hence, you resum these large logarithms into a b PDF via the DGLAP equation [5]

$$b(x,\mu^2) = \frac{\alpha_s(\mu^2)}{2\pi} \ln\left(\frac{\mu^2}{m_b^2}\right) \int_x^1 \frac{dz}{z} P_{bg}(z) g\left(\frac{x}{z},\mu^2\right). \tag{2}$$

This creates an improved perturbation series in which b degrees of freedom are an intrinsic part of proton structure. In NLO and NNLO calculations, careful subtraction of the resummed logarithms

Table 1. Inclusive LO and NLO t-channel single-top-quark cross section ($t + \bar{t}$) calculated for run II of the Tevatron ($\sqrt{S} = 1.96$ TeV) with various PDFs.

PDF	LO (pb)	NLO (pb)	Notes
CTEQ 4L/4M	2.26	2.41	6% deviation, known α_s bug
CTEQ 5L/5M1	2.08	2.07	< 0.5% (bug fixed)
CTEQ 6L1/6M	2.07	2.086	< 0.5%
CTEQ 6L1/6M	1.83	2.086	Scales set to m_t, 12% off as expected
CTEQ 14 llo/nlo	2.39	2.00	20% LO–NLO deviation!
HERAPDF 1.5 lo/nlo	1.965	1.798	9.3% deviation!
HERAPDF 2.0 lo/nlo	1.910	1.762	8.4% deviation, NLO 12% too small
NNPDF 3.0 lo/nlo	2.33	2.21	5.4% deviation, NLO 10% too big

leads to rapidly convergent series. However, any small mistake in either the calculation or PDF input (e.g., through faulty DGLAP evolution or poor fits) reintroduces these large logarithms, and formerly delicate cancellations that occur to enforce the equality of the inclusive cross sections between orders lead to large measurable deviations.

t-channel single-top-quark production is one of the most sensitive processes for testing the idea of heavy quark PDFs, the resummation framework for improved perturbation theory, and our mapping of the degrees of freedom of the proton onto universal parton distribution functions.

3 The failure of modern LO and NLO PDFs

Before updating the NNLO calculation to utilize the DDIS scales, I first updated the calculation of the inclusive t-channel cross section with modern PDFs using the public code ZTOP [6]. The ZTOP code was recently interfaced to the new LHAPDF [9] standard for accessing PDFs, and is the standard for LO and NLO t-channel production. For this study I focus on results for the Tevatron run II at 1.96 TeV, where a combined CDF and DØ analysis exists [10].

In the top half of Tab. 1 I show the results for inclusive single top production at the Tevatron calculated with older CTEQ PDFs [11–13] for which there are no NNLO fits. In order to be able to compare with Ref. [10], I use a top quark mass $m_t = 172.5$ GeV and the DDIS scales ($\mu_l = Q^2$, $\mu_h = Q^2 + m_t^2$). LO means a LO matrix element, with LO PDFs, and $\alpha_s(M_Z) = 0.130$. NLO means a NLO matrix element, with NLO PDFs, and $\alpha_s(M_Z) = 0.118$.

The CTEQ 4 PDF sets had a 6% disagreement between LO and NLO, however a bug in the running of α_s was discovered that accounted for the effect. The bug was fixed in CTEQ 5 with little change to the data in the PDF fits, and the LO and NLO agree within the Monte Carlo numerical integration. Everything continued to agree with CTEQ 6, and other distributions not shown, and the CTEQ 6 central value is identical to the combined Tevatron fit [10]. One observation from Tab. 1 is that choosing the top quark mass as a scale (as is done in Refs. [3, 4] and MCFM [14]) induces a large predictable 12% shift between orders. While the scale dependence formally decreases at higher orders, as it should, there is a shift with respect to the PDF extraction by a term proportional to $\ln(Q^2 + m_t^2/m_t^2)$. The DDIS scales are necessary here.

In the bottom half of Tab. 1 I show LO and NLO results for the inclusive cross section using modern PDFs, for which NNLO fits exist. Nothing agrees between LO and NLO, and the NLO calculations between PDF sets are inconsistent. Beginning with the latter point, the expected 90% confidence level NLO PDF uncertainty for this process is +8.8 − 7.3% as calculated with CTEQ 14 [15] PDFs — this is consistent with error calculations using CTEQ 6, HERAPDF, and NNPDF 3.0.

However, NNPDF [16] and HERAPDF [17] disagree with each other at NLO by 5σ, with CTEQ 14 splitting the difference! The Tevatron data [10] is not precise enough to rule out NNPDF and HERAPDF, but there is significant tension with the single top data.

While the NLO discrepancies are serious, the formal question addressed here — the consistency between LO and NLO – is a disaster. The CTEQ 14 PDFs, which are closest to the data at NLO, have a 20% deviation from LO when the difference should be zero. NNPDF 3.0 is shown as having a 5.4% deviation, but this is an artifact of choosing $m_t = 172.5$ GeV. If $m_t = 175$ GeV the discrepancy also grows to double-digits. Even HERAPDF 1.5 and 2 are off by almost 10%. No modern PDF gets back to the data. One might be tempted to ask these same questions at the LHC where there is more data. Unfortunately, all effects effects are suppressed due to accidental numerical cancellations that occur at 13 TeV and the typical x and Q^2 regions probed at the LHC.

4 What is wrong with the PDFs?

There are a large number of opportunities for errors to slip into any given set of calculations: from computer code bugs to the addition of inconsistent data. The first place to check for problems is to check the codes used for calculation. ZTOP has been continuously checked for over 13 years, and reproduces old results exactly. One bug that was found early in this comparison was, like CTEQ 4, LHAPDF 5 [18] miscalculated α_s when run in multisets mode.[1] This was corrected in LHAPDF 6. Another observation is that there are $< 0.1\%$ differences between PDFs returned from LHAPDF version of CTEQ 14, and the CTEQ interface to CTEQ 14. But be warned: in a NLO calculation, which has large numerical cancellations, it can take *millions* of Monte Carlo events to converge to the same result between the LHAPDF and CTEQ interfaces.

Another obvious possible solution to consider is that every PDF group has been sloppy in its LO fit. After all, the LO, NLO, and NNLO fits are independent, and there has not been a real emphasis on LO since 2002. This is something that needs to be examined by every group. It is difficult to trust the fitting procedures extended to higher orders if we cannot reproduce known results. However, it is logically possible that only NNLO (and maybe NLO) PDFs are valid, and for LO we must go back to using CTEQ 6L1.

Given that the disparities only arise in the newer PDF sets, the obvious questions to ask are what changed between 2002 and today, and why hasn't this been noticed before? The answer to the second question is that there has been a drought of LO PDFs for many years prior to the sets listed here. Hence, there was no opportunity to perform these checks with the newer fitting techniques. While calculations and fitting methods have evolved, the biggest change has been that the addition of jet data from the LHC, and final fits from HERA at DESY, have begun to heavily influence the fits of proton parton distribution functions.

Given the influx of new data sets in the fits I sought to determine what would happen if I removed the new data. NNPDF 3.0 includes sets that claim to fit the PDFs without LHC data (effectively no new jet data), or with HERA-only data. In the first part of Tab. 2 I recalculate at NLO using NNPDF 3.0 sets. The deviations do not change when LHC data is removed — which is curious, since jets data *should* change the NLO result; the gluon is best measured with jets data. A HERA-only fit by NNPDF seems even worse, which is strange given that HERA data is DIS-like. One caveat is that the NNPDF fit of HERA data is not consistent with the HERAPDF fit of HERA data shown in the second part of Tab. 2, so it should probably not be used for comparison. Unfortunately, there is no public LO version of the PDF fits with data sets removed. To truly make a comparison, the PDF groups need to provide the LO complement.

[1] Specifically, in LHAPDF 5 multiset mode, α_s was calculated for the first loaded PDF, and used for all other PDFs, even if the order should have been different.

Table 2. Inclusive LO and NLO t-channel single-top-quark cross section ($t + \bar{t}$) calculated for run II of the Tevatron ($\sqrt{S} = 1.96$ TeV) with variations from NNPDF and HERAPDF. Note: There are no LO variation sets — making direct comparison of LO to NLO difficult.

PDF	LO (pb)	NLO (pb)	Notes
NNPDF 3.0 lo/nlo	2.33	2.21	5% deviation
NNPDF 3.0 (no LHC)	—	2.22	5% deviation
NNPDF 3.0 (HERA)	—	2.10	11% deviation
HERAPDF 2.0 lo/nlo	1.910	1.762	8% deviation
HERAPDF 2.0 ("JETS")	—	1.830	4% — $+c$, dijets, α_s data

Turning to the HERAPDF fits, where we would have expected DIS data to map onto DDIS data, the story is more subtle. While the baseline HERAPDF 2.0 fit has an 8% deviation between orders, it does not include all of the relevant data. HERAPDF also performed an alternate fit called "JETS" that includes the charm final state (the most closely related to single-top), and jets data — which technically is differential DIS data. Both really should be included in the fit, and indeed the LO/NLO difference is cut in half in this case. Still missing is a LO version of the "JETS" fit.

There is a lot of room for simple mistakes to be causing the large discrepancies between orders, with less room to explain why at most one NLO calculation is correct. However, we should also be asking: is something deeper is occurring?

Hidden behind the fitting methods for proton structure is an assumption that the degrees of freedom in the proton map to the universal PDF shape each group uses in a clear way. There is good reason to believe this is not the case for the gluons that appear in t-channel single-top-quark production. The reason is that t-channel single-top involves pure $V - A$ interactions at all vertices in both production *and* decay. This is what generates the famous angular correlations between final state particles that was used to aid in discovery [1, 2, 8].

Current conservation means that only the left-handed polarization of the gluon is probed by t-channel single-top-quark.[2] DIS charged-current data that goes into PDF fits is also left-handed, though neutral-current data has a right-handed contribution. Jets data completely mix left- and right-handed gluons. Single-top sees a polarized gluon in an unpolarized proton. Could the numerical discrepancies we are seeing be a consequence of ignoring spin-dependent effects in the mapping of data onto PDF fit functions? One numerical hint comes from another color-singlet exchange process described at ISMD XLVI [19]: Higgs production sees 2–5% corrections from polarized gluons inside unpolarized protons. The large $\ln(m_t^2/m_b^2)$ terms which are losing their cancellations could easily boost this effect to 10–20%, while in most other processes the effect would be hidden. This is not a proof, but it is suggestive that an effect which does have to be there could be large enough to see.

5 Conclusions

Are PDFs still consistent with Tevatron data? Of course they are. The t-channel single-top-quark data from the Tevatron is not sufficient to place strong bounds on the b quark PDF. However, the calculations with modern PDFs that use significant amounts of LHC data as an input are much less in agreement than those that use older fits. Analytic constraints between LO and NLO PDFs are no longer satisfied, and NLO fits disagree with each other by up to 5σ. Something is clearly wrong, the question we now need to address is why is this failing?

[2]There is a negligible right-handed gluon contribution from the b propagator spin-flip that goes like $(m_b/m_t)^2$.

There are several avenues that should be followed to resolve the discrepancy discussed here. LO versions of fits with alternate data samples, such as removing LHC data or adding all HERA data, will help point toward which data sets might be driving the broken fits. Recalculation of the NNLO t-channel cross section using the DDIS scales will allow for validation of the NNLO fits. Will this be enough? It is possible the data is now so precise that we are seeing the breakdown of the way we map gluon degrees of freedom in the proton?

How might we use t-channel single-top-quark production as a foil for improving our PDFs in the future? One idea would be to use the analytic relationship between orders as a constraint on the fits. More speculatively, we may wish to consider fits are specialized to analytic constraints: e.g, fits of DIS data for DIS-like processes, fits of jets data for jets processes, etc. Ultimately, we should consider whether our data is now precise enough to consider alternate mappings of the gluon degrees of freedom in our fits that focus on these features. Perhaps we should be using spin-dependent PDFs even for unpolarized cross sections.

Acknowledgments

I would like to thank the ISMD organizers for an exceptional symposium and environment for lively discussions. This work was supported by the U.S. Department of Energy under award No. DE-SC0008347.

References

[1] T. Aaltonen *et al.* (CDF Collaboration), "First observation of electroweak single top-quark production," Phys. Rev. Lett. **103**, 092002 (2009) [arXiv:0903.0885 [hep-ex]].

[2] V. M. Abazov *et al.* (D0 Collaboration), "Observation of single top-quark production," Phys. Rev. Lett. **103**, 092001 (2009) [arXiv:0903.0850 [hep-ex]].

[3] M. Brucherseifer, F. Caola, and K. Melnikov, "On the NNLO QCD corrections to single-top production at the LHC," Phys. Lett. B **736**, 58 (2014) [arXiv:1404.7116 [hep-ph]].

[4] E. L. Berger, J. Gao, C.-P. Yuan, and H. X. Zhu, "NNLO QCD corrections to t-channel single top-quark production and decay," Phys. Rev. D **94**, 071501 (2016) [arXiv:1606.08463 [hep-ph]].

[5] T. Stelzer, Z. Sullivan, and S. Willenbrock, "Single-top-quark production via W–gluon fusion at next-to-leading order," Phys. Rev. D **56**, 5919 (1997) [hep-ph/9705398].

[6] Z. Sullivan, "Understanding single-top-quark production and jets at hadron colliders," Phys. Rev. D **70**, 114012 (2004) [hep-ph/0408049].

[7] B. W. Harris, E. Laenen, L. Phaf, Z. Sullivan, and S. Weinzierl, "Fully differential single-top-quark cross section in next-to-leading order QCD," Phys. Rev. D **66**, 054024 (2002) [hep-ph/0207055].

[8] Z. Sullivan, "Angular correlations in single-top-quark and Wjj production at next-to-leading order," Phys. Rev. D **72**, 094034 (2005) [hep-ph/0510224].

[9] A. Buckley, J. Ferrando, S. Lloyd, K. Nordström, B. Page, M. Rüfenacht, M. Schönherr, and G. Watt, "LHAPDF6: parton density access in the LHC precision era," Eur. Phys. J. C **75**, 132 (2015) [arXiv:1412.7420 [hep-ph]].

[10] T. A. Aaltonen *et al.* (CDF and D0 Collaborations), "Tevatron Combination of Single-Top-Quark Cross Sections and Determination of the Magnitude of the Cabibbo-Kobayashi-Maskawa Matrix Element $\mathbf{V_{tb}}$," Phys. Rev. Lett. **115**, 152003 (2015) [arXiv:1503.05027 [hep-ex]].

[11] H. L. Lai, J. Huston, S. Kuhlmann, F. I. Olness, J. F. Owens, D. E. Soper, W. K. Tung, and H. Weerts, "Improved parton distributions from global analysis of recent deep inelastic scattering and inclusive jet data," Phys. Rev. D **55**, 1280 (1997) [hep-ph/9606399].

[12] H. L. Lai, J. Huston, S. Kuhlmann, J. Morfin, F. Olness, J. F. Owens, J. Pumplin, and W. K. Tung (CTEQ Collaboration), "Global QCD analysis of parton structure of the nucleon: CTEQ5 parton distributions," Eur. Phys. J. C **12**, 375 (2000) [hep-ph/9903282].

[13] J. Pumplin, D. R. Stump, J. Huston, H. L. Lai, P. M. Nadolsky, and W. K. Tung, "New generation of parton distributions with uncertainties from global QCD analysis," J. High Energy Phys. **07** (2002) 012 [hep-ph/0201195].

[14] J. M. Campbell, R. K. Ellis and F. Tramontano, "Single top production and decay at next-to-leading order," Phys. Rev. D **70**, 094012 (2004) [hep-ph/0408158].

[15] S. Dulat *et al.*, "New parton distribution functions from a global analysis of quantum chromodynamics," Phys. Rev. D **93**, 033006 (2016) [arXiv:1506.07443 [hep-ph]].

[16] R. D. Ball *et al.* (NNPDF Collaboration), "Parton distributions for the LHC Run II," J. High Energy Phys. **04** (2015) 040 [arXiv:1410.8849 [hep-ph]].

[17] H. Abramowicz *et al.* (H1 and ZEUS Collaborations), "Combination of measurements of inclusive deep inelastic $e^{\pm}p$ scattering cross sections and QCD analysis of HERA data," Eur. Phys. J. C **75**, 580 (2015) [arXiv:1506.06042 [hep-ex]].

[18] M. R. Whalley, D. Bourilkov, and R. C. Group, "The Les Houches accord PDFs (LHAPDF) and LHAGLUE," hep-ph/0508110.

[19] D. Boer, "Multiple scattering aspects of gluon TMDs," presented at the XLVIth International Symposium on Multiparticle Dynamics, August 29–September 2, 2016, Jeju, South Korea.

The Weizsäcker-Williams distribution of linearly polarized gluons (and its fluctuations) at small x

Adrian Dumitru[1,2,3,★] and *Vladimir* Skokov[4,★★]

[1] *Department of Natural Sciences, Baruch College, CUNY, 17 Lexington Avenue, New York, NY 10010, USA*
[2] *The Graduate School and University Center, The City University of New York, 365 Fifth Avenue, New York, NY 10016, USA*
[3] *Physics Department, Brookhaven National Lab, Upton, NY 11973, USA*
[4] *RIKEN/BNL Research Center, Brookhaven National Laboratory, Upton, NY 11973, USA*

Abstract. The conventional and linearly polarized Weizsäcker-Williams gluon distributions at small x are defined from the two-point function of the gluon field in light-cone gauge. They appear in the cross section for dijet production in deep inelastic scattering at high energy. We determine these functions in the small-x limit from solutions of the JIMWLK evolution equations and show that they exhibit approximate geometric scaling. Also, we discuss the functional distributions of these WW gluon distributions over the JIMWLK ensemble at rapidity $Y \sim 1/\alpha_s$. These are determined by a 2d Liouville action for the logarithm of the covariant gauge function $g^2 \mathrm{tr}\, A^+(q)A^+(-q)$. For transverse momenta on the order of the saturation scale we observe large variations across configurations (evolution trajectories) of the linearly polarized distribution up to several times its average, and even to negative values.

1 Introduction

Dijet production in deep-inelastic $\gamma^* - A$ scattering at high energy can provide insight into the gluon fields of the nucleus in the regime of strong, non-linear fields [1]. At leading order a $q\bar{q}$ dijet is produced. Denote the average transverse momentum of the jets as $\vec{P} = (\vec{k}_1 - \vec{k}_2)/2$ and the transverse momentum imbalance as $q = \vec{k}_1 + \vec{k}_2$, where $\vec{k}_1$ and $\vec{k}_2$ are the transverse momenta of the two jets. In the "correlation limit" of roughly back to back jets [2] one has $P^2 \gg q^2$. In this limit the leading contribution (in powers of q^2/P^2) to the cross section can be obtained from Transverse Momentum Dependent (TMD) factorization. It predicts a distribution $xh^{(1)}(x, q^2)$ for linearly polarized gluons in an unpolarized target [3, 4] which gives rise to $\sim \cos 2\phi$ azimuthal anisotropies in dijet production [5–7], as well as in other processes [8–10]. ϕ is the angle between the transverse momentum vectors $\vec{P}$ and $\vec{q}$ (in a frame where neither the γ^* nor the hadronic target carry transverse momentum). The isotropic contribution to the dijet cross section is proportional to the conventional Weizsäcker-Williams (WW)

★e-mail: adrian.dumitru@baruch.cuny.edu
★★e-mail: vskokov@quark.phy.bnl.gov

© The Authors, published by EDP Sciences. This is an open access article distributed under the terms of the Creative Commons Attribution License 4.0 (http://creativecommons.org/licenses/by/4.0/).

gluon distribution $xG^{(1)}(x,q^2)$:

$$E_1E_2\frac{d\sigma^{\gamma_T^* A\to q\bar{q}X}}{d^3k_1d^3k_2d^2b} = \alpha_{em}e_q^2\alpha_s\delta\left(x_{\gamma^*}-1\right)z(1-z)\left(z^2+(1-z)^2\right)\frac{\epsilon_f^4+P^4}{(P^2+\epsilon_f^2)^4} \times\left[xG^{(1)}(x,q)-\frac{2\epsilon_f^2P^2}{\epsilon_f^4+P^4}\cos(2\phi)\,xh^{(1)}(x,q)\right], \quad (1)$$

$$E_1E_2\frac{d\sigma^{\gamma_L^* A\to q\bar{q}X}}{d^3k_1d^3k_2d^2b} = \alpha_{em}e_q^2\alpha_s\delta\left(x_{\gamma^*}-1\right)z^2(1-z)^2\frac{8\epsilon_f^2P^2}{(P^2+\epsilon_f^2)^4} \times\left[xG^{(1)}(x,q)+\cos(2\phi)\,xh^{(1)}(x,q)\right]. \quad (2)$$

z and $1-z$ are the momentum fractions of the quark and anti-quark, respectively, and $\epsilon_f^2 = z(1-z)Q^2$ (for massless quarks) with Q^2 the virtuality of the photon. Clearly, positivity of the cross section imposes the upper bound $|xh^{(1)}(x,q^2)| \leq xG^{(1)}(x,q^2)$. Note that even though P is taken to be the hard scale in the process, which can be greater than the saturation scale of the nucleus, that nevertheless the WW gluon distributions are probed at the much smaller momentum imbalance scale q. Therefore, the process can indeed provide information on these gluon distributions in the dense regime at $q < Q_s$. The WW gluon distributions also determine the divergence of the Chern-Simons current at the initial time in relativistic heavy-ion collisions [11] even though they are not the gluon distributions which enter the cross section for gluon production [12].

In the Color Glass Condensate (CGC) framework at small x the gluon fields are described by Wilson lines. They are path ordered exponentials in the strong color field of the target, and cross sections for different observables can be related to correlators of the Wilson lines. The Wilson line is a path ordered exponential of the covariant gauge field, whose largest component is A^+:

$$U(\vec{x}) = \mathbb{P}\exp\left\{ig\int dx^- A^+(x^-,\vec{x})\right\}. \quad (3)$$

The WW unintegrated gluon distribution [2, 12, 13], on the other hand, is defined in terms of the light cone gauge ($A^+=0$) field; it can be obtained by a gauge transformation

$$A^i(\vec{x}) = \frac{1}{ig}U^\dagger(\vec{x})\,\partial_i U(\vec{x})\,. \quad (4)$$

The trace (or the traceless part) of the two-point correlator of the light cone gauge field

$$xG^{ij}_{\rm WW}(x,\vec{q}) = \frac{1}{A_\perp}\frac{1}{2\pi}\left\langle A_a^i(\vec{q})A_a^j(-\vec{q})\right\rangle \quad (5)$$

defines $xG^{(1)}(x,q^2)$ and $xh^{(1)}(x,q^2)$ introduced above:

$$xG^{ij}_{\rm WW}(x,q^2) = \frac{1}{2}\delta^{ij}xG^{(1)}(x,q^2)+\frac{1}{2}\left(2\frac{q^iq^j}{q^2}-\delta^{ij}\right)xh^{(1)}(x,q^2)\,. \quad (6)$$

$A_\perp$ in eq. (5) denotes a transverse area over which the gluon distributions have been averaged over.

2 The WW gluon distributions at small x

These WW gluon distributions have been obtained at small x (high rapidity[1]) by a numerical solution of the JIMWLK evolution equations [14] in ref. [15], shown in fig. 1. At high transverse momentum

[1] $Y = \log x_0/x$ where x_0 determines the onset of small-x evolution; it is typically taken to be $x_0 = 10^{-2}$.

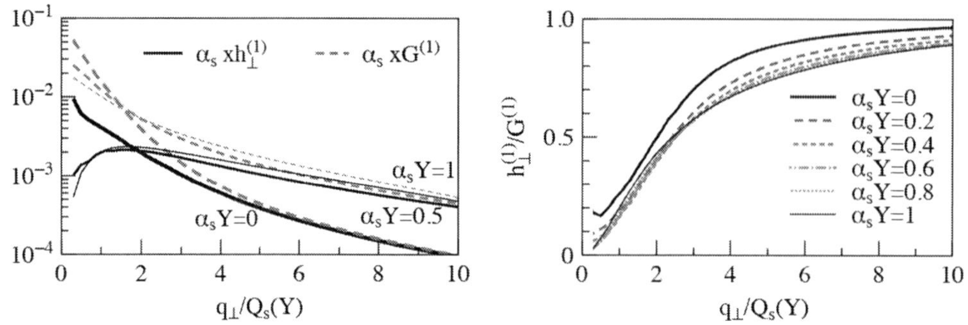

Figure 1. $xG^{(1)}(x,q^2)$ and $xh^{(1)}(x,q^2)$ WW gluon distributions versus transverse momentum q at different rapidities Y. $Q_s(Y)$ is the saturation momentum. The curves correspond to evolution at fixed α_s.

one finds that $xh^{(1)}(x,q^2) \to xG^{(1)}(x,q^2)$. This is easy to understand from the fact that in the dilute limit the classical light-cone gauge field is given by $A^i(q) = ig(q^i/q^2)\rho(q)$ where $g\rho(q)$ denotes the color charge density of the sources. For $A^i(q) \sim q^i$ eqs. (5,6) give $xh^{(1)}(x,q^2) = xG^{(1)}(x,q^2)$. Thus, the saturation of the above-mentioned bound on the distribution of linearly polarized gluons at high transverse momentum is a generic consequence of the dilute semi-classical field limit. On the other hand, at low q one has $xh^{(1)}(x,q^2)/xG^{(1)}(x,q^2) \approx 0$ implying that the angular dependence of the cross section (2) decreases. For more detailed predictions of $\langle \cos 2\phi \rangle$ obtained with the small-x WW gluon distributions see ref. [15].

The numerical solutions also indicate that the WW gluon distributions approach scaling functions $xh^{(1)}(x,q^2) = xh^{(1)}(q^2/Q_s^2(x))$, $xG^{(1)}(x,q^2) = xG^{(1)}(q^2/Q_s^2(x))$, at high rapidity. This is known as geometric scaling and has been discussed originally in the context of the dipole forward scattering amplitude (resp. the $\gamma^* - p$ total cross section) [16]. Geometric scaling of the WW distributions can be motivated from a Gaussian approximation to JIMWLK [17]. In this approximation, and also taking $N_c \gg 1$ for simplicity, they can be written in terms of the two-point function $\Gamma(r)$ of A^+ as [18]

$$xh^{(1)}(x,q^2) = \frac{4N_c}{\alpha_s(2\pi)^3}\int drr^3 J_2(qr)\left(1-[S(r^2)]^2\right)\frac{\Gamma''(r^2)}{\Gamma(r^2)} \tag{7}$$

$$xG^{(1)}(x,q^2) = \frac{4N_c}{\alpha_s(2\pi)^3}\int drrJ_0(qr)\left(1-[S(r^2)]^2\right)\left(\frac{\Gamma'(r^2)}{\Gamma(r^2)}+r^2\frac{\Gamma''(r^2)}{\Gamma(r^2)}\right). \tag{8}$$

$S(r^2) = \exp\left(-(1/2)C_F\Gamma(r^2)\right)$ is the S-matrix for a dipole of size r. At the JIMWLK fixed point $\Gamma(r^2)$, $r^2\Gamma'(r^2)$, and $r^4\Gamma''(r^2)$ are in fact functions of $r^2Q_s^2(x)$ only, rather than functions of both r and x. From the above expressions for $xh^{(1)}(x,q^2)$ and $xG^{(1)}(x,q^2)$ it follows that these functions then satisfy geometric scaling (also see ref. [19]).

3 The JIMWLK weight functional and the constraint effective potential for $g^2\mathrm{tr}\,|A^+(q)|^2$

Expectation values of observables at small x are computed by i) expressing the observable as a functional $O[A^+]$ of the covariant gauge field, ii) and averaging over the random semi-classical fields A^+ with the weight $W_Y[A^+]$:

$$\langle O\rangle = \frac{1}{Z}\int \mathcal{D}A^+(q)\, W_Y[A^+(q)]\, O[A^+(q)] \quad , \quad Z = \int \mathcal{D}A^+(q)\, W_Y[A^+(q)]\,. \tag{9}$$

Note that $A^+(q)$ is the soft semi-classical field which solves the Poisson equation, $A^+(q) = g\rho(q)/q^2$, with $g\rho(q)$ the random, effective color charge density that is the source of the soft gluon field. Hence, the above average over A^+ can also be written as an average over ρ.

The weight $W_Y[A^+(q)]$ for a given configuration $A^+(q)$ is determined by the solution of the JIMWLK functional RG equation [14]. The exact solution can only be obtained numerically. However, a non-local (in coordinate space) Gaussian mean-field approximation for $W_Y[\rho]$ has been proposed, see first reference in [17], which reproduces the proper gluon distribution $g^2\mathrm{tr}\, A^+(q)\, A^+(-q)$ both at small ($q^2 \ll Q_s^2$) as well as at high ($q^2 \gg Q_s^2$) transverse momentum:

$$W_G[A^+] = e^{-S_G[A^+]} \quad , \quad S_G[A^+] = \int \frac{\mathrm{d}^2 q}{(2\pi)^2}\, q^4 \frac{\mathrm{tr}\, A^+(q)\, A^+(-q)}{g^2 \mu^2(q^2)} \ . \tag{10}$$

For simplicity we restrict here to high transverse momentum where

$$\mu^2(q^2) \simeq \mu_0^2 \left(\frac{q^2}{Q_s^2}\right)^{1-\gamma} , \tag{11}$$

with $\gamma \simeq 0.64$ an anomalous dimension [20]. Q_s^2 and $\mu_0^2 \sim A^{1/3}$ are evaluated at the rapidity of interest but we will not spell out this dependence on Y explicitly.

The expectation value of $O[A^+]$ written in eq. (9) is an average over all configurations of $A^+(q)$. However, one may be interested in $\langle O \rangle$ evaluated over a subclass of configurations, for example those with a high (or low) number of gluons, or with an unusual transverse momentum distribution of gluons. To that end we introduce the constraint effective potential for $X(q) = g^2\mathrm{tr}\, |A^+(q)|^2$ by integrating over configurations at fixed $X(q)$ [21]:

$$Z = \int \mathcal{D}X(q)\, e^{-V_{\mathrm{eff}}[X(q)]} , \tag{12}$$

$$e^{-V_{\mathrm{eff}}[X(q)]} = \int \mathcal{D}A^+(q)\, W_Y[A^+(q)]\, \delta(X(q) - g^2\mathrm{tr}\, |A^+(q)|^2) \ . \tag{13}$$

For a Gaussian theory the integral over configurations at fixed $X(q)$ is easy to compute, and the resulting effective potential is [21]

$$V_{\mathrm{eff}}[X(q)] = \int \frac{\mathrm{d}^2 q}{(2\pi)^2} \left[\frac{q^4}{g^4 \mu^2(q)} X(q) - \frac{1}{2} A_\perp N_c^2 \log X(q)\right] , \tag{14}$$

$A_\perp$ is the transverse area over which

$$\mathrm{tr}\, |A^+(q)|^2 = \int_{A_\perp} \mathrm{d}^2 b \int \mathrm{d}^2 r\, e^{-i\vec{q}\cdot\vec{r}} A^+(\vec{b} - \vec{r}/2) A^+(\vec{b} + \vec{r}/2) \tag{15}$$

has been integrated. The stationary point of $V_{\mathrm{eff}}[X]$ determines the extremal gluon distribution

$$X_s(q) = \frac{1}{2} N_c^2 A_\perp \frac{g^4 \mu^2(q)}{q^4} \ . \tag{16}$$

$X_s(q)$ is the most likely gluon distribution function rather than the average. However, in the large-N_c limit it is equal to the expectation value of $\langle g^2\mathrm{tr}\, |A^+(q)|^2 \rangle$. Away from the extremal solution, the potential $V_{\mathrm{eff}}[X]$ provides insight into the distribution of functions $X(q)$ about the extremum. This distribution is determined by a "linear minus logarithmic" rather than by a polynomial potential.

It will be convenient for what follows to describe deviations from $X_s(q)$ by multiplying with $\eta(q)$ rather than by adding $\delta X(q)$. Hence, we introduce the function $\eta(q)$ through $X(q) = X_s(q)\,\eta(q)$. A fluctuation from the extremal field $X_s(q)$ has action

$$V_{\rm eff}[\eta(q)] = \frac{1}{2}N_c^2 A_\perp \int \frac{{\rm d}^2 q}{(2\pi)^2}\,[\eta(q) - 1 - \log\eta(q)]\ . \tag{17}$$

Note that $X(q) = g^2 {\rm tr}\,|A^+(q)|^2$ is a positive definite function and so is $\eta(q)$. We can therefore perform another field redefinition to introduce $\phi(q)$ through $e^{\phi(q)} = \eta(q)$ so that

$$V_{\rm eff}[\phi(q)] = \frac{1}{2}N_c^2 A_\perp \int \frac{{\rm d}^2 q}{(2\pi)^2}\left[e^{\phi(q)} - \phi(q) - 1\right]\ . \tag{18}$$

Thus, we found that it is a Liouville action in two dimensions which describes the distribution of $\log {\rm tr}\,|gA^+(q)|^2$ (relative to the average gluon distribution) in a Gaussian approximation to JIMWLK.

The action for the most likely distribution function $X_s(q)$ is of order $N_c^2 A^{1/3}$ (times zero, in dimensional regularization). So is the action for $X(q) = X_s(q)\eta(q)$ if $\eta(q) = O(1)$. Our discussion is restricted to the distribution of functions $X(q)$ which exhibit longitudinal coherence and are of order $N_c^2 A^{1/3}$. The small-x power counting assumes $g^4 A^{1/3} = O(1)$ [22], and so $X(q) \sim (A^{1/3})^0$ would correspond to a higher order correction in the coupling.

Knowing $V_{\rm eff}$ we can now evaluate the suppression probability for a modification of the gluon distribution such as

$$\eta(q) = 1 + \eta_0 \left(\frac{g^4\mu_0^2}{q^2}\right)^a \Theta\left(q^2 - \Lambda^2\right)\Theta\left(Q^2 - q^2\right)\ . \tag{19}$$

η_0 determines the amplitude of the distortion, Q^2 and $\Lambda^2 > Q_s^2$ determine its support, and the parameter a specifies the spectral shape. The action for such $\eta(q)$ when $Q^2 \gg \Lambda^2$ is

$$V_{\rm eff}[\eta(q)] \simeq \frac{1}{8\pi} N_c^2 A_\perp\, g^4\mu_0^2\,\eta_0 \times \begin{cases} \frac{1}{1-a}\left(\frac{Q^2}{g^4\mu_0^2}\right)^{1-a} & (a<1)\,, \\ \log\frac{Q^2}{\Lambda^2} & (a=1)\,, \\ \frac{1}{a-1}\left(\frac{g^4\mu_0^2}{\Lambda^2}\right)^{a-1} & (a>1)\,. \end{cases} \tag{20}$$

Hence, we find that a harder than average gluon distribution ($a < 0$) over $\Lambda < q < Q$ comes at a high price since $V_{\rm eff}[\eta(q)] \sim (Q^2/g^4\mu_0^2)^{1-a}$. On the other hand, gluon distributions which drop substantially faster than the most likely one (i.e. $X_s(q)$) correspond to $a \geq 1$, and such fluctuations can extend to high transverse momentum. For a more detailed discussion of the shape of the gluon distribution in the presence of a high (or low) gluon multiplicity "trigger" we refer to ref. [21].

4 The functional distribution of WW gluon distributions over the JIMWLK ensemble

In sec. 2 we discussed the WW gluon distributions averaged over the entire JIMWLK ensemble $W_Y[A+]$. In this section we discuss the distributions of these functions over the JIMWLK ensemble.

To obtain some basic analytic insight we write the expansion of $g^2 \mathrm{tr}\, A^i(q)A^j(-q)$ to fourth order in $A^+(q)$ obtained via eq. (4):

$$\begin{aligned}\delta^{ij}\mathrm{tr}\, A^i(q)A^j(-q) &= \frac{1}{2}q^2 A^{+a}(q)A^{+a}(-q)\\ +\frac{g^2}{8}f^{abe}f^{cde}\left(\frac{q^n q^m}{q^2}-\delta^{nm}\right)&\int\frac{\mathrm{d}^2k}{(2\pi)^2}k^n A^{+a}(q-k)A^{+b}(k)\int\frac{\mathrm{d}^2p}{(2\pi)^2}p^m A^{+c}(-q-p)A^{+d}(p)\,, \quad (21)\\ \left(2\frac{q^i q^j}{q^2}-\delta^{ij}\right)\mathrm{tr}\, A^i(q)A^j(-q) &= \frac{1}{2}q^2 A^{+a}(q)A^{+a}(-q)\\ -\frac{g^2}{8}f^{abe}f^{cde}\left(\frac{q^n q^m}{q^2}-\delta^{nm}\right)&\int\frac{\mathrm{d}^2k}{(2\pi)^2}k^n A^{+a}(q-k)A^{+b}(k)\int\frac{\mathrm{d}^2p}{(2\pi)^2}p^m A^{+c}(-q-p)A^{+d}(p)\,. \quad (22)\end{aligned}$$

In the weak-field limit the first term in these expansions dominates and the two WW gluon distributions are equal, configuration by configuration. The correction at fourth power in A^+ generates a "splitting". We can perform an average over a Gaussian ensemble by summing the two non-vanishing Wick contractions using

$$\left\langle A^{+a}(q)\,A^{+b}(k)\right\rangle = \delta^{ab}\,(2\pi)^2\delta(q+k)\frac{g^2\mu^2(q^2)}{q^4}\,. \quad (23)$$

At order N_c^2 this leads to

$$\delta^{ij}\,g^2\left\langle \mathrm{tr}\,A^i(q)A^j(-q)\right\rangle = \frac{1}{2}N_c^2 A_\perp\frac{g^4\mu^2(q)}{q^2} + N_c^3 A_\perp\frac{g^8}{4}\int\frac{\mathrm{d}^2k}{(2\pi)^2}k^2\left[1-(\hat{k}\cdot\hat{q})^2\right]\frac{\mu^2(q-k)}{(q-k)^4}\frac{\mu^2(k)}{k^4}\,. \quad (24)$$

The result for the average of eq. (22) is the same except that the sign of the second term is negative. Thus, one may wonder if the linearly polarized distribution could take negative values[2]. It is clear that at high q^2 the correction is $\sim 1/q^2$ power suppressed as compared to the leading contribution. This suppression ensures that $xh^{(1)}(x,q^2)$, averaged over all configurations, is a positive definite function (as seen in fig. 1).

Instead of averaging over all JIMWLK configurations we can use the approach from the previous section to integrate over all configurations at fixed $X(q) = g^2\mathrm{tr}\,|A^+(q)|^2$. To do so, instead of using eq. (23) we make the final integration over $X(q)$ explicit:

$$\left\langle A^{+a}(q)\,A^{+b}(k)\right\rangle = \delta^{ab}\,(2\pi)^2\delta(q+k)\frac{g^2\mu^2(q^2)}{q^4}\int\mathcal{D}X(\ell)\,e^{-V_{\mathrm{eff}}[X(\ell)]}\,\frac{X(q)}{X_s(q)}\,. \quad (25)$$

We can then rewrite eq. (24) as follows:

$$\begin{aligned}\delta^{ij}\,g^2\left\langle \mathrm{tr}\,A^i(q)A^j(-q)\right\rangle &= q^2\int\mathcal{D}X(\ell)\,e^{-V_{\mathrm{eff}}[X(\ell)]}\,X(q)\\ &+\frac{1}{N_c A_\perp}\int\frac{\mathrm{d}^2k}{(2\pi)^2}k^2\left[1-(\hat{k}\cdot\hat{q})^2\right]\int\mathcal{D}X(\ell)\,e^{-V_{\mathrm{eff}}[X(\ell)]}\,X(q-k)X(k)\,. \quad (26)\end{aligned}$$

As before, replacing the projector δ^{ij} by $2q^iq^j/q^2-\delta^{ij}$ reverses the sign of the last term. It is evident that for some functions $X(q)$ which contribute to the integral the correction in this last expression may be greater than the "leading" contribution. These configurations overcome the power suppression discussed above which arises at the saddle point of the integral; also, they give linearly polarized gluon distributions which are negative in some range of transverse momentum.

[2]This function does not have a "gluon density" / probability interpretation and so it needs not be positive definite.

We now show some numerical results obtained by Monte-Carlo sampling of the JIMWLK functional $W_Y[A^+]$ [21] for $N_c = 3$ colors and fixed α_s. We evaluate the WW gluon distributions on each configuration. They have to be integrated over a finite patch in impact parameter space greater than the inverse transverse momentum. We take

$$P^{ij} \int \mathrm{d}^2x\mathrm{d}^2y\, e^{-iq\cdot(x-y)} e^{-(x^2+y^2)/2R^2}\, g^2 \mathrm{tr}\, A^i(x)A^j(y)\,, \tag{27}$$

with $R = 2/Q_s(Y)$ and $q > Q_s(Y)$ the transverse momentum scale. P^{ij} denotes one of the two projectors mentioned above. This expression factorizes into two Fast Fourier Transforms which can be evaluated numerically very efficiently.

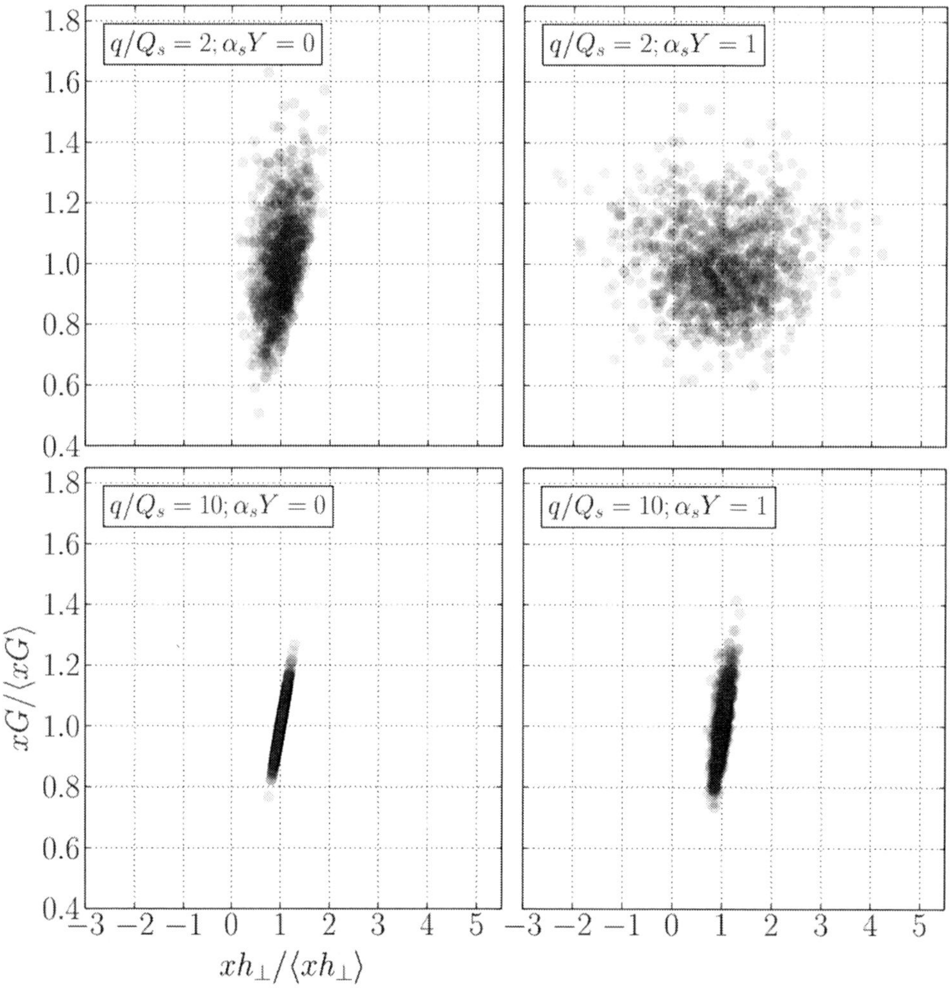

Figure 2. $xG^{(1)}(x, q^2)$ and $xh^{(1)}(x, q^2)$ WW gluon distributions for 1000 individual field configurations, evaluated at $q = 2Q_s(Y)$ or $q = 10Q_s(Y)$, respectively. Notice the different scales on the horizontal and vertical axes. Left: MV model initial condition; Right: JIMWLK ensemble at $\alpha_s Y = 1$.

In fig. 2 we show the $xG^{(1)}(x, q^2)$ and $xh^{(1)}(x, q^2)$ WW gluon distributions for individual configurations, relative to their average. For high transverse momentum far above $Q_s(Y)$ we observe, as expected, that the two functions are essentially equal, even for individual configurations. For $q = 2Q_s(Y)$ on the other hand the relative fluctuations of the linearly polarized distribution are much greater than those of the conventional WW distribution. For some configurations $xh^{(1)}(x, q^2)$ can take values up to several times its average while other evolution trajectories lead to negative values. This is an effect of evolution to small x since $xh^{(1)}(x, q^2) < 0$ at $q = 2Q_s(Y)$ does not occur at $Y = 0$ once in 10^4 configurations.

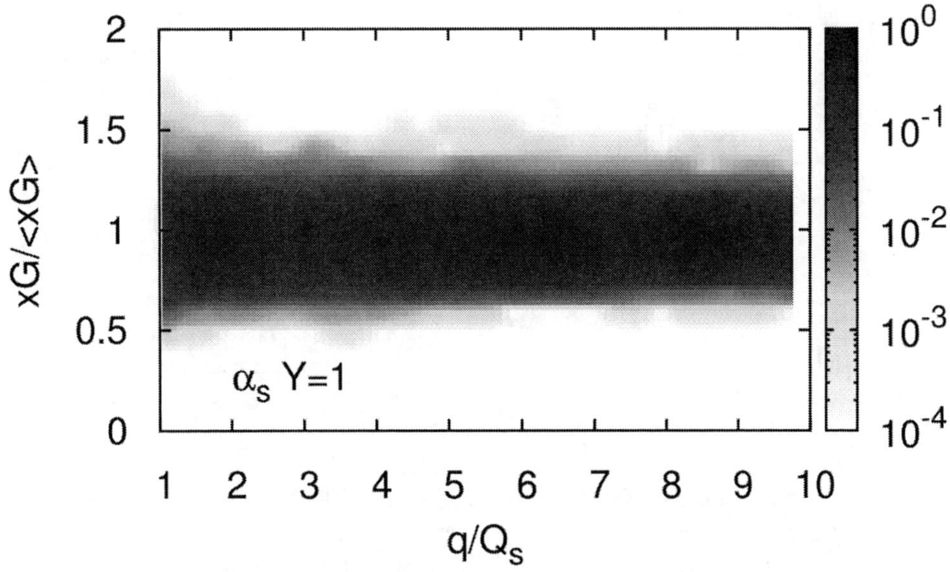

Figure 3. Distribution of functions $xG^{(1)}(x, q^2)/\langle xG^{(1)}(x, q^2)\rangle$ in the JIMWLK ensemble at $\alpha_s Y = 1$. The color coding indicates the probability for a particular function $xG^{(1)}(x, q^2)/\langle xG^{(1)}(x, q^2)\rangle$.

The functional distributions of $xG^{(1)}(x, q^2)$ and $xh^{(1)}(x, q^2)$ in the JIMWLK ensemble are shown in figs. 3 and 4, respectively. At high transverse momentum the distributions are strongly peaked about the most likely WW functions. On the other hand, when q is not very far above $Q_s(Y)$ the ensemble of linearly polarized WW gluon distribution functions is broad. At $\alpha_s Y \sim 1$ it includes non-positive definite functions as well as functions which take values several times their average.

Acknowledgements

A.D. thanks the organizers for the invitation to ISMD 2017; and gratefully acknowledges support by the DOE Office of Nuclear Physics through Grant No. DE-FG02-09ER41620, and from The City University of New York through the PSC-CUNY Research grant 60262-0048.

References

[1] E. C. Aschenauer *et al.*, arXiv:1708.01527 [nucl-ex].
[2] F. Dominguez, C. Marquet, B. W. Xiao and F. Yuan, Phys. Rev. D **83**, 105005 (2011)

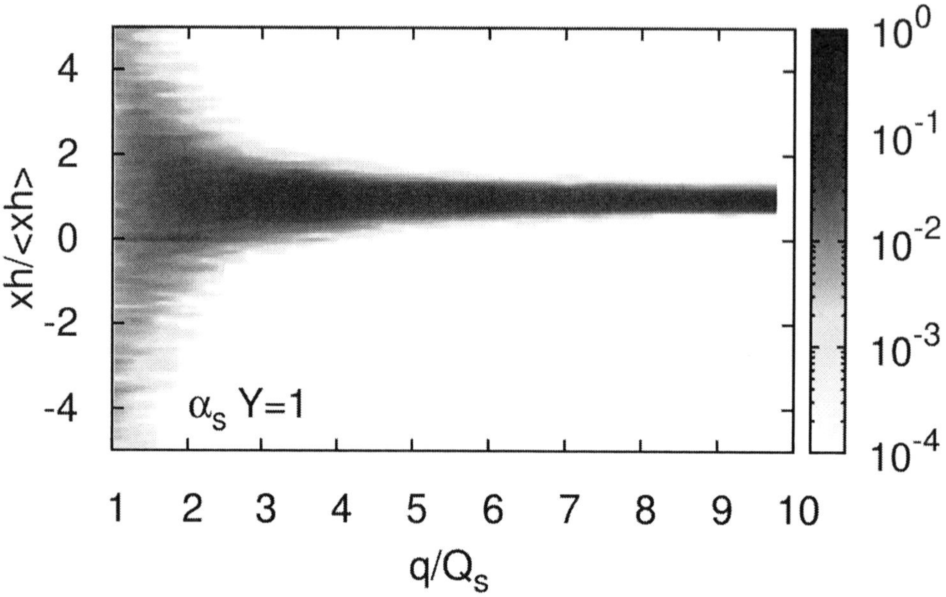

Figure 4. Distribution of functions $xh^{(1)}(x, q^2)/\langle xh^{(1)}(x, q^2)\rangle$ in the JIMWLK ensemble at $\alpha_s Y = 1$.

[3] P. J. Mulders and J. Rodrigues, Phys. Rev. D **63**, 094021 (2001)
[4] S. Meissner, A. Metz and K. Goeke, Phys. Rev. D **76**, 034002 (2007)
[5] D. Boer, P. J. Mulders and C. Pisano, Phys. Rev. D **80**, 094017 (2009)
[6] A. Metz and J. Zhou, Phys. Rev. D **84**, 051503 (2011)
[7] F. Dominguez, J. W. Qiu, B. W. Xiao and F. Yuan, Phys. Rev. D **85**, 045003 (2012)
[8] D. Boer, S. J. Brodsky, P. J. Mulders and C. Pisano, Phys. Rev. Lett. **106**, 132001 (2011)
[9] J. W. Qiu, M. Schlegel and W. Vogelsang, Phys. Rev. Lett. **107**, 062001 (2011)
[10] J. P. Lansberg, C. Pisano and M. Schlegel, Nucl. Phys. B **920**, 192 (2017);
J. P. Lansberg, C. Pisano, F. Scarpa and M. Schlegel, arXiv:1710.01684 [hep-ph].
[11] T. Lappi and S. Schlichting, arXiv:1708.08625 [hep-ph].
[12] D. Kharzeev, Y. V. Kovchegov and K. Tuchin, Phys. Rev. D **68**, 094013 (2003)
[13] F. Dominguez, J. W. Qiu, B. W. Xiao and F. Yuan, Phys. Rev. D **85**, 045003 (2012)
[14] J. Jalilian-Marian, A. Kovner, A. Leonidov and H. Weigert, Nucl. Phys. B **504**, 415 (1997); Phys. Rev. D **59**, 014014 (1998);
E. Iancu, A. Leonidov and L. D. McLerran, Phys. Lett. B **510**, 133 (2001); Nucl. Phys. A **692**, 583 (2001);
H. Weigert, Nucl. Phys. A **703**, 823 (2002)
[15] A. Dumitru, T. Lappi and V. Skokov, Phys. Rev. Lett. **115**, no. 25, 252301 (2015)
[16] A. M. Stasto, K. J. Golec-Biernat and J. Kwiecinski, Phys. Rev. Lett. **86**, 596 (2001)
[17] E. Iancu, K. Itakura and L. McLerran, Nucl. Phys. A **724**, 181 (2003);
J. Jalilian-Marian and Y. V. Kovchegov, Phys. Rev. D **70**, 114017 (2004);
H. Fujii, F. Gelis and R. Venugopalan, Nucl. Phys. A **780**, 146 (2006);
C. Marquet and H. Weigert, Nucl. Phys. A **843**, 68 (2010);

E. Iancu and D. N. Triantafyllopoulos, JHEP **1204**, 025 (2012)
[18] A. Dumitru and V. Skokov, Phys. Rev. D **94**, no. 1, 014030 (2016)
[19] C. Marquet, E. Petreska and C. Roiesnel, JHEP **1610**, 065 (2016)
[20] A. H. Mueller and D. N. Triantafyllopoulos, Nucl. Phys. B **640**, 331 (2002)
[21] A. Dumitru and V. Skokov, Phys. Rev. D **96**, 056029 (2017)
[22] Y. V. Kovchegov, Phys. Rev. D **61**, 074018 (2000)

System size, collision energy, and rapidity dependence of collective dynamics measured by the PHENIX experiment at RHIC

Seyoung Han[1,2,⋆] for the PHENIX collaborations

[1] *Ewha Womans University, Seoul, Republic of Korea*
[2] *RIKEN, Saitama, Japan*

Abstract. In high energy collisions, partons produced in the initial stage undergo multiple interactions and yield a collective motion as a whole. Recently, several questions have risen including how small the system can be for producing the collectivity, and how far in rapidity the collectivity extends. PHENIX has measured the particle flow in p + Au, d + Au, and ^{3}He + Au collisions over several energies as well as over wider rapidity range. In this work, we summarize our results on the particle flow and discuss the implications for collectivity in small collision systems.

1 Introduction

Many studies of the properties of the Quark Gluon Plasma (QGP) have been done. One of the key observations in heavy-ion collisions is that the created medium behaves like almost ideal fluid; we call it the strongly coupled QGP. The collectivity in heavy ion systems is multi-particle correlations with the large rapidity gap which originated from a common source. On the other hand, there also exist non-collectivity originated correlation such as jet, parton decays for example. In the di-hadron correlations v_n, the coefficient of Fourier harmonics is commonly attributed to hydrodynamic flow.

Until recently, the collective behavior was not expected in small systems. The observation of a long-range, near-side ridge in high multiplicity $p+p$ by the CMS experiment challenged this view [1]. PHENIX has also studied the collectivity in small collision systems. Previously these studies have been only on the p_T dependence in the most central events. More recent studies have expanded to all centralities to facilitate quantitative studies on the role of system size and non-flow effects, and also the rapidity η dependence to study the longitudinal structure of the observed correlations.

2 PHENIX detectors

Table 1 shows data sets of small collision systems taken by the PHENIX experiment. We collected data from various small collision systems, namely $p + p$, p + Au, d + Au, and ^{3}He + Au collisions at $\sqrt{s_{_{NN}}}$ = 200 GeV. Especially for d + Au collisions in 2016, we also have beam energy scan data in four collision energies [2]. With these data sets, we can study a variety of observables, such as the system size dependence and collision geometry.

⋆e-mail: seyoung_han@ewhain.net

© The Authors, published by EDP Sciences. This is an open access article distributed under the terms of the Creative Commons Attribution License 4.0 (http://creativecommons.org/licenses/by/4.0/).

Table 1. Small system data taken by PHENIX experiment.

	^{3}He + Au	d + Au	p + Au	$p + p$
200 GeV	✓	✓	✓	✓
62.4 GeV		✓		
39 GeV		✓		
19.6 GeV		✓		

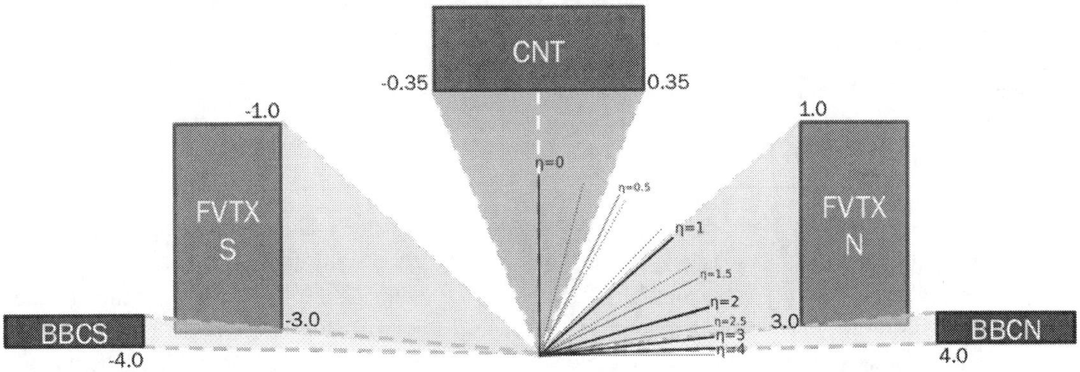

Figure 1. PHENIX detector coverage. The central arm (CNT) covers rapidity range of $-0.35 < \eta < 0.35$ with partial azimuthal coverage ($\Delta\phi = \pi$). The FVTX (forward silicon vertex tracker) and BBC (beam beam counter) are implemented in both south and north directions and cover $1 < |\eta| < 3$ and $3 < |\eta| < 4$ with full azimuthal coverage.

Figure 1 shows the rapidity coverage of a part of the PHENIX detector subsystems involved in the present analysis. The central arm (CNT) which covers mid-rapidity can measure charged particles and particle identification with a limited acceptance is also possible. The FVTX and BBC are the spectrometers at forward (positive, north) and backward (negative, south) rapidity. In the collectivity study of small systems, we use the beam beam counter (BBC) for triggering minimum bias and high multiplicity events, calculating event plane resolutions, and measuring two particle correlations. We also use the forward silicon vertex tracker (FVTX) to measure charged tracks and event planes. In the p + Au, d + Au and ^{3}He + Au runs, particle multiplicity is higher at the south arm because Au ions move to the south arm, which is the backward rapidity. Therefore, the BBC and FVTX located at backward rapidity are used for event plane determination.

The PHENIX detectors have an advantage in wide acceptance range which allows us to study correlations in various pseudo-rapidity (η) ranges. Figure 2 shows the two particle correlation functions of different $\Delta\eta$ gaps in 0–5% central d + Au collisions at $\sqrt{s_{NN}} = 200$ GeV [2]. $\Delta\eta$ gap increases from left to right. Since clear near-side ridge structure is shown in all combinations, we can expect non-zero v_2. In case of the correlation function between the central arm and the south FVTX, it shows the largest coefficient of the second order cosine modulation (c_2). It is possible to include a larger effect from intra-jet or di-jet (non-flow effects) due to the small $\Delta\eta$ gap. As $\Delta\eta$ increases the non-flow effects are expected to be decreased and c_2 in Figure 2 shows a decreasing trend. However, we can still see a near-side ridge structure even in the two particle correlation function in $6.2 < \Delta\eta < 7.8$, where the non-flow effects are expected to be negligible.

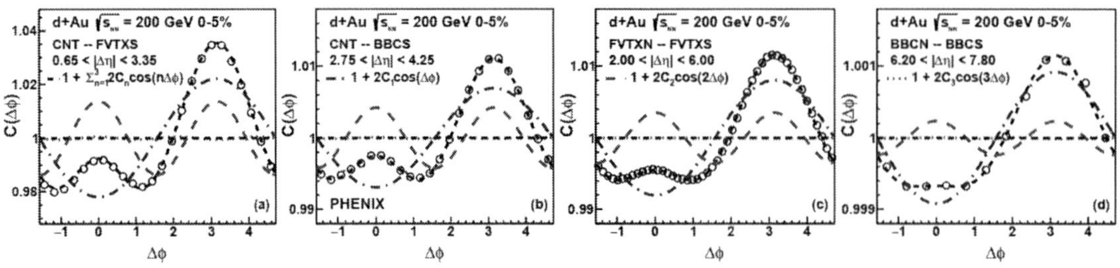

Figure 2. Two particle correlation functions in various $\Delta\eta$ gaps in d + Au collisions at $\sqrt{s_{NN}}$ = 200 GeV [2].

3 Analysis method

Two independent analysis methods are used for the collectivity study in small systems. One method is using event plane which is measured by the south FVTX or the south BBC.

The event plane resolution is calculated by comparing event planes from the three independent systems. We used the central arm, the south FVTX, and the south BBC for the resolution calculation and applied to the raw v_2. The raw v_2 of charged particles at the mid-rapidity is the amplitude of $\cos 2\phi$ modulation from the $\Delta\phi$ distribution between the charged particles in the central arm and event plane.

The other method is two particle correlation method (2PC) [3] which uses the ratio between the angular correlation of the particle pairs in a same event and mixed event with normalized by the total number of pairs of each. The Fourier coefficients of the second order cosine modulation (c_2) from the three detector combinations are used to get the v_2. We confirmed the consistency between these two methods. p + Au centrality dependence with function of p_T is the result by using the two particle correlation method.

4 System size dependence of v_2 and v_3

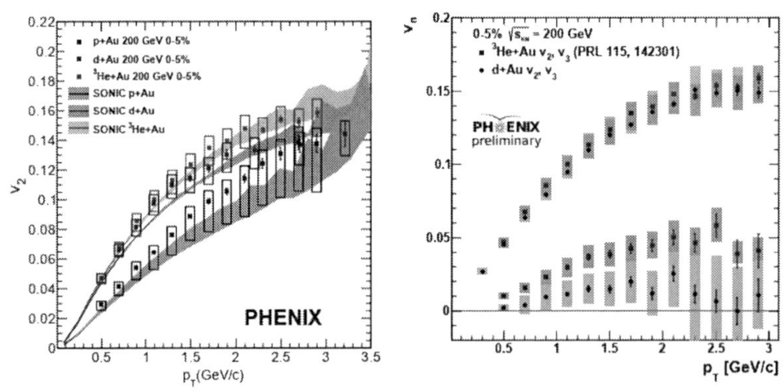

Figure 3. Collectivity measurement of charged hadrons at mid-rapidity as a function of p_T in 0–5% central events. (left) v_2 of p + Au, d + Au, and ^{3}He + Au collisions at $\sqrt{s_{NN}}$ = 200 GeV. (right) v_2 and v_3 of d + Au and ^{3}He + Au collisions at $\sqrt{s_{NN}}$ = 200 GeV [4].

PHENIX measured v_2 and v_3 of charged hadrons at the mid-rapidity as a function of p_T in 0–5% central events p + Au [4], d + Au [2], and ^{3}He + Au [5] collisions at $\sqrt{s_{NN}}$ = 200 GeV. In case of the v_2 shown in Figure 3, sizable v_2 observed in all three collision systems, and the v_2 measured in p + Au collisions is slightly smaller than the v_2 measured in d + Au and ^{3}He + Au collisions. The SONIC model [6] can describe the results in all three collision systems reasonably well.

The right plot of Figure 3 shows v_2 and v_3 of 0–5% of d + Au and ^{3}He + Au collisions at $\sqrt{s_{NN}}$ = 200 GeV. The measured v_2 in two collision systems are consistent, while the v_3 of ^{3}He + Au is larger than the v_3 of d + Au.

5 Beam energy scan of d + Au collisions

5.1 v_2 as a function of p_T in various centrality bins

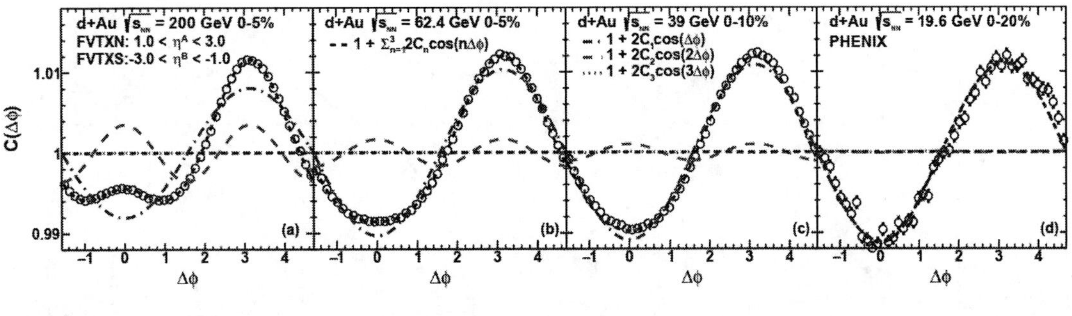

Figure 4. Two particle correlation between south and north FVTX in central d + Au collisions at various collision energies [2].

Figure 4 shows two particle correlations of charged hadrons measured in the south and north FVTX in d + Au collisions at various collision energies [2]. As the collision energy decreases, c_2, which is the second order coefficient of cosine modulation (shown as red dashed lines) also decreases.

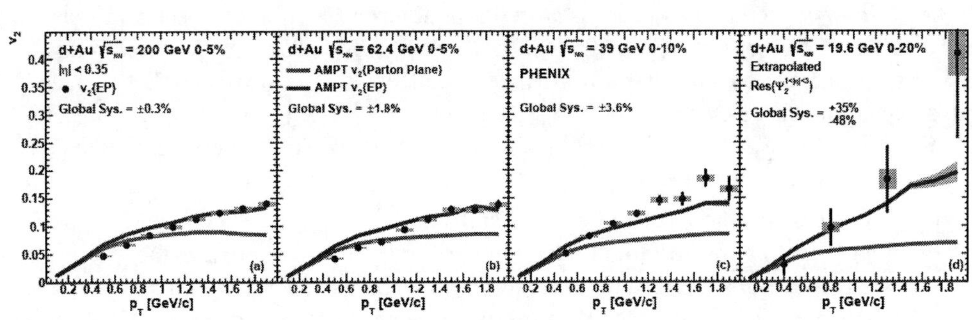

Figure 5. v_2 of charged hadrons as a function of p_T at mid-rapidity in central d + Au collisions at various collision energies.

For further understanding, the experimental data are compared with the AMPT [7] model v_2 calculations using both the parton plane and the event plane methods in Figure 5. The green curves (v_2{Parton Plane}) are the v_2 extracted by the parton plane method which is corresponding to initial collision geometry, and the blue curves ($v_2\{\mathrm{EP}\}$) are measured by using relative angular position to the event plane based on the produced particles in Monte-Carlo simulation. The AMPT $v_2\{\mathrm{EP}\}$ calculations follow the real PHENIX detectors acceptance at the four collision energies. The results of the AMPT $v_2\{\mathrm{EP}\}$ follow the increasing trend of the v_2 data as collision energy becomes lower. On the other hand, in case of the AMPT $v_2\{\text{Parton Plane}\}$, not only it is smaller than the data and the AMPT $v_2\{\mathrm{EP}\}$ but also it moderately decreases as the collision energy becomes lower. From these data comparisons with two sets of AMPT calculations, it is clear within the AMPT model that there are considerable fraction of the non-flow contribution in the measured v_2 especially dominated in the lower collision energy.

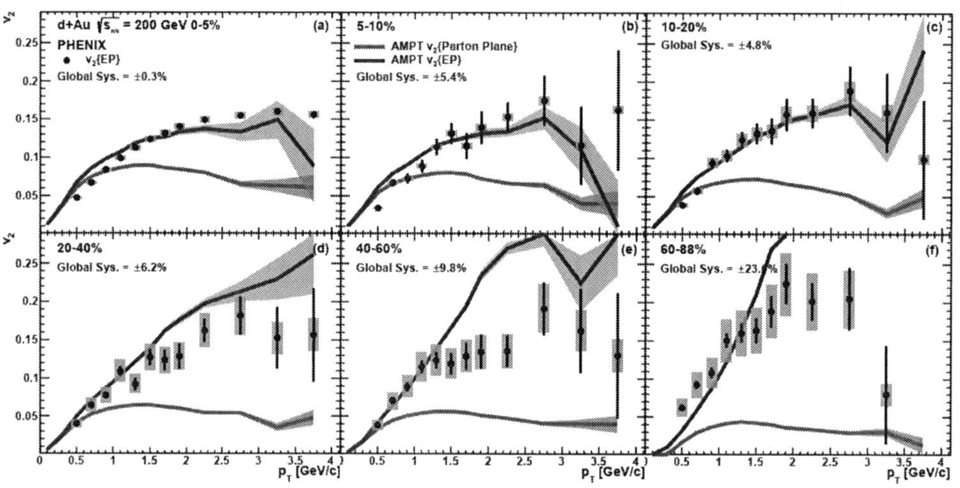

Figure 6. v_2 of charged hadrons as a function of p_T at mid-rapidity in various centrality classes of $d+\mathrm{Au}$ collisions at $\sqrt{s_{NN}} = 200$ GeV.

Figure 6 shows v_2 of charged hadrons at mid-rapidity as a function of p_T in a various centrality classes of $d+\mathrm{Au}$ collisions at $\sqrt{s_{NN}} = 200$ GeV. Non-zero v_2 is observed in all centrality bins, and measured value increases from central to peripheral collisions. The AMPT $v_2\{\mathrm{EP}\}$ agrees with the data in all centrality classes, and the AMPT $v_2\{\text{Parton Plane}\}$ gives lower value than the data. In case of the AMPT $v_2\{\text{Parton Plane}\}$, the amplitude decreases from central to peripheral collisions due to increasing non-flow contribution..

Figure 7 and 8 shows the results of $d+\mathrm{Au}$ collisions at $\sqrt{s_{NN}} = 62$ GeV and 39 GeV, respectively. From those two figures the amplitude of v_2 is slightly larger than that in the $\sqrt{s_{NN}} = 200$ GeV when comparing the results in the same centrality bins. The AMPT $v_2\{\mathrm{EP}\}$ agrees with the data at $\sqrt{s_{NN}} = 62$ GeV but gives lower value than the experimental data at $\sqrt{s_{NN}} = 39$ GeV especially at the peripheral events. It is similar with the $\sqrt{s_{NN}} = 200$ GeV results, the AMPT $v_2\{\text{Parton Plane}\}$ calculations always underestimate the data. The difference between the data and the AMPT $v_2\{\text{Parton Plane}\}$ is larger in high p_T and peripheral events due to a larger non-flow contribution.

In summary of the beam energy scan results, the measured v_2 is larger in higher p_T, peripheral collisions, and lower collision energy. The AMPT $v_2\{\mathrm{EP}\}$ reproduces general shape of p_T dependence in v_2. On the contrary, the AMPT $v_2\{\text{Parton Plane}\}$ calculations excluding possible non-flow effects therefore tend to underestimate the experimental data.

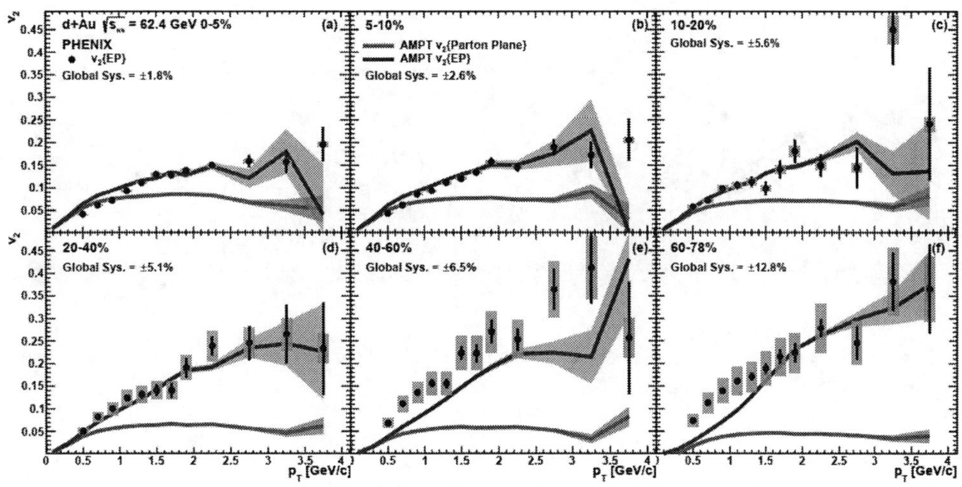

Figure 7. v_2 of charged hadrons as a function of p_T at mid-rapidity in various centrality classes of $d+$Au collisions at $\sqrt{s_{NN}} = 62$ GeV.

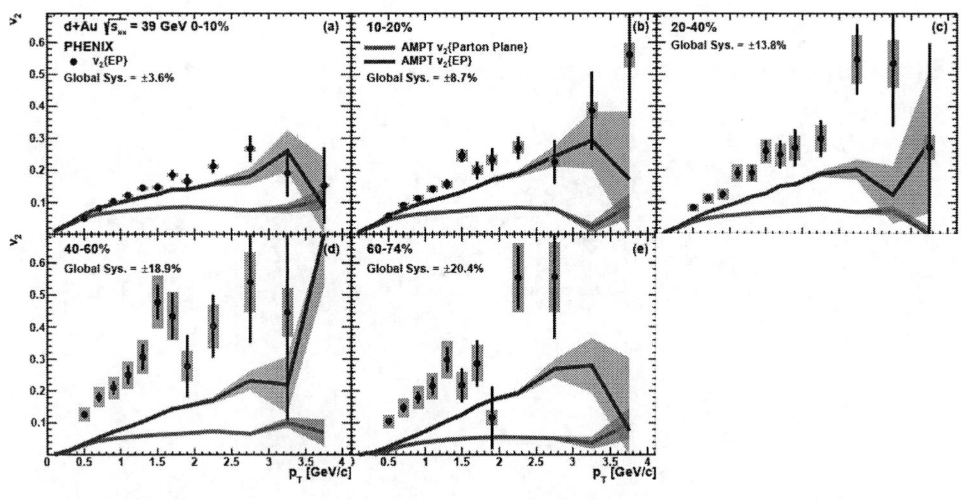

Figure 8. v_2 of charged hadrons as a function of p_T at mid-rapidity in various centrality classes of $d+$Au collisions at $\sqrt{s_{NN}} = 39$ GeV.

5.2 Rapidity dependence of v_2

Figure 9 shows v_2 of charged hadrons as a function of η in central $d+$Au collisions at $\sqrt{s_{NN}} = 200$ GeV (left), 62 GeV (middle), and 39 GeV (right) comparisons with the AMPT simulation. There is a clear energy dependence at the backward rapidity, which is the direction of the Au beam, whereas the rapidity dependence is weaker at the forward rapidity. In addition, the v_2 of the backward rapidity is larger than that of the forward rapidity at high collision energy, but this asymmetric shape quenches as the collision energy becomes smaller.

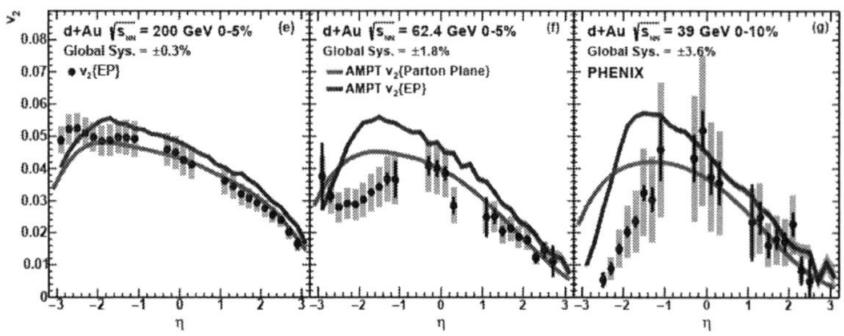

Figure 9. v_2 of charged hadrons as a function of η in central d + Au collisions at various collision energies and comparison with AMPT simulation.

The measured v_2 agrees with both AMPT $v_2\{\text{EP}\}$ and AMPT $v_2\{\text{Parton Plane}\}$ in all three collision energies at forward rapidity. Therefore at the backward rapidity the AMPT simulation reproduces the data at $\sqrt{s_{NN}}$ = 200 GeV, but it overestimates v_2 at $\sqrt{s_{NN}}$ = 62 GeV and 39 GeV while the AMPT $v_2\{\text{EP}\}$ and $v_2\{\text{Parton Plane}\}$ show a difference. This difference is possibly due to a larger non-flow contribution.

6 v_2 function of p_T in p + Au collisions

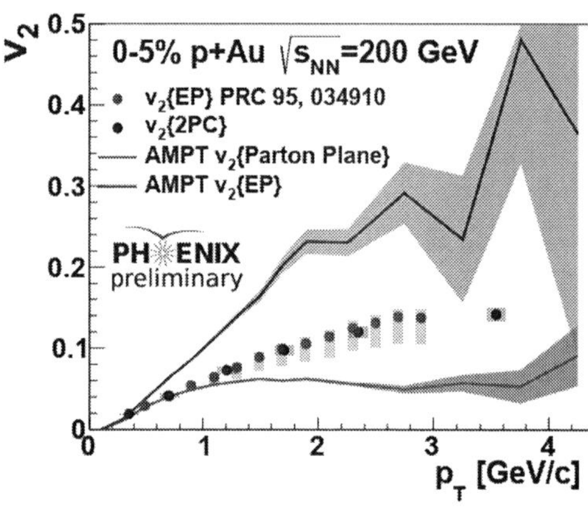

Figure 10. v_2 of charged hadrons as a function of p_T in the 0–5% centrality for p + Au collisions at $\sqrt{s_{NN}}$ = 200 GeV and comparison with AMPT simulation.

Figure 10 shows v_2 of charged hadrons as a function of p_T at mid-rapidity in 0–5% central events in p + Au $\sqrt{s_{NN}}$ = 200 GeV collisions. The new results of $v_2\{\text{2PC}\}$ with the three-sub detector method are consistent with the previous PHENIX results of $v_2\{\text{EP}\}$. In comparison with the AMPT simulation, the measured v_2 agree with the AMPT $v_2\{\text{Parton Plane}\}$ in $p_T < 1$ GeV/c, and the data in large p_T are laid in between the AMPT $v_2\{\text{EP}\}$ and $v_2\{\text{Parton Plane}\}$. The tendency between the data and AMPT simulation in p + Au collisions is different from that in the d + Au collisions, which shows a good agreement between the data and the AMPT $v_2\{\text{EP}\}$. In p + Au collisions, AMPT calculations

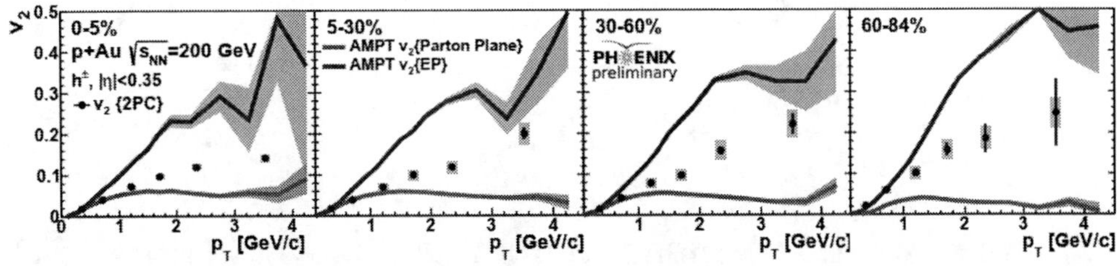

Figure 11. v_2 of charged hadrons as a function of p_T in various centrality classes of $p+\text{Au}$ collisions at $\sqrt{s_{NN}} = 200$ GeV and comparison with AMPT simulation.

overestimate $v_2\{\text{EP}\}$ in all centralities. Figure 11 shows $v_2\{\text{2PC}\}$ of charged hadrons as a function of p_T at mid-rapidity in various centrality classes of p + Au collisions at $\sqrt{s_{NN}} = 200$ GeV. The amplitude of v_2 in p + Au collisions is similar with the results in d + Au collisions, and the increasing trend of v_2 from central to peripheral collisions is shown both in p + Au and d + Au collisions. This trend is due to a larger non-flow contributions in peripheral collisions. The AMPT simulation shows a larger $v_2\{\text{EP}\}$ in p + Au than that in d + Au collisions, and this would be also related to the different non-flow contribution between the p + Au and d + Au collisions.

7 Conclusion

Collective-like behavior has been observed in small systems in PHENIX experiment. We measured v_2 and v_3 in d + Au and ^{3}He + Au collision systems. Data are well described by viscous hydro models such as SONIC [6]. In addition to the system size dependence, we studied v_2 dependencies on collision energy, centrality, rapidity, and transverse momentum. The present study indicates that the non-flow contributions in v_2 could be important in small collision systems. This contribution needs additional study to fully understand the nature of the observed correlations in small systems.

References

[1] CMS Collaboration, J. High Energy Phys. **09**, 091 (2010).
[2] PHENIX Collaboration, arXiv:1708.06983.
[3] S. Voloshin and Y. Zhang, Z. Phys. C **70**, 665 (1996).
[4] PHENIX Collaboration, Phys. Rev. C **95**, 034910 (2017).
[5] PHENIX Collaboration, Phys. Rev. Lett. **115**, 142301 (2015).
[6] M. Habich, J. L. Nagle, and P. Romatschke, Eur. Phys. J. C **75**, 15 (2015).
[7] Z.-W. Lin, C. M. Ko, B.-A. Li, B. Zhang, and S. Pal, Phys. Rev. C **72**, 064901 (2005).

Highlights of top quark measurements in hadronic final states at ATLAS

Serena Palazzo[1,2,*], on behalf of the ATLAS Collaboration

[1] *Università della Calabria*
[2] *INFN Cosenza*

Abstract. Measurements of inclusive and differential top quark production cross sections in hadronic final states, including hadronic τ decays, in proton-proton collisions with the ATLAS detector at the Large Hadron Collider are presented at center-of-mass energies of 7, 8 and 13 TeV. The inclusive cross section measurements reach high precision and are compared to other decay modes and the best available theoretical calculations. Differential measurements of the kinematic properties of top quark events are also discussed. These measurements use boosted top quarks, probing our understanding of top quark production in the TeV regime.

1 Introduction

The top quark was discovered by both the CDF [1] and DØ [2] collaborations in 1995 at the Tevatron collider. It is the most massive elementary particle known (m_t = 173.34 ± 0.27 (stat.) ± 0.71 (syst.) GeV [3]) and due to its very short lifetime, the top quark decays before hadronizing. This allows to experimentally test the properties of a bare quark. Top quark physics provides a sensitive probe to test the validity of the Standard Model (SM) and a tool to investigate the Higgs boson properties and to potentially discover physics beyond the SM. Some of the recent results achieved by the ATLAS collaboration [4], using data collected at a center-of-mass energy ($\sqrt{s}$) of 7, 8 and 13 TeV are here presented. A more complete list of results is available at the ATLAS public top quark results website [5].

2 Top quark pair production and decay

At leading order (LO), top quark pairs can be produced via two mechanisms: gluon-gluon fusion and $q\bar{q}$ annihilation. At the Tevatron collider, in $p\bar{p}$ collisions, the dominat process was the $q\bar{q}$ annihilation, whereas, at the LHC, in pp collisions, the dominant production process is the gluon-gluon fusion. The Feynman diagrams of both processes, at LO, are shown in Figure 1.

At next-to-leading order (NLO) the processes are a mixture of quark-gluon (qg) or antiquark-gluon ($\bar{q}$g). Some examples of the Feynman diagrams of these NLO processes are shown in Figure 2.

The top quark decays almost exclusively into a W boson and a bottom quark ($t \to W^+$ and $\bar{t} \to W^-$) since the CKM element V_{tb} is close to 1. The $t\bar{t}$ events are therefore classified according to the decays

*e-mail: serena.palazzo@cern.ch

© The Authors, published by EDP Sciences. This is an open access article distributed under the terms of the Creative Commons Attribution License 4.0 (http://creativecommons.org/licenses/by/4.0/).

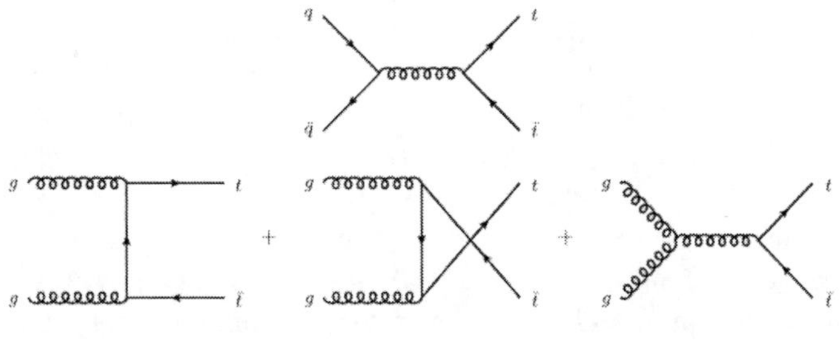

Figure 1. Top pairs production Feynman diagrams at LO. The two production processes are the gluon-gluon fusion and the $q\bar{q}$ annihilation.

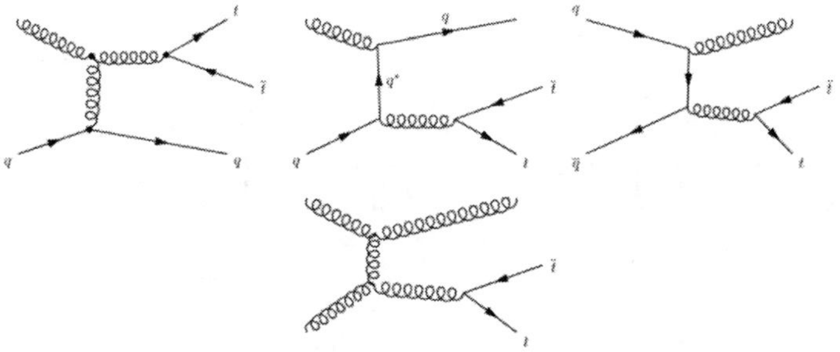

Figure 2. Examples of next-to-leading-order Feynman diagrams for top- antitop pair production via gluon fusion processes and the quark-anti-quark annihilation process.

mode of the two W bosons. When the two W bosons decay hadronically, the event is defined *fully hadronic*. This is the decay channel with the highest branching ratio ($\approx 44\%$) but it has also the largest background. When both the W bosons decay leptonically, the channel decay mode is called *di-lepton* and it has the lowest branching ratio ($\approx 10\%$) but it is the cleanest channel. Then, the last channel is the ℓ + jets, characterised by leptonic decay of one of the two W bosons and by the hadronic decay of the second W boson. This channel has a branching ratio of ($\approx 46\%$) and it is referred as *golden channel* because it has a moderate background.

3 $t\bar{t}$ cross-sections in the τ + jets channel at $\sqrt{s}$ of 7 and 8 TeV

The results of the top quark pairs cross sections in the τ+jets channel are here presented for the center-of-mass energies of $\sqrt{s}$ of 7 [6] and 8 TeV [7] with an integrated luminosity of 1.67 fb^{-1} and 20.1 fb^{-1}, respectively. The final states consist of a hadronically decaying τ lepton (τ_{had}) and jets corresponding to $t\bar{t} \rightarrow [b\tau_{\text{had}}\nu_\tau][bqq]$. Such an event topology corresponds to about 10% of the top quark pairs decays and this measurement is particularly important for charged Higgs boson production in top quark decays. In fact the existence of a charged Higgs boson would lead to an enhancement in the cross section for the considered top quark pair final state.

Starting with the measurement at 7 TeV, the candidate events are required to contain at least five jets with a transverse momentum (p_{T}) greater than 20 GeV. Two of these five jets must be originated from b-quarks. Concerning the other three jets, two of them must come from the hadronic decay of one of the two top quarks. The other jet, with high p_{T}, must be selected as τ candidate. This jet must have p_T greater than 40 GeV. The majority of τ decays are characterized by the presence of one or three charged hadrons in the final state. In this analysis, the number of tracks associated to a τ is used to separate the τ contribution from the background. The power of this separation is shown in Figure 3 where the distribution of the number of tracks is shown for the τ component (black line), the electron component (red dotted line) and the multijets events (blue line).

To extract the signal from the distribution of the number of tracks, the data sample is fitted with three probability density functions called also *templates*. These three templates are the *τ/electron* template which is obtained from $t\bar{t}$ simulated events; the remaining significant contributions are due to misidentified jets and are separated into two templates: the *gluon-jet* which describes the fakes jet coming from QCD multijet processes and the *quark-jet* template which describes the remaining processes (fakes-jet from $t\bar{t}$, W+jets and single-top quark events)

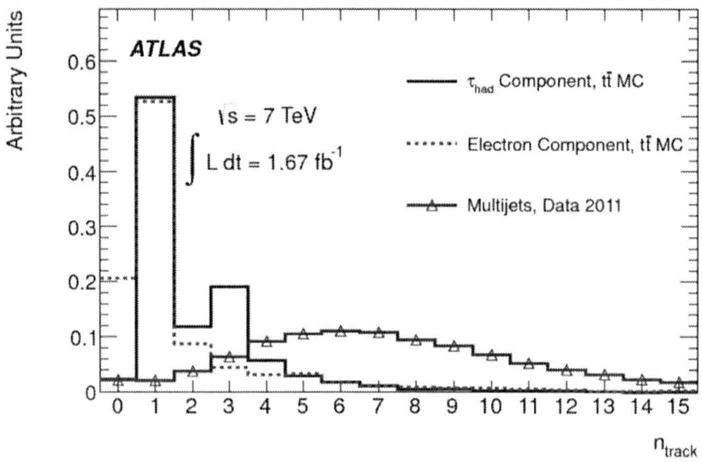

Figure 3. Distribution of the number of tracks for τ_{had} from Monte Carlo (MC) $t\bar{t}$ events (solid black line), electrons from MC $t\bar{t}$ events (dashed red line), and for jets from multijet events from data (blue triangles) [6].

A likelihood fit is used to extract the different contributions from the distribution of the number of tracks. The fit is shown in Figure 4 where are shown the events as a function of the number of tracks. To extract the number of signal events, predictions from simulation, are used to substract the background of W+jets and single-top quark events from the fitted number of τ/electron. This number is then scaled by the expected ratio that yield to the number of observed signal events from which the cross section is obtained. In this measurement, the leading uncertainties on the cross section come from the modeling of the initial and final state radiation (about 15%) and from the choice of the matrix element event generator (about 11%). The total systematic uncertainty on the cross section is 24%. Moving to the same analysis but using data collected at a center-of-mass-energy of 8 TeV, for this measurement four jets are required. At least two jets with missing transverse energy greater than 25 GeV and two jets originating from b-quark. Each event is also required to have at least one τ_{had} candidate which dacays into one or three charged particles.

The τ decays are identified as *single prong* (τ_1) when the τ decays to a single charged particle or *three prong* (τ_3) when it decays into three charged particle. To determine the number of signal events, two sources of background are estimated and subtracted: the first source of background containing real τ in the final state include single-top quark, W/Z+jets and diboson events; the other background

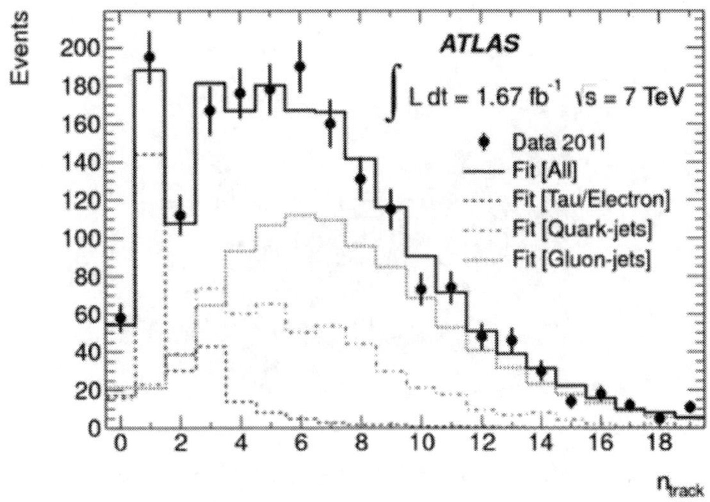

Figure 4. Distribution of the number of tracks for τ_{had} candidates after all selection cuts. The black point corresponds to the data while the solid black line is the fit result. Then the red dotted line correspond to the contribution from the τ/electron signal while the magenta and blue lines correspond to the quark-jet and gluon-jet background respectively [6].

source contains events where the τ in the final state is misidentified and is dominated by multijet processes. In order to determine the $t\bar{t}$ cross section the estimated backgrounds are substracted from the data. In Figure 5 the measured transverse momentum distribution is compared to the background and signal components and a good agreement is observed. The cross section is evaluated for both the τ_{had} that decays into one charged particle and for the τ_{had} that decays in three charged particles and then, the measurement is combined obtaining the total cross section $\sigma_{t\bar{t}}$ = 239 ± 4 (stat.) ± 28 (syst.) ± 5 (lumi.) pb. The total uncertainty on the measurement is 12%.

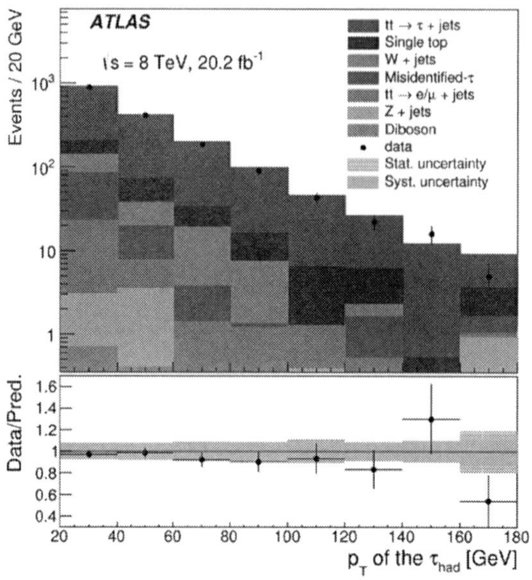

Figure 5. Distribution of the p_T of the τ_{had}. The measurements are compared to the predictions [7].

4 τ + lepton $t\bar{t}$ cross-section with $\sqrt{s}$ = 7 TeV

This analysis [8] uses a luminosity of 2.05 fb^{-1}. The final state has an electron (e) or a muon (μ) from the decay of the W bosons and an hadronically decaying of the τ lepton from the other W boson. This measurement is important because this channel is sensitive to the charged Higgs decays.

The event selection of this analysis is characterized by a primary vertex with at least five tracks, each one with a transverse momentum greater than 4 GeV. The additional conditions are: the presence of at least one τ candidate, only one isolated high p_T μ or e and a missing transverse energy greater than 30 GeV (to reduce the multijet background).

The τ identification methodology uses a *boosted decision tree discriminants* (BDT) to separate hadronic τ from jets. The BDT distributions could have a lepton and a τ candidate with the same sign, or a τ candidate and a lepton with opposite sign.

To estimate the background of this measurement, two different approaches are used. The first approach consists of a fit of the BDT distribution with a background and a signal template. The results of this method is reported on Table 1. The other method, used as a consistency check, is a matrix method based on a cut on the BDT output. The results of this method are shown in Table 2. As can be seen from the tables, the results obtained using the two methods, are in good agreement within the quoted uncertainties.

The cross section is derived from the difference between the same sign and the opposite sign events in the data sample with at least a b-tagged jet. The results are given separately for the *single prong* and *three prong* candidate events. The two results are then combined obtaining for the cross section the value $\sigma_{t\bar{t}}$ = 186 ± 13 (stat.) ± 20 (syst.) ± 7 (lumi.) pb, in agreement with the SM predictions. In Figures 6(a) and 6(b) are reported, respectively, the results of the BDT distributions of ℓ + τ events for the τ_1 type and the τ_3 type.

Table 1. Fit results using the background templates derived from 0 b-tag data and W+1jet. The MC columns give the expected τ signal, including also the contribution of non-$t\bar{t} \to \ell + \tau$ events and the expected number of $t\bar{t} \to \ell + \tau$ events after subtracting the contribution from non-$t\bar{t}$ events to the signal, assuming the theoretical $t\bar{t}$ cross section (164 pb) [8].

		Background template		MC	
		0 b-tag	W + 1jet	Signal	$t\bar{t}$
$\mu + \tau$	τ_1	490 ± 40	456 ± 32	432	388
	τ_3	135 ± 33	130 ± 50	126	116
$e + \tau$	τ_1	440 ± 50	430 ± 50	388	388
	τ_3	116 ± 32	120 ± 28	114	101
Combined	τ_1	930 ± 70	860 ± 50	820	726
	τ_3	260 ± 60	260 ± 40	239	217

Table 2. Matrix method, based on a cut on the BDT output [8].

		Background template	
		0 b-tag	W 1jet
$\mu + \tau$	τ_1	460 ± 60	440 ± 50
	τ_3	130 ± 40	105 ± 35
$e + \tau$	τ_1	420 ± 60	350 ± 50
	τ_3	140 ± 40	160 ± 40
Combined	τ_1	880 ± 70	800 ± 70
	τ_3	270 ± 60	260 ± 60

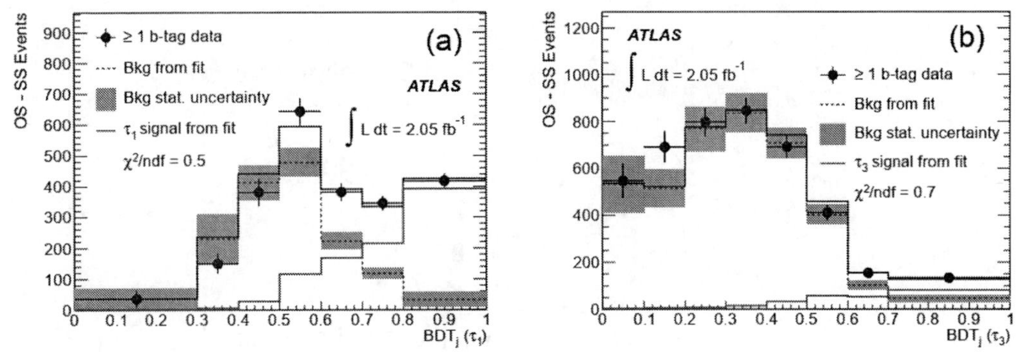

(a) (b)

Figure 6. BDT distributions of $\ell + \tau$ events for the τ_1 (a) type and the τ_3 (b) type. In the figures the τ_1 and τ_3 simulations are indicated by the red lines and the background by the dashed blue line. The blue band indicate the statistical uncertainty [8].

5 $t\bar{t}$ cross-section in the all-hadronic channel at $\sqrt{s}$ = 7 TeV

This analysis [9] uses 4.7 fb^{-1} of pp collision produced at the LHC and recorded by the ATLAS detector. The final state consists of the hadronically decay of the two W bosons and is characterized by a six-jet topology. This kind of event topology corresponds to 46% of $t\bar{t}$ decays, it has therefore a large branching ratio but also a large multijet background. This measurement is an important test of perturbative QCD and in addition allow to understand one of the major background to many new physics scenarios.

The events on this measurement are required to have at least a reconstructed primary vertex with five or more associated tracks. All jets are reconstructed with a jet vertex fraction smaller than 0.75 and p_T greater than 20 GeV. In addition at least five jets (out of which two should be b-tagged) are required with p_T greater than 55 GeV and $|\eta|$ smaller than 2.5. One more jet is required with p_T greater than 30 GeV and η less than 2.5. The dominant uncertainties are jet energy scale, b-tagging and and initial and final state radiation.

A kinematic fit has been performed to optimize the top quark mass reconstruction of $t\bar{t}$ events. This kinematic fit is based on a likelihood approach. The variable used to perform this fit and extract the cross section is the top quark mass distribution (m_t). In Figure 7 is shown the fit of this variable to the selected data.

The value of the measured cross-section is $\sigma_{t\bar{t}}$ = 168 ± 12 (stat.)$^{+60}_{-57}$(syst.) ± 7 (lumi.) pb and it is compatible with the SM predictions.

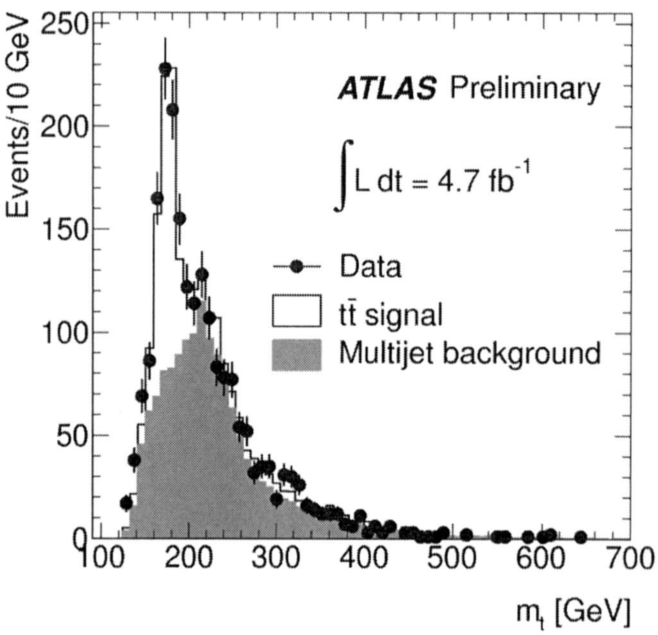

Figure 7. Fit of the m_t. The error bars of the data are only statistical. The $t\bar{t}$ signal in white and the multijet background in green are indicated as well [9].

6 $t\bar{t}$ differential cross-section in the all-hadronic channel at $\sqrt{s}$ = 13 TeV

This measurement [9] uses 14.7 fb^{-1} of data collected by the ATLAS detector in 2015 and in the first part of the 2016. In this analysis the final state consists of both W bosons decaying hadronically and are selected only top quark candidates with high p_T, the selected events contain therefore "fat jets". This measurement allows detailed studies of high p_T SM processes and searches of anomalies that could be signals of new physics. The event selection require for each event to have a primary vertex with five or more charged tracks. The reconstructed electron or muon should have p_T not greater than 25 GeV. The event should have at least 2 large-R jets with p_T greater than 350 GeV and the leading jet should have p_T greater than 500 GeV. Events are also required to have at least 2 small-R jets with p_T greater than 25 GeV and 2 small-R b-tagged jets each associated with one of the large-R jets. The largest background source came from multijets events. A data driven approach is employed to estimate the multijet background.

The differential cross section is obtained in a fiducial phase space at particle level using an unfolding tecnique that correct for detector effects. The differential cross sections have been obtained for different distributions. Just some of these distributions are here shown. In Figures 8 the hadronic top variables are shown: Figure 8(a) the normalised differential cross section of the leading p_T and in Figure 8(b) the normalised differential cross section of the second leading p_T. In general there is a good agreement between data and prediction within uncertainties.

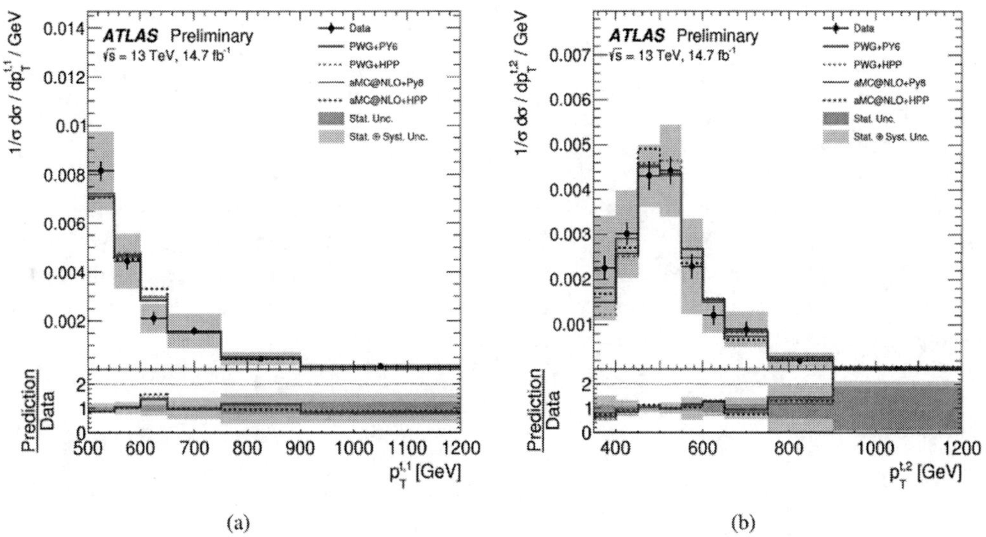

Figure 8. Normalised differential cross-section for the leading top quark p_T in Figure (a) and for the sub-leading top quark p_T in Figure (b) [10].

Two differential cross sections as a function of variables of the $t\bar{t}$ system are also shown in Figure 9. In Figure 9(a) the differential cross section for the $t\bar{t}$ system p_T is shown, while, in Figure 9(b) the differential cross section for the $t\bar{t}$ system mass is shown. The two distributions show a good agreement between data and predictions within the total uncertainties. In general the dominant uncertainties components are the large-R jets, the signal modeling and the b-tagging. The cross section measured is $\sigma_{fid} = 374 \pm 13\ (\mathrm{stat.})^{+111}_{-92}\ (\mathrm{syst.})$ pb and it is in agreement with the standard model predictions.

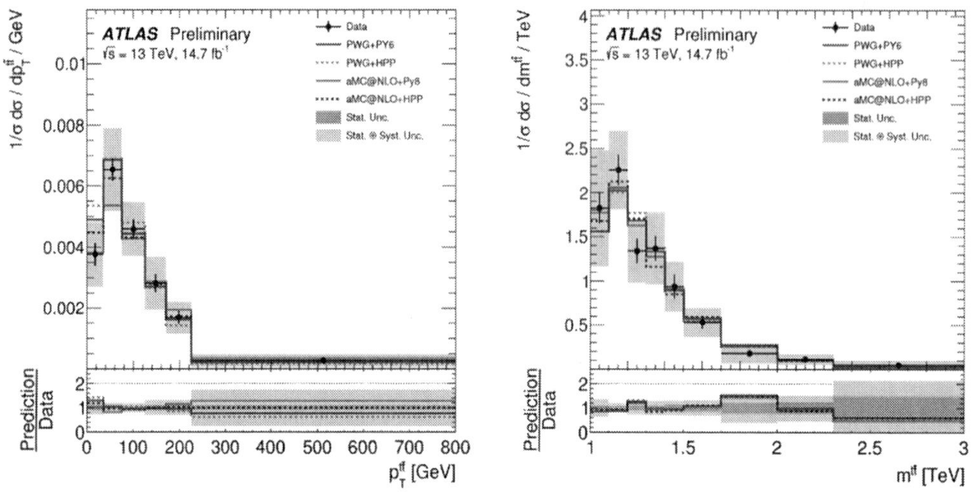

(a) (b)

Figure 9. Normalised differential cross-section for the leading top quark p_T in Figure (a) and for the sub-leading top quark p_T in Figure (b) [10].

7 Conclusions

High precision measurements of top quark inclusive and differential cross sections have been performed by the ATLAS experiment at 7, 8 and 13 TeV. The results shown are in good agreement with latest SM theory predictions. It has been shown that the ATLAS experiment is testing the SM at high precision with the cross section measurements. All the measurements will benefit from the incoming data samples which will allow to do more precise measurements.

References

[1] CDF Collaboration, Phys. Rev. Lett 8 (74) 2626 [hep-ex/9503002].

[2] D0 Collaboration, Phys. Rev. Lett. 8 (74) 2632 [hep-ex/9503003].

[3] ATLAS Collaboration, Measurement of the top quark mass in the $t\bar{t} \rightarrow$ dilepton channel from $\sqrt{s} = 8TeV$ ATLAS data, Phys. Lett. B 761 (2016) 350-371.

[4] ATLAS Collaboration, JINST 3 (2008) S08003.

[5] https://twiki.cern.ch/twiki/bin/view/AtlasPublic/TopPublicResults.

[6] ATLAS Collaboration, Measurement of the $t\bar{t}$ production cross section in the τ+jets channel using the ATLAS detector, Eur.Phys.J. C73 (2013).

[7] ATLAS Collaboration, Measurement of the $t\bar{t}$ production cross section in the τ + jets final state in pp collisions at $\sqrt{s} = 8$ TeV using the ATLAS detector, Phys. Rev. D 95, 072003.

[8] ATLAS Collaboration, Measurement of the top quark pair cross section with ATLAS in pp collisions at $\sqrt{s} = 7$ TeV using final states with an electron or a muon and a hadronically decaying τ lepton, Phys.Lett. B717 (2012) 89-108.

[9] ATLAS Collaboration, Measurement of the $t\bar{t}$ production cross section in the all-hadronic channel in ATLAS with $\sqrt{s} = 7$ TeV data, ATLAS-CONF-2012-031.

[10] ATLAS Collaboration, Measurements of $t\bar{t}$ differential cross-sections in the all-hadronic channel with the ATLAS detector using highly boosted top quarks in pp collisions at $\sqrt{s} = 13$ TeV, ATLAS-CONF-2016-100.

New results on low energy exclusive hadronic final states from BABAR

J. William Gary[1,★]

[1]*Department of Physics and Astronomy, University of California, Riverside, CA 92521, USA*

Abstract. The 3.6 standard deviation discrepancy between the standard model (SM) prediction for the muon anomalous magnetic moment $g_\mu - 2$ and the corresponding experimental measurement is one of the most persistent and intriguing potential signals in particle physics for physics beyond the SM. The largest uncertainty in the SM prediction for $g_\mu - 2$ arises from the uncertainty in the measured low energy inclusive $e^+e^- \to hadrons$ cross section. New results from the BABAR experiment at SLAC for the $e^+e^- \to \pi^+\pi^-\pi^0\pi^0$ and $e^+e^- \to KK\pi\pi$ cross sections are presented that significantly reduce this uncertainty. New BABAR results for other low energy exclusive hadronic processes are also discussed.

1 Introduction

The gyromagnetic ratio of the muon, g_μ, specifies the relationship between the muon's spin and its magnetic moment. The Dirac equation predicts $g_\mu = 2$ exactly. Radiative corrections alter this prediction to $g_\mu = 2(1 + a_\mu)$. The term $a_\mu = (g_\mu - 2)/2$ is known as the anomalous moment of the muon, or, for brevity, the "muon anomaly." The muon anomaly is much more sensitive to virtual heavy particle production in loops than the corresponding electron anomaly because the relative contributions of the virtual terms scale like $(m_\mu/m_e)^2 \approx 43,000$, where m_ℓ is the mass of lepton ℓ. Theory and experiment agree to high precision for the electron anomaly. However, for the muon anomaly, there is a tension between theory and experiment of about 3.6 standard deviations (see, for example, Ref. [1]). This long-standing discrepancy is considered to be one of the most intriguing and potentially most significant indications for physics beyond the standard model (SM).

In the SM the muon anomaly is expressed as $a_\mu = a_\mu^{QED} + a_\mu^{EW} + a_\mu^{had,LO} + a_\mu^{had,HO} + a_\mu^{had,LBLs}$, where "$QED$", "$EW$", "$had, LO$", "$had, HO$", and "$had, LBLs$" refer to the contributions from electromagnetic, weak, leading-order hadronic, higher-order hadronic, and hadronic light-by-light scattering terms, respectively. Example Feynman diagrams are shown in Fig. 1. Table 1 presents a summary of the relative size of the different terms, and their uncertainties. The electromagnetic term a_μ^{QED} is by far the largest term but has a small uncertainty. The second largest term, $a_\mu^{had,LO}$, from the leading-order hadronic contribution, has, in contrast, a large uncertainty. Summing the terms yields a total SM prediction of $a_\mu^{SM} = (1\,16\,591\,802 \pm 49) \times 10^{-11}$ [1], which lies $(287 \pm 80) \times 10^{-11}$ below the measured value of $a_\mu^{data} = (1\,16\,592\,091 \pm 63) \times 10^{-11}$ [3]. This difference corresponds to a discrepancy of about

★e-mail: bill.gary@ucr.edu

© The Authors, published by EDP Sciences. This is an open access article distributed under the terms of the Creative Commons Attribution License 4.0 (http://creativecommons.org/licenses/by/4.0/).

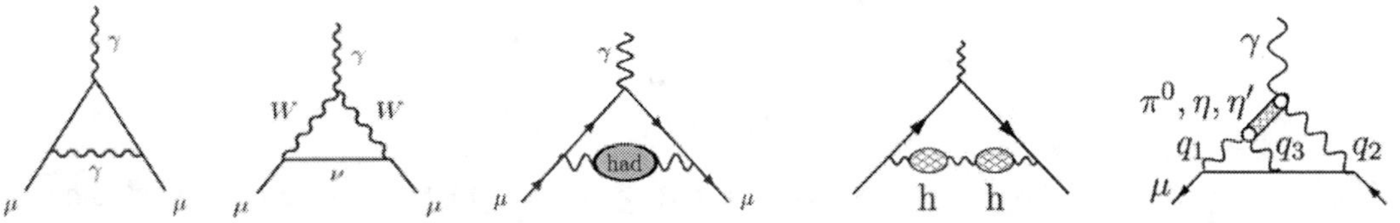

Figure 1. Feynman diagrams (from Ref. [2]) illustrating the contributions to the muon anomaly a_μ of (from left to right) electromagnetic, weak, leading-order hadronic, higher-order hadronic, and hadronic light-by-light scattering.

3.6 standard deviations. The overall uncertainty in the prediction, 49×10^{-11}, is dominated by the uncertainty of 42×10^{-11} in $a_\mu^{had,LO}$.

Table 1. SM results for the different contributions to the muon anomaly a_μ (from Ref. [1]).

a_μ term	SM prediction ($\times 10^{-11}$)
a_μ^{QED}	116584718.951 ± 0.080
a_μ^{EW}	153.6 ± 1.0
$a_\mu^{had,LO}$	6923 ± 42
$a_\mu^{had,HO}$	-98.4 ± 0.7
$a_\mu^{had,LbLs}$	105 ± 26
$a_\mu^{total,SM}$	116591802 ± 49

The longstanding tension in a_μ between experiment and theory could be an indication of new physics, such as from the supersymmetric process shown in Fig. 2 (left). The Muon $g-2$ experiment at Fermilab, with data collection scheduled to being in the Fall of 2017, expects to reduce the experimental uncertainty in a_μ by a factor of around 4 by 2019. The J-PARC E34 has a similar goal on a somewhat longer time scale. From the point of view of the SM prediction, the limiting factor is the uncertainty in the leading-order hadronic term $a_\mu^{had,LO}$.

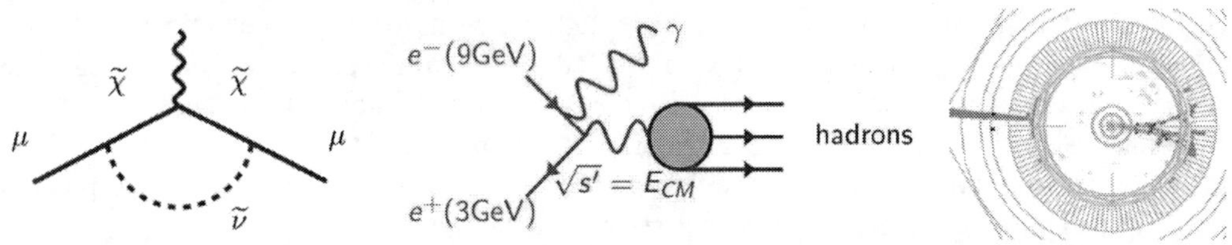

Figure 2. (left) Feynman diagram for a possible supersymmetric contribution to a_μ; (center) illustration of the ISR process; (right) example display of an ISR event, in which a recoiling hadronic system (on the right of the display) is opposite to a high energy ISR photon.

The relevant energy scales are too low for $a_\mu^{had,LO}$ to be calculated perturbatively. Lattice calculations [4] are not yet sufficiently precise. Instead, the most precise result for $a_\mu^{had,LO}$ comes from low energy $e^+e^- \to hadrons$ data, using the optical theorem and the following dispersion integral (see, for example, Ref. [5]):

$$a_\mu^{had,LO} = \left(\frac{m_\mu^2}{12\pi^3}\right) \int_{m_\pi^2}^{\infty} \frac{\hat{K}(s)}{s} \sigma_{e^+e^- \to hadrons}(s)\, ds, \tag{1}$$

where s is the square of the center-of-mass (CM) energy, $\hat{K}(s)$ is a kinematic factor with a magnitude on the order of unity, and $\sigma_{e^+e^- \to hadrons}$ is the inclusive $e^+e^- \to hadrons$ cross section. Because of the $1/s$ factor, low energy contributions ($\sqrt{s} < 2\,\text{GeV}$) dominate. At such low energies, the inclusive $e^+e^- \to hadrons$ cross section cannot be measured directly in a reliable manner because the simulations used to determine acceptance and reconstruction efficiencies are not reliable. Instead, one measures exclusive channels and sums their contributions, using isospin relations for unmeasured channels. Above around $\sqrt{s} = 1.8\,\text{GeV}$ one can start to use perturbative QCD, and also to directly measure the inclusive cross section. The BABAR experiment at SLAC has a longstanding program to measure exclusive cross sections for all possible hadronic states below $\sqrt{s} \approx 2.0\,\text{GeV}$ and to thereby contribute to the determination of the SM prediction for a_μ.

The BABAR experiment operated at the PEP-II asymmetric collider at SLAC between 1999 and 2008. Despite being shut down for almost 10 years, the data analysis is still active. There have been six BABAR papers submitted to journals so far in 2017, with a few others expected before the end of the year. The BABAR detector provided precise measurements of momentum and had excellent particle identification capabilities. Although it was primarily designed for the study of CP violation in B meson decays, its general purpose design made it suitable for a wide variety of other studies, such as those discussed here. The analyses presented here make use of the final BABAR data set collected in the vicinity of the $\Upsilon(4S)$ meson, corresponding to around $470\,\text{fb}^{-1}$ of data at $\sqrt{s} \approx 10.6\,\text{GeV}$.

The measurement of low energy hadronic cross sections is performed with the initial-state radiation (ISR) method. Thus events are required to contain a high-energy photon emitted by the incoming electron or positron, as illustrated in Fig. 2 (center), lowering the effective CM energy of the hadronic system. In the BABAR studies, the photon with the highest energy is assumed to be the ISR photon. It must have a CM energy larger than 3 GeV. The ISR events are easily recognizable, as illustrated by the example event display in Fig. 2 (right), and have high event acceptance. The rate of final-state photon radiation is negligible. The ISR method allows BABAR to access a wide range of energies in a single experiment: from threshold to around 5 GeV. Point-to-point systematic uncertainties are essentially eliminated, unlike measurements based on an energy scan, in which the e^+ and e^- beam energies are varied.

The study of intermediate resonance states in the hadronic systems is also important. Such studies shed light on the production process of hadrons and can be used to test theoretical models. From a practical point of view, knowledge of the intermediate resonance structure can be used to reduce systematic uncertainties in the acceptance and reconstruction efficiencies, since these terms depend on the intermediate state. For example, the angular distribution of particles depends on the intermediate state. Results from measurements of the intermediate resonance structure can be incorporated into Monte Carlo simulations, leading to a reduction in the uncertainties.

2 $e^+e^- \to \pi^+\pi^-\pi^0\pi^0$

The $e^+e^- \to \pi^+\pi^-\pi^0\pi^0$ cross section is one of the least known cross sections important for a_μ. The final BABAR results on this channel, presented in Ref. [6], are now available. These results supersede preliminary BABAR results from 2007 that were based on only around half the final BABAR data set. Exactly two charged-particle tracks, an ISR photon candidate, and at least four other photon candidates are required. A kinematic fit is made to the $e^+e^- \to \pi^+\pi^-\pi^0\pi^0$ signal hypothesis, constraining two 2γ combinations to the nominal π^0 mass. The combination of four photons that yields the smallest

$\chi^2_{4\pi\gamma}$ is retained. Signal candidates are required to have $\chi^2_{4\pi\gamma} < 30$. Background is subtracted using simulation normalized to data and a data sideband in $\chi^2_{4\pi\gamma}$. Around 150 000 signal events are obtained.

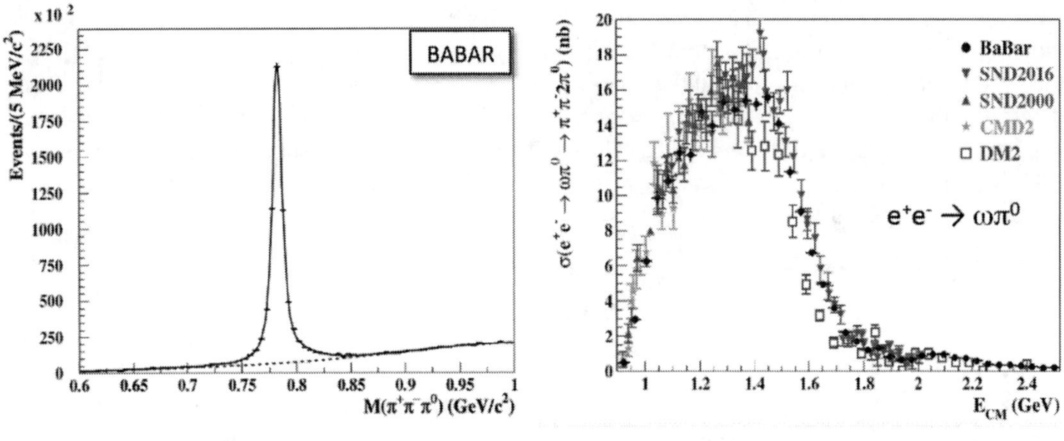

Figure 3. (left) The $\pi^+\pi^-\pi^0$ mass distribution in the selected sample of $e^+e^- \to \pi^+\pi^-\pi^0\pi^0$ events; (right) the $e^+e^- \to \omega\pi^0$ cross section.

A large fraction of the $e^+e^- \to \pi^+\pi^-\pi^0\pi^0$ cross section is found to be due to the $e^+e^- \to \omega\pi^0$ intermediate state with $\omega \to \pi^+\pi^-\pi^0$ decay. The $\pi^+\pi^-\pi^0$ mass distribution in the selected $e^+e^- \to \pi^+\pi^-\pi^0\pi^0$ events is shown in Fig. 3 (left). A prominent peak at the ω meson mass is seen. Our measurement of the $e^+e^- \to \omega\pi^0$ cross section is shown in Fig. 3 (right). The BABAR results are seen to be more precise than previous measurements, cover a wider energy range, and resolve discrepancies between the measurements of other experiments.

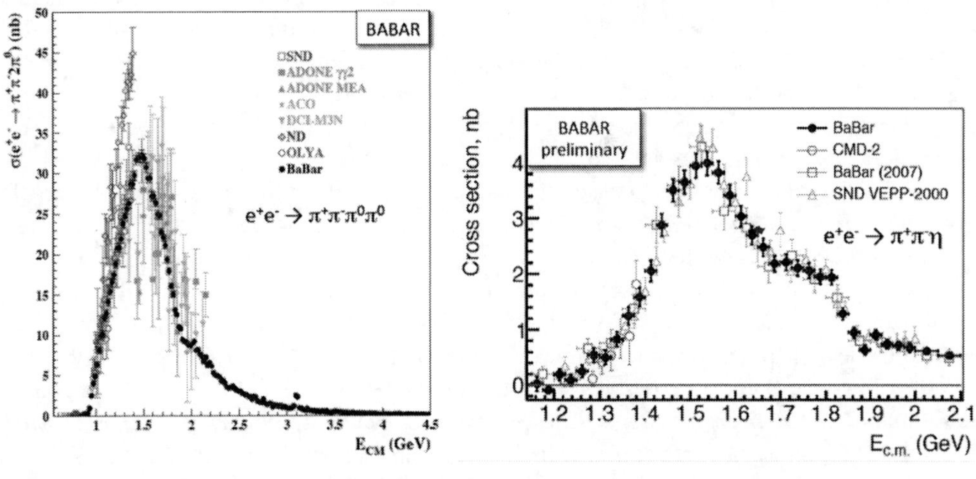

Figure 4. The (left) $e^+e^- \to \pi^+\pi^-\pi^0\pi^0$ and (right) $e^+e^- \to \pi^+\pi^-\eta$ cross sections.

Our result for the $e^+e^- \to \pi^+\pi^-\pi^0\pi^0$ cross section is shown in Fig. 4 (left). Again, the BABAR results are far more precise than previous results and cover a much wider energy range.

3 $e^+e^- \to \pi^+\pi^-\eta$ with $\eta \to \gamma\gamma$

A second new result from BABAR, this one still preliminary and presently without a publicly available document, is a measurement of the $e^+e^- \to \pi^+\pi^-\eta$ cross section, based on $\eta \to \gamma\gamma$ decays. This result complements and improves the precision of the BABAR result for $e^+e^- \to \pi^+\pi^-\eta$ from 2007 [7], which was based on around half the final BABAR data set and utilized $\eta \to \pi^+\pi^-\pi^0$ decays. For the new result and for all other studies presented here, similar analysis techniques to those used to determine the $e^+e^- \to \pi^+\pi^-\pi^0\pi^0$ cross section, discussed above, are employed. The result for the $e^+e^- \to \pi^+\pi^-\eta$ cross section, based on a selected data sample of around 8000 signal event candidates, is shown in Fig. 4 (right). The measurements are seen to agree with previous results and to improve the precision, especially for higher values of $\sqrt{s}$.

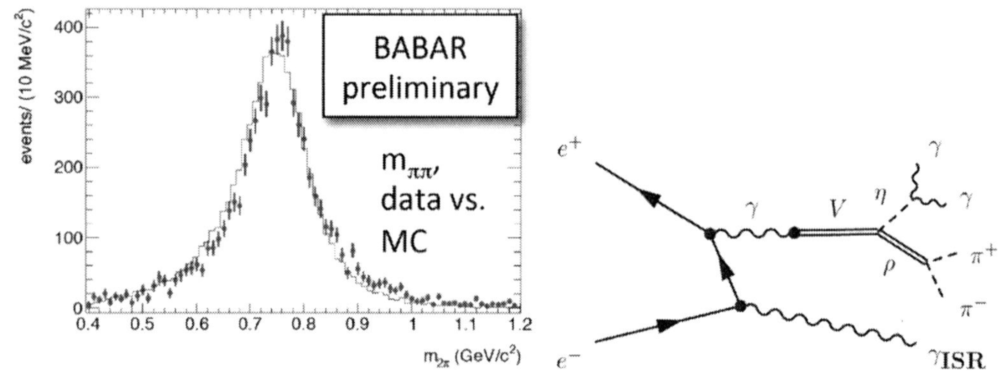

Figure 5. (left) The $\pi^+\pi^-$ mass distribution in the selected $e^+e^- \to \pi^+\pi^-\eta$ events; (right) illustration of the vector meson dominance model for $e^+e^- \to \pi^+\pi^-\eta$.

We observe the $e^+e^- \to \pi^+\pi^-\eta$ reaction to be dominated by the $\rho(770)\eta$ intermediate state, as illustrated in Fig. 5 (left). However, the $e^+e^- \to \pi^+\pi^-\eta$ cross section has a complicated structure, as seen from Fig. 4 (right). We can use the vector meson dominance model, based on production through the $\rho(770)$ and higher mass ρ states [see Fig. 5 (right)], to try to explain this structure. The coupling constants governing the decays are approximately real. Thus the phase differences between the different ρ meson excitations are taken to be zero and π only.

A model with two intermediate ρ mesons, in which the $\rho(1450)$ is included with a phase difference of zero relative to the $\rho(770)$, does not fit the data and is discarded. However, a model in which these two ρ mesons have a phase difference of π yields the dashed curve (model 1) in Fig. 6, which is seen to describe the data up to $\sqrt{s} \approx 1.7$ GeV. Adding a third resonance, the $\rho(1700)$, with either a relative phase of π or zero yields the red-solid and dotted curves in Fig. 6 (models 2 and 3, respectively). These two models each describe the data up to approximately 1.9 GeV. Adding the $\rho(2150)$ meson (model 4) allows the data to be described up to $\sqrt{s} \approx 2.2$ GeV, including the rather complicated structure in the 1.8 to 2.2 GeV region. However, even for model 4, the data lie above the prediction for $\sqrt{s} > 2.3$ GeV, implying that yet higher mass ρ resonances may be relevant for the $e^+e^- \to \pi^+\pi^-\eta$ process. A summary of the different models tested is given in Table 2.

4 $e^+e^- \to K_SK_L\pi^0$, $K_SK_L\eta$, and $K_SK_L\pi^0\pi^0$

BABAR recently published results on the cross sections for the reactions $e^+e^- \to K_SK_L\pi^0$, $K_SK_L\eta$, and $K_SK_L\pi^0\pi^0$ [8]. These results are presented in this section.

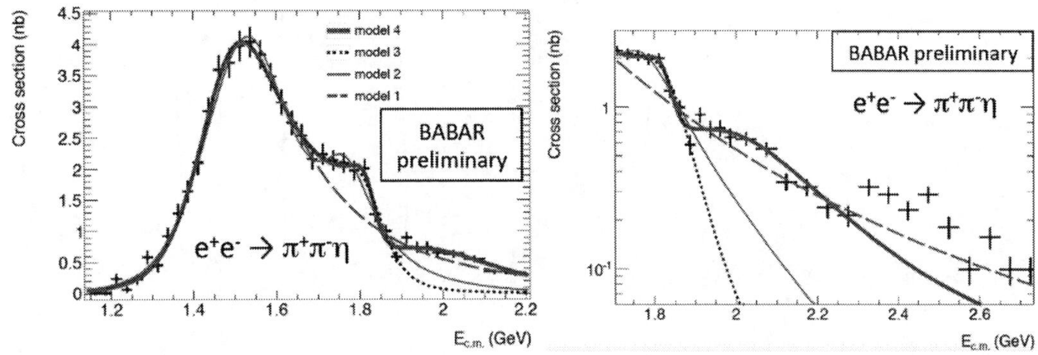

Figure 6. The measured $e^+e^- \to \pi^+\pi^-\eta$ cross section in comparison to predictions based on the vector meson dominance model: (left) for the full spectrum of measured results; (right) expanded view for $\sqrt{s} > 1.7\,\text{GeV}$.

Table 2. Summary of different variants of the vector meson dominance model used to attempt to describe the $e^+e^- \to \pi^+\pi^-\eta$ cross section data.

Model	Resonance model	Good fit for:
0	$\rho(770) + \rho(1450)$	Doesn't fit
1	$\rho(770) - \rho(1450)$	$\sqrt{s} < 1.7\,\text{GeV}$
2	$\rho(770) - \rho(1450) - \rho(1700)$	$\sqrt{s} < 1.9\,\text{GeV}$
3	$\rho(770) - \rho(1450) + \rho(1700)$	$\sqrt{s} < 1.9\,\text{GeV}$
4	$\rho(770) - \rho(1450) + \rho(1700) + \rho(2150)$	$\sqrt{s} < 2.2\,\text{GeV}$

Around 3700 signal events are obtained for $e^+e^- \to K_S K_L \pi^0$. The cross section is shown in Fig. 7 (left). This is the first measurement of this process. A peak is seen at the J/ψ meson mass. This is the first observation of $J/\psi \to K_S K_L \pi^0$ decays. (For all other reactions discussed in the remainder of this report, the same statement applies: we obtain the first observation of J/ψ decays to the presented channel.) The dominant intermediate state for $e^+e^- \to K_S K_L \pi^0$ is found to be $K^*(892)K$ [see Fig. 7 (right)]. Production through the $K^*(1430)K$ and $\phi(\to K_L K_S)\pi^0$ channels is also seen.

For the $e^+e^- \to K_S K_L \eta$ channel, we obtain 864 signal events. The measured cross section, which represents first measurement of this process, is shown in Fig. 8 (left). The process is dominated by the $e^+e^- \to \phi\eta$ intermediate resonant state, as illustrated in Fig. 8 (right).

Fig. 9 (left) shows the cross section for $e^+e^- \to K_S K_L \pi^0 \pi^0$, based on a sample of 392 signal events. This is the first measurement of this process. The reaction exhibits clear evidence for the $K^*(892)K\pi$ intermediate resonant state [9 (right)], although in this case the nonresonant component also is substantial.

5 $e^+e^- \to K_S K^+ \pi^- \pi^0$ and $K_S K^+ \pi^- \eta$

In another recent publication [9], we present the first measurements of the $e^+e^- \to K_S K^+ \pi^- \pi^0$ and $K_S K^+ \pi^- \eta$ cross sections.

The result for the $e^+e^- \to K_S K^+ \pi^- \pi^0$ channel, obtained from a selected sample of about 6400 signal events, is shown in Fig. 10 (left). A very large J/ψ signal is seen. The $K^*(892)K\pi$ and $K_S K^+ \rho(770)^-$ intermediate resonant states are dominant [Fig. 10 (center) and (right)]. A clear

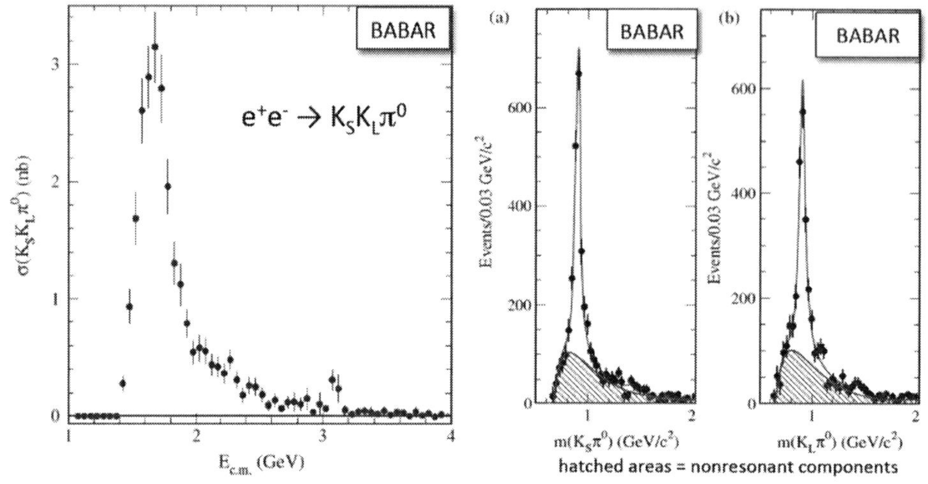

Figure 7. (left) The measured $e^+e^- \to K_S K_L \pi^0$ cross section; (right) the $K_S\pi^0$ and $K_L\pi^0$ mass distributions for the selected $e^+e^- \to K_S K_L \pi^0$ events.

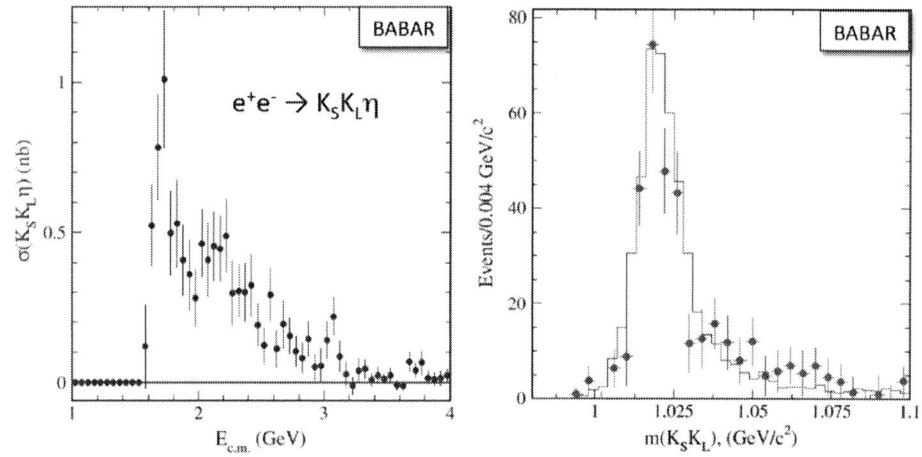

Figure 8. (left) The measured $e^+e^- \to K_S K_L \eta$ cross section; (right) the $K_S K_L$ mass distribution for the selected $e^+e^- \to K_S K_L \eta$ events.

$K^*(1430)K\pi$ signal is also seen. In addition, the double-resonant $K^*(892)K^*(892)$ intermediate channel is found to be responsible for around 15% of the total $e^+e^- \to K_S K^+\pi^-\pi^0$ cross section.

The result for the $e^+e^- \to K_S K^+\pi^-\eta$ cross section, based on a selected sample of 358 events, is shown in Fig. 11 (left). The results shown in Fig. 11 (center) and (right) demonstrate that much of this cross section arises from the resonant $e^+e^- \to K^*(892)K\eta$ process, particularly in the $K^*(892)^\pm \to K_S\pi^\pm$ channel.

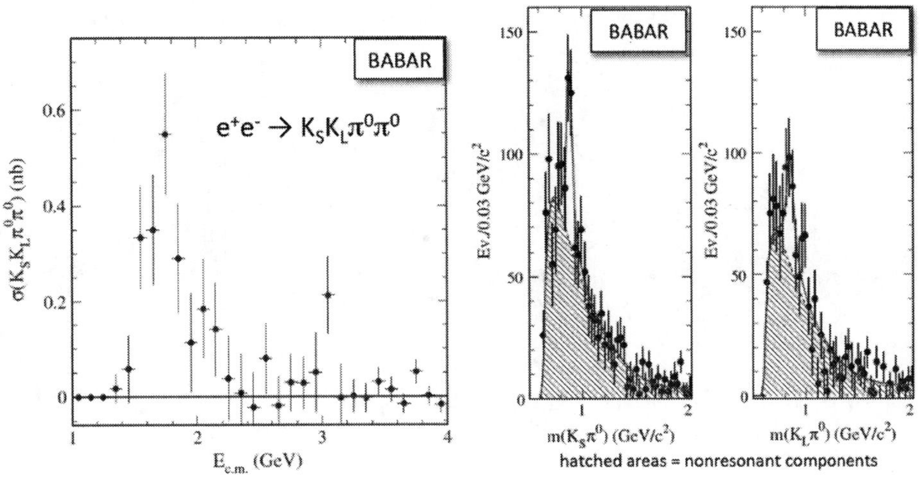

Figure 9. (left) The measured $e^+e^- \to K_S K_L \pi^0\pi^0$ cross section; (right) the $K_S\pi^0$ and $K_L\pi^0$ mass distributions for the selected $e^+e^- \to K_S K_L \pi^0\pi^0$ events.

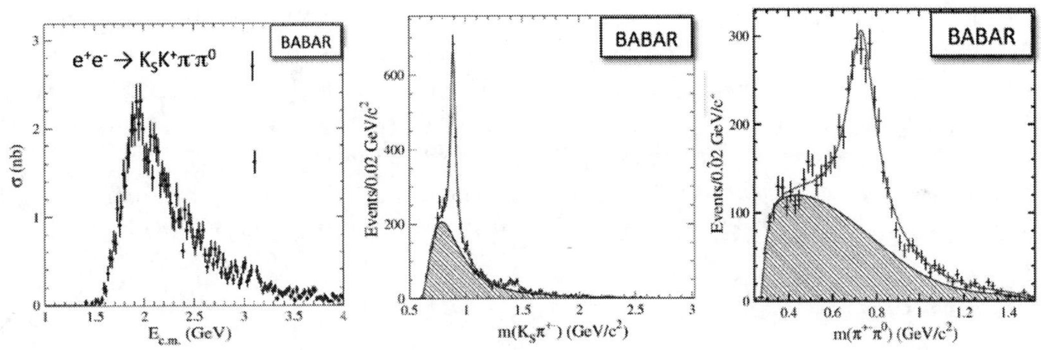

Figure 10. (left) The measured $e^+e^- \to K_S K^+\pi^-\pi^0$ cross section; the (center) $K_S\pi^\pm$ and (right) $\pi^\pm\pi^0$ mass distributions for the selected $e^+e^- \to K_S K^+\pi^-\pi^0$ events.

6 Implications for the muon anomaly a_μ

With respect to the muon anomaly a_μ, the new BABAR $\pi^+\pi^-\pi^0\pi^0$ cross section measurements yield $a_\mu^{\pi^+\pi^-\pi^0\pi^0} = (179 \pm 6) \times 10^{-11}$ (3.2% precision). The world average previously was $a_\mu^{\pi^+\pi^-\pi^0\pi^0} = (167 \pm 13) \times 10^{-11}$ (7.9% precision). Thus the new BABAR $\pi^+\pi^-\pi^0\pi^0$ data reduce the uncertainty in $a_\mu^{\pi^+\pi^-\pi^0\pi^0}$ by more than a factor of 2.

With the new results for $e^+e^- \to K_S K_L \pi^0$, $K_S K_L \pi^0\pi^0$, and $K_S K^+\pi^-\pi^0$, in conjunction with previous BABAR results published these past few years, BABAR has now measured all $e^+e^- \to KK\pi$ and $KK\pi\pi$ cross sections except those with a $K_L K_L$ pair. Thus, for the first time, the corresponding contributions to the muon anomaly, $a_\mu^{KK\pi}$ and $a_\mu^{KK\pi\pi}$, can be determined with no assumptions or reliance on isospin relations, except to assume the $K_L K_L$ rates to be the same as the $K_S K_S$ rates, which follows from CP symmetry. The $KK\pi$ channels comprise 12% of the total $e^+e^- \to$ *hadrons* cross section at $\sqrt{s} \approx 1.6$ GeV. The $KK\pi\pi$ channels comprise 25% of the total $e^+e^- \to$ *hadrons* cross

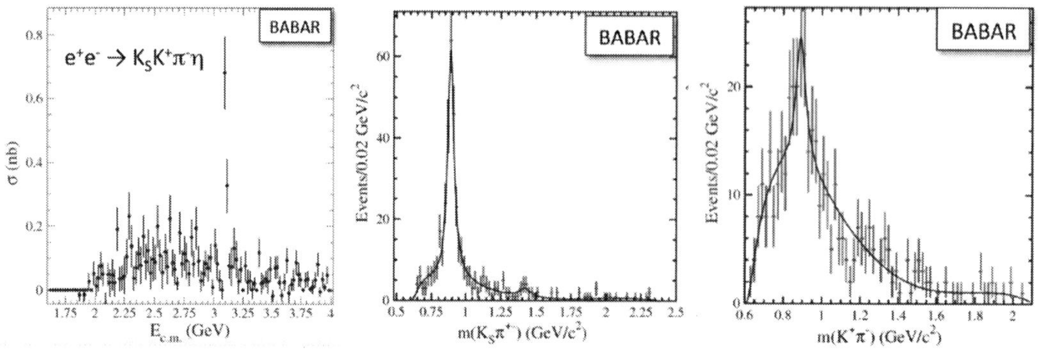

Figure 11. (left) The measured $e^+e^- \to K_S K^+\pi^-\eta$ cross section; the (center) $K_S\pi^\pm$ and (right) $K^+\pi^-$ mass distributions for the selected $e^+e^- \to K_S K^+\pi^-\eta$ events.

section at $\sqrt{s} \approx 2\,\text{GeV}$. Beyond their importance for the muon anomaly, the new BABAR $KK\pi$ and $KK\pi\pi$ data will therefore be important in upcoming tests of the perturbative QCD prediction for the total $e^+e^- \to hadrons$ cross section at $\sqrt{s} \approx 2\,\text{GeV}$.

Using the new BABAR data, the result for $a_\mu^{KK\pi}$ is $(24.5 \pm 1.5) \times 10^{-11}$. However, this represents only a small improvement in precision compared to the previous result: $a_\mu^{KK\pi} = (23.9 \pm 1.6) \times 10^{-11}$. In contrast, for $a_\mu^{KK\pi\pi}$, the improvement in precision is very substantial: the result based on the new BABAR data is $(8.5 \pm 0.5) \times 10^{-11}$ (6% precision) compared to the previous result $(13.5 \pm 3.9) \times 10^{-11}$ (30% precision). The previous $a_\mu^{KK\pi\pi}$ result primarily relied on isospin relations.

In the big picture, the overall uncertainty in the muon anomaly a_μ is reduced from 42×10^{-11} (Table 1) to 33×10^{-11} [10], representing a 20% improvement, because of the new BABAR data.

7 Summary

Low energy $e^+e^- \to hadrons$ cross section data currently provide the most accurate prediction for $a_\mu^{had,LO}$, which is the component of the of the muon anomalous magnetic moment a_μ associated with leading-order hadronic contributions. The uncertainty in $a_\mu^{had,LO}$ is the dominant uncertainty in the SM prediction for a_μ. The $e^+e^- \to hadrons$ data are also important because they yield information on hadron dynamics and allow tests of QCD, such as for the perturbative QCD prediction for the inclusive $e^+e^- \to hadrons$ cross section. The data can also be used to provide (for example) the first observations of certain J/ψ and $\psi(2S)$ branching fractions.

The new BABAR data reduce the uncertainty in the contribution to $a_\mu^{had,LO}$ from the $\pi^+\pi^-\pi^0\pi^0$ final state from 7 to 3%. For $KK\pi\pi$ states, the uncertainty is reduced from 30 to 6%. The overall uncertainty in $a_\mu^{had,LO}$ is reduced by about 20%. Future progress in reducing the uncertainty in $a_\mu^{had,LO}$ will need to come from a reduction in the systematic uncertainty in the all-important $e^+e^- \to \pi^+\pi^-$ channel. New results on $e^+e^- \to \pi^+\pi^-$ are expected from the BABAR and CMD3 experiments in the next year or so, which will hopefully resolve discrepancies that exist between current measurements.

References

[1] M. Davier, A. Hoecker, B, Malaescu, and Z. Zhang, Z. Phys. C **71**, 1515 (2011).
[2] F. Jegerlehner and A. Nyffeler, Phys. Rept. **477**, 1 (2009).

[3] E821 Collaboration, G. W. Bennett et al., Phys. Rev. D **73**, 072003 (2006).
[4] M. Della Morte, B. Jäger, A. Jüttner, and H. Wittig, JHEP **03**, 055 (2012).
[5] Particle Data Group, C. Patrignani et al., Chin. Phys. C, **40**, 100001 (2016) and 2017 update.
[6] BABAR Collaboration, J. P. Lees et al., arXiv:1709.01171, Submitted to Phys. Rev. **D**.
[7] BABAR Collaboration, J. P. Lees et al., Phys. Rev. D **76**, 092005 (2007).
[8] BABAR Collaboration, J. P. Lees et al., Phys. Rev. D **95**, 052001 (2017).
[9] BABAR Collaboration, J. P. Lees et al., Phys. Rev. D **95**, 092005 (2017).
[10] M. Davier, arXiv:1612.02743, Nucl. Part. Phys. Proc. **287–288** (2017) 70.

Systematic study of high p_T hadron production in small collision systems by the PHENIX experiment at RHIC

Takao Sakaguchi[1,*], *for the* PHENIX Collaboration

[1] *Brookhaven National Laboratory, Physics Department, Upton, NY 11973-5000, USA.*

Abstract. High p_T hadrons in the small systems of $d/^3$He+Au and p+Al/Au collisions have been measured at the midrapidity ($|\eta| < 0.35$) as well as at the forward and backward rapidities ($1.2 < |\eta| < 2.4$) at $\sqrt{s_{NN}}$ =200 GeV. A clear system and centrality ordering in the $R_{p/d/\mathrm{HeA}}$ was observed for midrapidity π^0 production. Hadron R_{cp} as a function of rapidity in forward and backward regions show an opposite trend to what HIJING++ predicted. Together with the difference of R_{cp} in p+Al/Au collisions, the observation can be understood consistently with the measured long-range two-particle correlations, hinting the possible medium creation similar to A+A collisions.

1 Introduction

The small collision systems such as p/d+A collisions have been considered a good laboratory to quantify cold nuclear matter effects, a necessary baseline for understanding the effects of the hot and dense medium produced in A+A collisions. This assumption has been confirmed by several measurements such as high transverse momentum (p_T) hadrons, jets, and direct photons, in the minimum bias d+Au collisions at $\sqrt{s_{NN}}$ =200 GeV [1–3]; the nuclear modification factors, $R_{d\mathrm{A}} \equiv (1/T_{d\mathrm{A}})(dN^{d\mathrm{A}}/dp_{\mathrm{T}}d\eta)/(d\sigma^{pp}/dp_{\mathrm{T}}d\eta)$, where the $T_{d\mathrm{A}}$ is the nuclear thickness function, are consistent with unity up to high p_T, and the slight deviation from the unity is consistent with the expectation by parton distribution functions such as EPS09 [4]. The observation of the ridge-like structure in the long-range azimuthal correlations in the central p+Pb collisions at $\sqrt{s_{NN}}$ =5.02 TeV at the LHC [5–7], however, called into question the view of such systems as consisting merely of cold nuclear matter. The study at the LHC was followed by the PHENIX experiment at RHIC, and a finite v_2 of hadrons in 0–5 % central d+Au collisions using both the two-particle angular correlation method and the event-plane method were observed [8–10]. These observations led the community to explore any phenomena found in A+A collisions, in p/d+A collisions.

The energy loss of hard scattered partons produced in the initial stage of the collisions, so-called jet quenching, has been one of the key observations in confirming the production of the QGP. The first evidence of this phenomena was found in the yield suppression of high transverse momentum (p_T) hadrons, the fragments of the hard scattered partons. The measurement of high p_T identified hadrons have been improved over the last decade, and reached to the level that a precise quantitative comparison of the data and theoretical models became realized [11, 12]. A recent study also found that the energy loss scales with the particle multiplicity ($dN_{\mathrm{ch}}/d\eta$) [13]. Obviously, the high p_T hadrons

* e-mail: takao@bnl.gov

© The Authors, published by EDP Sciences. This is an open access article distributed under the terms of the Creative Commons Attribution License 4.0 (http://creativecommons.org/licenses/by/4.0/).

will be a powerful tool to investigate the parton degree of freedom in the small systems like p/d+A collisions as well. In addition, a systematic study of the high p_T hadron spectra from small to large collision systems will be able to explore the onset of QGP as a function of the collision systems. In this paper, we show a new systematic measurement of the high p_T π^0 in $p/d/^3$He+Au collisions at midrapidity ($|\eta|$ <0.35) as well as the high p_T hadrons in p+Al/Au collisions at forward and backward rapidities ($1.2 < |\eta| < 2.4$), and discuss the hint of possible hot medium creation in connection with the long-range azimuthal correlation observed.

2 Measurement Method

The detector setup was the same as the one in the RHIC Year-2012 run as shown in Fig. 1. The

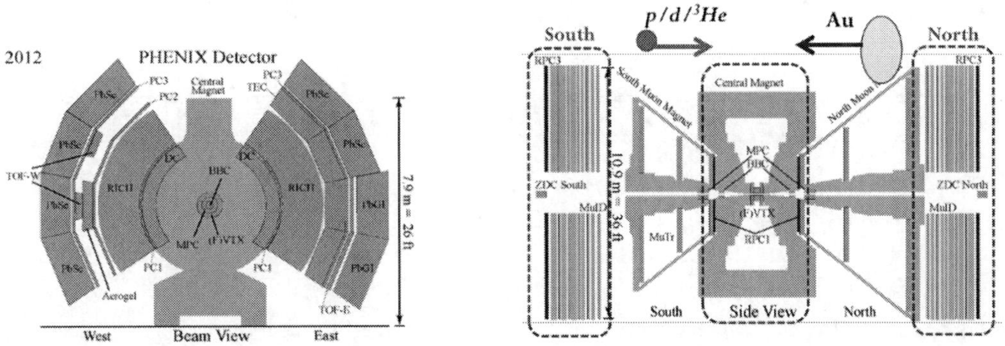

Figure 1. (Left) Beam view of the PHENIX detector in the RHIC Year-2012 run and later. (Right) Side view of the PHENIX detector with the beam direction of$p/d/^3$He and Au ions shown. The central and muon arms used for this analysis are shown in the squared dotted lines. Note that Al ions go to the same directions as Au ions.

detailed description of the PHENIX detector system can be found elsewhere [14]. The π^0's were reconstructed via $\pi^0 \rightarrow \gamma\gamma$, by primarily using a lead-scintillator sandwich type electromagnetic calorimeter (PbSc EMCal) in the midrapidity ($|\eta| < 0.35$). The threshold of cluster energy is set to 0.2 GeV and the photon clusters were selected using a shower shape cut. Then, an energy asymmetry cut of $\alpha = |E_1 - E_2|/(E_1 + E_2) < 0.8$ was applied on selecting pairs of photons from π^0 decay. The efficiency and acceptance of π^0's were estimated using a GEANT based detector simulation software. The hadrons in the forward and backward rapidity ($1.2 < |\eta| < 2.4$) were measured by a muon tracker (MuTr), a silicon vertex detector (FVTX), and a muon identification detector (MuID). The tracks that left signals in MuTr, FVTX and the first two sensitive layers of MuID are identified as high p_T hadrons.

This analysis used the events recorded by PHENIX in RHIC Year-14 (^{3}He+Au) and Year-15 (p+Al/Au) runs. We used the d+Au results from the literature published before [1]. The integrated luminosities recorded are 25 nb^{-1} (15 pb^{-1} pp-equivalent) for ^{3}He+Au collisions, 275 nb^{-1} (7.4 pb^{-1} pp-equivalent) for p+Al collisions, and 80 nb^{-1} (16 pb^{-1} pp-equivalent) for p+Au collisions. Three types of events were used in this analysis; the first type of events were triggered by the coincidence of signals from the two Beam-Beam counters (BBC) located at $3.1 < |\eta| < 3.9$ covering the full azimuth (minimum bias triggered events), the second type of events were triggered by a high energy tower hit in the EMCal coincided with a minimum bias trigger (ERT triggered events), and the third type of events were triggered by a high momentum track identified by a MuTr and MuID coincided with a minimum bias trigger (Muon-arm triggered events). The minimum bias trigger is not 100 % efficient to the inelastic collisions because of the limited acceptance and efficiency of the BBC. This inefficiency increases as the collision system becomes smaller. They were already studied in d+Au collisions by comparing the charge of the south BBC with a Glauber Monte Carlo simulation folded with a negative binomial distribution [15]. From this comparison, we determined that the trigger efficiency is 88 % for d+Au collisions. We followed the same method, and determined the trigger efficiency of p+Al, p+Au and ^{3}He+Au collisions also as 74 %, 84 % and 88 %, respectively. In case of 200 GeV Au+Au collisions, the efficiency was 94 % [12]. When dividing the events into centralities, an additional bias factor plays a role. The bias is originated from a correlation of particle yield in the rapidity range for the spectra measurements, and that in the rapidity range for the centrality measurement which is performed by the south BBC (-3.9< $|\eta|$ <-3.1). We estimated the bias factors for the ^{3}He+Au and p+Al/Au collisions by following the method applied for d+Au collisions [15], and determined as listed in Table 1. All the results (including minimum bias and centrality-dependent) shown

Table 1. Bias factors for centrality selections in small collision systems.

Cent (%)	0–5	5–10	10–20	0–20	20–40	40–60	40–72	60–84	60–88	0–100
p+Al	0.75	0.81	0.84	-	0.90	-	1.04	-	-	-
p+Au	0.86	0.90	0.94	0.90	0.98	1.02	-	1.00	-	0.86
d+Au	0.86	0.90	0.94	0.90	0.98	1.02	-	-	1.00	0.86
^{3}He+Au	0.86	0.90	0.94	0.90	0.98	1.02	-	-	1.00	0.86

in this paper are divided by these numbers after being corrected for acceptances and efficiencies. The numbers for 0–100 % will be used to obtain 0–100 % centrality result from the minimum bias data of the given collision system. For instance, in order to obtain 0–100 % result for d+Au collisions, the spectra for minimum bias events (0–88 % in this case) will be divided by 0.86.

3 Results

3.1 Mid-rapidity measurement

Figure 2 shows $R_{p/d/\mathrm{HeA}}$ of the π^0's for the minimum bias p+Au, d+Au, and ^{3}He+Au collisions at $\sqrt{s_{NN}}$ =200 GeV. $R_{p/d/\mathrm{HeA}}$ is defined as $R_{p/d/\mathrm{HeA}} \equiv (1/T_{p/d/\mathrm{HeA}})(dN^{p/d/\mathrm{HeA}}/dp_\mathrm{T}d\eta)/(d\sigma^{pp}/dp_\mathrm{T}d\eta)$, where the $T_{p/d/\mathrm{HeA}}$ is the nuclear thickness function. This measurement was performed at the midrapidity ($|\eta|$ <0.35). The magnitudes of the peaks seen at $p_\mathrm{T} \sim 5\,\mathrm{GeV}/c$, which are often called as Cronin-peak and understood as the consequence of the initial momentum broadening in the nucleus, are found to change as a function of the collision systems. As the projectile becomes lighter, the magnitude of the peak becomes larger. There is a common feature over the systems that some hint of suppression is seen at high p_T ($p_\mathrm{T} > 10\,\mathrm{GeV}/c$). The systematic uncertainties on data points are shown in boxes on them, the ones for $T_{p/d/\mathrm{HeA}}$ are shown as the boxes around unity at the right end of each plot, and the global uncertainty of 9.7 % which comes from $p + p$ luminosity normalization is not shown but is mentioned by texts. Note that both the number of participant nucleons (N_part) and the number of binary nucleon-nucleon collisions (N_coll) are different for each system. Figure 3 shows the centrality dependence of the π^0 $R_{p/d/\mathrm{HeA}}$ for p+Au, d+Au, and ^{3}He+Au collisions at $\sqrt{s_{NN}}$ =200 GeV. The similar feature for Cronin-peaks at $p_\mathrm{T} \sim 5\,\mathrm{GeV}/c$ is seen for all the systems and centralities. The slight suppression at higher p_T ($p_\mathrm{T} > 10\,\mathrm{GeV}/c$) tends to disappear and turns to an enhancement as going to more peripheral collisions for all the collision systems. The highest magnitude of the Cronin-peak is consistently seen for p+Au collisions and then d+Au and ^{3}He+Au collisions, following the

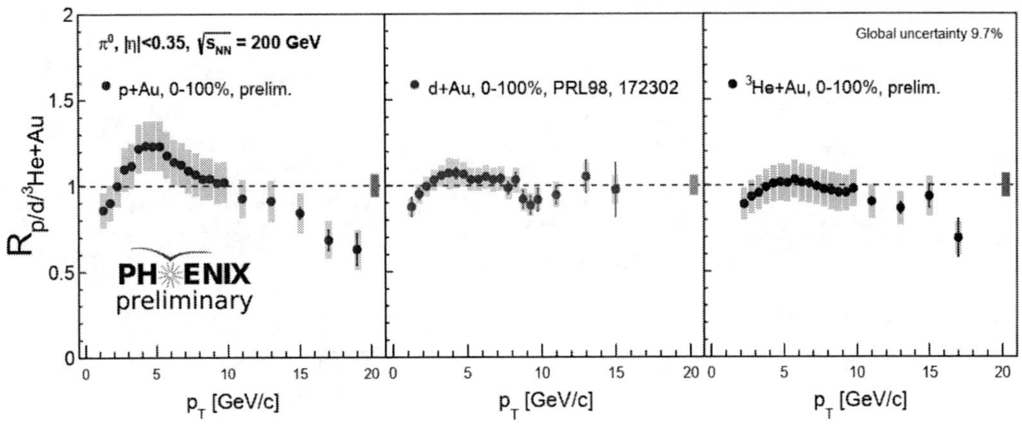

Figure 2. $R_{p/d/\mathrm{HeA}}$ of the π^0 for minimum bias p+Au, d+Au, and ^{3}He+Au collisions at $\sqrt{s_{NN}}$ =200 GeV (from left to right). Note that the N_{part} and N_{coll} are different for each system.

projectile order, except for the most peripheral bin. At high p_T the $R_{p/d/\mathrm{HeA}}$ of three collision systems tend to agree each other within the quoted uncertainties for all the centralities. Note that the same centrality doesn't mean the N_{part} and N_{coll} are same. In order to see the development of the suppression/enhancement at p_T ~5, 8 and 10 GeV/c, we have plotted the R_{AA} as a function of N_{part} and shown in Figure 4 (only d+Au and ^{3}He+Au collisions are shown here). For the lower N_{part} (N_{part} <~10), the $R_{d/\mathrm{HeA}}$ are different between two systems, while for high N_{part}, they tend to merge. The result can be compared with the R_{AA}'s from the peripheral Au+Au collisions, i.e., 60–70, 70–80, and 80–93 % centrality. In the previous publication, we measured the R_{AA} for p_T >5 GeV/c as ~0.78, ~ 0.87, and ~0.84, for N_{part} of 26.7±3.7, 13.7±2.5, and 5.6±0.8, respectively [12]. With these numbers, we found that the R_{AA}'s from the three collisions systems converge for N_{part} >~12, while a system ordering of R_{dAu}>R_{HeAu}>R_{AuAu} is observed for N_{part} <~12. We haven't calculated the Integrated $R_{p\mathrm{A}}$ for p+Au collisions, but if we take the numbers at p_T =5 GeV/c, the points are found to locate above d+Au points at N_{part} <~ 10.

We have made a comparison of the data with some models. A model including cold energy loss only (e.g. [16]) was found not to reproduce the system ordering of the Cronin-peak magnitude. The HIJING++ model [17] was found to give a reasonable description of the system ordering of the Cronin-peak as shown in Figure 5. It is, however, important to note that the peak positions are different from what were observed in the data; the data have the peaks at p_T ~ 5 GeV/c, while the model has the peaks at p_T ~1.5 GeV/c.

3.2 For(back)ward rapidity measurement

The HIJING++ model also gave a prediction of the rapidity dependence of the magnitude of the Cronin-peak in p+Au collisions as shown in Fig 6. The prediction says that the $R_{p\mathrm{A}}$ increases as going to p-going direction (positive rapidity), while decreases as going to Au-going direction (negative rapidity). This is mainly owing to the shadowing and anti-shadowing effect in the parton distribution function in p- and Au-going direction, respectively.

Figure 7 shows the centrality dependence of the R_{cp} of hadrons for 2< p_T <5 GeV/c as a function of rapidity in p+Au collisions at $\sqrt{s_{NN}}$ =200 GeV. The measurement was performed using the MuTr,

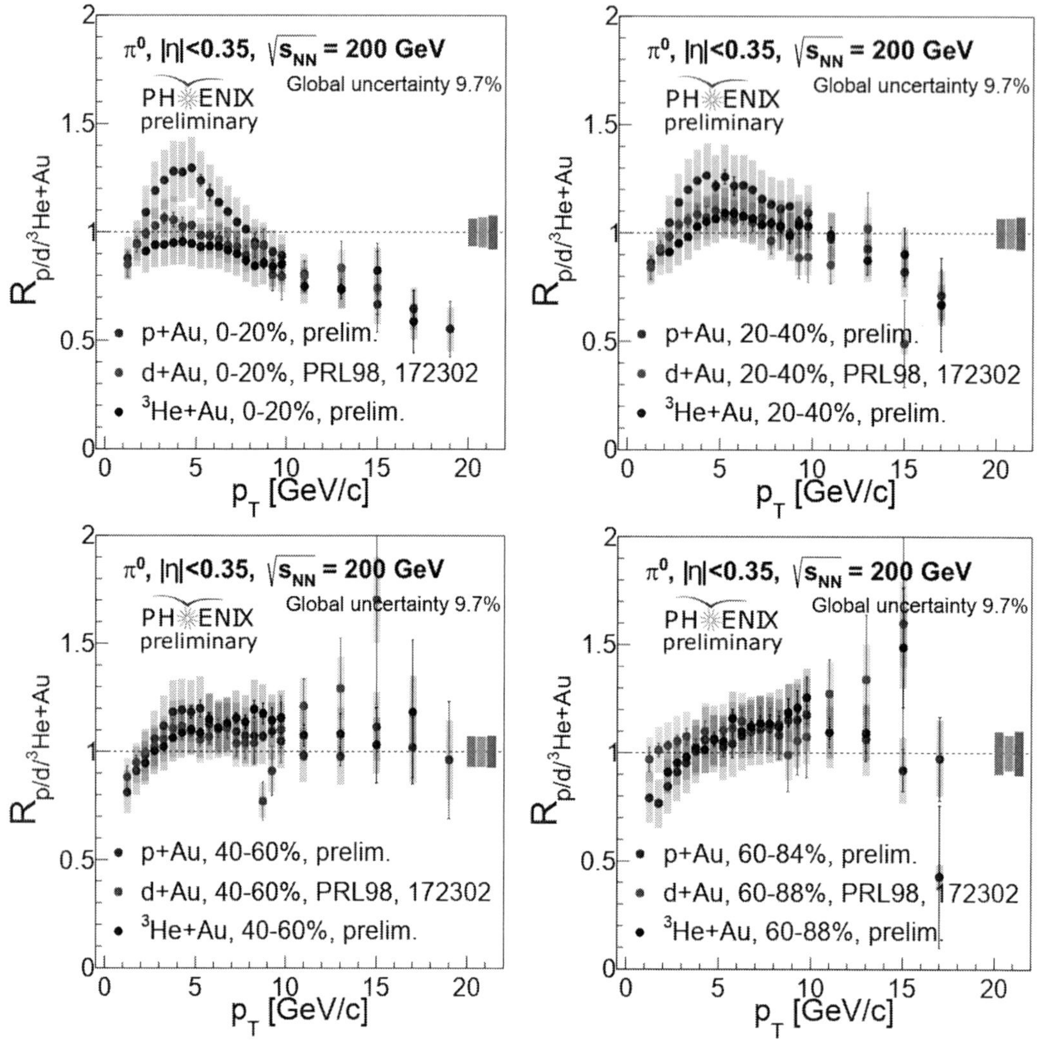

Figure 3. Centrality dependence of the π^0 $R_{p/d/\mathrm{HeA}}$ for p+Au, d+Au, and ^{3}He+Au collisions at $\sqrt{s_{NN}}$ =200 GeV.

FVTX and MuID as described in the analysis section. We don't show the measurement of midrapidity π^0 in the plots, but we confirmed that the π^0 $R_{p\mathrm{A}}$ points are located right on the linear interpolation of the for(back)ward measurements. It is seen that the rapidity dependence of the R_{cp} is apparently opposite to what was predicted by HINJING++. This calls additional/another physics mechanism to be played. Another observation is that the both enhancement and suppression as a function rapidity is highly centrality dependent, which is not expected from a parton distribution function. We have also measured the high p_{T} hadrons in for(back)ward rapidity in p+Al collisions in RHIC Year-2015 as shown in Figure 8. The result is compared with those in p+Au collisions. There are interesting features seen in the result. First, although the absolute magnitude are different, R_{cp} changes as a function of centrality both in p+Au and p+Al collisions and both in forward (p-going) and backward (Au/Al-going) rapidities; in more central collisions, R_{cp} is more suppressed in forward rapidity, and more enhanced in backward rapidity. The degree of suppression at the forward rapidity are similar

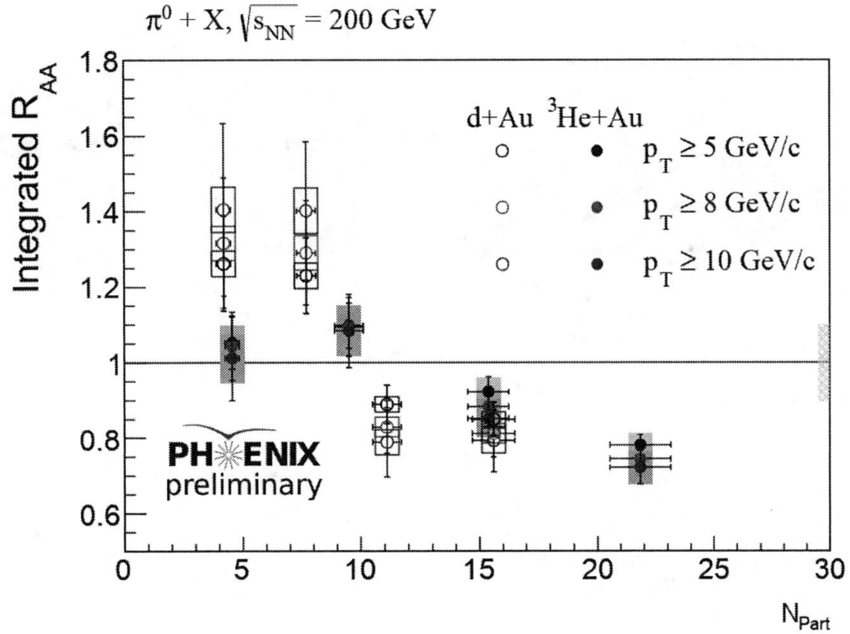

Figure 4. Integrated R_{AA} as a function of N_{part} for d+Au and ^{3}He+Au collisions at $\sqrt{s_{NN}}$ =200 GeV.

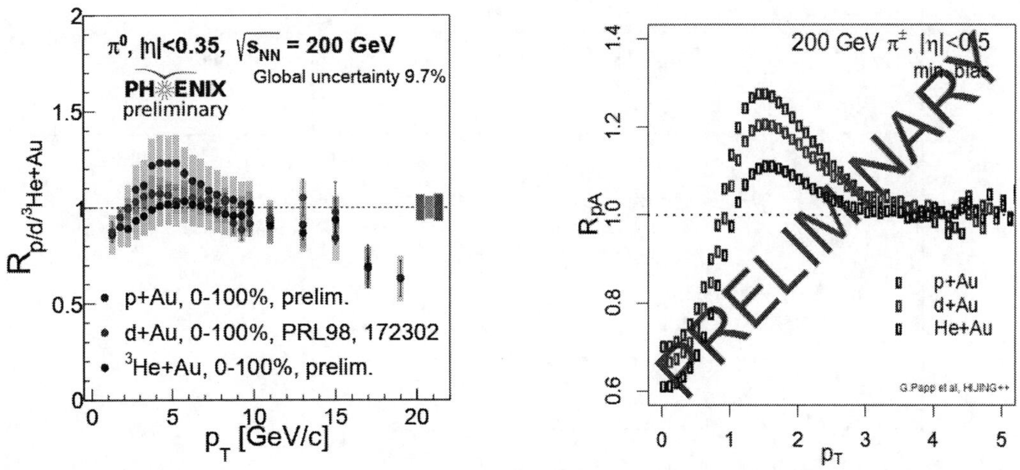

Figure 5. (Left) $R_{p/d/\mathrm{HeA}}$ for minimum bias $p/d/^3$He+Au collisions at $\sqrt{s_{NN}}$ =200 GeV. (Right) Calculation of the $R_{p/d/\mathrm{HeA}}$ by HIJING++ model.

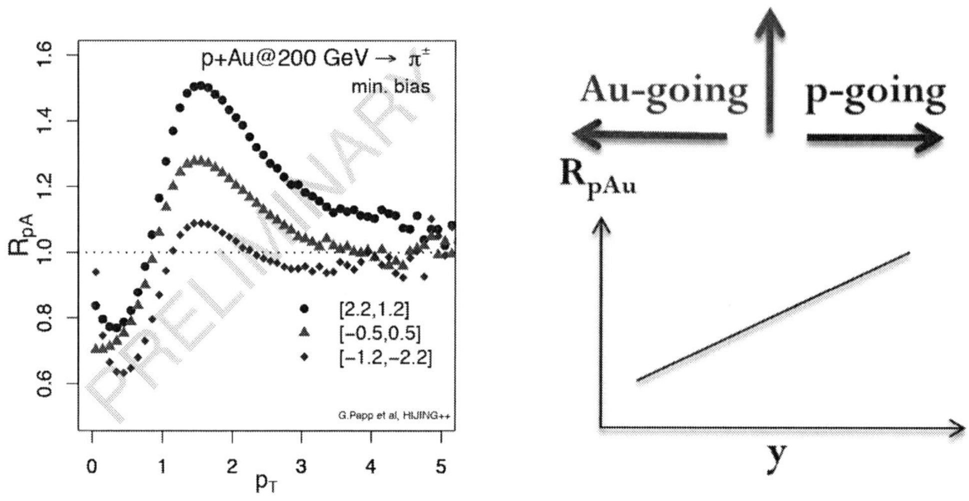

Figure 6. (Left) HIJING++ prediction of R_{pA} for pions for different rapidity regions in minimum bias p+Au collisions at $\sqrt{s_{NN}}$ =200 GeV. (Right) Schematic drawing of the overall trend of the R_{pA}.

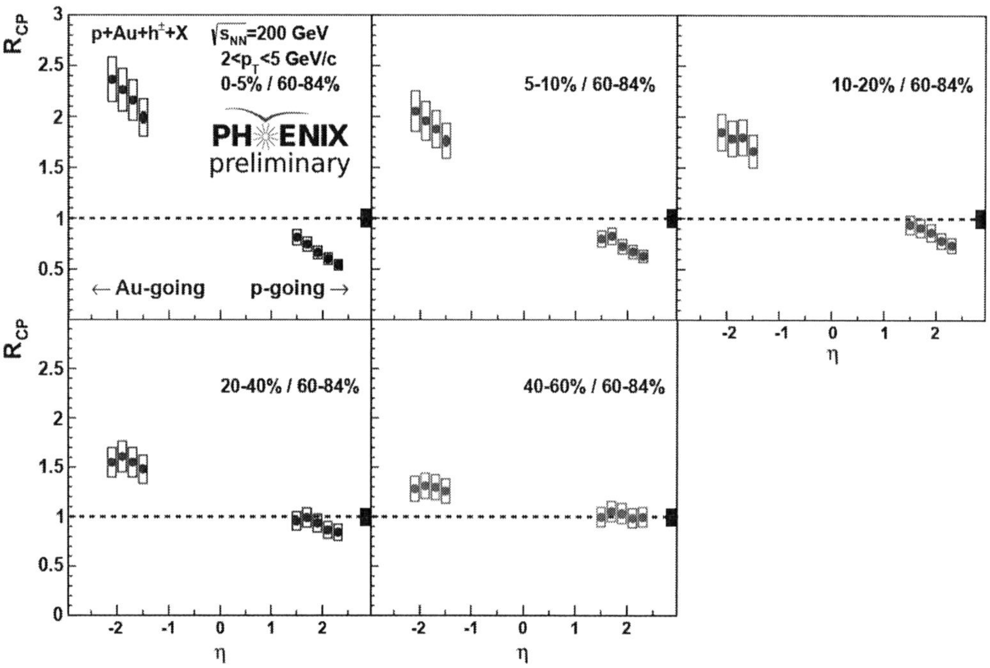

Figure 7. Hadron R_{cp} in p+Au collisions at for(back)ward rapidities as a function of centrality at $\sqrt{s_{NN}}$ =200 GeV. The p_T range is 2< p_T <5 GeV/c.

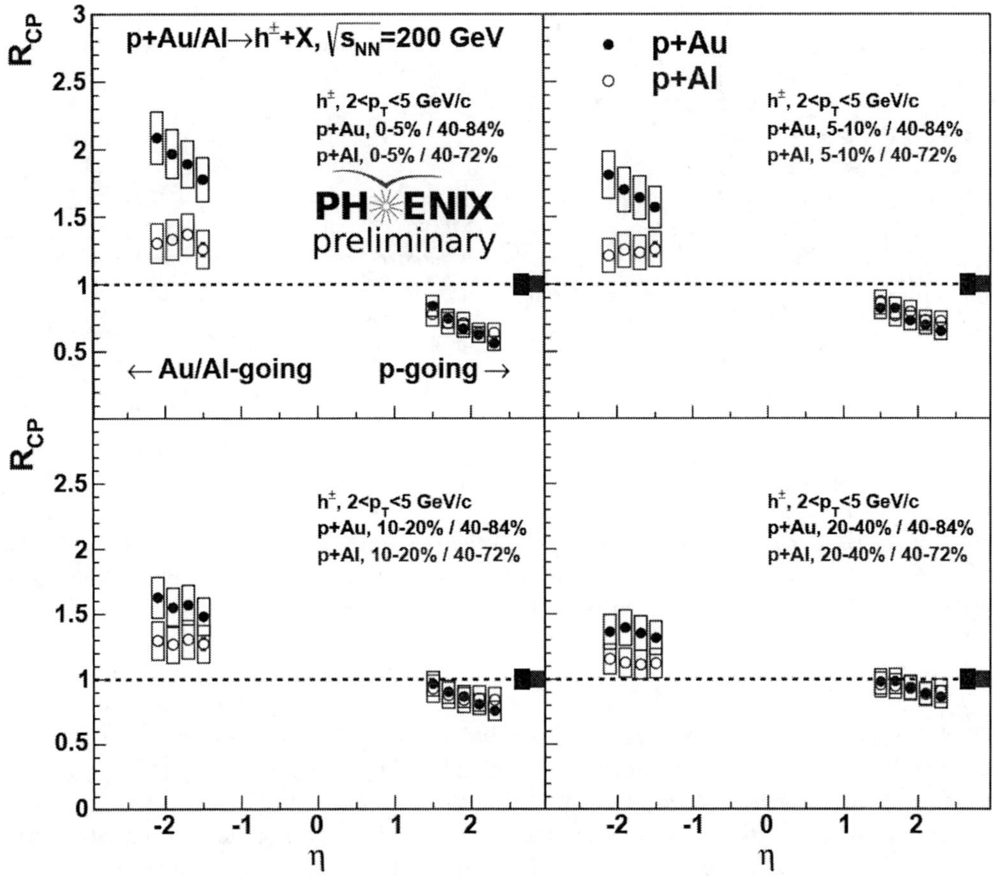

Figure 8. Hadron R_{cp} in p+Al collisions at for(back)ward rapidities as a function of centrality compared with those in p+Au collisions at $\sqrt{s_{NN}}$ =200 GeV. The p_T range is 2< p_T <5 GeV/c.

in p+Al/Au collisions. On the contrary, the degree of enhancement at the backward rapidity are very different as going to more central collisions.

3.3 Possible hot medium creation?

We have measured a flow-like signal in d+Au collisions through two-particle azimuthal angle correlation functions with a wide rapidity gap ($|\Delta\eta| \sim 3$) as mentioned in the introduction section. We observed a quadrupole component on top of a large dipole component when fitting the correlation functions with Fourier series. We recently measured a similar flow-like signal in p+Au collisions as well [18]. It is interesting to understand the different level of enhancement in R_{cp} at backward rapidity in p+Al/Au collisions in connection with the flow result. Figure 9 shows the two-particle correlation functions measured by PHENIX together with the hadron R_{cp}'s for 0–5 % centrality in p+Al/Au collisions. The two-particle correlation functions shown at the left panel are constructed by hadrons measured at CNT ($|\eta|$ <0.35) associated with the energy in a tower in the south MPC (MPCS, -3.7< η <-3.1), which is located in Au-going side. The functions shown at the right panel are constructed by

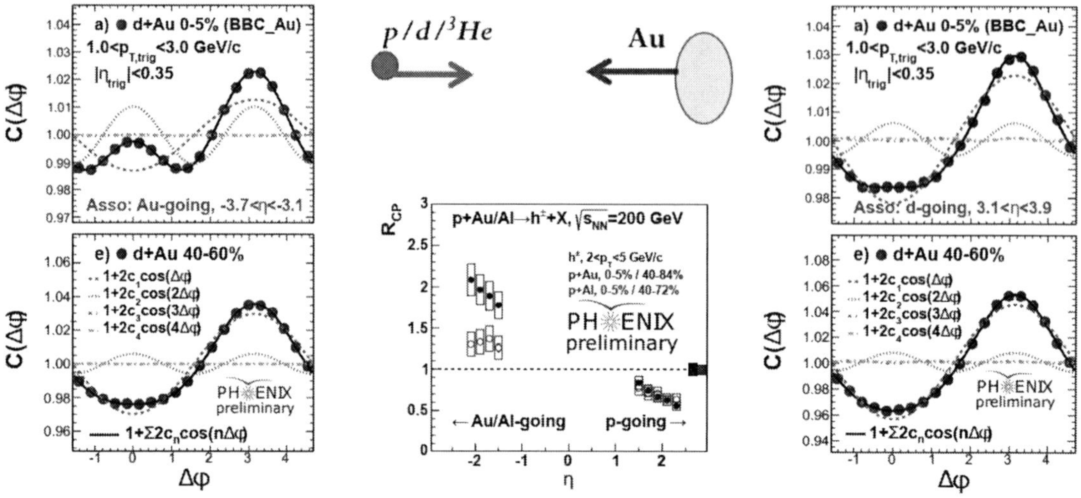

Figure 9. (Left and Right) Two-particle correlation functions measured in d+Au collisions at $\sqrt{s_{NN}}$ =200 GeV by PHENIX. The trigger p_T is 1.0< p_T <3.0 GeV/c and the centralities are 0–5 % and 40–60 %. (Middle) the R_{cp} measurement for 0–5 % centrality in p+Al/Au collisions at $\sqrt{s_{NN}}$ =200 GeV. Note that Al ions go to the same directions as Au ions.

hadrons measured at CNT ($|\eta|$ <0.35) associated with the energy in a tower in the north MPC (MPCN, 3.1< η <3.9), which is located in p/d-going side. The trigger p_T is 1.0< p_T <3.0 GeV/c and the centralities are 0–5 % and 40–60 %. On top of the dipole component in both cases, the quadrupole component is much prominent for Au-going side for 0–5 % centrality. For the 40–60 % centrality, both d-going and Au-going correlation functions look similar. The asymmetric feature of the ridge/flow can be understood in the context of the asymmetric pseudo-rapidity distribution of produced particles in the asymmetric collision systems [19]. In the middle panel, shown is the R_{cp} of the hadrons in forward and backward rapidities. The rapidity dependent increase of the R_{cp} is consistent with that of the ridge-like structure in the two-particle correlation functions. The possible scenario is that the enhancement of the R_{cp} is partly due to the the increase of the particle production by a possible hot medium created in the collisions. Both the centrality dependent and system size dependent increase of the R_{cp} in the backward rapidity may be understood as follows; (1) the possible hot medium creation is more prominent for more central collisions and (2) for larger collision systems, and (3) the medium is shifted towards Au-going direction for more central collisions.

4 Summary

High p_T hadrons in the small systems of $d/^3$He+Au and p+Al/Au collisions have been measured at the midrapidity ($|\eta|$ < 0.35) as well as at the forward and backward rapidities (1.2 < $|\eta|$ < 2.4) at $\sqrt{s_{NN}}$ =200 GeV. A clear system and centrality ordering in the $R_{p/d/\mathrm{HeA}}$ was observed for midrapidity π^0 production. Rapidity dependence of the Hadron R_{cp} in p+Au collisions show an opposite trend to what HIJING++ predicted. The R_{cp} in p+Au collisions was found higher than that of p+Al collisions at the backward rapidity, while the R_{cp} at forward rapidity shows the similar behavior. This result was compared with the two-particle long-range correlations measured in d+Au collisions that are similar to p+Au collisions, and found to be explained consistently with a possible hot medium creation similar to A+A collisions.

References

[1] S.S. Adler et al. (PHENIX), Phys. Rev. Lett. **98**, 172302 (2007), nucl-ex/0610036
[2] A. Adare et al. (PHENIX), Phys. Rev. **C87**, 054907 (2013), 1208.1234
[3] A. Adare et al. (PHENIX), Phys. Rev. Lett. **116**, 122301 (2016), 1509.04657
[4] K.J. Eskola, H. Paukkunen, C.A. Salgado, JHEP **04**, 065 (2009), 0902.4154
[5] S. Chatrchyan et al. (CMS), Phys. Lett. **B718**, 795 (2013), 1210.5482
[6] B. Abelev et al. (ALICE), Phys. Lett. **B719**, 29 (2013), 1212.2001
[7] G. Aad et al. (ATLAS), Phys. Rev. Lett. **110**, 182302 (2013), 1212.5198
[8] A. Adare et al. (PHENIX), Phys. Rev. Lett. **111**, 212301 (2013), 1303.1794
[9] A. Adare et al. (PHENIX), Phys. Rev. Lett. **114**, 192301 (2015), 1404.7461
[10] L. Adamczyk et al. (STAR), Phys. Lett. **B747**, 265 (2015), 1502.07652
[11] A. Adare et al. (PHENIX), Phys. Rev. **C77**, 064907 (2008), 0801.1665
[12] A. Adare et al. (PHENIX), Phys. Rev. **C87**, 034911 (2013), 1208.2254
[13] A. Adare et al. (PHENIX), Phys. Rev. **C93**, 024911 (2016), 1509.06735
[14] K. Adcox et al. (PHENIX), Nucl. Instrum. Meth. **A499**, 469 (2003)
[15] A. Adare et al. (PHENIX), Phys. Rev. **C90**, 034902 (2014), 1310.4793
[16] Y.T. Chien, A. Emerman, Z.B. Kang, G. Ovanesyan, I. Vitev, Phys. Rev. **D93**, 074030 (2016), 1509.02936
[17] G.G. Barnaföldi, G. Bíró, M. Gyulassy, S.M. Haranozó, P. Lévai, G. Ma, G. Papp, X.N. Wang, B.W. Zhang (2017), 1701.08496
[18] C. Aidala et al., Phys. Rev. **C95**, 034910 (2017), 1609.02894
[19] B.B. Back et al. (PHOBOS), Phys. Rev. **C72**, 031901 (2005), nucl-ex/0409021

Experimental Overview on Heavy Flavor Production in Heavy Ion Collisions.

Cesar L. da Silva[1,⋆]

[1]*Los Alamos National Lab - USA*

Abstract. The use of probes containing heavy quarks is one of the pillars for the study of medium formed in high energy nuclear collisions. The conceptual ideas formulated more than two decades ago, such as quark mass hierarchy of the energy that the probe lose in the media and color screening of bound heavy quarkonia states, have being challenged by the measurements performed at RHIC and LHC. A summary of the most recent experimental observations involving charm and bottom quarks in pp, pA, and AA collisions from collisions energies extending from $\sqrt{s_{NN}}$ =200 GeV to 8 TeV is presented. This manuscript also discuss possibilities of new measurements which can be at reach with increased statistics and detector upgrades.

1 Open heavy flavor: Charm and Bottom Quarks

The characterization of media formed in high energy collisions, such as the Quark-Gluon-Plasma (QGP), relies on the use of a probe which is produced and do not shower in the vacuum before the medium formation. Probes with different rest masses are required, and measurements may extend in a broad transverse momentum range. Charm(bottom)-quarks have formation times of 0.07(0.01) fm/c, much earlier than any prediction for QGP formation. The number of charm and bottom quarks, once created by a hard scattering, is preserved when crossing the medium [1–3]. Two main steps are needed to understand the heavy flavor (HF) probe before use it to make QGP tomography: 1) understand the production mechanisms and cross-sections in p+p collisions; and 2) understand how the HF sources are affected when the nucleon is in a nucleus.

Heavy flavor cross-section measurements reached a level of precision much better than the state-of-the-art perturbative Quantum Cromodynamics calculations, some examples can be found in [4–6]. Heavy flavor is mainly produced from gluons at RHIC and LHC. According to PYTHIA simulations of hard processes [7], heavy-quark pair creation and next-to-leading-order flavor excitation are the dominant production sources at RHIC energies ($\sqrt{s_{NN}}$ =200 GeV). Figure 1 shows how these processes produce different azimuthal correlations of muons from $c\bar{c}$ and $b\bar{b}$ pairs. As the energy increases at Tevatron and LHC, the gluon splitting contribution becomes dominant [8]. This is a fundamental difference between HF production at RHIC and LHC.

Initial state effects must be considered for HF production in collisions involving nucleus. Data from d+Au collisions taken at RHIC shows a significant enhancement of HF yield relative to p+p

⋆e-mail: cesar_luiz@lanl.gov

© The Authors, published by EDP Sciences. This is an open access article distributed under the terms of the Creative Commons Attribution License 4.0 (http://creativecommons.org/licenses/by/4.0/).

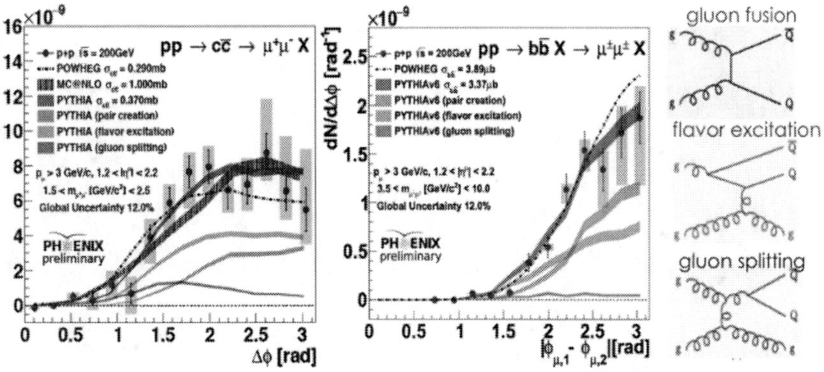

Figure 1: Azimuthal correlation of charm and bottom pairs from PHENIX at RHIC described by different heavy flavor production processes obtained from PYTHIA [7].

collisions at mid- and backward-rapidities [9, 10], corresponding to moderated to large fractional momentum x. Measurements from p+Pb collisions at LHC typically show no significant yield modification at mid- and backward rapidities [11, 12], however ALICE also reports a significant enhancement for the most central collisions. One explanation for these enhancements is incoherent multiple scattering of the initial gluon in the nucleus matter [13] and Cronin p_T broadening [14]. A significant HF suppression is observed at the forward direction of p+Pb collisions by LHCb [12], covering partons with small fractional momentum x. This suppression is attributed to shadowing [15], energy loss before the hard scattering [16], and gluon saturation [17, 18].

Heavy quarks lose energy ΔE when crossing the QGP medium through collisional scattering and gluon radiation [19]. Between 80-70% of high p_T mesons are consistently suppressed in central heavy ion collisions at RHIC and LHC, regardless its mass or collision energy [20]. Data from STAR (D-mesons) [1], PHENIX(c+b → e) [21], ALICE(D-mesons) [22] and CMS(D-meson, B-meson and B→J/ψ) [23, 24] indicate that quark mass dependency on the energy loss in the QGP ($\Delta E(u,d) > \Delta E(c) > \Delta E(b)$), caused by the suppression of gluon radiation in a cone $\Delta\theta < m_q/E$ [25] where m_q is the quark mass and E its energy, is apparently relevant only for $p_T \lesssim m_q$. Figure 2 compares the nuclear modification fraction R_{AA} of charm and bottom decays at RHIC using unfolding of electron decays on the left panel; D-mesons, B-meson and $B \to J/\psi$ at LHC are shown on the right panel. It is still not evident the maximum p_T where the meson suppression has a flavor dependency. Extra caution is also needed to interpret the data sets to account for the presence of initial state effects, such as the yield enhancements observed in d+Au collisions at RHIC.

Direct access to the initial heavy quark can be done with jets. CMS reported that b-jets with p_T >80 GeV/c have the same suppression as inclusive jets dominated by light quarks and gluons [26]. One possible explanation for the similar behavior between light and heavy quarks in the QGP medium is that heavy quarks are mainly produced in gluon splittings at LHC. Heavy quarks from these processes behave like gluons crossing the medium [27]. A preliminary study performed by CMS selects back-to-back b-jet pairs, minimizing the contribution of gluon splittings. The modification of the momentum imbalance between the b-jet pair in Pb+Pb collisions, relative to p+p collisions, still shows the same behavior as inclusive jets.

The second (elliptic flow) and third (triangular flow) Fourier components v_n of D-meson azimuthal asymmetry relative to the event reaction plane, measured at RHIC [28] and LHC [29, 30], indicate that charm quarks are strongly coupled to the medium with a significant diffusion coefficient. However,

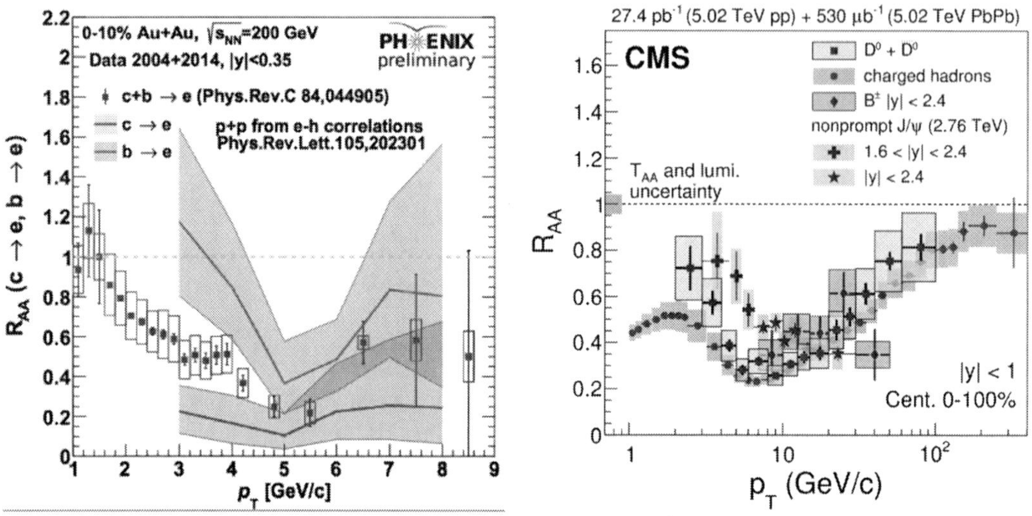

Figure 2: Nuclear modification factor of charm and bottom quarks versus transverse momentum measured by PHENIX at RHIC (left) and CMS at LHC (right).

significant v_2 have also been seen in many small system measurements, such as d+Au and p+Pb collisions.

2 Quarkonia

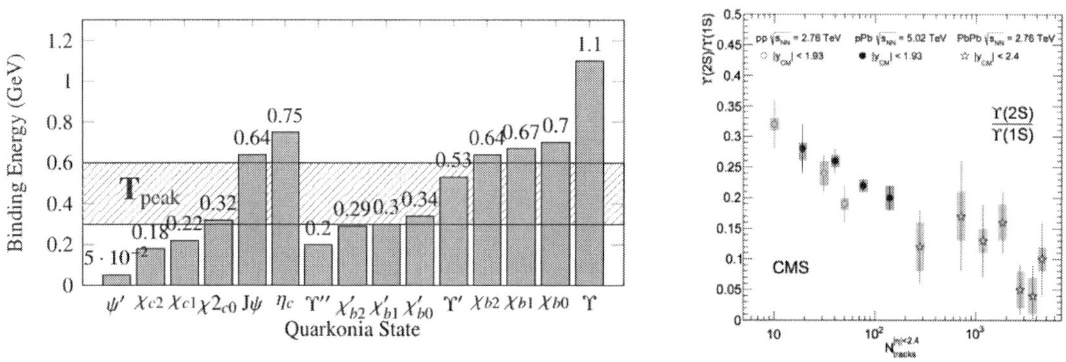

Figure 3: (left) Binding energy of several quarkonium states in vacuum [31] along with an estimation for the quark-gluon plasma peak temperature at RHIC [32] as hashed area. (right) Relative Υ(2S)/Υ(1S) yields versus particle multiplicity in p+p, p+Pb and Pb+Pb collisions measured by CMS [33].

Quarkonia states have a broad range of binding energies (Fig.3-left). The suppression (or not) of these different states depends on the local free energy of the medium. Recent experimental studies started to emphasize on the relative nuclear modifications between different quarkonia states, canceling out initial state effects. Examples are the suppression of the ψ(2S) relative to J/ψ in small system collisions at RHIC and LHC [34–36].

According to charm cross section measurements, each central Au+Au collision at RHIC produce around 20 $c\bar{c}$ pairs. In LHC this number is ten times larger. There is a significant probability for these charm quarks to recombine and form a new source of bound state charmonia in heavy ion collisions. PHENIX measurements indicate that J/ψ is less suppressed at mid-rapidity, where the number of $c\bar{c}$ pairs is the largest, compared to large rapidity [37]. Measurements in ALICE shows a nuclear modification similar to the large rapidity reported by PHENIX only at high p_T. The suppression is largely reduced towards low p_T [38]. These two observations can be evidences of charm recombination. A significant J/ψ elliptic flow v_2 is observed by ALICE [39], suggesting that the charm flow is transfered to J/ψ produced from coalescence.

Because of the much smaller number of $b\bar{b}$ pairs available per event, bottomonium states are expected to have small contribution from bottom recombination, even at LHC. A sequential suppression of Υ states has been observed by CMS in several collision systems. A pattern emerges when looking at Figure 3-right, which shows the ratio between Υ(2S)/Υ(1S) yields versus particle activity in the event at the same rapidity range of the measured Υ yields. The relative suppression of the weaker bounded 2S state increases with local particle density, regardless the collision system (p+p, p+Pb or Pb+Pb) [33]. This observation challenges the long standing picture of color screening of quarkonia states only in QGP medium [40].

3 Conclusions, Future Studies and Facilities

More than 15 years of data from RHIC and LHC shows that the QGP medium formed in high-energy nuclear collisions is opaque to high p_T hard probes, and the energy the hig-p_T probe lose in the medium does not depends on its constituent quark mass. The transition where the quark mass matters for the energy that the probe lose in the medium is still to be determined. Caution is also needed to correctly account for nuclear effects before the probe cross the medium. More studies are also needed for a conclusion on the diffusion coefficients obtained from charm flow measurements. The questions raised above drive the interest on low-p_T heavy quark measurements in the near future. New results will emerge from the data being analyzed from the recent high luminosity RHIC runs and LHC.

Higher harmonics $v_n > 2$ are starting to be available for D-mesons. Preliminary results on B-meson v_2 started to be available by CMS collaboration. The set of HF v_n measurements will help to elucidate how heavy quarks couple with the medium, and indicate how geometrical aspects of the collisions affect these results. Heavy quark correlations using semi-leptonics decays, mesons or jets have also a potential to select different production processes and even quantify the magnitude of collisional and radiative energy loss in QGP. Several detector upgrades are expected in LHC accessing a new level of precision in HF measurement. The sPHENIX upgrade at RHIC [41] will dedicates to jets and HF measurements, including D-, B-mesons and b-jets.

We are seeing a new wave of measurements now focusing in the excited state charmonia and bottomonium. Yield modifications relative to ground states reduces misinterpretations caused by initial state effects. Recent measurements of Υ indicate that states with weaker binding have a suppression relative to ground states which mainly depends on the local particle densities, regardless the collision system. New Υ measurements are expected from RHIC with different particle densities. Measurement of radiative decays of χ_C states is a promise from LHCb, which would greatly expands the understanding of charmonium breaking in heavy ion collisions.

References

[1] L. Adamczyk et al. (STAR Collaboration), Phys. Rev. Lett. **113**, 142301 (2014)
[2] A. Adare et al. (PHENIX), Phys. Rev. Lett. **98**, 172301 (2007)
[3] C. Aidala et al. (PHENIX Collaboration), Phys. Rev. C **96**, 064901 (2017)
[4] L. Adamczyk et al. (STAR Collaboration), Phys. Rev. D **86**, 072013 (2012)
[5] D. Acosta et al. (CDF), Phys. Rev. Lett. **91**, 241804 (2003)
[6] S. Acharya et al. (ALICE), Eur. Phys. J. **C77**, 550 (2017)
[7] T. Sjöstrand, S. Ask, J.R. Christiansen, R. Corke, N. Desai, P. Ilten, S. Mrenna, S. Prestel, C.O. Rasmussen, P.Z. Skands, Computer Physics Communications **191**, 159 (2015)
[8] V. Khachatryan et al. (CMS), JHEP **03**, 136 (2011)
[9] A. Adare et al. (PHENIX), Phys. Rev. Lett. **112**, 252301 (2014)
[10] A. Adare et al. (PHENIX), Phys. Rev. Lett. **109**, 242301 (2012)
[11] ALICE-PUBLIC-2017-008 (2017)
[12] R. Aaij et al. (LHCb), JHEP **10**, 090 (2017)
[13] Z.B. Kang, I. Vitev, E. Wang, H. Xing, C. Zhang, Phys. Lett. **B740**, 23 (2015)
[14] I. Vitev, Phys. Lett. **B562**, 36 (2003)
[15] N. Armesto, J. Phys. **G32**, R367 (2006)
[16] R. Neufeld, I. Vitev, B.W. Zhang, Physics Letters B **704**, 590 (2011)
[17] L.D. McLerran, R. Venugopalan, Phys. Rev. **D49**, 3352 (1994)
[18] F. Gelis, E. Iancu, J. Jalilian-Marian, R. Venugopalan, Ann. Rev. Nucl. Part. Sci. **60**, 463 (2010)
[19] J. Uphoff, O. Fochler, Z. Xu, C. Greiner, J. Phys. **G42**, 115106 (2015)
[20] A.M. Sirunyan et al. (CMS Collaboration), Phys. Rev. Lett. **119**, 152301 (2017)
[21] A. Adare et al. (PHENIX), Phys. Rev. **C93**, 034904 (2016)
[22] J. Adam et al. (ALICE), JHEP **03**, 081 (2016)
[23] A.M. Sirunyan et al. (CMS) (2017), `nucl-ex:1708.04962`
[24] V. Khachatryan et al. (CMS), Eur. Phys. J. **C77**, 252 (2017)
[25] Y. Dokshitzer, D. Kharzeev, Physics Letters B **519**, 199 (2001)
[26] S. Chatrchyan et al. (CMS Collaboration), Phys. Rev. Lett. **113**, 132301 (2014)
[27] J. Huang, Z.B. Kang, I. Vitev, Phys. Lett. **B726**, 251 (2013)
[28] L. Adamczyk et al. (STAR), Phys. Rev. Lett. **118**, 212301 (2017)
[29] S. Acharya et al. (ALICE) (2017), `nucl-ex:1707.01005`
[30] A.M. Sirunyan et al. (CMS) (2017), `nucl-ex:1708.03497`
[31] H. Satz, J. Phys. G **32**, R25 (2006)
[32] A. Adare et al. (PHENIX Collaboration), Phys. Rev. C **81**, 034911 (2010)
[33] S. Chatrchyan et al. (CMS), JHEP **04**, 103 (2014)
[34] A. Adare et al. (PHENIX Collaboration), Phys. Rev. C **95**, 034904 (2017)
[35] B.B. Abelev et al. (ALICE), JHEP **02**, 073 (2014)
[36] R. Aaij et al. (LHCb), JHEP **03**, 133 (2016)
[37] A. Adare et al. (PHENIX), Phys. Rev. Lett. **98**, 232301 (2007)
[38] J. Adam et al. (ALICE), Phys. Lett. **B766**, 212 (2017)
[39] E. Abbas et al. (ALICE), Phys. Rev. Lett. **111**, 162301 (2013)
[40] F. Karsch, M.T. Mehr, H. Satz, Z. Phys. C **37**, 617 (1987)
[41] A. Adare et al. (2015), `nucl-ex:1501.06197`

Bridging the two order parameters for the QCD deconfiment phase transition

Juan Pablo Carlomagno[1,2,⋆] and *Marcelo* Loewe[3,4,5,⋆⋆]

[1] *IFLP, CONICET – Dpto. de Física, Universidad Nacional de La Plata, C.C. 67, 1900 La Plata, Argentina*
[2] *CONICET, Rivadavia 1917, 1033 Buenos Aires, Argentina*
[3] *Instituto de Física, Pontificia Universidad Católica de Chile, Casilla 306, Santiago 22, Chile*
[4] *Centre for Theoretical and Mathematical Physics and Department of Physics, University of Cape Town, Rondebosch 7700, South Africa*
[5] *Centro Científico-Tecnológico de Valparaíso-CCTVAL, Universidad Técnica Federico Santa María, Casilla 110-V, Valparaíso, Chile*

Abstract. We study the relation between two order parameters for deconfinement, normally employed in the literature: the continuum threshold s_0, in the context of QCD Sum Rules, and the trace of the Polyakov loop Φ in the framework of a nonlocal SU(2) chiral quark model. We establish a bridge between both order parameters at finite temperature T and baryonic chemical potential μ. In our analysis, we also include the the chiral quark condensate, the order parameter for the chiral symmetry restoration.
We found that s_0 and Φ provide us with the same information for the deconfinement transition, both for the zero and finite chemical potential cases. At zero density, the critical temperatures for both quantities coincide exactly. This part of the analysis has been reinforced by the discussion of the corresponding susceptibilities and the static quark entropy behavior. At finite μ both order parameters provide evidence for the appearance of a quarkyonic phase.

1 Introduction

In QCD, when quarks are placed in a medium, the color charge is screened due to density and temperature effects [1]. If the density and/or the temperature increases beyond a certain critical value, one expects that the interactions between quarks will not be able to confine them inside a hadron, so that they are free to travel longer distances and deconfine. This transition from a confined to a deconfined phase is usually referred to as the deconfinement phase transition.

A separate phase transition is the realization of chiral symmetry, moving from a Nambu-Goldstone phase into a Wigner-Weyl phase. Based, on lattice QCD evidence [2] one expects these two phase transitions to take place at approximately the same temperature at zero chemical potential. At finite density these two transitions can arise at different critical temperatures. The result will be a quarkyonic phase, where the chiral symmetry is restored but the quarks and gluons remain confined.

⋆e-mail: carlomagno@fisica.unlp.edu.ar
⋆⋆e-mail: mloewe@fis.puc.cl

© The Authors, published by EDP Sciences. This is an open access article distributed under the terms of the Creative Commons Attribution License 4.0 (http://creativecommons.org/licenses/by/4.0/).

It has been customary to study the behavior of the trace of the Polyakov loop (PL) $\Phi(T,\mu)$ (order parameter for deconfinement phase transition) and quark anti-quark chiral condensate $\langle\bar{\psi}\psi\rangle(T,\mu)$ (chiral symmetry restoration), as function of temperature and chemical potential.

The goal of our discussion is to compare the Polyakov loop order parameter with a QCD deconfinement parameter [3], that corresponds to the squared energy threshold, $s_0(T,\mu)$, for the onset of perturbative QCD (PQCD) in hadronic spectral functions. For an actual general review see Ref. [4]. Around this energy, and at zero temperature, the resonance peaks in the spectrum dissapear or become very broad, approaching then the PQCD regime. With increasing temperature approaching the critical temperature for deconfinement, the spectral function should then be described entirely by PQCD.

When both T and μ are nonzero, lattice QCD simulations cannot be used, because of the sign problem in the fermionic determinant. Therefore, one need to resort either to mathematical constructions to overcome the above limitation, or to model calculations.

The two deconfinement order parameters mentioned before: $\Phi(T,\mu)$ and $s_0(T,\mu)$ can be used to realize a phenomenological description of the deconfinement transition at finite temperature and density.

The natural framework to determine s_0 has been that of QCD sum rules. This framework is based on the operator product expansion (OPE) of current correlators at short distances, extended beyond perturbation theory, and on Cauchy's theorem in the complex s-plane. The latter is usually referred to as quark-hadron duality. Vacuum expectation values of quark and gluon field operators effectively parametrize the effects of confinement. An extension of this method to finite temperature was first outlined in [3].

To analyze the role of the PL, we will concentrate on nonlocal Polyakov–Nambu–Jona-Lasinio (nlPNJL) models (see [5, 6] and references therein), in which quarks move in a background color field and interact through covariant nonlocal chirally symmetric four point couplings. These approaches, offer a common framework to study both the chiral restoration and deconfinement transitions. In fact, the nonlocal character of the interactions arises naturally in the context of several successful approaches to low-energy quark dynamics, and leads to a momentum dependence in the quark propagator that can be made consistent [7] with lattice results.

The aim of the present work is to study the relation between both order parameters for the deconfinement transition at finite temperature and chemical potential, Φ and s_0, using the thermal finite energy sum rules (FESR) with inputs obtained from nlPNJL models.

2 Finite energy sum rules

We begin by considering the (charged) axial-vector current correlator at $T = 0$

$$\Pi_{\mu\nu}(q^2) = i \int d^4x\, e^{iq\cdot x}\, \langle 0|T(A_\mu(x)A_\nu(0))|0\rangle = -g_{\mu\nu}\,\Pi_1(q^2) + q_\mu q_\nu \Pi_0(q^2)\,, \tag{1}$$

where $A_\mu(x) = :\bar{u}(x)\gamma_\mu\gamma_5 d(x):$ is the axial-vector current, $q_\mu = (\omega,\vec{q})$ is the four-momentum transfer, and the functions $\Pi_{0,1}(q^2)$ are free of kinematical singularities. Concentrating on the function $\Pi_0(q^2)$ and writing the OPE beyond perturbation theory in QCD , one of the two pillars of the sum rule method, one has

$$\Pi_0(q^2)|_{\rm QCD} = C_0\,\hat{I} + \sum_{N=1} C_{2N}(q^2,\mu^2)\langle\hat{O}_{2N}(\mu^2)\rangle\,, \tag{2}$$

where μ^2 is a renormalization scale. The Wilson coefficients C_N depend on the Lorentz indices and quantum numbers of the currents. Finally, the local gauge invariant operators $\hat{O}_N$, are built from the quark and gluon fields in the QCD Lagrangian. The vacuum expectation values of those operators

$\langle \hat{O}_{2N}(\mu^2) \rangle$, dubbed as condensates, parametrize nonperturbative effects and have to be extracted from experimental data or model calculations.

The second pillar of the QCD sum rules technique is Cauchy's theorem in the complex squared energy s-plane and this allows us to establish the following FESR. For details, we refer the reader to Ref. [4] and to the original article Ref. [6]

$$(-)^{N-1} C_{2N} \langle \hat{O}_{2N} \rangle = 4\pi^2 \int_0^{s_0} ds\, s^{N-1} \frac{1}{\pi} \mathrm{Im}\Pi_0(s)|_{\text{HAD}} - \frac{s_0^N}{N} \left[1 + O(\alpha_s)\right] \quad (N = 1, 2, \cdots). \tag{3}$$

For $N = 1$, the dimension $d = 2$ term in the OPE does not involve any condensate, as it is not possible to construct a gauge invariant operator of such a dimension from the quark and gluon fields. There is no evidence for such a term (at $T = 0$) from FESR analyses of experimental data on e^+e^- annihilation and τ decays into hadrons [8, 9]. At high temperatures, though, there seems to be evidence for some $d = 2$ term [10]. However, the analysis to be reported here is performed at lower values of T, so that we can safely ignore this contribution in the sequel.

The dimension $d = 4$ term, a renormalization group invariant quantity, is given by

$$C_4 \langle \hat{O}_4 \rangle = \frac{\pi}{6} \langle \alpha_s G^2 \rangle + 2\pi^2 (m_u + m_d) \langle \bar{q} q \rangle . \tag{4}$$

The extension of this program to finite temperature is fairly straightforward [3, 11, 12], with the Wilson coefficients in the OPE, Eq. (2), remaining independent of T at leading order in α_s, and the condensates developing a temperature dependence.

In the static limit ($\vec{q} \to 0$), to leading order in PQCD, and for $T \neq 0$ and $\mu \neq 0$ the function $\Pi_0(q^2)|_{\text{QCD}}$ in Eq. (1) becomes $\Pi_0(\omega^2, T, \mu)|_{\text{QCD}}$; to simplify the notation we shall omit the T and μ dependence in the sequel. A calculation of the spectral function in perturbative QCD, at finite temperature and finite density gives

$$\frac{1}{\pi} \mathrm{Im}\Pi_0(s)|_{\text{PQCD}} = \frac{1}{4\pi^2} \left[1 - \tilde{n}_+\left(\frac{\sqrt{s}}{2}\right) - \tilde{n}_-\left(\frac{\sqrt{s}}{2}\right)\right] - \frac{2}{\pi^2}\, T^2\, \delta(s) \left[\mathrm{Li}_2(-e^{\mu/T}) + \mathrm{Li}_2(-e^{-\mu/T})\right], \tag{5}$$

where $\mathrm{Li}_2(x)$ is the dilogarithm function, $s = \omega^2$, and $\tilde{n}_\pm(x) = (e^{(x\mp\mu)/T} + 1)^{-1}$ are the Fermi-Dirac thermal distributions for particles and antiparticles, respectively.

In the hadronic sector we assume pion-pole dominance of the hadronic spectral function, i.e. the continuum threshold s_0 to lie below the first radial excitation with mass $M_{\pi_1} \simeq 1300$ MeV.

We have, then

$$\frac{1}{\pi} \mathrm{Im}\Pi_0(s)|_{\text{HAD}} = 2\, f_\pi^2(T, \mu_B)\, \delta(s - m_\pi^2), \tag{6}$$

where $f_\pi(T, \mu_B)$ is the pion decay constant at finite T and μ, with $f_\pi(0, 0) = 92.21 \pm 0.14$ MeV [13]. Notice we will not include in our spectral function the first part of a_1 resonance obtained from the τ-decay data [14], since still there is no counterpart in the SU(2) nlPNJL model for the description of the hadronic vector resonance.

Turning to the FESR, Eq. (3), with $N = 1$ and no dimension $d = 2$ condensate, and using Eqs. (5) and (6) one finds

$$\int_0^{s_0(T,\mu)} ds \left[1 - \tilde{n}_+\left(\frac{\sqrt{s}}{2}\right) - \tilde{n}_-\left(\frac{\sqrt{s}}{2}\right)\right] = 8\pi^2 f_\pi^2(T, \mu) + 8T^2 \left[\mathrm{Li}_2(-e^{\mu/T}) + \mathrm{Li}_2(-e^{-\mu/T})\right]. \tag{7}$$

This is a transcendental equation determining $s_0(T,\mu)$ in terms of $f_\pi(T,\mu)$. The next thermal FESR at zero chemical potential, for completeness, is given by [14],

$$-C_4\langle\hat{O}_4\rangle(T) = 4\pi^2 \int_0^{s_0(T)} ds\, s\frac{1}{\pi}\mathrm{Im}\,\Pi_0(s)|_{\mathrm{HAD}} - \int_0^{s_0(T)} ds\, s\left[1 - 2n_F\left(\frac{\sqrt{s}}{2T}\right)\right], \tag{8}$$

where $n_F(x) = 1/(1+e^x)$ is the Fermi thermal function.

3 Thermodynamics at finite density in the PNJL model

We consider a nonlocal SU(2) chiral quark model that includes quark couplings to the color gauge fields [6]. The quark-antiquark currents include nonlocal covariant form factors $\mathcal{G}(z)$ and $\mathcal{F}(z)$ characterizing the corresponding four-fermion interactions. The scalar-isoscalar current will generate a momentum dependent quark mass in the quark propagator, while the "momentum" current will be responsible for a momentum dependent quark wave function renormalization (WFR) [7, 15, 16].

To proceed, we perform a bosonization of the theory, introducing bosonic fields $\sigma_{1,2}(x)$ and $\pi_a(x)$, and integrating out the quark fields. Details of this procedure can be found e.g. in Ref. [7].

In order to analyze the properties of meson fields it is necessary to consider the quadratic fluctuations in the Euclidean action:

$$S_E^{\mathrm{quad}} = \frac{1}{2}\int \frac{d^4p}{(2\pi)^4} \sum_M r_M\, G_M(p^2)\, \phi_M(p)\, \bar{\phi}_M(-p)\,, \tag{9}$$

where meson fluctuations $\delta\sigma_a$, $\delta\pi_a$ have been translated to a charged basis ϕ_M, being M the scalar and pseudoscalar mesons (σ, π^0, $\pi^\pm$) plus the σ_2 field, and G_M are the inverse dressed propagators. The coefficient r_M is 1 for charge eigenstates $M = \sigma_i, \pi^0$, and 2 for $M = \pi^+$. At finite temperature, the meson masses are obtained by solving $G_M(-m_M^2, 0) = 0$. The full expressions for the one-loop functions $G_M(q)$ can be found in Ref. [5, 7].

Following a standard procedure, we can finally identify the corresponding pion weak decay constant

$$f_\pi = \frac{m\, Z_\pi^{-1/2}}{m_\pi^2}\, F_0(-m_\pi^2)\,. \tag{10}$$

with

$$F_0(p^2) = 8\,N_c \int \frac{d^4q}{(2\pi)^4}\, g(q)\, \frac{Z(q^+)Z(q^-)}{D(q^+)D(q^-)}\, [q^+ \cdot q^- + M(q^+)M(q^-)] \tag{11}$$

where $q^\pm = q \pm p/2$ and $D(q) = q^2 + M^2(q)$, with $M(p)$ and $Z(p)$ defined as

$$M(p) = Z(p)\,[m + \bar{\sigma}_1 g(p)]\ , \quad Z(p) = [1 - \bar{\sigma}_2 f(p)]^{-1}\ ,$$

here $g(p)$ and $f(p)$ are the Fourier transforms of the form factors $\mathcal{G}(z)$ and $\mathcal{F}(z)$.

We extend the bosonized effective action to finite temperature T and chemical potential μ using the standard imaginary time formalism. Concerning the gauge fields, we assume that quarks move on a constant background field $\phi = A_4 = iA_0 = ig\,\delta_{\mu 0}\, G_a^\mu \lambda^a/2$, where G_a^μ are SU(3) color gauge fields. Then the traced Polyakov loop, which in the infinite quark mass limit can be taken as an order parameter of confinement, is given by $\Phi = \frac{1}{3}\mathrm{Tr}\,\exp(i\phi/T)$. We work in the so-called Polyakov gauge [17], where the matrix ϕ is given a diagonal representation $\phi = \phi_3\lambda_3 + \phi_8\lambda_8$.

With the constraint of ϕ_3 and ϕ_8 being real [18, 19], implies $\phi_8 = 0$, leaving only ϕ_3 as an independent variable, and therefore $\Phi = [2\cos(\phi_3/T) + 1]/3$.

Following the same prescriptions as in Refs. [5, 20, 21], the real part of $\Omega^{\rm MFA}$ at finite temperature T and chemical potential μ is given by

$$\Omega^{\rm MFA} = \Omega^{\rm reg} + \Omega^{\rm free} + \mathcal{U}(\Phi, T) + \Omega_0 \,, \tag{12}$$

where

$$\begin{aligned} \Omega^{\rm reg} &= -\, 4T \sum_{c,n} \int_{\vec{p}} \log \left[\frac{(\rho^c_{n,\vec{p}})^2 + M^2(\rho^c_{n,\vec{p}})}{Z^2(\rho^c_{n,\vec{p}})} \right] + \frac{\bar{\sigma}_1^2 + \kappa_p^2 \bar{\sigma}_2^2}{2\,G_S} \,, \\ \Omega^{\rm free} &= -\, 4T \int_{\vec{p}} \sum_{c,s=\pm 1} {\rm Re}\, \log \left[1 + \exp\left(-\frac{\epsilon_p + i s \phi_c}{T} \right) \right] , \end{aligned} \tag{13}$$

here $\bar{\sigma}_{1,2}$ are the mean field values of the scalar fields. We have also defined

$$\left(\rho^c_{n,\vec{p}}\right)^2 = \left[(2n+1)\pi T + \phi_c - \imath\mu\right]^2 + \vec{p}^{\,2} \,, \tag{14}$$

the sums over color indices run over $c = r, g, b$, with the color background fields components being $\phi_r = -\phi_g = \phi_3$, $\phi_b = 0$, and $\epsilon_p = \sqrt{\vec{p}^{\,2} + m^2}$.

One possible Ansatz for the Polyakov loop potential $\mathcal{U}(\Phi, T)$ is that based on the logarithmic expression of the Haar measure associated with the SU(3) color group integration [19]. Besides the logarithmic function, a widely used potential is that given by a polynomial function based on a Ginzburg-Landau Ansatz [22, 23]. The corresponding expressions can be found in Ref. [6].

Given the full form of the thermodynamical potential, the mean field values $\bar{\sigma}_{1,2}$ and ϕ_3 can be obtained as solutions of the coupled set of gap equations

$$\frac{\partial \Omega^{\rm MFA}_{\rm reg}}{(\partial\sigma_1, \partial\sigma_2, \partial\phi_3)} = 0 \,. \tag{15}$$

In order to fully specify the model under consideration, we proceed to fix the model parameters as well as the nonlocal form factors $g(q)$ and $f(q)$. We consider here Gaussian functions [6] which guarantee a fast ultraviolet convergence of the loop integrals. The values of the five free parameters can be found in Ref. [7].

Once the mean field values are obtained, the behavior of other relevant quantities as functions of the temperature and chemical potential can be determined. We concentrate, in particular, on the chiral quark condensate $\langle\bar{q}q\rangle = \partial\Omega^{\rm MFA}_{\rm reg}/\partial m$ and the traced Polyakov loop Φ, which will be taken as order parameters for the chiral restoration and deconfinement transitions, respectively. The associated susceptibilities will be defined as $\chi_{\rm ch} = \partial\langle\bar{q}q\rangle/\partial m$ and $\chi_{\rm PL} = d\Phi/dT$. In addition, in Ref [2] the deconfinement temperature, defined at the peak of the entropy of a static quark S_q is located at the same temperature, within errors, as the chiral susceptibility even at finite lattice spacing. This quantity defined as

$$S_q = -\frac{dF_q}{dT} \,, \tag{16}$$

with $F_q = -T \ln \Phi$, has the main advantage of being scheme independent.

In this work we define the deconfinement transition temperature, in the crossover region, with the peak of the Polyakov susceptibility $\chi_{\rm PL}$. In the region where the deconfinement is a first order phase transition we use the same prescription as Ref. [15], where the critical temperature is defined as the temperature where $\Phi = 0.4$.

4 Results

We begin our analysis studying the finite energy sum rules at zero density. In this scenario, when $\mu = 0$, the Eq. (7) becomes

$$8\pi^2 f_\pi^2(T) = \frac{4}{3}\pi^2 T^2 + \int_0^{s_0(T)} ds \left[1 - 2 n_F\left(\frac{\sqrt{s}}{2T}\right)\right], \tag{17}$$

where the pion decay constant at finite temperature and/or chemical potential is calculated using Eq. (10) and Eq. (11) as

$$F_0(p^2) = 8\,T \sum_{c,n} \int \frac{d^3\vec{q}}{(2\pi)^4}\, g(\rho^c_{n,\vec{q}})\, \frac{Z(\rho^{c\ +}_{n,\vec{q}})Z(\rho^{c\ -}_{n,\vec{q}})}{D(\rho^{c\ +}_{n,\vec{q}})D(\rho^{c\ -}_{n,\vec{q}})} \left[\rho^{c\ +}_{n,\vec{q}} \cdot \rho^{c\ -}_{n,\vec{q}} + M(\rho^{c\ +}_{n,\vec{q}})M(\rho^{c\ -}_{n,\vec{q}})\right] \tag{18}$$

where $\rho^{c\ \pm}_{n,\vec{q}} = \rho^c_{n,\vec{q}} \pm p/2$.

It is known that in local versions of the PNJL model, at zero chemical potential, the restoration of the chiral symmetry and the deconfinement transition take place at different temperatures (see e.g. Refs. [24, 25]), usually separated by approximate 20 MeV.

In Fig. 1 we plot the continuum threshold, the trace of the PL and the normalized quark condensate for the nonlocal (local) PNJL model in thick (thin) line, for the logarithmic and polynomial effective potentials. As we expected from previous results, in the local version both transitions do not occur simultaneously. In this scenario, the PQCD threshold vanishes at a critical temperature, $T_c^{s_0}$, located between the chiral critical temperature T_c^χ and the PL deconfinement temperature T_c^Φ (obtained through the corresponding susceptibilities).

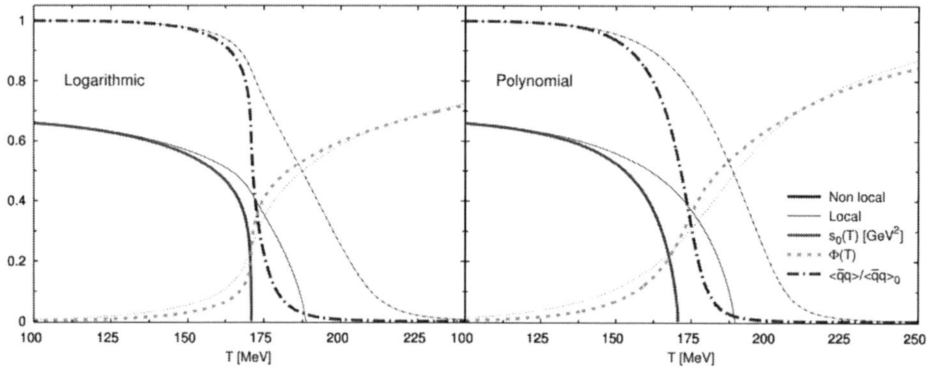

Figure 1. Continuum threshold (red solid line), trace of the Polyakov loop (green dashed line) and the normalized quark condensate (blue dotted line) as a function of the temperature for nonlocal (thick line) and local PNJL model (thin line).

In the case of the nonlocal PNJL model, for both effective potentials, s_0 and Φ have a similar critical temperature for the deconfinement transition of approximate $T_c \sim 170$ MeV. These temperatures are summarized in Table 1 and can be observed in Fig. 2, where we quote in the left (right) panel for $\mu = 0$, different normalized susceptibilities as functions of the temperature for the logarithmic (polynomial) PL potential.

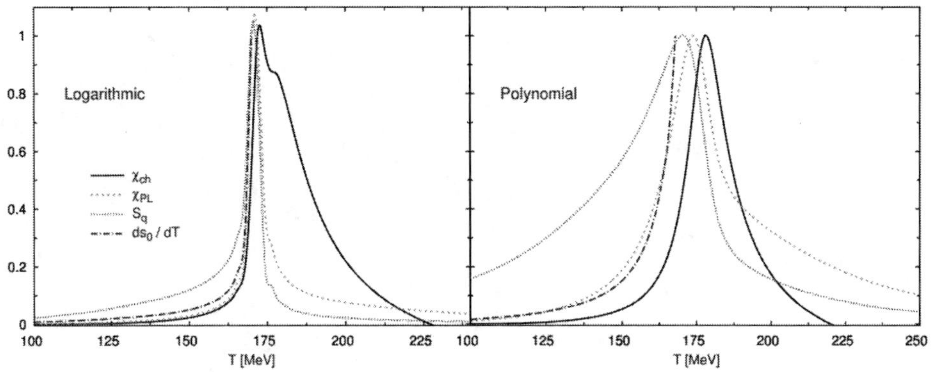

Figure 2. Normalized chiral susceptibility χ_{ch} (black solid line), Polyakov susceptibility $\chi_{\rm PL}$ (green dashed line), static entropy S_q (red dotted line) and ds_0/dT (blue dashed-dotted line) as a function of the temperature for logarithmic (left panel) and polynomial (right panel) PL potential.

	Logarithmic		Polynomial	
	Non local	Local	Non local	Local
T_c^χ [MeV]	171	205	176	201
T_c^Φ [MeV]	171	171	174	183
$T_c^{s_0}$ [MeV]	171	189	170	190

Table 1. Chiral critical temperatures T_c^χ and deconfinement temperatures T_c^Φ and $T_c^{s_0}$.

From lattice QCD calculations, at zero chemical potential, the chiral symmetry restoration and the deconfinement transition take place at the same critical temperature. This behavior was verified in nlPNJL models [5, 15, 26] and also obtained by finite energy sum rules [12]. We will now identify the relation between $s_0(T,\mu)$ and $\Phi(T,\mu)$, extending our previous discussion.

In Fig. 3 we plot, for the logarithmic Polyakov effective potential, the normalized quark condensate $\langle\bar{q}q\rangle/\langle\bar{q}q\rangle_0$, the trace of the PL Φ and the continuum threshold s_0 as functions of the temperature for three different values of chemical potential. In the middle panel we choose $\mu = 139$ MeV, which correspond to the critical end point chemical potential μ_{CEP}. For values of μ smaller than μ_{CEP}, the chiral restoration arises via a crossover transition. Beyond this critical density, a first order phase transition occurs. This value, together with the critical temperature $T_{CEP} = 161$ MeV determines the coordinates of the critical end point.

In the upper panel of Fig. 3, where $\mu = 100$ MeV, we see that the chiral and deconfinement transitions are crossovers occurring at the same critical temperature. The peak of the Polyakov susceptibility and the point where the continuum threshold vanishes occur at approximate the same temperature $T_c \sim 166$ MeV.

When μ becomes equal or higher than $\mu = 139$ MeV, the order parameter for the chiral symmetry restoration has a discontinuity signaling a first order phase transition. These gap in the quark condensate induces also a jump in the trace of the PL (see middle and lower panels in Fig. 3). The value of Φ at the discontinuity indicates that at this temperature the system remains confined but in a chiral symmetry restored state. This region is usually referred as the quarkyonic phase [27, 28].

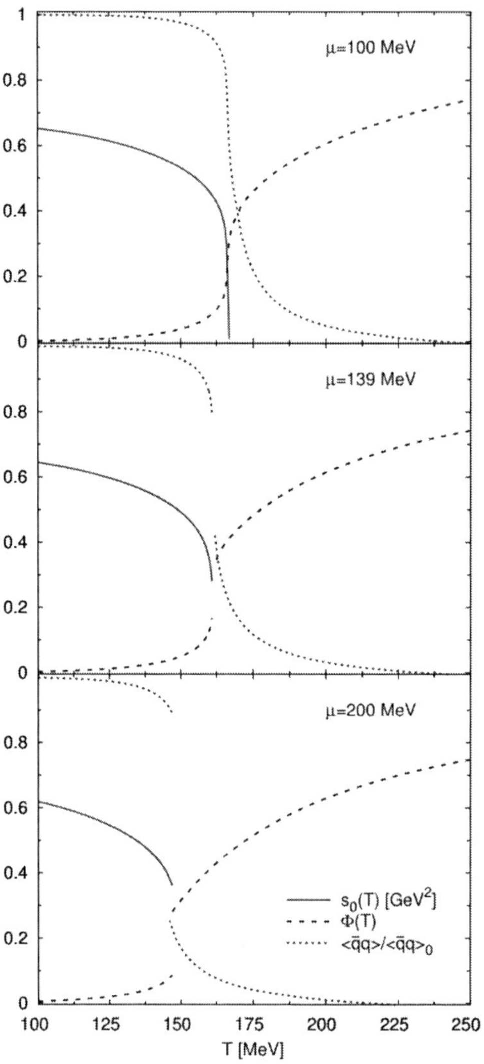

Figure 3. Continuum threshold (solid red line), trace of the Polyakov loop (black dashed lined) and the normalized quark condensate (blue dotted line) as a function of the temperature for the logarithmic effective potential.

We see in this way, that the Polyakov loop and the continuum threshold provide the same information. When the chiral symmetry is restored, s_0 and Φ show that we are still in a confined phase. This characterizes the occurrence of a quarkyonic phase.

5 Summary and conclusions

In this article we compare the behavior of two order parameters for the deconfinement transition: the continuum threshold and the trace of the Polyakov loop.

To accomplish this analysis, we use finite energy sum rules for the axial-vector current correlator.

On the other side, the Polyakov loop, is expected to vanish in the confined phase being different from zero in the deconfined phase.

By saturating the FESR with the pion pole in the spectral function, we used as an input the pion mass, the pion decay constant and the chiral quark condensate obtained from a nonlocal SU(2) Polyakov-NJL model with Gaussian form factors, establishing the connection between both approaches.

We determine, for the nlPNJL model, that the continuum threshold vanishes at the same temperature where the Polyakov susceptibility has its maximum value. In the case of the local PNJL, s_0 becomes zero between the critical temperature for the deconfinement transition, according to the Polyakov loop analysis, and the chiral restoration temperature.

At finite chemical potential, we find that for both deconfinement parameters, beyond the critical end point chemical potential, the system remains in its confined phase even when the chiral symmetry is restored. This is an evidence for the appearance of a quarkyonic phase.

We may conclude saying that our analysis gives strong support to the idea that both deconfinement parameters, in fact, provide the same kind of physical information.

Acknowledgements

This work has been partially funded by CONICET (Argentina) under Grant No. PIP 449; by the National University of La Plata (Argentina), Project No. X718; by FONDECYT (Chile), under grants No. 1130056, 1150171 and 1150847; and by Proyecto Basal (Chile) FB 0821.

References

[1] K. Fukushima and T. Hatsuda, Rept. Prog. Phys. **74**, 014001 (2011) doi:10.1088/0034-4885/74/1/014001 [arXiv:1005.4814 [hep-ph]].

[2] A. Bazavov, N. Brambilla, H.-T. Ding, P. Petreczky, H.-P. Schadler, A. Vairo and J. H. Weber, Phys. Rev. D **93**, no. 11, 114502 (2016) doi:10.1103/PhysRevD.93.114502 [arXiv:1603.06637 [hep-lat]].

[3] A. I. Bochkarev and M. E. Shaposhnikov, Nuovo Cim. A **92**, 17 (1986). doi:10.1007/BF02730424

[4] Alejandro Ayala, C. A. Dominguez and M. Loewe, Advances in High Energy Physics Volume 2017 (2017), Article ID 9291623, 24 pages https://doi.org/10.1155/2017/9291623 [hep-ph 1608.04284].

[5] J. P. Carlomagno, D. Gomez Dumm and N. N. Scoccola, Phys. Rev. D **88**, no. 7, 074034 (2013) doi:10.1103/PhysRevD.88.074034 [arXiv:1305.2969 [hep-ph]].

[6] J. P. Carlomagno and M. Loewe, Phys. Rev. D **95**, no. 3, 036003 (2017) doi:10.1103/PhysRevD.95.036003 [arXiv:1610.05429 [hep-ph]].

[7] S. Noguera and N. N. Scoccola, Phys. Rev. D **78**, 114002 (2008) doi:10.1103/PhysRevD.78.114002 [arXiv:0806.0818 [hep-ph]].

[8] C. A. Dominguez and K. Schilcher, Phys. Rev. D **61**, 114020 (2000) doi:10.1103/PhysRevD.61.114020 [hep-ph/9903483].

[9] C. A. Dominguez and K. Schilcher, JHEP **0701**, 093 (2007) doi:10.1088/1126-6708/2007/01/093 [hep-ph/0611347].

[10] E. Megias, E. Ruiz Arriola and L. L. Salcedo, Phys. Rev. D **81**, 096009 (2010) doi:10.1103/PhysRevD.81.096009 [arXiv:0912.0499 [hep-ph]].

[11] C. A. Dominguez and M. Loewe, Phys. Rev. D **52**, 3143 (1995) doi:10.1103/PhysRevD.52.3143 [hep-ph/9406213].

[12] Alejandro Ayala, A. Bashir, C. A. Dominguez, E. Gutierrez, M. Loewe and A. Raya, Phys. Rev. D **84**, 056004 (2011) doi:10.1103/PhysRevD.84.056004 [arXiv:1106.5155 [hep-ph]].

[13] K. A. Olive *et al.* [Particle Data Group Collaboration], Chin. Phys. C **38**, 090001 (2014). doi:10.1088/1674-1137/38/9/090001

[14] C. A. Dominguez, M. Loewe and Y. Zhang, Phys. Rev. D **86**, no. 3, 034030 (2012) Erratum: [Phys. Rev. D **90**, no. 3, 039903 (2014)] doi:10.1103/PhysRevD.90.039903, 10.1103/PhysRevD.86.034030 [arXiv:1205.3361 [hep-ph]].

[15] G. A. Contrera, M. Orsaria and N. N. Scoccola, Phys. Rev. D **82**, 054026 (2010) doi:10.1103/PhysRevD.82.054026 [arXiv:1006.4639 [hep-ph]].

[16] V. Pagura, D. Gomez Dumm and N. N. Scoccola, Phys. Lett. B **707**, 76 (2012) doi:10.1016/j.physletb.2011.11.064 [arXiv:1105.1739 [hep-ph]].

[17] D. Diakonov and M. Oswald, Phys. Rev. D **70**, 105016 (2004) doi:10.1103/PhysRevD.70.105016 [hep-ph/0403108].

[18] A. Dumitru, R. D. Pisarski and D. Zschiesche, Phys. Rev. D **72**, 065008 (2005) doi:10.1103/PhysRevD.72.065008 [hep-ph/0505256].

[19] S. Roessner, C. Ratti and W. Weise, Phys. Rev. D **75**, 034007 (2007) doi:10.1103/PhysRevD.75.034007 [hep-ph/0609281].

[20] D. Gomez Dumm and N. N. Scoccola, Phys. Rev. C **72**, 014909 (2005) doi:10.1103/PhysRevC.72.014909 [hep-ph/0410262].

[21] D. Gomez Dumm and N. N. Scoccola, Phys. Rev. D **65**, 074021 (2002) doi:10.1103/PhysRevD.65.074021 [hep-ph/0107251].

[22] O. Scavenius, A. Dumitru and J. T. Lenaghan, Phys. Rev. C **66**, 034903 (2002) doi:10.1103/PhysRevC.66.034903 [hep-ph/0201079].

[23] C. Ratti, M. A. Thaler and W. Weise, Phys. Rev. D **73**, 014019 (2006) doi:10.1103/PhysRevD.73.014019 [hep-ph/0506234].

[24] W. j. Fu, Z. Zhang and Y. x. Liu, Phys. Rev. D **77**, 014006 (2008) doi:10.1103/PhysRevD.77.014006 [arXiv:0711.0154 [hep-ph]].

[25] P. Costa, M. C. Ruivo, C. A. de Sousa, H. Hansen and W. M. Alberico, Phys. Rev. D **79**, 116003 (2009) doi:10.1103/PhysRevD.79.116003 [arXiv:0807.2134 [hep-ph]].

[26] J. P. Carlomagno, D. G. Dumm, V. Pagura and N. N. Scoccola, J. Phys. Conf. Ser. **630**, no. 1, 012049 (2015). doi:10.1088/1742-6596/630/1/012049

[27] L. McLerran and R. D. Pisarski, Nucl. Phys. A **796**, 83 (2007) doi:10.1016/j.nuclphysa.2007.08.013 [arXiv:0706.2191 [hep-ph]].

[28] H. Abuki, R. Anglani, R. Gatto, G. Nardulli and M. Ruggieri, Phys. Rev. D **78**, 034034 (2008) doi:10.1103/PhysRevD.78.034034 [arXiv:0805.1509 [hep-ph]].

Non-perturbative Approach to Equation of State and Collective Modes of the QGP

Shuai Y.F. Liu[1,*] and *Ralf* Rapp[1,**]

[1] *Cyclotron Institute and Department of Physics & Astronomy, Texas A&M University, College Station, TX 77843-3366, USA*

Abstract. We discuss a non-perturbative T-matrix approach to investigate the microscopic structure of the quark-gluon plasma (QGP). Utilizing an effective Hamiltonian which includes both light- and heavy-parton degrees of freedoms. The basic two-body interaction includes color-Coulomb and confining contributions in all available color channels, and is constrained by lattice-QCD data for the heavy-quark free energy. The in-medium T-matrices and parton spectral functions are computed selfconsistently with full account of off-shell properties encoded in large scattering widths. We apply the T-matrices to calculate the equation of state (EoS) for the QGP, including a ladder resummation of the Luttinger-Ward functional using a matrix-log technique to account for the dynamical formation of bound states. It turns out that the latter become the dominant degrees of freedom in the EoS at low QGP temperatures indicating a transition from parton to hadron degrees of freedom. The calculated spectral properties of one- and two-body states confirm this picture, where large parton scattering rates dissolve the parton quasiparticle structures while broad resonances start to form as the pseudocritical temperature is approached from above. Further calculations of transport coefficients reveal a small viscosity and heavy-quark diffusion coefficient.

1 Introduction

The theoretical study of the QCD phase diagram poses formidable challenges due to the strong force between quarks and gluons at intermediate and large distances. The QCD phase structure is believed to be tightly connected to two key phenomena of the standard model, namely hadronic mass generation and color confinement. In addition, it turned out that the strongly interacting fireball medium formed in ultra relativistic collisions of heavy nuclei possesses the smallest known ratio of viscosity to entropy density, giving rise to the notion of the strongly coupled quark-gluon plasma (sQGP) [1–3], a near-perfect liquid. The microscopic structure of the sQGP, including its prevalent degrees of freedom and its possible relation to the nearby phase transition(s) into hadronic matter, remains a forefront topic in contemporary research. In addition to the low viscosity inferred from fluid-dynamic models for the bulk evolution of the medium in heavy-ion collisions at RHIC and the LHC, the large modifications observed for the spectra and elliptic flow of heavy-flavor hadrons [4] have provided a more direct evidence of a frequent rescattering that (heavy) quarks undergo throughout the expansion

*e-mail: lshphy@gmail.com
**e-mail: rapp@comp.tamu.edu

© The Authors, published by EDP Sciences. This is an open access article distributed under the terms of the Creative Commons Attribution License 4.0 (http://creativecommons.org/licenses/by/4.0/).

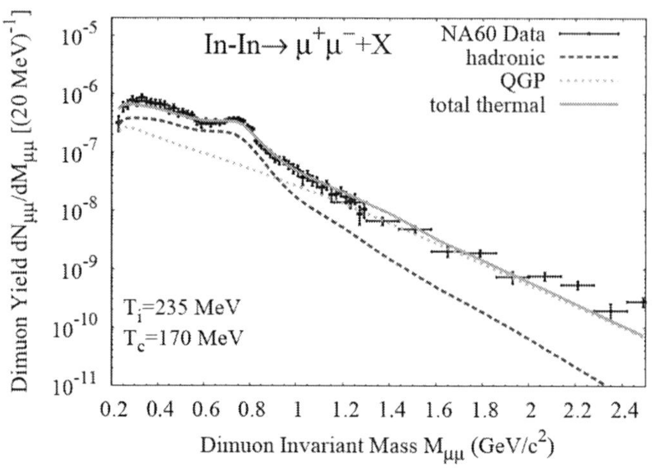

Figure 1. Dimuon excess spectra (after subtraction of background and the final-state hadron decay cocktail) [7] compared to theoretical calculations of thermal radiation from an expanding fireball with contributions from the QGP and hadronic phase with in-medium ρ and ω spectral functions; figure taken from Ref. [10].

of the nuclear fireball. At the same time, dilepton invariant-mass spectra emanating from the thermal radiation of the hot system [5–7] have revealed that the ρ-meson resonance peak, which dominates in the low-mass part of the electromagnetic spectral function in vacuum, melts due to strong rescattering in hot and dense hadronic matter [8], cf. Fig. 1. While this melting provides a signature of chiral symmetry restoration [9], it is also suggestive for a transition in the degrees of freedom in the medium, from well-defined hadronic states into a quasi-continuum of partons. The challenge remains how this melting can be seamlessly connected to a quark-gluon based description on the QGP side of the medium.

A microscopic description of the sQGP is generally expected to require nonperturbative ingredients, in terms of both the underlying interaction (beyond Color-Coulomb) and resummations of diagrams. The vicinity of the sQGP to the phase transition into hadrons suggests that (remnants of) a confining force and the emergence of bound states play an essential role. Computations of the heavy-quark (HQ) free energy in lattice QCD show that its long-distance limit stays above zero until temperatures of about 450 MeV [11], supporting to the presence of non-Coulombic contributions (the leading Coulombic contribution is negative). To capture these aspects, we have developed a thermodynamic T-matrix approach which resums the ladder series of an in-medium interaction kernel that includes the screened Coulomb and confining forces to characterize the interaction strength between the partons in the QGP, and allows for the dynamical formation of bound and resonance states. It provides a uniform treatment of light and heavy partons, can be constrained by pertinent results from lQCD and enables the calculation of spectral functions and transport coefficients for applications to heavy-ion phenomenology. We will briefly review the selfconsistent T-matrix approach in Sec. 2, discuss the lattice-QCD (lQCD) constraints on the underlying potential in Sec. 3, compute the equation of state (EoS) in Sec. 4, and evaluate the emerging spectral and transport properties in Secs. 5 and 6, respectively. Section 7 contains our conclusions and outlook.

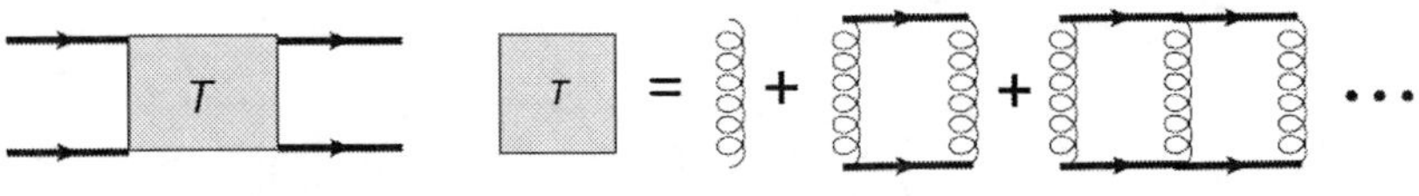

Figure 2. T-matrix resummation of ladder diagrams.

2 T-matrix approach

We start from an effective in-medium parton Hamiltonian,

$$H = \sum \varepsilon(\mathbf{p})\psi^\dagger(\mathbf{p})\psi(\mathbf{p}) + \frac{1}{2}\psi^\dagger(\frac{\mathbf{P}}{2} - \mathbf{p})\psi^\dagger(\frac{\mathbf{P}}{2} + \mathbf{p})V\psi(\frac{\mathbf{P}}{2} + \mathbf{p}')\psi(\frac{\mathbf{P}}{2} - \mathbf{p}') , \quad (1)$$

with single-particle energies $\varepsilon(\mathbf{p}) = \sqrt{M^2 + \mathbf{p}^2}$ containing the bare parton masses, M_i and a 2-body interaction kernel, V, where $\mathbf{p}, \mathbf{p}'$ and $\mathbf{P}$ denote the relative and total momenta of the pair. The sum is over momentum, spin, color, and particle species (3 light-quark flavors, gluons, charm and bottom quarks). Summing up the ladder diagrams generated from this Hamiltonian, we obtain the T-matrix equation [12–14],

$$T(z, \mathbf{p}, \mathbf{p}') = V(\mathbf{p}, \mathbf{p}') + \int_{-\infty}^{\infty} \frac{d^3\mathbf{k}}{(2\pi)^3} V(\mathbf{p}, \mathbf{k}) G^0_{(2)}(z, \mathbf{k}) T(z, \mathbf{k}, \mathbf{p}') \quad (2)$$

in the center-of-mass (CM) frame ($\mathbf{P}$=0) with external two-body energy z, cf. Fig. 2. In the Matsubara representation, the one- and uncorrelated two-body propagator take the form

$$G(i\omega_n, \mathbf{k}) = \frac{1}{[G^0(i\omega_n, \mathbf{k})]^{-1} - \Sigma(i\omega_n, \mathbf{k})} , \quad G^0_{(2)}(iE_n, \mathbf{k}) = -\beta^{-1} \sum_{\omega_n} G(iE_n - i\omega_n, \mathbf{k}) G(i\omega_n, \mathbf{k}) , \quad (3)$$

respectively, and $G^0 = 1/(i\omega_n - \varepsilon(\mathbf{k}))$ is the bare propagator. The selfenergy is obtained by closing T-matrix,

$$\Sigma(iw_n) = \int d\tilde{p}\ TG \equiv -\beta^{-1} \sum_{\nu_n} \int \frac{d^3\mathbf{p}}{(2\pi)^3} T(i\omega_n + i\nu_n) G(i\nu_n) , \quad (4)$$

which we evaluate using standard spectral representations. Equations (2) and (4) form a selfconsistency problem that we solve by numerical iteration, thereby satisfying thermodynamic conservation laws [15].

The key *inputs* to the Hamiltonian are the 2-body interaction kernel, V, and the bare parton masses, M. In the following two sections we discuss how we constrain them using lQCD data for the HQ free energy and the EoS.

3 Interaction kernel and HQ free energy

In Ref. [17] we have obtained a relation between the static HQ free energy and the potential within the T-matrix formalism, given by

$$F_{Q\bar{Q}}(r, \beta) = \frac{-1}{\beta} \ln\Big[\int_{-\infty}^{\infty} dE\, e^{-\beta E} \frac{-1}{\pi} \mathrm{Im}[\frac{1}{E + i\epsilon - \tilde{V}(r) - \Sigma_{Q\bar{Q}}(E + i\epsilon, r)}]\Big] . \quad (5)$$

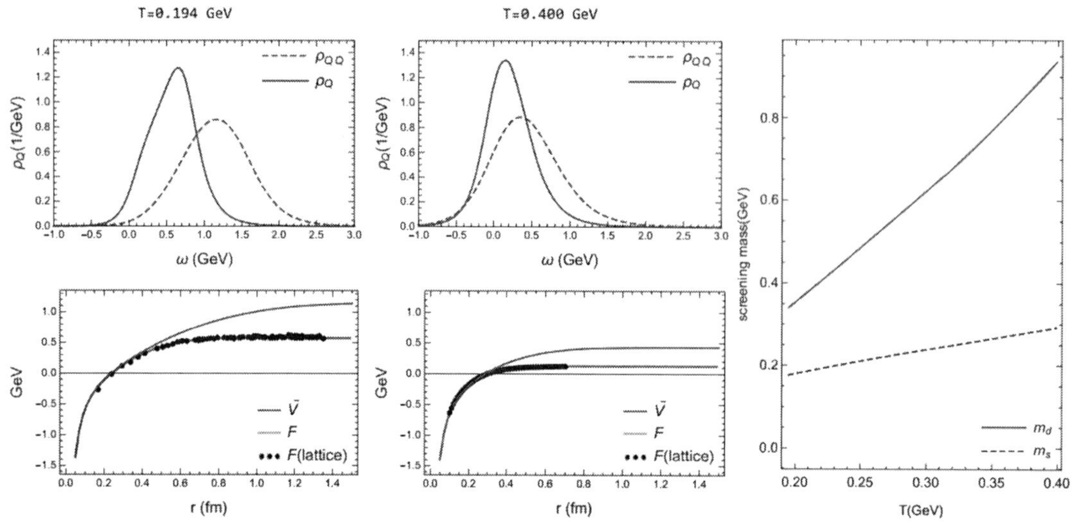

Figure 3. Our results for the selfconsistent fit to lQCD data [16] for the static HQ free energy and resulting potential, $\tilde{V}(r)$ (lower left and middle panels for two different temperatures) and the associated one- and uncorrelated two-body spectral functions, $\rho_Q(\omega,\infty)$ and $\rho_{Q\bar{Q}}(\omega,\infty)$, respectively (upper left and middle panels for the same two temperatures), for the strongly-coupled solution. The right column shows the fitted screening masses for Coulomb (solid line) and string (dashed line) interactions.

In the weak-coupling limit (or at low temperatures), where the imaginary part of the in-medium two-body selfenergy, $\Sigma_{Q\bar{Q}}$, is small, one readily recovers that the free energy coincides with the potential, $\tilde{V}$ (defined as containing a non-trivial infinite-distance limit, see Eq. (6) below). However, at strong coupling large imaginary part develop and lead to appreciable deviations between potential and free energy. As an ansatz for the potential, we employ a generalized in-medium (screened) Cornell potential [18],

$$\tilde{V}(r) = V_C(r) + V_S(r) + 2\Delta M_Q = -\frac{4}{3}\alpha_s\frac{e^{-m_d r}}{r} - \frac{\sigma e^{-m_s r-(c_b m_s r)^2}}{m_s} - \frac{4}{3}\alpha_s m_d + \sigma m_s, \tag{6}$$

including color-Coulomb (V_C) and string terms (V_S). Equation (6) is written in the color-singlet channel for static quarks, but in the calculations of the EoS reported in the next section we include all possible two-body color channels with appropriate Casimir factors and relativistic corrections [19]. The screening of the string term is constrained using $m_s = (c_s m_d^2\sigma/\alpha_s)^{1/4}$ [20] and also contains a quadratic term, $(c_b m_s r)^2$, to better mimic string breaking effects. The genuine two-body part is defined as $V(r) = V_C(r) + V_S(r)$, while the nonzero infinite-distance limit, $\tilde{V}(\infty)$, is related with the mass generated by self-interactions,

$$\Delta M_Q = -\frac{1}{2}\int dr\rho(r)V(r) = \frac{1}{2}(-\frac{4}{3}\alpha_s m_d + \sigma m_s) = \tilde{V}(\infty)/2\,, \tag{7}$$

corresponding to the classical static potential energy of a point charge, $\rho(r) = \delta(r)$, in its own potential; $\Sigma_{Q\bar{Q}}(z,r)$ denotes the analytical two-body selfenergy whose r-dependence includes interference effects [19, 21] which ensure that $\Sigma_{Q\bar{Q}}(z,r)$ vanishes for a color-singlet $Q\bar{Q}$ pair in the $r\to 0$ limit.

The solution for a fit of the underlying potential to the lQCD free-energy data turns out to not be unique. We therefore bracket the possible range of potentials by a "strongly-coupled solution" (SCS) and which maximally deviates from the free energy, and a "weakly-coupled solution" (WCS) which is as close as possible to the free energy (within our ability to find fits). In Fig. 3 we show the results for the fit to the free energy as well as the underlying potential and one- and uncorrelated two-body spectral functions of static quarks for the SCS. The potential is seen to significantly exceed the free energy at intermediate and large distances (especially at low temperatures), due to large imaginary parts of the selfenergies which in turn lead to broad the spectral functions. The latter are characterized by 1- and 2-body widths of near 1 and 2 GeV, respectively, which, remarkably, decrease from the lower to the higher temperature. While the screening mass for the Coulomb term increases quite strongly with temperature (roughly in accord with perturbative estimates, $m_d \sim gT$), the screening of the string term is rather weak. This leads to a long-range force which enables the static quark to interact with multiple thermal partons and thus significantly contributes to the large imaginary parts. In the WCS (not shown), the potential is much closer to the free energy and the imaginary parts are much smaller at low temperatures.

4 Equation of state and light-parton masses

Next we apply the Hamiltonian to calculate the EoS of the QGP using the Luttinger-Ward-Baym formalism [15, 22, 23] where the free energy of the system is given by

$$\Omega = \mp \sum \mathrm{Tr}\{\ln(-G^{-1}) + [(G^0)^{-1} - G^{-1}]G\} \pm \Phi \,, \tag{8}$$

where the upper (lower) signs corresponds to bosons (fermions). The "$\sum$Tr" includes a 3-momentum integral and summation over Matsubara frequencies (with factors of -$1/\beta$) and discrete quantum numbers. The Luttinger-Ward functional (LWF),

$$\Phi = \sum_{\nu=1}^{\infty} \Phi_\nu \,, \Phi_\nu = \sum \mathrm{Tr}\{\frac{1}{2\nu}\Sigma_\nu(G)G \,, \tag{9}$$

contains the 2-body interaction contribution to the EoS. It cannot be as straightforwardly resummed as the T-matrix due to additional $1/\nu$ factors for the ν-th order closed loop diagrams represented by closed selfenergies, $\Sigma_\nu(G)$, needed to eliminate double counting. In ladder approximation one has $\Sigma_\nu(G) = \int d\tilde{p}\, [VG^0_{(2)}VG^0_{(2)} \cdots V]G$, so that the LWF functional Φ can be expressed as

$$\Phi = \frac{1}{2}\sum \mathrm{Tr}\Big\{G\Big[V + \frac{1}{2}VG^0_{(2)}V + \ldots + \frac{1}{\nu}VG^0_{(2)}VG^0_{(2)} \ldots . V + \ldots\Big]G\Big\} \,. \tag{10}$$

The part in brackets, $[\cdots]$, has a structure similar to the T-matrix resummation,

$$T \;=\; V + VG^0_{(2)}V + \ldots + VG^0_{(2)}VG^0_{(2)} \ldots V + \ldots = \left[\sum_{\nu=0}^{\infty}\left(VG^0_{(2)}\right)^\nu\right] V = [1 - VG^0_{(2)}]^{-1}V \,, \tag{11}$$

except for the extra coefficients $1/\nu$. However, we can write

$$V + \frac{1}{2}VG^0_{(2)}V + \ldots + \frac{1}{\nu}VG^0_{(2)}VG^0_{(2)} \ldots V + \cdots = \left[\sum_{\nu=1}^{\infty}\frac{1}{\nu}\left(VG^0_{(2)}\right)^\nu\right][G^0_{(2)}]^{-1} = -\ln[1 - VG^0_{(2)}][G^0_{(2)}]^{-1} \equiv \mathrm{Log}\, T \tag{12}$$

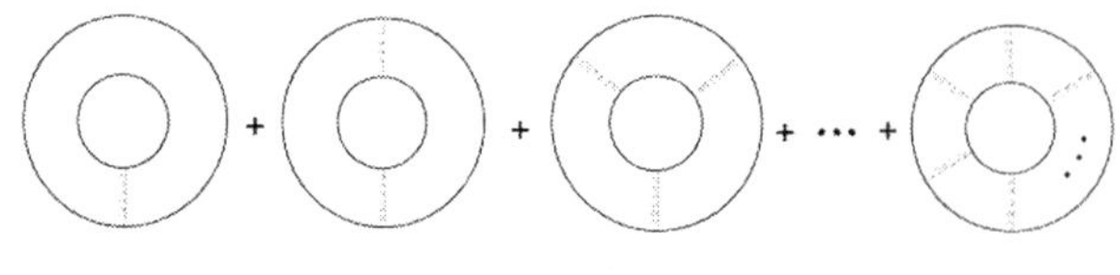

Figure 4. Examples of diagrams that are resummed by the generalized T-matrix approach for the EoS.

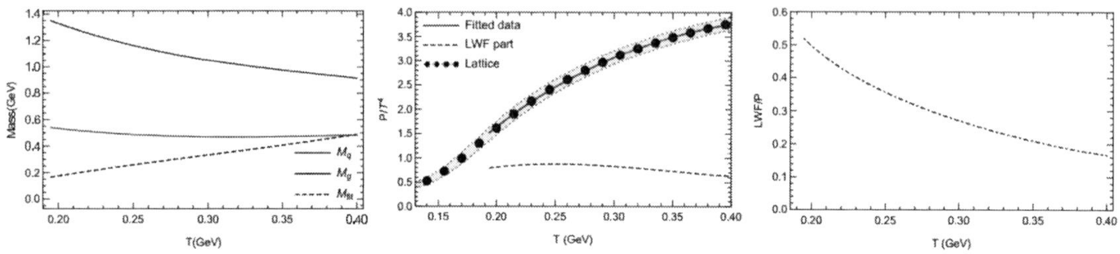

Figure 5. The fit results for the bare masses of the quarks and gluons in the Hamiltonian (left panel), the resulting fit to the lQCD data [11] for the QGP pressure (middle panel; solid line: total, dashed line: LWF contribution), and the relative contribution of the LWF part to the total pressure (right panel).

and to obtain a matrix-log representation of the skeleton series [20, 24], as a generalization of usual T-matrix resummation. The natural-base "Log" should be understood as a matrix operation in a discrete energy-momentum space together with other quantum numbers, defined through its power series [20, 24]. Closing one of the external lines of this quantity with a G, in resemblance of Eq. (4), we define $\text{Log}\,\Sigma \equiv \int d\tilde{p}\, \text{Log}\, T\, G$. The LWF is obtained by closing another external line as

$$\Phi = \frac{1}{2}\int d\tilde{p}\, G\, \text{Log}\Sigma,\ \Omega = \sum_j \mp d_j \int d\tilde{p}\Big\{ \ln(-G_j(\tilde{p})^{-1}) + [\Sigma_j(\tilde{p}) - \frac{1}{2}\text{Log}\Sigma_j(\tilde{p})]G_j(\tilde{p})\Big\}, \tag{13}$$

where j sums over the particle species. This procedure to sum the diagrams is illustrated in Fig. 4. The resummation of the LWF is critical to incorporate bound-state (or resonance) contributions to the pressure. The selfenergy of partons can then be obtained from the selfconsistent T-matrix equation via the functional derivative of this LWF, which explicitly illustrated the "conserving approximation" for the resulting 1- and 2-body spectral functions within this scheme [15, 23]. This, in turn, allows to probe transitions from parton to meson degrees of freedom, if any, in the system.

The resulting equation of state is shown in the middle panel of Fig. 5 where we utilized the bare light-parton masses figuring in the Hamiltonian as fit parameters, shown in the left panel in Fig. 5. We use the ansatz $M_q = M_{\text{fit}} + M_q^{\text{np}}$ and $M_g = 3/2M_{\text{fit}} + M_g^{\text{np}}$ with a fit parameter M_{fit} and additional mass contributions separately for quarks and gluons which follow from the nonperturbative part given by the infinite-distance limit of the potential, $\tilde{V}$, in the corresponding color channel. This ansatz allows the masses to smoothly transit from nonperturbative behavior at low temperature to perturbative behavior at high temperature. At low temperatures gluons are found to essentially decouple from the pressure due to their large masses. On the other hand, the LWF contribution grows with decreasing temperature, see right panel of Fig. 5, mostly driven by the formation of mesonic and diquark bound states / resonances. In this sense we find a gradual transition in the degrees of freedom from partonic to hadronic states as the temperature approaches T_c from above. This is different in the WCS, where the LWF does not exceed 15% of the total pressure and is rather constant with temperature.

5 Spectral properties of QGP

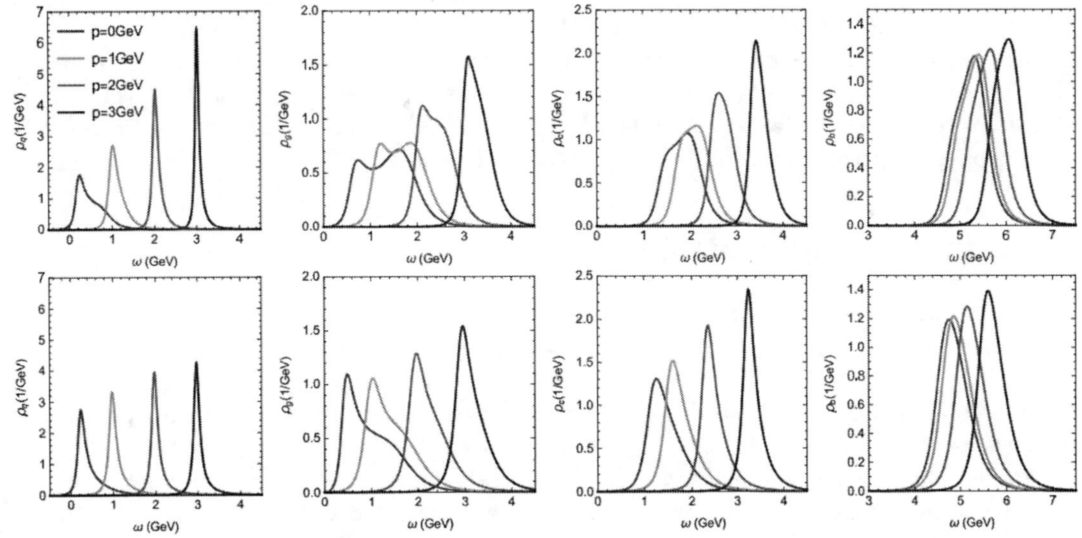

Figure 6. In-medium parton spectral functions from the selfconsistent solution in the SCS. From left to right, we show the light-quark (q), gluon (g), charm-quark (c) and bottom-quark (b) spectral functions for 4 different 3-momenta in each panel, and for temperatures T=0.194 GeV (upper row) and T=0.400 GeV (lower row). Note the melting of the low-momentum quark and gluon quasiparticles at low temperatures (with additional low-energy collective modes), while the charm and especially bottom quark remain good quasiparticles.

We proceed to inspect the spectral properties underlying our selfconsistent off-shell calculations of the EoS discussed in the previous section, cf. Fig. 6. Near T_c, the low-momentum light-parton spectral functions acquire widths of order 1 GeV, significantly exceeding their masses, and thus implying a melting of their quasiparticle structure. Consequently, their contribution to the EoS is suppressed, while at the same time the pertinent T-matrices develop broad mesonic bound states with appreciable strength (and a mass close to the vacuum ρ mass). These states therefore play a dual role of newly emerging degrees of freedom and providing large interaction strength for the partons, thus connecting the strong coupling nature of the QGP with a gradual transition to hadronic states. Furthermore, the

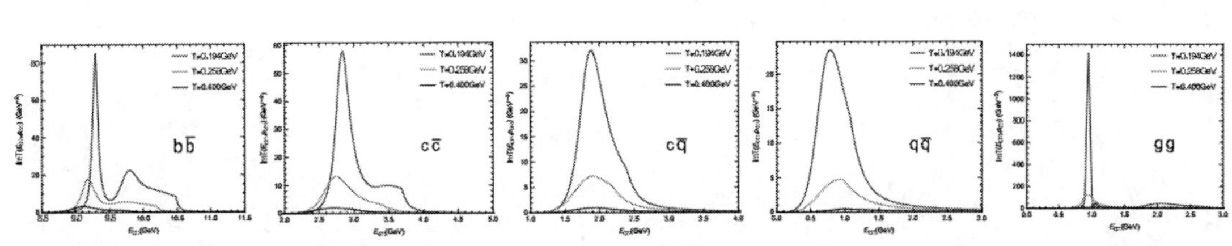

Figure 7. Imaginary part of the on-shell color-singlet S-wave T-matrices for different 3 different temperatures in each panel for: (from left to right) bottomonium ($b\bar{b}$), charmonium ($c\bar{c}$), D-meson ($c\bar{q}$), ρ-meson ($q\bar{q}$), and glueball (gg) channels. Note the decreasing y-axis scale for the first 4 panels.

attractive real part of the light-parton selfenergies leads to the appearance of low-energy collective modes in their spectral functions. These modes could play an important role for describing the quark susceptibilities where quasiparticle models face difficulties due to their large masses which suppress fluctuations. At high parton 3-momenta and higher temperature, the quasiparticle structures re-emerge – a manifestation of the weakening QCD force with increasing momentum transfer and illustrating the changing degrees of freedom as the QGP is probed with varying resolution. In the WCS (not shown), all partons remain well-defined quasiparticle at all temperatures and 3-momenta, while rather weak and sharp resonances appear in the T-matrices at low temperature.

6 Transport properties of QGP

Thus far the lQCD constraints imposed on our approach (HQ free energy, EoS and quarkonium correlator ratios (not explicitly discussed here)) cannot decisively distinguish between the SCS and WCS. As another constraint we have therefore investigated transport coefficients which figure in applications to heavy-ion phenomenology, specifically the HQ diffusion coefficient and the ratio of viscosity over entropy density.

For the HQ friction coefficient, we extend previous T-matrix calculations carried out for the free and internal energies as potential proxies [4, 13, 25]. In particular, due to the large widths in the SCS we account for off-shell effects based on the formalism described in Ref. [26]. Schematically, one has

$$A(p) = \left\langle (1 - \frac{\mathbf{p}\cdot\mathbf{p}'}{p^2})\rho_i\rho_i\rho_c \right\rangle , \tag{14}$$

where $\rho_{i(c)}$ are spectral functions for light partons (charm quark) and $\mathbf{p}$ ($\mathbf{p}'$) denotes the incoming (outgoing) charm-quark momentum. The spatial diffusion coefficient is defined as $\mathcal{D}_s = T/(A(0)M)$.

For the viscosity, η, a Kubo formula is employed using the leading-density energy-momentum tensor [27] with relativistic extension,

$$\eta = \lim_{\omega\to 0}\sum_i \frac{\pi d_i}{\omega}\int \frac{d^3\mathbf{p}d\lambda}{(2\pi)^3}\frac{p_x^2p_y^2}{\varepsilon_i^2(p)}\rho_i(\omega+\lambda,p)\rho_i(\lambda,p)[n_i(\lambda)-n_i(\omega+\lambda)] , \tag{15}$$

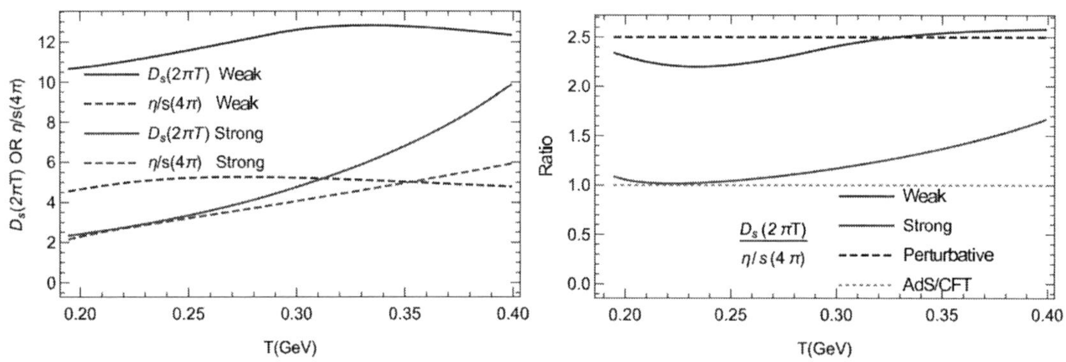

Figure 8. The transport coefficients $D_s(2\pi T)$ and $\eta/s(4\pi)$ of SCS and WCS (left); The ratio of $D_s(2\pi T)$ of $\eta/s(4\pi)$ of SCS and WCS (right).

where d_i and $n_i(\omega)$ are the partons' degeneracies and thermal distribution functions, respectively. The corrections from higher orders are expected to be small [28–31], which we have checked to remain valid within our approach.

The dimensionless quantities $\mathcal{D}_s(2\pi T)$ and $4\pi\eta/s$ characterize the interaction strength of the bulk medium (with smaller values indicating stronger coupling); they are shown in the left panel of Fig. 8. For the SCS both transport coefficients are within a factor of two of the conjectured quantum lower bounds of one, and increase with temperature indicating a transition to a more weakly coupled medium. On the contrary, for the WCS both transport coefficients are significantly larger and rather constant with temperature. Especially the HQ diffusion coefficient acquires magnitudes which do not compare well with current extractions from HF phenomenology in high-energy heavy-ion collisions [4].

In an attempt to better quantify the notions of "strongly" and "weakly" coupled media, one can inspect the ratio for the two dimensionless transport coefficients discussed above [32]. In particular, the ratio $r \equiv [2\pi T\mathcal{D}_s]/[4\pi\eta/s]$ is expected to be near one in the strong-coupling limit [33, 34], while perturbative estimates [35] appropriate for a weakly coupled system result in ~5/2. We plot this ratio in the right panel of Fig. 8 for both the SCS and WCS. Interestingly, for the SCS the ratio is around one for low temperatures, slowly increasing with temperature but still significantly below 5/2 at T=400 MeV. On the contrary, for the WCS the ratio is close to 5/2 characteristic for a weakly-coupled system even at low temperatures, with insignificant temperature dependence.

7 Conclusion

Unraveling the microscopic structure of the strongly coupled QCD medium in the vicinity of the transition from hadronic to partonic matter is one of the key challenges in the study of the QCD phase diagram. Toward this end, we have put forward a thermodynamic T-matrix formalism which includes basic nonperturbative ingredients that are expected to be relevant: For the underlying two-body interaction we account for remnants of the confining force encoded in a screened Cornell potential ansatz, while the T-matrix allows for a ladder resummation to account for dynamically formed bound states and retains the full off-shell properties of one- and two-body spectral functions. Our starting point is a relativistic Hamiltonian, whose in-medium two-body interaction is constrained by lattice-QCD data for the heavy-quark free energy and euclidean quarkonium correlators, while the light-parton masses in the kinetic term are fitted to the lattice equation of state. At this level, the solution to our set-up is not unique; however, a strongly coupled solution has the attractive features of melting light-parton degrees of freedom in connection with emerging hadronic bound states as the temperature approaches T_c from above. We furthermore evaluated transport properties of the QGP by computing the shear-viscosity-to entropy density ratio and the heavy-quark diffusion coefficient. In the strongly-coupled scenario, their values are within a factor of ~2 of the conjectured quantum lower bound, and slowly increase with temperature. The ratio of these two quantities corroborates that the QGP near T_c is a strongly-coupled medium near the quantum lower bound, which reiterates the necessity of including quantum off-shell effects in studying its properties. Future applications include the calculation of high-p_T parton transport, quark-number susceptibilities to explore the finite μ_q part of the phase diagram, and the explicit inclusion of condensate formation to treat chiral symmetry breaking and color superconductivity.

Acknowledgments
This work has been supported by the U.S. National Science Foundation under grant no. PHY-1614484.

References

[1] E. Shuryak, Rev. Mod. Phys. **89**, 035001 (2017)
[2] P. Braun-Munzinger, V. Koch, T. Schafer, J. Stachel, Phys. Rept. **621**, 76 (2016)
[3] U. Heinz, Int. J. Mod. Phys. A **30**, 1530011 (2015)
[4] F. Prino, R. Rapp, J. Phys. G **43**, 093002 (2016)
[5] R. Arnaldi et al. (NA60 Collaboration), Phys. Rev. Lett. **96**, 162302 (2006)
[6] I. Tserruya, Landolt-Bornstein **23**, 176 (2010)
[7] H.J. Specht (NA60 Collaboration), AIP Conf. Proc. **1322**, 1 (2010)
[8] R. Rapp, J. Wambach, H. van Hees, Landolt-Bornstein **23**, 134 (2010)
[9] P.M. Hohler, R. Rapp, Phys. Lett. B **731**, 103 (2014)
[10] R. Rapp, H. van Hees, Phys. Lett. B **753**, 586 (2016)
[11] A. Bazavov et al. (HotQCD Collaboration), Phys. Rev. D **90**, 094503 (2014)
[12] M. Mannarelli, R. Rapp, Phys. Rev. C **72**, 064905 (2005)
[13] F. Riek, R. Rapp, Phys. Rev. C **82**, 035201 (2010)
[14] F. Riek, R. Rapp, New J. Phys. **13**, 045007 (2011)
[15] G. Baym, L.P. Kadanoff, Phys. Rev. **124**, 287 (1961)
[16] A. Mocsy, P. Petreczky, M. Strickland, Int. J. Mod. Phys. A **28**, 1340012 (2013)
[17] S.Y.F. Liu, R. Rapp, Nucl. Phys. A **941**, 179 (2015)
[18] E. Megias, E. Ruiz Arriola, L. Salcedo, Phys. Rev. D **75**, 105019 (2007)
[19] S.Y.F. Liu, R. Rapp (2017), `arXiv:1711.03282`
[20] S.Y.F. Liu, R. Rapp (2016), `arXiv:1612.09138`
[21] M. Laine, O. Philipsen, P. Romatschke, M. Tassler, JHEP **0703**, 054 (2007)
[22] J.M. Luttinger, J.C. Ward, Phys. Rev. **118**, 1417 (1960)
[23] G. Baym, Phys. Rev. **127**, 1391 (1962)
[24] S.Y.F. Liu, R. Rapp, J. Phys. Conf. Ser. **779**, 012034 (2017)
[25] H. van Hees, M. Mannarelli, V. Greco, R. Rapp, Phys. Rev. Lett. **100**, 192301 (2008)
[26] P. Danielewicz, Annals Phys. **152**, 239 (1984)
[27] D.N. Zubarev, *Nonequilibrium statistical thermodynamics* (Plenum Publishing Corporation, 1974)
[28] M. Iwasaki, H. Ohnishi, T. Fukutome (2006), `hep-ph/0606192`
[29] M. Iwasaki, H. Ohnishi, T. Fukutome, J. Phys. G **35**, 035003 (2008)
[30] R. Lang, W. Weise, Eur. Phys. J. A **50**, 63 (2014)
[31] S. Ghosh, Int. J. Mod. Phys. A **29**, 1450054 (2014)
[32] R. Rapp, H. van Hees, in *Quark-Gluon Plasma 4*, edited by R. Hwa, X.N. Wang (World Scientific, Singapore, 2010), p. 111, `arXiv:0903.1096`
[33] P. Kovtun, D.T. Son, A.O. Starinets, Phys. Rev. Lett. **94**, 111601 (2005)
[34] S.S. Gubser, Phys. Rev. D **76**, 126003 (2007)
[35] P. Danielewicz, M. Gyulassy, Phys. Rev. D **31**, 53 (1985)

Experimental overview of correlations in small collision systems

A brief history

Soumya Mohapatra[1,*]

[1] *Department of Physics, Columbia University*

Abstract. The azimuthal anisotropies of particle yields observed in relativistic heavy ion collisions have been traditionally considered as a strong evidence of the formation of a deconfined quark-gluon plasma in these collisions. However multiple recent measurements in pp and p/D/He+A collisions show similar features as those observed in heavy ion collisions, indicating the possibility of the production of such a deconfined medium in smaller collision systems. This paper presents a comprehensive summary of such measurements in small systems. It includes measurements of identified and inclusive two-particle correlations in $\Delta\phi$ and $\Delta\eta$, with different procedures used to subtract the dijet contributions, as well as measurements of multi-particle cumulants $c_n\{2-8\}$. The traditional cumulant measurements confirm presence of collective phenomena in p+A collisions, but are biased by non-flow correlations and are not able to provide evidence for collectivity in pp collisions. To address this, a new subevent cumulant method that further suppresses the contribution non-flow effects was developed, whose measurements are also discussed.

1 Introduction

In relativistic heavy ion collisions, such as those at the RHIC and at the LHC, the colliding nuclei are heated to high enough temperatures such that the quarks and gluons that are ordinarily confined within nucleons get de-localized and a new state of matter called the quark-gluon plasma (QGP) is created. The medium expands anisotropically driven the pressure gradients between the medium and the outside vacuum, cooling down as it expands and finally hadronizes. The final particle yields in the azimuthal (ϕ) direction are typically parameterized as a Fourier series [1]:

$$dN/d\phi \propto (1 + 2\sum v_n \cos n(\phi - \Psi_n), \qquad (1)$$

where the v_n and Φ_n denote the magnitude and orientation of the single-particle anisotropies. Due to their hydrodynamic origin in heavy ion collisions, the v_n are often called flow harmonics.

Such azimuthal anisotropies were not expected to arise in proton-proton (pp) and proton-nucleus (p+A) collisions where the QGP was not expected to be formed. However in 2010 the CMS collaboration showed that in high-multiplicity pp collisions, two-particle correlations (2PCs) in relative azimuthal angle ($\Delta\phi$=$\phi^a - \phi^b$) and pseudorapidity separation ($\Delta\eta$=$\eta^a - \eta^b$), where the superscripts a and b label the two particles used in the correlation, showed the presence of a long range correlation structure along $\Delta\eta$ at $\Delta\phi$ ~0 [2]. This structure, commonly called "ridge", is also seen in heavy ion

*e-mail: soumya@cern.ch

© The Authors, published by EDP Sciences. This is an open access article distributed under the terms of the Creative Commons Attribution License 4.0 (http://creativecommons.org/licenses/by/4.0/).

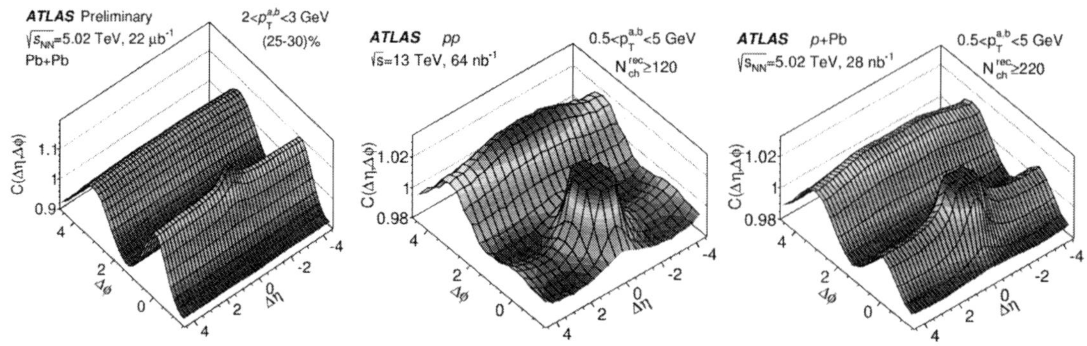

Figure 1. Two particle correlations ($C(\Delta\eta, \Delta\phi)$) in Pb+Pb(left), *pp* (center) and *p*+Pb (right) collisions. The Pb+Pb results are for 2< $p_T^{a,b}$ <3 GeV and for the 25–30% centrality interval. The *pp* results are for 0.5< $p_T^{a,b}$ < 5 GeV and for events having more than 120 reconstructed charged particles with p_T > 0.4 GeV. The *p*+Pb results are for 0.5< $p_T^{a,b}$ < 5 GeV and for events having more than 220 reconstructed charged particles with p_T > 0.4 GeV. The plots are from Refs. [4, 5]

collisions and arises there due to collective flow. Thus its presence in *pp* collisions indicated the possibility of collective effects being present in *pp* collisions. The left and middle panels of Figure 1 compare the ridge seen in Pb+Pb and *pp* collisions, respectively. The *pp* ridge is seen to be considerably weaker that the Pb+Pb ridge. The weakness of the ridge in *pp* collisions made it impossible to use the techniques that were developed for and used in analysis of 2PCs in A+A collisions. Further more, the *pp*-ridge could be explained by initial state models based on gluon saturation [3], and this was accepted by the heavy ion community as its origin.

2 Collectivity in *p*+A collisions

In 2012 the analysis of the first *p*+Pb collision data at the LHC from ATLAS, ALICE and CMS collaborations showed the presence of the ridge in *p*+Pb collisions as well [6–10]. Unlike the *pp* case, the long-range correlations in *p*+Pb were strong enough (see Figure 1 right panel), and the multiplicities were large enough, that the techniques used for analysis of 2PCs in A+A collisions could be applied. The 2PC analysis indicated that the long-range correlations were indeed arising from single-particle anisotropies, v_n, which could be extracted using the 2PCs [7]. The *p*+Pb-v_n were found to be remarkably similar to than seen in A+A collisions. In fact, after some trivial scalings, the v_2–v_4 in *p*+Pb had exactly the same p_T dependence as the v_n in Pb+Pb collisions [9]. This is demonstrated in Figure 2. The left panels show the comparison of v_2–v_4 as a function of p_T in Pb+Pb and *p*+Pb collisions. The centrality interval of 55–60% in the Pb+Pb data is chosen such that it approximately corresponds to the multiplicity range of the *p*+Pb data to which it is compared ($220 \leq N_{ch}^{rec} < 260$). The right panels of Figure 2, the Pb+Pb data are rescaled horizontally by a constant factor of 1.25, and the v_2 and v_4 are also down-scaled by an empirical factor of 0.66 to match the *p*+Pb data. The factor if 1.25 is to account for the difference in the $\langle p_T \rangle$ between the Pb+Pb and *p*+Pb data [11], while the empirical factor of 0.66 accounts for the differences in the collision geometry in the two systems.

Additionally, measurement of v_n via multi-particle cumulants and Lee-Yang Zeroes (LYZ) methods in *p*+Pb collisions also exhibited similarities with Pb+Pb collisions [10]. Figure 3 shows the v_2 obtained from 4, 6 and 8-particle cumulants ($v_2\{4\}$, $v_2\{6\}$ together with the v_2 from 2PCs ($v_2\{2, \Delta\eta > 2\}$)

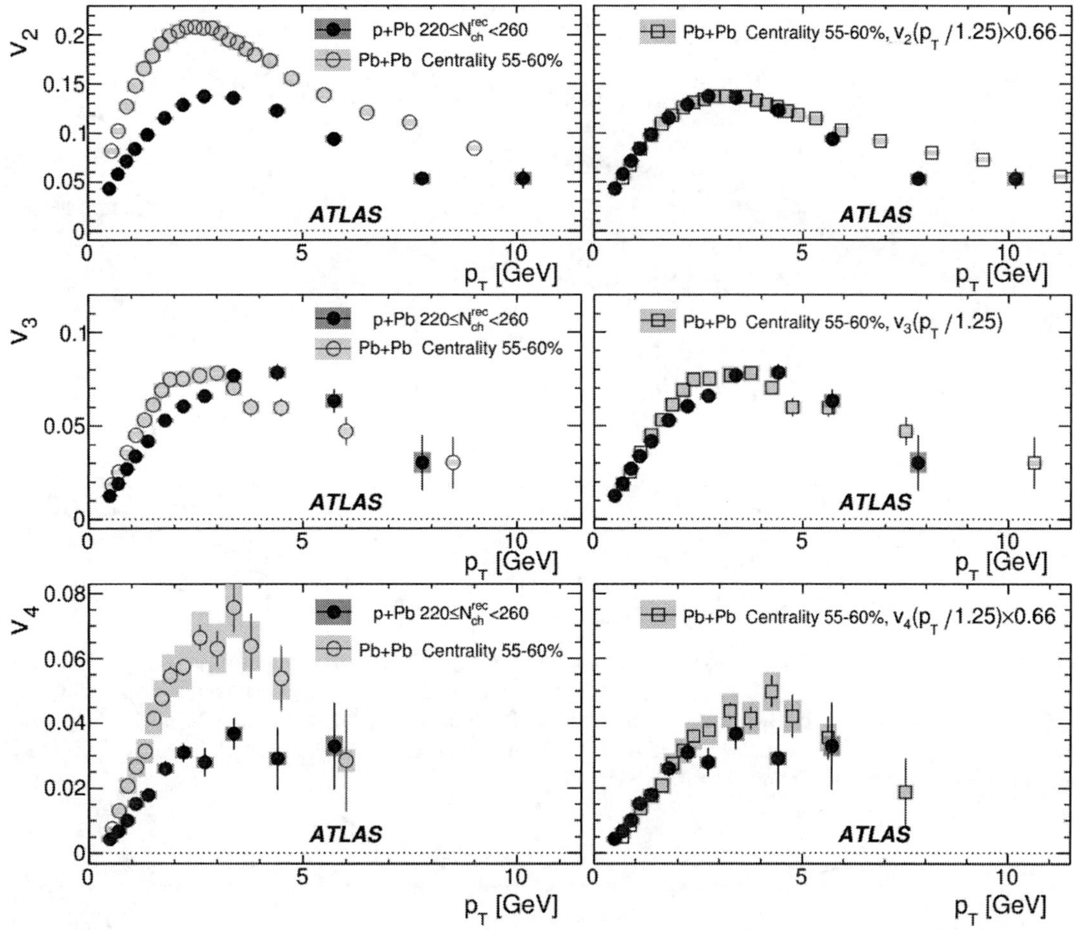

Figure 2. The coefficients v_2 (top row), v_3 (middle row) and v_4 (bottom row) as a function of p_T compared between p+Pb collisions with $220 \leq N_{ch}^{rec} < 260$ and Pb+Pb collisions in 55–60% centrality. The left column shows the original data with their statistical (error bars) and systematic uncertainties (shaded boxes). In the right column, the same Pb+Pb data are rescaled horizontally by a constant factor of 1.25, and the v_2 and v_4 are also down-scaled by an empirical factor of 0.66 to match the p+Pb data. The Figures are taken from Ref. [9].

in Pb+Pb (left) and p+Pb collisions (right). It is seen that in both systems, the v_2 obtained from the 4, 6 and 8 particle cumulants are identical within systematic uncertainties, and smaller than the 2PC-v_2. This ordering of the v_2 measured via these different methods is considered a characteristic of flow phenomena [10].

Further more, v_n measurements in other small-large collision systems, namely d+A and ^{3}He + A collisions, also demonstrated similarities with A+A collisions, and were well reproduced by hydrodynamic calculations [12–16]. A summary of such results from the PHENIX collaboration is shown in Figure 4. These and other measurements established with considerable certainty the presence of collective behavior in p/D/He+A collisions.

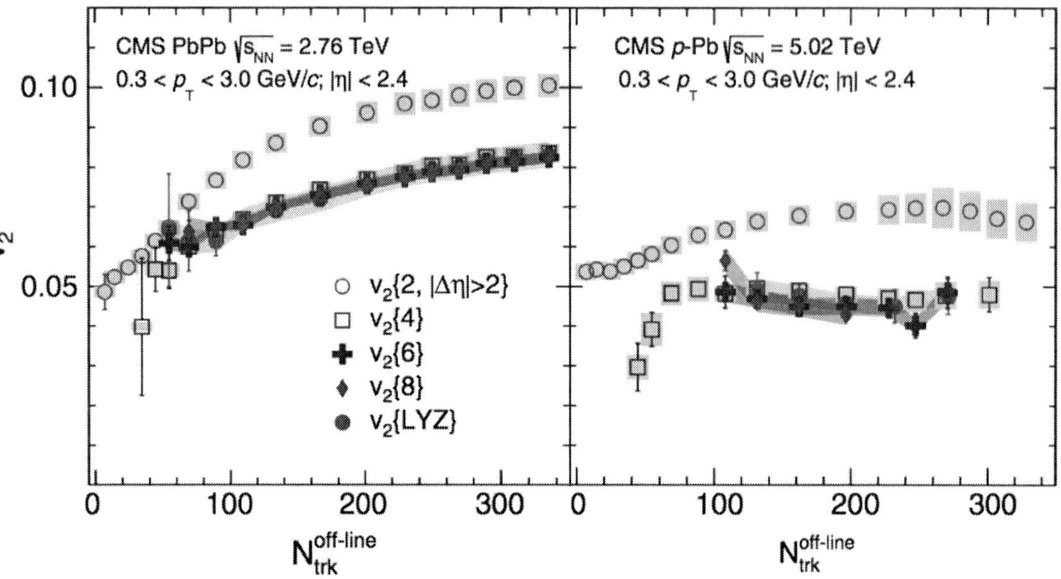

Figure 3. The multiplicity dependence of the v_2 obtained from 2PCs ($v_2\{2, \Delta\eta > 2\}$), from 4, 6 and 8-particle cumulants ($v_2\{4\}$, $v_2\{6\}$ and $v_2\{8\}$), and from Lee-Yang Zeroes (LYZ) methods. Results are for particles with $0.3 < p_T < 3$ GeV. The left panel corresponds to Pb+Pb collisions with $\sqrt{s_{NN}}$=2.76 TeV. The right panel corresponds to p+Pb collisions with $\sqrt{s_{NN}}$=5.02 TeV. The error bars and bands indicate statistical and systematic uncertainties, respectively. Figure taken from Ref. [10].

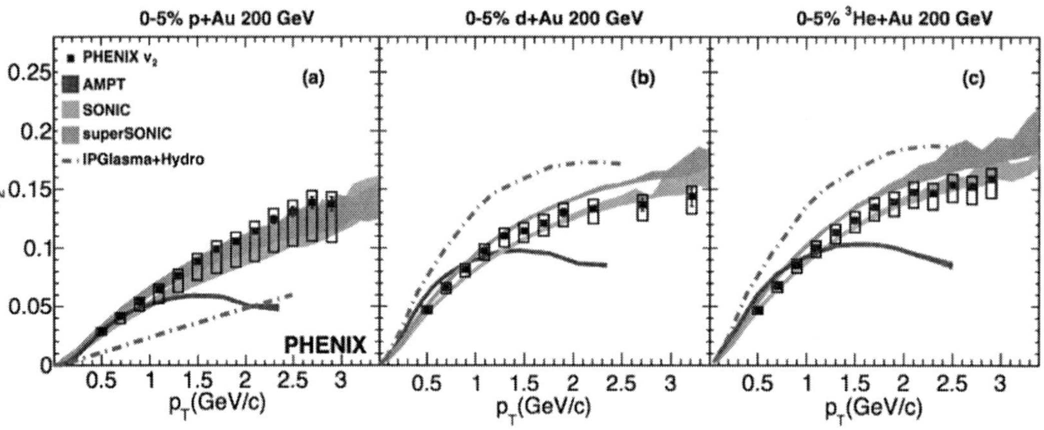

Figure 4. The p_T dependence of the v_2 measured in 0–5% most central p+Au (left), D+Au (center) and ^{3}He + Au collisions at $\sqrt{s_{NN}}$=200 GeV. Also shown on the plots are theoretical calculations from the AMPT, SONIC, superSONIC and IPGlasma+Hydro. Figure taken from [16].

3 Long-range correlations in pp collisions

These measurements thus necessitated revisiting the pp long-range correlations to perform a more detailed analysis along the lines of the p+Pb measurements and extract the v_n harmonics. However unlike the p+Pb measurements, the long-range correlation in pp collisions are much weaker compared to the other features seen in the 2PC, namely, the near-side jet peak at $\Delta\eta, \Delta\phi$ ~(0,0), and the away-side jet correlation at $\Delta\phi \sim \pi$. The near-side peak can be removed by focussing on the 1D correlation,$C(\Delta\phi)$, obtained by by projecting the long-range ($|\Delta\eta|$>2) part of the correlation on to the $\Delta\phi$-axis. This is shown in the left panel of Figure 5 for pp events at $\sqrt{s}$=13 TeV, that have more that 120 reconstructed charged particle tracks with p_{T}>0.4 GeV ($N_{\mathrm{ch}}^{\mathrm{rec}}$). While the $\Delta\eta$cut removes the near-side peak, the away-side jet peak still remains and dominates the 2PC, and must be somehow removed, prior to extracting the $v_n$from the correlation functions. This was recently done by the ATLAS collaboration who performed a "Template-fit" procedure to remove the contribution of the away-side jet to the 1D 2PC [5, 17]. In this procedure, the measured $C(\Delta\phi)$ distributions are assumed to result from a superposition of a "peripheral" $C(\Delta\phi)$ distribution, $C^{\mathrm{periph}}(\Delta\phi)$, scaled up by a multiplicative factor and a constant modulated by $\cos(n\Delta\phi)$ for n ≥2. The resulting template fit function,

$$C^{\mathrm{templ}}(\Delta\phi) = C^{\mathrm{ridge}}(\Delta\phi) + F\, C^{\mathrm{periph}}(\Delta\phi)\,, \tag{2}$$

where

$$C^{\mathrm{ridge}}(\Delta\phi) = G\left(1 + \sum_{n=2}^{\infty} 2v_{n,n}\cos(n\Delta\phi)\right), \tag{3}$$

has free parameters F and $v_{n,n}$. The parameter F is the multiplicative factor by which the $C^{\mathrm{periph}}(\Delta\phi)$ is scaled. The coefficient G, which represents the magnitude of the combinatoric component of $C^{\mathrm{ridge}}(\Delta\phi)$, is fixed by requiring that the integral of $C^{\mathrm{templ}}(\Delta\phi)$ be equal to the integral of the measured $C(\Delta\phi)$: $\int_0^{\pi} d\Delta\phi\ C^{\mathrm{templ}}(\Delta\phi) = \int_0^{\pi} d\Delta\phi\ C(\Delta\phi)$. In the ATLAS analysis, the $0 \le N_{\mathrm{ch}}^{\mathrm{rec}} < 20$ multiplicity interval is used to produce $C^{\mathrm{periph}}(\Delta\phi)$. The right panel of Figure 5 shows an example of the template fitting for the $N_{\mathrm{ch}}^{\mathrm{rec}}$>120 multiplicity interval. It is seen that the template describes the data quite well, reproducing both the near-side ridge as well as the narrowing of the away-side peak from low-$N_{\mathrm{ch}}^{\mathrm{rec}}$ to high-$N_{\mathrm{ch}}^{\mathrm{rec}}$ events.

Once the long-range correlation is extracted from the template fits, the single-particle anisotropies v_n can be extracted, following the standard procedure used in heavy ion analyses [1], where the Fourier coefficients $v_{n,n}$ of the 2PC (Eq. (3)) are related to the v_n as:

$$v_{n,n}(p_{\mathrm{T}}^{\mathrm{a}}, p_{\mathrm{T}}^{\mathrm{b}}) = v_n(p_{\mathrm{T}}^{\mathrm{a}}, p_{\mathrm{T}}^{\mathrm{b}}). \tag{4}$$

The left panel of Figure 6 shows the v_2 in 13 TeV and 5.02 TeV pp collisions obtained from the template fits. Also shown for comparison are the v_2 values in p+Pb collisions at $\sqrt{s_{\mathrm{NN}}}$=5.02 TeV. Very interestingly it is observed that the pp v_2 is independent of $N_{\mathrm{ch}}^{\mathrm{rec}}$, implying that the long range correlation is not a feature that is unique to high multiplicity events only, but in fact is present at all multiplicities. The p+Pb v_2 does show a dependence on $N_{\mathrm{ch}}^{\mathrm{rec}}$and in always larger than the pp v_2, possibly arising from geometrical effects that re present in p+Pb but not in pp collisions. However, at the lowest multiplicities the p+Pb v_2 values seem to approach the pp v_2 values. Another feature that is observed is that the pp v_2 values at 5.02 and 13 TeV are consistent with each other, implying a collision energy independence of the pp v_2 similar to what in observed in A+A collisions [4, 18]. The right panel of Figure 6 shows thev_2 as a function of p_{T}, for the same three datasets. As seen in the $N_{\mathrm{ch}}^{\mathrm{rec}}$ dependence, the v_2 values are consistent between the 5.02 and 13 TeV pp within the systematic uncertainties, over the measured p_{T} range. The pp v_2 values are smaller than the p+Pb v_2 across

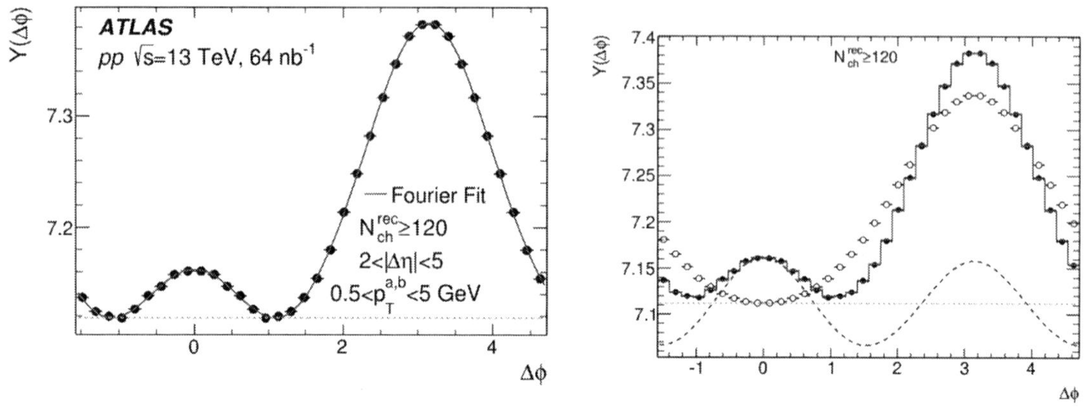

Figure 5. Left panel : The 1D two-particle correlation, in pp events at $\sqrt{s}$=13 TeV. The 1D correlations are obtained by projecting the long range ($\Delta\eta$>2) part of the 2PC to $\Delta\phi$. The plot is for events with more that 120 reconstructed tracks, and for $0.5 < p_T^a, p_T^b < 5.0$ GeV. The continuous line is drawn only to guide the eye. Right panel: Template fit to the 1D 2PC. The template fitting includes second-order, third-order and fourth-order harmonics. The solid points indicate the measured 2PC, the open points and curves show different components of the template (see legend). The Figures are taken from Ref. [5].

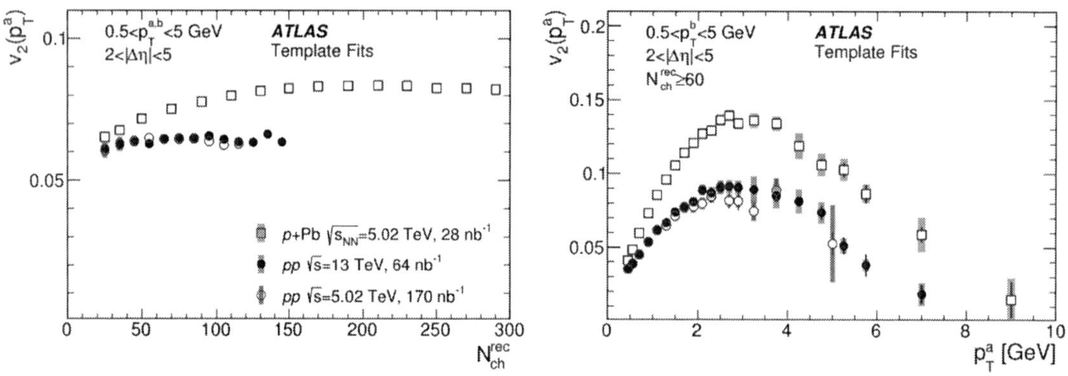

Figure 6. Left panels: comparison of the v_2 obtained from the template fitting procedure in the 13 TeV pp, 5.02 TeV pp, and 5.02 TeV p+Pb data, as a function of N_{ch}^{rec}. The results are for $0.5 < p_T^a, p_T^b < 5$ GeV. Right panels: the p_T dependence of the v_2 for the $N_{ch}^{rec} \geq 60$ multiplicity range. The error bars and shaded bands indicate statistical and systematic uncertainties, respectively. The Figures are taken from Ref. [5].

the whole p_T range. However the shapes of the $v_2(p_T)$ are quite similar between the pp and p+Pb collisions. This is seen in Figure 7 where the pp v_2 is scaled by an empirical factor of 1.51, choosen such that the maxima of the two $v_2(p_T)$ match. This similarity between the the v_2 in these two systems, and by extension to the v_2 in A+A (see Figure 2) again suggests similar origin for the v_n in the three systems.

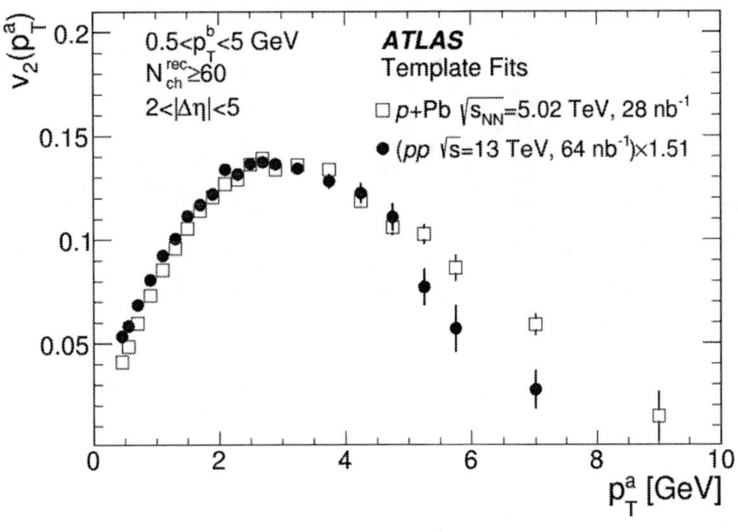

Figure 7. Comparison of the shapes of the $v_2(p_T)$ in 13 TeV pp and 5.02 TeV p+Pb data. The pp v_2 has been scaled by a factor of 1.51 along the y-axis in order to match the maximum of the v_2 in the two data sets. The results are for $N_{ch}^{rec} \geq 60$. The error bars indicate statistical uncertainties. The Figures are taken from Ref. [5].

Besides two-particle correlations, multi-particle cumulants [19] have also been extensively used in A+A and p+A collisions [20–22] to measure the v_n. The n-particle cumulant suppresses non-flow correlations involving n or fewer particles, and is very effective in involving non-flow azimuthal correlations in A+A and p+A collisions, as the non-flow correlations (such as those arising from jets, decays etc.) that only correlate few-particles among themselves. Thus it is natural to use the multi-particle cumulants in pp collisions to study azimuthal anisotropies. However, the multiplicities in pp collisions are so small that despite the suppression of such few particle correlations, they still affect the measured v_n. This inadequacy of the cumulants is measuring the v_n in pp collisions was recently demonstrated in Ref. [23], where the authors showed that the cumulants give significant and large non-zero values when analyzing MC events generated with the PYTHIA-8 [24] event generator. The authors of Ref. [23] proposed improved cumulant methods, called two-subevent and three-subevent cumulants, where the particles used in the cumulants are taken from two and three different η regions, respectively. This η separation, further suppresses jet-like correlations. Recently the ATLAS collaboration performed such subevent cumulant measurements in pp collisions [25]. Figure 8 shows the multiplicity dependence of the three-subevent $v_2\{4\}$ measured in 5.02 TeV and 13 TeV pp collisions (left and middle panels) and compares them to the $v_2\{4\}$ seen in 5.02 TeV p+Pb collisions (right panel). Also shown for comparison are the v_2 values obtained from the 2PCs. It is seen that the ordering v_2-2PC>$v_2\{4\}$ is observed in pp collisions just as it was observed in the p+Pb and Pb+Pb (Figure 3) collisions, again giving indications of collectivity in pp collisions.

4 Summary

The measurement of long-range correlations in p/D/He+A collisions and their similarity to A+A collisions is quite striking has established the presence of collective phenomena in such systems. Analysis of collectivity via measurement of the long-range correlation in pp collisions is considerably more

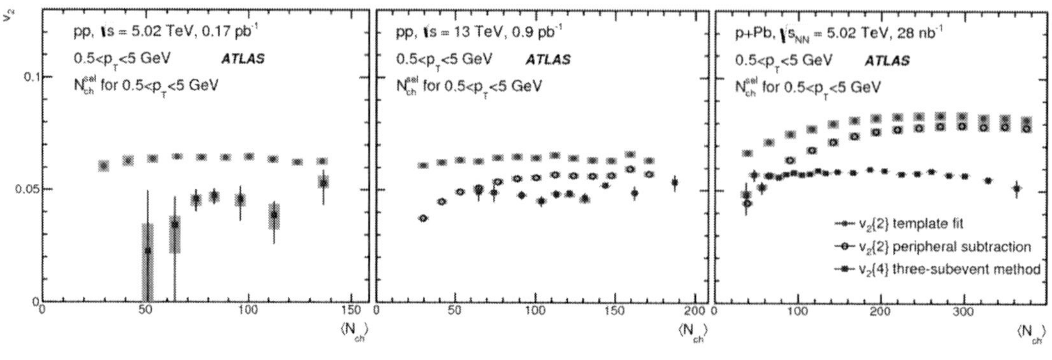

Figure 8. The $v_2\{4\}$ values calculated for charged particles with $0.5 < p_\mathrm{T} < 5$ GeV using the three-subevent method in 5.02 TeV pp (left panel), 13 TeV pp (middle panel) and 5.02 TeV p+Pb collisions (right panel). They are compared to the v_2 obtained from the 2PC analyses where the non-flow effects are removed by a template fit procedure (solid circles). Figure adapted from Ref. [25].

challenging. As described above, the most critical step in analyzing the long-range correlations in pp collisions is the removal of correlations arising from jets and dijets. The standard cumulants that work quite well in A+A collisions fail in pp collisions, as demonstrated in Ref. [25]. The subevent cumulants and template fitting procedure provide two effective methods of removing the jet and dijet correlations. However, even in these two methods, there might still be residual correlations from jets and dijets left over, so the results still must be taken with a pinch of salt. From the measurement, there are strong indications that global correlations - those that involve most of the particles produced in an event - are present in pp collisions. These correlations quantified by the Fourier coefficients v_n, demonstrate no dependence on the collision energy over $\sqrt{s}$ of 2.76–13 TeV, similar to the $\sqrt{s_\mathrm{NN}}$ independence of the v_n seen in A+A collisions. The correlations show no dependence on the multiplicity of the pp collision either, i.e. they are just as strong in low-multiplicity events as they are in high multiplicity events. The subevent cumulant measurements show that the $v_2\{2\}$ is larger than $v_2\{4\}$, similar to that seen in A+A collisions, indicating the presence of event-by-event v_2 fluctuations. Further more, the p_T dependence of the v_2 seen in pp collisions is qualitatively similar to that seen in p+A and A+A collisions, in fact up to an overall scaling the $v_2(p_\mathrm{T})$ look identical between pp, p+Pb and A+A collisions. All these observations give strong indications of collective behavior in pp collisions.

Acknowledgments

This work was supported by the US Department of Energy Office of Science, Office of Nuclear Physics under Award No. DE-FG02-86ER40281.

References

[1] ATLAS Collaboration, Phys. Rev. C **86**, 014907 (2012)

[2] CMS Collaboration, JHEP **1009**, 091 (2010)

[3] A. Dumitru, K. Dusling, F. Gelis, J. Jalilian-Marian, T. Lappi, R. Venugopalan, Phys. Lett. B **697**, 21 (2011)

[4] ATLAS Collaboration, ATLAS-CONF-2016-105 (2016)
[5] ATLAS Collaboration, Phys. Rev. **C96**, 024908 (2017), 1609.06213
[6] ALICE Collaboration, B. Abelev et al., Phys. Lett. B **719**, 29 (2013)
[7] ATLAS Collaboration, Phys. Rev. Lett. **110**, 182302 (2013)
[8] CMS Collaboration, Phys. Lett. B **724**, 213 (2013)
[9] ATLAS Collaboration, Phys. Rev. C **90**, 044906 (2014)
[10] CMS Collaboration, Phys. Rev. Lett. **115**, 012301 (2015)
[11] G. Başar, D. Teaney, Phys. Rev. C **90**, 054903 (2014)
[12] PHENIX Collaboration, A. Adare et al., Phys. Rev. Lett. **111**, 212301 (2013)
[13] PHENIX Collaboration, A. Adare et al., Phys. Rev. Lett. **114**, 192301 (2015)
[14] STAR Collaboration, L. Adamczyk et al., Phys. Lett. B **747**, 265 (2015)
[15] PHENIX Collaboration, A. Adare et al., Phys. Rev. Lett. **115**, 142301 (2015)
[16] PHENIX Collaboration, C. Aidala et al. , Phys. Rev. **C95**, 034910 (2017)
[17] ATLAS Collaboration, Phys. Rev. Lett. **116**, 172301 (2016)
[18] ALICE Collaboration, A. Jaroslav et al., Phys. Lett. **B762**, 376 (2016)
[19] N. Borghini, P.M. Dinh, J.Y. Ollitrault, Phys. Rev. **C63**, 054906 (2001)
[20] ATLAS Collaboration, Eur. Phys. J. C **74**, 3157 (2014)
[21] ATLAS Collaboration, Phys. Lett. B **725**, 60 (2013)
[22] ALICE Collaboration, Phys. Rev. Lett. **105**, 252302 (2010)
[23] J. Jia, M. Zhou, A. Trzupek, Phys. Rev. **C96**, 034906 (2017)
[24] T. Sjöstrand, S. Mrenna, P.Z. Skands, Comput. Phys. Commun. **178**, 852 (2008)
[25] ATLAS Collaboration (2017), arXiv 1708.03559

Recent ATLAS measurements of azimuthal anisotropies in pp and p+Pb collisions

Adam Trzupek[1,⋆] on behalf of the ATLAS Collaboration

[1] *Institute of Nuclear Physics PAS, ul. Radzikowskiego 152, 31-342 Kraków, Poland*

Abstract. The azimuthal anisotropies of particle yields observed in relativistic heavy-ion collisions are considered as an evidence of the formation on a deconfined Quark-Gluon Plasma produced in these collisions. Interestingly, recent measurements in pp and p+Pb systems from ATLAS and other experiments show similar features as those observed in A+A collisions, indicating the possibility of the production of such a deconfined medium in smaller collision systems. This report presents a summary of the recent ATLAS results on azimuthal anisotropies in pp collisions at 5.02 TeV and 13 TeV, p+Pb collisions at 5.02 TeV and 8.16 TeV as well as in peripheral 2.76 TeV Pb+Pb interactions. It includes measurements of two-particle correlations of charged particles as well as correlations of heavy flavor muons and charged particles in $\Delta\phi$ and $\Delta\eta$, with a template fitting procedure used to subtract the dijet contributions. Additionally, measurements of cumulants of multi-particle correlations, $c_n\{2\text{–}8\}$ are presented. The two-particle correlations and cumulants confirm a presence of collective phenomena in these collision systems, but the results on four-particle cumulants for pp collisions do not demonstrate a similar collective behaviour. However, the cumulant measurements in small collision systems can be biased by non-flow correlations. A novel subevent cumulant method that suppresses the contribution of non-flow effects was proposed recently by ATLAS allowing to measure significant azimuthal anisotropies in both pp and p+Pb collisions.

1 Introduction

One of the main signatures of the formation of strongly interacting Quark-Gluon Plasma (QGP) in heavy-ion collisions at the Relativistic Heavy Ion Collider (RHIC) and the Large Hadron Collider (LHC) is the large azimuthal anisotropy of produced particles [1, 2]. Interestingly, significant azimuthal anisotropy was also observed (at the LHC for the first time) in pp and p+Pb collisions [3, 4]. Since the first measurements, the collective phenomena in small systems are under extensive theoretical and experimental study, which significantly improves our understanding of the flow phenomena in heavy-ion collisions. However, there are still open questions such as what is the underlying mechanism for the observed correlations, what is the role of initial effects or whether the QGP exists in collisions involving light nuclei. In this report, the recent ATLAS [5] results on Fourier coefficients v_n of charged-particle azimuthal distributions, extracted from the two-particle correlations using a novel template fitting method, are shown for pp collisions at $\sqrt{s}$ = 5.02 TeV and 13 TeV and p+Pb collisions at $\sqrt{s_{\mathrm{NN}}}$ = 5.02 TeV [6] and $\sqrt{s_{\mathrm{NN}}}$ = 8.16 TeV [7]. The template fitting method was also used

⋆e-mail: adam.trzupek@ifj.edu.pl

© The Authors, published by EDP Sciences. This is an open access article distributed under the terms of the Creative Commons Attribution License 4.0 (http://creativecommons.org/licenses/by/4.0/).

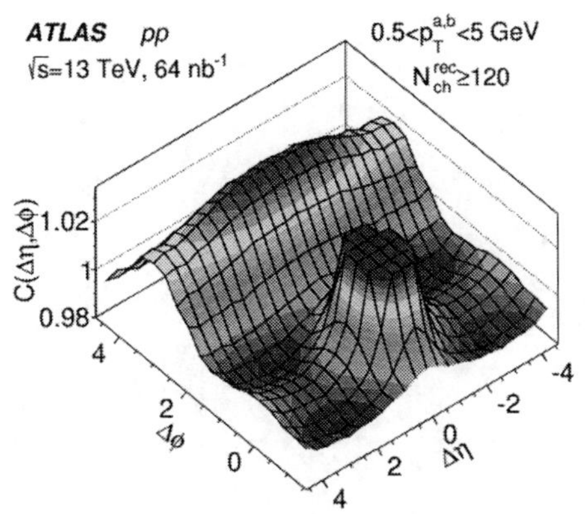

Figure 1. Two-particle correlation functions $C(\Delta\eta, \Delta\phi)$ in 13 TeV pp collisions corresponding to multiplicity range of $N_{\text{ch}}^{\text{rec}} \geq 120$. The plots are for charged particles of $0.5 < p_{\text{T}}^{a,b} < 5$ GeV. The distribution has been truncated to suppress the peak at $\Delta\eta = \Delta\phi = 0$ [4].

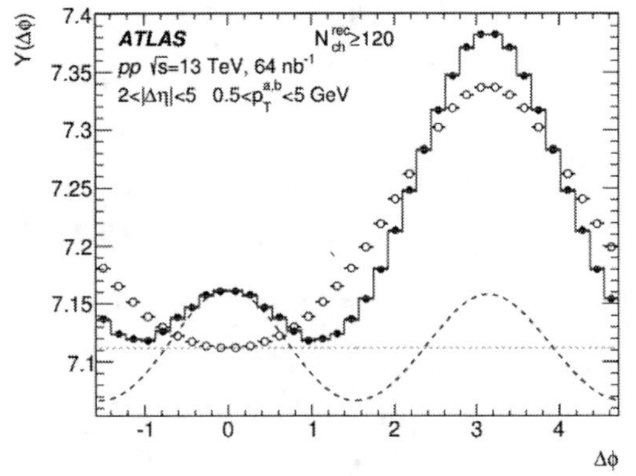

Figure 2. Template fits to the per-trigger particle yields $Y(\Delta\phi)$, in 13 TeV *pp* collisions for high multiplicity events of $N_{\text{ch}}^{\text{rec}} \geq 120$. The solid points indicate the measured $Y(\Delta\phi)$, the open points and curves show different components of the template fit [4].

to obtain v_2 of muons from heavy-flavor hadron decays in p+Pb collisions at $\sqrt{s_{\text{NN}}} = 8.16$ TeV [7]. For the small collision systems and additionally for low-multiplicity $\sqrt{s_{\text{NN}}} = 2.76$ TeV Pb+Pb collisions, the measurements of multi-particle cumulants and corresponding flow harmonics are also presented [8, 9].

2 Two-particle correlations

The two-particle correlation technique [3, 4] is deployed to measure the azimuthal anisotropy in small systems and is also commonly applied to probe collective phenomena in heavy-ion collisions [10–12]. Recently, a two-particle correlation analysis was performed by ATLAS [6, 7, 13] for *pp* collisions at $\sqrt{s} = 2.76$ TeV, 5.02 TeV and 13 TeV, and for p+Pb collisions at $\sqrt{s_{\text{NN}}} = 5.02$ TeV and 8.16 TeV. For low-multiplicity *pp* or p+Pb collisions, the correlation function measured in the relative pseudorapidity ($\Delta\eta$) and azimuthal angle ($\Delta\phi$) of two charged particles with transverse momenta p_{T}^{a} and p_{T}^{b}, is dominated by non-flow effects due to jet production, momentum conservation, resonance decays or Bose-Einstein correlations. The correlation function features a sharp peak centered at $(\Delta\phi, \Delta\eta) = (0, 0)$ and a broad, extended in $\Delta\eta$ structure at $\Delta\phi \approx \pi$. In high-multiplicity collisions, see Figure 1, an additional long-range structure in $\Delta\eta$ at $\Delta\phi \approx 0$, called "near-side ridge", is clearly visible. Also the correlation function at $\Delta\phi \approx \pi$ is broadened relative to low-multiplicity collisions, revealing a presence of the "away-side ridge". The strength of the ridge correlations is commonly quantified by the "per-trigger yield", $Y(\Delta\phi)$, which measures the average number of particle pairs associated with a trigger particle with p_{T}^{a} at the large pseudorapidity separation, $|\Delta\eta| > 2$, imposed to suppress the non-flow correlations. Using a template fitting function, the per-trigger yield is separated into two components: a scaled per-trigger yield for low-multiplicity interactions, $Y^{\text{periph}}(\Delta\phi)$, describing the back-to-back jet correlations, and an azimuthal modulation term describing the ridge, see Figure 2. In this approach, the non-flow correlations are described by the fit function, therefore, there is no need to perform the "peripheral subtraction" procedure on $Y(\Delta\phi)$ or $Y^{\text{periph}}(\Delta\phi)$, which has been commonly used in other analyses. In the peripheral subtraction procedure the non-flow effects are suppressed by setting the Y pedestal level by a zero-yield at minimum (ZYAM) [6]. The azimuthal modulation

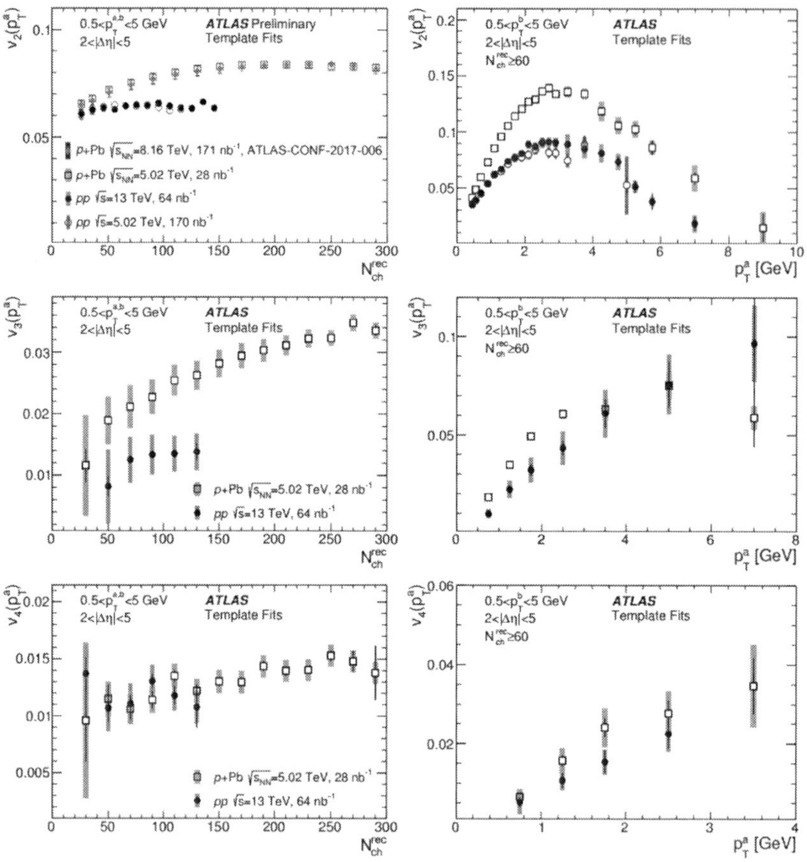

Figure 3. The $N_{\rm ch}^{\rm rec}$- and $p_{\rm T}$-dependences of the v_n in 13 TeV pp, 5.02 TeV pp, 5.02 TeV p+Pb and 8.16 TeV p+Pb collisions. Panels from top to bottom correspond to n=2, 3 and 4 respectively. The plots are for charged particles forming a pair with $0.5 < p_{\rm T} < 5$ GeV [6, 7].

amplitudes $2v_{n,n}$ obtained from the fitting procedure were found to factorize into a product of single particle flow harmonics $v_n \cdot v_n$, hence the flow harmonics are obtained from the formula: $v_n = \sqrt{v_{n,n}}$. A weak variation of $v_{n,n}$ on the applied pseudorapidity $|\Delta\eta|$ cut was observed. Figure 3 (the left column) shows measurements of v_2, v_3 and v_4 harmonics in 5.02 TeV and 13 TeV pp, and 5.02 TeV and 8.16 TeV p+Pb collisions as a function of the number of charged particles with $p_{\rm T} > 0.4$ GeV and $|\eta| < 2.5$, $N_{\rm ch}^{\rm rec}$. One can see that all three v_n harmonics in 5.02 TeV and 13 TeV pp data are $N_{\rm ch}^{\rm rec}$-independent, while the p+Pb v_2, v_3 and v_4 increase with increasing event multiplicity, except for v_2, which starts to be multiplicity independent for $N_{\rm ch}^{\rm rec} > 150$. Figure 3 also shows a comparison of the v_2 harmonic measured in p+Pb at $\sqrt{s_{\rm NN}} = 8.16$ TeV to those measured at $\sqrt{s_{\rm NN}} = 5.02$ TeV. The v_2 harmonics at the two collision energies are consistent within the systematic uncertainties.

Figure 3 (the right column) shows the $p_{\rm T}$ dependence of v_n in 5.02 and 13 TeV pp, and 5.02 TeV p+Pb collisions for $N_{\rm ch}^{\rm rec} \geq 60$. Similar v_2 harmonics are observed in pp collisions at both collision energies. With increasing transverse momentum, v_2 harmonics rise reaching a maximum near 3 GeV and then drop, reaching almost 0 at $p_{\rm T} \approx 7$ GeV. A more rapid increase of v_2 is observed in p+Pb collisions, but generally a similar trend is preserved in both systems. The v_3 harmonics in 13 TeV pp collisions increase with increasing $p_{\rm T}$ over the measured transverse momentum range. For p+Pb

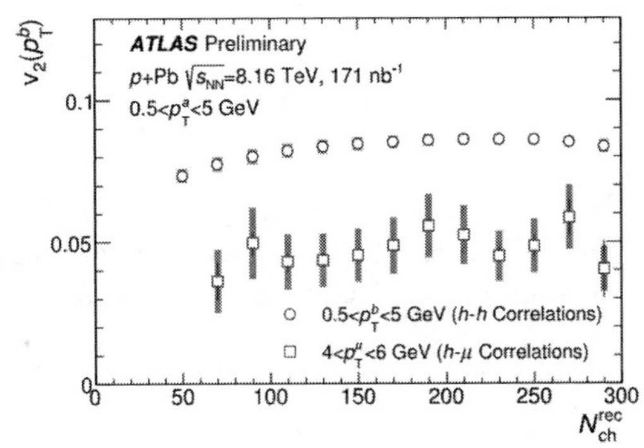

Figure 4. The v_2 values obtained from the template fits to $h-h$ correlations with $0.5 < p_T^{a,b} < 5$ GeV (circles) and to $h-\mu$ correlations with $4 < p_T^{\mu} < 6$ GeV (squares) [4].

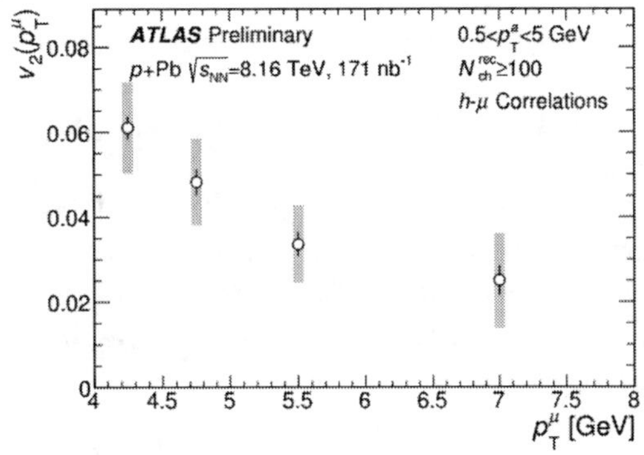

Figure 5. The p_T dependence of the muon-v_2 integrated over a broad multiplicity range of $N_{ch}^{rec} \geq 100$ [4].

collisions, larger v_3 values are measured than in pp for $p_T < 3$ GeV. At higher p_T, v_3 in p+Pb data saturates. The v_4 harmonics in 13 TeV pp and 5.02 TeV p+Pb collisions increase with increasing transverse momentum and larger v_4 values are measured in p+Pb than in pp collisions.

3 Correlations between muons and charged-particles

Charm and bottom quarks provide an important signature of the QGP formed in ultra-relativistic heavy ion collisions [14]. It is expected that at high transverse momenta, $p_T \gg 5$ GeV, heavy quarks produced in the early stage of the collision lose energy in the QGP with mass-dependent modifications [15]. At lower transverse momenta, $p_T \lesssim 5$ GeV, the quarks are expected to interact with the hot and dense medium acquiring an azimuthal anisotropy due to the collective expansion of the QGP. Measurements of heavy flavor quarks in A+A collisions were performed at RHIC and LHC [16–18]. A significant suppression of heavy quarks production due to energy loss was observed as well as a significant azimuthal anisotropy was measured.

Recently, ATLAS performed a measurement of the long-range correlations between reconstructed muons of $4 < p_T^{\mu} < 8$ GeV and inclusive charged particles in p+Pb collisions at $\sqrt{s} = 8.16$ TeV [7]. Majority of prompt muons at low transverse momenta, $p_T \lesssim 5$ GeV, results from decays of heavy-flavor hadrons containing charm or bottom quarks [19]. Therefore, if a QGP is formed in p+Pb collisions, its collective expansion may influence heavy-flavor hadron production [20], which can be probed by the measurement of two-particle correlations.

Based on a Monte Carlo simulation of the ATLAS detector [7], it was found that the sample of reconstructed muons with $4 < p_T < 8$ GeV is contaminated up to 45% by background muons originating from pion and kaon decays, muons produced in hadronic showers, and mis-associations of Inner Detector (ID) and Muon Spectrometer (MS) tracks. To lower the background fractions, the momentum imbalance, $\Delta p/p_{ID} = (p_{ID} - p_{MS})/p_{ID}$, is calculated for each muon, where p_{ID} and p_{MS} are the reconstructed muon momenta measured in the ID and MS, respectively. As the momentum of background muons measured in the MS is typically lower than the momentum measured in the ID, majority of background muons are characterised by the positive momentum imbalance, $\Delta p/p_{ID} > 0$. Therefore, only muons with $\Delta p/p_{ID} < 0$ are accepted for the correlation measurement. This requirement lowers the background fractions to 25% – 10% for the muon transverse momentum range from 4 GeV to 8 GeV.

The two-particle correlations of reconstructed muons and charged particles ($h - \mu$), obtained in a similar way as the two-particle correlations for inclusive charged particles ($h - h$), show a clear ridge structure in the high-multiplicity 8.16 TeV p+Pb collisions. To extract the muon-v_2, the two-particle correlation template fitting method, described in Section 2, is used. Figure 4 shows the measured muon-v_2 obtained from h–μ correlations of $4 < p_T^\mu < 6$ GeV and $0.5 < p_T^a < 5$ GeV as well as charged-particle v_2 obtained from $h - h$ correlations of $0.5 < p_T^{a,b} < 5$ GeV. Within systematic uncertainties, the muon-v_2 remains constant over the measured N_{ch}^{rec} range, and is smaller than v_2 for charged particles. However, it should be noted that different p_T ranges are used for both measurements. Figure 5 shows the p_T dependence of the muon-v_2 over transverse momentum range $4 < p_T^\mu < 8$ GeV measured for large multiplicity events of $N_{ch}^{rec} \geq 100$. A decrease of muon-v_2 is observed with increasing p_T^μ.

4 Multi-particle cumulants

A commonly used approach to measure flow harmonics, which efficiently suppresses the non-flow correlations, is the multi-particle cumulant method [21, 22]. In this method, the $2k$-particle azimuthal correlations, $corr_n\{2k\}$, are calculated and then used to obtain the multi-particle cumulants, $c_n\{2k\}$. By definition, $2k$-particle cumulants describe genuine $2k$-particle correlations, i.e. the correlations between fewer number of particles than $2k$, which include most of non-flow correlations, are subtracted. The cumulant method was used for systems including Pb+Pb, p+Pb and recently pp collisions [23–26]. ATLAS measured multi-particle cumulants in pp collisions at $\sqrt{s} = 5.02$ TeV and 13 TeV, and $\sqrt{s_{NN}} = 5.02$ TeV p+Pb and in low-multiplicity Pb+Pb collisions at $\sqrt{s_{NN}} = 2.76$ TeV [8]. To avoid event multiplicity fluctuations, which can mimic the collective-like effects [8, 24], cumulants are calculated in unit-size bins in the number of reference particles, M_{ref}, which is defined as the number of reconstructed charged particles with $|\eta| < 2.5$ and with p_T range: $0.3 < p_T < 3$ GeV or $0.5 < p_T < 5$ GeV. The cumulants are then calculated in broader, statistically significant multiplicity intervals by

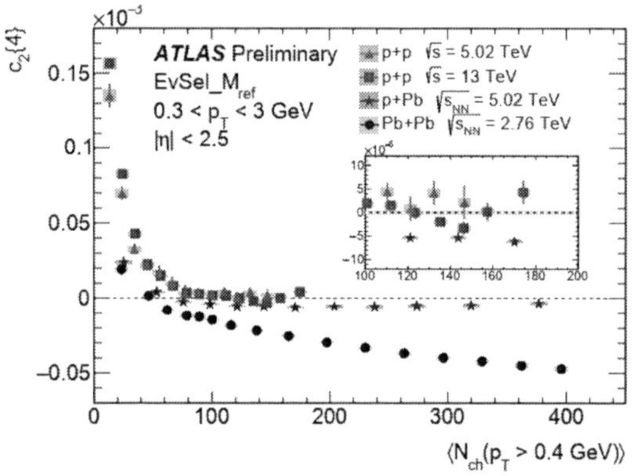

Figure 6. The second order cumulant $c_2\{4\}$ obtained from four-particle correlations as a function of $\langle N_{ch}(p_T > 0.4 \text{ GeV})\rangle$ for pp collisions at $\sqrt{s} = 5.02$ and 13 TeV, p+Pb collisions at $\sqrt{s_{NN}} = 5.02$ TeV and Pb+Pb collisions at $\sqrt{s_{NN}} = 2.76$ TeV [8].

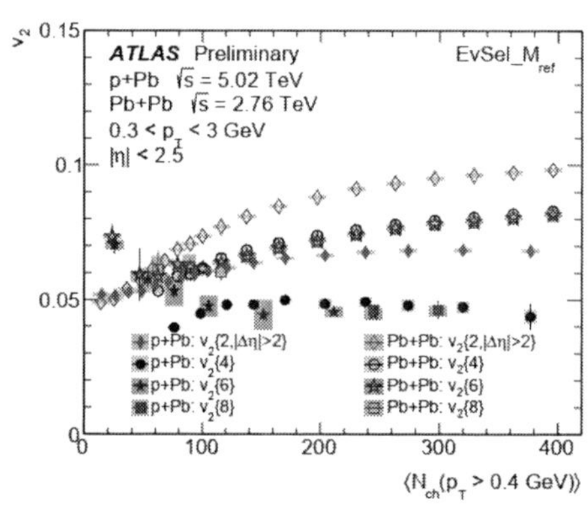

Figure 7. Comparison of $v_2\{2, |\Delta\eta| > 2\}$, $v_2\{4\}$, $v_2\{6\}$ and $v_2\{8\}$ as a function of $\langle N_{ch}(p_T > 0.4 \text{ GeV})\rangle$ for $\sqrt{s_{NN}} = 5.02$ TeV p+Pb collisions and $\sqrt{s_{NN}}$= 2.76 TeV Pb+Pb collisions. The results are presented for particles with $0.3 < p_T < 3$ GeV[8].

averaging with weights corresponding to the total number of $2k$-multiplets in M_{ref} bins and accounting for the event trigger efficiency [8, 22]. Results obtained for different collision systems or p_T-ranges are compared in a common event activity variable, $N_{\mathrm{ch}}(p_{\mathrm{T}} > 0.4)$, defined as the mean number of charged particles with $p_{\mathrm{T}} > 0.4$ GeV in events in the M_{ref} interval used for the cumulant calculations.

Figure 6 shows the four-particle cumulants measurement in pp collisions at $\sqrt{s} = 5.02$ and 13 TeV, p+Pb collisions at $\sqrt{s_{\mathrm{NN}}} = 5.02$ TeV and Pb+Pb collisions at $\sqrt{s_{\mathrm{NN}}}= 2.76$ TeV for $0.3 < p_{\mathrm{T}} < 3$ GeV. One can see that they follow the ordering $|c_2\{4\}|_{\mathrm{p+Pb}} < |c_2\{4\}|_{\mathrm{Pb+Pb}}$ for $N_{\mathrm{ch}}(p_{\mathrm{T}} > 0.4$ GeV$)>100$. Therefore, the magnitude of the corresponding elliptic flow, $\mathrm{v}_2\{4\} = \sqrt[4]{-c_2\{4\}}$, is larger for Pb+Pb collisions than for p+Pb events. For 5.02 TeV pp collisions, the $c_2\{4\}$ cumulants are positive or consistent with zero over the full range of particle multiplicities. For the 13 TeV pp data, the cumulants are positive over the large range of multiplicities, with the exception of N_{ch} from 130 to 150, where $c_2\{4\}$ is smaller than zero within 1–2 standard deviations. Therefore, these measurements of positive $c_2\{4\}$ cumulants in pp collisions do not allow to calculate the Fourier harmonics.

Figure 7 shows a comparison of the $\mathrm{v}_2\{2, |\Delta\eta| > 2\}$ harmonics calculated with the requirement of pseudorapidity separation $|\Delta\eta| > 2$, $\mathrm{v}_2\{4\}$, $\mathrm{v}_2\{6\}$ and $\mathrm{v}_2\{8\}$, for p+Pb and low-multiplicity Pb+Pb collisions. All derived v_2 harmonics have larger magnitudes in Pb+Pb collisions than in p+Pb collisions with the same multiplicity. For both systems for N_{ch} above 100, $\mathrm{v}_2\{2k\}$ are similar for $k = 2$, 3 and 4 while $\mathrm{v}_2\{2, |\Delta\eta| > 2\}$ is systematically larger than the harmonics $\mathrm{v}_2\{2k\}$ calculated with more than two-particle cumulants due to fluctuations in the initial-state geometry [28].

5 Subevent cumulant method

To further reduce the non-flow correlations, especially in pp collisions and low-multiplicity p+Pb collisions, ATLAS has developed an improved cumulant method based on the correlations between particles from different subevents separated in pseudorapidity [9]. In particular, correlating particles from two subevents constructed of particles with $\eta < 0$ (subevent a) and $\eta > 0$ (subevent b) suppresses non-flow contribution mainly from short-range correlations, like correlations within jets, while in the case of three-subevents correlating particles from subevent b of $|\eta| < 0.833$ with particles from subevents a of $\eta < -0.833$ and c of $\eta > 0.833$, additionally lowers contribution from dijets. The capability of the subevent cumulant method was verified with Pythia 8 event generator [27].

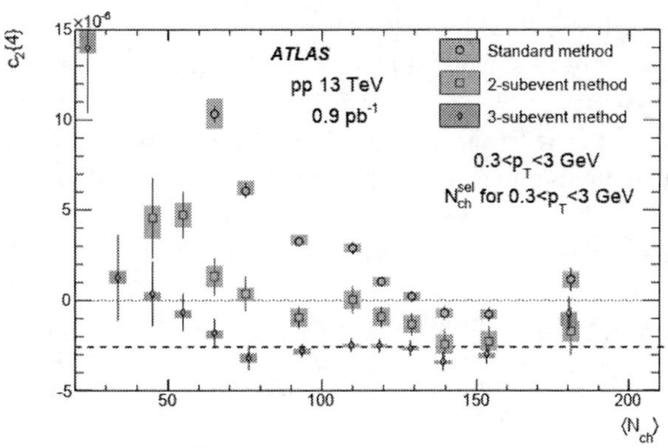

Figure 8. The $c_2\{4\}$ values calculated for charged particles with $0.3 < p_{\mathrm{T}} < 3$ GeV compared for the three cumulant methods from the 13 TeV pp data. The event averaging is performed for $N_{\mathrm{ch}}^{\mathrm{sel}}$ calculated for the same p_{T} range, which is then mapped to N_{ch}, the average number of charged particles with $p_{\mathrm{T}} > 0.4$ GeV. The dashed line indicates the $c_2\{4\}$ value corresponding to a 4% v_2 signal [9].

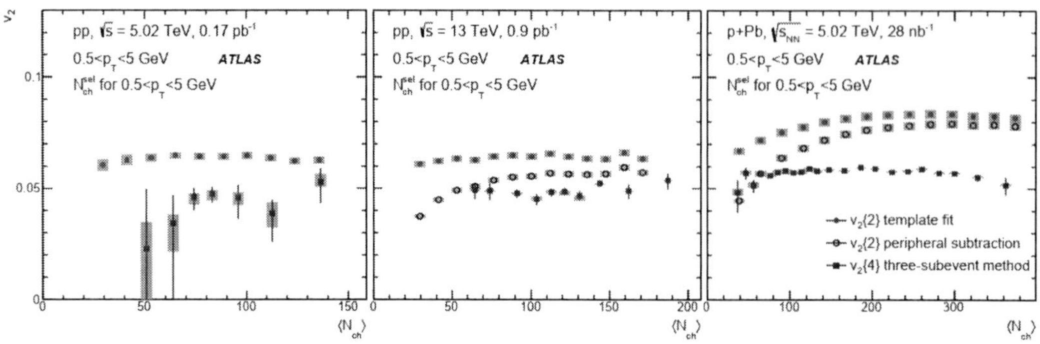

Figure 9. The $v_2\{4\}$ values calculated for charged particles with $0.3 < p_T < 3$ GeV using the three-subevent method in 5.02 TeV *pp* (left panel), 13 TeV *pp* (middle panel) and 5.02 TeV *p*+Pb collisions (right panel). They are compared to v_2 obtained from the two-particle correlation analyses where the non-flow effects are removed by a template fit procedure (solid circles) or with a fit after subtraction with a ZYAM assumption (peripheral subtraction, open circles) [9].

Figure 8 shows a comparison of the four-particle cumulant, $c_2\{4\}$, measured with the standard, two-subevent, and three-subevent methods for *pp* collisions at $\sqrt{s}$ =13 TeV. A clearly negative value of the $c_2\{4\}$ cumulant obtained with the two- and three-subevent methods is observed for $N_{ch} > 70$, which is consistent with the presence of collectivity in 13 TeV *pp* collisions. Results for 5.02 TeV *pp* and low multiplicity ($N_{ch} < 70$) *p*+Pb collisions at $\sqrt{s_{NN}}$ = 5.02 TeV show similar trends to the results for 13 TeV *pp* collisions, however, in *p*+Pb collisions for $N_{ch} > 100$, all three methods give comparable $c_2\{4\}$, suggesting that non-flow effects in *p*+Pb collisions are much smaller [9].

Negative values of $c_2\{4\}$ obtained from the subevent cumulant methods allow for calculations of flow harmonics. For the three-subevent cumulant method, the $v_2\{4\}$ for reference particles 0.3 $< p_T <$ 3 GeV in 5.02 TeV *pp*, 13 TeV *pp* and 5.02 TeV *p*+Pb collisions are presented in Figure 9. The results are compared to values of $v_2\{2\}$ obtained from the two-particle correlation template fitting method [10, 15]. For comparison, Figure 9 also shows $v_2\{2\}$ obtained from the two-particle correlation method which sets the pedestal level by a zero-yield at minimum (ZYAM) [6]. The $v_2\{4\}$ values are observed to be approximately independent of N_{ch} and lower than the two-particle correlation results for the three collision systems.

The ratio of the elliptic flow from the four-particle cumulant method and the two-particle correlations allows to estimate the number of independent sources contributing to the particle production, N_s [28]. Figure 10 shows the extracted values of N_s as a function of N_{ch} in 13 TeV *pp* and 5.02 *p*+Pb collisions, for charged particles with $0.3 < p_T < 3$ GeV, using the method presented in Ref. [28]. It is observed that the N_s value increases with N_{ch} in p+Pb collisions, reaching $N_s \approx 20$ at the highest multiplicity and the N_s values for 13 TeV *pp* and 5.02 TeV *p*+Pb collisions are similar in the range of overlapping N_{ch} multiplicities.

6 Summary

Using the novel, template fitting two-particle correlation method, v_n (n = 2, 3, 4) harmonics were measured as a function of multiplicity and transverse momentum in *pp* collisions at $\sqrt{s}$ = 5.02 TeV and 13 TeV as well as in *p*+Pb collisions at $\sqrt{s_{NN}}$ = 5.02 TeV and 8.16 TeV . The measurement shows that v_2 in *pp* collisions weakly depends on multiplicity and collision energy. The *p*+Pb v_2 values are larger than the *pp* v_2 for all multiplicities and are observed to increase with N_{ch}^{rec}. As a

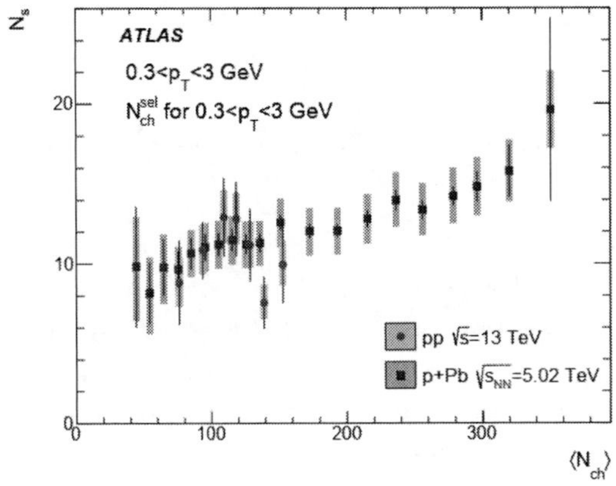

Figure 10. The number of sources inferred from $v_2\{2\}$ and $v_2\{4\}$ measurements via the model framework in 13 TeV *pp* and 5.02 TeV *p*+Pb collisions, for charged particles with $0.3 < p_T < 3$ GeV [9].

function of multiplicity, the v_2 harmonics in *p*+Pb at $\sqrt{s_{NN}} = 5.02$ TeV and 8.16 TeV are consistent within the measurement uncertainty. As a function of p_T, the *p*+Pb v_2 harmonics are larger than v_2 in *pp* collisions, but a similar dependence on the transverse momentum is seen, which also resembles the trend observed in Pb+Pb collisions.

The two-particle correlation template fitting method was also used to obtain v_2 of heavy flavor muons of $4 < p_T^{\mu} < 8$ GeV in *p*+Pb collisions at $\sqrt{s_{NN}} = 8.16$ TeV. The muon-v_2 harmonic is found to be constant as a function of multiplicity, while as a function of the transverse momentum a decrease of muon-v_2 is observed with increasing p_T.

Multi-particle cumulants were measured for *pp* at $\sqrt{s} = 5.02$ TeV and 13 TeV as well as for *p*+Pb at $\sqrt{s_{NN}} = 5.02$ TeV collisions and low-multiplicity Pb+Pb collisions at $\sqrt{s_{NN}} = 2.76$ TeV. The collective nature of multi-particle correlations is well confirmed for *p*+Pb and Pb+Pb collisions for charged-particle multiplicities above 100. The measured v_2 harmonics from multi-particle cumulants have larger values for Pb+Pb collisions as compared to *p*+Pb collisions and are similar for both systems at $N_{ch} > 100$ for $k = 2$, 3 and 4, while $v_2\{2, |\Delta\eta| > 2\}$ is systematically larger than the $v_2\{2k\}$ calculated with more than two-particle cumulants. This observation is consistent with models assuming fluctuation-driven initial-state anisotropies. For *pp* collisions, the four-particle cumulants are positive or consistent with zero over the full range of particle multiplicities, with the exception for 13 TeV *pp* collisions of N_{ch} from 130 to 150, where $c_2\{4\}$ is smaller than zero within 1–2 standard deviations. Therefore, these measurements in *pp* collisions, do not satisfy the requirement of being negative for the $v_2\{4\}$ harmonic calculation since are biased by non-flow contaminations.

Recently, ATLAS has proposed a novel sub-event method for the $c_2\{4\}$ cumulant measurement, in which particles from different sub-events separated in pseudorapidity are used for cumulant calculations. Monte Carlo simulations show that the subevent cumulant method efficiently suppresses non-flow effects. Clearly negative sign of the four-particle subevent cumulants, $c_2\{4\}$, allow for flow harmonics calculations both in *pp* and *p*+Pb collisions. The $v_2\{4\}$ harmonics in *pp* collisions at $\sqrt{s} =$ 5.02 TeV and 13 TeV and in *p*+Pb collisions at $\sqrt{s_{NN}} = 5.02$ TeV are observed to be approximately independent of N_{ch} and lower than the two-particle correlation results. These measurements allow studying the fluctuations in the initial-state geometry of small collision systems.

The measurements of two- and multi-particle correlations presented in this report significantly contribute to the understanding of collectivity in small systems and help to constrain the theoretical modelling.

This work was supported in part by the National Science Centre, Poland grant 2016/23/B/ST2/00702 and by PL-Grid Infrastructure.

References

[1] STAR Collaboration, Phys. Rev. Lett. 86 (2001) 402, arXiv: 0009011 [nucl-ex].
[2] ALICE Collaboration, Phys. Rev. Lett. 105 (2010) 252302, arXiv:1011.3914 [nucl-ex].
[3] CMS Collaboration, JHEP 1009 (2010) 091, arXiv:1009.4122 [hep-ex].
[4] ATLAS Collaboration, Phys. Rev. Lett. 110 (2013) 182302, arXiv:1212.5198 [hep-ex].
[5] ATLAS Collaboration, JINST 3 (2008) S08003.
[6] ATLAS Collaboration, PRC 96 (2017) 024908, arXiv:1609.06213 [nucl-ex].
[7] ATLAS Collaboration, ATLAS-CONF-2017-006, url: https://cds.cern.ch/record/2244808.
[8] ATLAS Collaboration, Eur. Phys. J. C 77 (2017) 428, arXiv:1705.04176 [hep-ex].
[9] ATLAS Collaboration, arXiv:1708.03559 [hep-ex].
[10] S. Wang et al., Phys. Rev. C 44 (1991) 1091.
[11] PHENIX Collaboration, Phys. Rev. Lett. 89 (2002) 212301.
[12] ATLAS Collaboration, Phys. Rev. C 86 (2012) 014907, arXiv:1203.3087 [hep-ex].
[13] ATLAS Collaboration, Phys. Rev. Lett. 116 (2016) 172301, arXiv:1509.04776 [hep-ex].
[14] S. Cao, G.-Y. Qin and S. A. Bass, Phys. Rev. C88 (2013) 044907, arXiv: 1308.0617 [nucl-th].
[15] P. B. Gossiaux and J. Aichelin, J. Phys. G36 (2009) 064028, arXiv: 0901.2462 [nucl-th].
[16] PHENIX Collaboration, A. Adare et al., Phys. Rev. C84 (2011) 044905, arXiv: 1005.1627 [nucl-ex].
[17] ALICE Collaboration, B. Abelev et al., Phys. Rev. Lett. 111 (2013) 102301, arXiv: 1305.2707 [nucl-ex].
[18] ATLAS Collaboration, ATLAS-CONF-2015-023, 2015, url: http://cds.cern.ch/record/2055674.
[19] ATLAS Collaboration, Phys. Lett. B 707 (2012) 438, arXiv:1109.0525 [hep-ex].
[20] G. D. Moore and D. Teaney, Phys. Rev. C71 (2005) 064904, arXiv:hep-ph/0412346.
[21] N. Borghini, P. M. Dinh and J. Y. Ollitrault, Phys. Rev. C 63 (2001) 054906.
[22] A. Bilandzic, R. Snellings and S. Voloshin, Phys. Rev. C 83 (2011) 044913, arXiv: 1010.0233 [nucl-ex].
[23] ATLAS Collaboration, Eur. Phys. J. C 74 (2014) 3157, arXiv: 1408.4342 [hep-ex].
[24] ALICE Collaboration, Phys. Rev. C 90 (2014) 054901, arXiv: 1406.2474 [nucl-ex].
[25] ATLAS Collaboration, Phys. Lett. B 725 (2013) 60, arXiv: 1303.2084 [hep-ex].
[26] CMS Collaboration, Phys. Lett. B 765 (2017) 193, arXiv: 1606.06198 [nucl-ex].
[27] J. Jia, M. Zhou, and A. Trzupek, Phys. Rev. C 96 (2017) 034906, arXiv:1701.03830 [nucl-th].
[28] L. Yan and J.-Y. Ollitrault, Phys. Rev. Lett. 112 (2014) 082301, arXiv:1312.6555 [nucl-th].

Collectivity and manifestations of minimum-bias jets in high-energy nuclear collisions

Thomas A. Trainor[1]*

[1] *CENPA 354290 University of Washington, Seattle, Washington, USA*

Abstract. Collectivity, as interpreted to mean flow of a dense medium in high-energy A-A collisions described by hydrodynamics, has been attributed to smaller collision systems – p-A and even p-p collisions – based on recent analysis of LHC data. However, alternative methods reveal that some data features attributed to flows are actually manifestations of minimum-bias (MB) jets. In this presentation I review the differential structure of single-particle p_t spectra from SPS to LHC energies in the context of a two-component (soft + hard) model (TCM) of hadron production. I relate the spectrum hard component to measured properties of isolated jets. I use the spectrum TCM to predict accurately the systematics of ensemble-mean $\bar{p}_t$ in p-p, p-A and A-A collision systems over a large energy interval. Detailed comparisons of the TCM with spectrum and correlation data suggest that MB jets play a dominant role in hadron production near midrapidity. Claimed flow phenomena are better explained as jet manifestations agreeing quantitatively with measured jet properties.

1 Introduction

Collectivity (interpreted to represent hydrodynamic flows [1]) has been attributed recently to smaller collision systems [2] based on certain correlation phenomena conventionally attributed to flows in larger A-A systems (e.g. v_2 data). But the term "collectivity" simply represents *any* correlated collection: correlations $\Rightarrow$ collectivity. For instance, dijet production represents a collective phenomenon that arguably dominates high-energy nuclear collisions. Several hadron production mechanisms may contribute to observed correlations. It should be our task to identify those mechanisms by comprehensive data analysis based on a variety of methods. The goal is then to extract *all* information from available particle data and interpret that information based on recognized physics principles. In this study measured $\bar{p}_t$ trends for three collision systems are compared to corresponding p_t spectra and measured jet properties.

2 Two-component (soft + hard) model or TCM for A-B collisions

The TCM is a comprehensive and accurate model of hadron production in A-B collisions near midrapidity inferred inductively from p_t spectra [3, 4] and angular correlations [5–7]. For instance, intensive charge density $\bar{\rho}_0$ can be separated into two components: $\bar{\rho}_0 = \bar{\rho}_s + \bar{\rho}_h$ (soft + hard). Based on

*e-mail: ttrainor99@gmail.com

© The Authors, published by EDP Sciences. This is an open access article distributed under the terms of the Creative Commons Attribution License 4.0 (http://creativecommons.org/licenses/by/4.0/).

comprehensive data analysis (e.g. [5, 8, 9]) the two components have been interpreted physically as follows: Soft component SC results from projectile-nucleon dissociation to *participant* low-x gluons $\sim \bar{\rho}_s \equiv n_s/\Delta\eta \propto \log(\sqrt{s}/10\text{ GeV})$. Hard component HC results from large-angle scattering of participant gluons which then fragment to a minimum-bias (MB) ensemble of dijets. Measured SC and HC (yields, spectra and correlations) are related in p-p collisions via $\bar{\rho}_h \equiv n_h/\Delta\eta \approx \alpha\bar{\rho}_s^2$ (*noneikonal* trend) with $\alpha \approx O(0.01)$. An eikonal trend (e.g. as in the Glauber model of A-A collisions) would instead follow $\bar{\rho}_h \propto \bar{\rho}_s^{4/3}$ [10]. The p-p $\bar{\rho}_h \approx \alpha\bar{\rho}_s^2$ trend is illustrated in Fig. 1 (a) as $n_h/n_s \approx 0.005\bar{\rho}_s$ (within $\Delta\eta = 1$) [3].

A TCM for hadron production in A-B collisions follows suite but requires additional geometry parameters defined within the Glauber model of collision geometry based on the eikonal approximation. $N_{part}/2$ is the number of participant nucleon pairs, N_{bin} is the number of N-N binary collisions, and $\nu \equiv 2N_{bin}/N_{part}$ is the mean number of binary collisions per participant pair. For instance, the event-ensemble mean of extensive total P_t (p_t integrated within some acceptance $\Delta\eta$) can be represented in TCM form as $\bar{P}_t = \bar{P}_{ts} + \bar{P}_{th}$ or

$$\bar{P}_t = (N_{part}/2)n_{sNN}\bar{p}_{tsNN} + N_{bin}n_{hNN}\bar{p}_{thNN};\quad \bar{P}_t/n_s = \bar{p}_{ts} + x(n_s)\nu(n_s)\bar{p}_{thNN}(n_s), \tag{1}$$

where $x \equiv \bar{\rho}_{hNN}/\bar{\rho}_{sNN} = n_{hNN}/n_{sNN}$ and the TCM incorporates factorization of N-N low-x gluon participants (noneikonal) and A-B nucleon participants (eikonal). Note that $\bar{p}_t$ SC $\bar{p}_{tsNN} \to \bar{p}_{ts} \approx 0.4$ GeV/c is observed to be universal for N-N collisions within A-B collisions.

3 Minimum-bias jet manifestations in p-p p_t spectra and correlations

Quantitative correspondence between measured jet properties and single-particle p_t spectra has been demonstrated previously for 200 GeV p-p and Au-Au spectra down to 0.5 GeV/c [8]. The demonstration relies on the observation that jet spectra and fragmentation functions (FFs) have simple universal forms (Gaussian and beta distribution) when plotted on rapidity variables. Some details are reviewed in this section (angular correlations are described below).

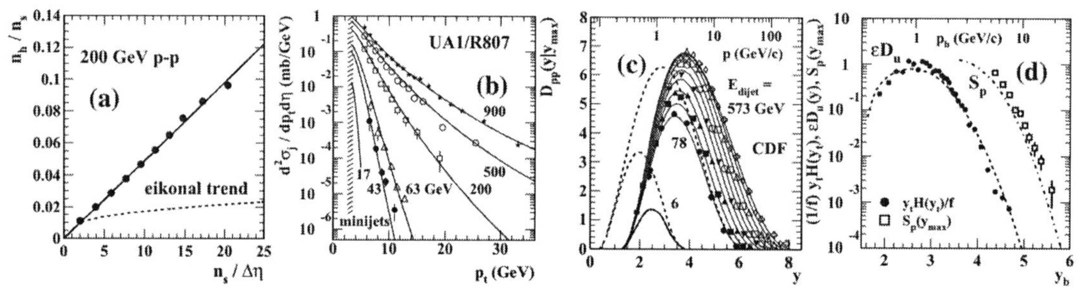

Figure 1: (a) Hard/soft ratio vs soft density $\bar{\rho}_s$. (b) Jet p_t spectra for several energies from the ISR and Sp$\bar{\text{p}}$S. Curves through data are from Ref. [9]. (c) Fragmentation functions from p-$\bar{p}$ collisions described by a simple parametrization from Ref. [9]. (d) Spectrum hard component for 200 GeV NSD p-p collisions (solid points) compared to a fragment distribution as predicted by Eq. (2) (dashed).

Figure 1 (b) shows jet spectra from the ISR (43 and 63 GeV points) and Sp$\bar{\text{p}}$S (remaining data points). The curves are derived from a universal parametrization of jet spectra applicable up to 13 TeV [9]. Panel (c) shows jet FFs from p-$\bar{p}$ collisions obtained by the CDF collaboration plotted on rapidity variable $y = \ln[(E + p)/m_\pi]$. The FFs show self-similar variation with jet energy making a simple parametrization possible. Schematically, the *fragment distribution* (FD) $P(p)$ describing the contribution to a p_t spectrum from MB dijets (spectrum HC) is the convolution of jet spectrum $P(E)$ for a given p-p collision energy and FF ensemble $P(p|E)$

$$P(p) \quad = \quad \int_{E_{min}}^{\infty} dE\, P(p|E)\, P(E), \tag{2}$$

with $E_{min} \approx 3$ GeV established by comparisons with hadron spectra [8, 11]. In panel (d) the dashed curve is $P(d)$. The solid points are the spectrum HC from 200 GeV NSD p-p collisions [3]. The open boxes are the 200 GeV UA1 jet spectrum appearing in panel (b). It is notable that the majority of jet fragments appear near 1 GeV/c. If E_{min} is reduced significantly below 3 GeV a large overestimate of the measured FD contribution below 1 GeV/c results, since the jet spectrum scales as $d\sigma_{jet}/dp_t \sim 1/p_t^6$ near that energy.

Two-particle angular correlations have been studied for 200 GeV p-p collisions [5–7] and Au-Au collisions [12] within $\Delta\eta = 2$. Measured quantity $\Delta\rho/\sqrt{\rho_{ref}}$ is essentially the number of correlated pairs per final-state hadron; η_Δ and ϕ_Δ are angle difference variables.

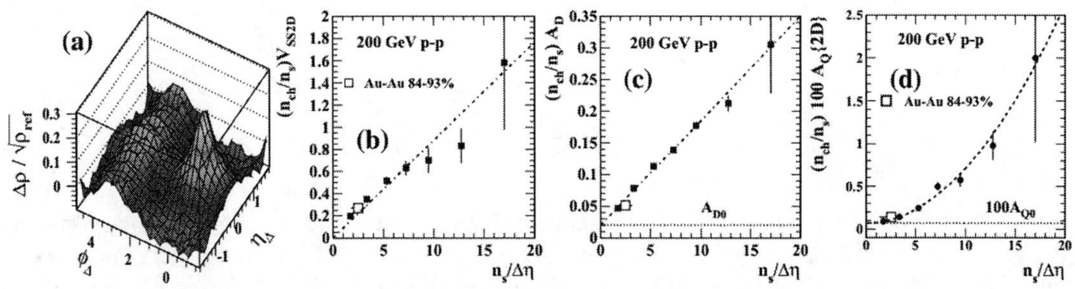

Figure 2: (a) 2D angular correlations for high-multiplicity 200 GeV p-p collisions. (b) SS 2D peak volume vs soft density $\bar{\rho}_s$. (c) AS 1D peak amplitude vs density $\bar{\rho}_s$. (d) Quadrupole amplitude A_Q vs $\bar{\rho}_s$. In (b-d) the vertical axis is proportional to number of correlated pairs divided by $\bar{\rho}_s$.

Figure 2 (a) shows angular correlations for high-multiplicity ($\approx 8\bar{\rho}_{0NSD}$) 200 GeV p-p collisions with no special p_t cuts. The main features are a SS (same-side on azimuth) 2D peak representing individual jets and an AS (away-side) 1D peak on ϕ_Δ representing back-to-back jet pairs. Superposed on the broader SS peak is a narrower 2D contribution from Bose-Einstein correlations. Just visible is a small SC contribution – a narrow 1D peak on η_Δ.

Figure 2 (b-d) shows systematic variation with n_{ch} or n_s of individual correlation amplitudes obtained by 2D model fits to correlation data [5]. The linear per-hadron trends in panels (b,c) indicate that the correlation hard component (dijets) follows a noneikonal trend: jet-correlated pairs $\propto \bar{\rho}_s^2$ as observed for the p-p spectrum HC. Panel (d) shows a quadratic trend on $\bar{\rho}_s$ for *nonjet quadrupole* component A_Q indicating that quadrupole correlated pairs are $\propto \bar{\rho}_s^3$. That trend is actually consistent with the quadrupole trend observed for A-A collisions. If $\bar{\rho}_s \sim$ low-x participant gluons and $\bar{\rho}_s^2 \sim$ participant-gluon binary collisions then $\bar{\rho}_s^3 \sim N_{part}N_{bin}$ for p-p collisions. The centrality trend observed for the nonjet quadrupole ($\bar{\rho}_0^2 v_2^2$) in Au-Au collisions is $N_{part}N_{bin}\epsilon_{opt}^2$ [13]. Absence of a ϵ_{opt}^2 factor for p-p collisions is consistent with the noneikonal $\bar{\rho}_h \propto \bar{\rho}_s^2$ trend: centrality is not relevant for p-p collisions.

4 p-p p_t spectrum TCM hard components

The illustrations above pertain to 200 GeV RHIC data as a testbed for MB dijet manifestations in spectra and angular correlations. In preparation for $\bar{p}_t$ analysis the full collision-energy and charge-multiplicity dependence of the spectrum hard component should be understood.

Figure 3 (a) shows the TCM hard components for *p*-*p* p_t spectra as a function of collision energy from $\sqrt{s}$ = 17 GeV to 13 TeV (curves of various line styles) [4] compared to data for two energies (points) [3, 14]. The TCM spectrum HC model is simple and relies on a few log($\sqrt{s}$) terms. The description of spectrum data is in all cases within systematic uncertainties of the data. Spectrum hard components show significant dependence on event multiplicity n_{ch} (e.g. multiple fine lines for the 200 GeV model). For the $\bar{p}_t$ study fiducial hard-component $\bar{p}_{th0}$ values in panel (b) are inferred from a *symmetric* spectrum HC model (for $\bar{\rho}_s \approx 2\bar{\rho}_{sNSD}$) [15].

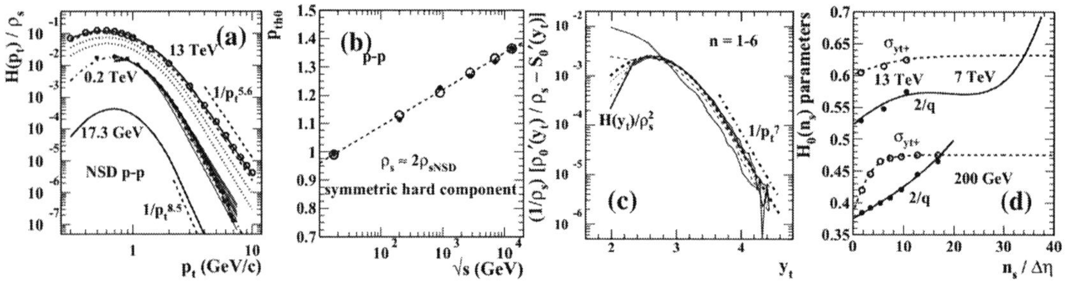

Figure 3: (a) *p*-*p* p_t spectrum TCM hard components for several collision energies (curves) compared to data (points). (b) Mean-p_t $\bar{p}_t$ hard components $\bar{p}_{th0}$ inferred from TCM model functions in panel (a). (c) p_t spectrum hard components for several multiplicity classes of 200 GeV *p*-*p* collisions. (d) Model parameters σ_{y_t+} and q for the TCM spectrum hard component, their variation with n_{ch}.

Figure 3 (c) shows spectrum hard components for several n_{ch} classes of 200 GeV *p*-*p* collisions [5]. The n_{ch} dependence of the shape has been known for more than ten years [3] and is attributed to bias of the underlying jet energy spectrum in response to the n_{ch} condition. The TCM is modified to accommodate those changes as follows: The hard-component model is a Gaussian plus exponential tail. The transition from Gaussian to exponential is determined by slope matching. The Gaussian widths are treated separately below and above the mode with widths σ_{y_t-} and σ_{y_t+} respectively. The exponential constant q is a separate parameter.

Figure 3 (d) shows evolution of spectrum HC model parameters σ_{y_t+} and q with n_{ch} at 200 GeV and 13 TeV. The trend for the width below the mode σ_{y_t-} is available in Ref. [4]. For 200 GeV the parameter trends are tightly constrained by spectrum data over the full n_{ch} interval relevant to $\bar{p}_t$ data. For 13 TeV the spectrum data span a more-limited n_{ch} interval. The $2/q$ trend (upper solid curve) has been extended by fitting $\bar{p}_t$ data in the present study.

The $\bar{p}_t$ analysis of Ref. [16] produced uncorrected $\bar{p}'_t$ values, where the prime denotes the effect on spectra of a lower-p_t acceptance cut at $p_{t,min}$ near 0.15 GeV/c. The p_t acceptance cut strongly affects the $\bar{p}_t$ SC but has negligible effect on the HC which is small below 0.5 GeV/c. The relation is $\bar{p}'_{ts} = \bar{p}_{ts}/\xi$ where ξ is the SC efficiency $\approx$ 0.78 for an *effective* p_t cutoff near 0.17 GeV/c. For universal $\bar{p}_{ts} \approx 0.4$ GeV/c uncorrected $\bar{p}'_{ts} \approx 0.51 \pm 0.2$ GeV/c.

5 p-p $\bar{p}_t$ TCM analysis

The TCM for $\bar{p}_t$ data from p-p collisions is constructed as follows [15]: The uncorrected yield n'_{ch} within angular acceptance $\Delta\eta$ relative to the corrected soft component n_s is $n'_{ch}/n_s = \xi + x(n_s)$ with $\xi = 0.76$ - 0.80 (depending on the effective $p_{t,min}$). For noneikonal p-p collisions $x(n_s, \sqrt{s}) \equiv n_h/n_s \approx \alpha(\sqrt{s})\bar{\rho}_s$ and $\alpha(\sqrt{s})$ is predicted from jet characteristics [4]. The TCM for uncorrected $\bar{p}'_t$ is the first of

$$\bar{p}'_t \equiv \frac{\bar{P}'_t}{n'_{ch}} \approx \frac{\bar{p}_{ts} + x(n_s, \sqrt{s})\,\bar{p}_{th}(n_s, \sqrt{s})}{\xi + x(n_s, \sqrt{s})}; \quad \frac{n'_{ch}}{n_s}\bar{p}'_t \approx \frac{\bar{P}_t}{n_s} = \bar{p}_{ts} + x(n_s, \sqrt{s})\,\bar{p}_{th}(n_s, \sqrt{s}), \tag{3}$$

where $\bar{p}_{ts} \approx 0.4$ GeV/c and $p_{th}(n_s, \sqrt{s})$ is the new information derived from p-p $\bar{p}_t$ data.

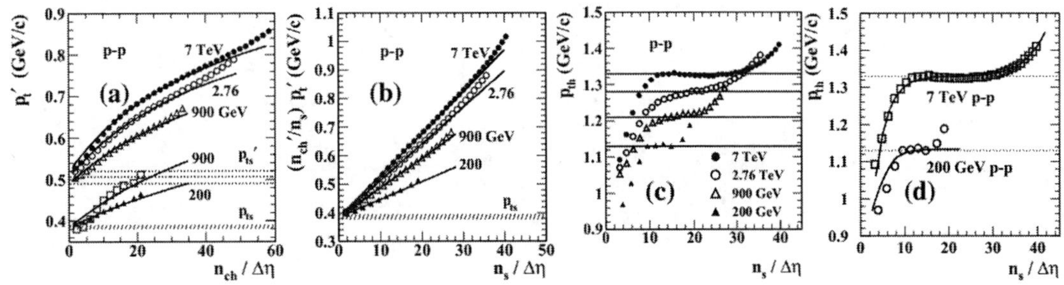

Figure 4: (a) Uncorrected $\bar{p}'_t$ data (points) from Ref. [16] vs the $\bar{p}_t$ TCM (solid curves). (b) Corrected $\bar{P}_t/n_s$ data (points) vs the TCM (solid lines). (c) $\bar{p}_{th}(n_s, \sqrt{s})$, trends vs $\bar{\rho}_s$ for four energies. (d) $\bar{p}_{th}(n_s, \sqrt{s})$, trends for 200 GeV and 7 TeV (points) compared to TCM trends from spectra (curves).

Figure 4 (a) shows uncorrected $\bar{p}'_t$ data from Ref. [16] (upper points) compared to nominally corrected data from STAR and UA1 (lower points). The $\bar{p}'_{ts}$ values for uncorrected data are in the range 0.49 - 0.525 GeV/c as expected (dotted lines). The solid curves are TCM trends assuming no n_{ch} variation of $\bar{p}_t$ hard component [held fixed at values $\bar{p}_{th0}$ from Fig. 3 (b)]. Figure 4 (b) shows corrected $\bar{P}_t/n_s$ data (points) according to the second of Eqs. (3). In that format the TCM is a straight line with slope $\alpha(\sqrt{s})\,\bar{p}_{th0}$ predicted from p_t spectra and jet data.

Figure 4 (c) shows $\bar{p}_{th}(n_s, \sqrt{s})$ data inferred from the uncorrected $\bar{p}'_t$ data according to Eqs. (3). The horizontal lines are the $\bar{p}_{th0}(\sqrt{s})$ values appearing in Fig. 3 (b). Clearly evident are substantial variations with n_s or $\bar{\rho}_s$. However, such variations are expected based on previous spectrum studies as in Refs. [3, 4]. Figure 4 (d) shows $\bar{p}_{th}(n_s)$ values for 200 GeV and 7 TeV (points) compared to a TCM including HC n_{ch} dependence according to Fig. 3 (d). The correspondence between TCM and $\bar{p}_t$ data is within data uncertainties. Accurate correspondence between p-p $\bar{p}_t$ data and the TCM compels the conclusion that there is a direct, quantitative correspondence among measured isolated-jet properties, p_t spectrum hard components and $\bar{p}_t$ data. p-p collisions are dominated by MB dijets and are noneikonal.

6 p-Pb $\bar{p}_t$ TCM analysis

For a TCM description of p-Pb collisions [15] the composite structure of the Pb nucleus must be incorporated by adding *mean participant pathlength* $\nu(n_s) \equiv 2N_{bin}/N_{part}$ as a geometry measure, so for instance $n'_{ch}/n_s \rightarrow \xi + x(n_s)\nu(n_s)$. The soft-component charge density factorizes as $\bar{\rho}_s \equiv (N_{part}/2)\bar{\rho}_{sNN}(n_s)$, similarly $\bar{\rho}_h \equiv N_{bin}\,\bar{\rho}_{hNN}(n_s)$ and hard/soft ratio $x(n_s) \rightarrow \bar{\rho}_{hNN}(n_s)/\bar{\rho}_{sNN}(n_s) \approx$

$\alpha\bar{\rho}_{sNN}(n_s)$ averaged over N-N collisions as for p-p collisions. Combining those relations gives $N_{part}(n_s)/2 = \alpha\bar{\rho}_s/x(n_s)$, and for p-A collisions $N_{part} \equiv N_{bin} + 1$. Thus, if a model for $x(n_s)$ is specified then N_{part}, N_{bin} and ν are also defined. As described below, $x(n_s)$ for more-peripheral p-A collisions is based on the noneikonal p-p trend and is then extrapolated to more-central collisions with $N_{part}/2 > 1$ based on $\bar{p}_t$ data. The TCM for p-Pb $\bar{p}'_t$ data is similar to Eqs. (3) but with additional $\nu(n_s)$ factors

$$\bar{p}'_t \equiv \frac{\bar{P}'_t}{n'_{ch}} \approx \frac{\bar{p}_{ts} + x(n_s)\nu(n_s)\,\bar{p}_{thNN}(n_s)}{\xi + x(n_s)\nu(n_s)}; \quad \frac{\bar{P}_t(n_s)}{n_s} = \bar{p}_{ts} + x(n_s)\nu(n_s)\,\bar{p}_{thNN}(n_s) \tag{4}$$

assuming $\bar{p}_{thNN}(n_s) \approx \bar{p}_{th0}$ (p-p values) for p-Pb collisions (no jet modification in p-A).

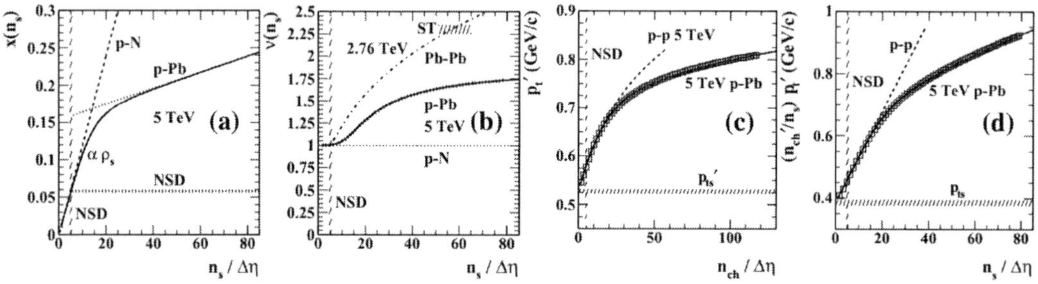

Figure 5: (a) TCM parameter $x(n_s)$ vs $\bar{\rho}_s$ for 5 TeV p-Pb collisions. (b) Corresponding $\nu(n_s)$ trend. (c) Uncorrected $\bar{p}'_t$ data (points) from Ref. [16] vs the $\bar{p}_t$ TCM (solid curve). (d) Corrected $\bar{P}_t/n_s$ data (points) vs the TCM (solid curve). Equivalent p-p trends are included for comparison.

Figure 5 (a) shows a $x(n_s)$ model for p-Pb collisions (solid curve): the observed p-p trend (dashed line) is smoothly transitioned to a second linear trend with reduced slope (dotted line). The additional p-Pb model parameters are $(\bar{\rho}_{s0}, m_0)$, a transition point and slope-reduction factor. Panel (b) shows the corresponding $\nu(n_s)$ trend (solid curve) defined by $x(n_s)$ as noted above. The dash-dotted curve is a Pb-Pb $\nu(n_s)$ trend for comparison.

Figure 5 (c) shows uncorrected $\bar{p}'_t$ data from Ref. [16] (points) with $\bar{p}'_{ts} \approx 0.525$ GeV/c. The solid curve is the TCM defined by Eq. (4) (first) with $\bar{\rho}_{s0} = 15$ and $m_0 = 0.1$ adjusted to accommodate $\bar{p}'_t$ data and $\bar{p}_{thNN}(n_s) \to \bar{p}_{th0} = 1.30$ GeV/c from Fig. 3 (b). The dashed curve is the p-p TCM for 5 TeV. Figure 5 (d) shows the same features for corrected $\bar{P}_t/n_s$ data with TCM described by Eq. (4) (second). This p-Pb exercise illustrates the transition from noneikonal p-p or p-N trend to eikonal p-A trend with increasing multiplicity. The value of $N_{part}/2$ remains near one (p-N only) up to the transition point but then slowly increases as greater p-A centrality becomes competitive to deliver larger overall charge multiplicity.

7 Pb-Pb $\bar{p}_t$ TCM analysis

For the Pb-Pb $\bar{p}_t$ TCM published yield measurements and a Glauber model based on the eikonal approximation can be used to determined some TCM model elements. Parameter $x(n_s)$ can be derived from *per-participant-pair* charge-density data based on the TCM

$$(2/N_{part})dn_{ch}/d\eta \approx (2/N_{part})\bar{\rho}_0(\nu) \quad = \quad \bar{\rho}_{sNN}[1 + x(\nu)\nu]. \tag{5}$$

Parameter $\nu(n_s)$ can be derived from the Glauber model in the form $\nu(n_s) \approx (N_{part}/2)^{1/3}$ [10] with $N_{part}/2 \approx \bar{\rho}_s/\bar{\rho}_{sNN}$ as in the previous section and $\bar{\rho}_{sNN} \approx \bar{\rho}_{sNSD}$ in both cases.

Figure 6 (a) shows per-participant-pair yield data from Ref. [17] (points). The dashed line is a Glauber linear superposition (GLS) model representing no jet modification. The upper bold dotted curve shows the expected effect of multiplicity fluctuations for more-central *A*-*A* collisions [18]. The solid curve is the TCM in Eq. (5) with $x(\nu)$ defined below. Panel (b) shows solution of Eq. (5) for parameter $x(\nu)$ (points) assuming $\bar{\rho}_{sNN} \to \bar{\rho}_{sNSD} \approx 4.6$ [4]. A *sharp transition* (ST), first observed for Au-Au jet-related 2D angular correlations [12], representing evidence of jet modification in the yield trend [8] is evident. The bold dotted curve is a simple TCM expression for $x(\nu)$ that starts from the NSD *p*-*p* value $x_{pp} \approx 0.045$, increases rapidly through the ST and maintains a saturation value ≈ 0.142 for more-central collisions, approximately threefold increase over the NSD value. The dashed curve is the equivalent for 200 GeV Au-Au collisions. The interval $\bar{\rho}_s = 10$ - 40 is not well-defined *a priori* by yield data so a TCM expression is required that includes an adjustable x_{pp} [15]

$$x(\nu) = x_{pp} + (0.142 - x_{pp})\{1 + \tanh[(\nu - 2.3)/0.5]\}/2, \tag{6}$$

with ST near $\nu = 2.3$.

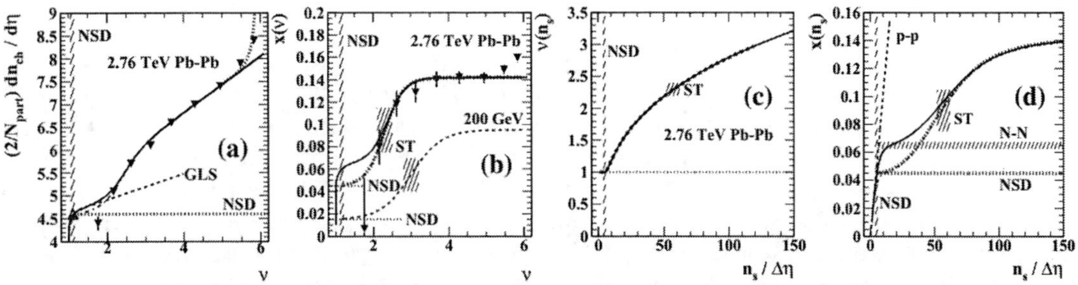

Figure 6: (a) Per-participant-pair charge density vs centrality measure ν (points) from Ref. [17] compared to the TCM (solid curve). (b) Data from panel (a) transformed to TCM parameter $x(\nu)$ compared to the TCM of Eq. (6) (solid, bold-dotted curves). (c) Glauber parameter $\nu(n_s)$ defined in the text. (d) Parameter $x(\nu)$ from panel (b) transformed to $x(n_s)$ based on $\nu(n_s)$ in panel (c).

Figure 6 (c) shows $\nu(n_s) \approx (\bar{\rho}_s/\bar{\rho}_{sNSD})^{1/3}$ with lower bound $\nu = 1$. Panel (d) shows $x(n_s) = x[\nu(n_s)]$ based on Eq. (6) and panel (c). For this model peripheral Pb-Pb follows the *p*-*p* trend (dashed line) for the most peripheral collisions, and more-central Pb-Pb shows the ST feature (jet modification). Yield data do not indicate where the transition from noneikonal *p*-*p* to eikonal Glauber trend occurs, requiring the adjustable x_{pp} parameter in Eq. (6).

Figure 7 (a) shows uncorrected $\bar{p}'_t$ data for 2.76 TeV Pb-Pb collisions from Ref. [16] (open squares). The uncorrected soft component is $\bar{p}'_{ts} \approx 0.51$ GeV/c. Data for *p*-*p* collisions are also shown (open circles). The Pb-Pb $\bar{p}_t$ TCM (solid curve) follows the *p*-Pb TCM in Eq. (4) (first) but with the $\nu(n_s)$ and $x(n_s)$ models in Fig. 6 (c,d). The dashed curve is the corresponding TCM for *p*-*p* data. The dash-dotted curve is a GLS model assuming no jet modification.

Figure 7 (b) shows product $x(n_s)\bar{p}_{thNN}(n_s) \approx \bar{P}_{thNN}/n_{sNN}$ (open squares) obtained from the data in panel (a) according to Eqs. (4) given the $\nu(n_s)$ model in Fig. 6 (c). The open circles are corresponding *p*-*p* data. The solid curve is the Pb-Pb TCM and the dashed curve is the *p*-*p* TCM. Note that the ST is clearly evident in this more-differential plot format, but not in panel (a). The Pb-Pb data follow the *p*-*p* trend closely up to and even beyond the NSD $\bar{\rho}_s$ value (vertical hatched band).

Figure 7 (c) shows $\bar{p}_{thNN}(n_s)$ data (points) obtained from panel (b) using the $x(n_s)$ model from Fig. 6 (d). The inferred $\bar{p}_{thNN}(n_s)$ values follow the *p*-*p* trend (bold dotted curve just visible) for peripheral Pb-Pb but then descend through the ST to a saturation value that is about 75% of the

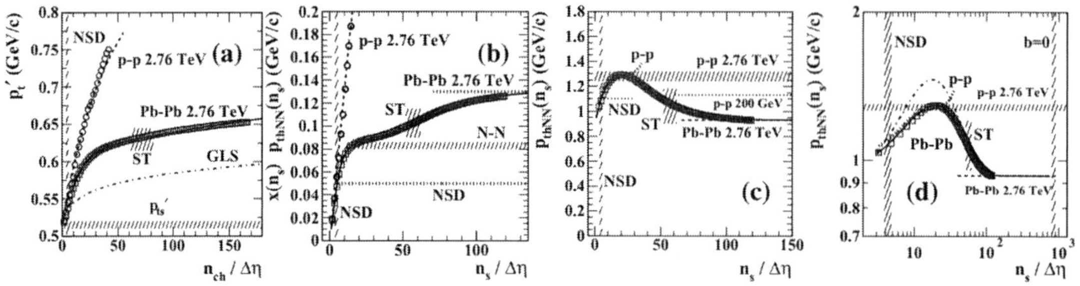

Figure 7: (a) Uncorrected $\bar{p}_t'$ data (points) from Ref. [16] vs the $\bar{p}_t$ TCM (solid curve). *p-p* data (open circles) and TCM (dashed curve) are included for comparison. The GLS curve represents the Pb-Pb trend for no jet modification. (b) Product $x(n_s)\ \bar{p}_{thNN}(n_s)$ obtained from the data in panel (a) by solving Eq. (4) using $\nu(n_s)$ in Fig. 6 (c). (c) $\bar{p}_{thNN}(n_s)$ obtained from data in panel (b) using $x(n_s)$ in Fig. 6 (d). (d) Panel (c) plotted in log-log format emphasizing the most-peripheral interval.

maximum. The constant saturation values for $x(n_s)$ and $\bar{p}_{thNN}(n_s)$ extend from just above the ST ($\bar{\rho}_s \approx 100$) to central collisions ($\bar{\rho}_s \approx 850$) suggesting that jets do not undergo further alteration within that interval. Panel (d) shows the same features in a log-log format providing better access to peripheral details and extending the plot to central Pb-Pb collisions. The correspondence between *p-p* (dotted curve) and Pb-Pb data (open squares) up to $\bar{\rho}_s \approx 20$ ($4 \times \bar{\rho}_{sNSD}$) is evident. The dash-dotted curve marks where the data would lie if the bold dotted curve in Fig. 6 (d) were used for $x(n_s)$. The data were actually transformed using the solid curve in that panel with $x_{pp} \to 0.065$ in Eq. (6), 1.45 times the NSD value. The solid curve through the $\bar{p}_{thNN}(n_s)$ data, representing new information obtained from Pb-Pb $\bar{p}_t$ data, is a parametrization that forms the third element of the Pb-Pb $\bar{p}_t$ TCM [15].

8 Comment on Glauber-model analysis of p-Pb collision centrality

The TCM for *p*-Pb $\bar{p}_t$ data described in Sec. 6 and Ref. [15] can be compared with a recent Glauber analysis of 5 TeV *p*-Pb centrality in Ref. [19]. For the Glauber analysis it is assumed that integrated charge multiplicity is proportional to Glauber N_{part}: "[n_{ch}] at mid-rapidity scales linearly with [N_{part}]." That assumption is equivalent to assuming that a probability (i.e. event-frequency) distribution on charge multiplicity – dP/dn_{ch} – is equivalent to a fractional cross-section distribution on N_{part} – $(1/\sigma_0)d\sigma/dN_{part}$ – inferred from a Glauber Monte Carlo.

Figure 8 (a) shows an event-frequency distribution (histogram) on V0A amplitude (charge multiplicity within a pseudorapidity interval in the Pb hemisphere) from Ref. [19]. The assumption $N_{part} \propto n_{ch}$ is equivalent to $dP/dn_{ch} \sim (1/\sigma_0)d\sigma/dN_{part}$ as noted above, and the fractional cross sections in the form $100(1 - \sigma/\sigma_0)$ noted in the panel and apparently obtained from a running integral of the histogram seem to be consistent with that statement.

Figure 8 (b) shows Glauber estimates of $(2/N_{part})dn_{ch}/d\eta$ from Ref. [19] (solid points) compared to the TCM equivalent for *p-p* (dashed) and *p*-Pb (solid) collisions from Ref. [15] as reported above. The large discrepancy is apparent. Given N_{part} and N_{bin} from the Glauber analysis corresponding values of ν are plotted in panel (c) (solid dots). The TMC equivalent is the solid curve, and the Glauber trend for Pb-Pb collisions is shown as the dashed curve. That ν for *p*-Pb collisions might anywhere greatly exceed that for Pb-Pb collisions is notable.

Figure 8 (d) shows fractional impact parameter in the form $1 - b/b_0$ (open circles) inferred from the fractional cross sections in panel (a) assuming that $\sigma/\sigma_0 = (b/b_0)^2$. The solid dots are the same

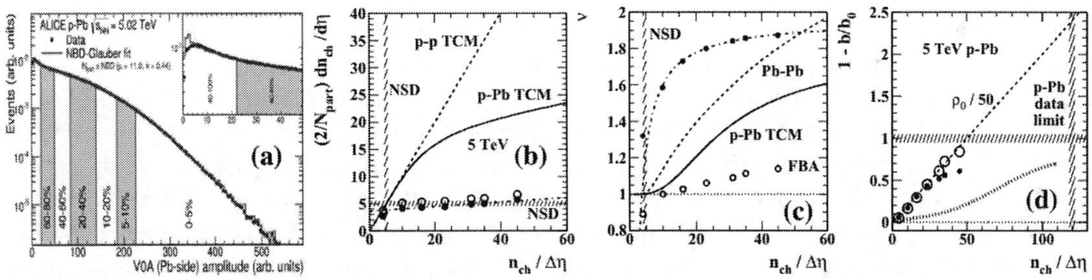

Figure 8: (a) Event-frequency distribution on charge multiplicity in a VOA detector with centrality intervals assigned by p-Pb Glauber analysis [19]. (b) Per-participant-pair p-Pb yields inferred from the Glauber analysis (solid points) and from the present study (solid curve). (c) Participant path length ν inferred from the Glauber analysis (solid points) and the present study (solid curve). Pb-Pb Glauber results (dashed) are included for comparison. (d) Fractional impact parameter in the form $1 - b/b_0$ inferred from the Glauber study (solid points) and indirectly from fractional cross sections in panel (a) (open circles) compared to a trend suggested by the present study (bold dotted curve).

quantity based on b values from the Glauber analysis and assuming that $b_0 \approx 7.1 + 0.85 \approx 8$ fm. Whereas the Glauber analysis in Ref. [19] reports 0-5% central p-Pb collisions for $\bar{\rho}_0 \approx 45$ the p-Pb $\bar{p}_t$ data from Ref. [16] (same collaboration) extend to $\bar{\rho}_0 \approx 120$, nearly three times larger. The bold dotted curve suggests a more realistic trend for b/b_0.

The basic problem with the analysis in Ref. [19] is the initial assumption $N_{part} \propto n_{ch}$ which is strongly violated for p-Pb collisions. Just as for p-p collisions in Sec. 5 initial increase of the n_{ch} condition is more easily met in p-Pb collisions by higher multiplicities in individual p-N collisions than by increasing p-Pb centrality. $N_{part}/2 \approx 1$ persists until $\bar{\rho}_{s0} \approx 15$, three times the NSD value. Only then does $N_{part}/2$ begin to increase above 1 significantly. Below that point $d\sigma/dn_{ch} \approx 0$ and p-Pb "centrality" remains fixed at 100%. The correct relation is

$$d\sigma/dn_{ch} = (dN_{part}/dn_{ch})d\sigma/dN_{part}, \tag{7}$$

where $d\sigma/dN_{part}$ may be obtained from a Glauber Monte Carlo but the Jacobian dN_{part}/dn_{ch} is determined by the p-Pb $\bar{p}_t$ TCM as described above. As noted, the Jacobian factor is essentially zero over a substantial n_{ch} interval below transition point $\bar{\rho}_{s0}$ in Sec. 6. Based on the assumption in Ref. [19] the Jacobian would have a fixed value near $2/n_{chNSD}$ for all n_{ch}.

9 Summary

Recently, claims have emerged that hydrodynamic flows must play a role in smaller collision systems (p-A and even p-p) based on observation of the same phenomena in those systems attributed to flows in more-central A-A systems. However, the argument could be reversed to conclude that if phenomena appear in collision systems so small and low-density that flows arising from particle rescattering are impossible then the same phenomena observed in more-central A-A may not represent flows. Recent $\bar{p}_t$ data may contribute to a resolution.

The two-component (soft + hard) model (TCM) of hadron production near midrapidity provides an accurate and comprehensive description of yield, spectrum and correlation data applicable to any A-B collision system. While one or more other mechanisms may contribute to hadron production in minor ways (e.g. a nonjet quadrupole component) projectile-nucleon dissociation (soft) and dijet production (hard) appear as the dominant processes in all cases.

From TCM analysis of $\bar{p}_t$ data described above for three successive collision systems the following may be concluded: (a) *p-p* $\bar{p}_{th}(n_s, \sqrt{s})$ hard-component trends agree with spectrum HC evolution and measured MB dijet properties. (b) *p-p* dijet production is *noneikonal*, *p-p* centrality is not relevant. (c) The *p*-Pb $\bar{p}_{th}(n_s)$ hard component establishes factorization of A-B Glauber and *N-N* noneikonal trends. (d) *p*-Pb $\bar{p}_t$ data confirm that MB dijets dominate $\bar{p}_t(n_s)$ variation. (e) In effect, MB dijets probe the centrality evolution of *p*-Pb collisions with n_{ch}. (f) Pb-Pb $\bar{p}_t$ data confirm that the Glauber model effectively describes more-central *A-A* collisions, but peripheral collisions accurately follow *p-p* (i.e. *N-N*) trends. (g) The Pb-Pb $\bar{p}_{thNN}(n_s)$ trend confirms that jets are *modified* quantitatively above the ST, but jets still dominate spectrum and correlation structure in more-central Pb-Pb collisions.

Thus, the likelihood that flows play any role in small collision systems appears negligible. Claimed novel phenomena may be better explained either as due to MB dijets or as a related QCD process (e.g. multipole radiation). The same argument may be extended even to the most-central *A-A* collisions at the highest energies. Before novelty is claimed the contribution of MB dijets and related QCD processes to data trends must first be well understood.

References

[1] P. Huovinen and P. V. Ruuskanen, Ann. Rev. Nucl. Part. Sci. **56**, 163 (2006).
[2] K. Dusling, W. Li and B. Schenke, Int. J. Mod. Phys. E **25**, no. 01, 1630002 (2016).
[3] J. Adams *et al.* (STAR Collaboration), Phys. Rev. D **74**, 032006 (2006).
[4] T. A. Trainor, J. Phys. G **44**, 075008 (2017).
[5] T. A. Trainor and D. J. Prindle, Phys. Rev. D **93**, 014031 (2016).
[6] R. J. Porter and T. A. Trainor (STAR Collaboration), J. Phys. Conf. Ser. **27**, 98 (2005).
[7] R. J. Porter and T. A. Trainor (STAR Collaboration), PoS **CFRNC2006**, 004 (2006).
[8] T. A. Trainor, Phys. Rev. C **80**, 044901 (2009).
[9] T. A. Trainor, Phys. Rev. D **89**, 094011 (2014).
[10] T. A. Trainor and D. J. Prindle, hep-ph/0411217.
[11] T. A. Trainor, J. Phys. G **42**, no. 8, 085105 (2015).
[12] G. Agakishiev, *et al.* (STAR Collaboration), Phys. Rev. C **86**, 064902 (2012).
[13] D. T. Kettler (STAR collaboration), Eur. Phys. J. C **62**, 175 (2009).
[14] J. Adam *et al.* (ALICE Collaboration), Phys. Lett. B **753**, 319 (2016).
[15] T. A. Trainor, arXiv:1708.09412.
[16] B. B. Abelev *et al.* (ALICE Collaboration), Phys. Lett. B **727**, 371 (2013).
[17] K. Aamodt *et al.* (ALICE Collaboration), Phys. Rev. Lett. **106**, 032301 (2011).
[18] T. A. Trainor, Phys. Rev. C **91**, 044905 (2015)
[19] J. Adam *et al.* (ALICE Collaboration), Phys. Rev. C **91**, no. 6, 064905 (2015).

Flow and correlation phenomena measurements in pp, pPb and PbPb collisions at CMS

Sandra S. Padula (on behalf of the CMS Collaboration)[1,⋆]

[1] *Universidade Estadual Paulista (UNESP), São Paulo, SP, Brazil*

Abstract. The quark-gluon plasma created in high energy collisions of large nuclei exhibits strong anisotropic collective behavior as a nearly perfect fluid, flowing with little frictional resistance or viscosity. It has been investigated extensively over the past years employing two or more particle correlations. An overview of collective flow and particle correlation measurements at CMS as a function of transverse momentum, pseudorapidity, event multiplicity, for both charged hadrons or identified particles will be presented. These results are compared among pp, pPb and PbPb systems and several aspects of their intriguing similarities are discussed.

1 Introduction

The state of deconfined quarks and gluons predicted by quantum chromodynamics, the quark-gluon plasma (QGP), was discovered at the Relativistic Heavy Ion Collider (RHIC) and found to be a strongly coupled medium, behaving as a nearly perfect fluid [1]. This finding was possible through the observation of anisotropic azimuthal correlations that can be associated to the collective properties of the hot and dense medium formed in such high energy collisions. Another important property observed at RHIC was the quench of the jets traversing the hot and dense medium, an effect which was more pronounced as the overlapping of the interacting nuclei increased from peripheral to head-on relativistic heavy ion collisions. At the Large Hadron Collider (LHC), the QGP has been further extensively investigated through a myriad of measurements performed over the recent years. In particular, the extension of the measurement of azimuthal correlations to particles with very high transverse momentum p_T, has added significant new insights into these observations, connecting both effects: in the low p_T region, the anisotropic azimuthal correlations are associated with the hydrodynamic collective flow, whereas in very high p_T realm, such correlations may reflect the path lengths followed by the quenched jet, reflecting the in-medium parton energy loss.

An additional remarkable discovery in AA collisions at RHIC energies was the "ridge-like" two-hadrons correlations, where particles very close in the azimuthal direction ($\Delta\phi \sim 0$) are correlated over a wide region in their relative pseudorapidity (up to $\Delta\eta \sim 4$). Amazingly, similar correlations were discovered in 2010 at the LHC in high multiplicity events produced in pp collisions at 7 TeV [2] and, some time later, also in pPb collisions, raising questions about the possible common nature of the systems formed in pp, pA and AA collisions.

⋆e-mail: sandra.padula@cern.ch
This material is based upon work supported by the So Paulo Research Foundation (FAPESP) under Grant No. 2013/01907-0.

© The Authors, published by EDP Sciences. This is an open access article distributed under the terms of the Creative Commons Attribution License 4.0 (http://creativecommons.org/licenses/by/4.0/).

Results related to the phenomena described above are reported in this paper. First, azimuthal correlations in *PbPb* collisions at $\sqrt{s_{NN}} = 5.02$ are presented, extending from low transverse momentum ($1 < p_T < 3$ GeV/c), up to very large p_T (≈ 100 GeV/c) [3]. The measurements of the corresponding Fourier coefficients are performed employing the scalar product method, by correlating charged tracks with the energy deposited in the hadronic forward calorimeters, as well as the multi-particle cumulant.

Next, measurements of two- and multi-particle angular correlations are discussed for high multiplicity events in pp collisions at $\sqrt{s} = 13$ TeV. Such correlations are investigated with unidentified charged particles ($h^{\pm}$), as well as reconstructed K_s^0 and Λ particles [4]. The results of v_2 and v_3 harmonics extracted from two-particle correlations are studied as functions of particle p_T and event multiplicity. The residual contribution of back-to-back jet correlations to the ridge phenomenon is estimated and removed by subtracting correlations obtained from very low multiplicity pp events. For high-multiplicity pp events, a mass ordering is observed for the v_2 values of charged hadrons (mostly pions), K_S^0, and $\Lambda/\bar{\Lambda}$, with lighter particle species exhibiting a stronger azimuthal anisotropy signal below $p_T \approx 2$ GeV/c. The v_2 signals are also extracted from four- and six-particle correlations using the cumulant method, on the search for a possible collective nature of the correlations. The pp results are directly compared to those found for pPb at 5 TeV and PbPb at 2.76 TeV collisions over a broad range of similar multiplicities, revealing similar observations in the three systems. These findings suggest a collective origin for the observed long-range correlations in high-multiplicity pp collisions.

Finally, a recent development is discussed, which allows for more deeply scrutinizing the final-state azimuthal anisotropy in different colliding systems by measuring the correlations between Fourier harmonic coefficients. In nucleus-nucleus collisions, these correlations have been shown to be strongly sensitive to initial-state fluctuations and to medium transport coefficients [5]. In what follows, the measurement of anisotropy harmonics (v_n, $n = 2-4$) and event-by-event correlations of different v_n, using data collected by the CMS experiment are presented for pp at 13 TeV, pPb at 5.02 and 8 TeV, and PbPb collisions at 5.02 TeV [6]. The v_n results are extracted via long-range ($\Delta\eta > 2$) two-particle correlations as a function of event multiplicity. Event-by-event correlations of v_2 versus v_3 and v_2 versus v_4 are measured using the four-particle Symmetric Cumulant (SC) method for these three colliding systems.

2 The CMS experiment

The CMS detector has nearly 4π solid angle acceptance and an extended coverage of more than ten units in pseudorapidity ($\eta = -\log[\tan(\theta/2)]$, with θ being the polar angle along the beam direction). Its large superconducting solenoid of 6 m internal diameter, with 3.8 T axial magnetic field, is surrounding the three-layer silicon pixel and the strip tracker, the crystal electromagnetic calorimeter, and a brass/scintillator hadron calorimeter. Muons are measured in gas-ionization detectors embedded in the steel return yoke. In addition to these detectors, CMS has extensive forward calorimetry, in particular two steel/quartz-fiber Cerenkov forward hadronic calorimeters (HF, $2.9 < |\eta| < 5.2$). A complete description of the CMS detector can be found in Ref.[7].

3 Anisotropy harmonics measured up to very high p_T

In non-central collisions, the eccentricities in the overlapping region of the two colliding nuclei result in spatial asymmetries. After the local thermal equilibrium is attained following the strong rescattering of the partons in the initial state, such asymmetries drive a collective anisotropic expansion, which is faster along the short axis of the almond shaped initial overlapping region due to the largest pressure gradient, resulting in the observed azimuthal anisotropies. The azimuthal dependence of the particle yield can be written in terms of an harmonic expansion, in which the Fourier coefficients v_2 and v_3 correspond to the elliptic and triangular flow, respectively. For characterizing the anisotropy of the azimuthal distribution of charged particles produced in PbPb collisions at $\sqrt{s_{NN}} = 5.02$ TeV these coefficients are measured with data collected by the CMS experiment. The scalar product method is employed for measuring the $v_2\{SP\}$ and $v_3\{SP\}$ anisotropy coefficients as a function of p_T up to 100 GeV/c, for several collision centrality intervals in the overall range 0-60%. The details about data samples, experimental procedures, the scalar product method and the systematic uncertainties, can be found in Ref. [3] . The results are shown in Fig. 1 left panel, with statistical and systematic uncertainties represented by bars and shaded boxes, respectively. In Fig. 1 left panel, the v_2 and v_3 results are compared to the CUJET3.0 [9] and SHEE [8] models for several centrality bins. An essencial difference between these two models is that in SHEE is that it includes initial-state geometry fluctuations (in addition to linear path-length dependence of parton energy loss), whereas CUJET3.0 uses a smooth hydrodynamic background (and uses perturbative quantum chromodynamics (pQCD) calculations to describe the hard parton interactions). Calculations from SHEE show a good agreement with the data, over the wide p_T and centrality ranges probed in this analysis.

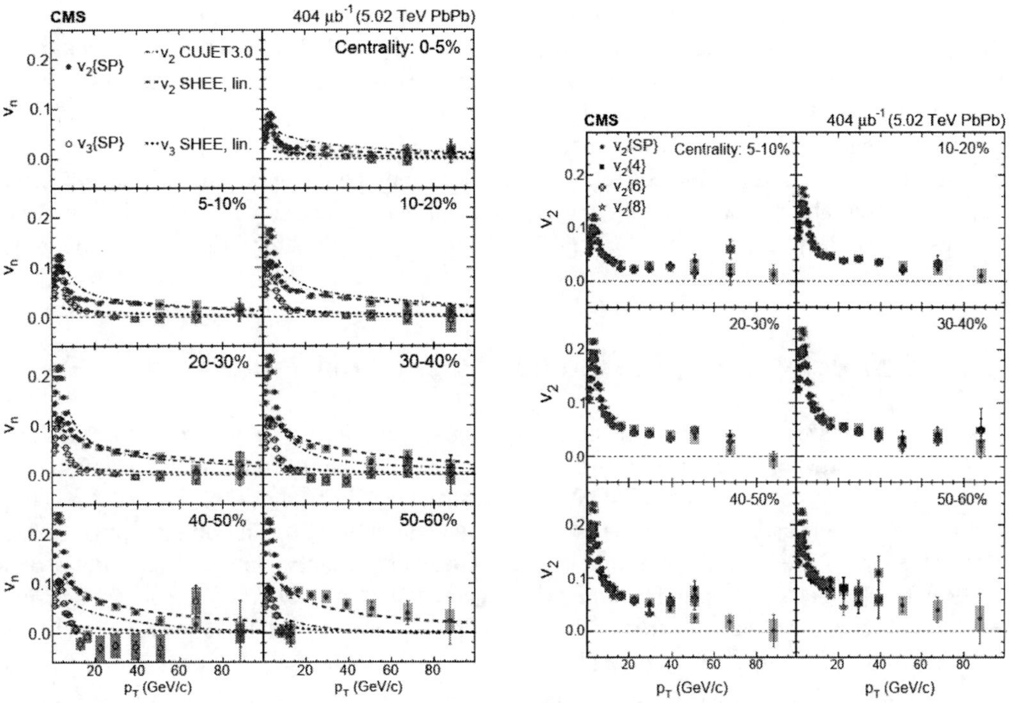

Figure 1. Left panel: Results of $v_2\{SP\}$ and $v_3\{SP\}$ measured in PbPb collisiosn at $\sqrt{s_{NN}} = 5.02$ TeV are shown in several centrality windows as a function of the transverse momentum p_T. Right panel: Results of $v_2\{SP\}$ and multi-particle cumulants $v_2\{4\}$, $v_2\{6\}$, $v_2\{8\}$ are shown as a function of p_T in six centrality ranges in the same collisions. Statistical and systematic uncertainties are represented, respectively, by bars and shaded boxes. The plots are from Ref. [3] .

It can be seen that, for all centralities, $v_2\{SP\}$ shows a rapid rise up to $p_T \approx 3$ GeV/c and then falls fast up to $p_T \approx 10$ GeV/c, decreasing more slowly for higher values of the transverse momentum, and remaining non-zero up to very high p_T. At low-p_T, $v_2\{SP\}$ increases from most central to mid-peripheral events (30-40%), and then decreases toward more peripheral ones, showing a similar p_T behavior in all measured centralities. The $v_3\{SP\}$ values are also shown in Fig. 1 left panel, exhibiting a similar behavior as that of $v_2\{SP\}$ for the same centrality ranges, although being practically centrality independent for low p_T. Besides, the measured $v_3\{SP\}$ shows non-zero values up to $p_T \approx 20$ GeV/c, which is consistent with picturing the system as being subjected to fluctuating initial conditions (the odd harmonics would be identically zero, by symmetry, if the initial conditions were described by nucleons distributed in a uniform and rotationally invariant way in both nuclei).

In order to investigate the origin of the non-zero anisotropy coefficients up to such high p_T and to explore its possible collective nature, the v_2 values are also measured by correlating 4, 6 and 8 particles. The Q-cumulant method [11] is used to measure $v_2\{4\}$, $v_2\{6\}$, $v_2\{8\}$, providing a fast, exact (free of approximations), and unbiased (no interference between different harmonics) evaluation of the cumulants. The method is also subjected to non-flow effects since, if M particles are produced in a collisions, direct k-particle correlations are of order $1/M^{k-1}$, so that they become smaller as k increases. Therefore, the 4th-, 6th-, 8th-order cumulant results are expected to be much less affected by non-flow contributions than the second-order one. The cumulants are expressed in terms of the moments of the magnitude of the corresponding flow vectors Q_n. The details of the technique and measurement are in Ref. [3] . The corresponding results are shown in Fig. 1 right panel, from which it can be seen that, in the range $p_T < 3$ GeV/c, $v_2\{SP\} > v_2\{4\} \approx v_2\{6\} \approx v_2\{8\}$, a result which is consistent with the expectations from hydrodynamics. For $p_T > 10$ GeV/c, $v_2\{SP\} \approx v_2\{m\}$, providing strong evidence that collective anisotropic particle emission in heavy ion collisions extends to very high p_T. Although the nature of the collectivity at high p_T may be related to the jet quenching phenomena, these observations of anisotropies at both low and high p_T are likely related to the initial state anisotropies and event-by-event fluctuations, posing a challenge to be jointly described by theoretical models.

4 Signs of collectivity also seen in pp and pPb collisions

The long-range two-particle correlation phenomena in high-multiplicity in small systems had its investigation further extended by measuring of two- and multi-particle azimuthal correlations with unidentified charged particles, as well as correlations of reconstructed V^0 particles (i.e., K^0_S and $\Lambda/\bar{\Lambda}$) in pp collision at $\sqrt{s_{NN}} = 5, 7$ and 13 TeV energies [4]. The technique used for investigating two-hadrons correlations defines, for each track multiplicity class, "trigger" particles as charged particles or V^0 candidates with $|\eta| < 2.4$, originating from the primary vertex within a given p_T^{trig} range. Particle pairs are then formed by associating each trigger particle with the remaining charged primary tracks with $|\eta| < 2.4$ and from a specified p_T^{assoc} interval (which can be either the same as or different from the p_T^{trig}). A pair is removed if the associated particle is the daughter of any trigger V^0 candidate (this contribution is negligible since associated particles are mostly primary tracks). The background to such signal combination is constructed by pairing the trigger particles in each event with the associated charged particles from 20 different randomly selected events in the same 0.5 cm wide longitudinal vertex position (z_{vtx}) range and from the same track multiplicity class. The same-event and mixed-event pair distributions are first calculated for each event, and then averaged over all the events within the track multiplicity class. Details of this analysis, containing the data samples, event selections, as well as a complete description of the technique used for studying dihadron correlations, and the estimate of corresponding systematic uncertainties can be found in Ref. [4] and in references therein.

The v_2 harmonics extracted from two-particle correlations are studied as functions of particle p_T and event multiplicity and are shown in Fig. 2. The left plot shows the measurements in the low

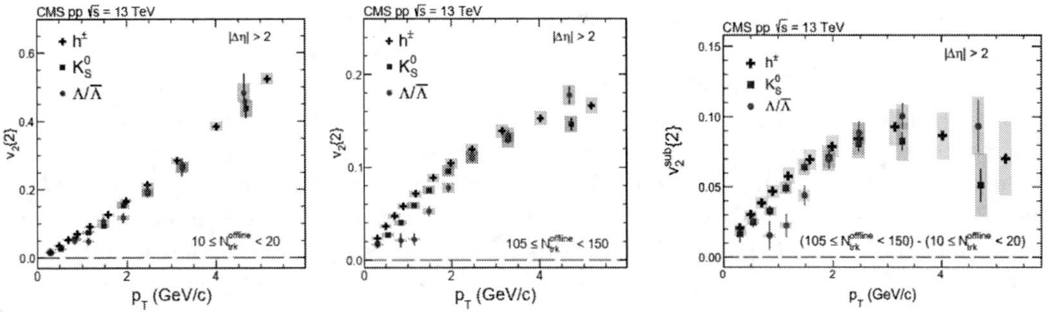

Figure 2. Results of $v_2\{2\}$ for inclusive charged particles, K_S^0 and $\Lambda/\bar{\Lambda}$ particles, are shown as a function of p_T in pp collisions at $\sqrt{s} = 13$ TeV, for the low multiplicity range $10 \leq N_{\mathrm{trk}}^{\mathrm{offline}} < 20$ (left), and for the high multiplicity $105 \leq N_{\mathrm{trk}}^{\mathrm{offline}} < 150$ (middle). The right plot shows $v_2^{\mathrm{sub}}\{2\}$ results of inclusive charged particles, K_S^0 and $\Lambda/\bar{\Lambda}$ after correcting for back-to-back jet correlations estimated from low-multiplicity data. Statistical and systematic uncertainties are represented, respectively, by bars and shaded boxes. The plots are from Ref. [4].

multiplicity range ($10 \leq N_{\mathrm{trk}}^{\mathrm{offline}} < 20$), which is associated mostly to jet correlations. The middle plot shows v_2 anisotropy coefficient in the high multiplicity range ($105 \leq N_{\mathrm{trk}}^{\mathrm{offline}} < 150$). The residual contribution to long-range correlations from back-to-back jets is estimated and removed by subtracting similar correlations obtained from very low multiplicity pp events. The resulting v_2 is also shown in Fig. 2 (right). No clear mass-ordering for low $N_{\mathrm{trk}}^{\mathrm{offline}}$ ranges is seen in Fig. 2. However, in the high multiplicity $N_{\mathrm{trk}}^{\mathrm{offline}}$ range, a clear mass-ordering is observed, which is indicative of the presence of radial flow. For the subtracted elliptic component $v_2^{\mathrm{sub}}\{2\}$, it can also be seen that for K_S^0, $v_2^{\mathrm{sub}}\{2\}$ is higher than for $\Lambda/\bar{\Lambda}$ at low p_T, but the order is reversed in the high-p_T range. It can also be seen that the values of $v_2^{\mathrm{sub}}\{2\}$ increase up to $p_T \approx 2-3$ GeV/c, and then decrease.

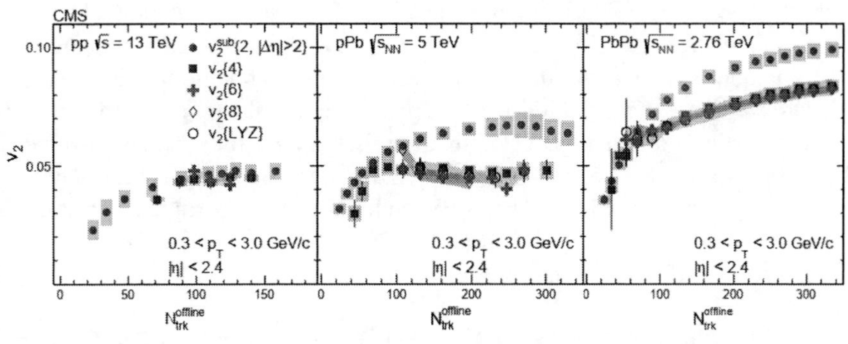

Figure 3. Results from $v_2^{\mathrm{sub}}\{2\}$ and $v_2\{4\}$, $v_2\{6\}$, $v_2\{8\}$ are shown as a function of $N_{\mathrm{trk}}^{\mathrm{offline}}$ for pp collisions in the high multiplicity realm, as well as for pPb and PbPb collisions at the LHC enegies, exhibiting a stricking similarity in the three systems. Shaded boxes represent systematic uncertainties. Statistical and systematic uncertainties are represented, respectively, by bars and shaded boxes. The plots are from Ref. [4].

The v_2 harmonics are also extracted using the multi-particle cumulant method, in order to search for a possible collective nature of the correlations. The pp data are directly compared to those found for pPb and PbPb systems over a broad range of similar track multiplicities [4]. The corresponding elliptic flow harmonics v_2 from this study are shown in Fig. 3, which were obtained with the cumulant method in high multiplicity pp collisions at 13 TeV (left), high multiplicity pPb collisions at 7 TeV (middle) and PbPb collisions at 2.76 TeV (right) energies in the nucleon-nucleon center-of-mass. First, the similarity of the results for these very different systems is indeed remarkable, differing mainly for a smaller ratio $v_2\{2\}/v_2\{4\}$ in the case of pp collisions. This smaller ratio perhaps suggests that there are fewer fluctuating sources, leading to less non-flow effects, in the initial conditions of pp collisions as compared to pPb and PbPb. Finally, the most striking result in the pp case, regarding the higher-order cumulants, with $v_2\{2\} \approx v_2\{4\} \approx v_2\{6\}$, clearly shows evidence of collective behavior in pp collisions, as expected from hydrodynamics, which had been already observed earlier in the pPb and PbPb cases. These observations pose an even stronger challenge to theoretical models for explaining the data.

5 Multiparticle azimuthal correlations with symmetric cumulants

In order to further explore the signs suggesting the observation of collectivity in the three systems and refine those measurements, a joint measurement of the anisotropy harmonics (v_n, n=2-4) and of event-by-event correlations of different v_n is performed in pp, pPb and PbPb collisions. The v_n results are extracted via long-range ($\Delta\eta > 2$) two-particle correlations as a function of event multiplicity. Event-by-event correlations of v_2 versus v_3 and v_2 versus v_4 are measured using the four-particle Symmetric Cumulant [5] method in all the three systems. Details on the analysis, containing the data samples, trigger and event selections, as well as a description of the techniques employed, and a complete estimate of corresponding systematic uncertainties can be found in can be found in [6].

For extracting the v_n ($n = 2 - 4$) coefficients in the long-range ($\Delta\eta > 2$) two-particle correlations measurements, as already performed in previous CMS papers, first the particle pair distribution is expressed as

$$\frac{dN_{\rm pair}}{d\Delta\phi} \propto 1 + \sum_n 2V_{n\Delta}\cos(n\Delta\phi), \tag{1}$$

where $V_{n\Delta}$ are the two-particle Fourier coefficients. The single-particle azimuthal anisotropy Fourier coefficients v_n can then be extracted as $v_n = \sqrt{V_{n\Delta}}$, assuming that factorization applies [12].

The SC technique was first introduced by the ALICE collaboration[5] in LHC measurements and is based on a 4-particle correlation calculations with cumulants. The difference between the standard cumulant calculation and the SC is that the first is used to compute diagonal v_n terms and the latter is used for measuring correlations between different coefficient orders. Therefore, to study the correlation between a harmonic n and m, one can build 2-and 4-particle correlator with: $\langle\langle 2\rangle\rangle_n = \langle\langle e^{i(n\phi_1 - m\phi_2)}\rangle\rangle \sim \langle v_n^2\rangle$ and $\langle\langle 4\rangle\rangle_{n,m} = \langle\langle e^{i(n\phi_1 + m\phi_2 - n\phi_3 - m\phi_4)}\rangle\rangle \sim \langle v_n^2 v_m^2\rangle$. The measured observable of the Scalar Product method, SC(n, m), is then defined as

$$\mathrm{SC}(n,m) = \langle\langle 4\rangle\rangle_{n,m} - \langle\langle 2\rangle\rangle_n \cdot \langle\langle 2\rangle\rangle_m = \langle v_n^2 v_m^2\rangle - \langle v_n^2\rangle\langle v_m^2\rangle. \tag{2}$$

Details of the procedure corresponding to the two methods ca be found in [6].

The measurements of v_2, v_3, and v_4 coefficients for $0.3 < p_T < 3$ GeV/c from long-range two-particle correlations are shown in Fig. 4, as a function of multiplicity in 13 TeV pp, 5.02 and 8.16 TeV pPb, and 5.02 TeV PbPb collisions. The contribution to v_n coefficients from back-to-back jet correlations are corrected by subtracting correlations from very low-multiplicity events ($v_n^{\rm sub}$), as done in Refs. [4, 13]. The v_n results before subtraction are also shown as lines in this figure. The effect of the low-multiplicity subtraction for $N_{\rm trk}^{\rm offline} > 200$ is very small in pPb and PbPb collisions. At

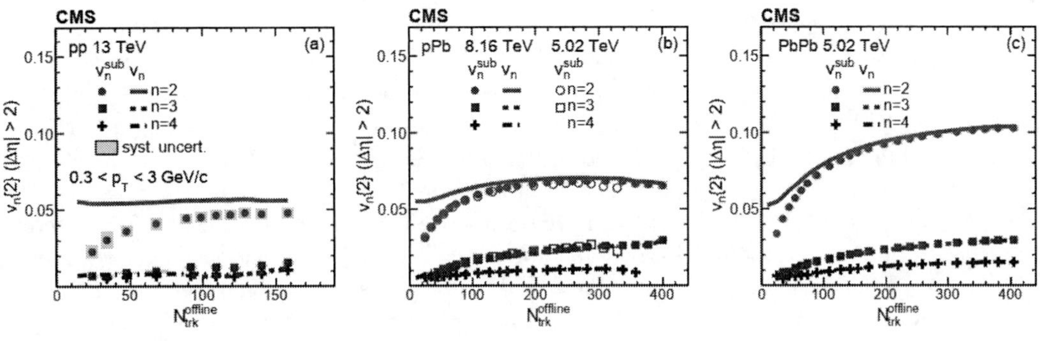

Figure 4. The v_2, v_3 [4], and v_4 coefficients from long-range two-particle correlations are shown as a function of $N_{\rm trk}^{\rm offline}$ in 13 TeV pp (a), 5.02 TeV citeCMS-2-4-cumm and 8.16 YeV pPb (b), and 5.02 TeV PbPb collisions (c). The results corrected by low-multiplicity subtraction are denoted as $v_n^{\rm sub}$. The lines show the v_n results before subtraction of low-multiplicity correlations. The gray boxes represent systematic uncertainties. The plots are from Ref. [6].

low multiplicities, however, this correction is significant, in particular for pp collisions where dijet correlations are expected to be the main source of correlations. Due to the higher collision energy and integrated luminosity, the new 8.16 TeV pPb results allowed for extending the measurements of the v_n coefficients to higher multiplicities as compared to the 5.02 TeV pPb data. Still from Fig. 4, it can be seen that the v_2 coefficient increases with $N_{\rm trk}^{\rm offline}$, saturating for values of this variable above 200. It can also be seen that v_4 has finite values in all three systems, being about 50% smaller than the v_3 coefficients for $N_{\rm trk}^{\rm offline} > 100$, mainly visible in the pPb and PbPb results.

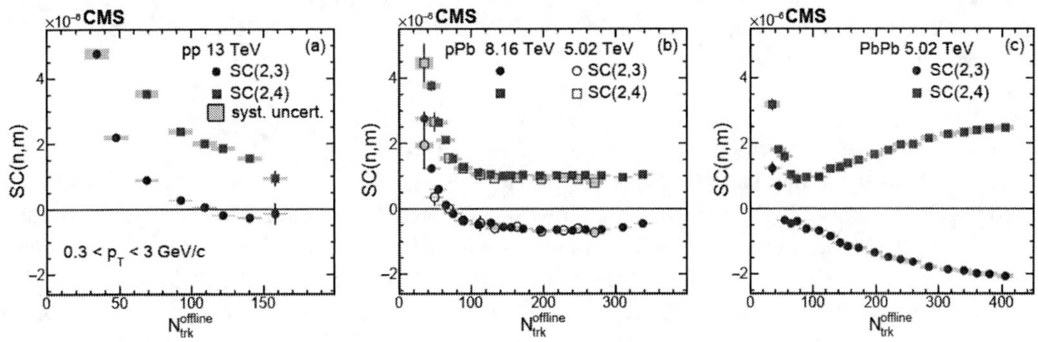

Figure 5. The SCs for the second and third coefficient (red points) and the second and fourth coefficient (blue points) as a function of $N_{\rm trk}^{\rm offline}$ in 13 TeV pp (a), 5.02 TeV and 8.16 TeV pPb (b), and 5.02 TeV PbPb collisions (c). The gray boxes represent systematic uncertainties. The plots are from Ref. [6].

Results from symmetric cumulants allow to further study the correlations of different v_n coefficients. The measurements of SC(2,3) and SC(2,4), for $0.3 < p_T < 3$ GeV/c from four-particle correlations, are shown in Fig. 5, as a function of multiplicity in 13 TeV pp, 5.02 and 8.16 TeV pPb, and 5.02 TeV PbPb. From this figure, it can be seen that both SC(2,3) and SC(2,4) decrease as $N_{\rm trk}^{\rm offline}$

increases in pp collisions. The SC(2, 4) values always remain positive, while an indication of a transition to negative values for SC(2, 3) can be envisaged for $N_{\rm trk}^{\rm offline} > 110$, although the measurement is not precise enough to allow for a firm conclusion. For pPb and PbPb data at sufficiently high multiplicities (e.g., $N_{\rm trk}^{\rm offline} > 60$), clear negative values of SC(2, 3) are observed, while SC(2, 4) values are positive in all the multiplicity range. Besides, it can be seen that the PbPb data are consistent with the results reported at $\sqrt{s_{_{NN}}}$ = 2.76 TeV [5]. In lower multiplicity ranges ($N_{\rm trk}^{\rm offline} < 100$) for all three systems, both SC(2, 3) and SC(2, 4) have positive values, which increase as $N_{\rm trk}^{\rm offline}$ decreases. Nevertheless, in the low multiplicity region, short-range few-body correlations, such as in jets, are likely to have a dominant contribution, which needs to be properly taken into account. Besides, the jet contribution at low $N_{\rm trk}^{\rm offline}$ might be different in pp, pPb and PbPb, and perhaps could account for the slightly different behaviors of the SCs observed in the data in this multiplicity range.

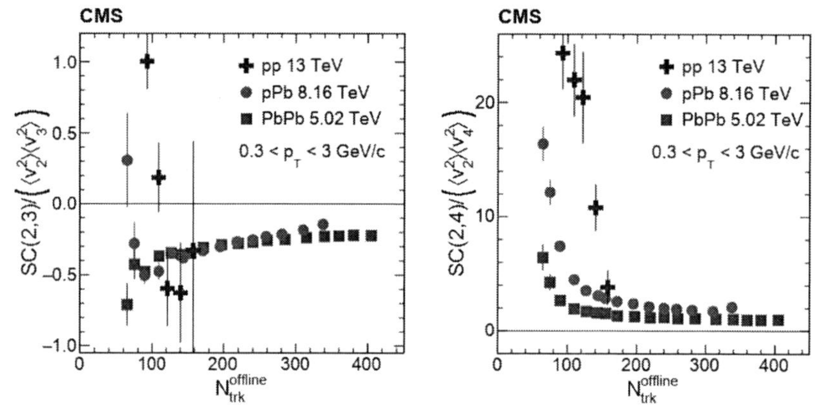

Figure 6. The SCs for the second and third coefficients (left) and the second and fourth coefficients (right) normalized by $\langle(v_2^{\rm sub})^2\rangle\langle(v_3^{\rm sub})^2\rangle$ and $\langle(v_2^{\rm sub})^2\rangle\langle(v_4^{\rm sub})^2\rangle$ from two-particle correlations. The track p_T range is $0.3 < p_T < 3$ GeV/c and the results are shown as a function of $N_{\rm trk}^{\rm offline}$ in 13TeV pp, 8.16TeV pPb, and 5.02 TeV PbPb collisions. The plots are from Ref. [6].

In Fig. 5, the absolute magnitudes of SC(2, 3) and SC(2, 4) are seen to be larger in PbPb as compared to pPb systems at high multiplicities. However, this may be related to the different magnitude of v_n coefficients as seen in Fig. 4. Therefore, to investigate intrinsic correlations between v_n coefficients and compare with different collision systems in a more quantitative way, SC(2, 3) and SC(2, 4) are normalized by $\langle(v_2^{\rm sub})^2\rangle\langle(v_3^{\rm sub})^2\rangle$ and $\langle(v_2^{\rm sub})^2\rangle\langle(v_4^{\rm sub})^2\rangle$, respectively, based on the v_n values from two-particle correlations in Fig. 4. One must note that the two-particle correlation v_n with a rapidity gap is used for the normalization, therefore, the results could be affected by the event-plane decorrelation as measured in [15] within a few percent. Nevertheless, all systems would be affected consistently, so that the conclusions from the results would remain unchanged. The resulting normalized SCs in all three colliding systems are shown in Fig. 6. The normalized SC(2, 3) values are found to be very similar between pPb and PbPb systems at high multiplicities. Taken together with the v_n results in Fig. 4, these measurements strongly suggest a unified origin to explain the collective behavior observed in large and small hadronic collisions. It is interesting to note that this common behavior may even apply to pp collisions for $N_{\rm trk}^{\rm offline} > 120$, where SC(2, 3) tends to converge to a unified value for all three systems, even though the statistical uncertainties are still too large to draw a firm conclusion. The SC(2, 4), on the other hand, shows a clear dependence on the system size with a larger value for smaller systems. Complementary, the observed difference between SC(2, 4) values in pPb and PbPb collisions, however, may point to a different contribution of initial-state fluctuations or transport properties of the medium such as the shear viscosity to entropy ratio [14].

6 Summary and conclusions

Several measurements with data collected with the CMS experiment and related to azimuthal anisotropy correlations were reported here. The first, with charged particles produced in PbPb collisions at $\sqrt{s_{_{NN}}}$ = 5.02 TeV, in which v_2 and v_3 coefficients were determined for different collision centralities, extending over the widest transverse momentum range studied to date ($1 < p_T < 100$ GeV/c). Positive v_2 values were found up to $p_T \approx 70$ GeV/c, while for v_3, the values are consistent with zero for $p_T > 20$ GeV/c. For $p_T < 3$ GeV/c, it is observed that $v_2\{SP\} > v_2\{4\} \approx v_2\{6\} \approx v_2\{8\}$, which is consistent with a collective behavior arising from the hydrodynamic expansion of a quark-gluon plasma and such similarity of $v_2\{4\}$, $v_2\{6\}$, and $v_2\{8\}$ at high p_T suggests that v_2 originates from the path-length dependence of parton energy loss associated with an asymmetric initial collision geometry.

Data collected in pp collisions at $\sqrt{s}$ = 13 TeV have been used to measure two- and multi-particle azimuthal correlations with unidentified charged particles ($h^{\pm}$), as well as reconstructed K^0_s and Λ particles, exploring the correlation data over a broad particle multiplicity range. The v_2 and v_3 Fourier coefficients are extracted from long-range two-particle correlations, from which contributions from back-to-back jet correlations, estimated using low-multiplicity data, are subtracted. In the higher-multiplicity region, a particle species dependence of v_2 is observed with and without correcting for jet correlations, a similar behavior to what had been observed for identified particles produced in pPb and AA collisions at RHIC and the LHC. Even more striking is the observation that the v_2 values obtained with 2-, 4-, and 6-particle correlations at $\sqrt{s}$ = 13 TeV, are found to be comparable within uncertainties, providing strong evidence in support of a collective origin for the observed long-range correlations in high-multiplicity pp collisions.

The v_2, v_3, and v_4 Fourier coefficients are extracted from long-range two-particle correlations for 8.16 TeV pPb collisions and compared to those in 13 TeV pp, 5.02 TeV pPb and 5.02 TeV PbPb results. Using a four-particle cumulant technique, correlations between different coefficients are obtained. A negative (positive) correlation is observed between v_2 and v_3 (v_4) in pPb collisions, similarly to what is observed in PbPb systems. Normalized correlation coefficients for v_2 and v_3 are found to be quantitatively similar between pPb and PbPb, whereas the results for v_2 and v_4 are larger in pPb than in PbPb. The corresponding result in pp collisions shows a similar trend at high multiplicity as in pPb and PbPb, however, the statistical uncertainties do not allow for a firm conclusion.

The set of measurements discussed above shows results which provide evidence of collectivity in pp, pPb and PbPb systems alike, suggesting that the observed behavior in data may have a similar origin. This represent a considerable challenge to theoretical models, imposing significant constraints in their tentative to explain the experimental findings.

References

[1] M. Gyulassy and L. McLerran, *New forms of QCD matter discovered at RHIC*, Nucl. Phys. **A750** (2005) 30.

[2] CMS Collaboration, *Observation of Long-Range Near-Side Angular Correlations in Proton-Proton Collisions at the LHC*, JHEP **09** (2010) 091, doi:10.1007/JHEP09(2010)091, arXiv:1009.4122.

[3] CMS Collaboration, *High-p_T track v_n harmonics in PbPb collisions at 5.02 TeV*, submitted to Phys. Lett. **B**, arXiv:1702.00630 [hep-ex].

[4] CMS Collaboration, *Evidence for collectivity in pp collisions at the LHC*, Phys. Lett. B 765 (2017) 193, arXiv:1606.06198.

[5] ALICE Collaboration, *Correlated event-by-event fluctuations of flow harmonics in PbPb collisions at $\sqrt{s_{NN}}$ = 2.76 TeV*, Phys. Rev. Lett. **117** (2016) 182301, arXiv:1604.07663.

[6] CMS Collaboration, *Observation of correlated azimuthal anisotropy Fourier harmonics in pp and pPb collisions at the LHC*, submitted to Phys. Rev. Lett., arXiv:1709.09189 [hep-ex].

[7] CMS Collaboration, *The CMS experiment at the CERN LHC*, JINST **0803** (2008) S08004.

[8] J. Noronha-Hostler, B. Betz, J. Noronha, and M. Gyulassy, *Event-by-Event Hydrodynamics + Jet Energy Loss: A Solution to the RAA $\otimes$ v_2 Puzzle*, Phys. Rev. Lett. **116** (2016) 252301, doi:10.1103/PhysRevLett.116.252301, arXiv:1602.03788.

[9] J. Xu, J. Liao, and M. Gyulassy, *Bridging soft-hard transport properties of Quark-Gluon Plasmas with CUJET3.0*, JHEP **02** (2016) 169, doi:10.1007/JHEP02(2016)169, arXiv:1508.00552.

[10] M. Luzum and J.-Y. Ollitrault, *Eliminating experimental bias in anisotropic-flow measurements of high-energy nuclear collision*, Phys. Rev. **C87** (2013), n. 4, 044907, doi:10.1103/PhysRevC.87.044907, arXiv:1209.2323.

[11] A. Bilandzic et al., *Generic framework for anisotropic flow analyses with multiparticle azimuthal correlations*, Phys. Rev. **C89** (2014), no. 6, 064904, doi:10.1103/PhysRevC.89.064904, arXiv:1312.3572.

[12] S. Voloshin and Y. Zhang, *Flow study in relativistic nuclear collisions by Fourier expansion of azimuthal particle distributions*, Z. Phys. **C 70** (1996) 665, doi:10.1007/s002880050141, arXiv:hep-ph/9407282.

[13] CMS Collaboration, *Multiplicity and transverse momentum dependence of two- and four-particle correlations in pPb and PbPb collisions*, Phys. Lett. **B 724** (2013) 213, doi:10.1016/j.physletb.2013.06.028, arXiv:1305.0609.

[14] K. Welsh, J. Singer, and U. W. Heinz, *Initial state fluctuations in collisions between light and heavy ions*, Phys. Rev. **C 94** (2016) 024919, doi:10.1103/PhysRevC.94.024919, arXiv:1605.09418.

[15] CMS Collaboration, *Evidence for transverse momentum and pseudorapidity dependent event plane fluctuations in PbPb and pPb collisions*, Phys. Rev. **C 92** (2015) 034911, doi:10.1103/PhysRevC.92.034911, arXiv:1503.01692.

EPJ Web of Conferences **172**, 05006 (2018) https://doi.org/10.1051/epjconf/201817205006
ISMD 2017

Vector boson tagged jets and jet substructure

Ivan Vitev[1,⋆]

[1] *Los Alamos National Laboratory, Theoretical Division, Mail Stop B283
Los Alamos, NM 87545, USA*

Abstract. In these proceedings, we report on recent results related to vector boson-tagged jet production in heavy ion collisions and the related modification of jet substructure, such as jet shapes and jet momentum sharing distributions. Z^0-tagging and γ-tagging of jets provides new opportunities to study parton shower formation and propagation in the quark-gluon plasma and has been argued to provide tight constrains on the energy loss of reconstructed jets. We present theoretical predictions for isolated photon-tagged and electroweak boson-tagged jet production in Pb+Pb collisions at $\sqrt{s_{NN}}$ = 5.02 TeV at the LHC, addressing the modification of their transverse momentum and transverse momentum imbalance distributions. Comparison to recent ATLAS and CMS experimental measurements is performed that can shed light on the medium-induced radiative corrections and energy dissipation due to collisional processes of predominantly quark-initiated jets. The modification of parton splitting functions in the QGP further implies that the substructure of jets in heavy ion collisions may differ significantly from the corresponding substructure in proton-proton collisions. Two such observables and the implication of tagging on their evaluation is also discussed.

1 Introduction

In ultrarelativistic collisions of heavy nuclei a new deconfined state of matter, the quark-gluon plasma (QGP), is expected to be formed. The attenuation of the production rate of high transverse momentum particles in nucleus-nucleus (A+A) relative to the binary collision-scaled proton-proton (p+p) reactions has been proposed a long time ago as a signature of its formation [1]. The quenching of hadrons opposite a direct photon was studied [2] shortly after – a precursor of the modern tagged reconstructed jet observables [3, 4]. The production of a vector boson in association with jets is a powerful channel to probe the fundamental properties of Quantum Chromodynamics (QCD). Vector boson-tagged jets are also very well suited to studying the effects of the QGP. The tagging bosons escape the region of the hot dense medium unaffected and can be used to constrain the energy of the away-side parton shower. Previous studies of vector boson tagged jet production in heavy ion collisions have been performed in the framework of perturbative QCD [5, 6]. A Boltzmann transport model [7], an event generator JEWEL [8] and a hybrid strong/weak coupling model [9] have also been used. On the experimental side, isolated γ-tagged and Z^0-tagged jets results in Pb+Pb collisions at $\sqrt{s_{NN}}$ = 5.02 TeV have become available at the LHC from the ATLAS and CMS collaborations [10–12]. These new measurements call for improved theoretical calculations [13] to interpret the experimental data.

⋆e-mail: ivitev@lanl.gov

© The Authors, published by EDP Sciences. This is an open access article distributed under the terms of the Creative Commons Attribution License 4.0 (http://creativecommons.org/licenses/by/4.0/).

Vector-boson tagged jet cross section modification cannot be understood in isolation. The emergence of in-medium parton branching, qualitatively different from the one which determines the jet properties in $e^+ + e^-$, $e^- + p$, and $p + p$ collisions, also implies non-trivial modification of jet substructure observables. These include jet shapes, jet fragmentation functions and jet momentum sharing distributions. In these proceedings we also report on calculation of two of the above [14, 15] in a theoretical framework that uses the same fundamental input – the in-medium splitting functions.

2 Vector boson tagged jets

Let us now turn to the new theoretical results on V+jet modification at the LHC.

2.1 Proton collisions

Before we proceed to heavy ion collisions, it is important to validate the perturbative calculations in the simpler proton-proton reactions. We use Pythia 8 [16], which is a popular high energy phenomenology event generator. It utilizes leading-order perturbative QCD matrix elements+parton shower, combined with the Lund string model for hadronization an can can describe well the main properties of a p+p event structure. This is shown in the left panel of figure 1 in comparison to Z^0+jet measurements made by the CMS collaboration at $\sqrt{s}$ = 7 TeV [17]. We also find good agreement with γ-tagged jet measurements, see [18].

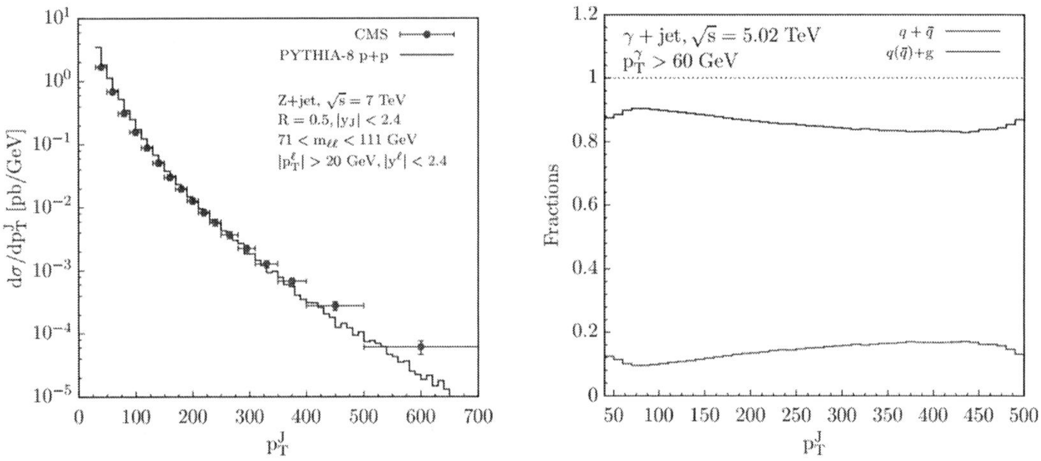

Figure 1. Left panel: Pythia 8 simulations are compared to CMS measurements of Z^0+jet production in p+p collisions at the LHC at $\sqrt{s}$ = 7 TeV. Results are shown differentially versus the jet p_T^J. Right panel: the fractional contributions of different subprocesses to isolated-γ+jet production cross sections in p+p collisions at $\sqrt{s}$ = 5.02.

At leading order two dominant channel for V+jet are implemented in Pythia, namely $q+\bar{q} \to V+g$ and $q(\bar{q}) + g \to V + q(\bar{q})$. We have checked that the $g + g \to V + g$ channel contribution to the cross section is negligible. The utility of vector boson tagging beyond constraining the recoil jet energy is that it can be used for flavor selection. We have shown this in the right panel of figure 1 on the example of photon-tagged jets at $\sqrt{s}$ = 5.1 TeV. Results indicate that γ tagging or Z^0 tagging preferentially selects quark jets and at high p_T the sample has ~ 80% purity. The accumulated statistics allows us to use the fully differential $d\sigma/dp_T^V dp_T^J$ to high transverse momenta of the jet and the vector boson.

2.2 Heavy ion collisions collisions

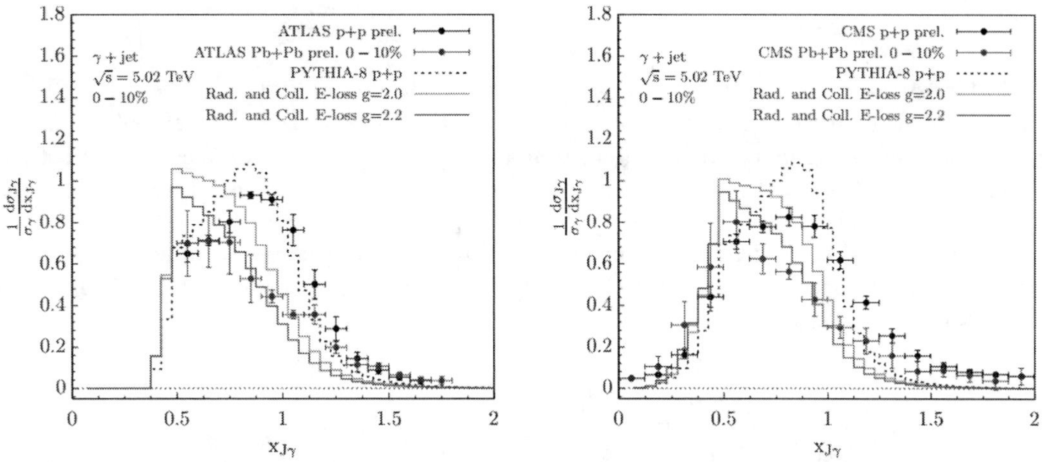

Figure 2. Left panel: The isolated photon-tagged jet asymmetry distributions are shown and compared to ATLAS data in central Pb+Pb collisions at the LHC [12]. The transverse momenta for the isolated photon and the jet are $p_T^\gamma > 60$ GeV and $p_T^J > 30$ GeV, respectively. The jet radius parameter is $R = 0.4$. Right panel: same comparison but for CMS data [11] with jet radius $R = 0.3$.

We start from the baseline p+p cross sections obtained from Pythia. However, in heavy ion collisions we take into account the reaction geometry, the hydrodynamic expansion of the QGP and radiative and collisional energy losses. At an impact parameter $|\mathbf{b}_\perp|$ we evaluate the cross sections as

$$\frac{d\sigma^{AA}(|\mathbf{b}_\perp|)}{dp_T^V dp_T^J} = \int d^2\mathbf{s}_\perp T_A\left(\mathbf{s}_\perp - \frac{\mathbf{b}_\perp}{2}\right) T_A\left(\mathbf{s}_\perp + \frac{\mathbf{b}_\perp}{2}\right) \sum_{q,g} \int_0^1 d\epsilon \frac{P_{q,g}(\epsilon; s_\perp, |\mathbf{b}_\perp|)}{1 - f_{q,g}^{\text{loss}}(R; s_\perp, |\mathbf{b}_\perp|)\,\epsilon}$$
$$\times \frac{d\sigma_{q,g}^{NN}\left(p_T^V, p_T^J/\{1 - f_{q,g}^{\text{loss}}(R; s_\perp, |\mathbf{b}_\perp|)\,\epsilon\}\right)}{dp_T^J dp_T^V}. \quad (1)$$

Here hard production follows the binary collision density $\propto T_A(\mathbf{s}_\perp - \mathbf{b}_\perp/2)\,T_A(\mathbf{s}_\perp + \mathbf{b}_\perp/2)$ and we use an optical Glauber model with inelastic nucleon-nucleon scattering cross sections $\sigma_{\text{in}} = 70$ mb at $\sqrt{s_{NN}} = 5.02$ TeV. In equation 1 $P_{q,g}(\epsilon)$ is the probability that a fraction ϵ of the parent parton is redistributed through medium-induced multiple gluon emission. We employ the soft gluon emission limit of the of the SCET$_\text{G}$ [19, 20] splitting functions [21, 22] used to predict inclusive light hadron attenuation in heavy ion collisions [23, 24]. We note that the spectrum coincides with the original energy loss derivation of [25]. Of course, only the fraction of the energy that is redistributed outside of the jet cone leads to jet cross section suppression. We evaluate this fraction as follows

$$f_{q,g}^{\text{loss}}(R; \text{rad} + \text{coll}) = 1 - \left(\int_0^R dr \int_{\omega_{\min}}^E d\omega \frac{dN_{q,g}^g(\omega, r)}{d\omega dr}\right) \Bigg/ \left(\int_0^{R_{\max}} dr \int_0^E d\omega \frac{dN_{q,g}^g(\omega, r)}{d\omega dr}\right). \quad (2)$$

Equation 2 treats radiative and collisional energy losses on the same footing since for convenience we have express this total collisional energy loss as an integral over the spectrum of the medium induced gluons

$$\Delta E^{\rm coll}_{q,g}({\rm tot.}) = \sum_{i=1}^{N^{\rm tot.\ partons}_{q,g}} \int_{z_i}^{\infty} \frac{d\Delta E^{\rm coll}_i}{d\Delta z} d\Delta z\,, \quad \Delta E^{\rm coll}_{q,g}({\rm tot.}) = \int_0^{\omega_{\rm min}} d\omega \int_0^{R_{\rm max}} dr\, \omega \frac{dN^g_{q,g}(\omega, r)}{d\omega dr}. \qquad (3)$$

To lowest leading order in perturbative QCD, the transverse momentum of the vector boson is balanced by the transverse momentum of the jet, $p_T^V = p_T^J$. Higher order processes and the development of parton showers introduce deviations from this equality. The exact differential distribution of $d\sigma/dp_T^V dp_T^J$ is also affected by the jet reconstruction algorithm, jet radius choice, experimental cuts, and detector resolution. Nevertheless, the quenching and shift of this distribution to smaller values of p_T^J are currently the best proxies for jet energy loss. To study these effects one introduces the transverse momentum imbalance $\mathrm{x_{JV}} = \mathrm{p_T^J/p_T^V}$. distribution can be obtained from the double differential distribution of V+jet cross section

$$\frac{d\sigma}{d\mathrm{x_{JV}}} = \int_{p_T^{J,\rm min}}^{p_T^{J,\rm max}} dp_T^J \frac{p_T^J}{\mathrm{x_{JV}^2}} \frac{d\sigma(p_T^V = p_T^J/\mathrm{x_{JV}}, p_T^J)}{dp_T^V dp_T^J}\,, \qquad (4)$$

where $p_T^{J,\rm min}$ and $p_T^{J,\rm max}$ are matched to the desired cuts of the experimental measurements.

In figure 2 we plot the normalized momentum imbalance distributions for the γ+jet final state in p+p and Pb+Pb collisions at the LHC. Comparison of the calculations to the ATLAS measurements [12] are shown in the left panel. Comparison of the calculations to the CMS measurements [11] are shown in the right panel. The black dashed histogram shows Pythia 8 simulations for the the $\mathrm{x_{JZ}}$ distribution in p+p collisions. The black solid points represent the ATLAS and CMS results in the elementary nucleon-nucleon collisions. It can be seen that the $\mathrm{x_{JZ}}$ distribution from Pythia 8 simulation is narrower than the one measured by the experiments. Part of the reason is that experimental measurements are not unfolded for detector resolution effects. This can be seen on the example of Z^0+jet momentum imbalance by CMS in Ref. for the p+p reference [10]. We applied the same smearing functions that experiment applies to Monte Carlo simulations, to our differential p_T distributions for p+p and Pb+Pb collisions. We got broader $\mathrm{x_{JZ}}$ distributions and better agreement between the curves to the data points in both p+p and Pb+Pb collisions. In figure 2 we also give results of our theoretical calculations in Pb+Pb collisions. The green and magenta histograms correspond to jet-medium coupling strengths $g = 2.0$ and $g = 2.2$, respectively, values that have worked well in describing the light inclusive hadron [23, 24] and jet suppression data [18] at the LHC. Our results include both medium-induced radiative energy loss and the parton showers energy dissipation of through collisional processes in the QGP. We see the downshift of $\mathrm{x_{JV}}$, which quantitatively agrees with measurements in terms of the difference between p+p and Pb+Pb collisions.

Another often studied observable sensitive to nuclear modification effects in V+jet systems is $\mathrm{I_{AA}}$. It is defined as ratio of the differential cross section for tagged jets in A+A collisions to the binary collision scaled p+p result. We present $\mathrm{I_{AA}}$ in isolated γ+jet production at the LHC in 0 – 30% Pb+Pb collisions in figure 3. By comparing to CMS experimental data, we find that our theoretical results agree with data for a wide kinematic range. In each p_T^γ window, the jet energy loss effects are shown in four cases and coded with different colors - two different jet-medium coupling strengths and simulations where collisional energy loss effects are either included or excluded. Figure 3 illustrates that there is kinematic sensitivity of $\mathrm{I_{AA}}$ and the largest suppression is observed along the diagonal region of the transverse momenta of the trigger γ and the recoil jet. The reason for this behavior in $p_T^\gamma \approx p_T^J$ arises from the steeper falling cross section in that region. The cross section is suppressed in the region $p_T^J > p_T^\gamma$ and is. enhanced for $p_T^J < p_T^\gamma$. This is characteristic of in-medium tagged-jets as discussed in previous works [5, 6].

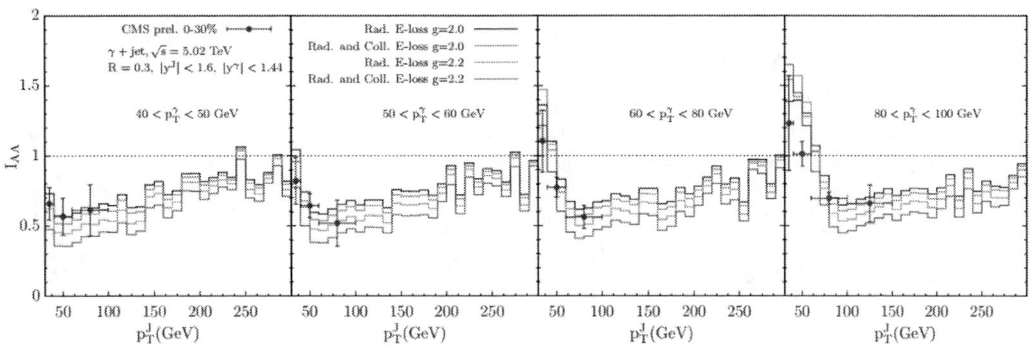

Figure 3. Theoretical results for I_{AA} compared to CMS data with several different transverse momentum cuts. Shown are four different scenarios for energy loss effects with and without collisional energy loss and with couplings between the jet and the medium $g = 2.0$ and $g = 2.2$

Table 1. Theoretical predictions for the difference of the average x_{JZ} between p+p and Pb+Pb $0-30\%$ central collisions. The center of mass energy is chosen to be $\sqrt{s} = 5.02$ TeV, the transverse momentum cut for the recoil jet is $p_T^J > 30$ GeV, and data is from CMS [10].

	$\Delta\langle x_{JZ}\rangle$			
p_T^Z (GeV)	$40-50$	$50-60$	$60-80$	$80-120$
CMS [10]	0.061±0.059	0.123±0.051	0.124±0.052	0.068±0.042
Rad. + Coll. $g = 2.0$	0.022	0.050	0.075	0.086
Rad. + Coll. $g = 2.2$	0.024	0.058	0.093	0.119

To better quantify the shift toward smaller values of the x_{JV} distribution, we introduce the mean value of x_{JV} and the difference between proton and heavy ion collisions

$$\langle x_{JV}\rangle = \left(\int dx_{JV} x_{JV} \frac{d\sigma}{dx_{JV}}\right) \Big/ \left(\int dx_{JV} \frac{d\sigma}{dx_{JV}}\right), \quad \Delta\langle x_{JV}\rangle = \langle x_{JV}\rangle_{pp} - \langle x_{JV}\rangle_{PbPb}. \tag{5}$$

The calculated downshifts of the x_{JV} distribution is consistent with the experimental data [10] within the measurement uncertainties for different p_T^Z cuts. More importantly, when the downshift is recast into energy loss of the jet recoiling against the Z^0 boson, we find that this experimentally determined jet energy loss corresponds within 15% accuracy to the theoretically computed energy redistributed out of the jet cone due to in-medium interactions of the parton shower

3 Jet substructure

We now turn to the question of jet substructure in heavy ion collisions.

3.1 Momentum sharing distributions

A new jet substructure observable, called the groomed momentum sharing, was recently proposed and studied using the soft drop jet grooming procedure [26, 27]. It is sensitive to the hard branching in the jet formation and is controlled by the leading-order Altarelli-Parisi splitting functions. Having reconstructed a jet of radius R using the anti-k_T algorithm, one can recluster the jet using the

Cambridge/Aachen algorithm and goe through the branching history, dropping the soft branch until $z_{cut} < \min(p_{T_1}, p_{T_2})/(p_{T_1} + p_{T_2}) \equiv z_g$. One can also demand that the angular separation between the two branches is greater than an angular scale Δ, $\Delta < \Delta R_{12} \equiv r_g$. By selecting the groomed jet radius r_g, one gains access to the momentum sharing distribution $p(z_g)$ at different splitting angles, and can study the angular distribution ρz_g. We evaluate

$$p_i(z_g) = \frac{\int_{k_\Delta}^{k_R} dk_\perp \overline{\mathcal{P}}_i(z_g, k_\perp)}{\int_{z_{cut}}^{1/2} dx \int_{k_\Delta}^{k_R} dk_\perp \overline{\mathcal{P}}_i(x, k_\perp)} \quad \text{where } k_\Delta = \omega x(1-x)\tan\frac{\Delta}{2},\ k_R = \omega x(1-x)\tan\frac{R}{2}. \tag{6}$$

The symmetrized splitting functions are $\overline{\mathcal{P}}_i(x, k_\perp) = \sum_{j,l}\left[\mathcal{P}_{i\to j,l}(x, k_\perp) + \mathcal{P}_{i\to j,l}(1-x, k_\perp)\right]$ and in the presence of a QGP, $\mathcal{P}_{i\to jl}(x, k_\perp) = \mathcal{P}^{vac}_{i\to jl}(x, k_\perp) + \mathcal{P}^{med}_{i\to jl}(x, k_\perp)$. The medium-induced component were calculated using soft-collinear effective theory with Glauber gluon interactions (SCET$_G$) [19–22].

Analytic and numerical considerations show that for $x < 1/2$ in Eq. (6) medium-induced component goes as $1/x^2$ [21]. We thus expect that the momentum sharing distribution will be enhanced at small z_g and suppressed near $z_g = 1/2$. Numerical results [15] are shown in the left panel of figure 4 and compared to preliminary CMS data. The parameters for the groomed soft-dropped jets are $\beta = 0$, $z_{cut} = 0.1$ and $\Delta R_{12} > 0.1$. We estimate the theoretical uncertainty by varying the jet-medium coupling $g = 2.0 \pm 0.2$. Analogous calculations in the soft gluon emission limit have also been presented [28, 29]. The finalized CMS data [30] changed significantly and is now in excellent agreement with the theoretical predictions of [15]. This signifies the need for reliable pQCD theory to interpret heavy ion data and motivates further studies, such as the groomed jet radius distribution. The anticipated modification is shown in the right panel of figure 4 for several different p_T bins. The peak of this distribution reflects the medium enhancement of large-angle splitting for hard branching processes.

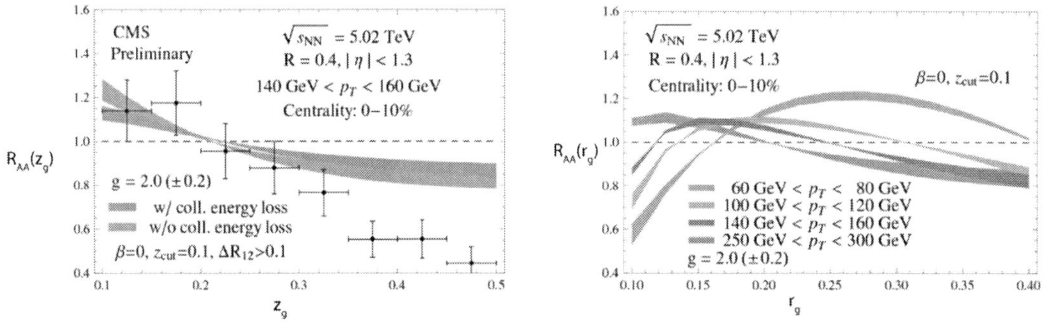

Figure 4. Left panel: theoretical calculations compared to preliminary CMS data for the ratio of momentum sharing distributions of inclusive anti-k_T $R = 0.4$ jets in central Pb+Pb and p+p collisions at $\sqrt{s_{NN}} = 5.02$ TeV. Purple and red bands correspond to calculations with and without the implementation of collisional energy loss. Right panel: the groomed jet radius modification of inclusive jets in proton-proton and central lead-lead collisions at $\sqrt{s_{NN}} = 5.02$ TeV. We show results for four p_T bins with 60 GeV $< p_T <$ 80 GeV , 100 GeV $< p_T <$ 120 GeV, 140 GeV $< p_T <$ 160 GeV and 250 GeV $< p_T <$ 300 GeV.

3.2 Jet shapes

A complementary observable, in fact the first one that has been quantitatively studied in relation to inclusive jet modification in heavy ion collisions [31], is the jet shape. The jet shape [32] describes the transverse energy profile inside a jet of size R, reconstructed using a jet algorithm. The integral jet shape, defined as the fraction of the transverse energy E_T of the jet within a subcone of size r, reads

$$\Psi_J(r) = \frac{\sum_{i,\, d_{i\hat{n}}<r} E_T^i}{\sum_{i,\, d_{i\hat{n}}<R} E_T^i}, \quad \text{where } d_{i\hat{n}} = \sqrt{(\eta_i - \eta_{jet})^2 + (\phi_i - \phi_{jet})^2} \tag{7}$$

In heavy ion collisions, the differential jet shape $\rho(r) = \frac{d}{dr}\Psi(r)$ is more sensitive to in-medium modification, which can be presented as $M_\rho(r) = \rho^{AA}(r)/\rho^{pp}(r)$.

A robust theory of jets in high energy nuclear collisions should describe, within its regions of applicability, the modification of the jet shapes in addition to the attenuation of the inclusive and dijet cross sections [33]. We have taken the opportunity to address this key issue in [14], building upon an improved description of $\rho^{pp}(r)$ [34] in the framework of SCET [35]. Our results are shown in figure 5, where we impose cuts on the jet transverse momentum $p_T > 100$ GeV and pseudorapidity $0.3 < |\eta| < 2.0$ of jets. We take the coupling between the jet and the QGP to be $g = 2$ and the theoretical uncertainty is estimated by varying the jet energy scales $\frac{1}{2}\mu_{j_R} < \mu < 2\mu_{j_R}$ in the calculations. Several contributions to the jet shape modification are shown. The blue band corresponds to the initial state cold nuclear matter effects. The effect of quenching is added in the red band, which enhances the fraction of quark jets that are narrower than the gluon jets. The green band correspond to the full calculation the green band is the full calculation and adds the event-by-event broadening. It is evident that the jet shape modification is highly nontrivial and we find that it compares well to the data from CMS [36] with jet radius $R = 0.3$.

We also present predictions for the modification of jet shapes in Pb+Pb collisions at $\sqrt{s_{\rm NN}} \approx 5.1$ TeV, focusing on the differences between inclusive jets and tagged jets. The results are given in the right panel of figure 5. While $M_\rho(r)$ for jet shapes at $\sqrt{s_{\rm NN}} \approx 5.10$ TeV is similar to the one at $\sqrt{s_{\rm NN}} = 2.76$ TeV, the modification of γ-tagged jet shapes is more pronounced. Photon-tagged jet shapes also exhibit a different modification pattern. Since photon tagging selects primarily quark jets, see previous section, the quenching does not play a significant role and the enhancement toward the periphery of jets is more pronounced. Other examples of jet shape calculations include [37, 38].

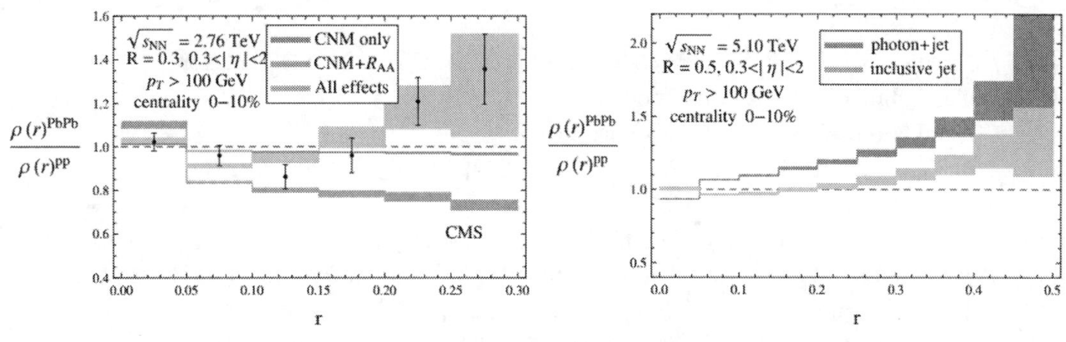

Figure 5. Left panel: the modification of differential jet shapes of inclusive jets in Pb+Pb central collisions at $\sqrt{s_{\rm NN}} = 2.76$ TeV at the LHC Different contributions are shown and compared to CMS data. Right panel: theoretical predictions for the differential jet shape modification of inclusive and photon-tagged jets with $R = 0.3$ in Pb+Pb collisions at $\sqrt{s_{\rm NN}} \approx 5.1$ TeV at the LHC.

4 Conclusions

To summarize, in these proceedings we reported results on vector boson-tagged jet production [13] and jet substructure [14, 15] in heavy ion collisions. We presented a new study of γ-tagged or Z^0-tagged jets in Pb+Pb reactions at the LHC a center-of-mass energy per nucleon pair of 5.02 TeV. Good description of the mean momentum imbalance shift $\Delta\langle x_{JV}\rangle$ and the tagged jet nuclear modification factor I_{AA} can be achieved in a framework that accounts for the QGP-induced radiative and collisional dissipation of the parton shower energy outside of the jet cone. At the heart of such quenching effects for reconstructed jets is the in-medium parton branching – qualitatively and quantitatively different from the standard Altarelli-Parisi splittings in the vacuum. New techniques, such as the study of the subjet momentum sharing distributions, have recently emerged and can probe precisely this parton dynamics. We also presented the first calculation of this subjet momentum sharing distribution in heavy ion collisions. The validation of the theoretical predictions by recent experimental measurements complements our earlier results on the modification of the jet shape and corroborates the perturbative picture of parton shower modification in the medium. Finally, we found that the modification of γ-tagged and Z^0-tagged jet substructure can differ substantially from the corresponding substructure modification of inclusive jets. Future experimental results for such observables can significantly enhance our understanding of in-medium QCD dynamics.

References

[1] X.N. Wang, M. Gyulassy, Phys. Rev. Lett. **68**, 1480 (1992), hep-ph/9605213
[2] X.N. Wang, Z. Huang, I. Sarcevic, Phys. Rev. Lett. **77**, 231 (1996), hep-ph/9605213
[3] D.K. Srivastava, C. Gale, T.C. Awes, Phys. Rev. **C67**, 054904 (2003), nucl-th/0212081
[4] R.B. Neufeld, I. Vitev, B.W. Zhang, Phys. Rev. **C83**, 034902 (2011), 1006.2389
[5] W. Dai, I. Vitev, B.W. Zhang, Phys. Rev. Lett. **110**, 142001 (2013), 1207.5177
[6] R.B. Neufeld, I. Vitev, Phys. Rev. Lett. **108**, 242001 (2012), 1202.5556
[7] X.N. Wang, Y. Zhu, Phys. Rev. Lett. **111**, 062301 (2013), 1302.5874
[8] R. Kunnawalkam Elayavalli, K.C. Zapp, Eur. Phys. J. **C76**, 695 (2016), 1608.03099
[9] J. Casalderrey-Solana, D.C. Gulhan, J.G. Milhano, D. Pablos, K. Rajagopal (2015), 1508.00815
[10] A.M. Sirunyan et al. (CMS), Phys. Rev. Lett. **119**, 082301 (2017), 1702.01060
[11] Collaboration (CMS), Analisys note (2016)
[12] Collaboration (ATLAS), Analysis note (2016)
[13] Z.B. Kang, I. Vitev, H. Xing, Phys. Rev. **C96**, 014912 (2017), 1702.07276
[14] Y.T. Chien, I. Vitev, JHEP **05**, 023 (2016), 1509.07257
[15] Y.T. Chien, I. Vitev, Phys. Rev. Lett. **119**, 112301 (2017), 1608.07283
[16] T. Sjostrand, S. Mrenna, P.Z. Skands, Comput.Phys.Commun. **178**, 852 (2008), 0710.3820
[17] V. Khachatryan et al. (CMS), Phys. Rev. **D91**, 052008 (2015), 1408.3104
[18] Z.B. Kang, F. Ringer, I. Vitev, Phys. Lett. **B769**, 242 (2017), 1701.05839
[19] A. Idilbi, A. Majumder, Phys.Rev. **D80**, 054022 (2009), 0808.1087
[20] G. Ovanesyan, I. Vitev, JHEP **1106**, 080 (2011), 1103.1074
[21] G. Ovanesyan, I. Vitev, Phys.Lett. **B706**, 371 (2012), 1109.5619
[22] M. Fickinger, G. Ovanesyan, I. Vitev, JHEP **1307**, 059 (2013), 1304.3497
[23] Z.B. Kang, R. Lashof-Regas, G. Ovanesyan, P. Saad, I. Vitev, Phys. Rev. Lett. **114**, 092002 (2015), 1405.2612

[24] Y.T. Chien, A. Emerman, Z.B. Kang, G. Ovanesyan, I. Vitev, Phys. Rev. **D93**, 074030 (2016), 1509.02936
[25] M. Gyulassy, P. Levai, I. Vitev, Phys. Rev. Lett. **85**, 5535 (2000), nucl-th/0005032
[26] A.J. Larkoski, S. Marzani, G. Soyez, J. Thaler, JHEP **05**, 146 (2014), 1402.2657
[27] M. Dasgupta, A. Fregoso, S. Marzani, G.P. Salam, JHEP **09**, 029 (2013), 1307.0007
[28] Y. Mehtar-Tani, K. Tywoniuk, JHEP **04**, 125 (2017), 1610.08930
[29] N.B. Chang, S. Cao, G.Y. Qin (2017), 1707.03767
[30] A.M. Sirunyan et al. (CMS) (2017), 1708.09429
[31] I. Vitev, S. Wicks, B.W. Zhang, JHEP **0811**, 093 (2008), 0810.2807
[32] S.D. Ellis, Z. Kunszt, D.E. Soper, Phys.Rev.Lett. **69**, 3615 (1992), hep-ph/9208249
[33] Y. He, I. Vitev, B.W. Zhang, Phys.Lett. **B713**, 224 (2012), 1105.2566
[34] Y.T. Chien, I. Vitev, JHEP **12**, 061 (2014), 1405.4293
[35] C.W. Bauer, S. Fleming, D. Pirjol, I.W. Stewart, Phys.Rev. **D63**, 114020 (2001), hep-ph/0011336
[36] S. Chatrchyan et al. (CMS), Phys.Lett. **B730**, 243 (2014), 1310.0878
[37] J. Brewer, K. Rajagopal, A. Sadofyev, W. Van Der Schee (2017), 1710.03237
[38] Y. Tachibana, N.B. Chang, G.Y. Qin, Phys. Rev. **C95**, 044909 (2017), 1701.07951

 https://doi.org/10.1051/epjconf/201817205007

Collectivity in small systems - Initial state vs. final state effects

Moritz Greif[1], *Carsten* Greiner[1], *Björn* Schenke[2], *Sören* Schlichting[3,⋆], and *Zhe* Xu[4]

[1] *Institut für Theoretische Physik, Johann Wolfgang Goethe-Universität, Max-von-Laue-Str. 1, D-60438 Frankfurt am Main, Germany*
[2] *Physics Department, Brookhaven National Laboratory, Upton, NY 11973, USA*
[3] *Department of Physics, University of Washington, Seattle, WA 98195-1560, USA*
[4] *Department of Physics, Tsinghua University and Collaborative Innovation Center of Quantum Matter, Beijing 100084, China*

Abstract. Observations of long rang azimuthal correlations in small collision systems (p+p/A) have triggered an enormous excitement in the heavy-ion community. However, it is presently unclear to what extent the experimentally observed correlations should be attributed to initial state momentum correlations and/or the final state response to the initial state geometry. We discuss how a consistent theoretical description of the non-equilibrium dynamics is important to address both effects within a unified framework and present first results from weakly coupled non-equilibrium simulations in [1] to quantify the relative importance of initial state and final state effects based on theoretical calculations.

1 Introduction

Experimental measurements of long. range (in rel. rapidity $\Delta\eta$) azimuthal correlations (in rel. angle $\Delta\phi$) in high-energy proton-proton (p+p) and proton-nucleus (p+A) collisions, have revealed many interesting features that call for a deeper theoretical understanding [2]. Most strikingly, many features observed in these "small system" are qualitatively similar to previous observations in nucleus-nucleus (A+A) collisions, including e.g. the transverse momentum and hadron-species dependence of these correlations [3, 4] as well as the fact that the observed correlations are "collective" in the sense that many particles are correlated with each other [5–7]. However, there are also some important differences between the experimental results in proton-proton/nucleus and nucleus-nucleus collisions, including e.g. the fact that (so far) no evidence of jet-quenching phenomena has been reported in small systems.

Based on the apparent similarities and differences between small (p+p/A) and large (A+A) systems, different theoretical explanations have been explored which attribute the observed correlations either to initial state momentum correlations (see e.g. [8–13]) or the final state response to the event geometry (see e.g. [14–19]). Since the format of this proceeding is not suitable to provide an exhaustive review of the current state of the art of these calculations,[1] we will focus on the key challenge of

⋆e-mail: sslng@uw.edu
[1] We refer the interested reader to [2, 20] for detailed reviews on these subjects.

© The Authors, published by EDP Sciences. This is an open access article distributed under the terms of the Creative Commons Attribution License 4.0 (http://creativecommons.org/licenses/by/4.0/).

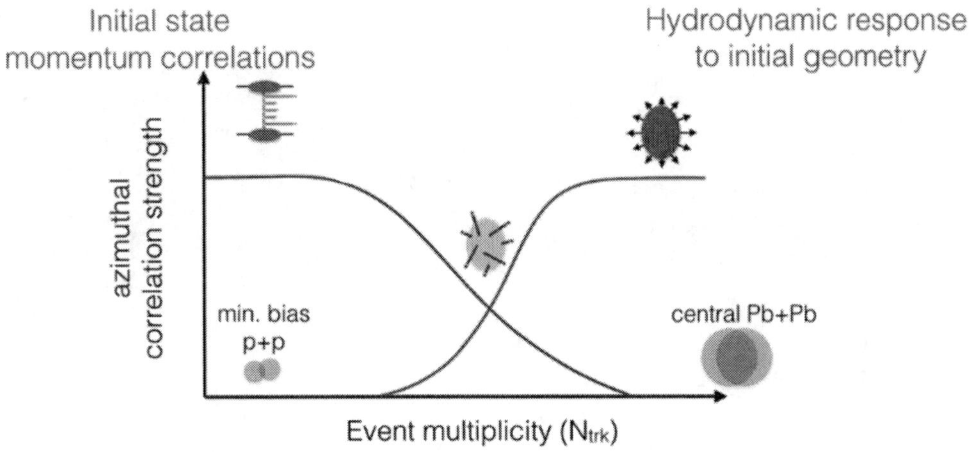

Figure 1. Illustration of the expected relative strength of different sources of long range azimuthal correlations: (blue) initial state momentum correlations (red) final state response to initial state geometry. Beyond the asymptotic limits ($N_{trk} \rightarrow 0$ and $N_{trk} \rightarrow \infty$), one expects long range azimuthal correlations to be sensitive to the non-equilibrium dynamics of the medium. Fig. from [20, 21]

quantifying the relative importance of these two effects. Based on a short outline of the general expectations in Sec. 2.1, we present first results addressing this questions within a weak coupling approach to the underlying non-equilibrium dynamics in Sec. 2.2, followed by some concluding remarks and perspectives for future studies in Sec. 3.

2 Quantifying the importance initial state vs. final state effects

2.1 General expectations

Generally speaking, it is important to take into account both initial state and final state effects in order to obtain a robust theoretical description across a wide kinematic range all the way from low multiplicity p+p ($dN_{p+p}^{min.bias}/dy \sim 5$) to high-multiplicity A+A collisions ($dN_{Pb+Pb}^{0-5\%}/dy \sim 1500$). While one naturally expects initial state effects to dominate correlations in low mult. p+p collisions, where the opacity of the medium is small and final state interactions are subleading, the opacity of the medium in high mult. A+A collisions is very large such that even highly energetic jets loose a substantial amount of energy and azimuthal correlations between low momentum particles emerge pre-dominantly due to the hydrodynamic response of the medium to the initial state geometry. However, as illustrated in Fig. 1, this leaves a large range of possibilities where for the typical degrees of freedom the opacity of the medium is neither particularly small to be ignored nor particularly large for the medium to be described hydrodynamically. Hence one expects experimental observables to become sensitive to the non-equilibrium dynamics of the medium, and new theoretical developments are needed to describe long. range azimuthal correlations in this regime.

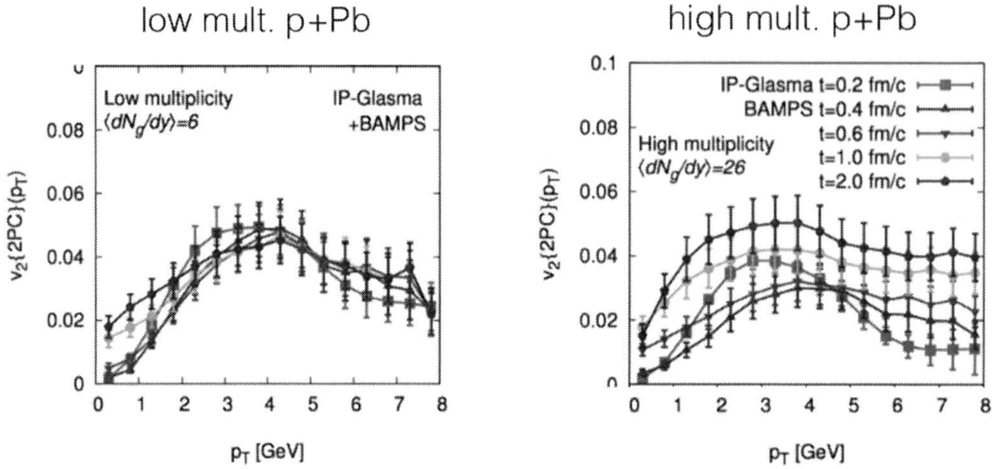

Figure 2. Comparison of the time evolution of the (two-particle) azimuthal correlation coefficient v_2^{gluon} in low. multiplicty (left) and high multiplicity (right) p+Pb collisions. Fig. from [1]

2.2 Non-equilibrium description of initial state & final state effects

Based on a weak-coupling picture, a consistent theoretical description of the non-equilibrium dynamics in high-energy collisions [22–24] can be achieved by matching a classical-statistical lattice description of particle production and early time dynamics (see e.g. [23]) to an effective kinetic description of the subsequent non-equilibrium dynamics (see e.g. [24]). Event-by-event simulations of the non-equilibrium dynamics of high-energy p+p,p+A and A+A collisions, can then be performed within this framework by

1) Simulating particle production and early time dynamics ($\tau < 0.2 fm/c$) based on the IP-Glasma model [25, 26]
2) Extracting the phase space distribution of gluons $dN_g/d\eta d^2\mathbf{x}dyd^2\mathbf{p}$, which includes all relevant information about initial state momentum correlations[2] as well the initial state geometry
3) Simulating final state re-scattering dynamics in a parton cascade (BAMPS) [30–32] based on pQCD matrix elements for $2 <-> 2$ and $2 <-> 3$ interactions [33, 34]

and details of the implementation are discussed in [1]. By investigating the time evolution of azimuthal correlations of partons over the course of the non-equilibrium evolution, one can then start to assess the relative importance of initial state and final state effects from a theoretical perspective.

Some results from [1] for azimuthal correlations in p+Pb collisions are summarized in Fig. 2, where the time evolution of the gluon $v_2(p_T)$ in low multiplicity (left) and high multiplicity (right) events is compared. Despite the fact that sizable initial state momentum correlations are present at $t = 0.2 fm/c$ in both high and low multipiclity events (also see [28, 29]), one observes a strong modification of the correlations in high multiplicity events $(\left(dN_g/dy\right)/\langle dN_g/dy\rangle > 2.5)$, where most

[2]Event-by-event the initial state momentum correlations due to Bose enhancement of small-x gluons (Glasma graphs) factorize into products of (azimuthally anisotropic) single particle distribution, i.e. $d^2N_{2g}/d^2\mathbf{p}_1d^2\mathbf{p}_2 = dN_g/d^2\mathbf{p}_1 \times dN_g/d^2\mathbf{p}_2$ [27–29] such that the associated one body density $dN_g/d\eta d^2\mathbf{x}dyd^2\mathbf{p}$ contains all relevant information about such correlations.

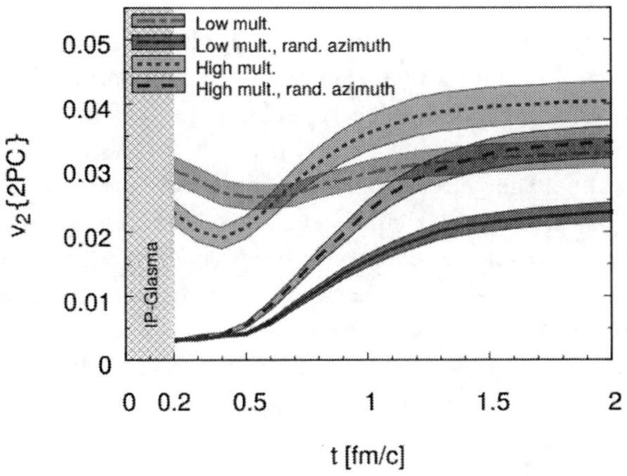

Figure 3. Evolution of the p_T integrated (gluon) v_2 for low (green) and high (red) multiplicity p+Pb events. Effect of initial state correlations on v_2 can be inferred from comparison of full simulations with "rand. azimuth" curves, where initial state momentum correlations have been artificially removed. Fig. from [1]

of the initial state correlations get destroyed on a time scale $\sim 1fm/c$ while new final state induced correlations are build up simultaneously in response to the system geometry.[3] Conversely, in low multiplicity events $(0.5 < \left(dN_g/dy\right)/\langle dN_g/dy\rangle < 1)$ final state re-scattering only affects correlations at very low momenta $p_T \lesssim 2\text{GeV}$; while initial state momentum correlations at higher momenta $p_T \gtrsim 2\text{GeV}$ persist throughout the evolution. Based on the discussion in Sec. 2.1 it is natural that the differences between the two event classes can be attributed to a larger number of large angle scatterings in high multiplicity events. However, it is at least somewhat surprising that even in high multiplicity events an average number of ≈ 1 large angle scattering is sufficient to build up a geometric response, although similar observations have been reported previously in [35].

Even though the relative importance of initial state and final state effects depends strongly on the momenta under consideration, an approximate measure of the different effects can be obtained based on the p_T integrated v_2 which is depicted in Fig. 3. One finds that the time evolution of v_2 can be characterized by three distinct stages, where for $t \lesssim 0.2fm/c$ correlations are dominated by initial state, before for $0.2fm/c \lesssim t \lesssim 0.5fm/c$ scatterings partially destroy initial state correlations until for $0.5 \lesssim t \lesssim 1.0fm/c$ new correlations build up in response to event geometry. By comparing the results of full event-by-event simulations, with those where initial state correlations have been artificially removed by randomizing the azimuthal angles of the initial state gluons ("rand. azimuth"), one can attempt to further disentangle initial state and final state contributions. Even though geometric response ultimately dominates v_2 at low p_T, the results in Fig. 3 indicate that there are still sizable effects of initial state correlations (even on the p_T integrated v_2) on the oder of $\sim 25\%$ for high multiplicity and $\sim 50\%$ for low multiplicity events.

3 Conclusions & Outlook

Experimental observations of pronounced azimuthal correlations in small systems have started to challenge our current understanding of the space time evolution of high-energy collisions. So far two lines of theoretical explanations have been developed, which attribute the observed correlations either to intrinsic momentum correlations in the initial state or a collective expansion of the medium driven by final state interactions, while typically neglecting the other effect. Despite a variety of

[3] A careful analysis reveals that the final state induced correlations are strongly correlated with the event geometry [1].

successful phenomenological applications in both frameworks, both of these approaches fall short in consistently describing the underlying dynamics over a wide range of multiplicities, where one expects a gradual change from an initial state dominated to a final state dominated scenario. Hence it is important to develop new theoretical tools, which can properly address the features of the non-equilibrium dynamics interpolating between these two extremes. So far, first calculations in this spirit [1] indicate that both initial state and final state effects can be quantitatively important for two-particle correlations in p+Pb collisions. Based on such a consistent theoretical description of the non-equilibrium evolution, it would also be interesting to explore in more detail to what extent specific observables, such as e.g. the existence of (flow-like) higher order cumulants of v_2 at low momenta or the survival of (jet-like) back-to-back correlations a higher momenta, are indicative of the dominance of one of the underlying physics mechanisms. However, these studies are still in their infancies and further theoretical progress will be required to clarify these points.

Acknowledgements: We would like to thank Kevin Dusling, Tuomas Lappi, Aleksas Mazeliauskas, Jean-Francois Paquet, Derek Teaney, Prithwish Tribedy and Raju Venugopalan for insightful discussions and fruitful collaborations related to this work. S.S. and B.S. acknowledge support by the U.S. Department of Energy, Office of Science, Office of Nuclear Physics under Grant DE-FG02-97ER41014 (S.S.) and DE-SC0012704 (B.S.). M.G., and C.G. acknowledge support by the Deutsche Forschungsgemeinschaft (DFG) through the grant CRC-TR 211 "Strong-interaction matter under extreme conditions". Z.X. was supported by the National Natural Science Foundation of China under Grants No. 11575092 and No. 11335005, and the Major State Basic Research Development Program in China under Grants No. 2014CB845400 and No. 2015CB856903. Numerical calculations used the resources of the Center for Scientific Computing (CSC) Frankfurt and the National Energy Research Scientific Computing Center, a DOE Office of Science User Facility supported by the Office of Science of the U.S. Department of Energy under Contract No. DE-AC02-05CH11231.

References

[1] M. Greif, C. Greiner, B. Schenke, S. Schlichting and Z. Xu, arXiv:1708.02076 [hep-ph].

[2] K. Dusling, W. Li and B. Schenke, Int. J. Mod. Phys. E **25**, no. 01, 1630002 (2016) doi:10.1142/S0218301316300022 [arXiv:1509.07939 [nucl-ex]].

[3] B. B. Abelev *et al.* [ALICE Collaboration], Phys. Lett. B **726**, 164 (2013) doi:10.1016/j.physletb.2013.08.024 [arXiv:1307.3237 [nucl-ex]].

[4] V. Khachatryan *et al.* [CMS Collaboration], Phys. Lett. B **742**, 200 (2015) doi:10.1016/j.physletb.2015.01.034 [arXiv:1409.3392 [nucl-ex]].

[5] S. Chatrchyan *et al.* [CMS Collaboration], Phys. Lett. B **724**, 213 (2013) doi:10.1016/j.physletb.2013.06.028 [arXiv:1305.0609 [nucl-ex]].

[6] V. Khachatryan *et al.* [CMS Collaboration], Phys. Rev. Lett. **115**, no. 1, 012301 (2015) doi:10.1103/PhysRevLett.115.012301 [arXiv:1502.05382 [nucl-ex]].

[7] G. Aad *et al.* [ATLAS Collaboration], Phys. Lett. B **725**, 60 (2013) doi:10.1016/j.physletb.2013.06.057 [arXiv:1303.2084 [hep-ex]].

[8] K. Dusling and R. Venugopalan, Phys. Rev. D **87**, no. 5, 051502 (2013) doi:10.1103/PhysRevD.87.051502 [arXiv:1210.3890 [hep-ph]].

[9] K. Dusling and R. Venugopalan, Phys. Rev. D **87**, no. 9, 094034 (2013) doi:10.1103/PhysRevD.87.094034 [arXiv:1302.7018 [hep-ph]].

[10] A. Dumitru, L. McLerran and V. Skokov, Phys. Lett. B **743**, 134 (2015) doi:10.1016/j.physletb.2015.02.046 [arXiv:1410.4844 [hep-ph]].

[11] K. Dusling, P. Tribedy and R. Venugopalan, Phys. Rev. D **93**, no. 1, 014034 (2016) doi:10.1103/PhysRevD.93.014034 [arXiv:1509.04410 [hep-ph]].

[12] B. Schenke, S. Schlichting, P. Tribedy and R. Venugopalan, Phys. Rev. Lett. **117**, no. 16, 162301 (2016) doi:10.1103/PhysRevLett.117.162301 [arXiv:1607.02496 [hep-ph]].

[13] K. Dusling, M. Mace and R. Venugopalan, arXiv:1706.06260 [hep-ph].

[14] P. Bozek, Phys. Rev. C **85**, 014911 (2012) doi:10.1103/PhysRevC.85.014911 [arXiv:1112.0915 [hep-ph]].

[15] G. Y. Qin and B. Müller, Phys. Rev. C **89**, no. 4, 044902 (2014) doi:10.1103/PhysRevC.89.044902 [arXiv:1306.3439 [nucl-th]].

[16] K. Werner, M. Bleicher, B. Guiot, I. Karpenko and T. Pierog, Phys. Rev. Lett. **112**, no. 23, 232301 (2014) doi:10.1103/PhysRevLett.112.232301 [arXiv:1307.4379 [nucl-th]].

[17] B. Schenke and R. Venugopalan, Phys. Rev. Lett. **113**, 102301 (2014) doi:10.1103/PhysRevLett.113.102301 [arXiv:1405.3605 [nucl-th]].

[18] H. Mäntysaari, B. Schenke, C. Shen and P. Tribedy, Phys. Lett. B **772**, 681 (2017) doi:10.1016/j.physletb.2017.07.038 [arXiv:1705.03177 [nucl-th]].

[19] R. D. Weller and P. Romatschke, Phys. Lett. B **774**, 351 (2017) doi:10.1016/j.physletb.2017.09.077 [arXiv:1701.07145 [nucl-th]].

[20] S. Schlichting and P. Tribedy, Adv. High Energy Phys. **2016**, 8460349 (2016) doi:10.1155/2016/8460349 [arXiv:1611.00329 [hep-ph]].

[21] S. Schlichting, Nucl. Phys. A **956**, 216 (2016) doi:10.1016/j.nuclphysa.2016.02.026 [arXiv:1601.01177 [hep-ph]].

[22] R. Baier, A. H. Mueller, D. Schiff and D. T. Son, Phys. Lett. B **502**, 51 (2001) doi:10.1016/S0370-2693(01)00191-5 [hep-ph/0009237].

[23] J. Berges, K. Boguslavski, S. Schlichting and R. Venugopalan, Phys. Rev. D **89**, no. 11, 114007 (2014) doi:10.1103/PhysRevD.89.114007 [arXiv:1311.3005 [hep-ph]].

[24] A. Kurkela and Y. Zhu, Phys. Rev. Lett. **115**, no. 18, 182301 (2015) doi:10.1103/PhysRevLett.115.182301 [arXiv:1506.06647 [hep-ph]].

[25] B. Schenke, P. Tribedy and R. Venugopalan, Phys. Rev. Lett. **108**, 252301 (2012) doi:10.1103/PhysRevLett.108.252301 [arXiv:1202.6646 [nucl-th]].

[26] B. Schenke, P. Tribedy and R. Venugopalan, Phys. Rev. C **86**, 034908 (2012) doi:10.1103/PhysRevC.86.034908 [arXiv:1206.6805 [hep-ph]].

[27] K. Dusling, F. Gelis, T. Lappi and R. Venugopalan, Nucl. Phys. A **836**, 159 (2010) doi:10.1016/j.nuclphysa.2009.12.044 [arXiv:0911.2720 [hep-ph]].

[28] T. Lappi, S. Srednyak and R. Venugopalan, JHEP **1001**, 066 (2010) doi:10.1007/JHEP01(2010)066 [arXiv:0911.2068 [hep-ph]].

[29] B. Schenke, S. Schlichting and R. Venugopalan, Phys. Lett. B **747**, 76 (2015) doi:10.1016/j.physletb.2015.05.051 [arXiv:1502.01331 [hep-ph]].

[30] Z. Xu and C. Greiner, Phys. Rev. C **71**, 064901 (2005) doi:10.1103/PhysRevC.71.064901 [hep-ph/0406278].

[31] Z. Xu and C. Greiner, Phys. Rev. C **76**, 024911 (2007) doi:10.1103/PhysRevC.76.024911 [hep-ph/0703233].

[32] Z. Xu, K. Zhou, P. Zhuang and C. Greiner, Phys. Rev. Lett. **114**, no. 18, 182301 (2015) doi:10.1103/PhysRevLett.114.182301 [arXiv:1410.5616 [hep-ph]].

[33] O. Fochler, J. Uphoff, Z. Xu and C. Greiner, Phys. Rev. D **88**, no. 1, 014018 (2013) doi:10.1103/PhysRevD.88.014018 [arXiv:1302.5250 [hep-ph]].

[34] J. Uphoff, F. Senzel, O. Fochler, C. Wesp, Z. Xu and C. Greiner, Phys. Rev. Lett. **114**, no. 11, 112301 (2015) doi:10.1103/PhysRevLett.114.112301 [arXiv:1401.1364 [hep-ph]].

[35] L. He, T. Edmonds, Z. W. Lin, F. Liu, D. Molnar and F. Wang, Phys. Lett. B **753**, 506 (2016) doi:10.1016/j.physletb.2015.12.051 [arXiv:1502.05572 [nucl-th]].

Forward backward multiplicity correlations: a tool to study properties of the proton-proton collisions

Edgar Dominguez-Rosas[1,*] and *Eleazar* Cuautle Flores[1,**]

[1] *Instituto de Ciencias Nucleares, Universidad Nacional Autónoma de México. Apartado Postal 70-543, Ciudad de México 04510, México*

Abstract. Forward-backward multiplicity correlations have been used to study hadron production mechanisms in electron-positron, proton-proton and more recently in lead-lead collisions. The experimental results on this correlations and its comparison to different models reveals an incomplete agreement. In this work, we present an study of forward backward multiplicity correlations in proton-proton collisions using PYTHIA event generator, at LHC energies. Detailed analysis is presented in the case of soft and hard QCD processes, incorporating color reconnection model as part of hadronization mechanism and multiple parton interactions effects in the correlations. Our results and its comparison to available experimental data suggest that this kind of correlations are great tools to characterize the events and gives the possibility to disentangle phenomena in hard and soft QCD processes.

1 Introduction

The Forward Backward (F − B) multiplicity correlations had been studied long time ago, on different colliding system. One of the first results in p+p collisions in ISR at $\sqrt{s} = 52.6$ MeV/c [1] where was found positive values for the correlations, five years later there were results of $\mathrm{p} + \bar{\mathrm{p}}$ collisions at $\sqrt{s} = 570$ MeV/c [2] in SPS, again positive values of the correlation were found, and a possible dependence on the energy was reported. Following this kind of correlations, results on e^+e^- collisions at $\sqrt{s} = 29$ GeV [3] were published, the reported results shows no F − B correlations, this result was interpreted as a consequence of the independent fragmentation of the jets produced in the collision. Few years later, results of the E735 collaboration at Tevatron [4] in p+p collision at $\sqrt{s} \approx 1$ TeV confirmed the dependence of the correlation with the energy.
Results on heavy ions at $\sqrt{s_{NN}} = 200$ GeV, published by the STAR Collaboration at RHIC [5], found strong correlations in case of Au+Au collisions for centrality 0 − 10%, meanwhile for p+p collisions was found very low correlation. Recently at the LHC were published similar results in p+p collision by ATLAS [6] and ALICE [7] collaboration at $\sqrt{s} = 0.9, 2.7, 9$ TeV, they observed strong correlations. Their analysis was more detailed investigating the p_{T} dependence of azimuthal and pseudorapidity distributions of F − B multiplicity correlation.

*e-mail: edgar.dominguez.rosas@cern.ch
**e-mail: ecuautle@nucleares.unam.mx

© The Authors, published by EDP Sciences. This is an open access article distributed under the terms of the Creative Commons Attribution License 4.0 (http://creativecommons.org/licenses/by/4.0/).

In the present work we analyze F – B multiplicity correlations and possible effects on it due to final state interactions in the hadronization mechanism, those like color reconnection (CR) as well as from the number of multiple parton interactions (nMPI), these phenomenological models are successfully used to describe kinematic variables like average transverse momentum of charged hadrons and multiplicity distributions, hence its importance to investigate the effects on F – B multiplicity correlations.

The F – B multiplicity correlation strength is defined as correlation coefficient of forward and backward multiplicity, where n_F and n_B are the numbers of charged particles falling into the forward and backward pseudorapidity intervals of $\delta\eta$, respectively:

$$b_{Corr} = \frac{\langle n_F n_B \rangle - \langle n_F \rangle \langle n_B \rangle}{\langle n_F^2 \rangle - \langle n_F \rangle^2}, \tag{1}$$

this strength of the correlation factor, $b_{\rm corr}$, is really sensitive to the changes of multiplicity, for instance, for ranges of multiplicity is possible to get negative values of $b_{\rm corr}$, this results are consequences of the definition, but those results have no physical meaning.

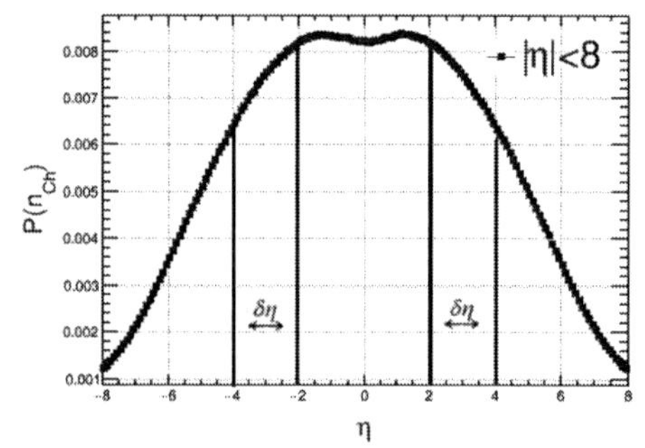

Figure 1: Pseudorapidity distribution illustrating $\delta\eta$ in forward-backward region

In the Fig. 1 shows the pseudorapidity distribution, where is illustrated the $\delta\eta$ variable which is the interval of the multiplicity to measure the correlation. This variable allow to study the out coming hadrons of the collisions, those are expected to come from the collisions if we analyze central pseudorapidity regions, while for larger values of η we expect to see hadrons coming from the fragmentation region, and also from colliding beam.

2 Forward-Backward multiplicity correlation with PYTHIA

The analysis of the correlation is realized with Monte Carlo event generator, PYTHIA 8.2 [8], taking into account soft and hard QCD process, with a sample of 25 billions of events for each set of studies, selecting charged primary hadrons with transverse momentum in the range $0.3 < p_{\rm T} < 1.5$ GeV/c and pseudorapidity range $|\eta| < 1$. These values are taken to be able to compare our results with those presented by the ALICE experiment [7].

Pseudorapidity distribution is the first variable to be understood since any change on it, will bring consequences on the correlations. It is important to quantify the effects of soft and hard QCD processes in an independent way. The left panel of the Fig. 2 shows these two distributions generated in p+p collisions in a width η range showing a higher number of charged hadrons produced in the central region for hard processes while the opposite is observed for large η values (fragmentation region). The differences could be understood in terms of the kind of processes occurred in the collisions, it means, hard processes mean harder collision between interaction partons, producing jets in transverse direction respect to that of the beam. Right panel of the figure shows the transverse momentum distribution for soft and hard QCD processes, from which one sees a spectrum hardest for hard QCD processes, also it shows a spectrum for central rapidity compared to fragmentation

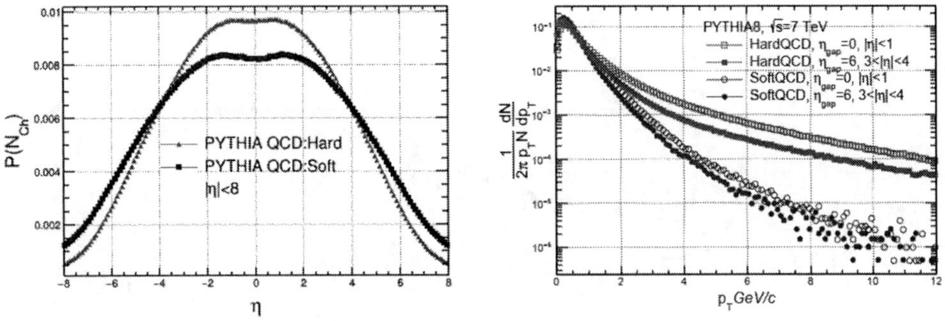

Figure 2: Pseudorapidity distributions (left) for soft and hard QCD processes and transverse momentum distributions (right) at $\sqrt{s} = 7$ TeV

regions, these differences produce direct effects on the behavior of the F − B multiplicity correlations.

Figure 3 shows b_{corr} as a function of $\delta\eta$: for central ($0 < |\eta| < 1$) and fragmentation ($3 < |\eta| < 4$) region, from here one can see the evolution of the strength correlation parameter, it shows a higher values for the soft processes respect to the hard ones. It is important to observe a faster saturation of the b_{corr} than those calculated for soft processes. Comparisons short to long range correlations are shown in the bottom of the figure, where almost a scaling behavior of the b_{corr} for soft processes is observed while for the hard processes it increases faster in the fragmentation region.

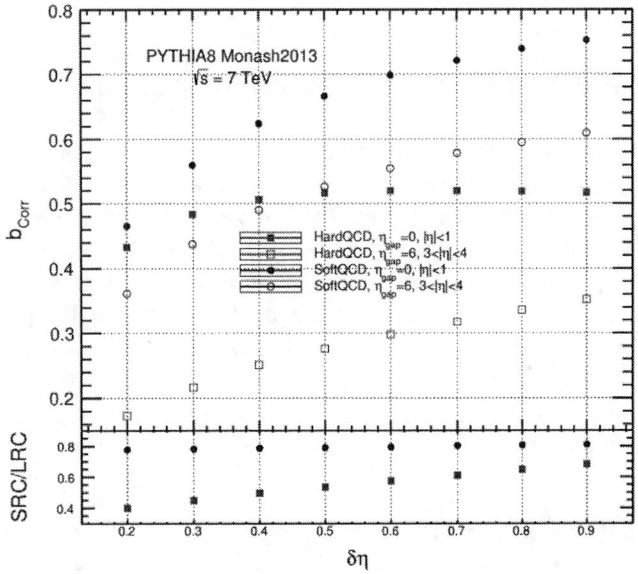

Figure 3: The correlation factor for p+p collision at $\sqrt{s} = 7$ TeV, for soft and hard QCD processes in central and fragmentation pseudorapidity regions

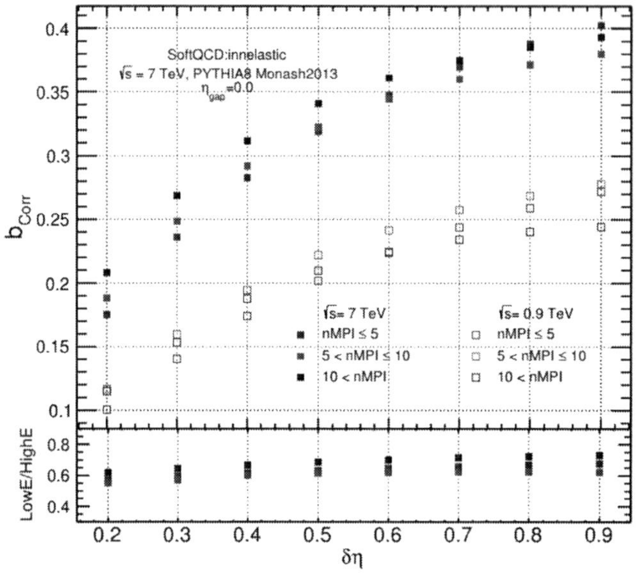

Figure 4: The correlation factor in terms of ranges of nMPI for p+p collision at $\sqrt{s} = 0.9, 7$ TeV.

3 F – B Correlations and multiple parton interactions

The multiple parton interactions is gaining interest since from the experimental point of view there are evidences of their existence and from theoretical side, there are calculations [9] which have been included in PYTHIA event generator. Figure 4 shows the correlations strength for two energies and for each one them, different ranges in the number of multiple parton interactions. The behavior in both cases is a continuous growth when the nMPI increase, however, comparing the correlation at low to high energy, one sees an almost flat behavior, it means, that F – B multiplicity correlation shows a scaling with the energy. Combined effects on b_{corr} coming from energy and nMPI could be understood from the relationship between these two parameters, since nMPI is modeled in PYTHIA with the damping parameter p_{0T} [8] where this parameter is function of p_T, and this last one depends directly on the collision energy.

4 F – B correlations and color reconnection

The CR has been proposed as an alternative mechanism to produce flow like effects in proton-proton collisions [10] where the direct variable that changes with the CR is the multiplicity distribution, it decreases when CR increases. CR has been taken to describe successfully the average transverse momentum of charge hadrons [11]. Figure 5 shows a reduction of the multiplicity when CR is included, and this will produce some effect on the strength in the F – B correlation multiplicity.

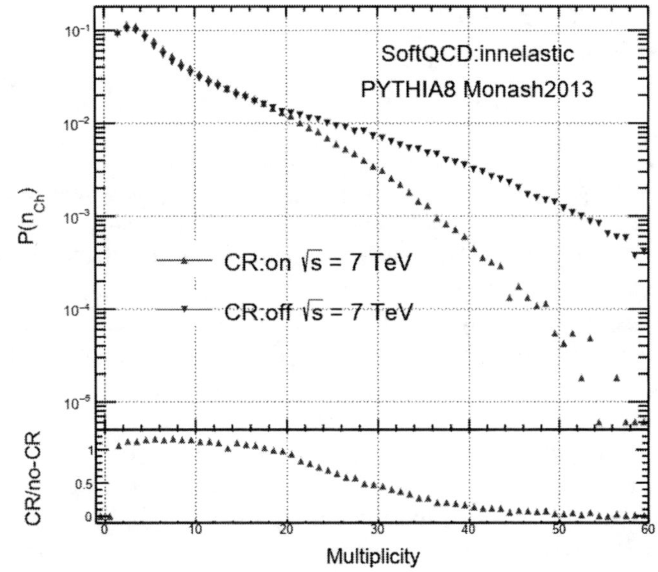

Figure 5: Multiplicity distribution of charge hadrons with and without color reconnection for p + p collision at $\sqrt{s}$ = 7 TeV.

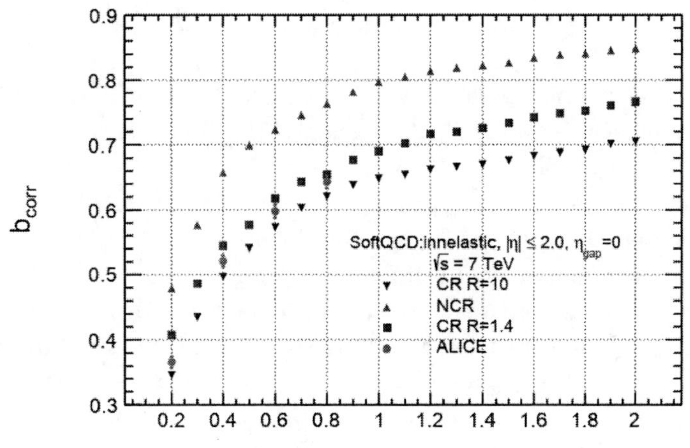

Figure 6: The correlation factor for Soft QCD processes in p + p collisions at $\sqrt{s}$ = 0.9, 7 TeV. Comparison is done among different strength of CR and the experimental ALICE data.

Figure 6 shows the b_{corr} as a function of the pseudorapidity width $\delta\eta$ at 7 TeV for different values of the strength of the CR. It is observed that higher values of the CR reduce the multiplicity and also the b_{corr}. The possible range of values of CR strength is from 1.4 to 10, and more precise value could be extracted from a fit of the experimental data which are shown in the same figure. However the CR is not the only possible effect present in the $b_{\rm corr}$, then it is useful to make a more detailed analysis on the F − B correlation multiplicity.

5 Conclusions

In the present work we have compared the charged hadron production in the forward and backward, in central and fragmentation pseudorapidity regions. The analysis was done in p+p collision at $\sqrt{s}$ = 0.9 and 7 TeV using PYTHIA event generator, with the effects of the number of multiple parton interactions and different strength values of color reconnection, splitting the events into soft and hard QCD processes. The analysis of F − B multiplicity correlations shows a weak dependence with the number of multiple parton interactions while the energy produce almost scaling behavior of the correlations. However, when we took soft QCD processes a scaling between long and short range correlations is observed, meanwhile for hard processes a faster saturation of the $b_{\rm corr}$ at central rapidity respect to the fragmentation region is observed. The color reconnection, which reduces the multiplicity also reduce the strength of the correlations, a combination of both numbers of multiple parton interactions and color reconnection could bring us an insight more detailed of hadron production mechanism in p+p and other colliding systems.

Acknowledge

Partial support was received by DGAPA-PAPIIT and CONACYT projects IN109817 and 280362 respectively.

References

[1] British-French-Scandinavian Collaboration, M. Albrow *et al.*,Nucl. Phys. **B145** 305 (1978).
[2] UA5 Collaboration, K. Alpgard *et al.*, Phys. Lett. **B123** 361 (1983).
[3] M. Derrick, K. Gan, P. Kooijman, J. Loos, B. Musgrave, *et al.*,Phys. Rev. **D34** 3304 (1986).
[4] E735 Collaboration, T. Alexopoulos, *et al.*, Phys. Lett. **B353** 155 (1995).
[5] STAR Collaboration, B. Abelev et al., Phys. Rev. Lett. **103** 172301 (2009).
[6] ATLAS Collaboration, G. Aad *et al.*,JHEP **07** 019 (2012).
[7] ALICE Collaboration, Jaroslav Adam, *et al*, JHEP **1505** 097 (2015).
[8] T. Sjöstrand, S. Ask, J. R. Christiansen, R. Corke, N. Desai, P. Ilten, S. Mrenna, S. Prestel, C. O. Rasmussen, P. Z. Skands, Computer Physics Communications **191** 159 (2015.)
[9] Torbjörn Tjöstran and Maria van Zijl, Phys. Rev. **D36**, 2019 (1987).
[10] A. Ortiz Velasquez, P. Christiansen, E. Cuautle Flores, I. A. Maldonado Cervantes, and G. Paić, Phys. Rev. Lett. **111**, 042001 (2013).
[11] ALICE Collaboration,B. Abelev Phys. Lett. **B727** 371-380 (2013).

Medium response to jets in heavy ion collisions

Yasuki Tachibana[1,2,⋆]

[1] *Institute of Particle Physics and Key Laboratory of Quark and Lepton Physics (MOE), Central China Normal University, Wuhan, Hubei 430079, China*
[2] *Department of Physics and Astronomy, Wayne State University, Detroit, Michigan 48201, USA*

Abstract. A short overview on recent progress in studies of medium response to jet quenching in heavy ion collisions is presented. We show the typical features of medium response and give comment on their connection to jet observables by introducing the work done by the author and collaborators as an example.

1 Introduction

The deconfined state of quarks and gluons, so-called quark-gluon plasma (QGP), has been created and explored experimentally in heavy ion collisions at the Relativistic Heavy Ion Collider (RHIC) and the Large Hadron Collider (LHC) [1]. In particular, the observation of the large anisotropic flow patterns indicates the collective behavior of the bulk QGP, which implies the strong interactions among the constituents. Various relativistic hydrodynamic models have been developed and very successful in explaining the space-time evolution of the bulk QGP [2–10]. As another remarkable feature of the strongly coupled QGP, jet quenching, namely the large energy loss of jets, has also been revealed in heavy ion collisions [11, 12]. The jets are collimated sprays of particles resulting from the QCD showering of high-p_T partons produced in early stage hard scatterings. Under the existence of the QGP medium, partons in jets collide and exchange the energy and momentum with the medium constituents via the strong interactions during the penetration in the medium. In addition, the successive collisions inside the medium cause bremsstrahlung in QCD and induce additional partonic showering in the jets (medium-induced radiation). Through the collisions and medium-induced radiations, the structure of the jets is significantly modified, and their modification patterns provide the information of the interactions between jets and QGP medium.

Medium response, the effect of the wake generated by the jet propagation in the medium, is another possible contribution to the medium modification of the jet structures and has recently been focused [13–19]. The energy and momentum deposited by the jets are supposed to excite the medium, and induce flows following the jets. The jet-induced flow carries the deposited energy and momentum and enhances the hadron emission from the medium around the jet axis direction. Since the medium response happens only together with the jet propagation, these enhanced hadrons correlated with the jet propagation cannot be subtracted with background and detected with the fragments from the energetic jet shower.

⋆e-mail: yasuki.tachibana@wayne.edu

© The Authors, published by EDP Sciences. This is an open access article distributed under the terms of the Creative Commons Attribution License 4.0 (http://creativecommons.org/licenses/by/4.0/).

2 Motivations

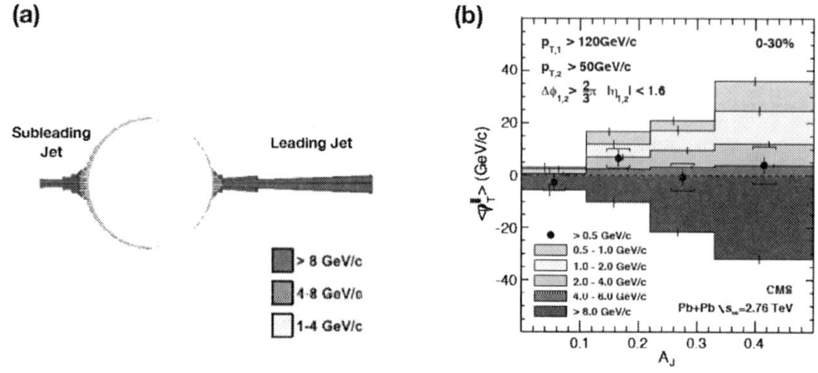

Figure 1. (a) Azimuthal angle distribution of transverse momentum in the transverse plane in a dijet event at LHC observed by CMS Collaboration. Each color in histgrams represents the contribution in a transverse-momentum range. Figure is adapted from the slides of Ref. [20]. (b) Missing transverse momentum $\langle \not{p}_{\mathrm{T}}^{\parallel} \rangle = \langle -\sum_i p_{\mathrm{T}}^i \cos(\phi_i - \phi_{\mathrm{Leading\ Jet}}) \rangle$ as a function of dijet asymmetry ratio $A_J = \frac{p_{\mathrm{T},1}-p_{\mathrm{T},2}}{p_{\mathrm{T},1}+p_{\mathrm{T},2}}$ measured for dijet events in PbPb collisions at LHC by CMS Collaboration [21]. The solid black circles show the total $\langle \not{p}_{\mathrm{T}}^{\parallel} \rangle$ and each color in histograms represents the contribution in a transverse-momentum range. Figure is adapted from Ref. [21].

Detailed measurements of fully reconstructed jets and their correlated particles provide us information on a lot of physics over a wide range of energy-momentum scales involved in jet quenching phenomenon. For example, CMS Collaboration observed dijet events with large $p_{\mathrm{T}}^{\mathrm{jet}}$-imbalance in PbPb collisions at the LHC [21]. Figure 1 (a) is the azimuthal angle-distribution of the transverse momentum in a dijet event. One can see two peaks; the taller peak on the right side for the leading jet and the shorter peak on the left side for the subleading jet. As a consequence of the jet quenching effect, the transverse momentum of the leading jet $p_{\mathrm{T},1}$ and the subleading jet $p_{\mathrm{T},2}$ are uneven and the large asymmetry ($p_{\mathrm{T},1} > p_{\mathrm{T},2}$) can be seen. However, at large angles from the jet axes, a huge enhancement of low-p_{T} particles is observed together with the jets. When the contribution of these low-p_{T} particles is taken into account, the whole transverse momentum in a dijet event turns out to be balanced. This transverse-momentum balance is confirmed for any asymmetric dijet events [Fig. 1 (b)] (further details can be found in Ref. [21]), and shows evidence of momentum conservation in dijet events in heavy ion collisions. The enhanced low-p_{T} particles originate from the energy and momentum of the jets redistributed through the jet quenching effect, and can naturally be considered as a consequence of the medium response because their transverse momenta are sufficiently low [15, 22, 23]. Accordingly, to study the full picture of jet quenching phenomenon it is necessary to consider the effect of medium response.

Medium response effect may give a manifestation of the QGP's fluidity associated with jets and information about the bulk QGP properties, e.g. sound velocity, and transport coefficients. For the QGP fluid, the energetic partons in jets can be considered fast-moving sources depositing the energy and momentum, which are supposed to excite the conical shockwave, so-called Mach cone, as a hydrodynamic medium response [24–28]. The shockwave front propagates toward a specific angle from the direction of the moving source which is determined by the sound velocity of the fluid. The shape of the Mach cone can also be affected by the finite viscosities of the QGP fluid, e.g., the shockwave front can be smeared by the shear viscosity [29–36]. Furthermore, medium response involves the

QGP transport properties related to how part of the jet energy and momentum relaxes into the hydro mode in the QGP medium. This relaxation process has not been studied well but is very important to define the scale separation between the hard jet part and the soft fluid part, and to obtain the space-time profile of the energy-momentum deposition into the QGP fluid [29, 37, 38].

3 Modelings for medium response in recent studies

Recently, various models for numerical simulation studies with inclusion of the medium response effect have been developed and used to show the contribution of the effect on the jet sub-structure observables. Some event-by-event models for in-medium jet evolution include the medium response as recoil partons. In this approach, when they consider a two-particle (t-channel) scattering between a jet parton and a medium constituent, a sample of parton from the medium is taken with the thermal distribution. Then after the scattering, the scattered parton from the thermal medium is called the recoil parton and tracked as part of the jet partons. Jet Evolution With Energy Loss (JEWEL) [18, 19, 39, 40], the jet-quenching Monte-Carlo code based on the BDMPS-Z formalism [41–45], has the option to include the recoil partons traveling freely in the medium without any further interaction. In the recent version of Linear Boltzmann Transport (LBT) model [13, 46–48], the recoil partons is implemented together with the in-medium partonic radiation based on the higher-twist formalism [49, 50]. In LBT model, the recoil partons also take part in the interactions with the medium constituents, and the appearance of the Mach cone-like structure that can also be seen in jet studies with Boltzmann Approach of MultiParton Scatterings (BAMPS) model [32, 33] or A Multi-Phase Transport (AMPT) model [23] is shown.

When part of the jet energy and momentum is thermalized and deposited into the QGP fluid, the dynamics of the medium response can be described by relativistic hydrodynamics. To study the hydrodynamic medium response, the hydrodynamic equation with source term,

$$\partial_\mu T^{\mu\nu}_{\rm fluid}(x) = J^\nu(x), \tag{1}$$

is commonly used [16, 17, 22, 29–31, 51–62]. Here, $T^{\mu\nu}_{\rm fluid}(x)$ is the energy-momentum tensor for the medium fluid, and $J^\nu(x)$ is the source term representing the space-time profile of the energy and momentum deposited by jets. In the recent studies of full jet shower with hydrodynamic medium response, Eq. (1) is solved while being coupled to jet transport equations; the source term in Eq. (1) is calculated from the solutions of the jet transport equations based on the energy-momentum conservation for the combined system of the jets and the QGP fluid. It should be noted that the hydrodynamic simulation with the jet source term must be performed in full (3+1)-dimensional space time for a complete theoretical treatment, since the jet-induced flow, e.g. Mach cone, always has a three-dimensional structure and violates the boost invariance that the fluid profile is supposed to have as an approximate symmetry around midrapidity. The hadrons from the medium fluid with medium response are obtained by using Copper-Frye formula [63], which is most commonly used in hydrodynamic models for heavy ion collisions. After the background subtraction, the hadrons from the medium fluid are summed to the final-state particles calculated from the jet transport equation. Then, the jet including the contribution of the hydrodynamic medium response is obtained. An example of this approach is the model used in Ref. [16] (we refer it as Jet Shower Transport + Hydro model below). In this model, it is assumed that all partons scattered by the jet partons instantaneously rethermalize in the medium fluid. As another example, Coupled LBT Hydro model [17] is a full event-by-event model employing LBT Model for the jet evolution calculation. In this model, they assume that all the jet partons in LBT simulation with energy below a certain cutoff energy are completely absorbed in the medium fluid, and deposit their energy and momentum. In this framework, the contribution of both hydrodynamic medium response and recoil partons can be handled. Incidentally, most recently, it is suggested to use Eq. (1) not only for the medium response to jets but also for the early space-time evolution of the QGP fluid that gradually forms while receiving the energy and momentum deposited by the matters in the pre-thermalized (or pre-hydronized) state [64, 65].

In addition, also the latest version of Hybrid Strong/Weak Coupling model [15, 66] contains a simplified description for the medium response. The lost energy and momentum of the jet calculated by $\mathcal{N} = 4$ supersymmetric Yang-Mills theory [67] are stored into a boost-invariant Bjorken fluid as a perturbation, and then the hadrons enhanced by the perturbation are obtained by using the linear expansion of Cooper Frye formula. In this approach, the space-time evolution of the jet-induced flow is not described.

4 Full jet study with Jet Shower Transport + Hydro model

Here we briefly introduce the full jet study by Jet Shower Transport + Hydro model [16] and show the typical features of medium response contribution in jet observables. Some calculations with other models show different results but mostly they are the consequences of the competition with the effects other than the medium response.

4.1 Jet Shower Transport + Hydro model

In Jet Shower Transport + Hydro model, the relativistic hydrodynamic equations with source term (1) is numerically solved in (3 + 1) dimensional space-time to study the space-time evolution of the medium with energy-momentum deposition from the jet. Neglecting the small viscosity of the medium, the medium is modeled as an ideal fluid here. The initial profile of the medium fluid for central PbPb collisions at 2.76A TeV is generated by the optical Glauber model with the impact parameter $b = 0$. The jet shower evolution is described in terms of the energy and transverse momentum distributions of the partons in the jet shower, $f_i(\omega_i, k_{i\perp}^2, t) = dN_i(\omega_i, k_{i\perp}^2, t)/d\omega_i dk_{i\perp}^2$, by solving the transport equation [68],

$$\frac{df_i}{dt} = \left(\hat{e}_i\frac{\partial f_i}{\partial\omega_i} + \frac{1}{4}\hat{q}_i\nabla_{k_{i\perp}}^2 f_i\right) + \sum_j\int d\omega_j dk_{j\perp}^2\left[\frac{d\tilde{\Gamma}_{j\to i}(\omega_i, k_{i\perp}^2|\omega_j, k_{j\perp}^2)}{d\omega_i dk_{i\perp}^2 dt}f_j - \frac{d\tilde{\Gamma}_{i\to j}(\omega_j, k_{j\perp}^2|\omega_i, k_{i\perp}^2)}{d\omega_j dk_{j\perp}^2 dt}f_i\right]. \quad (2)$$

The first two terms describe the energy-momentum changes via the scatterings with the medium constituents in the longitudinal direction (collisional energy loss), and in the transverse direction (transverse momentum broadening), respectively. The last two terms account for the medium-induced radiations. In this framework, all the strengths of the medium effects are determined solely by the jet quenching parameter for quarks $\hat{q}_q$ (see Ref. [68] for the details of the model for the jet shower evolution, and Ref. [16] for the choice of the parameters).

Assuming that the energy and momentum transferred to the medium partons from the jet are quickly thermalized, the source term $J^\nu(x)$ in Eq. (1) is constructed by using the solution of Eq. (2) as follows:

$$J^\nu(x) = -\sum_i\int\frac{d\omega_i dk_{i\perp}^2 d\phi_i}{2\pi}k_i^\nu\left[\left(\hat{e}_i\frac{\partial}{\partial\omega_i} + \frac{1}{4}\hat{q}_i\nabla_{k_\perp}^2\right)f_i(\omega_i, k_{i\perp}^2, t)\right]\delta^{(3)}\left(\boldsymbol{x} - \boldsymbol{x}_0^{\mathrm{jet}} - \frac{\boldsymbol{k}_i}{\omega_i}t\right). \quad (3)$$

Here, the jets are generated by the Glauber model+Pythia [69, 70], and assumed to be created in the transverse plane at $\eta_s = 0$. The initial proper time of the hydrodynamic medium evolution and jet-medium interaction is set as $\tau_0 = 0.6$ fm/c. Hereafter we refer to the contribution of particles in jet shower described by Eq. (2) as shower part, and to that of particles coming from the hydrodynamic medium response as hydro part.

4.2 Results

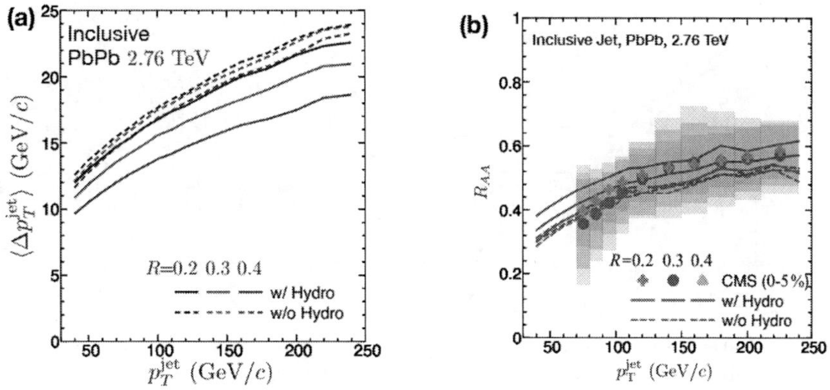

Figure 2. (a) Total $p_{\rm T}^{\rm jet}$ loss as a function of initial $p_{\rm T}^{\rm jet}$, and (b) nuclear modification factor for inclusive jets in central PbPb collisions at 2.76A TeV with the jet-cone sizes $R = 0.2, 0.3$, and 0.4 [16]. The solid lines show the results for jets with hydro part, and the dashed lines shows those for jets without hydro part. Experimental data are measured by the CMS Collaboration [71]. The colored shaded boxes denote the systematic uncertainties.

The jets in the final state are contributed from two parts: the shower part and the hydro part. The shower part loses energy due to three types of mechanism:

1. The collisional energy loss reduces the energy of all the jet shower partons.
2. The transverse momentum broadening tends to kick the jet shower partons out of the jet cone.
3. Medium-induced radiated partons outside the jet cone carry away the energy from the jet shower.

On the other hand, some of the particles from the hydro part carring the energy deposited by the jets via the mechanisms 1 and 2 are emitted inside the jet cone and detected as part of the jet. In this way, the lost energy from the shower part is partially recovered by the hydro part. To see this effect, the mean value of the total transverse momentum loss $\langle \Delta p_{\rm T,init}^{\rm jet} \rangle = \langle p_{\rm T,init}^{\rm jet} - p_{\rm T,final}^{\rm jet} \rangle$ for inclusive jets as a function of $p_{\rm T,init}^{\rm jet}$ is shown in Fig. 2 (a). One can see the feature that the total $p_{\rm T}^{\rm jet}$ loss with the hydro part is smaller than that without the hydro part due to the partial energy-momentum recovery mentioned above. Also, the jet cone size dependence becomes stronger when the hydro part is included. This is because more energy and momentum carried by the hydro part are recovered by the increasing of jet cone size. These effects of the hydro part can be seen also in the nuclear modification factor for single inclusive jet spectrum $R_{\rm AA} = (1/\langle N_{\rm coll} \rangle)(d^2N_{\rm jet}^{\rm AA}/d\eta_p dp_{\rm T}^{\rm jet})/(d^2N_{\rm jet}^{\rm pp}/d\eta_p dp_{\rm T}^{\rm jet})$ shown in Fig. 2 (b).

Figure 3 shows the jet shape function for leading and subleading jets in dijet events. The jet shape function is defined as

$$\rho(r) \quad = \quad \frac{1}{N_{\rm jet}} \sum_{\rm jet} \left[\frac{1}{p_T^{\rm jet}} \frac{\sum_{\rm trk \in (r-\delta r/2, r+\delta r/2)} p_T^{\rm trk}}{\delta r} \right], \tag{4}$$

where $r = \sqrt{(\eta_{\rm p} - \eta_{\rm jet})^2 + (\phi - \phi_{\rm jet})^2}$. In the small-$r$ region, the hydro part does not affect much the jet shape. However, the contribution of the hydro part becomes larger by increasing the value of r,

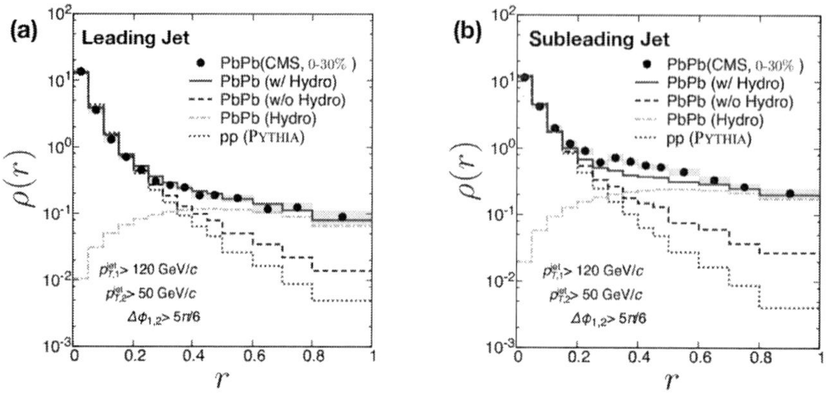

Figure 3. Jet shape functions for (a) leading and (b) subleading jet in dijet events at 2.76A TeV [16]. The dijet events are selected with the leading jet, $p_{T,1}^{\mathrm{jet}} > 120$ GeV/c, the subleading jet, $p_{T,2}^{\mathrm{jet}} > 50$ GeV/c, and the relative azimuth angle between the leading and subleading jets, $\Delta\phi_{1,2} > 5\pi/6$. The solid lines and dashed lines show the results for jets with and without hydro part in central PbPb collisions, respectively. The dashed dotted lines are the hydro part contribution in the jets with hydro part. The dotted lines are the results for p+p collisions obtained by Pythia. Experimental data are measured by the CMS Collaboration [71] (the shaded boxes indicate the systematic uncertainties).

and finally in the very large-r region ($r \gtrsim 0.5$), the contribution of the hydro part dominates the jet shape function. The result including the significant broadening by the hydro part provides a good description of the experimental data from the CMS collaboration [71] in the whole region ($0 \leq r \leq 1$), and indicates that jet-induced flow can spread quite widely around the jet axis. This means that the particles emitted at large angle regions with respect to the jet axis direction provide a good opportunity to study the medium response.

5 Conclusion, comments and outlook

Medium response to jets in heavy ion collisions appears as the excitation of the QGP medium caused by the jet energy-momentum deposition. The jet-correlated hadrons emitted by the excited medium can be detected as part of the jet and affect the energy and structures of jets in the final state. In the full jet study with Jet Shower Transport + Hydro model, the sizable contribution of the energy-momentum recovered by the medium response has been seen. The hadrons from the medium response are supposed to have small p_{T} compared with the hadrons fragmented from the hard intrinsic jet shower, and modify the hadron p_{T} distribution in jets. In Ref. [17], the γ-triggered jet fragmentation function and γ-hadron correlation as a function of $\xi = -\ln p_{\mathrm{T}}^{h}/p_{\mathrm{T}}^{\gamma}$ are calculated at RHIC energy by Coupled LBT Hydro model, and the significant enhancement of soft (large-ξ) hadrons due to the jet-induced medium excitation is shown. Also in a study with JEWEL [19], they show that the soft hadrons from the recoil partons modify the z_g-distribution in jets. The energy and momentum deposited in the medium are redistributed to the wide-angle region around the jet axis direction by the medium response. As a result, increase of the jet-cone size dependence in jet energy loss and the strong enhancement of jet shape at large-r have been seen in the results from Jet Shower Transport + Hydro model. The large number of soft particles at large angles from the jet axis due to the medium response effect is shown in the calculations of missing transverse momentum from various models such as hydrodynamics with source term [22], Hybrid Strong/Weak Coupling model [15], and AMPT [23] as well.

Medium response may provide new approaches to extract the QGP's transport properties from jet events. The shear viscosity dependence of the jet-induced flow pattern and the resulting jet-hadron correlation have been studied in several approaches based on linearized hydrodynamics [29–31, 34–36], or BSMPS model calculation [32, 33]. For further detailed study to connect them to realistic observables in heavy ion collisions, full $(3+1)$-dimensional viscous hydrodynamic model simulations would need to be done. While we have mostly focused on the dynamics of the deposited energy and momentum after their relaxation into the hydro mode, the detailed mechanism of the relaxation process is not fully understood but needed to be considered when obtaining the finite structure of the source term in Eq. (1). It is one of the essential problems of the jet study in heavy ion collisions, which defines the separation between the hard intrinsic jet scale and the soft medium fluid scale.

Thanks to a wealth of the experimental data and their sophisticated analyses, observables containing rich information of jet structures have become available. Medium response is an essential ingredient for an integrated theoretical framework for the jet study which should be developed for the interpretation of the experimental data from various aspects in order to achieve a comprehensive understanding of the jet quenching phenomenon.

6 Acknowledgements

The author would like to thank N.-B. Chang, T. Hirano, G.-Y. Qin, and X.-N. Wang for discussions on this topic. The author was supported in part by the Natural Science Foundation of China (NSFC) under Grants No. 11375072 and No. 11405066, by the Chinese Ministery of Science and Technology under Grant No. 2014DFG02050, and by the Major State Basic Research Development Program in China under Grant No. 2014CB845404.

References

[1] K. Yagi, T. Hatsuda, Y. Miake, Camb. Monogr. Part. Phys. Nucl. Phys. Cosmol. **23**, 1 (2005)
[2] U.W. Heinz, P.F. Kolb, Nucl. Phys. **A702**, 269 (2002)
[3] T.D. Lee, Nucl. Phys. **A750**, 1 (2005)
[4] M. Gyulassy, L. McLerran, Nucl. Phys. **A750**, 30 (2005)
[5] E.V. Shuryak, Nucl. Phys. **A750**, 64 (2005)
[6] T. Hirano, M. Gyulassy, Nucl. Phys. **A769**, 71 (2006)
[7] B. Schenke, S. Jeon, C. Gale, Phys. Rev. Lett. **106**, 042301 (2011)
[8] Z. Qiu, U.W. Heinz, Phys. Rev. **C84**, 024911 (2011)
[9] Z. Qiu, C. Shen, U. Heinz, Phys. Lett. **B707**, 151 (2012)
[10] C. Gale, S. Jeon, B. Schenke, P. Tribedy, R. Venugopalan, Phys. Rev. Lett. **110**, 012302 (2013)
[11] J.P. Blaizot, Y. Mehtar-Tani, Int. J. Mod. Phys. **E24**, 1530012 (2015)
[12] G.Y. Qin, X.N. Wang, Int. J. Mod. Phys. **E24**, 1530014 (2015)
[13] X.N. Wang, Y. Zhu, Nucl. Phys. **A932**, 447 (2014)
[14] Y. He, T. Luo, X.N. Wang, Y. Zhu, Phys. Rev. **C91**, 054908 (2015)
[15] J. Casalderrey-Solana, D. Gulhan, G. Milhano, D. Pablos, K. Rajagopal, JHEP **03**, 135 (2017)
[16] Y. Tachibana, N.B. Chang, G.Y. Qin, Phys. Rev. **C95**, 044909 (2017)
[17] W. Chen, S. Cao, T. Luo, L.G. Pang, X.N. Wang (2017)
[18] R. Kunnawalkam Elayavalli, K.C. Zapp, JHEP **07**, 141 (2017)
[19] J.G. Milhano, U.A. Wiedemann, K.C. Zapp, `arXiv:1707.04142`
[20] C. Roland, J.Phys. **G38**, 124020 (2011)

[21] S. Chatrchyan et al. (CMS Collaboration), Phys. Rev. **C84**, 024906 (2011)
[22] Y. Tachibana, T. Hirano, Phys. Rev. **C90**, 021902 (2014)
[23] Z. Gao, G.L. Ma, G.Y. Qin, H.Z. Zhang, `arXiv:1612.02548`
[24] H. Stoecker, Nucl. Phys. **A750**, 121 (2005)
[25] J. Casalderrey-Solana, E.V. Shuryak, D. Teaney, J.Phys.Conf.Ser. **27**, 22 (2005)
[26] J. Ruppert, B. Muller, Phys. Lett. **B618**, 123 (2005)
[27] L.M. Satarov, H. Stoecker, I.N. Mishustin, Phys. Lett. **B627**, 64 (2005)
[28] T. Renk, J. Ruppert, Phys. Rev. **C73**, 011901 (2006)
[29] R.B. Neufeld, Phys. Rev. **C79**, 054909 (2009)
[30] R.B. Neufeld, T. Renk, Phys. Rev. **C82**, 044903 (2010)
[31] R.B. Neufeld, I. Vitev, Phys. Rev. **C86**, 024905 (2012)
[32] I. Bouras, A. El, O. Fochler, H. Niemi, Z. Xu et al., Phys. Lett. **B710**, 641 (2012)
[33] I. Bouras, B. Betz, Z. Xu, C. Greiner, Phys. Rev. **C90**, 024904 (2014)
[34] A. Ayala, J.D. Castaño-Yepes, I. Dominguez, M.E. Tejeda-Yeomans, Phys. Rev. **C92**, 024910 (2015)
[35] A. Ayala, I. Dominguez, J. Jalilian-Marian, M.E. Tejeda-Yeomans, Phys. Rev. **C94**, 024913 (2016)
[36] L. Yan, S. Jeon, C. Gale, `arXiv:1707.09519`
[37] T. Renk, Phys. Rev. **C88**, 044905 (2013)
[38] E. Iancu, B. Wu, JHEP **10**, 155 (2015)
[39] K.C. Zapp, F. Krauss, U.A. Wiedemann, JHEP **1303**, 080 (2013)
[40] K.C. Zapp, Eur. Phys. J. **C74**, 2762 (2014)
[41] R. Baier, Y.L. Dokshitzer, A.H. Mueller, S. Peigne, D. Schiff, Nucl. Phys. **B483**, 291 (1997)
[42] B.G. Zakharov, JETP Lett. **63**, 952 (1996)
[43] R. Baier, Y.L. Dokshitzer, A.H. Mueller, S. Peigne, D. Schiff, Nucl. Phys. **B484**, 265 (1997)
[44] B.G. Zakharov, JETP Lett. **65**, 615 (1997)
[45] R. Baier, Y.L. Dokshitzer, A.H. Mueller, D. Schiff, Nucl. Phys. **B531**, 403 (1998)
[46] T. Luo, Y. He, E. Wang, X.N. Wang, Nucl. Phys. **A956**, 585 (2016)
[47] S. Cao, T. Luo, G.Y. Qin, X.N. Wang, Phys. Rev. **C94**, 014909 (2016)
[48] S. Cao, T. Luo, G.Y. Qin, X.N. Wang, `arXiv:1703.00822`
[49] X.N. Wang, X.f. Guo, Nucl. Phys. **A696**, 788 (2001)
[50] A. Majumder, Phys. Rev. **D85**, 014023 (2012)
[51] A.K. Chaudhuri, U. Heinz, Phys. Rev. Lett. **97**, 062301 (2006)
[52] A.K. Chaudhuri, Phys. Rev. **C75**, 057902 (2007)
[53] A.K. Chaudhuri, Phys. Rev. **C77**, 027901 (2008)
[54] R.B. Neufeld, B. Muller, J. Ruppert, Phys. Rev. **C78**, 041901 (2008)
[55] R.B. Neufeld, Phys. Rev. **D78**, 085015 (2008)
[56] R.B. Neufeld, B. Muller, Phys. Rev. Lett. **103**, 042301 (2009)
[57] G.Y. Qin, A. Majumder, H. Song, U. Heinz, Phys. Rev. Lett. **103**, 152303 (2009)
[58] B. Betz, J. Noronha, G. Torrieri, M. Gyulassy, D.H. Rischke, Phys. Rev. Lett. **105**, 222301 (2010)
[59] R.P.G. Andrade, J. Noronha, G.S. Denicol, Phys. Rev. **C90**, 024914 (2014)
[60] S. Floerchinger, K.C. Zapp, Eur. Phys. J. **C74**, 3189 (2014)
[61] M. Schulc, B. Tomášik, Phys. Rev. **C90**, 064910 (2014)

[62] Y. Tachibana, T. Hirano, Phys. Rev. **C93**, 054907 (2016)
[63] F. Cooper, G. Frye, Phys. Rev. **D10**, 186 (1974)
[64] M. Okai, K. Kawaguchi, Y. Tachibana, T. Hirano, Phys. Rev. **C95**, 054914 (2017)
[65] C. Shen, B. Schenke, arXiv:1710.00881
[66] Z. Hulcher, D. Pablos, K. Rajagopal, arXiv:1707.05245
[67] P.M. Chesler, K. Rajagopal, Phys. Rev. **D90**, 025033 (2014)
[68] N.B. Chang, G.Y. Qin, Phys. Rev. **C94**, 024902 (2016)
[69] M.L. Miller, K. Reygers, S.J. Sanders, P. Steinberg, Ann. Rev. Nucl. Part. Sci. **57**, 205 (2007)
[70] T. Sjostrand, S. Mrenna, P.Z. Skands, Comput. Phys. Commun. **178**, 852 (2008)
[71] V. Khachatryan et al. (CMS Collaboration), Phys. Rev. **C96**, 015202 (2017)

Measurements of jets in heavy ion collisions

Christine Nattrass[1,⋆]

[1] *University of Tennessee, Knoxville, TN, USA-37996*

Abstract. The Quark Gluon Plasma (QGP) is created in high energy heavy ion collisions at the Relativistic Heavy Ion Collider (RHIC) and the Large Hadron Collider (LHC). This medium is transparent to electromagnetic probes but nearly opaque to colored probes. Hard partons produced early in the collision fragment and hadronize into a collimated spray of particles called a jet. The partons lose energy as they traverse the medium, a process called jet quenching. Most of the lost energy is still correlated with the parent parton, contributing to particle production at larger angles and lower momenta relative to the parent parton than in proton-proton collisions. This partonic energy loss can be measured through several observables, each of which give different insights into the degree and mechanism of energy loss. The measurements to date are summarized and the path forward is discussed.

1 Introduction

A hot, dense liquid of quarks and gluons called the Quark Gluon Plasma (QGP) is formed when heavy ions collide at relativistic speeds. High energy partons scattered early in the collision can be used as probes of the medium. In high energy collisions where a QGP is not formed, such as p+p collisions, these hard scattered partons will fragment into a collimated spray of particles called a jet. Partons traversing a QGP can lose energy either radiatively or through collisions with medium partons. These interactions lead to jets which are broader on average than those in p+p collisions, with fewer particles carrying a large fraction of the jet's energy, $z = p_T/E_{jet}$, and more particles at lower z.

This picture of partonic energy loss, concurrent with broadening of the jet and softening of the fragmentation function, has been qualitatively confirmed by several measurements [1]. Recently a number of novel observables, mostly used in particle physics for distinguishing between quark and gluon jets, have been investigated as a means for measuring the modifications of the struture of jets through medium interactions in greater detail.

While there are extensive measurements, the constraints of the properties of the QGP from measurements of jets are largely informed by only a handful of measurements of the modifications of single particle spectra [2]. As the field works to measure jet observables to higher precision, it is worth remembering some of the lessons from particle physics about the complications of measurements of jets. The background in heavy ion collisions is very large and the experimental techniques may lead to a number of biases in the results.

⋆e-mail: christine.nattrass@utk.edu

© The Authors, published by EDP Sciences. This is an open access article distributed under the terms of the Creative Commons Attribution License 4.0 (http://creativecommons.org/licenses/by/4.0/).

2 Measurements of jets

In $p+p$ collisions, hard partons scattered early in the collision fragment into collimated sprays of particles called jets. As high energy partons traverse the medium, they lose energy though emission of gluons and collisions with medium partons. This lost energy retains most of its spatial correlations with the parent parton, so jet quenching will result in a suppression of final state hadrons carrying a large fraction of the parton's momentum and an enhancement in particles carrying a small fraction of the parton's momentum. On average the final state hadrons will be somewhat less correlated with the parent parton's momentum than before medium interactions, resulting in a broadening of the distribution of jet constituents.

2.1 Energy loss

The most straightforward way to measure partonic energy loss is through the measurement of the nuclear modification factor

$$R_{AA} = \frac{\sigma^{NN}}{\langle N_{coll}\rangle} \frac{d^2N_{AA}/dp_T d\eta}{d^2\sigma_{pp}/dp_T d\eta} \quad (1)$$

where σ^{NN} is the integrated nucleon-nucleon cross section, $\langle N_{coll}\rangle$ is the average number of binary nucleon-nucleon collisions for a given range of impact parameter, N_{AA} is the number of particles, $p_{\rm T}$ is the transverse momentum, η is the pseudorapidity, and σ_{pp} is the inelastic cross section in $p+p$ collisions. If nucleus-nucleus collisions were simply a superposition of nucleon-nucleon collisions, R_{AA} would be approximately one because the high $p_{\rm T}$ particle cross section would scale with the number of binary collisions.

The R_{AA} has been measured for several different final state hadrons, ranging in mass from pions [3, 4] to B mesons [5], and the production rate of all of these colored probes are suppressed at high momenta in central heavy ion collisions. In contrast, the medium is nearly transparent to electroweak probes. The R_{AA} of reconstructed jets is also substantially suppressed [6–8]. Even heavy quarks are suppressed at comparable levels to light hadrons, as shown by measurements of single final state particles [5, 9, 10] and reconstructed jets [11–13]. This energy loss has also been seen using dihadron correlations [14–16], hadron-jet correlations [17, 18], the azimuthal anisotropies of jets [19, 20], and di-jet energy asymmetries [21–23].

2.2 Fragmentation

The most direct observations of the broadening of the distribution of jet constituents and the softening of those constituents is through measurements of the fragmentation function. At leading order, isolated direct photons are produced in inverse Compton scattering, q+g $\rightarrow$ q+γ. The energy of the outgoing quark jet will equal the energy of the photon. Fragmentation functions can be measured through this process and demonstrate both a suppression of high $z = p_T/E_{jet}$ hadrons and an enhancement of low z processes [24]. Measurements of fragmentation functions using reconstructed jets also show this suppression [25–27]. As for energy loss, modified fragmentation in heavy ion collisions has been observed through high momentum di-hadron correlations [28], jet-hadron correlations [29, 30], and measurements of di-jet energy asymmetries [23].

Partonic energy loss is expected to be path length dependent. The path length can be varied using measurements of jets relative to the reaction plane. Studies of jet-hadron correlations with the angle of the jet relative to the reaction plane varied found no path length dependence within uncertainties [31]. This can be explained by theoretical models which include jet-by-jet fluctuations in the amount of energy lost [32], meaning that while on average jets traveling out-of-plane will lose more energy and be more modified by interactions with the medium than those traveling in-plane because they see more medium, a particular jet going out-of-plane may lose less energy than a particular jet going in-plane. This indicates that fluctuations such as these can have a large impact on observables and may be more important than the path length dependence of energy loss in some cases.

2.3 Jet structure and new observables

Recent studies have begun investigating new observables to see if they may be more sensitive to medium modifications of jets in the medium than traditional observables. Many of these observables are used in particle physics for separating quark and gluon jets. The jet girth, g, is given by

$$g = \sum_i \frac{p_T^i}{p_T^{jet}} |r_i|, \tag{2}$$

where r_i is the angular distance between the jet axis and particle i. The dispersion is given by

$$\mathrm{p_T^D} = \sqrt{\Sigma_i p_{T,i}^2} / \Sigma_i p_{T,i} \tag{3}$$

where $p_{T,i}$ is the momentum of a constituent and the sum is over the constituents. The girth and the dispersion both are measures of the width of the jet. Measurements by ALICE indicate that, if anything, surviving jets in $Pb+Pb$ collisions are narrower than in $p+p$ collisions [33]. This apparent contradiction with the observations of modified fragmenation can be explained by a bias in the surviving jets reconstructed by a jet finder. Jets which are reconstructed at high momenta may be more likely to be unmodified or to have been quark jets, which are narrower. It is important to note that nearly every observable is biased towards a particular selection of jets, so it is necessary to look at multiple observables to get a complete picture of partonic energy loss.

$LeSub$, the difference between the momenta of the leading and subleading jet constituents, has also been measured in $Pb+Pb$ collisions and is consistent with expectations for both unmodified jets and jets which have interacted with the medium [33]. This observable is robust to a high background but may not be sensitive. Jet grooming, a technique to measure the substructure of jets, is also under investigation [34, 35], as are measurements of the jet mass [36, 37].

3 The path forward

Several important steps were proposed in [1] in order to determine the properties of the medium with higher precision from measurements of jets. The main improvements required are to understand the biases in measurements, make more quantitative measurements, make more quantitative comparisons to theory, and to come to an agreement on the treatment of background in heavy ion collisions. Each of these points is summarized below.

3.1 Understand bias

The techniques used for the suppression and subtraction of background impose biases in the results. Even the use of a jet finder may bias the sample towards less modified jets. This is unavoidable because of the large backgrounds in central heavy ion collisions, but the discussions of these techniques should not be relegated to the method section. Attempts to correct for these biases often use jets from PYTHIA embedded in a heavy ion collision. Jets in heavy ion collisions fragment differently so such corrections may skew the results. Rather than trying to correct for such biases, the bias should be considered part of the definition of the measurement. Theoretical calculations should apply the same techniques in order to be comparable to the data.

3.2 An agreement on the treatment of background

The history of the definition of jets in particle physics provides some useful lessons for measurements in heavy ion collisions. After years of different definitions of jets in theory and experiment, the Snowmass Accord postulated that a good jet finding algorithm must be both infrared and colinear safe, meaning that the same jet would be found in the presence of either soft radiation (which cannot be calculated to high precision perturbatively) and if a leading parton had a colinear splitting [38]. Seeded cone algorithms were replaced by sequential algorithms [39]. The definition of the jet is a combination of the jet finding algorithm and the resolution parameter, which limits the size of the jet.

There are several experimental approaches to the treatment of background. For example, ALICE explicitly requires a jet to have a constituent of at least 5 GeV/c [6] and ATLAS requires jets to be matched to a track jet consisting of high momentum tracks [7]. In the limit of high energy jets, these algorithms appear to give consistent results, but at lower energies they may not. An example of this is the tension between ALICE, ATLAS, and CMS measurements of jet R_{AA} [1]. Since these background subtraction and suppression algorithms impose a bias, a similar approach to that in particle physics is probably necessary, that the background suppression and subtraction techniques be considered part of the definition of a jet.

Experimentally, the operational distinction between the jet signal and the background is based on whether or not final state hadrons retain some short range correlations with the parent parton, typically $R = \sqrt{\Delta\eta^2 + \Delta\phi^2} \leq 1$. Theoretically, it is less clear what should be considered a background particle. We therefore called for a meeting in [1] to discuss the definition of the background and come to an agreement between theorists and experimentalists on the definition and treatment of the background.

3.3 Make more quantitative comparisons to theory

The best constraint of the properties of the medium to date comes from the JET collaboration [2]. In this study, the partonic energy loss of a hard parton traveling through a static brick of QGP was calculated using several different theoretical models for partonic energy loss. These calculations were systematically compared to measurements of single hadron R_{AA} in order to constrain $\hat{q}$, the partonic energy loss in the medium squared divided by the path length. While this is a monumental achievement, theoretical developments allowing more accurate calculations of partonic energy loss and comparisons to other observables would likely constrain the properties of the medium better.

The JETSCAPE collaboration is working on developing realistic event generators combining realistic hydrodynamical simulations with realistic partonic energy loss models. The final stage of the JETSCAPE project will be to use these models to do a Bayesian analysis in order to constrain the properties of the QGP. Such analyses have been done incorporating measurements of single hadron spectra and azimuthal anisotropies [40, 41]. These Bayesian analyses require detailed accounting of systematic uncertainties and their correlations with each other and possibly with other experiments. This detailed accounting of systematic uncertainties is not currently the norm in the field, but without it we cannot optimize our constraints of QGP properties.

3.4 Make more differential measurements

Since the field to date has not had a full simulation of jets in a hydrodynamical fluid, it has not yet been possible to apply experimental methods to a model to see which observables are least sensitive to the background and most sensitive to the properties of the QGP. In the era of JETSCAPE such studies will be possible. It may turn out that some of the new observables to measure jet structure are the most sensitive, however, it is still unclear. It is important that the field do these detailed studies to find the best observables and to stop focusing on observables which turn out to be insensitive to the properties of the QGP. Measurements of jets by the sPHENIX experiment will also be important for constraining the properties of the medium at RHIC energies.

4 Conclusions

Jets are used to measure partonic energy loss, a process which can lead to constraints on the properties of the QGP. An impressive array of measurements have demonstrated that the theoretical picture of partonic energy loss in the medium is correct, however, very few of these measurements have yet informed quantitative constraints on the properties of the QGP. Forthcoming theoretical developments will allow the incorporation of jet observables into Bayesian analyses so that they can be used to constrain QGP properties. However, to do this successfully requires an agreement between theorists and experimentalists on the appropriate treatment of the background. It furthermore requires an understanding of measurement biases and the sensitivities of various observables to medium properties. It will also require careful reporting of systematic uncertainties and their correlations. We will only be able to make precision constraints of the properties of the medium by doing these things.

References

[1] M. Connors, C. Nattrass, R. Reed, S. Salur (2017), 1705.01974
[2] K.M. Burke et al. (JET), Phys.Rev. **C90**, 014909 (2014)
[3] A. Adare et al. (PHENIX), Phys. Rev. Lett. **101**, 232301 (2008)
[4] J. Adam et al. (ALICE), Phys. Rev. **C93**, 034913 (2016)
[5] A.M. Sirunyan et al. (CMS), Phys. Rev. Lett. **119**, 152301 (2017), 1705.04727
[6] J. Adam et al. (ALICE), Phys.Lett. **B746**, 1 (2015)
[7] G. Aad et al. (ATLAS), Phys.Rev.Lett. **114**, 072302 (2015)
[8] V. Khachatryan et al. (CMS), Phys. Rev. **C96**, 015202 (2017), 1609.05383
[9] A. Adare et al. (PHENIX Collaboration), Phys.Rev. **C84**, 044905 (2011)
[10] J. Adam et al. (ALICE), JHEP **03**, 081 (2016)
[11] S. Chatrchyan et al. (CMS), Phys. Rev. Lett. **113**, 132301 (2014), [Erratum: Phys. Rev. Lett.115,no.2,029903(2015)]
[12] V. Khachatryan et al. (CMS), Phys. Lett. **B754**, 59 (2016)
[13] A.M. Sirunyan et al. (CMS), Phys. Lett. **B772**, 306 (2017)
[14] C. Nattrass, N. Sharma, J. Mazer, M. Stuart, A. Bejnood, Phys. Rev. **C94**, 011901 (2016)
[15] J. Adams et al. (STAR), Phys.Rev.Lett. **97**, 162301 (2006)
[16] K. Aamodt et al. (ALICE), Phys.Rev.Lett. **108**, 092301 (2012)
[17] J. Adam et al. (ALICE), JHEP **09**, 170 (2015)
[18] L. Adamczyk et al. (STAR) (2017), 1702.01108
[19] J. Adam et al. (ALICE), Phys. Lett. **B753**, 511 (2016)
[20] G. Aad et al. (ATLAS), Phys.Rev.Lett. **111**, 152301 (2013)
[21] G. Aad et al. (ATLAS), Phys.Rev.Lett. **105**, 252303 (2010)
[22] S. Chatrchyan et al. (CMS), Phys. Rev. **C84**, 024906 (2011)
[23] L. Adamczyk et al. (STAR), Phys. Rev. Lett. **119**, 062301 (2017), 1609.03878
[24] A. Adare et al. (PHENIX), Phys.Rev. **D82**, 072001 (2010)
[25] G. Aad et al. (ATLAS), Phys.Lett. **B739**, 320 (2014)

[26] S. Chatrchyan et al. (CMS), Phys.Rev. **C90**, 024908 (2014)
[27] S. Chatrchyan et al. (CMS), JHEP **1210**, 087 (2012)
[28] G. Agakishiev et al. (STAR), Phys.Rev. **C85**, 014903 (2012)
[29] V. Khachatryan et al. (CMS), JHEP **02**, 156 (2016)
[30] L. Adamczyk et al. (STAR), Phys.Rev.Lett. **112**, 122301 (2014)
[31] J. Mazer (ALICE), J. Phys. Conf. Ser. **832**, 012010 (2017), `1703.09287`
[32] K.C. Zapp, Phys.Lett. **B735**, 157 (2014)
[33] L. Cunqueiro (ALICE), Nucl. Phys. **A956**, 593 (2016)
[34] Y.j. Lee, These proceedings (2017)
[35] A.M. Sirunyan et al. (CMS) (2017), `1708.09429`
[36] A. Majumder, J. Putschke, Phys. Rev. **C93**, 054909 (2016)
[37] S. Acharya et al. (ALICE) (2017), `1702.00804`
[38] J.E. Huth et al., *Toward a standardization of jet definitions*, in *1990 DPF Summer Study on High-energy Physics: Research Directions for the Decade (Snowmass 90) Snowmass, Colorado, June 25-July 13, 1990* (1990), pp. 0134–136, `http://lss.fnal.gov/cgi-bin/find_paper.pl?conf-90-249`
[39] G.P. Salam, Eur.Phys.J. **C67**, 637 (2010)
[40] J. Novak, K. Novak, S. Pratt, J. Vredevoogd, C. Coleman-Smith, R. Wolpert, Phys. Rev. **C89**, 034917 (2014)
[41] J.E. Bernhard, J.S. Moreland, S.A. Bass, J. Liu, U. Heinz, Phys. Rev. **C94**, 024907 (2016)

Photoproduction of Upsilon states in ultraperipheral collisions at the LHC

Magno Machado[1,*]

[1]*High Energy Physics Phenomenology Group, GFPAE IF-UFRGS*
Caixa Postal 15051, CEP 91501-970, Porto Alegre, RS, Brazil

Abstract. The exclusive photoproduction of upsilon states is investigated in the context of ultra-peripheral collisions at the LHC energies. Predictions are presented for their production in proton-proton, proton-nucleus and nucleus-nucleus collisions at the energies available at the LHC run 2. The theoretical framework considered in the analysis is the light-cone color dipole formalism, which includes consistently parton saturation effects and nuclear shadowing corrections. For PbPb collisions we consider both the coherent and incoherent contributions.

1 Introduction

The exclusive photoproduction of heavy vector mesons is a kind of diffractive process where, besides a soft scale characterized by the hadron size, there is clearly a hard scale (mesons mass m_V) that allows to analyse the reaction from the perturbative QCD point of view. This advantage creates ways to investigate the pomeron exchange which could lead to a better understanding of this object in terms of QCD. Its investigation in ultraperipheral collisions (UPCs) [1] is an essential tool to understand the low-x physics and also to investigate the gluon density in this regime. In the UPC case, the exclusive photoproduction dominates the process through the emission of quasi-virtual photons which interact with the target. The theoretical approaches addressing such reactions include vector meson dominance (VDM), perturbative QCD calculations, k_T-factorization approach and color dipole formalism. The photon-target interaction amplitude, when considering the light-cone dipole formalism [2], can be written as a convolution between the photon-meson wave functions overlap and the elementary dipole-target cross section [3]. In addition, the process considered here is quasi-elastic, $Q^2 \approx 0$, and in the region of small-x, the gluon density may increase to the point where gluon fusion, gg $\rightarrow$ g, becomes significant. This kind of fusion produces nonlinear effects in the evolution equations. For instance, at mid-rapidity the typical value of Bjorken variable is $x = \frac{m_\Upsilon}{\sqrt{s_{AA}}} \simeq 10^{-3}$ for PbPb collisions at the LHC. Differently from DGLAP equations which is a linear equation, dipole models incorporate linear and non-linear effects[4]. In pQCD the exclusive meson photoproduction has a differential cross-section $\propto [xg(x, Q^2)]^2$, where the gluon distribution functions do not take into account the effects of saturation since they are evolved by DGLAP equations. The inclusion of parton saturation and nuclear effects are crucial in describing even the experimental observations in heavy-ion collisions at RHIC (for instance, see Ref. [5]). Within the color dipole approach one can introduce information on dynamics

*e-mail: magnus@if.ufrgs.br

© The Authors, published by EDP Sciences. This is an open access article distributed under the terms of the Creative Commons Attribution License 4.0 (http://creativecommons.org/licenses/by/4.0/).

beyond the leading logarithmic QCD approach for meson production and computing predictions for the radially excited states is a reasonably easy task [3]. From the experimental point of view, the considered process is quite clear due to the presence of two rapidity gaps. The rapidity gap describes a region between the axis beam and the decay of meson ($e^+ + e^-, \mu^- + \mu^+$) free of particles. In this way, the experimental observation of this kind of quasi-diffractive process is facilitated. In addition, the absorption corrections in this case is not strong as in the corresponding final state in proton-proton collisions.

In this contribution we summarize the results obtained in Refs. [6, 7], where we investigate the exclusive production of $\Upsilon(1S)$ and its radially excited states $\Upsilon(2S)$ and $\Upsilon(3S)$ in PbPb collisions for LHC energy. In Ref. [7] the coherent photoproduction of Υ states at various energies in pp, pPb PbPb collisions at the LHC has been considered. Those calculations were carried out in the theoretical framework of the color light-cone dipole formalism [2] and focused only on the coherent channel where the initial state particles remain intact after interaction. It was shown that the corresponding predictions describe correctly the experimental results from LHCb Collaboration [8] for Υ photoproduction in pp collisions. Those data were obtained for typically large rapidities and the x-values to be covered are increasingly smaller for forward rapidities. We roughly get $x = \frac{m_\Upsilon}{\sqrt{s_{AA}}}e^{-y} \simeq 8 \times 10^{-5}$ at $y = 3$ and it is clear that gluon dynamics is being probed at extremely low-x and low perturbative scales $\mu^2 \simeq 20$ GeV^2. This kinematical range is in the limit of application of usual pQCD and saturation approach should be relevant. For nuclear targets, the nuclear saturation scale $Q^2_{sat,A} \simeq cA^{1/3}Q^2_{sat,p}$ (with $c \simeq 0.3$) reaches 2 GeV^2 in those cases [9]. In Ref. [6], a detailed study of the incoherent cross section for the upsilon states has been done. This is quite important, as it was pointed out that incoherent diffraction probes the fluctuations in the interactions strengths of multiparton Fock states in the nuclear wavefunctions [10]. The connection between incoherent diffraction and fluctuations is a quite rich subject and the pioneering works are found in Refs. [11]. Recently, the topic is very active and we call attention to the following works [12, 13].

This contribution is organized as follows. In the next section we give the main theoretical information to obtain the rapidity distribution of coherent and incoherent production of $\Upsilon(1S, 2S, 3S)$ states in hadron-hadron and heavy ion collisions for the energies available at the LHC. The approach considered here is quite robust due to the successful description of experimental results measured by LHCb Collaboration [8] for $\Upsilon(1S)$ in pp collisions. In the section 3 we present the phenomenological calculations, discuss the main theoretical uncertainties and a comparison with other approaches is performed. For instance, we compare the present calculation to the predictions available using the STARlight Monte Carlo [14, 15]. Finally, we show the main conclusions.

2 Theoretical Framework

The exclusive meson photoproduction in nucleus-nucleus collisions can be factorized in terms of the equivalent flux of photons of the nucleus projectile and photon-target production cross section [1]. In UPCs there is the absence of strong interactions between the projectile particle and the target. In this case the reaction is characterized by impact parameter $> 2\,R_A$ and as the interaction is ultra-relativistic and purely electromagnetic, one can use the Weizsäcker-Williams approximation [1]. The photon energy spectrum, $dN_\gamma^A/d\omega$, which depends on the photon energy ω, is well known [1]. The rapidity distribution for Υ states photoproduction in AA collisions can be written down as,

$$\frac{d\sigma}{dy}(AA \to A \otimes \Upsilon(nS) \otimes Y) = \omega \frac{dN_\gamma^A}{d\omega}\, \sigma(\gamma A \to \Upsilon(nS)\, Y) + (y \to -y)\,, \tag{1}$$

where the photon flux in nucleus is denoted by $dN_\gamma^A/d\omega$ and $Y = A$ (coherent case) or $Y = A^*$ (incoherent case). The symbol $\otimes$ denotes the large rapidity gap between the produced meson and the final states nucleus.

Similar calculation can be carried out in proton-proton and proton-nucleus interactions. For highly energetic protons as those produced by the LHC beams the photon flux can be approximated by [1, 15]

$$\frac{dN_\gamma^p}{d\omega} = \frac{\alpha_{em}}{2\pi\omega}\left[1+\left(1-\frac{2\omega}{\sqrt{s}}\right)^2\right]\left[\ln\Omega - \frac{11}{6}+\frac{3}{\Omega}-\frac{3}{2\Omega^2}+\frac{1}{3\Omega^3}\right] \quad (2)$$

where $\Omega = 1 + 0.71\ \mathrm{GeV}^2/Q_{min}^2$ e $Q_{min}^2 = (\omega/\gamma_L)^2$, with $\gamma_L = \sqrt{s}/\left(2m_p\right)$. On the other hand, for nucleus having charge Z, the photon flux is approximately given by [1, 15]

$$\frac{dN_\gamma^A}{d\omega} = \frac{2Z^2\alpha_{em}}{\pi}\left[\xi K_0(\xi)K_1(\xi) - \frac{\xi^2}{2}\left(K_1^2(\xi) - K_0^2(\xi)\right)\right], \quad (3)$$

where $\xi = \omega(R_p + R_A)/\gamma_L$ for pA collisions and $\xi = \omega(2R_A)/\gamma_L$ for AA collisions. We will discuss about the rapidity distribution for those collisions in next section.

The produced state with mass m_V has rapidity $y \simeq \ln(2\omega/m_V)$ and the square of the γA centre-of-mass energy is given by $W_{\gamma A}^2 \simeq 2\omega\sqrt{s}$. The photon-Pomeron interaction will be described within the light-cone dipole frame, where the probing projectile fluctuates into a quark-antiquark pair with transverse separation r (and momentum fraction z) long after the interaction, which then scatters off the hadron. The cross section for exclusive photoproduction of Υ states off a nucleon target is given by,

$$\sigma(\gamma p \to \Upsilon\, p) = \frac{\left|\sum_{h,\bar{h}}\int dz\, d^2r\, \Psi^\gamma_{h,\bar{h}}\sigma_{dip}(x,r)\,\Psi^{V*}_{h,\bar{h}}\right|^2}{16\pi B_\Upsilon}, \quad (4)$$

where Ψ^γ and Ψ^V are the light-cone wavefunction of the photon and of the vector meson ($V = \Upsilon$), respectively. The dipole-proton cross section is denoted by $\sigma_{dip}(x,r)$ and the diffractive slope parameter by B_V. In this context, we are implicitly assuming that the proton shape is Gaussian and that the impact parameter dependence factorizes out from the dipole-nucleon scattering amplitude. Here, we consider the energy dependence of the slope using the Regge motivated expression [7].

The exclusive photoproduction off nuclei for coherent and incoherent processes can be simply computed in high energies where the large coherence length $l_c \gg R_A$ is fairly valid. The expressions for both cases are given by [16],

$$\sigma(\gamma A \to \Upsilon A) = \int d^2b\left|\sum_{h,\bar{h}}\int dz\, d^2r\, \Psi^\gamma_{h,\bar{h}}\Psi^{V*}_{h,\bar{h}}\left[1-\exp\left(-\frac{1}{2}\sigma_{dip}(x,r)T_A(b)\right)\right]\right|^2,$$

$$\sigma(\gamma A \to \Upsilon A^*) = \int d^2b\,\frac{T_A(b)}{16\pi\, B_V}\left|\sum_{h,\bar{h}}\int dz\, d^2r\, \Psi^\gamma_{h,\bar{h}}\Psi^{V*}_{h,\bar{h}}\sigma_{dip}(x,r)\exp\left[-\frac{1}{2}\sigma_{dip}(x,r)T_A(b)\right]\right|^2,$$

where $T_A(b) = \int dz\rho_A(b,z)$ is the nuclear thickness function. In the numerical evaluations, we have considered the boosted Gaussian wavefunction and several phenomenological saturation models, which encode the main properties of the saturation approaches. Accordingly, the cross sections above include both the skwedness and real part of amplitude corrections. Namely, we multiply the result above by $K^2 = R_g^2(1+\beta^2)$, where $\beta = \tan(\pi\lambda_{ef}/2)$ is the ratio of real to imaginary parts of the scattering amplitude and R_g incorporates the off-forward correction (see [7] for details). The effective power on energy, λ_{ef} is determined for each case. In order to take into account the threshold correction for the dipole cross section, we have multiplied them by a factor $(1-x)^7$.

Finally, we set the parameters and phenomenological models to be considered in next section. For the slope parameter it was considered the energy dependency from the Regge phenomenology [7],

$$B_\Upsilon = b_{el}^\Upsilon + 2\alpha'\log\left(\frac{W_{\gamma A}^2}{W_0^2}\right) \tag{5}$$

with $\alpha' = 0.164$ GeV^{-2}, $W_0 = 95$ GeV, $b_{el}^{\Upsilon(1S)} = 3.68$ GeV^{-2}, $b_{el}^{\Upsilon(2S)} = 3.61$ GeV^{-2} and $b_{el}^{\Upsilon(3S)} = 3.57$ GeV^{-2}. It will be taken into account only for the incoherent cross section. For the meson wavefuntion, we will use the Boosted-Gaussian model [17] because it can be applied in a systematic way for excited states. The corresponding function is given by [18],

$$\phi_{nS}(r,z) = \left[\sum_{k=0}^{n-1} \alpha_{nS,k} R_{nS}^2 \hat{\mathcal{D}}^{2k}(r,z)\right] G_{nS}(r,z), \tag{6}$$

with $\alpha_{nS,0} = 1$. The operator $\hat{\mathcal{D}}^2(r,z)$ is defined by

$$\hat{\mathcal{D}}^2(r,z) = \frac{m_f^2 - (\frac{1}{r}\partial_r + \partial_r^2)}{4z(1-z)} - m_f^2, \tag{7}$$

and it acts on the following generatrix function

$$G_{nS}(r,z) = \mathcal{N}_{nS}\, z(1-z)\, \exp\left(-\frac{m_f^2\mathcal{R}_{nS}^2}{8z(1-z)} - \frac{2z(1-z)r^2}{\mathcal{R}_{nS}^2} + \frac{m_f^2\mathcal{R}_{nS}^2}{2}\right). \tag{8}$$

The main physical quantity is the dipole scattering cross section. We consider the following phenomenological models in our analysis: GBW [19], CGC [20] and BCGC [21]. The GBW model is defined by the eikonal shape for the dipole cross section,

$$\sigma_{q\bar{q}}^{GBW}(x,r) = \sigma_0\left(1 - e^{-r^2Q_s^2(x)/4}\right), \tag{9}$$

where $\sigma_0 = 2\pi R^2$ is a constant and $Q_s^2(x) = (x_0/x)^\lambda$ GeV^2 denotes the saturation scale. We also consider the CGC model [20], based in the Color Glass Condensate framework, in which gluon saturation effects are incorporated via an approximate solution of the Balitsky-Kovchegov equation [4]. The expression for the CGC model is given by,

$$\sigma_{q\bar{q}}^{CGC}(x,r) = \sigma_0 \begin{cases} \mathcal{N}_0\left(\frac{rQ_s}{2}\right)^{\gamma_{eff}(x,r)} & : rQ_s \le 2 \\ 1 - e^{-Aln^2(BrQ_s)} & : rQ_s > 2 \end{cases}$$

where $\gamma_{eff}(x,r) = 2\,(\gamma_s + (1/\kappa\lambda\ \ln(1/x))ln(2/rQ_s))$ is the effective anomalous dimension and one has the constant $\kappa = 9.9$.

In order to investigate the theoretical uncertainty associated to the models for the dipole cross section, we use the original values (OLD label) of parameters for the fits including the charm contribution. That is, for GBW-OLD we follow Ref. [19], for CGC-OLD Ref. [22] is considered and bCGC-OLD refers to Ref. [21]. The bCGC model uses the same functional form of Eq. (10) and replaces the saturation scale in the following way: $Q_s^2(x) \to Q_s(x,b)^2 = (x_0/x)^\lambda \exp[-b^2/(2\gamma_s B_{CGC})]$. Moreover, we consider the updated version of those models, GBW-NEW [13], CGC-NEW [23] and

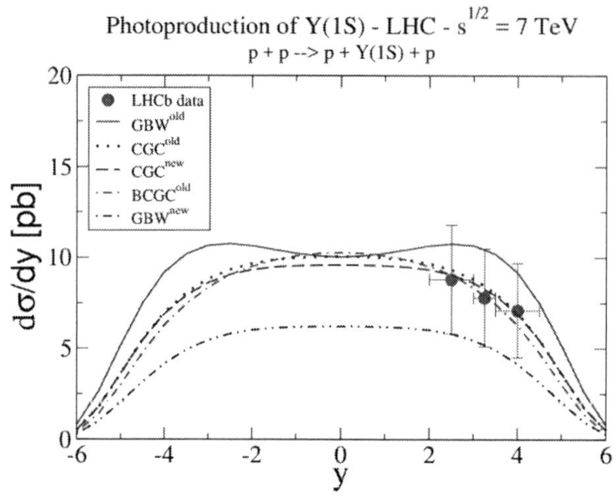

Figure 1. Rapidity distribution for the $\Upsilon(1S)$ state in pp collisions at $\sqrt{s} = 7$ TeV using boosted gaussian model (BG) for wavefunction. Theoretical uncertainty here comes from distinct models for dipole cross section. Data from LHCb Collaboration[8]

bCGC-NEW [23], respectively. A comment is in order here: the GBW-NEW parametrization is very different from other color dipole fits, as it includes energy evolution of the subnucleonic shape of the proton and it can potentially significantly affect the incoherent cross section. In particular, GBW-OLD and GBW-NEW are qualitatively very different and GBW-NEW was not fitted to all F_2 small-x data (DESY-HERA) as discussed in details in Ref. [13].

3 Results and discussions

Let us start the analysis by computing the theoretical predictions for the proton-proton case. In pp collisions, the rapidity distribution of vector meson V is given by

$$\frac{d\sigma}{dy}(p+p \to p+p+V) \;=\; S^2_{gap}\left[\omega \frac{dN^p_\gamma(\omega)}{d\omega}\sigma(\gamma p \to V+p) + (y \to -y)\right], \tag{10}$$

where the rapidity of the produced meson is related to the photon energy by $y \simeq \ln(2\omega/m_V)$. The parameter S^2_{gap} quantifies the absorptive corrections[24], and in such a process one has the presence of large rapidity gap between the produced meson and the final state protons. In the present calculation we will take $S^2_{gap} \approx 0.8 - 0.9$. Finally, the notation $(y \to -y)$ indicates the symmetry target-projectile in the pp collision. In Figure 1, it is presented the results for photoproduction $\Upsilon(1S)$ in pp collisions considering the different models presented in the last section. The relative normalization and the overall behavior on rapidity are fairly reproduced by all the models in the forward region in comparison to the experimental results from LHCb Collaboration [8]. Given the present level of the experimental uncertainties it is not possible to make definitive statements about the precision of the distinct models investigated. The theoretical uncertainty reaches a factor two considering the same wave-function and distinct dipole cross sections. Predictions for the upsilon states in pp collisions at 13 TeV and pA collisions at 8.2 TeV are presented in Ref. [7]. The general trend follows the one predicted at lower energy, including the theoretical uncertainty and overall behavior. The cross sections are quite sizeable and the relative contribution of the radial excited states compared to the lowest state follows the pattern shown at $y = 0$. Namely, $\sigma[\Upsilon(2S)]/\sigma[\Upsilon(1S)] \simeq 0.17$ and $\sigma[\Upsilon(3S)]/\sigma[\Upsilon(1S)] \simeq 0.083$. Our predictions for the Υ state ratios are lower that those predicted by STARlight Monte Carlo, as presented in Ref. [14]. The origin can be the fact that the different states are obtained from an extrapolation of HERA-DATA and using a fixed ratio for the distinct states in [14]. In our case, the evolution on energy is dynamically generated by parton saturation approach models and the meson wavefuntions have non-trivial behavior on the overlap function.

The calculation discussed above can be directly compared to another theoretical approaches. For instance, the present result is consistent with the NLO pQCD analysis done in Ref. [25], which also compares the outcoming predictions to the LHCb data. It is verified that the theoretical uncertainty from color dipole approach is somewhat smaller than the presented there (particularly, LO and NLO pQCD predictions present large disparities). Moreover, it is worth to note that the present work updates the analysis done in Ref. [18], where the photoproduction of Υ states in pp collisions has been done for the first time in the scope of color dipole approach, also used in [26] for $\Upsilon(1S)$. The focus here is to provide predictions for the next (and current) runs of LHC.

In Ref. [7] predictions for upsilon states in pA ultra-peripheral collisions were presented. In particular, in proton-lead collisions if the quarkonium rapidity, y, is positive in the nucleus beam direction its rapidity distribution reads as [1]:

$$\frac{d\sigma}{dy}(Pb + p \rightarrow Pb + p + V) \;\; = \;\; \frac{dN_{\gamma}^{Pb}(y)}{d\omega}\sigma_{\gamma p \rightarrow V+p}(y) + \frac{dN_{\gamma}^{p}(-y)}{d\omega}\sigma_{\gamma Pb \rightarrow V+Pb}(-y),$$

where $\frac{dN_{\gamma}(y)}{d\omega}$ is the corresponding photon flux . The case for the inverse beam direction is straightforward. We use the Weiszäcker-Williams method to calculate the flux of photons from a charge Z nucleus as referred in the previous section. In the numerical calculation we disregard the contribution coming from the photon flux related to the proton source. The integrated cross section (in the range $2 \leq y \leq 4.5$) is predicted to be 243.41 ± 20.37 pb and 4.97 ± 0.48 pb, respectively. The error includes the theoretical uncertainty related to the model of the dipole cross section. When considering the higher energy of $\sqrt{s} = 8.2$ TeV the values found are 340.51 ± 36.76 pb and 7.14 ± 0.85 pb. Moreover, for the lowest state $\Upsilon(1S)$ one has 2.92 ± 1.43 pb at $\sqrt{s} = 5.02$ TeV and 5.45 ± 2.04 pb at $\sqrt{s} = 8.2$ TeV. In Figure 2 the rapidity distribution is shown for the $\Upsilon(1S)$ state at $\sqrt{s} = 8.2$ TeV. In addition, we compare our results to recent CMS Collaboration data [27], where the theoretical predictions are consistent with measurements within the errors. The model deviation is more intense on the very forward rapidity and the theoretical uncertainty reaches a factor two similarly to the pp case. The rapidity distribution exhibits directly the influence of the x-dependence of the dipole cross section in the interval $0 \leq y \leq 4$, and specially for rapidities around $y = 3$. The shapes are similar for the lowest state and its radial excited states (see Ref. [7]). Therefore, it is feasible that a consideration of Υ photoproduction in this rapidity interval offers potential in discriminating models of dipole cross sections. It is timely having a measurement of Υ production at mid-rapidity, as the corresponding cross section in the ψ case it is a challenge when considering the color dipole approaches. The predictions can be directly compared to the work in Ref. [30], where the perturbative two-gluon exchange formalism has been considered. We have not verified a second peak in the rapidity distribution as presented in [30]. The reason is that the dynamics embedded in the color dipole approach considered here corresponds to strong shadowing corrections. The ratio $\sigma[\Upsilon(2S)]/\sigma[\Upsilon(1S)]$ and $\sigma[\Upsilon(3S)]/\sigma[\Upsilon(1S)]$ are still of same order of magnitude compared to the proton-proton case.

Concerning the theoretical predictions for the coherent process for PbPb collisions at 5.5 TeV, we use the Glauber-Gribov approach presented in previous sectiona and disregard any absorptive corrections. In Figure 3, it is presented the results for photoproduction of Υ states, including its radial

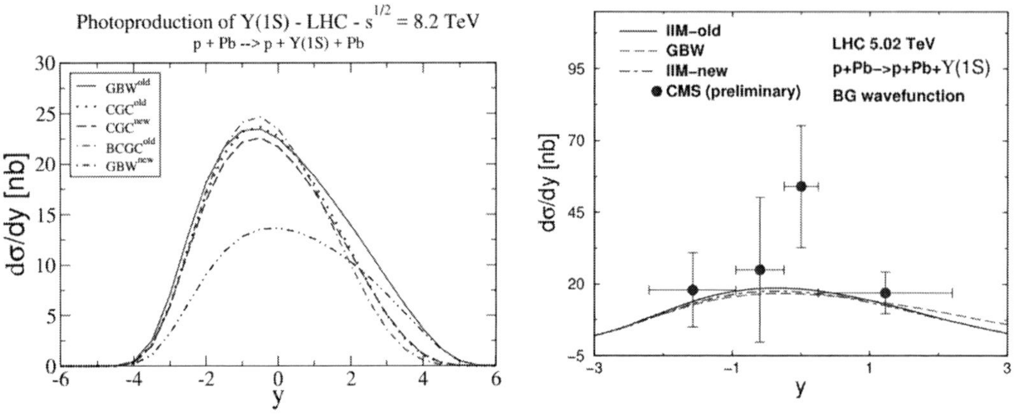

Figure 2. Rapidity distribution for the vector meson state $\Upsilon(1S)$ in pA collisions at $\sqrt{s}$ = 8.2 TeV (left panel) and at 5.02 TeV (right panel). Data from CMS Collabotation [27].

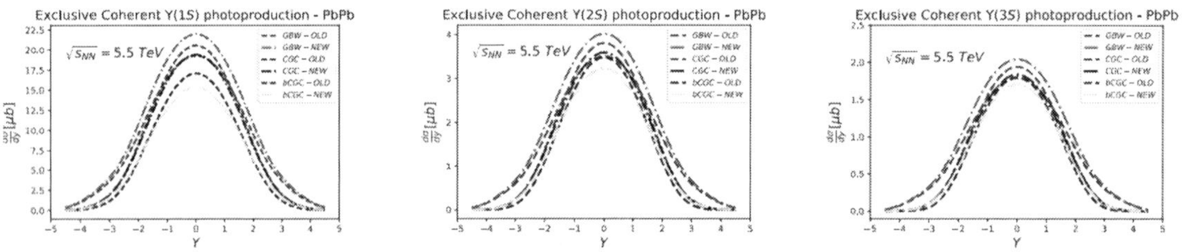

Figure 3. Exclusive coherent photoproduction of $\Upsilon(1S, 2S, 3S)$ in PbPb at $\sqrt{s}$ = 5.5 TeV for GBW, CGC and bCGC dipole models.

excitations, taking into account the different models presented in the last section. The theoretical uncertainty is relatively large, being of order 15 % for the $1S$ state (similar for the remaining $2S$ and $3S$ states). We could have an additional uncertainty related to the vector meson wave function, however in Ref. [28] it was shown that this is not the case for Υ states (the overall theoretical uncertainty is within the experimental error bars in pp case [8]). For the main contribution, we have $d\sigma_{coh}/dy\,(y = 0) = 18.5 \pm 3.5\ \mu$b for $\Upsilon(1S)$. The relative contribution of the excited states compared to the bound states is $\Upsilon(1S)/\Upsilon(2S)/\Upsilon(3S) = 1/0.17/0.09$. We see that the relative normalization and the overall behavior is changed mostly at mid-rapidity when comparing the old and updated versions of the dipole cross sections (the deviation at large rapidities is less evident). Notice that the LHCb data for upsilon production in pp collisions is reproduced by all the models in the forward region [8] as shown in Ref. [7]. Therefore,the current level of the experimental uncertainties does not allow us to make definitive statements about the precision of the models considered. For sake of completeness, we present the integrated cross sections considering distinct cuts on rapidity. In Table 1, we present the results for the full rapidity coverage, $-4 < y < 4$, and forward rapidities, $2 \leq y \leq 4.5$. In both tables 1 and 2, we present only the updated versions of the dipole cross sections.

We now focus on the incoherent reaction, $PbPb \rightarrow Pb\,\Upsilon\,Pb^*$. This is a new contribution to the literature concerning the upsilon production. The rapidity distribution is shown in Fig. 4 using the same notation as the previous figure. As already known, the incoherent cross section is smaller

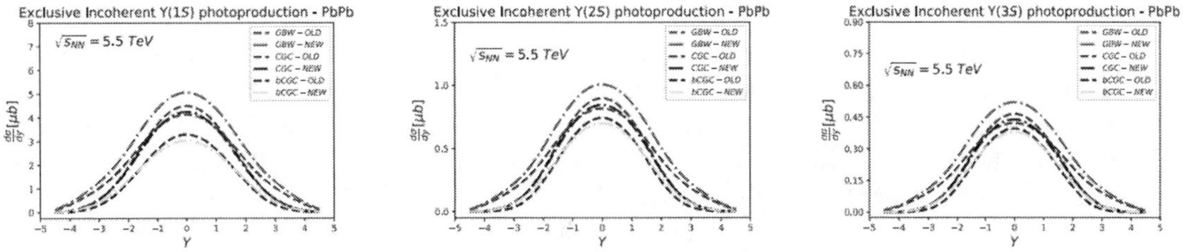

Figure 4. Exclusive incoherent photoproduction of $\Upsilon(1S, 2S, 3S)$ in PbPb at $\sqrt{s} = 5.5$ TeV for GBW, CGC and bCGC dipole models.

Table 1. Integrated cross section (in units of μb) for coherent reactions, $PbPb \rightarrow Pb\,\Upsilon\,Pb$, for full rapidity coverage (and forward rapidities). Here, we consider the updated versions of dipole cross sections.

process: PbPb	$\sqrt{s} = 5.5$ TeV	$\vert y\vert \leq 4$ $(2 \leq y \leq 4.5)$	
$\Upsilon(nS)$	GBW	CGC	b-CGC
$\Upsilon(1S)$	163.7 (60.8)	171.9 (63.8)	143 (53.1)
$\Upsilon(2S)$	20.3 (7.8)	22.0 (8.2)	20.5 (7.7)
$\Upsilon(3S)$	10.3 (3.9)	11.9 (4.3)	10.9 (4.1)

Table 2. Integrated cross section (in units of μb) for incoherent reactions, $PbPb \rightarrow Pb\,\Upsilon\,Pb^*$, for full rapidity coverage (and forward rapidities).Here, we consider the updated versions of dipole cross sections.

process: PbPb	$\sqrt{s} = 5.5$ TeV	$\vert y\vert \leq 4$ $(2 \leq y \leq 4.5)$	
$\Upsilon(nS)$	GBW	CGC	b-CGC
$\Upsilon(1S)$	61.2 (25.6)	58.5 (24.4)	44.5 (18.6)
$\Upsilon(2S)$	8.2 (3.4)	8.0 (3.4)	6.9 (2.9)
$\Upsilon(3S)$	4.2 (1.8)	4.2 (1.8)	3.8 (1.6)

than the coherent one. The typical ratio is $(d\sigma_{inc}/dy)/(d\sigma_{coh}/dy) \simeq 0.2$. For instance, we obtain $d\sigma_{inc}/dy\,(y = 0) = 3.75 \pm 1.25$ μb for $\Upsilon(1S)$. The theoretical uncertainty seems to be larger than in the coherent case. The integrated cross sections are shown in Table 2 in the rapidity ranges $|y| \leq 4$ and $2 \leq y \leq 4.5$.

The calculations performed above can be compared to another theoretical approaches available in the literature. Let start comparing them to the STARlight Monte Carlo [14]. For the coherent production, the predictions for the Υ state ratios are lower that STARlight results. As discussed in Ref. [7], the possible origin comes from the extrapolation of HERA-DATA and taking a fixed ratio for the distinct states in the Monte Carlo, whereas in current case the evolution on energy is dynamically generated by the parton saturation approach models and mostly by the meson wavefuntions for the radially excited states. The nuclear effects are also computed in a different way in the two formalisms. In STARLight, the nuclear shadowing is calculated using vector meson dominance (VDM) plus Glauber model for hadronic collisions. In our case, shadowing comes from the multiple scatterings of color dipoles and is described by the Glauber-Gribov approach. We verified that our results are also smaller that in Ref. [14], which can be related to more shadowing in color dipole models compared to the VDM+Glauber approach. In Ref. [30] only the coherent contribution was computed and the theoretical uncertainty we have found in the color dipole approach is comparable to perturbative QCD formalism. Concerning similar dipole calculations, more recent investigations are available in Refs. [28, 31]. In Ref. [31] only the coherent $\Upsilon(1S)$ production has been considered at 5.02 TeV. The results are smaller than ours and the main reason is the wavefunction chosen (Light Cone Gaussian wavefunction which gives smaller overall normalization compared to Boosted Gaussian one). The authors in [31] did not investigated the theoretical uncertainty associated to the wavefunction and dipole cross sections as well (only the uncertainty coming from one model for dipole cross section was addressed). In Ref. [28] the theoretical uncertainty for the coherent and incoherent cross section was investigated. However, predictions for higher energies in PbPb collisions were not presented and only the $\Upsilon(1S)$ state was considered (the results are consistent with ours in that case). Finally, we did not consider photonuclear breakup in the present study. We will consider them in future analysis as they are important and the distinct channels have been measured for ρ and J/ψ photoproduction in UPCs [32, 33]. This sort of analysis was recently done in Ref.[34], where the coherent $\Upsilon(1S)$ production was considered using a pQCD model with NLO accuracy. An important point discussed in [34] is that the large y region gives the dominant contribution for 0nXn and XnXn channel and they probe larger photon-target centre-of-mass energy than the case without neutron tagging.

4 Summary

We presented the predictions of rapidity distribution and integrated cross sections for the $\Upsilon(1S, 2S, 3S)$ states for the LHC energies, including discussions for pp and pA modes and fosusing mainly on PbPb collisions. The rapidity intervals used in total cross section were selected to match with the rapidity coverage of LHCb and ALICE detectors both covering $2 \leq y \leq 4.5$. The main contribution is the computation of the incoherent cross section within the color dipole approach and Glauber-Gribov treatment of nuclear shadowing. The cross section for the excited states are also calculated in a consistent formalism where the wavefunction of $2S$ and $3S$ states are theoretically well constrained. The usual procedure in the literature involves only an extrapolation of DESY-HERA production ratios to the LHC energies. Our calculations are directly comparable to the STARLight calculation, where distinct procedures are involved in the computation of nuclear shadowing (VDM plus Glauber model versus color dipole plus Glauber-Gribov approach) and how to obtain the incoherent cross section. The color dipole model is in agreement with measurements of $\Upsilon(1S)$ state in pp and pA collisions.

5 Acknowledgments

I would like to thank the organizers of XLVII International Symposium on Multiparticle Dynamics (ISMD2017) for warm hospitality and for an exciting conference.

References

[1] G. Baur, K. Hencken, D. Trautmann, S. Sadovsky, Y. Kharlov, Phys. Rep. **364**, 359 (2002); C. A. Bertulani, S. R. Klein and J. Nystrand, Ann. Rev. Nucl. Part. Sci. **55**, 271 (2005).

[2] N. N. Nikolaev, B. G. Zakharov, Phys. Lett. B **332**, 184 (1994); Z. Phys. C **64**, 631 (1994).

[3] J. Nemchik, N. N. Nikolaev, E. Predazzi and B. G. Zakharov, Phys. Lett. B **374**, 199 (1996).

[4] F. Gelis, E. Iancu, J. Jalilian-Marian and R. Venugopalan, Ann. Rev. Nucl. Part. Sci. **60**, 463 (2010); H. Weigert, Prog. Part. Nucl. Phys. **55**, 461 (2005); J. Jalilian-Marian and Y. V. Kovchegov, Prog. Part. Nucl. Phys. **56**, 104 (2006); A.L. Ayala, M.B. Gay Ducati and E.M. Levin, Nucl.Phys. B **493**, 305 (1997).

[5] A. Dainese, in Proceedings of the 38th International Symposium of Multiparticle Dynamics (ISMD2008): Hamburg, Germany. September 15-20 2008, p.118-124, arXiv:0902.0377 [hep-ph].
[6] M. B. Gay Ducati, F. Köpp and M.V. T. Machado. Phys. Rev. D **96**, 054001 (2017).
[7] M. B. Gay Ducati, F. Köpp, M.V. T. Machado, and S. Martins. Phys. Rev. D **94**, 094023 (2016).
[8] R. Aaij et al. [LHCb Collaboration], JHEP **1509**, 084 (2015).
[9] K. Dusling , F. Gelis, T. Lappi and R. Venugopalan, Nucl. Phys. A **836**, 159 (2010) .
[10] T. Lappi, H. Mäntysaari, R. Venugopalan, Phys. Rev. Lett. **114**, 082301 (2015). H. Mäntysaari and B. Schenke, Phys. Rev. Lett. **117**, 052301 (2016).
[11] L. Frankfurt, G.A. Miller and M. Strikman, Phys. Rev. Lett. **71**, 2859 (1993); A. Caldwell and H. Kowalski, Phys. Rev. C **81**, 025203 (2010).
[12] Kirill Tuchin, Phys. Rev. C **79**, 055206 (2009); T. Lappi and H. Mäntysaari, Phys. Rev. C **83** 065202 (2011) ; T. Toll and T. Ullrich, Phys. Rev. C **87**, 024913 (2013) ; H. Mäntysaari, B. Schenke, C. Shen and P. Tribedy, Phys. Lett. B **772**, 681 (2017).
[13] J. Cepila, J.G. Contreras, J. D. Tapia Takaki, Phys. Lett. B **766**, 186 (2017).
[14] S.R. Klein, J. Nystrand, J. Seger, Y. Gorbunov and J. Butterworth, Comput. Phys. Commun. **212**, 258 (2017).
[15] S. Klein and J. Nystrand, Phys. Rev. C **60**, 014903 (1999).
[16] B. Z. Kopeliovich and B. G. Zakharov, Phys. Rev. D **44**, 3466 (1991); Yu.P.Ivanov, J.Huefner, B.Z.Kopeliovich and A.V.Tarasov, AIP Conf.Proc. **660**, 283 (2003).
[17] J. Nemchik, N. N. Nikolaev, E. Predazzi and B. G. Zakharov, Z. Phys. C **75**, 71 (1997).
[18] B.E. Cox, J. R. Forshaw and R. Sandapen, JHEP **0906** , 034 (2009).
[19] K. Golec-Biernat and M. Wüsthoff, Phys. Rev. D **59**, 014017 (1998).
[20] E. Iancu, K. Itakura, S. Munier, Phys. Lett. B **590**, 199 (2004).
[21] H. Kowalski, L. Motyka and G. Watt, Phys. Rev. D **74**, 074016 (2006).
[22] G. Soyez, Phys. Lett. B **655**, 32 (2007).
[23] A.H. Rezaeain and I. Schmidt, Phys. Rev. D **88**, 074016 (2013).
[24] W. Schäfer, A. Szczurek, Phys. Rev. D **76**, 094014 (2007); A. Rybarska, W. Schäfer, A. Szczurek, Phys. Lett. B **668** , 126 (2008).
[25] S.P. Jones, A.D. Martin, M.G. Ryskin and T. Teubner, J. Phys. G: Nucl. Part. Phys. **43**, 035002 (2016).
[26] V.P Gonçalves, B.D. Moreira and F.S. Navarra, Phys. Lett. B **742**, 172 (2015); V.P Gonçalves, B.D. Moreira, F.S. Navarra and D. Spiering, Phys. Rev. D **94**, 014009 (2016); G. Gil da Silveira, V.P Gonçalves and M.M. Jaime, Phys.Rev. D **95** 034020 (2017).
[27] R. Chudasama *et al.* [CMS Collaboration], PoS ICPAQGP **2015**, 042 (2017), [arXiv:1607.00786 [hep-ex]].
[28] G. Sampaio dos Santos and M.V.T. Machado, J. Phys. G **42**, 105001 (2015).
[29] N . Armesto,A. H. Rezaeian, Phys.Rev. D **90**, 054003 (2014).
[30] A. Adeluyi and A. Nguyen, Phys. Rev. C **87**, 027901 (2013).
[31] V.P. Gonçalves, B.D. Moreira and F.S. Navarra, Phys. Rev. D **95**, 054011 (2017).
[32] J. Adam et al. [ALICE Collaboration], JHEP **1509**, 095 (2015).
[33] V. Khachatryan et al. [CMS Collaboration], Phys. Lett. B **396**, 772 (2017).
[34] V. Guzey, E. Kryshen and M. Zhalov, Phys. Rev. C **93**, 055206 (2016).

Probing Anomalous Quartic Interactions at the LHC with proton tagging

Cristian Baldenegro[1,⋆]

[1] *University of Kansas*

Abstract. One of the main goals of the Large Hadron Collider is to find signatures of physics Beyond the Standard Model of particle physics. One way to do this is by studying with high precision the interactions of the Standard Model. In this talk, we address the discovery potential of New Physics in the exclusive channel $pp \to pXp$ which relies on the general purpose detectors at the Large Hadron Collider and their respective forward proton detector stations, located at about $\sim$ 210 m w.r.t. the interaction point. These reactions are highly sensitive to quartic electroweak gauge interactions. As a proof of concept, we discuss the exclusive diphoton production at high diphoton invariant mass. We quote sensitivities on the anomalous $\gamma\gamma\gamma\gamma$ coupling for an integrated luminosity of 300 fb^{1} at the center-of-mass energy of 14 TeV. We also discuss the discovery potential of $3\gamma Z$ anomalous quartic gauge coupling by measuring the $pp \to p(\gamma\gamma \to Z\gamma)p$ reaction.

1 Introduction

One of the primary goals of the Large Hadron Collider (LHC) is the search of signatures of Beyond the Standard Model (BSM) physics. Despite the numerous efforts to find BSM physics at the LHC, none have been successful. Nonetheless, they have ruled out many extension models and have set stronger bounds for their parameter space [1–7]. While these searches are very powerful in specific cases, they end up being highly model-dependent. It is absolutely necessary to exhaust every possibility of New Physics signatures with the best sensitivities accessible at the LHC.

Anomalous quartic gauge interactions between the photons, $W^{\pm}$, and Z^{0} bosons can be probed with very high precision at the LHC by measuring central exclusive reactions $pp \to pXp$, where $X = \gamma\gamma, Z\gamma, ZZ, \ell\bar{\ell}$, among others. The system X is reconstructed in the central detector, like ATLAS or CMS, while the outgoing protons, which remain intact after the exchange of quanta carrying the quantum numbers of the vacuum, can be reconstructed with dedicated forward proton detectors at about ± 210 meters from the nominal interaction point (See Fig. 1). The CMS-TOTEM Precision Proton Spectrometer (CT-PPS) and the ATLAS Forward Physics (AFP) experiments are equipped and operational for the standard high luminosity fills at the LHC to study exclusive and semi-exclusive reactions. The full reconstruction of the final state gives three main advantages: i) due to the kinematic constraint, the number of background events stemming from the simultaneous secondary interactions at the LHC, the pile-up events, is negligible; ii) the production of a massive system X is dominated by photon-exchange events and; iii) systematic uncertainties related to proton dissociation are absent in these reactions, since protons remain intact.

⋆e-mail: cbaldenegro@ku.edu

© The Authors, published by EDP Sciences. This is an open access article distributed under the terms of the Creative Commons Attribution License 4.0 (http://creativecommons.org/licenses/by/4.0/).

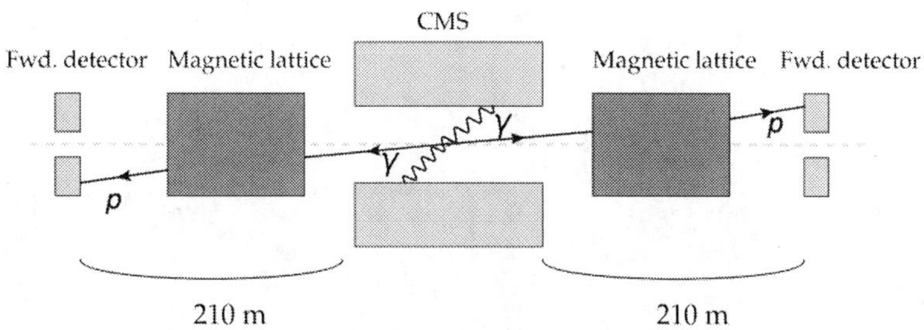

Figure 1. CMS-TOTEM Precision Proton Spectrometer schematic diagram. The $pp \rightarrow p\gamma\gamma p$ reaction is displayed as an example. The diphoton is measured in CMS and the outgoing intact protons with small fractional momentum loss are measured with the forward detectors at $\sim$ 210 m w.r.t. the nominal interaction point of CMS. The intact protons are transported through the accelerator's magnetic lattice (in blue), which acts as an effective spectrometer on the intact protons. The final state is reconstructed completely. Distances in this diagram are not to scale.

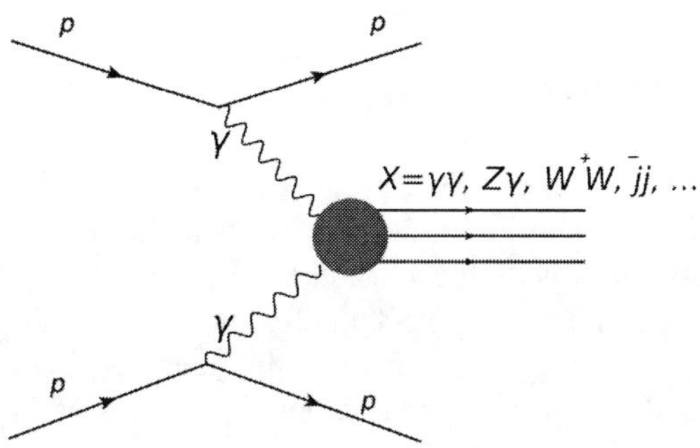

Figure 2. Schematic diagram of $pp \rightarrow p(\gamma\gamma \rightarrow X)p$ reaction. The photon-produced system X is measured in the central detector, the outgoing intact protons are tagged using very forward detectors.

Our focus is on photon-induced processes in pp collisions at high invariant masses, as in Fig. 2. These processes are the most sensitive to the anomalous quartic couplings between the SM vector gauge bosons. Pure gauge interactions can appear due to loops of new real or virtual charged particles. This happens, for instance, in composite Higgs models which require the existence of new charged particles of spin $\frac{1}{2}$ and 1 [8]. New particles with exotic electric charges can also appear in warped extra-dimension models with custodial symmetry [9]. Pure gauge interactions can also be induced via the exchange a resonant neutral state of spin 0 or 2. A prime example would be the Kaluza-Klein [11, 12] graviton.

The CT-PPS experiment has published the first physics results for an integrated luminosity of 10 fb^{-1} by measuring the semi-elastic production of $\mu^+\mu^-$ pairs using the proton tag method [10]. This first measurement proves that the chain of central and forward detection at the standard high luminosity fills works: from trigger, calibration, alignment and reconstruction has the potential of providing interesting physics results, necessary for future New Physics searches. The AFP experiment started taking data in the Summer of 2017, being an independent experiment probing the same reactions as in CT-PPS at high luminosity fills.[1]

[1]Private communication with members of the ATLAS collaboration.

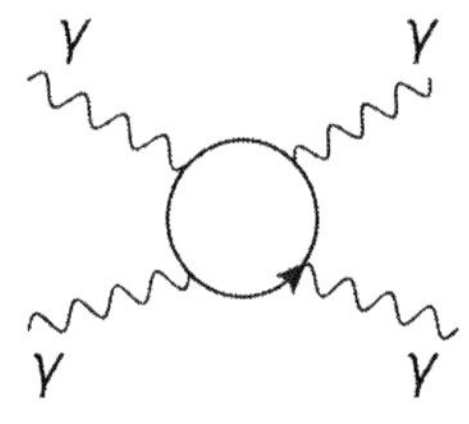

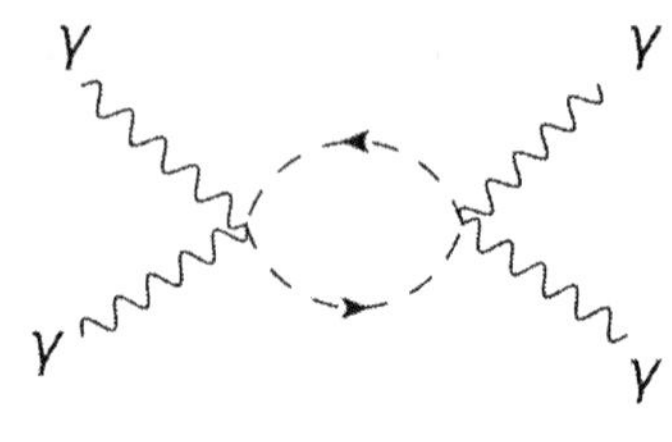

Figure 3. Typical diagrams illustrating electrically charged particles contributing to light-by-light scattering at the lowest order in perturbation theory. (a) The box contribution contains a loop that can involve the SM fermions and W-bosons. *Any* electrically charged BSM particle contributes to this process. (b) Bubble contribution from virtual charged scalars. Only BSM particles contribute to this diagram, since there are no scalars in QED. t-channel and u-channel diagrams are not shown here.

2 Light-by-light scattering ($\gamma\gamma \to \gamma\gamma$)

Light-by-light scattering ($\gamma\gamma \to \gamma\gamma$) was first predicted to exist in QED in 1933 [13]. It was first shown that the leading-order contribution to this process was finite in Ref. [14], but it was not until 2017 that it was first directly observed. Indeed, the first direct evidence of light-by-light scattering was found by the ATLAS Collaboration in Pb-Pb collisions [15], setting an important precedent in photon physics at the LHC. [2] The diphoton mass in this measurement ranges from 10 to 25 GeV. Complementary to the ATLAS Pb-Pb measurement, the pp process will probe diphoton masses from $\sim$ 300 GeV to 2 TeV with high transverse momenta photons and with proton tagging. As explained below, this gives sensitivity to regions where the anomalous interaction effects are more important. The Muon g-2 experiment [17] is also sensitive to the four photon coupling, as it modifies directly the QED vertex of the muon. While the precision in this experiment is already exceptional, and the measurement is expected to improve even more at Fermilab, the sensitivity to BSM particle properties in CT-PPS or AFP is expected to be superior [18].

Any new electrically charged particles would contribute to the scattering of light-by-light ($\gamma\gamma \to \gamma\gamma$) via loop diagrams. Light-by-light scattering amplitudes are fully characterized by the mass, spin and electric charge of the particle in the loop [18]. As a result, the search for charged particles can be performed in a fully model-independent way.

As described in [18], using kinematic information from tagged protons can constrain background events by roughly two orders of magnitude. In Fig. 4, we show the leading SM processes which result in two photons and two intact protons in the final state [19–22]. Fig. 2 corresponds to the exclusive QCD diphoton production via colorless two-gluon exchange, while Fig. 2 shows the photon exchange. In Fig 2 we see that quark, lepton and $W^{\pm}$ loops are considered in order to calculate the correct SM cross section for diphoton production. The two-gluon exchange process is dominant at low invariant masses whereas the photon induced ones dominate at higher diphoton masses [23]. This leads to a strong conclusion: if one observes two photons with high invariant mass in the central detector and two intact protons in the forward detectors -and the kinematics are properly balanced between the forward and central detectors- we have high confidence this is due to a photon-induced process.

[2]Already limits have been set on the non-linear Born-Infeld extension of QED [16]. This imposed a lower bound on the mass of $\sim$ 11 TeV of the corresponding magnetic monopole, the strongest to this date.

In the assumption of a new physics mass scale Λ heavier than the experimentally accessible energy E, all new physics manifestations can be described using an effective Lagrangian valid for $\Lambda \gg E$. Among these operators, the pure photon dimension-eight operators

$$\mathcal{L}_{4\gamma} = \zeta_1^{4\gamma} F_{\mu\nu}F^{\mu\nu}F_{\rho\sigma}F^{\rho\sigma} + \zeta_2^{4\gamma} F_{\mu\nu}F^{\nu\rho}F_{\rho\lambda}F^{\lambda\mu} \tag{1}$$

can induce the $\gamma\gamma\gamma\gamma$ interaction. The unpolarized differential cross section induced by the EFT Lagrangian reads,

$$\frac{d\sigma_{\gamma\gamma\to\gamma\gamma}}{d\Omega} = \frac{1}{16\pi^2 s}(s^2 + t^2 + st)^2\left[48\zeta_1^2 + 40\zeta_1\zeta_2 + 11\zeta_2^2\right] \tag{2}$$

Imposing unitarity on the S-wave from the EFT amplitudes, we find the bound

$$\zeta_1, \zeta_2 < (10^{-12} - 10^{-11})\mathrm{GeV}^{-4} \tag{3}$$

In Fig. 2, we see the diphoton invariant mass distribution for signal and backgrounds. As mentioned previously, the signal appears at high diphoton mass. We also see that the dominant background, by several orders of magnitude, at high invariant mass is the $\gamma\gamma$+ pile-up contribution, which is due to the overlap of non-diffractive diphoton production and the presence of two intact protons due to the secondary interactions, commonly known as pile-up interactions. This background can be largely suppressed by applying the rapidity and mass matching criteria (see Fig. 7) between the forward proton detectors information and the central detector information. [3]. The signal selection efficiency exceeds 70%.

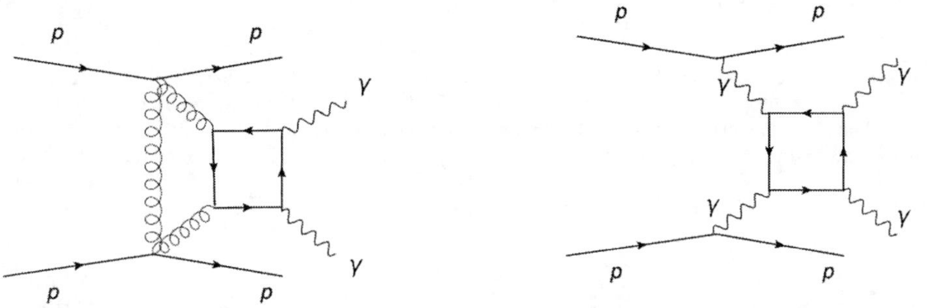

Figure 4. Exclusive $pp \to p\gamma\gamma p$ processes in the SM at the lowest order in perturbation theory. (a) Two-gluon exchange with a box of virtual charged particles leading to a $\gamma\gamma$ final state, and (b) the well-known light-by-light scattering process with two leading intact protons.

3 $\gamma\gamma\gamma Z$ coupling

The $\gamma\gamma\gamma Z$ interaction is induced at one-loop level in the SM via loops of fermions and $W^\pm$ bosons. Loops of heavy particles charged under $SU(2)_L\times U(1)_Y$ contribute to the $\gamma\gamma\gamma Z$ couplings. These loop contributions only depend on the mass and quantum numbers of the particle in the loop. This anomalous interaction results in the $\gamma\gamma \to \gamma Z$ scattering and generates the rare SM decay $Z \to \gamma\gamma\gamma$.

[3]The fractional momentum loss of the intact protons ξ is measured by the proton taggers. The diphoton mass can be reconstructed with them via $m_{\gamma\gamma} = \sqrt{\xi_1\xi_2 s}$ (often called missing diproton mass, since it comes from the intact protons) and the diphoton rapidity with $y_{\gamma\gamma} = 1/2\log(\xi_1/\xi_2)$. This is valid for every central exclusive reaction.

Cut / Process	Signal (full)	Signal with (without) f.f (EFT)	Excl.	DPE	DY, di-jet + pile up	$\gamma\gamma$ + pile up
$[0.015 < \xi_{1,2} < 0.15,$ $p_{\mathrm{T1,(2)}} > 200, (100)$ GeV$]$	65	18 (187)	0.13	0.2	1.6	2968
$m_{\gamma\gamma} > 600$ GeV	64	17 (186)	0.10	0	0.2	1023
$[p_{\mathrm{T2}}/p_{\mathrm{T1}} > 0.95,$ $\lvert\Delta\phi\rvert > \pi - 0.01]$	64	17 (186)	0.10	0	0	80.2
$\sqrt{\xi_1\xi_2 s} = m_{\gamma\gamma} \pm 3\%$	61	16 (175)	0.09	0	0	2.8
$\lvert y_{\gamma\gamma} - y_{pp}\rvert < 0.03$	60	12 (169)	0.09	0	0	0

Figure 5. Number of signal events for a given coupling value and background events after various selections for an integrated luminosity of 300 fb^{-1} and $\mu = 50$ at $\sqrt{s} = 14$ TeV. At least one converted photon is required. Excl. stands for exclusive backgrounds and DPE for double pomeron exchange backgrounds.

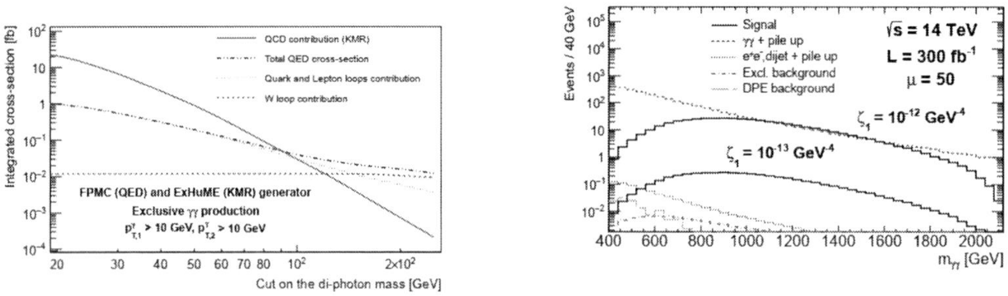

Figure 6. (a) SM exclusive diphoton cross-section as a function of the diphoton inv. mass cut. Considered events involve two intact protons in the forward proton detectors and photons with $p_{\mathrm{T}} > 10$. The QCD contribution (two-gluon exchange in magenta) dominates at low masses, while QED (diphoton production dashed lines) dominates at higher masses. The QED production contains two pieces: lepton/fermion loops (dotted line) and W boson loops (dashed-dotted line). (b) Diphoton invariant mass distribution for the signal ($\zeta_1 = 10^{-12}$, 10^{-13} GeV^{-4}) and for the backgrounds (dominated by $\gamma\gamma$ with protons from pile-up), requesting two protons in the forward detectors and two photons of $p_{\mathrm{T}} > 50$ GeV.

If a $\gamma\gamma\gamma\gamma$ signal is observed, the $pp \to p(\gamma\gamma \to \gamma Z)p$ channel would provide a clear test whether or not the underlying exchanged particle is universally coupled to the gauge kinetic terms.

In the presence of New Physics with a mass scale Λ heavier than the experimentally accessible energy E, all New Physics manifestations can be described using an effective Lagrangian valid for $\Lambda \gg E$. In this low-energy EFT, the $\gamma\gamma\gamma Z$ interactions are described by two dimension-eight $\gamma\gamma\gamma Z$ operators

$$\mathcal{L}_{\gamma\gamma\gamma Z} = \zeta \mathcal{O}^{\gamma Z} + \tilde{\zeta}\tilde{\mathcal{O}}^{\gamma Z} = \zeta F^{\mu\nu}F_{\mu\nu}F^{\rho\sigma}Z_{\rho\sigma} + \tilde{\zeta}F^{\mu\nu}\tilde{F}_{\mu\nu}F^{\rho\sigma}\tilde{Z}_{\rho\sigma}\,, \tag{4}$$

with $\tilde{F}^{\mu\nu} = \frac{1}{2}\epsilon^{\mu\nu\rho\sigma}F_{\rho\sigma}$. The $\mathcal{O}_2^{\gamma Z} = F^{\mu\nu}F_{\nu\rho}F^{\rho\sigma}Z_{\sigma\mu}$ operator can sometimes be encountered in the literature. It is related to the above basis via the identity $4\mathcal{O}_2^{\gamma Z} = 2\mathcal{O}^{\gamma Z} + \tilde{\mathcal{O}}^{\gamma Z}$. The $\mathcal{O}^{\gamma Z}, \tilde{\mathcal{O}}^{\gamma Z}$ provides a somewhat clearer mapping onto the properties of the underlying physics.

We find the unpolarized differential cross section to be

$$\frac{d\sigma^{\mathrm{NP}}_{\gamma\gamma\to\gamma Z}}{d\Omega} = \frac{\beta}{16\pi^2 s}\left[(3\zeta^2 + 3\tilde{\zeta}^2 - 2\zeta\tilde{\zeta})(st + tu + us)^2 - 4(\zeta^2 + \tilde{\zeta}^2 - \zeta\tilde{\zeta})^2 m_Z^2 stu\right], \tag{5}$$

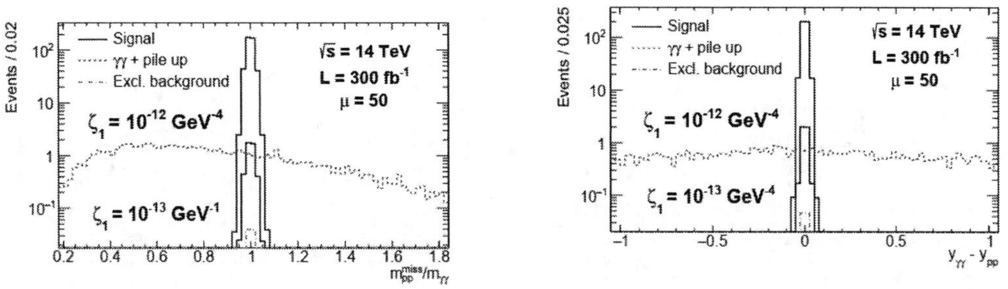

Figure 7. Missing diproton mass $m_{pp} = \sqrt{\xi_1 \xi_2 s}$ to central mass ratio distribution (Top) and rapidity difference distribution (Bottom) in the $p\gamma\gamma p$ channel for the signal and background within the fractional momentum loss acceptance $0.015 < \xi_{1,2} < 0.15$ considering two different coupling values after applying the required cuts on p_T, invariant mass $m_{\gamma\gamma}$, p_T ratios and angle separation. The integrated luminosity is 300 fb^{-1} and the average pile-up is $\mu = 50$. Exclusive processes peak at unity and zero respectively.

where s, t, and u are the usual Mandelstam variables and $\beta = 1 - m_Z^2/s$ for the γZ final state.

By imposing unitarity on the S-wave of the EFT amplitudes and neglecting the Z boson mass one finds the conditions

$$|\zeta + \tilde{\zeta}|s^2 < 4\pi\,, \quad |\zeta - \tilde{\zeta}|s^2 < \frac{12\pi}{5}\,. \tag{6}$$

The discovery potential of this anomalous coupling via the exclusive production $pp \to p(\gamma\gamma \to Z\gamma)p$ has been addressed in detail in [24]. The EFT approach is used in the study, and conclusions to scenarios where loops of new particles charged under $SU(2)_L \times U(1)_Y$ or neutral resonance production. Two dimension-8 operators contribute to the anomalous quartic interaction $3\gamma Z$ with couplings ζ, $\tilde{\zeta}$. Just as in the exclusive diphoton production, the only significant background is the overlap of non-diffractive events with γZ final states and the simultaneous intact protons coming from the uncorrelated single diffractive and double diffractive events. This background can be reduced by comparing the forward and central measurements, as seen in Fig. 8 for the $pp \to p(\gamma\gamma \to Z\gamma)p \to p(\gamma\gamma \to \text{Dijet} + \gamma)p$ case.

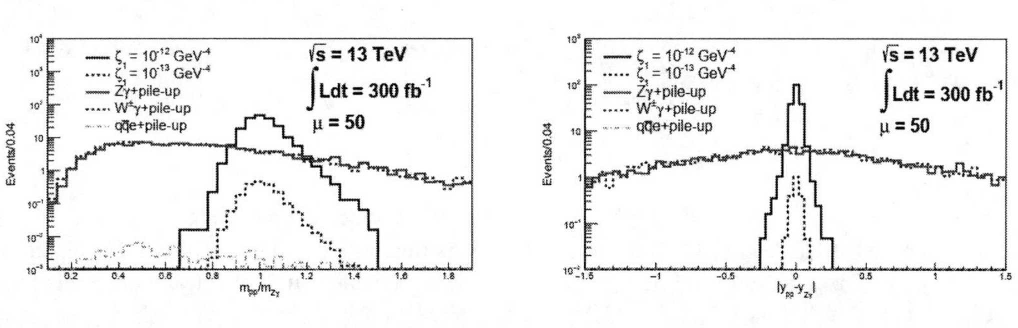

Figure 8. Missing diproton mass $m_{pp} = \sqrt{\xi_1 \xi_2 s}$ to central mass ratio distribution (Top) and rapidity difference distribution (Bottom) in the $jj\gamma$ channel for the signal and background within the acceptance $0.015 < \xi_{1,2} < 0.15$ considering two different coupling values after applying the requirement on p_T, invariant mass $m_{Z\gamma}$, p_T ratios and angle separation. The integrated luminosity is 300 fb^{-1} and the average pile-up is $\mu = 50$. The signal width is due to a combined effect of the reconstructed jet energy low resolution ($\approx 15\%$) and the $\xi_{1,2}$ resolution from the proton detectors. The asymmetry on the $m_{pp}/m_{Z\gamma}$ distribution is due to the resolution on the jet energy.

Coupling (GeV^{-4})	ζ ($\tilde{\zeta}=0$)		$\zeta=\tilde{\zeta}$	
Luminosity	300 fb^{-1}		300 fb^{-1}	
Pile-up (μ)	50		50	
Channels	5 σ	95% CL	5 σ	95% CL
$\ell\bar{\ell}\gamma$	$2.8\cdot10^{-13}$	$1.8\cdot10^{-13}$	$2.5\cdot10^{-13}$	$1.5\cdot10^{-13}$
$jj\gamma$	$2.3\cdot10^{-13}$	$1.5\cdot10^{-13}$	$2\cdot10^{-13}$	$1.3\cdot10^{-13}$
$jj\gamma\bigoplus\ell\bar{\ell}\gamma$	$1.93\cdot10^{-13}$	$1.2\cdot10^{-13}$	$1.7\cdot10^{-13}$	$1\cdot10^{-13}$

Table 1. 5σ discovery and 95% CL exclusion limits on ζ, $\tilde{\zeta}$. Sensitivities are given for 300 fb^{-1} for μ = 50 pile-up interactions. The last row corresponds to the search of $Z\gamma$ production in both channels with their respective selection cuts.

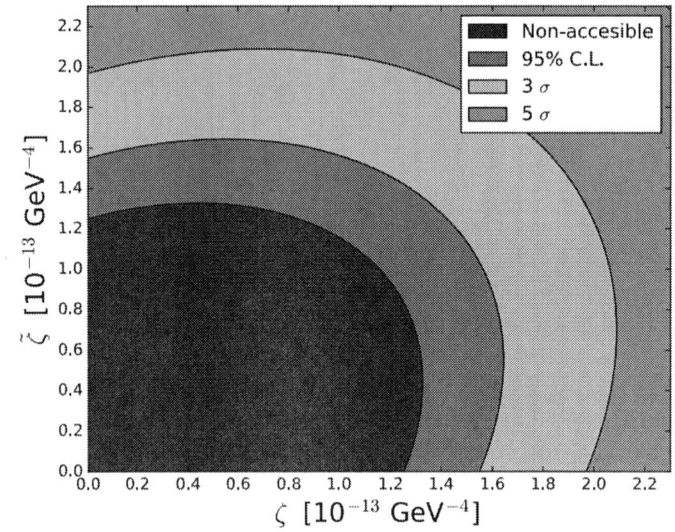

Figure 9. Sensitivity in the $\zeta-\tilde{\zeta}$ plane for 300 fb^{-1} and μ = 50. Orange, green and light blue can be probed at 5 σ, 3σ, and 95% C.L. using proton tagging at the LHC.

The exclusive channel allows the possibility of studying exclusive $Z\gamma$ production with the Z decaying leptonically or hadronically. The signature $\ell\bar{\ell}\gamma$ is much cleaner, but generates vastly fewer events than $q\bar{q}\gamma$. The sensitivity on the anomalous coupling considering both decay channels at 13 TeV, with 300 fb^{-1} of data, is $1.7\cdot10^{-13}$ GeV^{-4} at 5 σ and $1\cdot10^{-13}$ GeV^{-4} at 95% CL.

The strongest bound on the branching ratio $\mathcal{B}(Z\to\gamma\gamma\gamma)$ was found by the ATLAS Collaboration at 8 TeV [25]: $\mathcal{B}(Z\to\gamma\gamma\gamma)<2.2\cdot10^{-6}$. This bound translates to a limit on the anomalous couplings

$$\left[\zeta^2+\tilde{\zeta}^2-\frac{1}{2}\zeta\tilde{\zeta}\right]^{1/2}<1.3\cdot10^{-9}\ \mathrm{GeV}^{-4}\ (95\%\ \mathrm{CL}). \qquad (7)$$

Imagining that the same search is done at 13 TeV with 300 fb^{-1}, we roughly expect this bound to improve by an order of magnitude. This remains far away from the expected sensitivities obtained in the exclusive channel at the same luminosity by roughly three orders of magnitude.

4 Conclusions

The exclusive channel provides an excellent test ground to probe quartic electroweak gauge boson interactions with cross-sections at the order of fractions of fb. Search of quartic interactions in the exclusive channel allows to study New Physics in a model-independent. The couplings can be characterized as a function of the mass, spin and charge under $SU(2)_L \times U(1)_Y$ of the new particles contributing via loops in $\gamma\gamma \to \gamma\gamma$ and $\gamma\gamma \to \gamma Z$ scattering.

The CT-PPS experiment has published the first results for an integrated luminosity of 10 fb^{-1} for the semi-elastic production of $\mu^+\mu^-$ pairs at the LHC using the proton tagging method. This first measurement proves that the chain of exclusive and forward detection works at the standard high luminosity fills, which goes from trigger, calibration, alignment and reconstruction can provide interesting physics results, necessary for future BSM searches. The ATLAS Forward Physics project started taking data in the Summer of 2017, being an independent experiment probing the same reactions as in CT-PPS.

References

[1] CMS Collaboration. Search for signatures of extra dimensions in the diphoton mass spectrum at the Large Hadron Collider. Phys. Rev. Lett., 108:111801, 2012.

[2] CMS Collaboration. Search for quark compositeness in dijet angular distributions from pp collisions at $\sqrt{s} = 7$ TeV. JHEP, 05:055, 2012.

[3] CMS Collaboration. Search for large extra dimensions in dimuon and dielectron events in pp collisions at $\sqrt{s} = 7$ TeV. Phys. Lett., B711:15–34, 2012.

[4] ATLAS Collaboration. Search for pair production of heavy vector-like quarks decaying to high-pT W bosons and b quarks in the lepton-plus-jets final state in pp collisions at $\sqrt{s}$=13 TeV with the ATLAS detector. 2017.

[5] ATLAS Collaboration. A search for resonances decaying into a Higgs boson and a new particle X in the XH qqbb final state with the ATLAS detector. 2017.

[6] CMS Collaboration. Search for low mass vector resonances decaying into quark-antiquark pairs in proton-proton collisions at $\sqrt{s} = 13$ TeV. 2017.

[7] CMS Collaboration. Search for dark matter produced with an energetic jet or a hadronically decaying W or Z boson at $\sqrt{s} = 13$ TeV. JHEP, 07:014, 2017.

[8] K. Agashe, R. Contino, and A. Pomarol. The Minimal composite Higgs model. Nucl. Phys., B719:165–187, 2005.

[9] K. Agashe, A. Delgado, M.J. May, and R. Sundrum. RS1, custodial isospin and precision tests. JHEP, 08:050, 2003.

[10] CMS and TOTEM collaborations, Evidence for proton-tagged, central semi-exclusive production of high-mass muon pairs at 13 TeV with the CMS-TOTEM Precision Proton Spectrometer, CMS-PAS-PPS-17-001, 2017.

[11] Th. Kaluza. Zum Unitatsproblem in der Physik. Sitzungsber. Preuss. Akad. Wiss. Berlin (Math. Phys.), page 966, 1921.

[12] O. Klein. Quantentheorie und funfdimensionale Relativitatstheorie. Zeitschrift fur Physik, 37(12):895–906, Dec 1926.

[13] O. Halpern. Scattering Processes Produced by Electrons in Negative Energy States. Phys. Rev., 44:855.2–856, 1933.

[14] H. Euler and B. Kockel. Ueber die Streuung von Licht an Licht nach der Diracschen Theorie. Naturwiss., 23:246, 1935.

[15] ATLAS Collaboration. Measurement of exclusive W + W production and search for exclusive Higgs boson production in pp collisions at $\sqrt{s} = 8$ TeV using the ATLAS detector. Phys. Rev., D94(3):032011, 2016.

[16] J. Ellis, N.E. Mavromatos, and T. You. Light-by-Light Scattering Constraint on Born-Infeld Theory. Phys. Rev. Lett., 118(26):261802, 2017.

[17] G. Venanzoni. The Fermilab Muon g 2 Experiment. PoS, EPS-HEP2015:568, 2015.

[18] S. Fichet, G. von Gersdorff, B. Lenzi, C. Royon, and M. Saimpert. Light-by-light scattering with intact protons at the LHC: from Standard Model to New Physics. JHEP, 02:165, 2015.

[19] E. Chapon, C. Royon, and O. Kepka. Anomalous quartic W +W , ZZ, and trilinear W +W couplings in two-photon processes at high luminosity at the LHC. Phys. Rev., D81:074003, 2010.

[20] O. Kepka and C. Royon. Anomalous W W coupling in photon-induced processes using forward detectors at the LHC. Phys. Rev., D78:073005, 2008.

[21] S. Fichet, G. von Gersdorff, and C. Royon. Measuring the Diphoton Coupling of a 750 GeV Resonance. Phys. Rev. Lett., 116(23):231801, 2016.

[22] S. Fichet, G. von Gersdorff, and C. Royon. Scattering light by light at 750 GeV at the LHC. Phys. Rev., D93(7):075031, 2016.

[23] V. A. Khoze, A. D. Martin, and M. G. Ryskin. Prospects for new physics observations in diffractive processes at the LHC and Tevatron. Eur. Phys. J., C23:311–327, 2002.

[24] C. Baldenegro, S. Fichet, Gero von Gersdorff, and C. Royon. Probing the anomalous Z coupling at the LHC with proton tagging. JHEP, 06:142, 2017.

[25] ATLAS Collaboration. Search for new phenomena in events with at least three photons collected in pp collisions at $\sqrt{s} = 8$ TeV with the ATLAS detector. Eur. Phys. J., C76(4):210, 2016.

3 parton production at DIS at small x

Martin Hentschinski[1,*]

[1] *Departamento de Actuaria, Física y Matemáticas, Universidad de las Americas Puebla, Santa Catarina Martir, 72820, Puebla, Mexico*

Abstract. We use the spinor helicity formalism to calculate the cross section for production of three partons of a given polarization in Deep Inelastic Scattering (DIS) off proton and nucleus targets at small Bjorken x. The target proton or nucleus is treated as a classical color field (shock wave) from which the produced partons scatter multiple times. The resulting expressions are used to study azimuthal angular correlations between produced partons in order to probe the gluon structure of the target hadron or nucleus as well as to study energy loss in DIS reactions.

1 Introduction

The perturbative Regge limit of Quantum Chromodynamics (QCD) – which is characterized by the presence of a hard scale Q^2 far bigger than the QCD confinement scale and high center of mass energy s, such that $x = Q^2/s \ll 1$ – continues to provide challenges to those who wish to explore it. While Balitsky-Fadin-Kuraev-Lipatov (BFKL) evolution [1] is known to yield the correct perturbative high energy evolution equation, if the full resummation program is fully perturbative, there exist various theoretical arguments, which suggests that at some – yet unknown – value of energy, non-linear effects will complement BFKL evolution and tame the observed (and by BFKL predicted) rise of the perturbative gluon and bring it finally to hold. A dynamical mechanism which achieves such perturbative unitarity of QCD cross-sections at low x is known under the term 'gluon saturation' and has been originally proposed by Gribov, Levin and Ryskin [2]. The framework of the Color Glass Condensate (CGC) provides on the other hand an effective action approach to gluon saturation [3]. Within this framework, the high energy hadron or nucleus is described as a a weakly-coupled yet non-perturbative system of gluons with high occupancy number, characterized by a semi-hard scale Q_s, called the saturation scale. As a result one obtains a non-linear generalization of the BFKL evolution equation [4–6], which includes multiple-scattering effects; the latter provide the necessary slow down and saturation of the growth with energy, see [7] and references therein

While the above mentioned non-linear dynamics is of high interest and many efforts are dedicated towards its exploration, one finds at the same time that at current collider energies, linear low x evolution provides a pretty good description of data, see for instance the study [8] of exclusive vector meson production, where the energy dependence results directly of the NLO BFKL fit of [9]. To reveal saturation effects (or rather their onset) at current energies, it is therefore necessary to study specially

*e-mail: martin.hentschinski@udlap.mx

© The Authors, published by EDP Sciences. This is an open access article distributed under the terms of the Creative Commons Attribution License 4.0 (http://creativecommons.org/licenses/by/4.0/).

designed observables, which provide increased sensitivity to the presence of high gluon densities. In the following we give some details on a calculation of such an observables as well as a first numerical simulation. The hope is that this observable will provide deeper inside into the question, whether we already experience the on-set of non-linear corrections or not. For more details we refer the interested reader to the original publication [12, 13].

2 3 parton production in DIS at low x

Since the CGC-framework assumes high gluon densities in the probed target, the gluonic field of the hadron/nucleus latter can be no longer treated perturbatively. The resulting formalism represents therefore the DIS scattering process in the low x limit as the scattering of the virtual photon on a strong background color field; this field then essentially represents the target, together with a suitable averaging procedure over field configurations. Defining light-cone vectors $n, \bar{n}$ through the four momenta of virtual photon and target, *i.e.* $n \sim p$ and $n \cdot \bar{n} = 1$ with $v^- \equiv n \cdot v, v^+ \equiv \bar{n} \cdot v$ for a generic four vector v, theory calculations are performed using quark and gluon propagators in the presence of the background field $A^{+,c}$. One finds for the propagators

$$\tilde{S}_F(p,q) = (2\pi)^d \delta^{(d)}(p-q)\tilde{S}_F^{(0)}(p) + \tilde{S}_F^{(0)}(p)\tau_f(p,q)\tilde{S}_F^{(0)}(q),$$
$$\tilde{G}_{\mu\nu}(p,q) = (2\pi)^d \delta^{(d)}(p-q)\tilde{G}_{\mu\nu}^{(0)}(p) + \tilde{G}_{\mu\alpha}^{(0)}(p)\tau_g(p,q)\tilde{G}_{\alpha\nu}^{(0)}(q), \tag{1}$$

with the free propagators

$$\tilde{S}_F^{(0)}(p) = \frac{i\not{p}+m}{p^2-m^2+i0}, \qquad \tilde{G}_{\mu\nu}^{(0)}(p) = \frac{id_{\mu\nu}(p)}{p^2+i0}, \qquad d_{\mu\nu}(p) = -g_{\mu\nu} + \frac{n_\mu^- p_\nu + p_\mu n_\nu^-}{n^- \cdot p}, \tag{2}$$

and the interaction terms

$$\tau_f(p,q) = 2\pi\delta(p^- - q^-)\not{n}^- \int d^{d-2}z e^{-iz\cdot(p-q)} \left\{\theta(p^-)[V(z)-1] - \theta(-p^-)[V^\dagger(z)-1]\right\},$$
$$\tau_g(p,q) = -2\pi\delta(p^- - q^-)2p^- \int d^{d-2}z e^{-iz\cdot(p-q)} \cdot \left\{\theta(p^-)[U(z)-1] - \theta(-p^-)[U^\dagger(z)-1]\right\}. \tag{3}$$

The latter contain Wilson lines in the fundamental (V) and adjoint (U) representation which resum the interaction with the background field with

$$V(z) \equiv V_{ij}(z) \equiv \mathrm{P}\exp ig \int_{-\infty}^{\infty} dx^- A^{+,c}(x^-, z)t^c \tag{4}$$

and $U^{ba}(z_t) = 2\mathrm{tr}(t^a V(z_t) t^b V^\dagger(z_t))$; P denotes path ordering of the gluon field in the exponent while t^c is a SU(N_c) generator in the fundamental representation.

A well known possibility to increase gluon densities and therefore the sensitivity to saturation is to consider furthermore scattering on large nuclei. In such a nuclear environment, gluon densities are expected to be enhanced by a factor $A^{\frac{1}{3}}$, with A the mass number of the nucleon. This sensitivity to high density effects is then furthermore enhanced, if one considers exclusive final states: while for inclusive observables the presence of high gluon densities leads mainly to a modification of the x dependence of gluon densities [11], exclusive production processes can provide matrix elements which differ for high gluon densities from their low density counter parts. They are therefore more

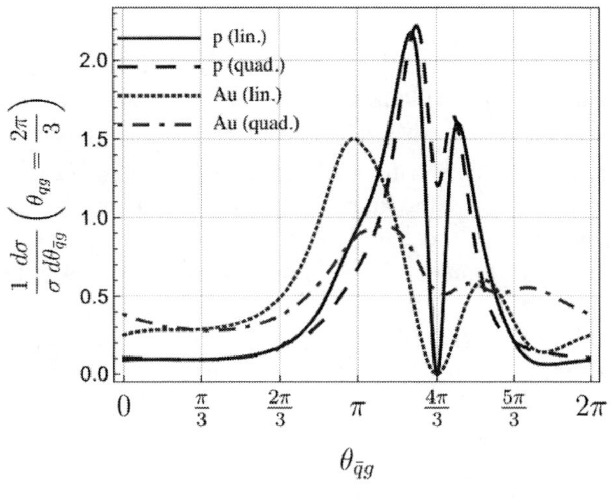

Figure 1. In the Mercedes-Benz star configuration, all transverse momenta have equal absolute value, with relative angle $\Delta = 2\pi/3$. We show the normalized cross-section for three parton production in the presence of large gluon densities against $\Delta\theta_{\bar{q}g}$ with $\Delta_{qg} = 2\pi/3$ for proton and gold up to linear and quadratic order in $N^{(2)}$. For the study the photon momentum fractions of the final state particles have been fixed to $z_1 = z_2 = 0.2$, $z_3 = 0.6$ while $|\boldsymbol{p}_1| = |\boldsymbol{p}_2| = |\boldsymbol{p}_3| = 2$ GeV and $Q^2 = 9$ GeV^2 (plot originally published in [12]).

sensitive to effects related to high gluon densities and provide a tool for their exploration. In the following we study the process

$$\gamma^*(q) + \text{target}(p) \to q(p_1) + \bar{q}(p_2) + g(p_3) + X, \tag{5}$$

where we refer for details to [12, 13]. This observable has the big advantage that it depends on the gluon distribution in the target in a highly non-linear way and therefore gives hope to reveal the presence of such non-linear effects. The results of a first numerical study are shown in Fig. 1. With the differential cross-section for the process Eq. (5) expressed in terms of dipole and quadrupole operators of Wilson lines Eq. (4),

$$S^{(2)}_{(\boldsymbol{x}_1\boldsymbol{x}_2)} \equiv \frac{1}{N_c}\text{tr}\left[V(\boldsymbol{x}_1)V^\dagger(\boldsymbol{x}_2)\right], \qquad S^{(4)}_{(\boldsymbol{x}_1\boldsymbol{x}_2\boldsymbol{x}_3\boldsymbol{x}_4)} \equiv \frac{1}{N_c}\text{tr}\left[V(\boldsymbol{x}_1)V^\dagger(\boldsymbol{x}_2)V(\boldsymbol{x}_3)V^\dagger(\boldsymbol{x}_4)\right], \tag{6}$$

this study uses the large N_c and Gaussian approximation [14] to express the quadrupole in terms of the dipole operator. For the dipole profile itself a model motivated by recent rcBK[1] fits to HERA data [15] has been used,

$$S^{(2)}_{(\boldsymbol{x}_1\boldsymbol{x}_2)} = \int d^2\boldsymbol{l}\, e^{-i\boldsymbol{l}\cdot\boldsymbol{x}_{12}}\, \Phi(\boldsymbol{l}^2) = 2\left(\frac{Q_0|\boldsymbol{x}_{12}|}{2}\right)^{\rho-1}\frac{K_{\rho-1}(Q_0|\boldsymbol{x}_{12}|)}{\Gamma(\rho-1)},$$
$$\text{where} \quad \Phi(\boldsymbol{l}^2) = \frac{\rho-1}{Q_0^2\pi}\left(\frac{Q_0^2}{Q_0^2+\boldsymbol{l}^2}\right)^\rho, \tag{7}$$

and Q_0 is a scale proportional to the saturation scale. To justify the dilute limit associated with the Gaussian approximation, we study the cross-section at large photon virtuality $Q^2 = 9$ GeV^2

[1]The running coupling Balitsky-Kovchegov equation (rcBK) provides a non-linear generalization of the BFKL equation as discussed above, see [5, 15] for details

and expand the differential cross-section up to quadratic order in $N^{(2)} = 1 - S^{(2)}$. We further use $\rho = 2.3$ and $Q_0^{\text{proton}} = 0.69$ GeV, as motivated by inclusive DIS fits at $x = 0.2 \times 10^{-3}$. Since we want to search for effects due to large gluon densities, we further simulate scattering on a gold nucleus through the re-scaling $Q_0^{\text{Au}} = A^{1/6} \cdot Q_0^{\text{proton}} = 1.67$ GeV. At linear order in $N^{(2)}$, the cross-section is proportional to the Fourier transform of the dipole, $\Phi\left((\boldsymbol{p}_1 + \boldsymbol{p}_2 + \boldsymbol{p}_3)^2\right)$ and therefore gives direct access to the gluon distribution in the target. At this order the cross-section is – up to the detailed shape and x-dependence of the dipole factor Eq. (7) – identical to a corresponding BFKL calculation. The configuration dominant in the 'collinear' limit corresponds then to the vanishing of the sum of transverse momenta $\boldsymbol{p}_1 + \boldsymbol{p}_2 + \boldsymbol{p}_3 = 0$, *i.e.* the transverse momentum transfer between projectile and target is zero. To study deviations from this configuration we first chose $|\boldsymbol{p}_1| = |\boldsymbol{p}_2| = |\boldsymbol{p}_3|$. For this choice the 'collinear' limit $\boldsymbol{p}_1 + \boldsymbol{p}_2 + \boldsymbol{p}_3 = 0$ is then given by the Mercedes-Benz star configuration, see Fig. 1. We observe vanishing of the partonic cross section at these 'collinear' configurations, Fig. 1, with a strong double peak. This vanishing is explained by the vanishing of the partonic matrix element at leading order in $N^{(2)}$ for zero momentum transfer between projectile and target, which itself is expected from gauge invariance. For a complete study at hadronic level, we expect the double peak to turn into a single peak. While in the case of the proton we find (besides the double-peak structure) mainly a smearing of the collinear peak at $\Delta_{\bar{q}g} = 4\pi/3$, the effect of a larger gluon saturation scale for a gold nucleus can be clearly seen: while the shape of the linear gold curve is similar to the proton curve, it is wider and depleted with respect to the former. Including further quadratic terms in $N^{(2)}$, we start to probe effects due to high gluon densities at the level of the matrix element beyond the mere modification of Eq. (7) and beyond the corresponding BFKL matrix element. In the case of the proton, these corrections are small, as expected for $x = 0.2 \cdot 10^{-3}$ and a relatively large photon virtuality $Q^2 = 9$ GeV^2. For a highly saturated gold nucleus, they are on the other hand sizable. While this apparently puts doubts on the reliability of the current expansion up to quadratic order in $N^{(2)}$, it clearly demonstrates the sensitivity of this processes to high and saturated gluon densities and identifies it as suitable tool in searching for such effects already at currently accessible collider energies.

3 Conclusions

In this contribution, we provided a short overview on the use of the 3 parton production process in Deep Inelastic Scattering to pin down the presence of high density effects. Those effects in turn would then allow a further exploration of the dynamics associated with gluon saturation.

Acknowledgments

I am grateful for support by the symposium organizers and the Red Mexicana Científica y Tecnológica para ALICE LHC.

References

[1] L. N. Lipatov, Sov. J. Nucl. Phys. **23** (1976) 338, E. A. Kuraev, L. N. Lipatov, V. S. Fadin, Phys. Lett. B **60** (1975) 50, Sov. Phys. JETP **44** (1976) 443, Sov. Phys. JETP **45** (1977) 199. I. I. Balitsky, L. N. Lipatov, Sov. J. Nucl. Phys. **28** (1978) 822; V. S. Fadin, L. N. Lipatov, Phys. Lett. B **429** (1998) 127, M. Ciafaloni, G. Camici, Phys. Lett. B **430** (1998) 349.

[2] L. V. Gribov, E. M. Levin and M. G. Ryskin, Phys. Rept. **100**, 1 (1983).

[3] L. D. McLerran and R. Venugopalan, Phys. Rev. D **49**, 2233 (1994), Phys. Rev. D **49**, 3352 (1994).

[4] J. Jalilian-Marian, A. Kovner, L. D. McLerran and H. Weigert, Phys. Rev. D **55**, 5414 (1997); J. Jalilian-Marian, A. Kovner, A. Leonidov and H. Weigert, Nucl. Phys. B **504**, 415 (1997), Phys. Rev. D **59**, 014014 (1999), Phys. Rev. D **59**, 014015 (1999), Phys. Rev. D **59**, 034007 (1999), A. Kovner, J. G. Milhano and H. Weigert, Phys. Rev. D **62**, 114005 (2000); A. Kovner and J. G. Milhano, Phys. Rev. D **61**, 014012 (2000); E. Iancu, A. Leonidov and L. D. McLerran, Nucl. Phys. A **692**, 583 (2001), Phys. Lett. B **510**, 133 (2001); E. Ferreiro, E. Iancu, A. Leonidov and L. McLerran, Nucl. Phys. A **703**, 489 (2002); H. Weigert, Nucl. Phys. **A703**, 823-860 (2002); J. -P. Blaizot, E. Iancu, H. Weigert, Nucl. Phys. **A713**, 441-469 (2003).

[5] I. Balitsky, Nucl. Phys. B **463**, 99 (1996); Y. V. Kovchegov, Phys. Rev. D **60**, 034008 (1999), Phys. Rev. D **61**, 074018 (2000).

[6] Y. V. Kovchegov and E. Levin, Nucl. Phys. B **577**, 221 (2000) [hep-ph/9911523]; M. Hentschinski, H. Weigert and A. Schafer, Phys. Rev. D **73**, 051501 (2006) [hep-ph/0509272]; E. Levin and A. Prygarin, Phys. Rev. C **78**, 065202 (2008) [arXiv:0804.4747 [hep-ph]].

[7] E. Iancu and R. Venugopalan, In *Hwa, R.C. (ed.) et al.: Quark gluon plasma* 249-3363; H. Weigert, Prog. Part. Nucl. Phys. **55**, 461 (2005); J. Jalilian-Marian, Y. V. Kovchegov, Prog. Part. Nucl. Phys. **56**, 104-231 (2006); F. Gelis, E. Iancu, J. Jalilian-Marian, R. Venugopalan, Ann. Rev. Nucl. Part. Sci. **60**, 463-489 (2010); J. L. Albacete and C. Marquet, Prog. Part. Nucl. Phys. **76**, 1 (2014).

[8] I. Bautista, A. Fernandez Tellez and M. Hentschinski, Phys. Rev. D **94** (2016) no.5, 054002 doi:10.1103/PhysRevD.94.054002 [arXiv:1607.05203 [hep-ph]].

[9] M. Hentschinski, A. Sabio Vera and C. Salas, Phys. Rev. Lett. **110** (2013) no.4, 041601 [arXiv:1209.1353 [hep-ph]].

[10] M. Hentschinski, A. Sabio Vera and C. Salas, Phys. Rev. D **87** (2013) no.7, 076005 [arXiv:1301.5283 [hep-ph]].

[11] J. Bartels and K. Kutak, Eur. Phys. J. C **53**, 533 (2008) [arXiv:0710.3060 [hep-ph]]; L. Motyka and M. Sadzikowski, Acta Phys. Polon. B **45**, no. 11, 2079 (2014) [arXiv:1411.7774 [hep-ph]].

[12] A. Ayala, M. Hentschinski, J. Jalilian-Marian and M. E. Tejeda-Yeomans, arXiv:1604.08526 [hep-ph].

[13] A. Ayala, M. Hentschinski, J. Jalilian-Marian and M. E. Tejeda-Yeomans, Nucl. Phys. B **920** (2017) 232 doi:10.1016/j.nuclphysb.2017.03.028 [arXiv:1701.07143 [hep-ph]].

[14] F. Dominguez, C. Marquet, B. -W. Xiao and F. Yuan, Phys. Rev. D **83**, 105005 (2011).

[15] J. L. Albacete, N. Armesto, J. G. Milhano, P. Quiroga-Arias and C. A. Salgado, Eur. Phys. J. C **71**, 1705 (2011).

Elastic and inelastic diffraction at the LHC

László Jenkovszky[1,*] and *István* Szanyi[2,**]

[1]*Bogolyubov Institue for Theoretical Physics, Nat. Ac. Sc. of Ukraine, Kiev*
[2]*Uzhgorod National University, Uzhgorod*

Abstract. New phenomena discovered at the LHC in nearly forward scattering are discussed. The properties of elastic and inelastic diffraction are related due to Regge factorisation and pomeron dominance.

1 Introduction

At the LHC, in the nearly forward direction for the first time one can see the isolated pomeron, uncontaminated by secondary (non-leading) reggeon contributions. Elastic proton-proton-proton scattering, single and double proton diffraction dissociation are related by factorization, as shown in Fig. 1. The input pomeron is a double pole in the j plane, lying on a non-linear trajectory. Deviation from the exponential diffraction cone are studied in details. Predictions for future experiments are presented.

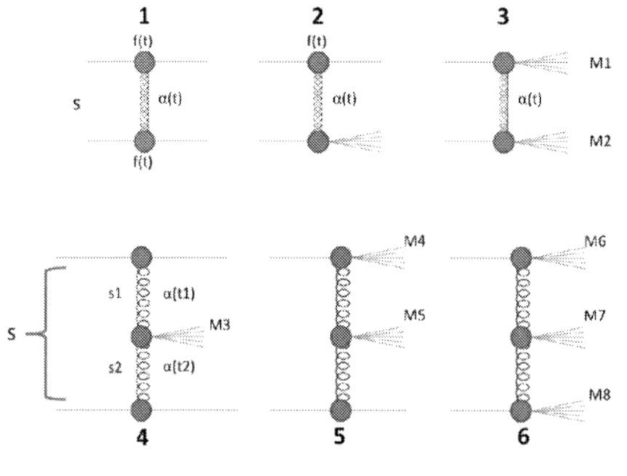

Figure 1. Diagrams for elastic scattering and diffraction dissociation (single, double and central).

*e-mail: jenk@bitp.kiev.ua
**e-mail: sz.istvan03@gmail.com

© The Authors, published by EDP Sciences. This is an open access article distributed under the terms of the Creative Commons Attribution License 4.0 (http://creativecommons.org/licenses/by/4.0/).

2 Elastic scattering

The scattering amplitude is a sum of four terms, two asymptotic (pomeron (P) and odderon (O)) and two non-asymptotic ones (secondary Regge pole contributions). P and f have positive C-parity, thus entering in the scattering amplitude with the same sign in pp and $\bar{p}p$ scattering, while the odderon and ω have negative C-parity, thus entering pp and $\bar{p}p$ scattering with opposite signs, as shown below:

$$A(s,t)^{\bar{p}p}_{pp} = A_P(s,t) + A_f(s,t) \pm [A_\omega(s,t) + A_O(s,t)], \tag{1}$$

where the symbols P, f, O, ω stand for the relevant Regge-pole amplitudes and the super(sub)script, evidently, indicate $\bar{p}p(pp)$ scattering with the relevant choice of the signs in the sum (1). This sum can be extended by adding more reggeons, whose role may become increasingly important towards lower energies; their contribution can be effectively absorbed by f and ω.

We treat the odderon, C-odd counterpart of the pomeron on equal footing, differing by its $C-$parity and the values of its parameters (to be fitted to the data).

The main subject of our study is the pomeron, and it is a double pole, or DP [1, 2] lying on a non-linear trajectory, whose intercept is slightly above one. This choice is motivated by the unique properties of the DP: it produces logarithmically rising total cross sections at unit pomeron intercept. By letting $\alpha_P(0) > 1$, we allow for a faster rise of the total cross section [1], although the intercept is about half that in the DL model [3] since the double pole (or dipole) itself drives the rise in energy. Due to its geometric form (see below) the DP reproduces itself against unitarity (eikonal) corrections. As a consequence, these corrections are small, and one can use the model at the "Born level" without complicated (and ambiguous) unitarity (rescattering) corrections. DP combines the properties of Regge poles and of the geometric approach.

Regge trajectories are non-linear complex functions. In a limited range and with limited precision, they can be approximated by linear trajectories (which is a common practice, reasonable when non-linear effects can be neglected). This non-linearity is manifest e.g. as the "break" i.e. a change the slope $\Delta B \approx 2$ GeV$^{-}2$ around $t \approx -0.1$ GeV^2 and at large $|t|$, beyond the second maximum, $|t| > 2$ GeV^2, where the cross section flattens and the trajectories are expected to slowdown logarithmically.

A simple mechanism of the diffractive dip-bump structure combining geometrical features and Regge behaviour was suggested in Ref. [2]. In that model, the dip is generated by the pomeron contribution. The relevant pomeron is a double pole arises from the interference between this dipole with a simple one, it is accompanied by. The dip-bump in the model shows correct dynamics, that is it develops from a shoulder, progressively deepening in the ISR energy region. As energy increases further, the dip is filled by the odderon contribution. At low energies the contribution from non-leading, "secondary" reggeons is also present.

The dipole pomeron produces logarithmically rising total cross sections and nearly constant ratio of σ_{el}/σ_{tot} at unit pomeron intercept, $\alpha_P(0) = 1$. While a mild, logarithmic increase of σ_{tot} does not contradict the data, the rise of the ratio σ_{el}/σ_{tot} beyond the SPS energies requires a supercritical DP intercept, $\alpha_P(0) = 1 + \delta$, where δ is a small parameter $\alpha_P(0) \approx 0.05$. Thus DP is about "twice softer" then that of Donnachie-Landshoff's model [3], in which $\alpha_P(0) \approx 0.08$.

We use the normalization where

$$\frac{d\sigma}{dt} = \frac{\pi}{s^2}|A(s,t)|^2 \text{ and } \sigma_{tot} = \frac{4\pi}{s}\Im m A(s,t)\Big|_{t=0}. \tag{2}$$

Neglecting spin dependence, the invariant proton(antiproton)-proton elastic scattering amplitude is that of Eq. (1). The secondary reggeons are parametrized in a standard way, with linear Regge

[1] A supercritical pomeron trajectory, $\alpha_P(0) > 1$ in the DP is required by the observed rise of the ratio σ_{el}/σ_{tot}, or, equivalently, departure form geometrical scaling.

trajectories and exponential residua, where R denotes f or ω - the principal non-leading contributions to pp or $\bar{p}p$ scattering:

$$A_R(s,t) = a_R e^{-i\pi\alpha_R(t)/2} e^{b_R t} \left(s/s_0\right)^{\alpha_R(t)}, \tag{3}$$

with $\alpha_f(t) = 0.70 + 0.84t$ and $\alpha_\omega(t) = 0.43 + 0.93t$.

The pomeron is a dipole in the j–plane

$$A_P(s,t) = \frac{d}{d\alpha_P}\left[e^{-i\pi\alpha_P/2} G(\alpha_P)\left(s/s_0\right)^{\alpha_P}\right] = \tag{4}$$

$$e^{-i\pi\alpha_P(t)/2}\left(s/s_0\right)^{\alpha_P(t)}\left[G'(\alpha_P) + \left(L - i\pi/2\right)G(\alpha_P)\right].$$

Since the first term in squared brackets determines the shape of the cone, one fixes

$$G'(\alpha_P) = -a_P e^{b_P[\alpha_P - 1]}, \tag{5}$$

where $G(\alpha_P)$ is recovered by integration, and, as a consequence, the pomeron amplitude Eq. (4) can be rewritten in the following "geometrical" form

$$A_P(s,t) = i\frac{a_P}{b_P}\frac{s}{s_0}[r_1^2(s)e^{r_1^2(s)[\alpha_P-1]} - \varepsilon_P r_2^2(s)e^{r_2^2(s)[\alpha_P-1]}], \tag{6}$$

where $r_1^2(s) = b_P + L - i\pi/2$, $r_2^2(s) = L - i\pi/2$, $L \equiv \ln(s/s_0)$. Typical choices for the pomeron trajectory are

$$\alpha_P \equiv \alpha_P(t) = 1 + \delta_P + \alpha_{1P}t, \tag{TR.1}$$

$$\alpha_P \equiv \alpha_P(t) = 1 + \delta_P + \alpha_{1P}t - \alpha_{2P}\left(\sqrt{4\alpha_{3P}^2 - t} - 2\alpha_{3P}\right), \tag{TR.2}$$

$$\alpha_P \equiv \alpha_P(t) = 1 + \delta_P - \alpha_{1P}\ln(1 - \alpha_{2P}t). \tag{TR.3}$$

The main features of the nonlinear trajectories are: 1) presence of a threshold singularity required by t–channel unitarity and responsible for the change of the slope in the exponential cone (the so-called "break") near $t = -0.1$ GeV^2 [4], and 2) logarithmic asymptotic behaviour providing for a power fall-off of the cross sections in the "hard" region. The combination of theses properties, however is not unique.

2.1 Diffraction minimum (dip-bump)

Figure 2 (a) shows the pp and $\bar{p}p$ total elastic scattering cross section calculated from the model (values of parameters see in [5]). Figure 2 (b) shows the ratio of the real to imaginary part of the forward amplitude. The model with a linear trajectory describes reasonably the data a wide range of collision energies for pp and $\bar{p}p$. Figures 3 (a,b,c) show the fitted $\bar{p}p$ and pp differential cross sections and predictions for three different center of mass energies are shown. The yellow area exhibits the statistical uncertainty on the calculations, described earlier. Calculations are characterized by an approximately exponential fall-off in the range $0 < |t| < 8$ GeV^2, with the slope changing around $-t \approx 0.6$ GeV^2. The dip moved towards lower momentum transfer and became almost filled by the odderon.

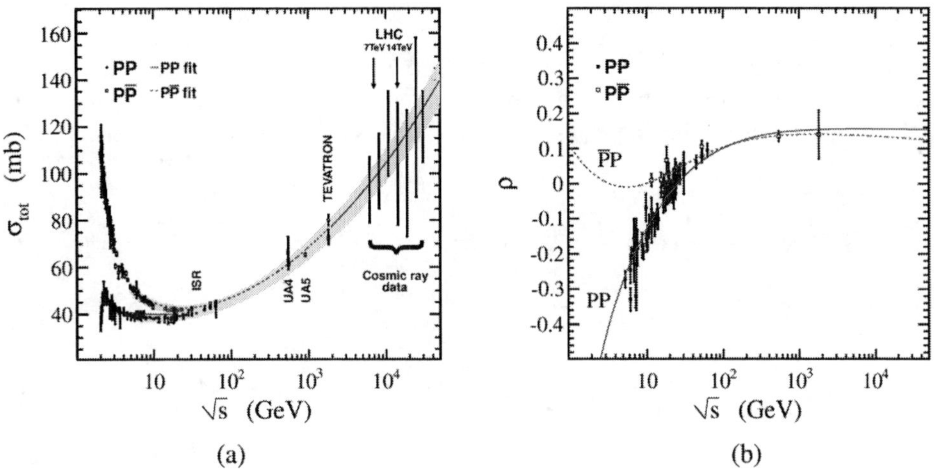

Figure 2. (a) Total pp and $\bar{p}p$ cross sections calculated from the model, Eqs. (1-6, TR.1), and fitted to the data in the range $\sqrt{s} = 5 - 30$ TeV and 5 GeV — 1.8 TeV, respectively. (b) Ratio of the real to imaginary part of the forward amplitude for pp and $\bar{p}p$, calculated from the model and fitted to the data.

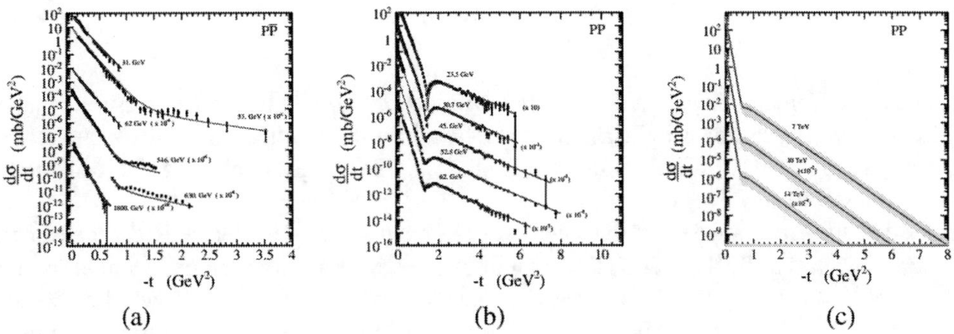

Figure 3. (a) $\bar{p}p$ differential cross sections calculated from the model, Eqs. (1-6, TR.1), and fitted to the data, and fitted to the data in the range $-t = 0.1 - 8$ GeV^2. (b) pp differential cross sections calculated from model and fitted to the data. (c) predictions for the pp differential cross section calculated from the model for three different LHC energies.

2.2 Low-$|t|$ "break" and proton shape

The diffraction cone of high-energy elastic hadron scattering deviates from a purely exponential dependence on t due to two structures clearly visible in proton-proton scattering: the so-called "break" (in fact, a smooth curve with a concave curvature) near $t = -0.1$ GeV^2, whose position is nearly independent of energy and the prominent "dip" – diffraction minimum, moving slowly (logarithmically) with s towards smaller values of $|t|$, where s and t are the Mandelstam variables. The physics of these two phenomena are quite different. As illustrated in Fig. 4, the "break" is likely a reflection due to the "pion cloud", which controls the "static size" of nucleon. This effect, first observed in 1972 at the ISR, was interpreted [4, 6, 7] as the manifestation of t-channel unitarity, generating a two-pion loop in the cross channel, Fig. 5, and was referred to by Bronzan [8] as the "fine structure" of the pomeron. The dip (diffraction minimum), on the other hand, is generally accepted as a consequence of s-channel unitarity or absorption corrections to the scattering amplitude. As such, dip location reflects the "size" of proton, moving towards smaller $|t|$ values as the total cross section increases with energy.

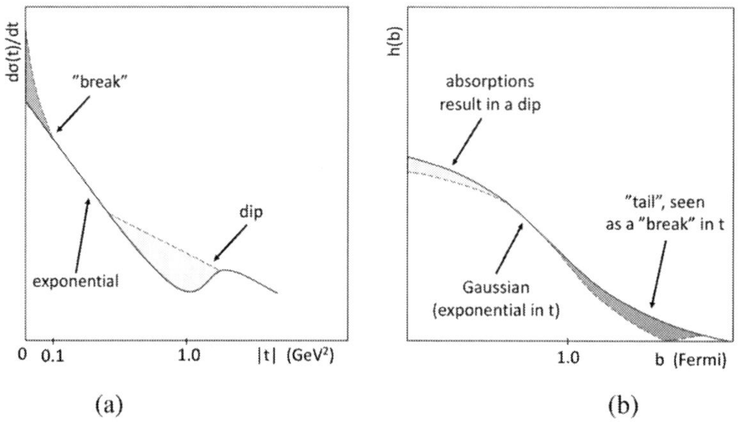

Figure 4. Schematic (qualitative) view of the "break", followed by the diffraction minimum ("dip"), shown both as function in t and its Fourier transform (impact parameter representation), in b.

The departure from the linear exponential behavior was confirmed by recent measurements by the TOTEM Collaboration at the CERN LHC, first at 8 TeV (with a significance greater then 7σ) [9] and subsequently at 13 TeV [10]. At the ISR the "break" was illustrated by plotting the local slope $B(t)$ for several t-bins at fixed values of s.

At the LHC the effect appears of the same order of magnitude and is located near the same value of t. Different from the ISR [11], TOTEM quantifies the deviation from the exponential by normalizing the measured cross section to a linear exponential form, (see Eq. (9) below). For the sake of completeness we will exhibit this "break effect" both in the normalized form and for $B(t)$.

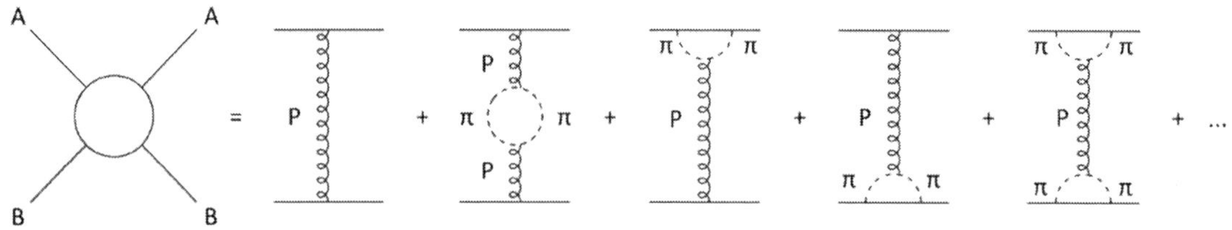

Figure 5. Diagram for elastic scattering with t-channel exchange containing a branch point at $t = 4m_\pi^2$.

The new LHC data from TOTEM at 8 TeV confirm the conclusions [4, 6, 7] about the nature of the break and call for a more detailed analysis and better understanding of this phenomenon. The new data triggered intense theoretical work in this direction [12–14], but many issues still remain open. Although the curvature for $B(t)$, both at the ISR and the LHC is concave, a convex behavior cannot be excluded in other reactions and/or new energies. While the departure from a linear exponential was studied in details both at the ISR and LHC energies, an interpolation between the two is desirable to clarify the uniqueness of the phenomenon. This is a challenge for the theory, and it can be done within Regge-pole models. Below we do so by adopting a simple Regge-pole model, with a pomeron and two secondary reggeons, f and ω exchanges. The odderon may also be present. However its contribution at low $|t|$ is too small to be identified unambiguously.

Having identified [4, 6, 7] the observed "break" with a two-pion exchange effect, we investigate further two aspects of the phenomenon, namely: 1) to what extent is the "break" observed recently at the LHC a "recurrence" of that seen at the ISR (universality)? 2) what is the relative weight of the Regge residue (vertex) compared to the trajectory (propagator) in producing the "break"? We answer these questions by means of a detailed fit to the elastic proton-proton scattering data in the relevant kinematic range $0.05 < -t < 0.35$ GeV^2 ranging between the ISR energies ($23.5 \leq \sqrt{s} \leq 62.5$ GeV), and those available at the LHC.

As shown by Barut and Zwanziger [15], t-channel unitarity constrains the Regge trajectories near the threshold, $t \to t_0$ by

$$\Im\, \alpha(t) \sim (t - t_0)^{\Re\, \alpha(t_0)+1/2}, \tag{7}$$

where t_0 is the lightest threshold, $4m_\pi^2$ in the case of the vacuum quantum numbers (pomeron or f meson). Since the asymptotic behavior of the trajectories is constrained by dual models with Mandelstam analyticity by square-root (modulus $\ln t$): $|\ \frac{\alpha(t)}{\sqrt{t}\ln t}\ |_{t\to\infty} \leq$ const, (see Ref. [4] and references therein), for practical reasons it is convenient to approximate, for the region of t in question, the trajectory as a sum of square roots. Higher thresholds, indispensable in the trajectory, may be approximated by their power expansion, *i.e.* by a linear term, matching the threshold behavior with the asymptotic.

At the ISR, the proton-proton differential cross section was measured at $\sqrt{s}$ = $23.5, 30.7, 44.7, 52.8$ and 62.5 GeV [16]. In all the above energies the differential cross section changes its slope near $-t = 0.1$ GeV^2. By using a simple Regge-pole model we have mapped the "break" fitted at the ISR onto the LHC TOTEM 8 TeV data. The simple Regge-pole model is constructed by a leading supercritical pomeron and two secondary, f and ω contributions.

The detailed results of fits and the parameters see in Ref [17]. To demonstrate the important features more clearly, we show the results of the mapping in higher resolution in Fig. 6 and Fig. 7. In Fig. 6, we exhibit the shape of local slopes, defined by

$$B(s,t) = \frac{d}{dt}\ln(d\sigma/dt)\,, \tag{8}$$

for our fit at six s values. To better demonstrate the goodness of our fit and to anticipate the comparison with the LHC data, we also present in Fig. 7 the ISR data in normalized form as used by TOTEM [9]:

$$R = \frac{d\sigma/dt}{d\sigma/dt_{ref}} - 1, \tag{9}$$

where $d\sigma/dt_{ref} = Ae^{Bt}$, with A and B constants determined from a fit to the experimental data.

Both Fig. 6 and Fig. 7 re-confirm the earlier finding that the "break" can be attributed the presence of two-pion branch cuts in the Regge parametrization. The effect of pion cloud manifest itself most importantly through the Regge residue, rather then the pomeron trajectory.

3 Diffraction dissociation (SD and DD)

In most of the papers SD is calculated from the triple Regge limit of an inclusive reaction, as shown in Fig. 8.

In that limit, the double diffraction cross section can be written as

$$\frac{d^2\sigma}{dtdM_x^2} = \frac{G_{13,2}^{PP,P(t)}}{16\pi^2 s_0^2}\left(\frac{s}{s_0}\right)^{2\alpha_P(t)-2}\left(\frac{M^2}{s_0}\right)^{\alpha_P(0)-\alpha_P(t)}.$$

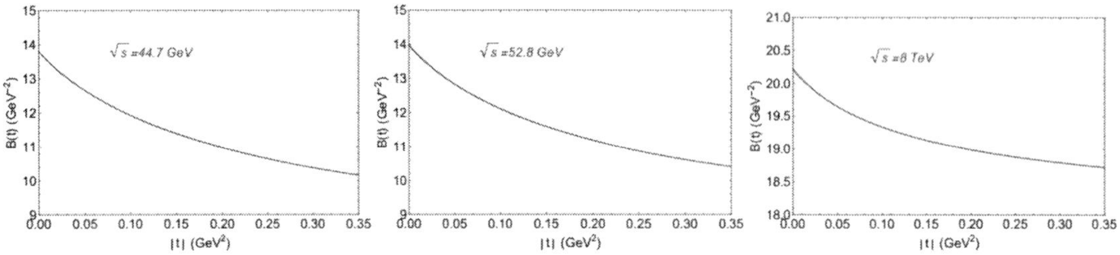

a) 44.7 GeV b) 52.8 GeV c) 8 TeV

Figure 6. Local slopes.

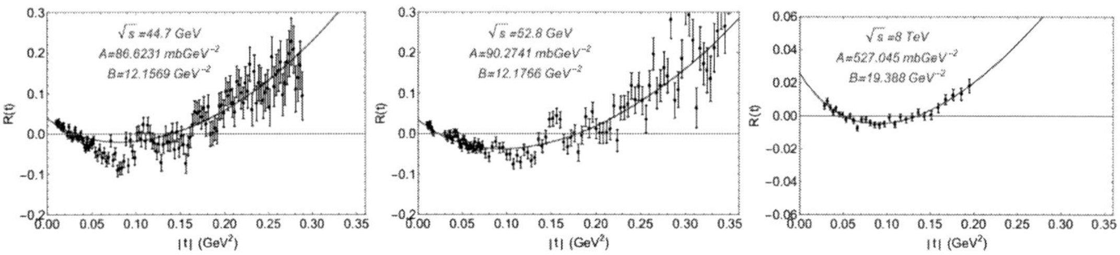

a) 44.7 GeV b) 52.8 GeV c) 8 TeV

Figure 7. $R(t)$ ratios.

This approach has two shortcomings. The first one is that it leaves outside the small-M^2 resonance region. The second one is connected with the fact that whatever the pomeron, the (partial) SD cross section overshoots the total one, thus obviously conflicting with unitarity. Various ways of resolving this deficiency are known from the literature, including the vanishing (decoupling) of the triple pomeron coupling, but none of them can be considered completely satisfactory.

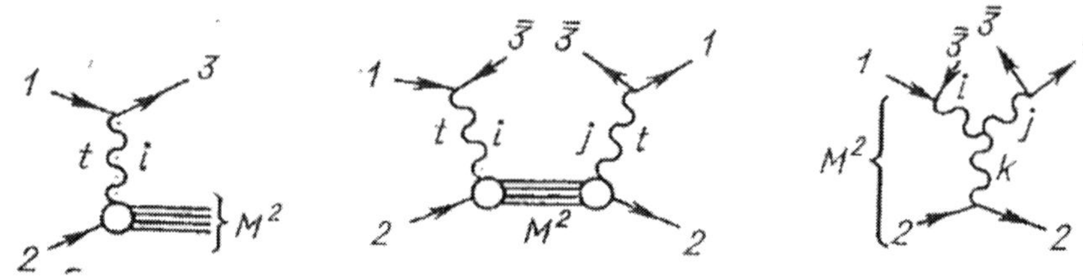

Figure 8. From SD to the triple Regge limit.

We instead follow the idea [18] according to which the reggeon (here, the pomeron) is similar to the photon and that the reggeon-nucleon interaction is similar to deep-inelastic photon-nucleon scattering (DIS), with the replacement $-Q^2 = q^2 \to t$ and $s = W^2 \to M_x^2$. There is an obvious difference between the two: while the C parity of the photon is negative, it is positive for the pomeron. We believe that while the dynamics is essentially invariant under the change of C, the difference between the two being accounted for by the proper choice of the parameters. Furthermore, while Jaroszewicz and Landshoff [14], in their pomeron-nucleon DIS structure function (SF) (or Pp total cross section) use the Regge asymptotic limit, we include also the low missing mass, resonance behaviour.

It is evident that Regge factorization is essential in both approaches (triple Regge and the present one). It is feasible when Regge singularities are isolated poles. While the pre-LHC data require the inclusion of secondary reggeons, at the LHC we are in the fortunate situation of a single pomeron exchange (pomeron dominance) in the t channel in single and double diffraction (not necessarily so in central diffraction, to be treated elsewhere). Secondary Regge pole exchanges will appear however, in our dual-Regge treatment of Pp scattering (see below), not to be confused with the the t channel of pp. This new situation makes diffraction at the LHC unique in the sense that for the first time Regge-factorization is directly applicable. We make full use of it.

Diffraction is limited both in the missing mass (coherence), $\xi \lesssim 0.05$ and in t ("soft" collisions). There is a transition region in t from "soft" to "hard" collisions, with a possible dip-bump structure between the two. To be sure, in our analysis we leave outside these interesting but controversial points, concentrating on the "first" cone with clear exponential behaviour.

With the advent of the LHC, diffraction, elastic and inelastic scattering entered a new area, where it can be seen uncontaminated by non-diffraction events. In terms of the Regge-pole theory this means, that the scattering amplitude is completely determined by a pomeron exchange, and in a simple-pole approximation, Regge factorization holds and it is of practical use!

3.1 A compilation of basic formulae

This subsection contains a compilation of the main formulae used in the calculations and fits to the data.

The single diffraction (SD) dissociation cross section is:

$$2 \cdot \frac{d^2\sigma_{SD}}{dtdM_x^2} = F_p{}^2(t) F_{inel}{}^2(t, M_x^2) \left(\frac{s}{M_x^2}\right)^{2(\alpha(t)-1)} . \qquad (10)$$

Double diffraction (DD) dissociation cross section:

$$\frac{d^3\sigma_{DD}}{dtdM_1^2dM_2^2} = N_{DD} F_{inel}{}^2(t, M_1^2) F_{inel}{}^2(t, M_2^2) \left(\frac{ss_0}{M_1^2M_2^2}\right)^{2(\alpha(t)-1)} . \qquad (11)$$

with the norm $N_{DD} = \frac{1}{4A_{el}}$, and the inelastic vertex:

$$F_{inel}{}^2(t, M_x^2) = A_{res} \frac{1}{M_x^4} \sigma_T^{Pp}(M_i^2, t) + C_{bg}\sigma_{Bg}, \qquad (12)$$

where the pomeron-proton total cross section is the sum N^* resonances and the Roper resonance, with a relevant norm factor R (we remove the t dependent $f_{res}(t)$ out of the sum):

$$\sigma_T^{Pp}(M_x^2, t) = R \frac{[f_{res}(t)]^2 \cdot M_{Roper}\left(\frac{\Gamma_{Roper}}{2}\right)}{\left(M_x^2 - M_{Roper}^2\right)^2 + \left(\frac{\Gamma_{Roper}}{2}\right)^2} + [f_{res}(t)]^4 \sum_{n=1,3} \frac{\mathcal{I}m\,\alpha(M_x^2)}{(2n + 0.5 - \mathcal{R}e\,\alpha(M_x^2))^2 + (\mathcal{I}m\,\alpha(M_x^2))^2}, \qquad (13)$$

and the background corresponding to non-resonance contributions:

$$\sigma_{Bg} = \frac{f_{bg}(t)}{\frac{1}{(M_x^2-(m_p+m_\pi)^2)^\epsilon} + (M_x^2)^\eta}, \qquad (14)$$

The integrated cross sections are calculated as:

$$\frac{d\sigma_{SD}}{dt} = \int_{M_1^2}^{M_2^2} \frac{d^2\sigma_{SD}}{dtdM_x^2} dM_x^2 \tag{15}$$

for the case of SD and:

$$\frac{d\sigma_{DD}}{dt} = \int\int_{f(M_{x_1}^2, M_{x_2}^2)} \frac{d^3\sigma_{SD}}{dtdM_{x_1}^2 dM_{x_2}^2} dM_{x_1}^2 dM_{x_2}^2 \tag{16}$$

for the case of DD.
We calculate also the fully integrated cross sections:

$$\sigma_{SD} = \int_0^1 dt \int_{M_{th}^2}^{0.05s} dM_x^2 \frac{d^2\sigma_{SD}}{dtdM_x^2}, \tag{17}$$

$$\sigma_{DD} = \int_0^1 dt \int\int_{\Delta\eta>3} dM_{x_1}^2 dM_{x_2}^2 \frac{d^3\sigma_{DD}}{dtdM_{x_1}^2 dM_{x_2}^2} \tag{18}$$

and

$$\frac{d^2\sigma_{DD}}{dM_{x_1}^2 dM_{x_2}^2} = \int_0^1 \frac{d^3\sigma_{DD}}{dtdM_{x_1}^2 dM_{x_2}^2} dt. \tag{19}$$

3.2 Fitting procedure

The model contains 12 parameters, a large part of which is fixed either by their standard values (e.g. those of Regge trajectories, except for the pomeron slope, whose slope exceeds the "standard" value to meet the SD data) or are set close to the previous fits [18].

The elastic data and Regge theory fix the parameters $s_0, \alpha(0), \alpha', A_{el}, b_{el}$.

The data at larger $|t|$, with the dip-bump structure and subsequent flattening of the cross section, both in elastic scattering and in SD may indicate the onset of new physics and the transition to hard scattering, implying a non-exponential residue and/or a non-linear pomeron trajectory, that goes beyond the present study.

Single diffraction dissociation (SD) is an important pillar in our fitting procedure. The following parameters were fitted to the SD data: A_{res}, C_{bg}, R, b_{res}, b_{bg}, ς, η. As input data we use double differential cross sections $\frac{d^2\sigma_{SD}}{dtdM^2}$ versus M_x^2 at $|t| = 0.05$ GeV^2; b)single differential cross sections $\frac{d\sigma_{SD}}{dt}$ vs. t; c) fully integrated cross sections versus energy $\sqrt{s}$.

Central exclusive diffraction (diagram 6 in Fig. 1 was calculted recently in Ref.[19]).

4 Conclusions

Below is a brief summary of our results:

At the LHC, in the diffraction cone region ($t < 1$ GeV^2) proton-proton scattering is dominated (over 95%) by pomeron exchange (quantified in Ref. [5]). This enables full use of factorized Regge-pole models. Contributions from non-leading (secondary) trajectories can (and should be) included in the extension of the model to low energies, e.g. below those of the SPS.

ACKNOWLEDGMENTS

L.J. thanks Alexandro Ayala and his team for they hospitality and support at the Symposium.

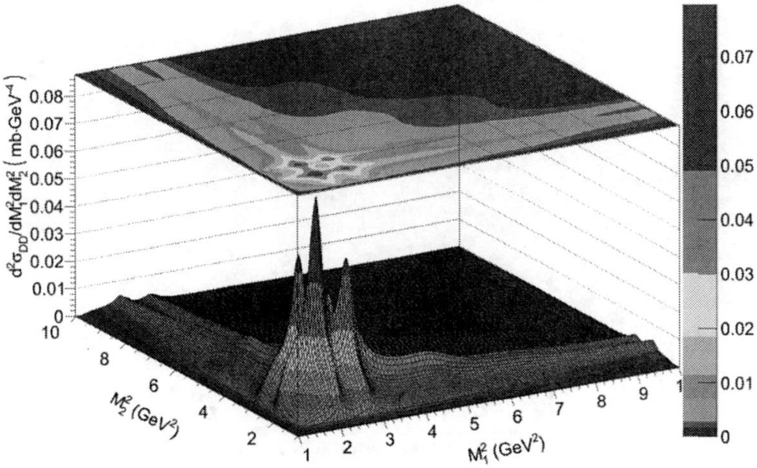

Figure 9. Double differential DD cross section as function of M_1^2 and M_2^2, integrated in t;, see Eq. (11).

References

[1] R. Fiore *et al.* Int. J. Mod. Phys., A24, 2551-2559 (2009); arXiv:hep-ph/0812.0539.

[2] L.L. Jenkovszky and A.N. Vall, ITP Preprint, Kiev 1974; Czech. J. Phys. B, **26**, 447 (1976).

[3] S. Donnachie and P. Landshoff, Phys. Lett., **B 123**, 345 (1983); Nucl. Phys. **267**, 690 (1985).

[4] G. Cohen-Tannoudji *et al.*, Nuov. Cim., **5**, 957 (1972).

[5] L.L. Jenkovszky, A.I. Lengyel, D.I. Lontkovsky, Int. J. Mod. Phys. A, **26**, 4755 (2011); arXiv:1105.1202.

[6] C.-I Tan and D.M. Tow, Phys. Letters B. **53B**, 452 (1975).

[7] U. Sukhatme, Chung-I Tan, and Tran Thanh Van, Z. Phys. C, Particles and Fields, **1**, 95 (1979).

[8] J.B. Bronzan, *Fine structure of the Pomeron*, in Symposium on the Pomeron, ANL report, ANL/HEP 7327, p.34, (Argonne National Laboratory, 1973).

[9] TOTEM Collab. (G. Antchev *et al.*), Nucl. Phys., **B 899** 527 527 (2015); arXiv:1503.08111.

[10] M. Deile, *Elastic and Total Cross-Section Measurements by TOTEM*, Contribution at the 17. Conference on Elastic and Diffractive Scattering, (Prague, Czech Republic, 2017).

[11] B. Barbiellini *et. al.*, Phys. Lett. **B39**, 663 (1972).

[12] L. Jenkovszky, I. Szanyi, Phys. Part. Nuclei Lett., **14**, 687 (2017); arXiv:1701.01269.

[13] D.A. Fagundes, L. Jenkovszky, E.Q. Miranda, G. Pancheri, P.V.R.G. Silva, Int. J. Mod. Phys. A **31**, 1645022 (2016); arXiv:1509.02197.

[14] L. Jenkovszky and A. Lengyel, Acta Phys. Pol., **B 46**, 863 (2015); arXiv:1410.4106.

[15] A.O. Barut and D.E. Zwanziger, Phys. Rev. **127**, 974 (1962).

[16] http://durpdg.dur.ac.uk/review/pp2/exphtml/CERN-ISR.shtml

[17] L. Jenkovszky, I. Szanyi and C.-I Tan. *Shape of Proton and the Pion Cloud*, arXiv:1710.10527.

[18] R. Fiore *et al.* Phys. ReV D **83**, 056014 (2011); arXiv:056014; R. Fiore *et al.*, Yad. Fizika, **77** 1535 (2014); arXiv:1011.0664.

[19] R. Fiore, L. Jenkovszky and R. Schicker, EPJ C **76(1)** 1-10 (2016); arXiv:1512.04977.

EPJ Web of Conferences **172**, 06005 (2018)
ISMD 2017

https://doi.org/10.1051/epjconf/201817206005

Soft diffraction at LHC

Jan Kašpar[1,2,3,⋆]

[1] *INFN - Sezione di Pisa, Pisa, Italy*
[2] *CERN, Geneva, Switzerland*
[3] *Institute of Physics of the ASCR, Prague, Czech Republic*

Abstract. This contribution reviews and compares various LHC results on soft diffraction, in particular elastic scattering, total, inelastic and elastic cross-section, single and double diffraction.

1 Introduction

The aim of this contribution is to compare soft-diffraction results across the LHC experiments, to identify and understand differences and trends that occur in the data. There are measurements from ALICE [1], ATLAS [2], ATLAS-ALFA [3], CMS [4], LHCb [5] and TOTEM [6]. The focus is put on elastic scattering, total, inelastic and elastic cross-section, single and double diffraction. Other interesting topics, e.g. central diffraction, are not covered only due to length restrictions to this contribution.

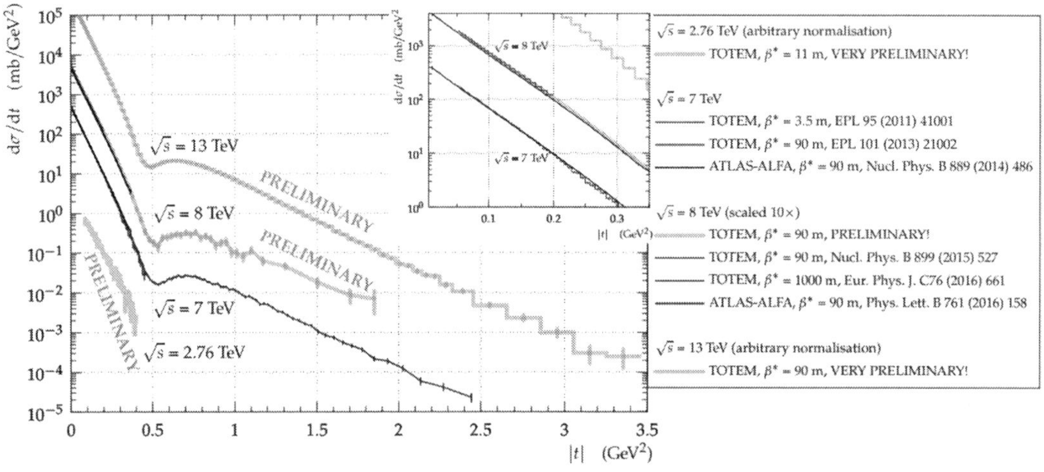

Figure 1. A summary of elastic scattering differential cross-section measurements at LHC. The red and blue lines refer to measurements by TOTEM and ATLAS ALFA, respectively. The grey lines correspond to preliminary and yet unpublished data by TOTEM. Uncertainties shown are statistical only.

⋆e-mail: jan.kaspar@cern.ch

© The Authors, published by EDP Sciences. This is an open access article distributed under the terms of the Creative Commons Attribution License 4.0 (http://creativecommons.org/licenses/by/4.0/).

2 Elastic scattering

At the LHC there are two experiments equipped with forward-proton taggers needed to measure elastic scattering: ATLAS-ALFA and TOTEM. Their measurements are summarised in Figure 1. Different regions of four-momentum transfer squared, t, probe different physics regimes of elastic scattering. At very low $|t|$, the interference between Coulomb and nuclear amplitudes (CNI) takes place. At larger values of $|t|$ up to $\approx 0.4\,\text{GeV}^2$, the differential cross-section presents almost exponential decrease sometimes called diffraction cone. This region is followed by a dip and a bump which are often explained by an interference of different amplitude components. At high values of $|t|$, where the cross-section exhibits a smooth and uniform decrease, one may expect a transition from non-perturbative to perturbative regime.

The inset of Figure 1 compares the measurements of ATLAS-ALFA and TOTEM which are in general in good agreement, especially for what the shape is concerned. At $\sqrt{s} = 8\,\text{TeV}$, the measurement by ATLAS-ALFA is a little lower than the one by TOTEM which will be further discussed in Section 3.

Figure 1 confirms the phenomenon of "diffraction cone shrinkage" also at LHC energies: the position of the dip moves to lower $|t|$ values with increasing energy. At $\sqrt{s} = 7\,\text{TeV}$ the dip occurs at $|t| \approx 0.53\,\text{GeV}^2$ while the preliminary data at 13 TeV indicate it at about $|t| \approx 0.48\,\text{GeV}^2$.

The $\sqrt{s} = 13\,\text{TeV}$ TOTEM cross-section at $|t| \gtrsim 1\,\text{GeV}^2$ presents a uniform decrease. Although this measurement is preliminary, the yet-to-be applied corrections are unlikely to create any structures. This observation contradicts predictions of many theoretical/phenomenological models where secondary dips or oscillations appear (see e.g. [7]). The TOTEM measurement, in turn, excludes all such models. In fact, it rules out all models built on the optical analogy – where multiple dips or oscillations are inevitable. On the other hand, the measurement may be favouring a hypothesis that large $|t|$ elastic scattering is dominated by a perturbative QCD amplitude, e.g. triple gluon exchange proposed by Donnachie and Landshoff [8].

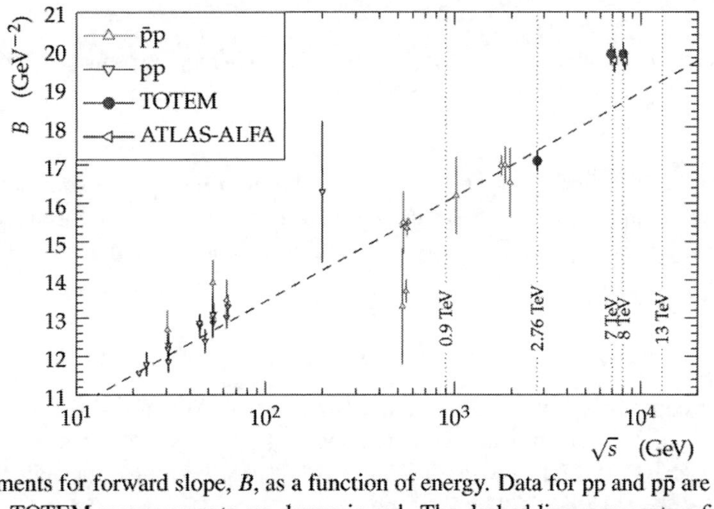

Figure 2. Measurements for forward slope, B, as a function of energy. Data for pp and $p\bar{p}$ are drawn in blue and green, respectively. TOTEM measurements are drawn in red. The dashed line represents a fit of pre-LHC data [9] by TOTEM.

Figure 2 summarises measurements of forward slope, B:

$$B = \frac{\mathrm{d}}{\mathrm{d}t} \log \frac{\mathrm{d}\sigma}{\mathrm{d}t}\bigg|_{t=0} . \tag{1}$$

All measurements for $\sqrt{s} < 3\,\text{TeV}$ are compatible with a fit linear in $\log s$, see the dashed line. However, this trend is not respected by $\sqrt{s} = 7$ and $8\,\text{TeV}$ measurements by ATLAS-ALFA [3, 10] and TOTEM [11, 12] (well compatible between the two experiments). This observation may be indicating an onset of a new scattering regime between $\sqrt{s} = 2.76$ and $7\,\text{TeV}$. Preliminary $13\,\text{TeV}$ results by TOTEM support this hypothesis further constraining the break point towards $\sqrt{s} \approx 3\,\text{TeV}$.

The diffraction cone presents differential cross-section with approximately exponential decrease. To visualise small deviations from this leading behaviour Figure 3 shows the relative difference of measured $\mathrm{d}\sigma/\mathrm{d}t$ from a reference exponential. All measurements, by ATLAS-ALFA [3, 10] and by TOTEM [11, 13], give a qualitatively similar picture at all available energies. For $|t| \lesssim 0.2\,\text{GeV}^2$ one can observe deviations with a "wide U" shape and a significance of more than $7\,\sigma$ in case of TOTEM $\sqrt{s} = 8\,\text{TeV}$ measurement [13]. For $|t| \gtrsim 0.2\,\text{GeV}^2$ the slope increases, thus the differential cross-section falls steeper towards the dip.

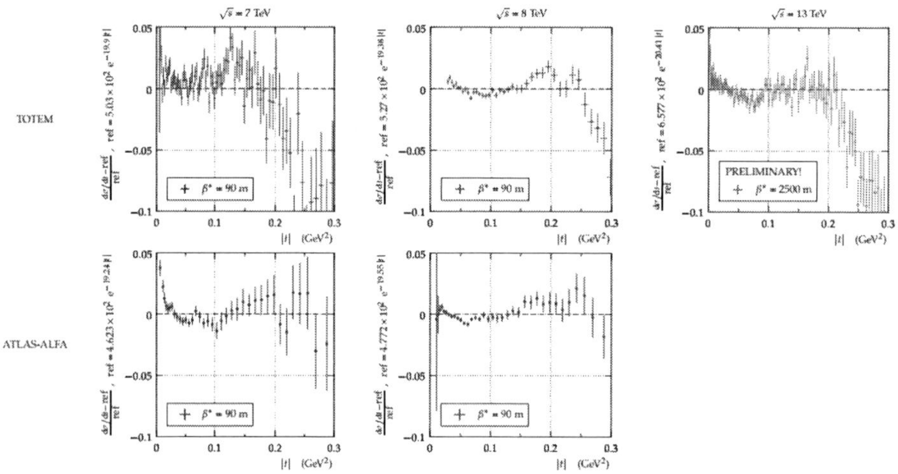

Figure 3. Details of the diffraction cone: relative deviation from a reference exponential. Top row: measurements by TOTEM, bottom row: measurements by ATLAS-ALFA.

Since the Coulomb scattering amplitude is known from QED, the Coulomb-nuclear interference occurring at very low $|t|$ enables to determine the phase of the nuclear amplitude. This is typically quantified in terms of the ρ parameter, the ratio of real to imaginary part of the nuclear scattering amplitude:

$$\rho = \left.\frac{\Re \mathcal{A}^{\mathrm{N}}}{\Im \mathcal{A}^{\mathrm{N}}}\right|_{t=0} . \tag{2}$$

Figure 4 shows a compilation of ρ measurements vs. energy. ATLAS-ALFA has not published any result yet, but $\sqrt{s} = 8\,\text{TeV}$ results are expected soon. TOTEM's measurement at this energy [16] lies a little below the COMPETE extrapolation [15] but is still compatible with it due to the large uncertainty. At $\sqrt{s} = 13\,\text{TeV}$ TOTEM has collected a larger-statistics sample and is aiming at ρ uncertainty about 0.01 which should be sufficient to probe the compatibility with the COMPETE models. This is interesting since deviations may hint on the existence of the Odderon (e.g. [17, 18]), the even-under-crossing partner of the Pomeron. ρ is particularly sensitive parameter since the real part of the Pomeron-exchange amplitude is expected small and therefore the contribution from the Odderon may be significant albeit it is generally dominated by the Pomeron. The other Odderon discovery channels in elastic scattering are either not available or seem excluded at the LHC. The

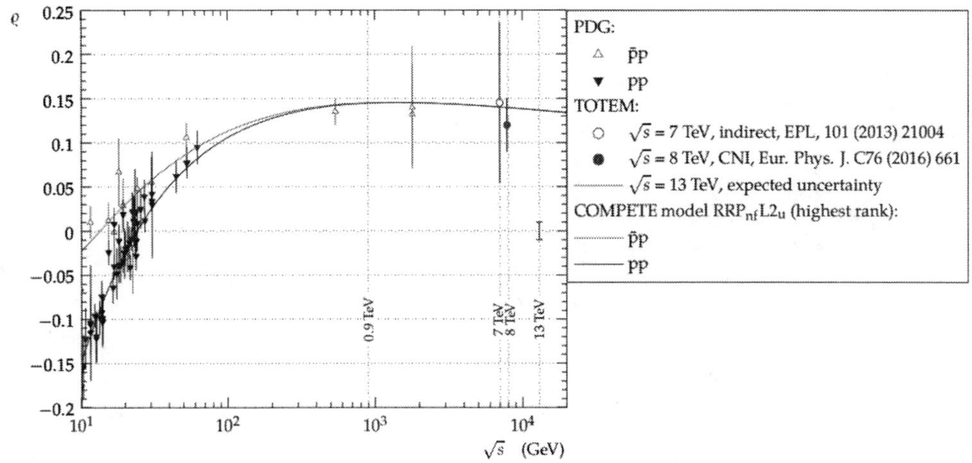

Figure 4. ρ measurements plotted as a function of energy. The pre-LHC data are taken from collection [14]. The solid lines represent pre-LHC fits by the COMPETE Collaboration [15]. Green and blue colours refer to p$\bar{\text{p}}$ and pp scattering, respectively; TOTEM measurements are drawn in red.

accelerator does not provide polarised beams to study polarisation effects. It can neither provide a collision energy similar to the previous accurate p$\bar{\text{p}}$ measurement to compare dips/shoulders in pp and p$\bar{\text{p}}$ scattering. Finally, the preliminary TOTEM 13 TeV data seem to suggest that high $|t|$ scattering is dominated by a perturbative-QCD amplitude, thus hiding potential Odderon signals.

3 Cross-sections

Figure 5 shows a compilation of total, inelastic and elastic cross-sections. The pre-LHC data are taken from [14], an additional cosmic-ray measurement is by the Auger Collaboration [19]. ATLAS-ALFA [3, 10] and TOTEM [12, 20, 21] have contributed by several measurements of all three cross-sections. Additional inelastic cross-section measurements comprise results from ALICE [22], ATLAS [23, 24], CMS [25, 26] and LHCb [27]. $\sqrt{s}$ = 13 TeV results from TOTEM are expected soon, ATLAS-ALFA has collected data at this energy, too. Furthermore, these two experiments aim at measurements at $\sqrt{s}$ = 0.9 TeV possibly even at the end of 2017. This will constitute 5 sampling points and thus strong constraint for the energy dependence of the cross-section. The presently available data are compatible with $\log^2 s$ growth of all the cross-sections.

It is interesting to compare the total cross-section measurements by ATLAS-ALFA and TOTEM. While the $\sqrt{s}$ = 7 TeV points are fully compatible, there is about 2 σ tension at 8 TeV. It can be traced down to the normalisation (thus to luminosity or inefficiency determination) of the differential cross-section, cf. Figure 1, inset. The two measurements therefore give quite different trends: while the TOTEM points are compatible with the COMPETE fit, the ATLAS-ALFA points give extrapolation to 13 TeV about 10 mb lower. Consequently, the soon-available 13 TeV results will provide important insight for this comparison.

The comparison is even more complicated for the inelastic cross-section given the large variety of results. One can observe that while ALICE and TOTEM results are compatible with the dashed-dotted line, all others lie below. This can be understood by studying the systematic uncertainties related to each measurement. Those performed with central detectors only necessarily need to correct for events escaping the apparatus in the forward direction. This correction is sizeable and strongly

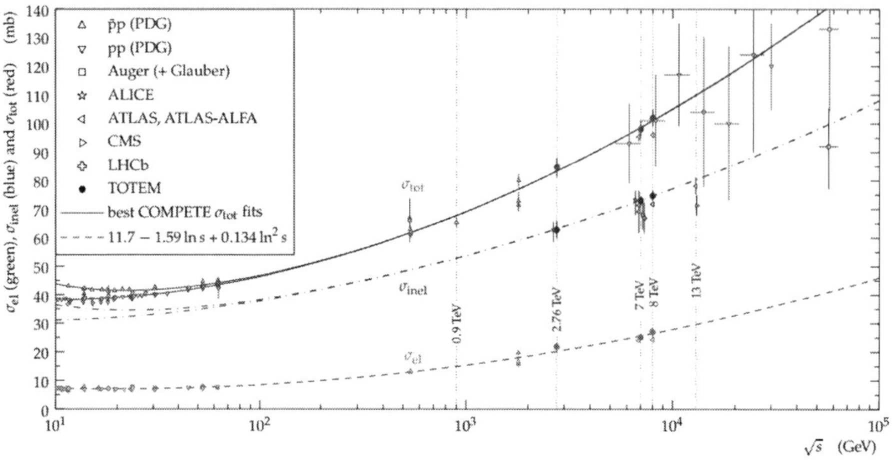

Figure 5. Total (red), inelastic (blue) and elastic (green) cross-section measurements plotted as a function of energy. For references see text. The solid lines represent the pre-LHC fits by COMPETE [15]. The dashed line shows a σ_{el} fit by TOTEM, including its 7 and 8 TeV and PDG [14] points. The dashed-dotted lines corresponds to the difference between the σ_{tot} fits by COMPETE and the σ_{el} fit by TOTEM.

model dependent. For example, the CMS analysis [26] takes a "democratic" approach and uses a mean correction over a handful of Monte-Carlo predictions. In another example, the ATLAS analysis [24] first tunes and compares various models to data and for the final correction selects only those which describe the data well. This procedure tends to select models predicting larger corrections. In other words, naive selection of Monte-Carlo predictions can lead to underestimation of the inelastic cross-section. This conclusion is supported also by σ_{inel} measurements by experiments including forward-proton taggers: ATLAS-ALFA and TOTEM can determine the inelastic cross-section by $\sigma_{tot} - \sigma_{el}$ and the results favour larger values. Finally, at $\sqrt{s} = 7$ TeV TOTEM determined the inelastic cross-section by 3 different methods [20], yielding consistent results compatible with the dash-dotted line.

4 Single diffraction

ALICE has published [22] a single-diffraction measurement based on counting events with single-sided topology. The surviving protons were not tagged in this analysis and only integrated cross-sections are presented, see Figure 7.

CMS has published [28] a measurement based on counting events with rapidity gaps on one side of the detector. Also in this analysis the forward protons were not identified. The cross-section was presented as a function of ξ, the fractional momentum loss: showing a gentle decrease of the cross-section with ξ, where none of the considered Monte-Carlo generators was able describe all the features in detail.

TOTEM has preliminary results [29] from an analysis at $\sqrt{s} = 7$ TeV where the surviving protons are tagged. As illustrated in Figure 6, the analysis is performed in 3 diffractive-mass (or ξ) bins. They are defined by the acceptance of TOTEM subdetectors and the relation between the rapidity gap, Δy, and ξ: $\Delta y \approx -\log \xi$. For each ξ bin, the results are presented as a function of the proton four-momentum transfer squared, t. The distributions are compatible with exponentials where the slope, B, decreases with increasing ξ. Summing the preliminary per-bin cross-sections yields (6.5 ± 1.3) mb.

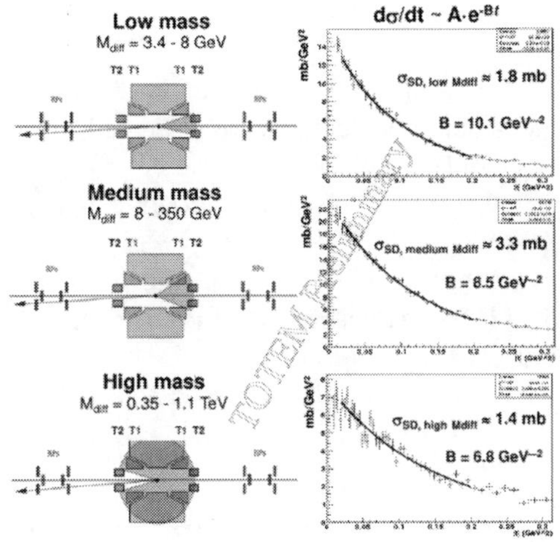

Figure 6. Illustration of TOTEM single diffraction measurement, adapted from material by V. Avati and F .Oljemark. The relative uncertainty of cross-sections is estimated to 20 % and 15 % for the slope *B*.

This can be complemented with a cross-section of (2.6 ± 2.2) mb which corresponds to events with all particles more forward than the T2 acceptance [30]. As this cross-section is expected to be dominated by single diffraction, one can estimate $\sigma_{SD} = (9.1 \pm 2.9)$ mb. This value corresponds to cut-off $\xi \lesssim 0.022$ and is represented by the green point in Figure 7.

There is also an ongoing combined CMS+TOTEM analysis at $\sqrt{s} = 8$ TeV which combines the benefits of a central detector and forward-proton taggers. This analysis aims at event classification as non-diffractive, single-diffractive and double-diffractive.

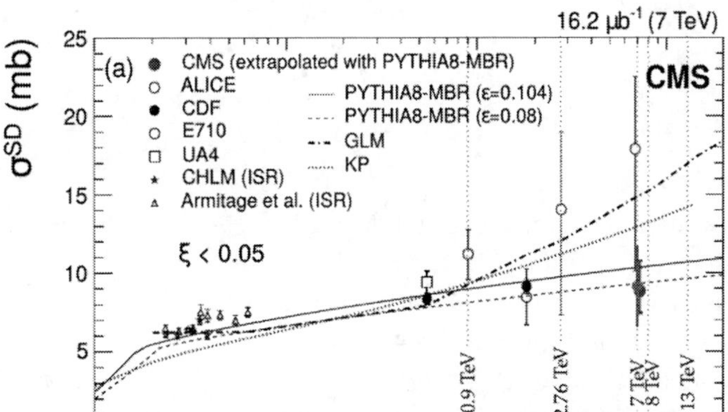

Figure 7. Single-diffractive cross-section as a function of energy. Plot adapted from [28]. The ALICE data are from [22]. The green point corresponds to a compilation of TOTEM data, see text.

ATLAS published a study of events with rapidity gaps [31] at $\sqrt{s} = 7$ TeV where, however, no attempt was made to disentangle non-diffractive, single-diffractive and double-diffractive components.

5 Double diffraction

ALICE has published [22] a double-diffraction measurement based on counting events with central rapidity gap $\Delta\eta > 3$. Only integrated cross-sections are presented, see Figure 8.

CMS has published [28] a similar measurement, also counting events with central rapidity gap $\Delta\eta > 3$. The results are presented as a function of the rapidity gap, featuring decreasing cross-section with increasing gap size.

TOTEM made a measurement at $\sqrt{s} = 7\,\text{TeV}$ [32] however constrained to a narrow fiducial region where background could be controlled and subtracted. Consequently, no attempt was made to extrapolate the results to the full phase space.

As mentioned in Section 4, the combined CMS + TOTEM analysis is expected to provide double-diffraction cross-section at $\sqrt{s} = 8\,\text{TeV}$.

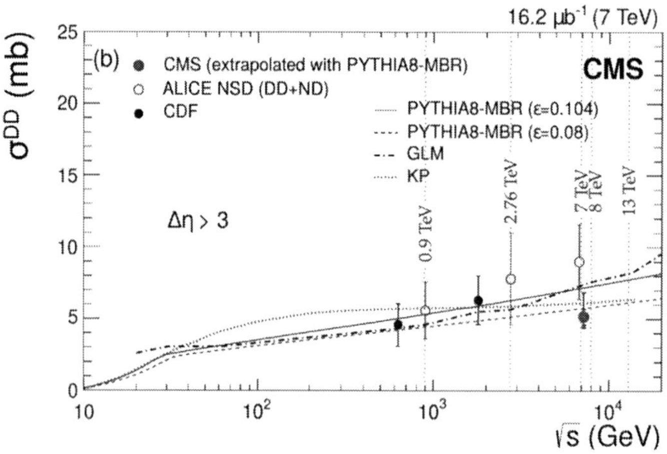

Figure 8. Double-diffractive cross-section as a function of energy. Plot adapted from [28]. The ALICE data are from [22].

References

[1] ALICE Collaboration, J. Instrum. **3**, S08002 (2008)

[2] ATLAS Collaboration, J. Instrum. **3**, S08003 (2008)

[3] ATLAS Collaboration, Nucl. Phys. B **889**, 486–548 (2014)

[4] CMS Collaboration, J. Instrum. **3**, S08004 (2008)

[5] CMS Collaboration, J. Instrum. **3**, S08005 (2008)

[6] G. Anelli et al. (TOTEM Collaboration), J. Instrum. **3**, S08007 (2008)

[7] J. Kašpar, Comput. Phys. Commun. **185**, 1081–1084 (2014); http://elegent.hepforge.org/

[8] A. Donnachie, P. V. Landshoff, Phys. Lett. B **387**, 637–641 (1996)

[9] ISR (CR Collaboration), Phys. Lett. **B62**, 460 (1976); ISR (ACHGT Collaboration), Phys. Lett. **B39**, 663 (1972); ISR (R-211), Nucl. Phys. **B262**, 689 (1985); ISR (R-210), Phys. Lett. **B115**, 495 (1982); UA1, Phys. Lett. **B147**, 385 (1984); UA4, Phys. Lett. **B127**, 472 (1983) and Phys. Lett. **B198**, 583 (1987); UA4/2, Phys. Lett. **B316**, 448 (1993); CDF, Phys. Rev. **D50**, 5518 (1994); E710, Phys. Rev. Lett. **68**, 2433 (1992) and Nuovo Cimento **A106**, 123 (1992); D0, D0 Note 6056-CONF; pp2pp, Phys. Lett. **B579**, 245 (2004)

 https://doi.org/10.1051/epjconf/201817206005

[10] ATLAS Collaboration, Phys. Lett. B **761**, 158–178 (2016)
[11] TOTEM Collaboration, Europhys. Lett. **101**, 21002 (2013)
[12] TOTEM Collaboration, Phys. Rev. Lett. **111**, 012001 (2013)
[13] TOTEM Collaboration, Nucl. Phys. B **899**, 527–546 (2015)
[14] K. Nakamura et al. (Particle Data Group), J. Phys. G **37**, 075021 (2010)
[15] J. R. Cudell et al. (COMPETE Collaboration), Phys. Rev. Lett. **89**, 201801 (2002)
[16] TOTEM Collaboration, Eur. Phys. J. C **76**, 661 (2016)
[17] C. Ewerz, The Odderon in Quantum Chromodynamics, https://arxiv.org/abs/hep-ph/0306137
[18] R. Avila, P. Gauron and B. Nicolescu, Eur. Phys. J. C **49**, 581 (2007)
[19] P. Abreu et al. (Pierre Auger Collaboration), Phys. Rev. Lett. **109**, 062002 (2012)
[20] TOTEM Collaboration, Europhys. Lett. **101**, 21004 (2013)
[21] TOTEM Collaboration, Total cross-section measurement at $\sqrt{s} = 2.76\,\text{TeV}$, publication in preparation
[22] ALICE Collaboration, Eur. Phys. J. **C73**, 2456 (2013)
[23] ATLAS Collaboration, Nature Commun. **2**, 463 (2011)
[24] ATLAS Collaboration, Phys. Rev. Lett. **117**, 182002 (2016)
[25] CMS Collaboration, CMS-PAS-FWD-11-001, https://cds.cern.ch/record/1373466
[26] CMS Collaboration, CMS-PAS-FSQ-15-005, https://cds.cern.ch/record/2145896
[27] LHCb collaboration, J. High Energ. Phys. **02**, 129 (2015)
[28] CMS Collaboration, CMS-PAS-FSQ-12-005; arXiv:1503.08689
[29] F. Oljemark and K. Österberg on behalf of TOTEM collaboration, Studies of Soft Single Diffraction with TOTEM at the LHC at 7 TeV, Contribution to LHC students poster session 2013, https://indico.cern.ch/event/238337/
[30] TOTEM Collaboration, Europhys. Lett. **101**, 21003 (2013)
[31] ATLAS Collaboration, Eur. Phys. J. C **72**, 1926 (2012)
[32] TOTEM Collaboration, Phys. Rev. Lett. **111**, 262001 (2013)

Pomeron Physics at the LHC

Federico Deganutti[1,2,⋆], *David* Gordo Gomez[1,3,⋆⋆], *Timothy* Raben[1,⋆⋆⋆], and *Christophe* Royon[1,⋆⋆⋆⋆]

[1]*Department of Physics and Astronomy, University of Kansas, 1082 Malott, 1251 Wescoe Hall Dr., Lawrence, KS 66045-758*
[2]*Dipartimento di Fisica ed Astronomia, Università di Firenze, Piazza di San Marco, 4, 50121 Firenze FI, Italia*
[3]*Instituto de Física Teórica UAM/CSIC, Nicolás Cabrera 13-15, Campus de Cantoblanco UAM, 28049 Madrid, España & Universidad Autónoma de Madrid, E-28049 Madrid, España*

Abstract. We present current and ongoing research aimed at identifying Pomeron effects at the LHC in both the weak and strongly coupled regimes of QCD.

1 Introduction

Hadron collisions in many kinematic regimes display Regge behavior where cross sections grow as a power of the center of mass energy. Regge Theory grew out of pre-QCD S-matrix theory. Amplitudes are seen as unitary, Lorentz invariant functions of analytic momenta.[1] Poles in scattering amplitudes represent particle exchange or bound state production.[2] Using a partial wave analysis, the dominant contribution to simple amplitudes is the exchange of an *entire trajectory* of particles: Pomeron exchange. Single Pomeron exchange is predictive of a power-law rise in the total cross section[3] $\sigma_{tot} \sim s^{\alpha_0 - 1}$ Since the beginning of Pomeron physics, fits to p-p total cross section data have shown some power-law behavior.[4] In these proceedings we review progress in using Pomeron exchange to describe LHC data.

Weak Coupling Pomeron

Balitsky, Fadin, Kuraev, Lipatov (BFKL) asked, and provided the first important answers to, the question: what happens in the Regge limit of QCD? Large logarithms get in the way

⋆e-mail: fedeganutti@ku.edu
⋆⋆e-mail: david.gordo@csic.e
⋆⋆⋆e-mail: timothy.raben@ku.edu **Speaker**
⋆⋆⋆⋆e-mail: christophe.royon@ku.edu

[1]The goal was to describe scattering experiments without a detailed description of an underlying microscopic theory.

[2]Generically amplitudes also have cuts corresponding to multiparticle production and non-linear interactions. The physics of cuts much less well understood and we omit any detailed discussion

[3]Asymptotically, as $s \to \infty$, cross-sections are likely subject to the Froissart bound indicating a maximal growth $\sigma_{tot} \sim log^2(s)$.

[4]The contribution responsible for the empirically observed σ_{tot} rise is referred to as the *soft Pomeron*. Although recent σ_{tot} measurements show a deviation from a strict power-law, the soft Pomeron approach is still used to fit data and for modeling.[1]

© The Authors, published by EDP Sciences. This is an open access article distributed under the terms of the Creative Commons Attribution License 4.0 (http://creativecommons.org/licenses/by/4.0/).

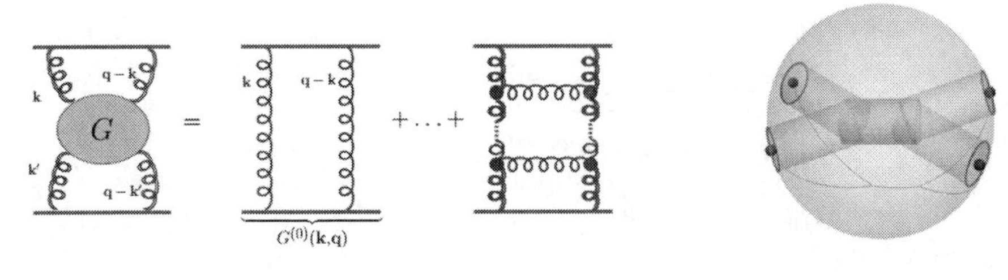

(a) (b)

Figure 1. (a) Ladder diagram for the BFKL weak coupling Pomeron. (b) Cartoon of a 2-to-2 scattering process using the AdS/CFT applicable for the BPST strong coupling Pomeron.

of usual perturabtion theory; a resummation of terms $\alpha_s\, log(s)$ to all orders is necessary. Scattering amplitudes are dominated by the t-channel exchanged of an infinite ladder of reggized gluons (See Figure1(a)) leading to Regge behavior (power law growth of cross sections with energy). The BFKL equation, an integral equation for Green's function in Mellin space, is used to described this exchange kernel. The solution can be written

$$G(\mathbf{k},\mathbf{k}',\mathbf{q},Y) = \int_{-i\infty}^{+i\infty} \frac{d\omega}{2\pi i} e^{Y\omega} f_\omega(\mathbf{k},\mathbf{k}',\mathbf{q}) \rightarrow \int_{-i\infty}^{+i\infty} \frac{d\omega}{2\pi i} e^{Y\omega} \sum_{n\in\mathcal{Z}} \int_{\frac{1}{2}-i\infty}^{\frac{1}{2}+i\infty} \frac{d\gamma}{2\pi\, i} \frac{E_{\gamma,n}(k)E^*_{\gamma,n}(k')}{\omega - \bar{\alpha}_s\chi(\gamma,n)}, \quad (1)$$

where in the leading log (LL) approximation: $\chi(\gamma,n) = 2\psi(1) - \psi(\gamma + \frac{|n|}{2}) - \psi(1-\gamma+\frac{|n|}{2})$ and $\omega_0 = \frac{4\alpha_s N_c}{\pi} ln(2)$.

Strong Coupling Pomeron

Pomeron phenomenology is applicable over a wide range of kinematics and, particularly if one focuses on small momentum transfers, involves non-perturbative physics. The most promising description of a unified soft/hard Pomeron framework comes from the AdS/CFT duality conjecture. In this framework, a $\mathcal{N}=4$ susy Yang-Mills theory is dual to a string theory in a higher dimensional curved space. The utility of such a duality is far beyond the scope here, but we mention a few pertinent properties: the duality is strong/weak so that a perturbative, weakly coupled string theroy calculation can give insights into strongly coupled Yang-Mills; the gluon sector of the $\mathcal{N}=4$ theory is similar to that of QCD; and the extra dimensions and new fields become natural mechanisms for modeling the onset of confinement and saturation.

Within the AdS/CFT duality, the Pomeron has been identified with the Regge trajectory of the graviton[5]. (Otherwise known as the BPST Pomeron[2]) Scattering amplitudes computed via Feynman diagram like approach (See Figure1(b) for a pictorial.) with the leading Regge amplitudes written as a convolution of wavefunctions and Reggeon propagators over AdS space: $A \sim \psi_1(z)\psi_2(z) * \chi(z,z',s,t) * \psi_3(z')\psi_4(z')$. Here s and t are Mandelstam like invariants and z is an AdS coordinate. Reggeon propagators are reminiscent of the weak coupling partonic description

$$\chi_R \sim \int dj(\alpha'\hat{s})^j(1+\cos(-i\pi j))G_j(t,z,z') \quad (2)$$

[5]The BPST Pomeron naturally connects a hard and soft regime and has diffusive behavior similar to the BFKL Pomeron.

Operator dimensions of AdS Reggeons have an anomalous part and admit a non-trivial convergent expansion in terms of spin, coupling, and twist. These $\Delta - J$ curves can be calculated to high order using a mix of conformal, string, and integrability techniques. Minimizing these curves gives Reggeon intercepts.[3]

Applications of the BPST Pomeron have included deep inelastic total cross section measurements from HERA[4–6], vector meson production[7] and diffractive Higgs production [8] at the LHC, to predict glueball masses [9], and to predict cross sections of non-diffractive central η production [10].

2 Dijet Processes

Here we review previous efforts, and describe our program in progress, for observing BFKL effects at the current LHC collider kinematics[6]. We are focused on exclusive and inclusive dijet production as in Figure 2.

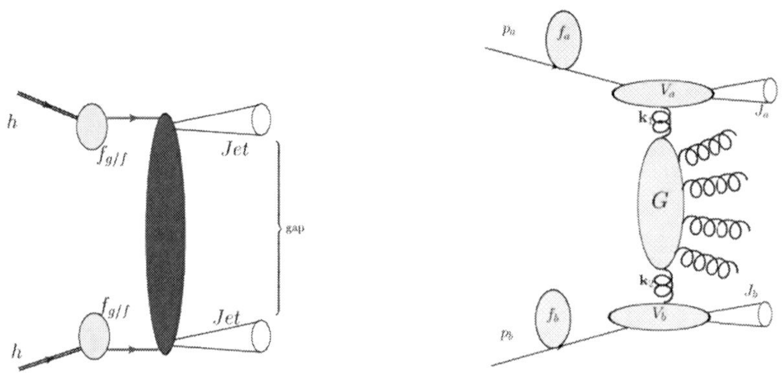

(a) (b)

Figure 2. (a) Mueller-Tang jet-gap-jet process. (b) Mueller-Navelet inclusive dijet process.

2.1 Mueller-Navelet Jets

In the Mueller-Navelet (M-N)[11] jet process all the radiation is treated inclusively except for the two jets farthest apart in rapidity. The link between this semi-inclusive process and the purely elastic amplitude of the Pomeron exchange is provided by the optical theorem. Theoretically, this is one of the simplest jet processes sensitive to BFKL effects as the outer most dijet can be thought of coming from 2-to-2 forward ($t = 0$) parton scattering. The BFKL equation greatly simplifies in this regime making both analytic and numerical computations much more tractable; the BFKL gluon ladder is seen as an interference diagram with real emissions that contribute to the cross section instead of being part of the virtual corrections to the 2 to 2 scattering amplitude.

BFKL dynamics are predicted to manifest as an increase in the decorrelation of the momentum of the tagged jets. The configuration of perfect correlation, valid at tree level, is incrisingly altered by the addition of any other emission. In the forward Regge limit, the additional real emissions are enhanced by the large log s. The larger is the rapidity separation between the outer most tagged jets, the wider is the rapidity interval that can be spanned by the ordered gluon emission, which generates the powers of logarithms: M-N jets with a large rapidity separation should be very sensitive to BFKL effects. The jet azimuthal (de)correlation

[6]This kinematical regime is traditionally dominated by DGLAP evolution.

can be quantified considering the Fourier modes of the average cosine of the angular azimuthal difference between the tagged jets $C_n/C_m = \langle\cos(n(\phi_{J_1} - \phi_{J_2} - \pi))\rangle / \langle\cos(m(\phi_{J_1} - \phi_{J_2} - \pi))\rangle$. Thanks to the high energy factorization, the coefficients C_n are given by a convolution between the probe dependent jet vertices and the universal gluon Green function:

$$\frac{C_m(|\mathbf{k}_{J_1}|, |\mathbf{k}_{J_2}|, Y)}{dY} = \int dx_1 dx_2 f(x_1) f(x_2) dy_{J_1} dy_{J_2} \delta(y_{J_1} - y_{J_2} - Y) d\phi_{J_1} d\phi_{J_2}$$
$$\times \cos(m(\phi_{J_1} - \phi_{J_2} - \pi)) \int d\mathbf{k}_1 \mathbf{k}_2 V(\mathbf{k}_{J_1}, x_{J_1}, \mathbf{k}_1) G(\mathbf{k}_1, \mathbf{k}_2, \hat{s}) V(\mathbf{k}_{J_2}, x_{J_2}, \mathbf{k}_2). \quad (3)$$

Here Y is the rapidity difference between the outgoing jets.[7] The remaining finite part is called the jet vertex $V(\mathbf{k}_{J_1}, x_{J_1}, \mathbf{k}_1)$. The jet vertices, as for the gluon Green function, depend on the approximation order and can be defined unambiguously in the BFKL approach. They have been proven to be infrared safe up to the next-to-leading order.

A test of the M-N BFKL predictions was carried out in 2013 by the CMS collaboration[12][8]. Events with at least two emerging jets with with transverse energy $|\mathbf{k}_J| > 35$ GeV and $|y_J| < 4.7$ entered the analysis. Jets were reconstructed from the energy deposition in the calorimeter and clustered with the anti-k_T algorithm[14, 15] with a distance parameter $R = 0.5$. Results were compared to the BFKL analytical predictions at NLL order[9] and the predictions of several Monte Carlo generators like PYTHIA[18] and HERWIG++[19] which are based on the DGLAP evolution.

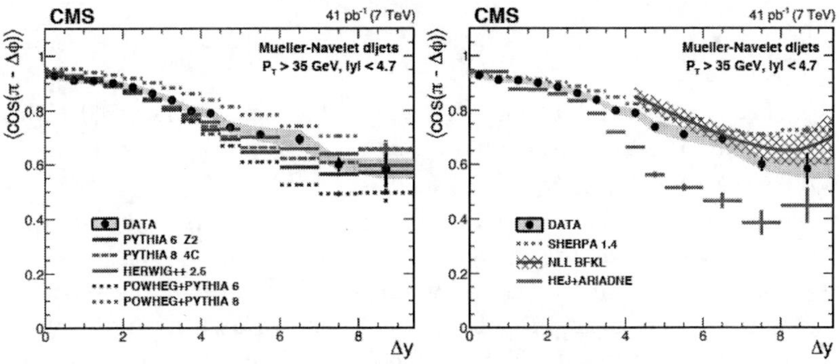

Figure 3. From [12], average $\cos(\Delta\phi)$, i.e. C_1/C_0, as a function of the rapidity difference Y compared to DGLAP MC generators on the left and to the analytic NLL BFKL calculations at partonic level on the right. Similar analysis holds for most other ratios C_m/C_n.

Concerning the BFKL predictions, few conclusions were drawn: the analytic NLL calculation predicts a decorrelation above the observed data for the three conformal moments in the whole range of rapidities, except for the data points at $Y \simeq 9$. A better agreement with data was observed for the ratios C_n/C_m $(m, n \neq 0)$ along the whole rapidity range. The

[7]Sometimes in the literature Δy is used.

[8]Similar types of analyses have been done with dijet inclusive processes at electron-ion colliders. For a recent example see[13].

[9]The NLO vertex found in the literature, and extensively used for theoretical analyses, is not in fully compatible with the M-N prescription that is used in experimental analyses. The inconsistency can be corrected, and the difference can amount up to ~ 4–10% change. Although important, it does not qualitatively alter the conclusions drawn in the analyses of CMS data. The corresponding paper is still unpublished[16], but an outline can be found in[17].

DGLAP-based MC generator HERWIG++ reproduces the experimental data with satisfactory accuracy for all the observables.

The analysis confirms that contamination from DGLAP evolution dijets affects substantially all the observables containing the total cross-section C_0 and suggests that a description of such observables based on the BFKL approach alone is perhaps not feasible at the current initial energies; the kinematic domain explored in the CMS analysis seems to lie in a transition region between the DGLAP and BFKL regimes for angular decorrelation. This opens the path for alternative or complementary strategies to better disentangle BFKL effects on top of the standard M-N angular decorrelation observables.

2.2 New Observables for Inclusive Dijet Production

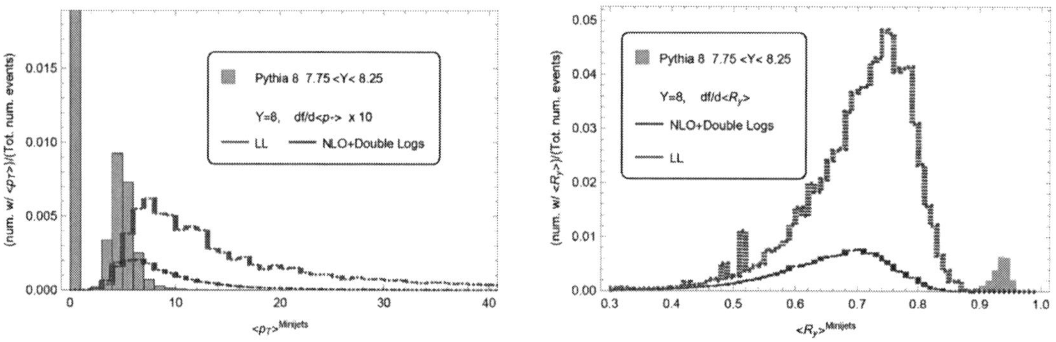

(a) (b)

Figure 4. Preliminary comparison of mini-jet p_T and $\mathcal{R}_y$ between BFKLex (red curves) and DGLAP-based PYTHIA8 (gold). PYTHIA results have been rescaled by one order of magnitude for ease of comparison. BFKLex data is from [20].

Since the geometry of the CMS detector limits the highest accessible rapidity, BFKL effects cannot be enhanced at LHC via increasing the center-of-mass energies. However, observables could be more sensitive to the differences between DGLAP and BFKL evolutions. A promising candidate is to consider the same Mueller-Navelet definition with an additional requirement that excludes events with an insufficient number of distinguished jets in between the tagged jets: counting mini-jet radiation. In fact, the BFKL ordered emission is expected to give rise to a larger mini-jet multiplicity compared to the collinear emissions re-summed in the DGLAP picture.

In addition to multiplicity, other observables were recently proposed [21, 22] using this mini-jet radiation

$$\langle p_T \rangle = \frac{1}{N} \sum_i |k_i|, \qquad \langle \theta \rangle = \frac{1}{N} \sum_i \theta_i, \qquad \langle \mathcal{R}_y \rangle = \frac{1}{N+1} \sum \frac{y_i}{y_{i-1}}, \tag{4}$$

reflecting the average transverse momentum, emission angle, and rapidity ratio of the mini-jets respectively. Initial investigations showed that these observables could be simulated using the BFKLex MC to give non trivial predictions. In Figure 4, results of a comparison between DGLAP based PYTHIA8 and the BFKLex for the average transverse momenta and rapidity ratio are plotted. These preliminary results[23] are extremely encouraging: the BFKL average p_T has a very heavy tail and the BFKL $\mathcal{R}_y$ is peaked differently from the DGLAP expectation.

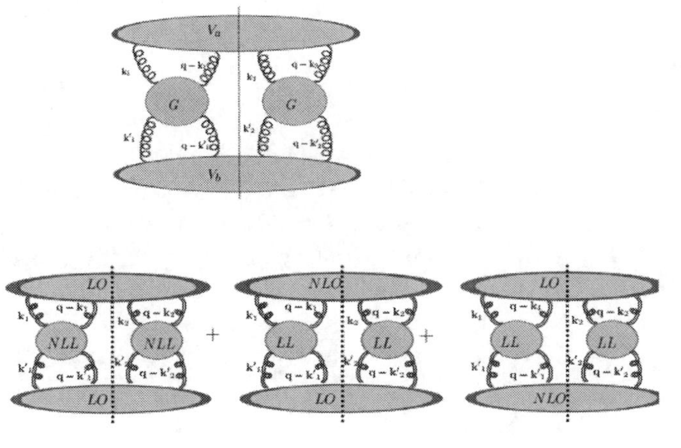

Figure 5. (Top) Schematic representation of the partonic elastic scattering via color-singlet exchange. G denotes the Green function and V the impact factors. (Bottom) Combinations of jet vertex and Green's function order that contribute to the full NLO M-T process.

However, there are still challenges left to pursue. The data in Figure 4 has cuts for the M-N outer jets similar to that of CMS, $|p_T| > 35$ GeV. However, the mini-jets are simulated with a rather modest $|k_i| > 1$ GeV; increasing to the sensitivity of CMS will require much longer runs. Secondly the overall normalization of the BFKLex is *not* fixed. This is primarily an effect of not incorporating PDF effects in the MC; BFKLex only uses the BFKL kernel. This issue can be circumvented either by an including PDF effects in BFKLex or normalizing the distributions in a region where there is a small overall rapidity gap and small p_T difference between the outer most jets; a region where both DGLAP and BFKL effects might be suppressed and the predictions should agree. The consideration of new observables, like higher moment companions of $\mathcal{R}_y$ might also alleviate these issues.

2.3 Mueller-Tang Jets

The process of using large rapidity gaps to isolate and measure high energy asymptotic behavior goes back to [24–26]. The properties of the BFKL hard Pomeron at finite momentum transfer can be investigated at hadron colliders by looking for highly exclusive processes where a dijet is separated by a large rapidity interval devoid of radiation [27]. These dijet events are known as Mueller-Tang dijet events or jet-gap-jet events [28]. At the parton level the simplest configuration consists of 2-to-2 elastic parton scattering where, at large s, the non-forward elastic amplitude is dominated by the exchange of a Pomeron. No real emission is allowed in the internal rapidity region suppressing DGLAP evolution[10]. Although this is a much cleaner experimental signal of BFKL physics, handling the non-forward amplitude is a more difficult task theoretically.

An observable that gained large popularity is the differential cross section ratio of jet-gap-jet events to inclusive dijet events: $R = d\sigma^{JGJ}/d\sigma^{dijet}$. In the BFKL approach the jet-gap-jet partonic cross section can be computed as[11]

$$\frac{d\hat{\sigma}}{dJ_1 dJ_2 d\mathbf{q}} = \int d\mathbf{k}_{1,2} d\mathbf{k}'_{1,2} V_a(\mathbf{k}_1, \mathbf{k}_2, J_1, \mathbf{q}) \mathcal{G}(\mathbf{k}_1, \mathbf{k}'_1, \mathbf{q}, Y) \times \mathcal{G}(\mathbf{k}_2, \mathbf{k}'_2, \mathbf{q}, Y) V_b(\mathbf{k}'_1, \mathbf{k}'_2, J_2, \mathbf{q}), \tag{5}$$

[10]This can be seen in Figures7(a) and 8(a) in the large excess of BFKL events at zero multiplicity.

[11]The observable R contains the gap survival probability $\mathcal{S}$. This has an intrinsic non-perturbative nature [29–33] In most applications it has been assumed to be an empirically determined constant depending only on the center of mass energy, although recent results have shown it should have some kinematic dependence[34, 35].

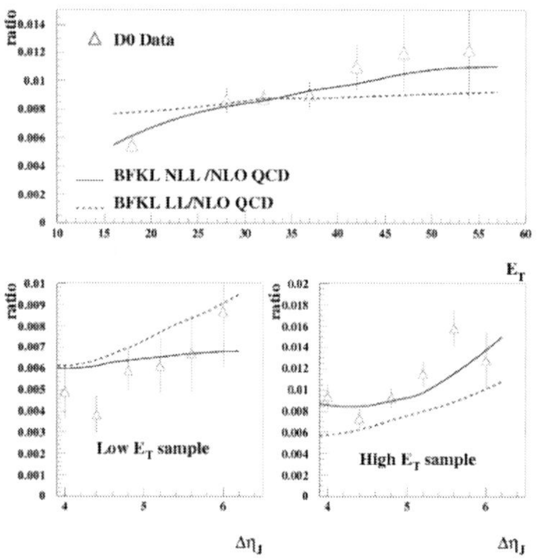

Figure 6. Comparisons between the D0 measurements of the jet-gap-jet event ratio using the NLL-BFKL kernel (solid line), LL-BFKL (dashed line) predictions with respect to the 2-to-2 NLO DGLAP prediction. Taken from [36]. The NLL calculation is in fair agreement with the data while the LL one leads to a worse description.

where $J = \{\mathbf{k}_J, x_J\}$ collects the variables that specify the jets. It is represented schematically in Fig. 5. The non-forward gluon Green's function and the jet vertices depend on the approximation order[12].

To identify M-T events, the two hardest jets resulting from the color-singlet exchange are selected which: are strongly correlated in their azimuthal angular separation; are balanced in transverse momenta; and lie on opposite hemispheres of the central detector ($\eta_1 \cdot \eta_2 < 0$). Experimentally, the rapidity gap is defined by means of the absence of particle activity in a fixed rapidity region in the central detector, for instance, at the Tevatron they counted the number of calorimeter towers with energy $E > 0.2$ GeV in the fixed rapidity region $|\eta| < 1$ and defined the rapidity gap events as the ones falling in the lowest multiplicity. It is possible to obtain a high-purity sample of jet-gap-jet events with this definition; for a judicious choice of jet momentum, rapidity, and size gap-the DGLAP evolution is highly suppressed.

M-T jets were investigated at the Tevatron at $\sqrt{s} = 1.8$ TeV[37, 38] and by the CMS collaboration at 7 TeV[39][13]. The rapidity gap condition was applied only on the central rapidity region $|\eta| < 1$ and only on charged particles with $p_T > 0.2$ GeV. The ratio of dijet events with rapidity gap to the total of dijet events has been measured as a function of the second leading jet p_T and the rapidity difference $\Delta\eta$. Even in this regime, the BFKL can be used to compute the partonic elastic amplitude. The implementation in a Monte Carlo is necessary in order to make a comparison with data; cross section measurements are sensitive to the jet size and the observed rapidity separation between the leading jets is smaller due to the soft radiation from the edge of the jets. Underlying event effects, which are estimated to be small, with the latest MC tuning should also be incorporated in the simulation. A phenomenological study was done of data released by the D0 Collaboration[36] to the BFKL prediction by implementing the NLL BFKL kernel into HERWIG6.5. (See Figure6). Importantly it was found that improving the BFKL kernel to NLL order and including a large number of conformal spin contributions are both necessary to obtain good agreement with data. In the CMS analysis summary, the results are compared only with the LL BFKL predictions (See Fig. 7(b)). These plots hint the need of higher-order corrections.

[12] If the jet vertices are kept at leading order the general expression can be *greatly* simplified. This fact has been extensively used in the phenomenological analyses.

[13] This is the only experimental analysis coming from the LHC to date, though there is an ongoing 13 TeV analysis according to private communications with members of the CMS collaboration.

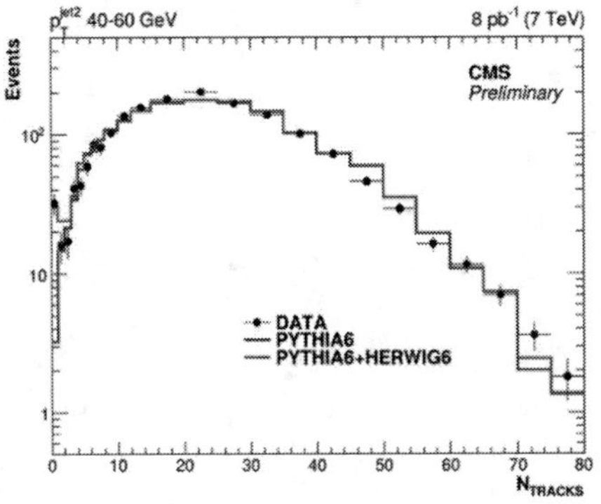

(a)

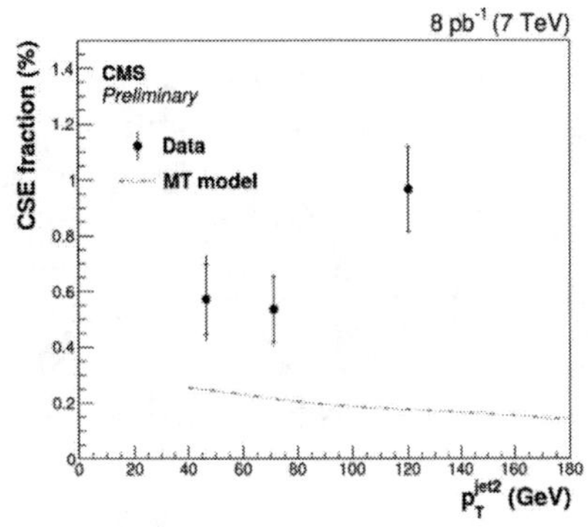

(b)

Figure 7. From [39]: (a) Number of charged particle tracks $N_{\rm Tracks}$ for 40 GeV$< p_{\rm T}^{\rm sublead} <$ 60 GeV with $|\eta^{1,2}| > 1.5$ and $\eta^1 \cdot \eta^2 < 0$ in the fixed rapidity interval $|\eta| < 1$ of CMS. The red curve combines the LL M-T dijet and PYTHIA6 NLO dijet events. The excess at low multiplicities can be explained due to the color-singlet exchange. (b) The fraction of color-singlet exchange $f_{\rm CSE}$ dijet events as a function of the subleading $p_{\rm T}$ measured for $\sqrt{s}$=7 TeV, compared to the LL BFKL predictions for $N_{\rm Tracks} = 0$. A large numeric and qualitative disagreement between the two can be seen.

The hard pomeron exchange embedded into HERWIG6.5 lacks the high order jet vertex correction to complete the NLL order. Nonetheless, it is expected that once the normalization is fixed for a given data set [14] the NLL BFKL is able to describe reasonably well the Tevatron data.

The charged multiplicity distribution measured by CMS in Fig. 7(a) is well described by DGLAP models (PYTHIA6) except at zero multiplicity, where we expect to see the jet-gap-jet events. The excess is consistent with the prediction of the HERWIG6.5 event generator using the color-singlet exchange process, even if the LL approximation BFKL kernel is used. There is a hint that the Mueller-Tang jets is an excellent venue to extract the signal of a BFKL Pomeron, even if the event generation was done with the LL approximation kernel and the leading conformal spin. Refining the theoretical prediction up to the full NLL BFKL order is an essential step toward this goal to show that the hard color-singlet exchange gives the correct differential distributions.

2.4 Jet-Gap-Jet Predictions at NLO

Until now the description of M-T jets has focused on incorporating the NLL BFKL kernel, considering higher conformal spin corrections, and understanding soft events that can affect the rapidity gap signature. [36, 40–42]. The path forward[15] is clear: add the NLO jet vertex[16] to the NLL BFKL kernel and perform a full NLO analysis of LHC data. Still, the full NLO

[14] The rescattering of proton remnants can destroy the gap resulting in a suppression factor that can affect at least the overall normalization.

[15] A group of Uppsala University has studied the jet-gap-jet process since the time of the Tevatron [43, 44]. They have developed the HARDCOL package, a modified version of the PYTHIA6 event generator, where they embedded the solution of the non-forward LL BFKL equation for the jet vertices and kernel. This implementation considers also the multi parton-parton interactions with the latest tuning in PYTHIA6 based on LHC data at 7 TeV. In addition, they used the Soft Color Interaction model for color rearrangements in the final state through soft gluon exchanges, since such rearrangements can have large effects on rapidity gaps. This approach is complimentary to that proposed here.

[16] Recently the NLO jet vertices were calculated[45, 46] using Lipatov's effective action. Although the vertices were shown to be finite, there is a careful cancellation of soft and collinear divergences between the

calculation is numerically very non-trivial. The convolutions over the transverse momenta in Eqn. 5 can be computed analytically if the vertices are kept at the LL approximation. At NLO, they have to be solved numerically utilizing Monte Carlo integration techniques. Similarly to the M-N process, the corrections are expected to be large and important to accurately predict BFKL physics, though the calculation is absolutely necessary to verify this. In [36], to speed-up computation time, the amplitude is fit to an ad-hoc parameterization and fed into HERWIG6.5 in place of standard QCD 2-to-2 process. More recently a variety of parameterizations was tested[47] and preliminary results show that the best fit parameterization can be physically motivated by considering asymptotics of the NLL kernel. Figure 8 was generated using this parameterization. Jets are reconstructed using the anti-kt algorithm with a distance parameter of $R = 0.4$ and the selection cuts described in the CDF, D0 and CMS analyses are applied to them. A definition of the rapidity gap similar to that used at the Tevatron and the LHC is used. While HERWIG6.5 counts with underlying event effects, the modeling does not take into account LHC-era results.

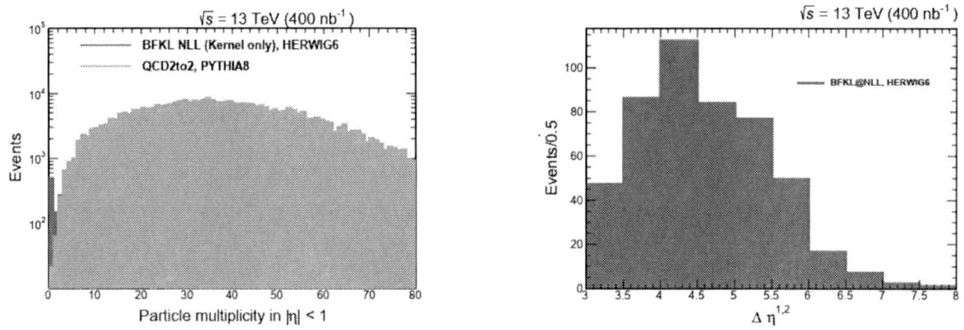

(a) (b)

Figure 8. (a) Particle multiplicity in the fixed rapidity region $|\eta| < 1$. Stable particles pass the selection cut $p_T > 0.2$ GeV. A high-purity Mueller-Tang dijets sample can be extracted in the lowest multiplicity bin. (b) Rapidity separation between the two hardest jets in the lowest multiplicity bin.

2.5 Inclusive Central Production

The BFKL Pomeron, BPST Pomeron, and QCD as extremely short distances exhibit conformal symmetry[17]. However, conformal symmetry has historically eschewed a description of scattering process because of the inability to define asymptotic states. Using the AdS/CFT, it was recently shown[48, 49] that effects due to conformal properties can be exhibited in LHC scattering. The central idea, extended in these works, is that instead of considering the scattering of asymptotic states, the emphasis should be on the flow of IR safe quantities. In addition, by utilizing generalized optical theorems, cross sections can be expressed as a discontinuity over the appropriate amplitude. For example, the differential cross section for inclusive central production can be computed via the AdS/CFT

$$\frac{1}{\sigma_{tot}}\frac{d^3\sigma_{ab\to X}}{d\boldsymbol{p}_\perp^2 dy} \sim \frac{1}{2i\,s} Disc_{M^2>0}[\mathcal{T}_{abc'\to a'b'c}] \sim \frac{A}{(\boldsymbol{p}_\perp + C)^B} \tag{6}$$

real and virtual corrections. In addition, the jet definitions involve complicated non-analytic pieces to prevent gap contamination. Both of these require careful numerical treatments.

[17]It is important to note that they are different symmetries.

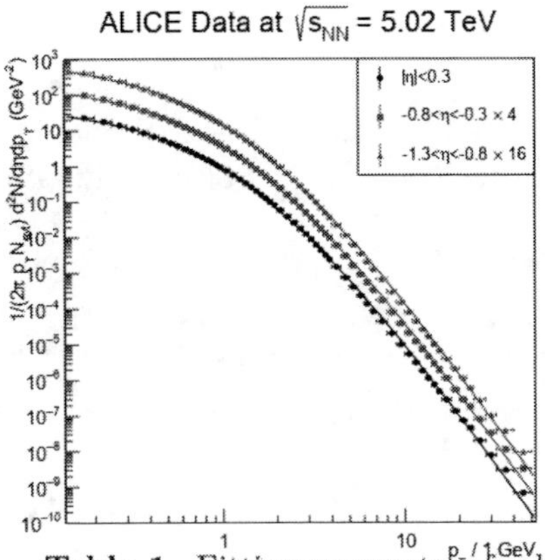

Figure 9. Fits of the ansatz, Eq.(6), to the ALICE $\sqrt{s_{NN}} = 5.02$TeV data set for the inclusive production of charged particles in the central regions.

Table 1. Fitting parameters for holographic description of inclusive central production in p-X scattering.

Dataset	A/10 (GeV^{-2})	B	C/(1 GeV)
ALICE 5.02 TeV, $\lvert\eta\rvert < 0.3$	38.48 ± 8.26	7.23 ± 0.09	1.32 ± 0.04
ALICE 5.02 TeV, $-0.8 < \eta < -0.3$	37.60 ± 7.97	7.22 ± 0.08	1.30 ± 0.04
ALICE 5.02 TeV, $-1.3 < \eta < -0.8$	43.00 ± 9.29	7.30 ± 0.09	1.31 ± 0.04
ATLAS 8 TeV	4.46 ± 2.60	7.03 ± 0.264	1.07 ± 0.123
ATLAS 13 TeV	5.77 ± 3.38	6.96 ± 0.265	1.12 ± 0.126

Using this ansatz, data from $\sqrt{s_{NN}} = 5.02$ TeV p-pb collisions at ALICE, and 8 and 13 TeV p-p collisions at ATLAS were analyzed (See Figure 1). Results show conformal behavior near $B = 2\Delta = 8$ regardless of confinement model used. A lower cutoff $C \sim \Lambda_{qcd}$ is consistent with the onset of confinement effects. Further discussion of deviations from exact conformal behavior can be found in [48].

References

[1] M.J. Menon, P.V.R.G. Silva, J. Phys. **G40**, 125001 (2013), [Erratum: J. Phys.G41,019501(2014)], 1305.2947

[2] R.C. Brower, J. Polchinski, M.J. Strassler, C.I. Tan, JHEP **12**, 005 (2007), hep-th/0603115

[3] R.C. Brower, M.S. Costa, M. Djurić, T. Raben, C.I. Tan, JHEP **02**, 104 (2015), 1409.2730

[4] R.C. Brower, M. Djuric, I. Sarcevic, C.I. Tan, JHEP **11**, 051 (2010), 1007.2259

[5] R.C. Brower, M. Djurić, T. Raben, C.I. Tan, PoS **DIS2015**, 089 (2015), 1508.05063

[6] R. Brower, R.C. Brower, M. Djurić, T. Raben, C.I. Tan (2014), pp. 173–168, 1412.3443

[7] M.S. Costa, M. Djurić, N. Evans, JHEP **09**, 084 (2013), 1307.0009

[8] R.C. Brower, M. Djuric, C.I. Tan, JHEP **09**, 097 (2012), 1202.4953

[9] A. Ballon-Bayona, R. Carcassés Quevedo, M.S. Costa, M. Djurić, Phys. Rev. **D93**, 035005 (2016), 1508.00008

[10] N. Anderson, S. Domokos, N. Mann, Phys. Rev. **D96**, 046002 (2017), 1612.07457

[11] A.H. Mueller, H. Navelet, Nucl. Phys. **B282**, 727 (1987)

[12] CMS-Collaboration, Journal of High Energy Physics **2016**, 139 (2016)

[13] V. Andreev et al. (H1), JHEP **05**, 056 (2015), 1502.01683

[14] M. Cacciari, G.P. Salam, G. Soyez, Journal of High Energy Physics **2008**, 063 (2008)

[15] S. Catani, Y.L. Dokshitzer, M.H. Seymour, B.R. Webber, Nucl. Phys. **B406**, 187 (1993)

[16] D. Colferai, F. Deganutti, unpublished (2018)

[17] D. Colferai, F. Deganutti, A. Niccoli (2017), 1709.10370
[18] T. Sjöstrand, S. Mrenna, P. Skands (2007), 0710.3820v1
[19] M. Bahr, S. Gieseke, M. Gigg, D. Grellscheid, K. Hamilton, O. Latunde-Dada, S. Platzer, P. Richardson, M. Seymour, A. Sherstnev et al., Eur. Phys. J. **C58**, 639 (2008), 0803.0883
[20] G. Chachamis, A. Sabio Vera (2016), hep=ph/1610.01334
[21] G. Chachamis, A. Sabio Vera, JHEP **02**, 064 (2016), 1512.03603
[22] G. Chachamis, A. Sabio Vera, PoS **DIS2016**, 179 (2016)
[23] G. Chachamas, F. Deganutti, D. Gordo Gomez, T. Raben, C. Royon (2018), To Appear
[24] J.D. Bjorken, Physical Review D **47**, 101 (1993)
[25] Y.L. Dokshitzer, S.I. Troian, V.A. Khoze, Sov. J. Nucl. Phys. **46**, 712 (1987), [Yad. Fiz.46,1220(1987)]
[26] Y. Dokshitzer, V. Khoze, T. Sjöstrand, Physics Letters B **274**, 116 (1992)
[27] A.H. Mueller, W.K. Tang, Phys. Lett. **B284**, 123 (1992)
[28] A.H. Mueller, W.K. Tang, Physics Letters B **284**, 123 (1992)
[29] E. Levin, A.D. Martin, M.G. Ryskin (1998), hep-ph/9812352v1
[30] E. Levin (1999), hep-ph/9912402v1
[31] E. Levin, S. Tapia (2012), 1204.0675v1
[32] M.G. Ryskin, A.D. Martin, V.A. Khoze (2008), 0812.2413v2
[33] A. Chuinard, C. Royon, R. Staszewski, JHEP **04**, 092 (2016), 1510.04218
[34] M. Luszczak, R. Maciula, A. Szczurek, M. Trzebinski, JHEP **02**, 089 (2017), 1606.06528
[35] I. Babiarz, R. Staszewski, A. Szczurek (2017), 1704.00546v1
[36] O. Kepka, C. Marquet, C. Royon, Phys. Rev. **D83**, 034036 (2011), 1012.3849
[37] D. Collaboration (1998), hep-ex/9809016v1
[38] CDF-Collaboration, Phys. Rev. Lett. **80**, 1156 (1998)
[39] CMS-Collaboration, CMS-PAS-FSQ-12-001 (2015)
[40] C. Marquet, C. Royon, M. Trzebiński, R. Žlebčík, Phys. Rev. **D87**, 034010 (2013), 1212.2059
[41] L. Motyka, A.D. Martin, M.G. Ryskin, Phys. Lett. **B524**, 107 (2002), hep-ph/0110273
[42] M. Hentschinski, J.D. Madrigal Martínez, B. Murdaca, A. Sabio Vera, Phys. Lett. **B735**, 168 (2014), 1404.2937
[43] R. Enberg, G. Ingelman, L. Motyka (2001), hep-ph/0111090v1
[44] A. Ekstedt, R. Enberg, G. Ingelman (2017), 1703.10919
[45] M. Hentschinski, J.D. Madrigal, B. Murdaca, A.S. Vera, Nucl.Phys.B **08**, 010 (2014), 1406.5625v1
[46] M. Hentschinski, J.D.M. Martínez, B. Murdaca, A.S. Vera, Nucl.Phys.B **10**, 026 (2014), 1409.6704v1
[47] M. Trzebinski, P. Swierska, Acta Physica Polonica B **46**, 2415 (2015), 1504.06271v2
[48] R. Nally, T.G. Raben, C.I. Tan (2017), 1702.05502
[49] R. Nally, T. Raben, C.I. Tan (2017), 1711.02727

Exclusive diffractive results from ATLAS, CMS, LHCb, TOTEM at the LHC

Christophe Royon[1,*]

[1] *Department of Physics and Astronomy, The University of Kansas, Lawrence, USA*
On behalf of the ATLAS, CMS, LHCb and TOTEM collaborations

Abstract. In this report, we describe the most recent on exclusive diffraction from the ATLAS, CMS, LHCb, TOTEM experiments at the LHC

1 Introduction: Definition of exclusive diffraction

We will start this report by defining what we call "Exclusive" diffraction. The first left diagram of Fig. 1 corresponds to Double Pomeron Exchanges in inclusive diffraction. In this event, both protons are intact in the final state and two pomerons are exchanged. Gluons are extracted from each pomeron in order to produce jets. Some energy is lost in pomeron remnants. The three other diagrams in Fig. 1 are exclusive in the sense that the full energy is used to produce the diffractive object. In other word, there is no energy loss in pomeron remnants. The second diagram corresponds to exclusive diffraction [1], the third one to photon exchanges and the last one to photon pomeron exchanges that produce vector mesons. In the following, we will discuss the measurements at the LHC corresponding to these different diagrams.

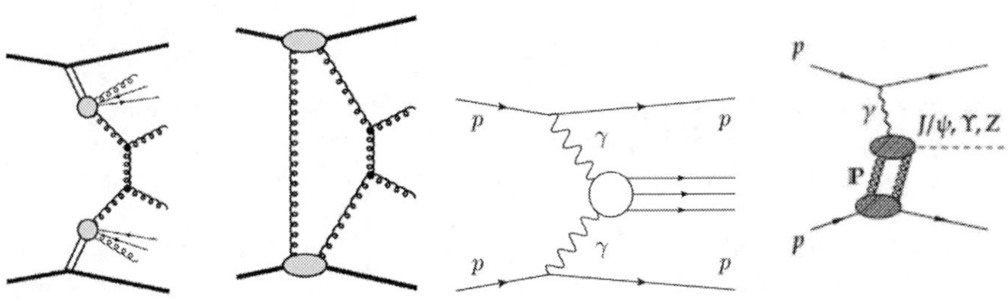

Figure 1. Diagrams of inclusive and exclusive diffractive processes.

2 Low mass exclusive diffraction

In this section, we will focus on low mass exclusive diffraction results from LHCb and CMS, namely on vector meson and pion production.

*e-mail: christophe.royon@ku.edu

© The Authors, published by EDP Sciences. This is an open access article distributed under the terms of the Creative Commons Attribution License 4.0 (http://creativecommons.org/licenses/by/4.0/).

2.1 Measurement of central exclusive production in LHCb

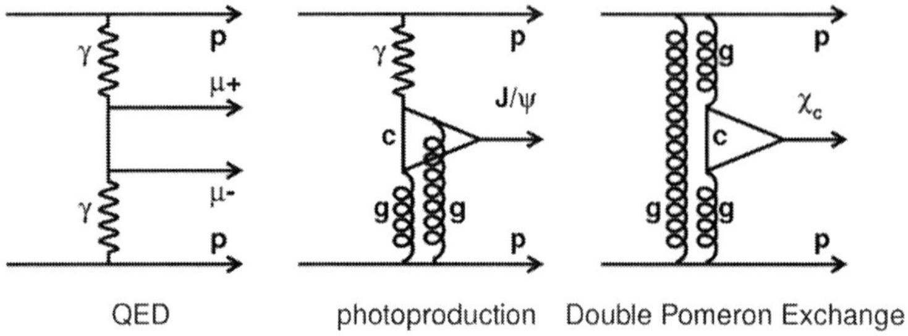

Figure 2. Central exclusive production in LHCb.

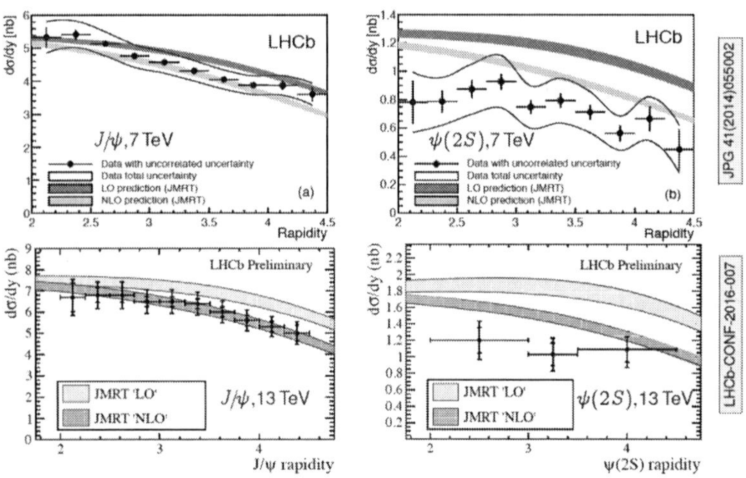

Figure 3. Measurement of the J/Ψ and $\Psi(2S)$ cross sections as a function of rapidity using 7 and 13 TeV data. The 13 TeV data include the new HERSCHEL detector..

In Fig. 2, we display three diagrams that correspond to exclusive productions in LHCb. The first left diagram corresponds to the QED production of di-muon, the middle one the photoproduction of J/Ψ, and the right one the inclusive Double Pomeron Exchange production of χ_C mesons. The exclusive production of J/Ψ is strongly sensitive to the gluon distribution in the proton since it appears as a square dependence in the amplitude. The signal is to measure the J/Ψ in the central part of LHCb and the background is due to inclusive diffractive processes where the pomeron remnants are not detected since exclusive processes are selected via the rapidity gap method. The issue is the gap detection since, for some events, the pomeron remnants are outside detector acceptance.

In order to improve the coverage at high rapidities, the LHC collaboration installed a new detector called HERSCHEL (High Rapidity Shower Counter for LHCb). HERSCHEL is installed on both sides of LHCb and allows a coverage in pseudo rapidity of $-10.0 < \eta < -3.5$ and $5.0 < \eta < 10.0$ which leads to a coverage of LHCb of $-10.0 < \eta < -1.5$ and $2.0 < \eta < 10.0$. Herschel was used

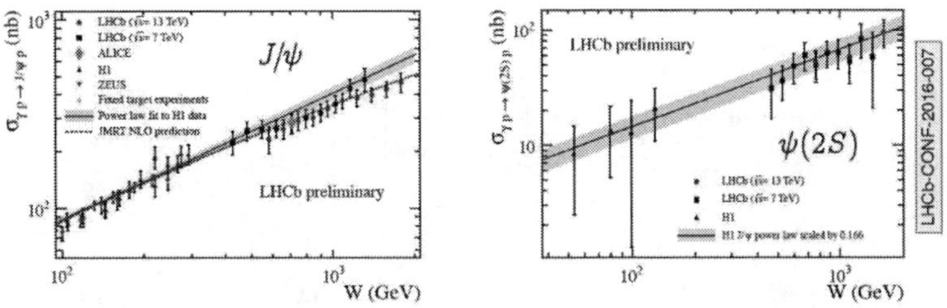

Figure 4. Measurement of the $\gamma p \to J/\Psi p$ and $\gamma p \to \Psi(2S)p$ cross sections as a function of renergy using 7 and 13 TeV data. Data cannot be described by a simple power law.

to select exclusive data at 13 TeV, and the differential cross sections for J/Ψ and $\Psi(2S)$ compared between 7 and 13 TeV data are shown in Fig. 3 [2]. The results are in agreement with NLO calculations and the best PDF parametrisation is from JMRT [3].

This measurement allow LHCb to compare with previous measurements performed by the H1 and ZEUS collaborations at HERA using the following simple relation between HERA and LHCb measurements

$$\frac{d\sigma}{dy_{pp\to pVp}} = r(y)\left[k_+ \frac{dn}{dk_+}\sigma_{\gamma p\to Vp}(W_+) + k_- \frac{dn}{dk_-}\sigma_{\gamma p\to Vp}(W_-)\right] \tag{1}$$

where the $pp \to pVp$ is measured by LHCb, $\sigma(W_-)$ (resp. $\sigma(W_+)$) at HERA, and W_+ i(resp. W_-) s extracted from the previous formula. The results are shown in Fig. 4 [2] and a simple power law does not lead to a good description of data.

2.2 Measurement of exclusive pion production in CMS

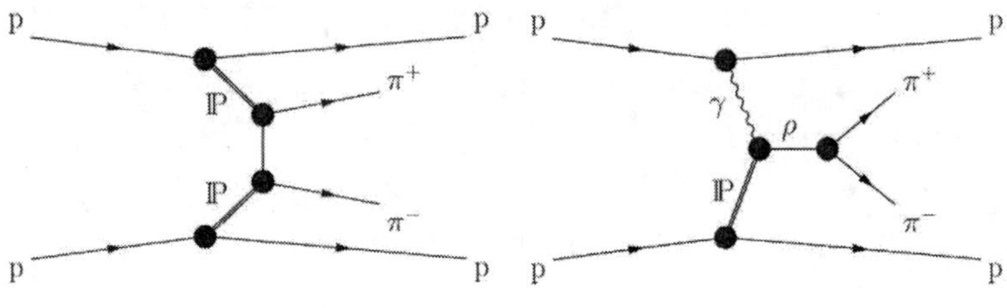

Figure 5. Leading diagrams leading to exclusive pion production.

In this section, we will describe the measurement of exclusive pion production by the CMS collaboration [4]. The two diagrams that are dominant at low masses are shown in Fig. 5. The left diagram corresponds to double pomeron exchange and is implemented in the DIME Monte carlo [5] whereas the right diagram corresponds to γ-pomeron exchanges producing a ρ meson that decays into a pion pair as implemented in the STARLIGHT Monte Carlo [6]. The measurement was performed using low pile up runs since the production cross section is high and not much accumulated luminosity is needed. The same measurement in a high pile up environment would be challenging because of the low p_T track requirement. The experimental signature is to request two opposite tracks from the same

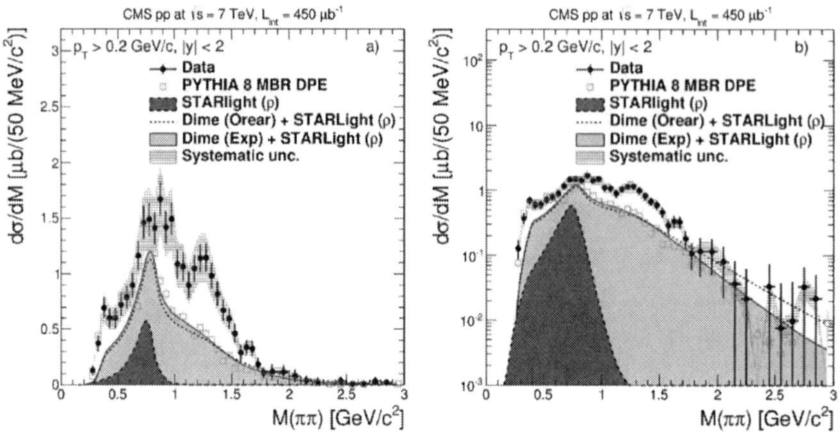

Figure 6. Measurement of exclusive pion cross section by the CMS collaboration as a function of the di-pion mass in linear and logarithmic scales.

primary vertex and no additional signal in the calorimeter outside the two pions with $p_T(\pi) > 0.2$ GeV and $|y(\pi)| < 2$. The background is computed using same sign events directly in data.

The measurement of the di-pion cross section is shown in Fig. 6 in both linear and logarithmic scales [4]. We see a discrepancy between the measurements and the sum of the DIME and STARLIGHT Monte Carlo expectations both in shape and normalization. Both Monte Carlos do not contain proton dissociation and it might be a reason for most of the discrepancy (since the measurements require the existence of a rapidity gap in the CMS detector, it is impossible to distinguish experimentally between the cases when the protons are intact or dissociate). The exclusive pion cross section was measured to be $\sigma_{\pi^+\pi^-} = 26.5 \pm 0.3$ (stat) ± 5.0 (syst) ± 1.1 (lumi) μb.

3 Exclusive measurements at high mass and photon exchange processes

In this section, we will describe the results on exclusive production at high mass together with the ones on photon exclusive exchange processes.

3.1 ATLAS and CMS results on exclusive WW production

ATLAS and CMS measured the exclusive production of a pair of W bosons $pp \to pWWp$ using the rapidity gap technique. This measurement is motivated by the search for quartic anomalous $\gamma\gamma WW$ couplings defined as additional terms in the SM lagrangian

$$\mathcal{L}_6^0 \sim \frac{-e^2}{8}\frac{a_0^W}{\Lambda^2}F_{\mu\nu}F^{\mu\nu}W^{+\alpha}W_\alpha^- - \frac{e^2}{16\cos^2(\theta_W)}\frac{a_0^Z}{\Lambda^2}F_{\mu\nu}F^{\mu\nu}Z^\alpha Z_\alpha$$

$$\mathcal{L}_6^C \sim \frac{-e^2}{16}\frac{a_C^W}{\Lambda^2}F_{\mu\alpha}F^{\mu\beta}(W^{+\alpha}W_\beta^- + W^{-\alpha}W_\beta^+) - \frac{e^2}{16\cos^2(\theta_W)}\frac{a_C^Z}{\Lambda^2}F_{\mu\alpha}F^{\mu\beta}Z^\alpha Z_\beta$$

All parameters are equal to 0 in the SM.

This search has to be performed using the highest possible integrated luminosity at the LHC since exclusive WW events are rare which means at high pile up at the LHC. The CMS and ATLAS collaborations used respectively 19.7 fb^{-1} and 20.2 fb^{-1} at a center-of-mass of 8 TeV (and CMS an additional 5.05 fb^{-1} at 7 TeV). The exclusivity selection requires opposite sign electron and muon originating from the common primary vertex and no extra track from that vertex. In addition, $M_{e\mu} >$ 20 GeV to avoid the low mass resonance region and $P_T^{e\mu} > 30$ GeV to remove Drell-Yan and $\gamma \to \tau\tau$ events.

CMS and ATLAS measured the exclusive WW production and obtained $\sigma(pp \to pWWp \to p\mu ep) = 2.2^{+3.3}_{-2.0}$ fb at 7 TeV (SM: 4.0 ± 0.7 fb), $\sigma(pp \to pWWp \to p\mu ep) = 10.8^{+5.1}_{-4.1}$ fb at 8 TeV (SM: 6.2 ± 0.5 fb) for CMS after correction for proton dissociation, and $\sigma = 6.9 \pm 2.2(stat) \pm 1.4(syst)$ fb) for ATLAS (SM: 4.4 ± 0.3 fb). The observed significance for 7 and 8 TeV combination: is 3.4 σ for CMS and 3.0 σ for ATLAS [7]

Since data are in agreement with the standard model, it is possible to obtain the most stringent limits to date on $\gamma\gamma WW$ quartic anomalous coupling and the results are given in Fig. 7. It is worth noticing that these limits can potentially be improved by about two orders of magnitude measuring the intact protons in the final state using the CT-PPS and AFP detectors [8].

Dimension-6 AQGC parameter	7 TeV ($\times 10^{-4}$ GeV^{-2})	8 TeV ($\times 10^{-4}$ GeV^{-2})	7+8 TeV ($\times 10^{-4}$ GeV^{-2})
$a_0^W/\Lambda^2(\Lambda_{cutoff} = 500\text{GeV})$	$-1.5 < a_0^W/\Lambda^2 < 1.5$	$-1.1 < a_0^W/\Lambda^2 < 1.0$	$-0.9 < a_0^W/\Lambda^2 < 0.9$
$a_C^W/\Lambda^2(\Lambda_{cutoff} = 500\text{GeV})$	$-5 < a_C^W/\Lambda^2 < 5$	$-4.2 < a_C^W/\Lambda^2 < 3.4$	$-3.6 < a_C^W/\Lambda^2 < 3.0$

Coupling	Λ_{cutoff}	Observed allowed range [GeV $^{-2}$]	Expected allowed range [GeV $^{-2}$]
a_0^W/Λ^2	500 GeV	$[-0.96 \times 10^{-4}, 0.93 \times 10^{-4}]$	$[-0.90 \times 10^{-4}, 0.87 \times 10^{-4}]$
a_C^W/Λ^2	500 GeV	$[-3.5 \times 10^{-4}, 3.3 \times 10^{-4}]$	$[-3.3 \times 10^{-4}, 3.1 \times 10^{-4}]$
a_0^W/Λ^2	∞	$[-1.7 \times 10^{-6}, 1.7 \times 10^{-6}]$	$[-1.5 \times 10^{-6}, 1.6 \times 10^{-6}]$
a_C^W/Λ^2	∞	$[-6.4 \times 10^{-6}, 6.3 \times 10^{-6}]$	$[-5.9 \times 10^{-6}, 5.8 \times 10^{-6}]$

Figure 7. Limits on $\gamma\gamma WW$ quartic anomalous couplings by the CMS and ATLAS collaborations..

3.2 Exclusive di-muon production and the AFP and CT-PPS detectors

The Atlas Forward Proton [9] (AFP) and CMS-TOTEM Precision Proton Spectrometer [10] (CT-PPS) detectors were installed recently in order to detect and measure intact protons in the final state that leads to a possible better identification of exclusive events. The LHC magnets bend the scattered protons outside the beam envelope. The roman pots detectors are located at about 210-220 m from the center part of the CMS or ATLAS detectors and cover a region in diffractive mass between typically 350 and 2000 GeV depending on the beam lattice. The CT-PPS detector started taking data in 2016 and could accumulate about 15 fb^{-1} in 2016. Both CT-PPS and AFP have been taking data continuously in 2017.

The ATLAS and CMS collaborations measured in 2016 the exclusive di-muon production with two different strategies. The corresponding diagrams are shown in Fig. 8. The LHC is turned into a $\gamma\gamma$ collider and the flux of quasi-real photons is computed using the Equivalent Photon Approximation. ATLAS did not have the possibility of detecting intact protons at the time of this measurement, and they selected exclusive $\mu\mu$ production using the rapidity gap technique. The exclusivity requirement in presence of pile up is that zero additional track exists within 1 mm of the $\mu^+\mu^-$ vertex. From the rapidity gap technique, it is impossible to know if the protons are dissociate or not and the measurement corresponds to the three diagrams shown in Fig. 8. The ATLAS collaboration fitted the $\mu\mu$

acoplanarity distribution at 13 TeV using 3.2 fb^{-1} of data in order to extract the fiducial cross section for $p_T^{\mu} > 6$ GeV ($12 < m_{\mu\mu} < 30$ GeV) and for $p_T^{\mu} > 10$ GeV ($30 < m_{\mu\mu} < 70$ GeV) and found $\sigma^{excl.fid}_{\gamma\gamma\rightarrow\mu\mu} = 3.12 \pm 0.07(stat) \pm 0.10(syst)$ pb [11]. The comparison with the Superchic Monte Carlo [12] indicates an insufficient suppression in Monte Carlo due to absorptive effects.

The CT-PPS collaboration measured exclusive di-muon production by tagging one intact proton in the final state. This is the first time the semi-exclusive di-muon processes are measured with proton tag. In Fig. 8, the two left diagrams correspond to the signal whereas the rightmost diagram is part of the background. The reason that only one proton is requested to be tagged is that less than one event is expected for double tagged events with about 15 fb^{-1} of data due to the mass acceptance above about 350 GeV for the forward proton detectors. The main background is due du Drell-Yan di-muon production with the intact proton originating from pile up events. This background is estimated using Drell-Yan Z events in data and extrapolating from the Z peak region to our exclusive di-muon signal region. 17 events are found with protons in the CT-PPS acceptance and 12 show a less than 2σ matching between the values of ξ computed using the TOTEM roman pots and using the di-muon measured in CMS as shown in Fig. 9 [13].

The correlation between the di-muon mass and rapidity is shown in Fig. 10. We see that there is no event with two protons tagged due to the acceptance in mass of the TOTEM roman pots. The highest di-muon mass events is at 341 GeV. This analysis shows that CT-PPS is now working perfectly and that we have a good understanding of triggers, data taking, forward detector alignment...

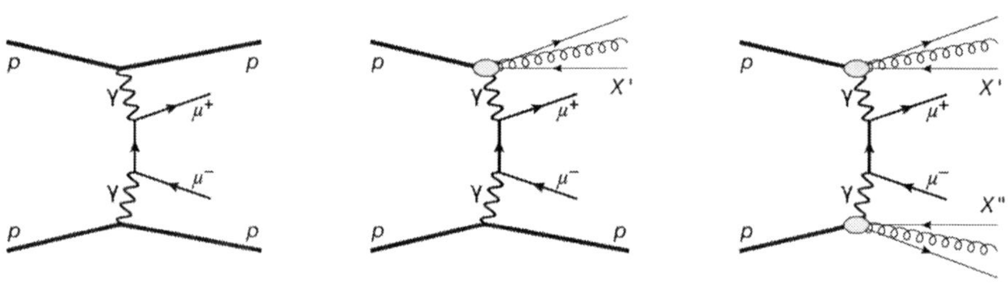

Figure 8. Diagrans leading to di-muon production via photon exchanges..

3.3 Exclusive di-photon production via photon exchanges

The ATLAS collaboration recently measured exclusive di-photon production in *PbPb* production at the LHC. Exclusive di-muons can be produced via photon exchanges and the cross section is enhanced by a factor Z^4 with respect to the pp case. In 480 μb^{-1} of *PbPb* data at $\sqrt{s}$ =5.02 TeV, 13 events were observed for an estimated background of 2.6±0.7. For photon $E_T > 3$ GeV, $|\eta| < 2.4$, $M_{\gamma\gamma} > 6$ GeV, $p_T^{\gamma\gamma} < 2$ GeV:, ATLAS measured $\sigma = 70 \pm 24(stat) \pm 17(syst)$ nb in agreement with Standard Model expectations [14].

Using CT-PPS and AFP, it is also possible to search for $\gamma\gamma\gamma\gamma$ anomalous couplings in a very clean way. Within the acceptance of the CT-PPS and AFP detectors during standard high luminosity runs at the LHC (basically for a di-photon mass above 350 GeV), it is possible to show that the exclusive production of di-photons is completely dominated by photon exchange processes and gluon exchanges can be neglected. Since the signal only shows two photons and two intact protons in the final state, we measure all final state particles. That allows us to obtain a negligible background for 300 fb^{-1} at the LHC. The basic idea is to compare the proton missing mass and the di-photon mass as

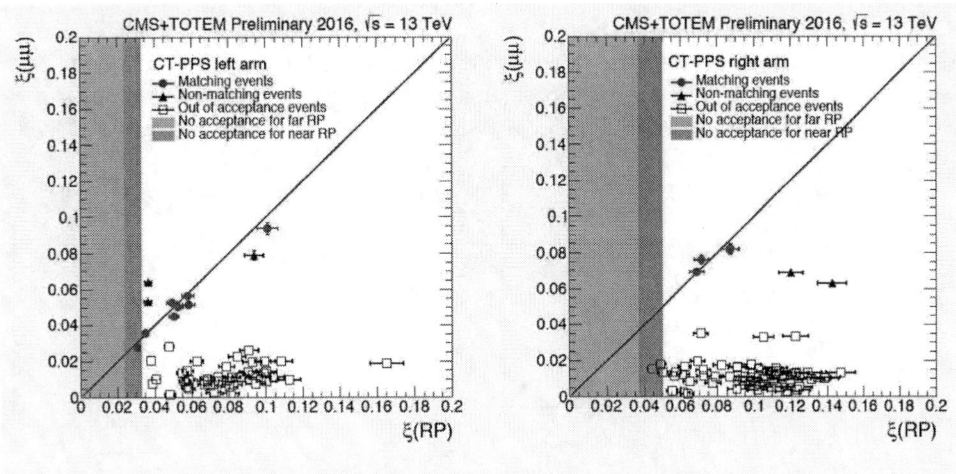

Figure 9. Correlation between the ξ values computed using the TOTEM roman pots and the di-muon measured in CMS. The 12 semi-exclusive events are indicated in red. The left (right) plot displays the left (right) arm of TOTEM.

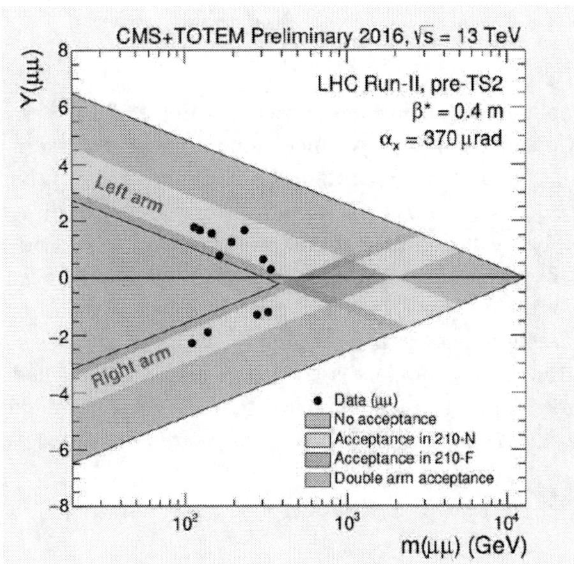

Figure 10. Correlation between the di-muon mass and rapidity showing the semi-exclusive di-muon events on the left and right arms of TOTEM.

shown in Fig. 11, left [15]. The signal peaks around 1.0 and the gaussian width is due to the detector resolution whereas the pile-up background leads to a much flatter distribution since the two protons are not related with the two photons. The same requirement can be performed using the difference in rapidity between the di-photon and di-proton systems. This leads to a sensitivity better by two orders of magnitude to $\gamma\gamma\gamma\gamma$ anomalous couplings compared to other methods at the LHC without detecting intact protons [15]. It is possible to show that the gain of two orders of magnitude is also valid for $\gamma\gamma WW$ and $\gamma\gamma ZZ$ anomalous couplings whereas the gain reaches three orders of magnitude for $\gamma\gamma\gamma Z$ anomalous couplings [8, 17]. The search for anomalous couplings with tagged protons is now being pursued by both CMS-TOTEM and ATLAS collaborations.

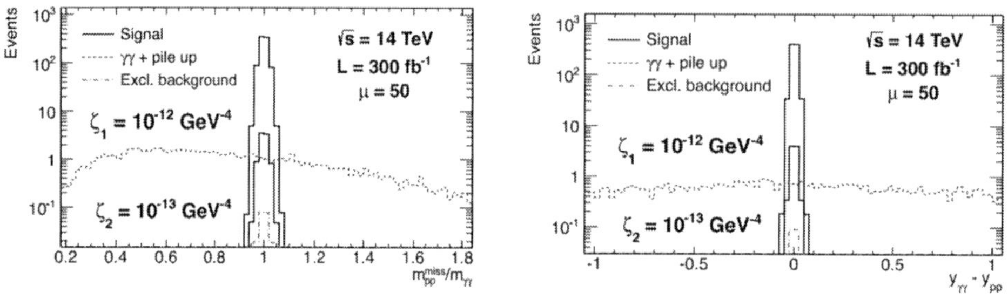

Figure 11. Left: Ratio between the proton missing mass and di-photon mass for exclusive di-photon signal events and background. Right: Difference between the di-photon and di-proton rapidity for exclusive di-photon signal and background.

4 Conclusion

We discussed many complementary measurements concerning exclusive production at the LHC from the ATLAS, CMS, LHCb and TOTEM experiments either using the "rapidity gap" technique or the proton tags. The LHCb collaboration measured the J/Ψ and $\Psi(2S)$ production and the preferred model is JMRT at NLO. The CMS exclusive pion measurement is in disagreement with theoretical expectations which is probably due to the fact that proton dissociation is not included in models. The exclusive production of W pair production was observed and led to the best limits to date on $\gamma\gamma WW$ anomalous couplings. For the first time, the CMS and TOTEM collaborations observed the exclusive production of di-muons at high mass with a proton tag. Last but not least, AFP and CT-PPS open a completely new search for quartic anomalous couplings between γ, W and Z bosons with a gain in reach by two or three orders of magnitude compared to more standard methods at the LHC, reaching some of the values predicted by extra-dimension or composite Higgs models.

References

[1] V. Khoze, A. Martin, M. Ryskin, Eur. Phys. J. C**23** (2002) 311.
[2] LHCb Coll., LHCb-CONF-2016-007; LHCb Coll., J. Phys. G **41** (2014) 055002.
[3] P. Jones, A. Martin, G. Ryskin, T. Teubner, JHEP **11** (2013) 085.
[4] CMS Coll., ArXiv:1706.08310.
[5] L. Harland Lang, V. A. Khoze, M. G. Rykin, ArXiv:1312.4553.
[6] S. R. Klein, J. Nystrand, J. Seger, Y. Gorbunv, J. Butterworth, Comp. Phys. Comm. **212** (2017) 258.
[7] CMS Coll. JHEP08 (2016) 119; ATLAS Coll., Phys Rev D94 (2016) 032011.
[8] E. Chapon, O. Kepka, C. Royon, Phys. Rev. D**81** (2010) 074003.
[9] ATLAS Coll., CERN-LHCC-2011-012.

[10] CT-PPS project, see https://cds.cern.ch/record/1753795; CMS and TOTEM coll., CERN-LHCC-2014-021.

[11] ATLAS Coll., preprint ArXiv:1708.04503.

[12] L. Harland Lang, V.A. Khoze, M.G. Ryskin, preprint ArXiv:1508.02718.

[13] CMS and TOTEM Coll., CMS-PAS-PPS-17001.

[14] ATLAS Coll., Nature Physics **13** (2017) 852.

[15] G. von Gersdorff, S. Fichet, C. Royon, Phys. Rev. Lett. **116** (2016) no 23, 231801; G. von Gersdorff, S. Fichet, C. Royon, Phys. Rev. D**93** (2016) no 7, 075031; G. von Gersdorff, S. Fichet, B. Lenzi, C. Royon, M. Saimpert, JHEP **1502** (2015) 165; G. von Gersdorff, S. Fichet, O. Kepka, B. Lenzi, C. Royon, M. Saimpert, Phys. Rev. D**89** (2014) 114004.

[16] M. Boonekamp et al., preprint ArXiv:1102.2631.

[17] C. Baldenegro, G. von Gersdorff, S. Fichet, C. Royon, JHEP **1706** (2017) 142; O. Kepka, C. Royon, Phys. Rev. D**78** (2008) 073005.

Measurement of the cosmic ray spectrum and chemical composition in the 10^{15}-10^{18} eV energy range

Andrea Chiavassa[1,⋆]

[1]*Dipartimento di Fisica dell'Universitá degli Studi di Torino & INFN*
Via Pietro Giuria 1, 10125, Torino, Italy

Abstract. Cosmic ray in the $10^{15}-10^{18}$ eV energy range can only be detected with ground based experiments, sampling Extensive Air Showers (EAS) particles. The interest in this energetic interval is related to the search of the knee of the iron component of cosmic ray and to the study of the transition between galactic and extra-galactic primaries. The energy and mass calibration of these arrays can only be performed with complete EAS simulations as no sources are available for an absolute calibration. The systematic error on the energy assignment can be estimated around $30 \pm 10\%$.
The all particle spectrum measured in this energy range is more structured than previously thought, showing some faint features: a hardening slightly above 10^{16} eV and a steepening below 10^{17} eV. The studies of the primary chemical composition are quickly evolving towards the measurements of the primary spectra of different mass groups: up to now we are able to separate (on a event by event basis) light and heavy primaries. Above the knee a steepening of the heavy primary spectrum and a hardening of the light ones have been detected.

1 Introduction

Experiments studying primary cosmic rays in the $10^{15}-10^{18}$ eV energy range are ground based arrays, typical active surface $\sim 4 \times 10^4 m^2 < \Sigma < 10^6 m^2$, detecting either the Cherenkov light emitted during Extensive Air Showers (EAS) development (TUNKA-133[1], TALE[2]) either the electromagnetic and muon EAS components (KASCADE-Grande[3], Ice Top[4]). Pros and cons of both techniques are well known and can be briefly summarized: Cherenkov arrays are limited by statistical problems (their duty cycle being $\sim 10-15\%$) but can detect the light emitted during the whole shower development, thus performing an almost calorimetric measurement of the primary energy and are sensible to the depth of the shower maximum. While surface arrays (100% duty cycle) detecting the electromagnetic and muon EAS components are mainly limited by the systematic errors on their energy and mass calibrations, that are necessarily based on a full EAS simulation. Recently experiments based on the detection of the radio signals emitted during EAS developments are giving first physics results like the LOFAR ones about primary chemical composition[5].

Lately in spite of these limitations important improvements have been obtained, mainly because of the increased resolution in the detection of the shower parameters, reaching, with experiments covering very large active surfaces, the level of 15-20% on the shower size (i.e. the number of electrons

⋆e-mail: achavas@to.infn.it

© The Authors, published by EDP Sciences. This is an open access article distributed under the terms of the Creative Commons Attribution License 4.0 (http://creativecommons.org/licenses/by/4.0/).

at observation level) and on the total muon number. In parallel to the increased array resolution the hadronic interaction models have evolved after the results of the LHC experiments, covering an energy range reaching the knee energy, even if collider experiments data cover a kinematic range different from the one relevant for EAS development.

In section 2 I will discuss the main sources of systematic errors in the energy and mass calibration and finally in section 3 I will review recent results on the measurement of the cosmic ray spectrum and chemical composition in the $10^{15} - 10^{18}$ eV energy range.

2 Systematic errors in the energy and mass calibration

Experiments operating in the $10^{15} - 10^{18}$ eV energy range cannot be calibrated with known sources, therefore we must use full EAS simulation introducing systematic errors due to arbitrary choices that must be made running simulation codes.

Surface arrays evaluate the energy of the primary particle measuring the number of particles (either electrons or muon) at observation level, while cherenkov experiments determine the photon's one. The correlation between these experimental parameters and the cosmic ray energy is obtained by means of EAS simulations.

The center of mass energy of the interaction between the incoming cosmic ray and atmospheric nuclei is similar (in the considered energy range) to the one of LHC proton-proton interactions, but LHC experiments give precise informations in the central pseudo-rapidity region, while EAS develop in the very forward one. Therefore the experimental results of LHC experiments are in the energy range of interest but in a different kinematic region; the high energy hadronic interaction models used in the EAS simulation codes are tuned to these data.

The so-called "post-LHC" hadronic interaction models more widely used in the EAS simulation are: EPOS-LHC[6], QGSJetII-04[7] and SIBYLL2.3c[8]. The EAS description we obtain from all of them is not reproducing all the experimental data (for instance the muon development in atmosphere[9]) therefore we still don't have a completely reliable and exact EAS simulation. Nevertheless the differences between models (in this energy range) have been reduced with respect to the situation before the LHC data taking. For instance the behavior of the particle number versus the pseudo-rapidity: all interaction models give similar results in the central region, while their prediction are different in the forward one.

Running EAS simulation codes (the more widely diffused in the community is CORSIKA[10], that can be considered as the reference one) we must choose an hadronic interaction model. We can evaluate the differences in the energy assignments, for the same value of the experimental parameter, running the EAS simulation with the same input values and different hadronic interaction models. This difference was estimated around 20% by the KASCADE-Grande experiment [11] for the pre-LHC models.

The EAS parameters used to evaluate the energy of the primary particle depend also on its mass. EAS experiments cannot (because of the shower development fluctuations) measure the mass of the particle originating the shower, that is therefore a free parameter of the EAS simulations. To derive the so called "all particle" cosmic ray spectrum we can estimate the, energy dependent, mean value of the primary mass[12, 13] and use this value to calculate the calibration curve correlating the experimental observable and the primary energy. The models, describing the mean value of the cosmic ray mass dependence on energy, are usually constructed starting from the abundance and the spectral slopes of each element measured at low energy by direct experiments. Then different hypothesis are made about the change of slope of the single element spectra.

Concluding the energy calibration of EAS experiments (around knee energies) can only be performed by means of a complete EAS simulation, systematic errors, roughly about $\sigma_{sys}(E)/E \sim$

$30 \pm 10\%$, are introduced by the (arbitrary) choice of the hadronic interaction model and of the primary particle mass.

3 All particle spectrum measurements

Figure 1 shows a summary[1, 2, 4, 11, 14–21] of the all particle spectrum measurement. In this plot the results obtained by different experiments, operating at different heights above sea level, with different techniques and using different hadronic interaction models for the absolute energy calibration, have been collected.

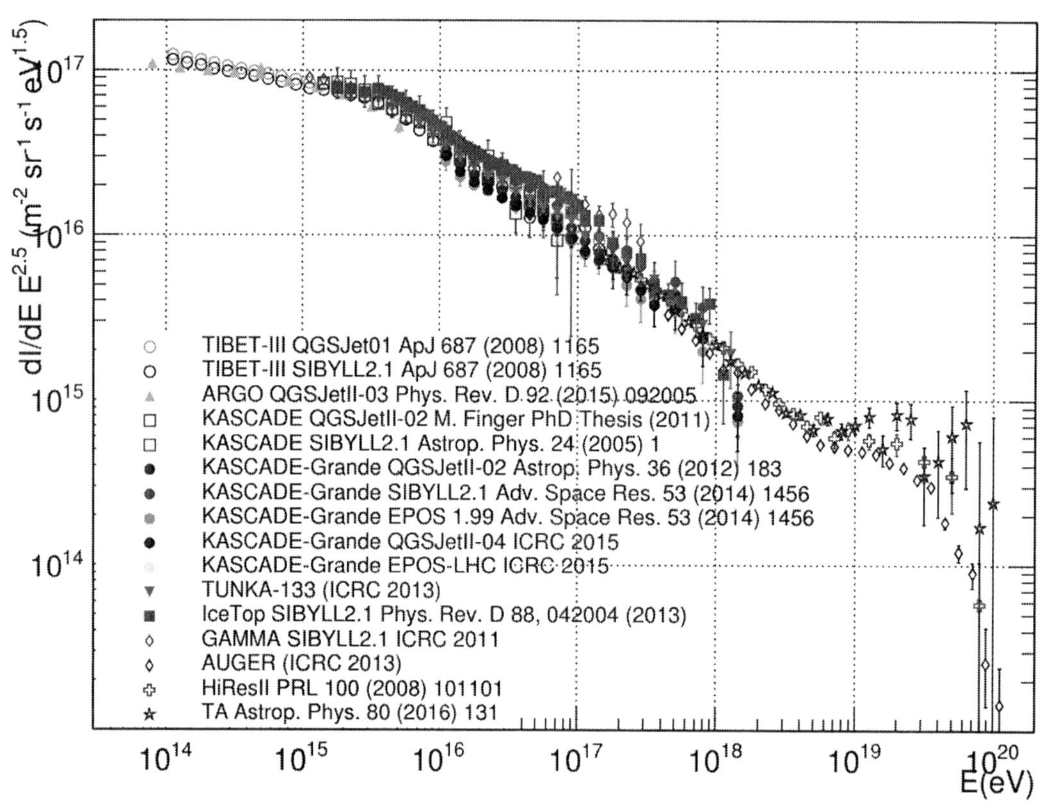

Figure 1. Compilation of all particle spectrum measurements.

Already from this plot covering large energy and flux ranges we can see that the primary spectrum shows more features than the two well known and established ones called the "knee" and the "ankle". Other two faint features (a hardening slightly above 10^{16} eV and a steepening around 10^{17} eV) have been observed in the all particle spectrum. These features have been claimed by the KASCADE-Grande experiment[18] and then confirmed by Ice Top[4], TUNKA-133[1] and TALE[2]. The spectral shapes agree quite well, the differences in the corresponding energies are smaller than the systematic error due to the energy calibration. In figure 2 the same marker is used for the results of a single experiment, while the same color is used for data calibrated with the same hadronic interaction model. We can appreciate that differences between markers with the same color are smaller than those between the same marker with different colors. Thus showing that differences between measurements can be mainly ascribed to the data energy calibration.

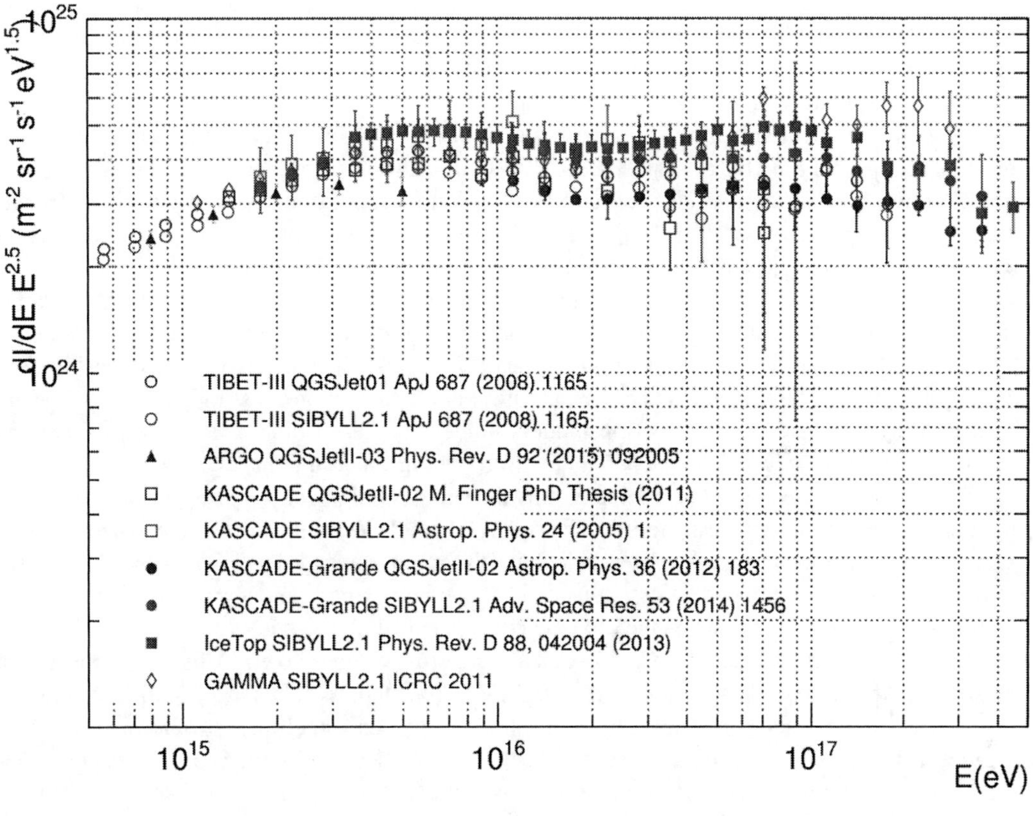

Figure 2. All particle spectra measurements around the knee. Results from the same experiment but calibrated with different hadronic interaction models are shown by the same marker with different colors. While different markers with the same colors represent the results of different experiments calibrated with the same hadronic interaction model.

The origin of these spectral features are object of discussion in the community (as an example see[22]) and can be partly explained by containment inside magnetic fields (either in the acceleration region or during the primary particles propagation). It is important to check whether these features can be ascribed to a particular primary component and therefore all analysis techniques aiming to the measurement of the spectra of different mass groups are very important.

4 Measurement of the cosmic ray mass group spectra

The last generation of surface EAS arrays has reached a resolution allowing studies separating events in mass groups: light (represented by Hydrogen and Helium), intermediate (Carbon or Oxygen) and heavy (Silicon and Iron). We can divide analysis techniques in two families: event by event classification and statistical methods.

Statistical analyses are performed either unfolding, from the number of events measured in each bin of the N_e vs N_μ plots, the spectra of the primary mass groups[17], or by machine learning approach like the neural network technique[23]. Both techniques allow the separation of four or even five primary mass groups, but their results heavily depend on the hadronic interaction model used in the EAS simulation. The spectra of the different mass groups, published by the KASCADE[17], KASCADE-Grande[24] and Ice Top[23] experiments, show sudden and correlated flux jumps, limited to very small energy intervals, in the spectra of different primary components indicating that these features are artificially introduced by the method.

Event by event classifications are performed defining experimental observables depending on the mass of the primary particle. The KASCADE collaboration introduced a selection method based on the ratio of the muon and electron numbers converted (applying the constant intensity cut technique) to a reference value of the zenith angle (i.e. to the same atmospheric depth)[25]. This method has then been used by the KASCADE-Grande experiment at higher energies ([26] and [27]). While the ARGO-YBJ experiment separated light and heavy primaries using a parameter obtained combining the number of particles measured by the surface detector with those describing the shape of the Cherenkov light images obtained by a prototype of the LHAASO Cherenkov telescopes[15].

Event by event classification allows the separation of two mass groups (light and heavy primaries). The choice of the hadronic interaction models influences the absolute flux values, all experiments claim that the event selection is not artificially introducing spectral features.

The KASCADE experiment published[25] the differential spectra, not energy calibrated, of the light and heavy primaries mass groups. The light elements spectrum shows a clear steepening, at the same value of the experimental observable (i.e. the muon density at $45.5m$ from the shower core) also the all particle one shows a (less enhanced) steepening. From the integral flux above this spectral feature we can argue that this change of slope corresponds with the knee.

The main results of the KASCADE-Grande experiment are shown in figure 3: the steepening observed in the all particle spectrum at 8×10^{16} eV is detected, with enhanced statistical significance, in the spectrum of heavy primary[26]. This feature was expected at a similar energy by models attributing the knee of the primary spectrum (at $E \sim 2-4 \times 10^{15}$ eV) to light primaries and explaining the knee as a feature due to containment inside magnetic fields. A hardening of the light primaries is observed in the all particle spectrum at $10^{17.08}$ eV[27], this feature is not visible in the all particle spectrum because this component is not the dominant one at these energies.

The ARGO-YBJ experiment published the spectrum of the light primaries showing a very sharp knee-like feature around $7 \pm 2.3 \pm 0.7 \times 10^{14}$ eV[15]. This feature is hardly conceivable with the KASCADE results, either showing an unexpected spectral feature or an incorrect monte-carlo description of the EAS development. ARGO-YBJ and KASCADE were located at very different heights above sea level (4300 m and 130 m), therefore showers were measured at different development stages. The all particle spectrum published by the ARGO-YBJ is not showing a steepening at the same energy, even if the fraction of light primaries below the knee is very high: this is not fully understood yet.

In figure 4 the EAS experiments measurements of the light primaries spectrum are shown together with the results of the CREAM[28] balloon experiment.

Recently it has been shown that the detection of the radio emission during EAS development is sensible to the atmospheric depth of the shower maximum. The LOFAR collaboration published

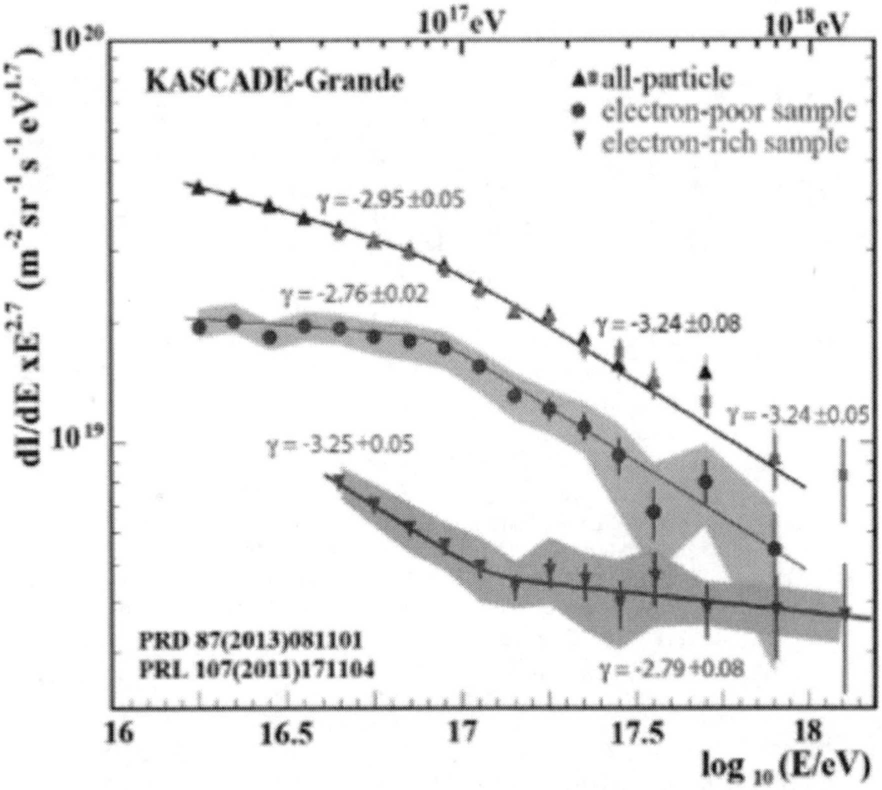

Figure 3. All particle spectrum measured by the KASCADE-Grande collaboration shown together with the spectra of the light and heavy primary components (from[27]).

interesting results[5] obtained with a good X_{max} precision but with small statistics. These results, interpreted with post-LHC hadronic interaction models, indicate a chemical composition dominated by light elements even above the knee of the spectrum. In the LOFAR publication a difference with respect to the results of the traditional surface arrays is claimed, in my opinion this difference can be mainly attributed to the hadronic interaction models used in the EAS simulation: all published results by traditional surface arrays are calibrated by pre-LHC hadronic interaction models and we know that post-LHC models (predicting an higher μ number) foresee an increase of the light elements fraction.

5 Concluding Remarks

The high resolution reached by last generation EAS experiments and the developments in the EAS simulation, due to the increased knowledge of hadronic interactions, brought relevant improvements in the study of primary cosmic rays in the 10^{15}-10^{18} eV energy range. The main limitation of current experiments is due to the absolute energy and mass calibration: a detailed study of their implications is necessary to reach a deeper understanding of this energy range. In my opinion it is important to evaluate not only the effects due to high energy hadronic models but also those due to the analysis strategies followed by experiments.

We can summarize the major recent experimental achievements:

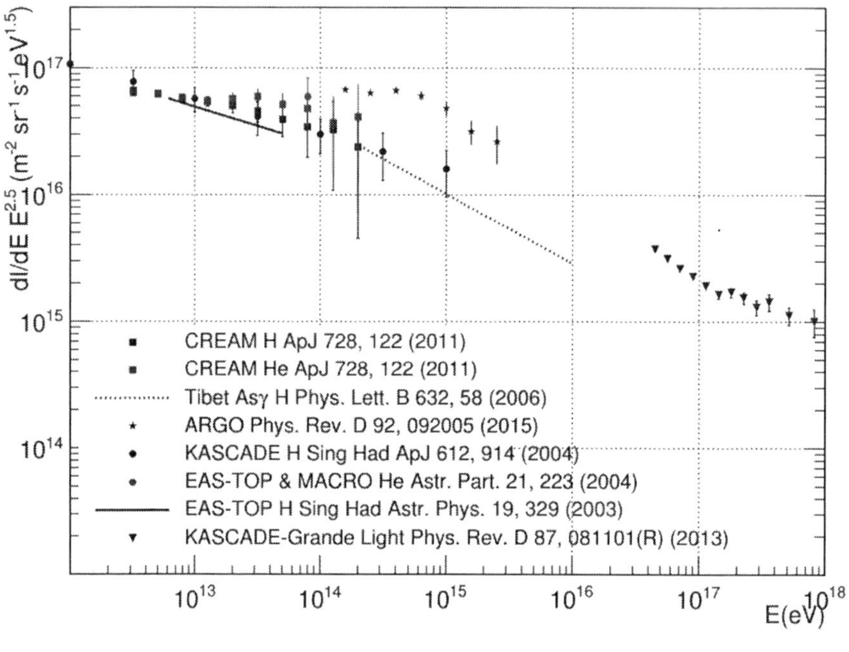

Figure 4. H and He spectra measured by the CREAM[28] experiment compared with light primaries spectra measured by EAS experiments. The KASCADE results[25] are not shown in this plot as they are presented in a not energy calibrated form.

- the primary all particle spectrum above the knee cannot be described by a single slope power law, faint but significant structures have been identified.
- The knee of the primary spectrum, from the KASCADE and KASCADE-Grande results, is attributed to the light primary component, but high altitude experiments (e.g. TIBET-III and ARGO-YBJ) published results that are not in full agreement with this claim.
- A knee-like feature in the spectrum of the heavy primary component and a hardening of the spectrum of light primaries have been observed.

Future experiments will improve the situation if they will explain the differences between current results (mainly those about the light primaries spectrum) and will measure the anisotropies of the arrival directions for different mass groups. To obtain these results future experiments must combine high resolution ($< 15\%$ on both electron and muon numbers), high statistics ($\sim$ km^2 effective areas) and large dynamic range (at least 2-3 orders of magnitude in energy).

References

[1] V.V. Prosin et al., Nucl. Instr. and Meth. A **756**, 94 (2014).
[2] R.U. Abbasi et al., Astroparticle Physics. **80**, 131 (2016).

[3] W.D. Apel et al., Nucl. Instr. and Meth. A **620**, 202 (2010).
[4] M.G. Aartsen et al., Phys. Rev. D **88**, 042004 (2013).
[5] S. Buitnik et al., Nature **531**, 70 (2016).
[6] T. Pierog at al., Phys. Rev. C **92**, 034906 (2015).
[7] S. Ostapchenko, Phys. Rev. D **83**, 014018 (2011).
[8] F. Riehn et al., EPJ Web Conf. **99**, 12001 (2015).
[9] W.D. Apel et al., Astroparticle Physics **95**, 25 (2017).
[10] D. Heck et al., Report FZKA 6019, Forschungszentrum Karlsruhe (1998).
[11] W.D. Apel et al., Advances in Space Research **53**, 1456 (2014).
[12] J.R. Höerandel, Astropart. Phys. **19**, 193 (2003).
[13] T.K. Gaisser et al., Astroparticle Physics **35(12)**, 801 (2012).
[14] M. Amenomori et al., ApJ **678**, 1165 (2008).
[15] B. Bartoli et al., Phys. Rev. D **92**, 092005 (2015).
[16] M. Finger, PhD Thesis, Karlsruhe University (2011).
[17] T. Antoni et al., Astropart. Phys. **24**, 1 (2005).
[18] W.D. Apel et al., Astropart. Phys. **36**, 183 (2012).
[19] A.P. Garkaya et al., J. Phys. G: Nucl. Part. Phys. **35**, 115201 (2008).
[20] A. Aab et al., arXiv:13075059v1, (2013).
[21] R.U. Abbasi et al., Phys. Rev. Lett. **100**, 101101 (2008).
[22] T.K. Gaisser et al., Front. Phys. **8(6)**, 748 (2013).
[23] K. Rawlins for the IceCube collaboration, Journal of Physics (conference series) **718**, 052033 (2016).
[24] W.D. Apel et al., Astropart. Phys. **47**, 54 (2013).
[25] T. Antoni et al., Astropart. Phys. **16**, 373 (2002).
[26] W.D. Apel et al., Phys. Rev. Lett. **107**, 171104 (2011).
[27] W.D. Apel et al., Phys. Rev. D **87**, 081101(R) (2013).
[28] Y.S. Yoon et al., ApJ **728**, 122 (2011).

Neutron stars velocities and magnetic fields

Daryel Manreza Paret[1,2,⋆], *A.* Perez Martinez[3,⋆⋆], *Alejandro.* Ayala[1], *G.* Piccinelli[4], and *A.* Sanchez[5]

[1] *Instituto de Ciencias Nucleares, Universidad Nacional Autónoma de México, Apartado Postal 70-543, CdMx 04510, México*
[2] *Facultad de Física, Universidad de la Habana, La Habana, Cuba*
[3] *Instituto de Cibernética Matemática y Física, ICIMAF, La Habana, Cuba*
[4] *Centro Tecnológico, FES Aragón-UNAM, México*
[5] *Facultad de Ciencias, UNAM, México*

Abstract. We study a model that explain neutron stars velocities due to the anisotropic emission of neutrinos. Strong magnetic fields present in neutron stars are the source of the anisotropy in the system. To compute the velocity of the neutron star we model its core as composed by strange quark matter and analice the properties of a magnetized quark gas at finite temperature and density. Specifically we have obtained the electron polarization and the specific heat of magnetized fermions as a functions of the temperature, chemical potential and magnetic field which allow us to study the velocity of the neutron star as a function of these parameters.

1 Introduction

Neutron stars exhibit an odd large distribution of velocities. In a reported study [1] of the data from the proper motion of 233 pulsars, velocities as high as 1000km s^{-1} are obtained . These velocities are very high compared with typical velocities of other stars, so that the natural question of how to explain it arise. There are several ideas that involve magnetic fields to explain these velocities:

1. Topological currents [2].
2. Electromagnetic rocket effect: Electromagnetic radiation from an off-centered rotating magnetic dipole imparts a kick to the pulsar along its spin axis [3].
3. Neutrino–Magnetic Field Driven Kicks: Asymmetric neutrino emission induced by strong magnetic fields [4].

We have investigated the third option which can be summarized in the following ideas

- Neutrino emissivity from the process $d \longrightarrow u + e + \bar{\nu}_e, \;\; u + e \longrightarrow d + \nu_{e},$.
- The polarisation of the electron spin will fix the neutrino emission in one direction along the magnetic field.

⋆e-mail: dmanreza@gmail.com
⋆⋆e-mail: aurora@icimaf.cu

© The Authors, published by EDP Sciences. This is an open access article distributed under the terms of the Creative Commons Attribution License 4.0 (http://creativecommons.org/licenses/by/4.0/).

- The neutrinos work as a propulsion mechanism for the neutron star.

In order to compute the pulsar velocity we use momentum conservation of the neutron star by the emission of neutrinos

$$dv = \frac{\chi_e}{M_{NS}}\frac{4}{3}\pi R^3 \epsilon dt, \tag{1}$$

where χ_e is the electron spin polarization which is equal to the fraction of neutrinos emitted in the direction of the magnetic fiel, M_{NS} and R are the mass and radius of the neutron star respectively and ϵ is the neutrino emissivity.

If the change in the emissivity is due to the change in temperature, we can use the cooling equation due to neutrino emission

$$-\epsilon = \frac{du}{dt} = \frac{du}{dT}\frac{dT}{dt} = C_v\frac{dT}{dt}, \tag{2}$$

therefor $C_v dT = -\epsilon dt$ and

$$dv = -\frac{\chi_e}{M_{NS}}\frac{4}{3}\pi R^3 C_v dT, \tag{3}$$

$$v = -\frac{1}{M_{NS}}\frac{4}{3}\pi R^3 \int_{T_i}^{T_f} \chi_e C_v dT. \tag{4}$$

The work is organized as follows: In Sec. 2 we analyze the properties of a relativistic fermion gas in a magnetic field and we give exact expressions for the polarization and specific heat of charged fermions as a function of temperature, chemical potential and magnetic field. Numerical results are presented in Sec. 3. We finally conclude and discuss our findings in Sec. 4.

2 Thermodynamics of magnetized fermions

The one-loop thermodynamical potential of a magnetized Fermi gas is given by [5]

$$\Omega_f(B,\mu,T) = -\frac{e_f d_f B}{2\pi^2}\int_0^\infty dp_3 \sum_{l=0}^{\infty}(2-\delta_{l0})\left[E_{lf} + \frac{1}{\beta}\ln\left(1+e^{-\beta(E_{lf}-\mu)}\right)\left(1+e^{-\beta(E_{lf}+\mu)}\right)\right], \tag{5}$$

where

$$E_{lf} = \sqrt{p_3^2 + 2|e_f|Bl + m_f^2}, \tag{6}$$

is the energy spectrum of charged particles in a constant magnetic field, f stands for fermion, β is the inverse of the temperature, $l = \nu + 1/2 + s/2$ is the Landau level quantum number. $\nu = 0, 1, 2, \ldots$, and $s = \pm 1$. The s values correspond to the projection of the spin operator eigenvalues along the direction of the magnetic field.

From the thermodynamical potential Eq. (5) we can compute the specific heat as

$$C_{Vf} = T\left(\frac{\partial S_f}{\partial T}\right)_{V,N}, \tag{7}$$

where the entropy is given by $S_f = -(\partial\Omega_f/\partial T)_{V,N}$. So that we obtain

$$C_{Vf} = \frac{e_f d_f B}{2\pi^2}\int_0^\infty dp_3 \sum_{l=0}^{\infty}(2-\delta_{l0})\frac{(E_{lf}-\mu_f)^2}{2T^2[1+\cosh\frac{E_{lf}-\mu_f}{T}]}. \tag{8}$$

We can also compute the electron polarization of charged fermions in a magnetic field which is defined as

$$\chi = \frac{n_- - n_+}{n_- + n_+}, \tag{9}$$

where $n_\pm$ are the number densities of fermions with spin parallel ($s = +1$) and anti–parallel ($s = -1$) to the magnetic field direction.

The general expression for the particle number density can be obtained from Eq. (5) as

$$n = -\left(\frac{\partial \Omega}{\partial \mu}\right)_V = \frac{dm^3}{2\pi^2} b \sum_{l=0}^{\infty} (2 - \delta_{l0}) \int_0^{\infty} dx_3 \frac{1}{e^{\frac{m}{T}\left(\sqrt{x_3^2+1+2lb}-x\right)} + 1}, \tag{10}$$

where $x_3 = p_3/m$ is the dimensionless momentum and $b = B/B_c^f$ ($B_c^f = m_f^2/e_f$). In such a way, $n_\pm$ are given by

$$n_- = \frac{dm^3}{2\pi^2} b \sum_{\nu=0}^{\infty} \int_0^{\infty} dx_3 \frac{1}{e^{\frac{m}{T}\left(\sqrt{x_3^2+1+2\nu b}-x\right)} + 1}, \tag{11}$$

$$n_+ = \frac{dm^3}{2\pi^2} b \sum_{\nu=1}^{\infty} \int_0^{\infty} dx_3 \frac{1}{e^{\frac{m}{T}\left(\sqrt{x_3^2+1+2\nu b}-x\right)} + 1}, \tag{12}$$

where we have used the relation between l, ν and s, changing the summation over l to the summation over ν (it is important to notice that the change is $\sum_{l=0}^{\infty}(2 - \delta_{l0}) \to \sum_{s=\pm1} \sum_{\nu=0}^{\infty}$).

From Eqs. (9), (10) and (11), we can numerically compute the dependence of χ on the parameters B, T, and μ from the following expression

$$\chi = \left\{ 1 + \frac{2 \sum\limits_{\nu=1}^{\infty} \int_0^{\infty} dx_3 \frac{1}{e^{\frac{m}{T}\left(\sqrt{x_3^2+1+2\nu b}-x\right)}+1}}{\int_0^{\infty} dx_3 \frac{1}{e^{\frac{m}{T}\left(\sqrt{x_3^2+1}-x\right)}+1}} \right\}^{-1}. \tag{13}$$

3 Numerical results

For the numerical analysis of the velocity, we can rewrite Eq. (4) as

$$v = -803.925 \, \frac{\text{km}}{\text{s}} \left(\frac{1.4 M_\odot}{M_{NS}}\right) \left(\frac{R}{10\,\text{km}}\right)^3 \left(\frac{I}{\text{MeV fm}^{-3}}\right), \tag{14}$$

where

$$I = \int_{T_i}^{T_f} \chi_e C_v dT. \tag{15}$$

In order to solve the integral in Eq. (15), we need to know the dependence of χ and C_V on the magnetic field, the chemical potential and temperature inside the neutron star. In this way, we impose the conditions that exist inside the core of neutron stars which are determined by the beta decay equilibrium equations between the quark species

$$d \to u + e + \bar{\nu}_e, \quad u + e \to d + \nu_e, \tag{16a}$$

$$s \to u + e + \bar{\nu}_e, \quad u + d \to u + s. \tag{16b}$$

Eqns. (17) along whit charge neutrality and baryon number conservation are the stellar equilibrium conditions that give us how the chemical potentials depends on temperature. To impose these conditions we have to solve the system of equations

$$\mu_u + \mu_e - \mu_d = 0 \ , \ \mu_d - \mu_s = 0, \quad (17a)$$

$$2n_u - n_d - n_s - 3n_e = 0, \quad (17b)$$

$$n_u + n_d + n_s - 3n_B = 0, \quad (17c)$$

where the n_f are given by Eq. (10)

From Eq. (13) we can study how the electron polarization varies as a function of the temperature, chemical potential and magnetic field. In the left panel of Fig. (1) shows the behavior of the polarization as a function of the magnetic field for different values of the temperature and a fixed chemical potential of $x = 10$. Notice that, with the increase of the magnetic field, the polarization grows, while the effect of increasing the temperature is to inhibit the increase of the polarization. Also, notice that for low temperatures, the typical effect of Landau number transitions analogue to the Haas-van Alphen oscillations in the magnetization, can be appreciated. The right panel of Fig. (1) shows the polarization as a function of the temperature for different values of the magnetic field an a fixed chemical potential of $x = 10$. The effect of increasing the temperature is to diminish the polarization, while the increase of the magnetic field increases the polarization.

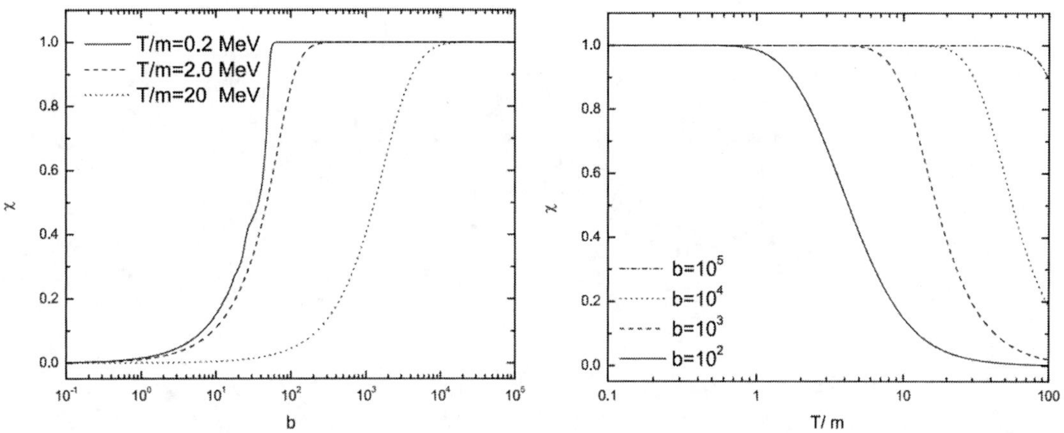

Figure 1. Left panel: polarization χ as a function of the magnetic field and temperature, for a fixed chemical potential of $x = 10$. Right panel: polarization χ as function of the temperature for several values of the magnetic field and fixed chemical potential of $x = 10$.

Fig. (2) shows the polarization as a function of the chemical potential for different values of the magnetic field and two values of temperature. An increase of the chemical potential leads to a decrease in the polarization.

In Fig. (3) we show the specific heat as a function of the temperature for a fixed value of the magnetic field and baryon density. We compare our numerical result whit previous results.

In the left panel of Fig. (3) we show the behaviour of the velocity as a function of the neutron star radius for different values of the magnetic field and a fixed central density, notice that for higher values of the magnetic field, the neutron star can reach higher velocities for small radius values while for low values of the magnetic field the star would requiere larger radius to reach velocities of the order of $v_{\mathrm{kick}} \sim 1000$ km s^{-1}. In the right panel of Fig. (3) we analice how the velocity is affected

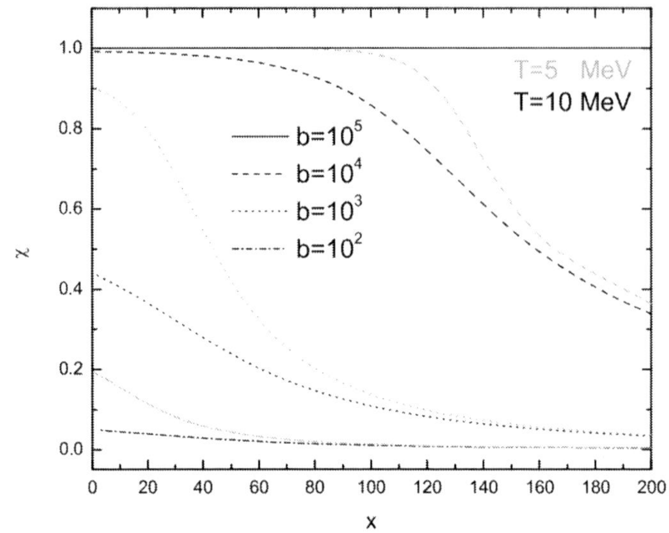

Figure 2. Polarization χ as function of the chemical potential for several values of the magnetic field and two fixed values of the temperature.

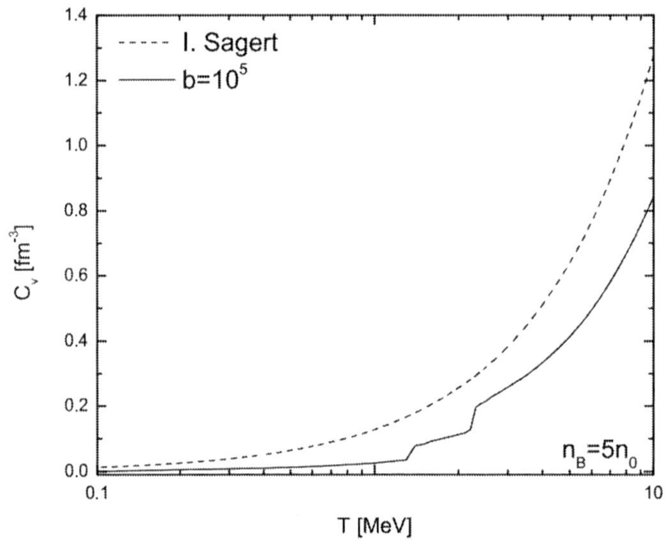

Figure 3. Heat capacity as function of the temperature for a fixed magnetic field and baryon density.

when the baryon density changes, an increase in the baryon density implies that the stars can reach higher velocities for the same value of magnetic fields an radius.

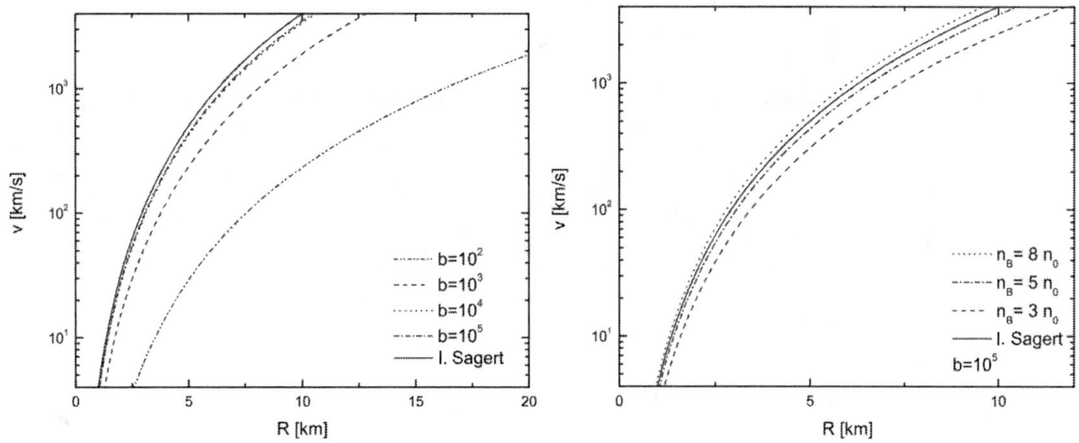

Figure 4. Left panel: velocity as a function of the neutron star radius for different values of the magnetic field and a fixed central density of $n_B = 5\,n_0$. Right panel: velocity as a function of the neutron star radius for different values of the central density and a fixed magnetic field.

4 Conclusions

We have studied a model that can explain the observations of large velocities found on neutron stars. Our model is based on the magnetic induced anisotropic emission of neutrinos as a mechanism of propulsion. We have computed the exact numerical expressions for the electron polarization and the specific heat of a magnetized quark gas as a function of the temperature, chemical potential and magnetic field.

The velocity of NS was computed for SQM in presence of a magnetic field in stellar equilibrium. We have obtained kick velocities $v_{\rm kick} \sim 1000$ km s^{-1} for different values of magnetic fields and star radius. We also have studied the dependence of the kick velocity with the central densities of the star obtaining that when the central density increases the stars can reach higher velocities for the same value of magnetic fields.

Acknowledgement

D.M.P. and A.P.M. have been supported by the grant CB0407 and acknowledge the receipt of the grant from the Abdus Salam International Centre for Theoretical Physics, Trieste, Italy. D.M.P. has been also supported by a DGAPA-UNAM fellowship. Support for this work has been received in part by UNAM-DGAPA-PAPIIT GrantNo. IN101515, IA107017, IN117817 and by Consejo Nacional de Ciencia y Tecnología Grant No. 256494.

References

[1] G. Hobbs, D.R. Lorimer, A.G. Lyne, M. Kramer, MNRAS **360**, 974 (2005), astro-ph/0504584
[2] J. Charbonneau, A. Zhitnitsky, JCAP **8**, 010 (2010), 0903.4450
[3] D. Lai, *Physics of neutron star kicks*, in *Astrophysics and Space Science Library*, edited by K.S. Cheng, H.F. Chau, K.L. Chan, K.C. Leung (2000), Vol. 254 of *Astrophysics and Space Science Library*, p. 127, astro-ph/9912522

[4] I. Sagert, J. Schaffner-Bielich, Astron. Astrophys. (2007), [Astron. Astrophys.489,281(2008)], 0708.2352
[5] E.J. Ferrer, V. de La Incera, J.P. Keith, I. Portillo, P.L. Springsteen, Physical Review C **82**, 065802 (2010), 1009.3521

Tests of the SIBYLL 2.3 high-energy hadronic interaction model using the KASCADE-Grande muon data

J.C. Arteaga-Velázquez[1,⋆], D. Rivera-Rangel[1,⋆⋆], W.D. Apel[2], K. Bekk[2], M. Bertaina[3], J. Blümer[2,4,14], H. Bozdog[2], I.M. Brancus[5], E. Cantoni[3,6,15], A. Chiavassa[3], F. Cossavella[4,16], K. Daumiller[2], V. de Souza[7], F. Di Pierro[3], P. Doll[2], R. Engel[2], D. Fuhrmann[8,17], A. Gherghel-Lascu[5], H.J. Gils[2], R. Glasstetter[8], C. Grupen[9], A. Haungs[2,18,⋆⋆⋆], D. Heck[2], J.R. Hörandel[10], T. Huege[2], K.H. Kampert[8], D. Kang[2], H.O. Klages[2], K. Link[4], P. Łuczak[11], H.J. Mathes[2], H.J. Mayer[2], J. Milke[2], B. Mitrica[5], C. Morello[6], J. Oehlschläger[2], S. Ostapchenko[12], T. Pierog[2], H. Rebel[2], M. Roth[2], H. Schieler[2], S. Schoo[2], F.G. Schröder[2], O. Sima[13],, G. Toma[5], G.C. Trinchero[6], H. Ulrich[2], A. Weindl[2], J. Wochele[2], and J. Zabierowski[11]

[1] *Institute of Physics and Mathematics, Universidad Michoacana de San Nicolás de Hidalgo, Morelia, Mexico*
[2] *Institut für Kernphysik, KIT - Karlsruher Institut für Technologie, Germany*
[3] *Departimento di Fisica, Università degli Studi di Torino, Italy*
[4] *Institut für Experimentelle Teilchenphysik, KIT - Karlsruher Institut für Technologie, Germany*
[5] *Horia Hulubei National Institute of Physics and Nuclear Engineering, Bucharest, Romania*
[6] *Osservatorio Astrofisico di Torino, INAF Torino, Italy*
[7] *Universidade de São Paulo, Instituto de Física de São Carlos, Brasil*
[8] *Fachbereich Physik, Universität Wuppertal, Germany*
[9] *Department of Physics, Siegen University, Germany*
[10] *Department of Astrophysics, Radboud University Nijmegen, The Netherlands*
[11] *National Centre for Nuclear Research, Department of Astrophysics, Lodz, Poland*
[12] *Frankfurt Institute for Advanced Studies (FIAS), Frankfurt am Main, Germany*
[13] *Department of Physics, University of Bucharest, Bucharest, Romania*
[14] *Now: Head of Division V at KIT - Karlsruher Institut für Technologie, Germany*
[15] *Now at: Istituto Nazionale di Ricerca Metrologica, INRIM, Torino, Italy*
[16] *Now at: DLR Oberpfaffenhofen, Germany*
[17] *Now at: University of Duisburg-Essen, Duisburg, Germany*
[18] *Spokesperson KASCADE-Grande*

Abstract. The KASCADE-Grande observatory was a ground-based air shower array dedicated to study the energy and composition of cosmic rays in the energy interval $E = 1$ PeV -1 EeV. The experiment consisted of different detector systems which allowed the simultaneous measurement of distinct components of air showers (EAS), such as the muon content. In this contribution, we study the total muon number and the lateral density distribution of muons in EAS detected by KASCADE-Grande as a function of the zenith angle and the total number of charged particles. The attenuation length of the muon content of EAS is also measured. The results are compared with the predictions of the SIBYLL 2.3 high-energy hadronic interaction model.

⋆e-mail: arteaga@ifm.umich.mx
⋆⋆e-mail: drivera@ifm.umich.mx
⋆⋆⋆e-mail: andreas.haungs@kit.edu

© The Authors, published by EDP Sciences. This is an open access article distributed under the terms of the Creative Commons Attribution License 4.0 (http://creativecommons.org/licenses/by/4.0/).

1 Introduction

Hadronic interaction models are an important element in the analysis of data from extensive air showers (EAS) induced by cosmic rays in the atmosphere. However, they are subject to theoretical and experimental uncertainties, which may hamper cosmic-ray studies. Recently, a lot of progress has been done towards the reduction of such uncertainties motivated, in part, by the results of the measurements of the LHC [1]. In this regard, one of the models that has been updated is SIBYLL [2].

In this work, muon predictions of the post-LHC model SIBYLL 2.3 will be tested using the data of the KASCADE-Grande observatory [3]. Shower muons are employed for these tests since they are very sensitive to the early hadronic processes occurring in the shower [4]. This way a failure of the model to describe the observed muon data of EAS may imply a problem in the description of the hadronic physics of air showers.

2 The KASCADE-Grande observatory

KASCADE-Grande was a ground-based air-shower observatory dedicated to investigate the energy spectrum, the elemental composition and the arrival direction of cosmic rays in the energy range from 10^{15} and 10^{18} eV [3]. The instrument was located at the Karlsruhe Institute of Technology, Campus North (110 m a.s.l.) in Karlsruhe, Germany and consisted of several detector systems aimed to measure with high precision different components and properties of the EAS (c.f. fig. 1, left). For instance, the shower size or total number of charged particles (electrons plus muons) ($E_{th} > 3\,\text{MeV}$), the shower core position at ground and the angle of incidence of the event were estimated from the data collected with the Grande array, which was composed by 37 scintillator detectors separated by a mean distance of 137 m and distributed over a surface of $0.5\,\text{km}^2$. Meanwhile the total number of muons ($E_{th} > 230\,\text{MeV}$) was obtained from local muon density measurements performed with 192 shielded detectors belonging to a $200 \times 200\,\text{m}^2$ array called KASCADE [3]. Systematic uncertainties for N_{ch} and N_μ were estimated as described in [3] and [5], respectively. They were found to be $\leq 15\%$ and $\leq 25\%$ for each case.

3 Experimental data and MC simulations

To perform the present studies, we have considered measured events collected during the full data acquisition period of the experiment, i.e., from December 2003 up to November 2012. On the other hand, in order to diminish the effect of systematic uncertainties in the results of our analyses, we applied several selection cuts to the data. In particular, we considered only events that were measured during stable data acquisition runs with no hardware problems. In addition, we included events that passed successfully the full reconstruction chain [3]. Besides, the EAS cores were required to be located within the limits of a central area of $2.2 \times 10^5\,\text{m}^2$ inside the KASCADE-Grande array and within radial distances in the interval $R = [100\,\text{m}, 600\,\text{m}]$ measured from the center of the KASCADE array. On the other hand, we rejected events with $\theta \geq 40^\circ$. Finally, low energy events, which have poor reconstructed muon numbers, were removed from the data sample by requiring showers that activated more than 11 Grande stations and had a high N_μ number (specifically $\gtrsim 3 \times 10^4$ and $\gtrsim 4 \times 10^4$ for data with $\theta \lesssim 30^\circ$ and $30^\circ \lesssim \theta < 40^\circ$, respectively). Using the above selection cuts, we got an experimental sample with about 1.24×10^7 events and an effective time of $\Delta t = 1.6 \times 10^8$ s.

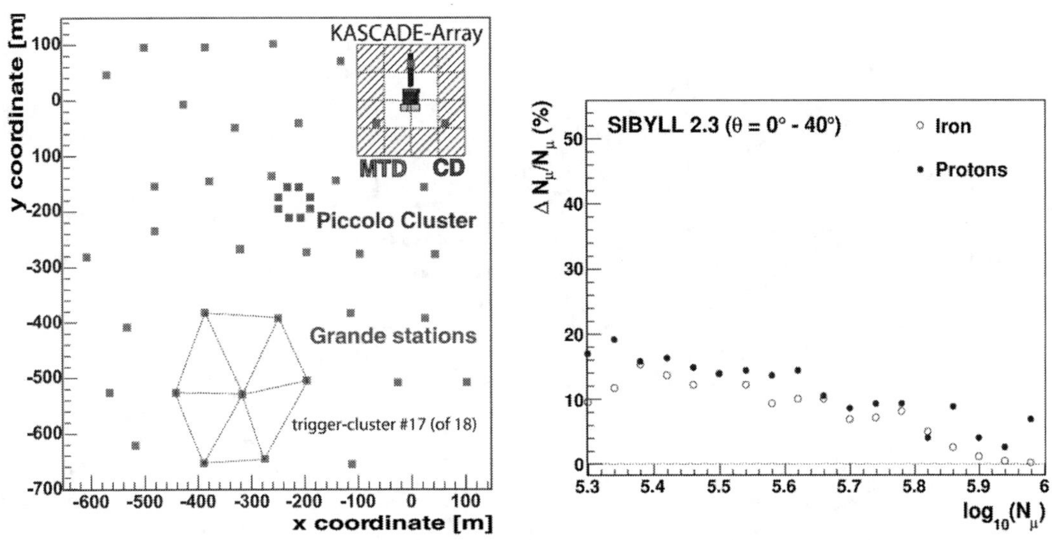

Figure 1. Left: The KASCADE-Grande experiment [5]. Small squares represent the Grande stations. The KASCADE array is seen at the upper right hand of the figure. Close to the center of KASCADE, the Muon tracking detector (MTD) and the Central detector (CD) are shown. KASCADE detectors are arranged in 16 clusters (big squares). The outer 12 clusters contain the 192 shielded plastic scintillator stations employed for the muon measurements. One example of a Grande-trigger is shown. It consists of 6 activated stations in hexagonal configuration plus a central one. **Right**: Systematic error (reconstructed minus true parameter) of the total number of muons in EAS measured with KASCADE-Grande as a function of the estimated shower muon content. The error was estimated using simulations with SIBYLL 2.3.

On the other hand, the MC simulations were built using SIBYLL 2.3 [2] and Fluka 2011.2 [6] as high- and low-energy ($E_h \leq 200\,\text{GeV}$) hadronic interaction models, respectively. The production and development of the shower were simulated with CORSIKA 7.5 [7], while the response of the detector to the passage of the shower, with a GEANT 3.21 [8] based program. MC events were generated for zenith angles $\theta < 42°$ and the energy interval from $E = 10^{14}$ eV to 3×10^{18} eV using an E^{γ} primary spectrum with spectral index $\gamma = -2$. For the analysis, the MC events were weighted in order to simulate an E^{-3} power-law spectrum, which is in better agreement with the experimental data (see, for example, [9]). MC data samples were produced for five primary nuclei: H, He, C, Si and Fe, each of them with roughly the same number of events. In total, $\sim 1.28 \times 10^7$ MC showers were generated. All of them were processed with the same reconstruction algorithm employed with the experimental events [3]. In addition, for the analysis we have applied the same selection cuts on the MC sample that were used with the measured data.

According to MC simulations the maximum efficiency is reached for $\log_{10}(E/\text{GeV}) = 7.2 \pm 0.3$ and $\log_{10}(N_\mu) = 5.3 \pm 0.3$ depending of the primary particle and the arrival direction of the EAS. On the other hand, the mean shower core and angular resolutions were found to be ≤ 11 m and $\leq 0.7°$, respectively, for our selected data sample. Finally, the bias in the muon size resulted to be smaller than 20% for $\log_{10}(N_\mu) > 5.3$ (see fig. 1, right). We have improved the accuracy of N_μ, for both MC and measured data, by correcting the reconstructed muon number for systematic biases using the correction function defined in [9]. This way, the final bias on N_μ becomes smaller than 8%.

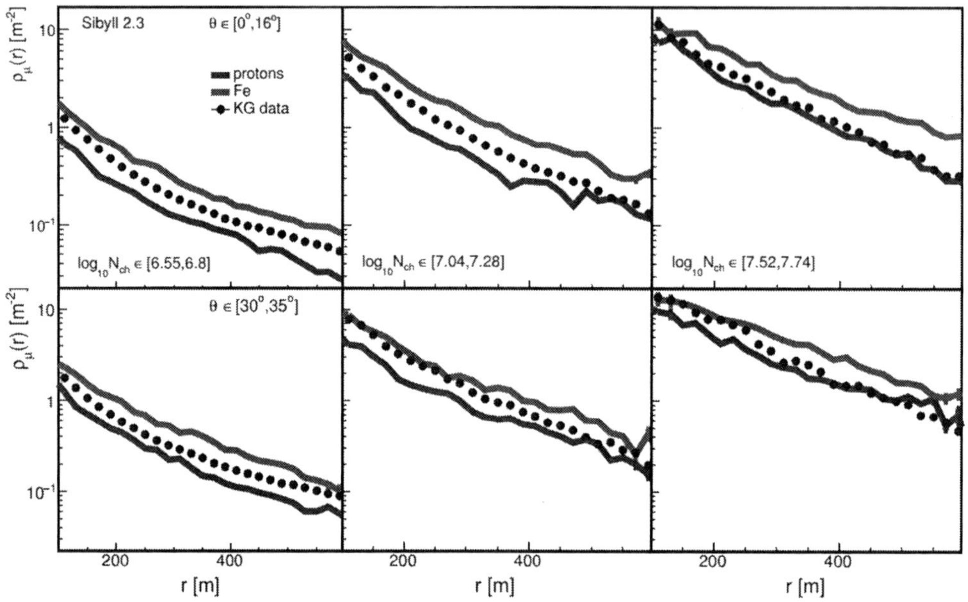

Figure 2. Mean muon lateral density distributions of EAS measured with KASCADE-Grande (points) compared with the predictions of SIBYLL 2.3 for iron (red upper lines) and proton (lower blue lines) primaries [10]. Statistical error bars in measured data are smaller than the size of the markers. From MC simulations the energy scale of the above data spans the interval from $E \sim 10^{7.2}$ GeV to 10^{9} GeV. The primary energy was estimated event-by-event from both N_{ch} and the corrected N_{μ} using a formula derived from SIBYLL 2.3 simulations (following the procedure described in [9]).

4 Testing the predictions of SIBYLL 2.3 for shower muons

4.1 Muon radial density distributions

The aim of this analysis is to compare the predictions of SIBYLL 2.3 and the KASCADE-Grande measurements for the mean muon lateral distributions at the shower plane, $\rho_{\mu}(r)$, at different zenith angles and $\log_{10}(N_{ch})$ intervals. For this purpose, we divided our data in five bins of θ, each of them with equal acceptance: $[0°, 16.71°]$, $[16.71°, 23.99°]$, $[23.99°, 29.86°]$, $[29.86°, 35.09°]$ and $[35.09°, 40°]$. Next, we subdivided these data sets in bins of $\log_{10}(N_{ch})$: $[6.0, 6.55]$, $[6.55, 6.8]$, $[6.8, 7.04]$, $[7.04, 7.28]$, $[7.28, 7.52]$, $[7, 52, 7.74]$ and $[7.74, 8.0]$. Then, for each θ and shower size intervals we built the mean measured $\rho_{\mu}(r)$ distributions (neglecting corrections due to atmospheric absorption when projecting on the shower disk plane). Finally, we compared these experimental distributions against the corresponding predictions from SIBYLL 2.3 for pure protons and iron nuclei, respectively. In fig. 2 we show a plot with results for two zenith angle intervals and three distinct shower size ranges. In particular, we observe that the measured distributions are within the MC predictions. In general, that holds for vertical and inclined showers with $\log_{10}(N_{ch}) < 7.74$. For higher values of the total number of charged particles, we found some deviations, but here statistical fluctuations avoid us to drive a conclusion.

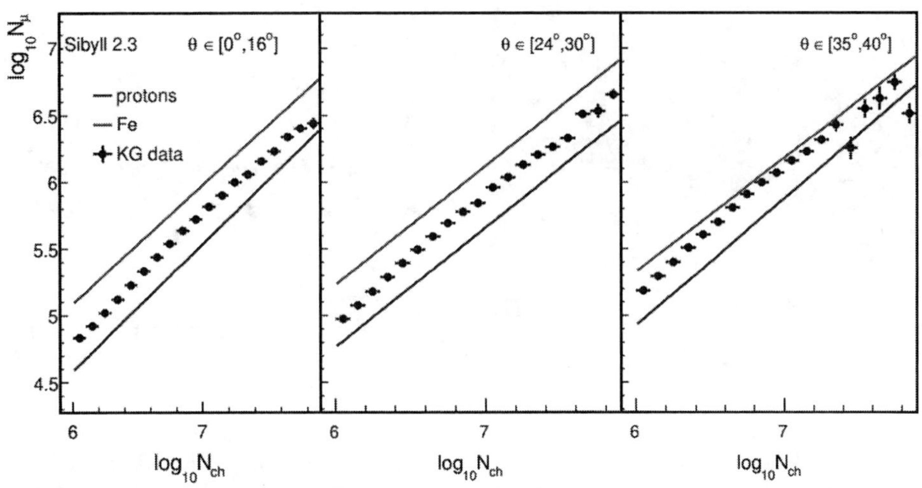

Figure 3. Logarithm of the corrected muon size versus the logarithm of the total number of charged particles in air showers as measured with the KASCADE-Grande detector (black points). Measurements are confronted with predictions of SIBYLL 2.3 for iron nuclei (upper red lines) and protons (lower blue lines) [10]. Vertical error bars represent the errors on the mean. Data is shown for the interval $N_{ch} = 10^6 - 10^{7.9}$, which roughly corresponds to mean primary energies between $E \sim 10^{6.8}$ GeV to $10^{9.1}$ GeV for the measured data. The energies were estimated event-by-event from both N_{ch} and the corrected N_μ using a formula derived from SIBYLL 2.3 simulations (following the procedure described in [9]).

4.2 The total muon number

In this study, we compared the measurements for the corrected muon number as a function of the shower size with the expectations from the SIBYL 2.3 model for hydrogen and iron nuclei for the θ ranges described in subsection 4.1. The results for three selected zenith angle intervals are presented in fig. 3. In general, we do not observe any deviation of the measured data from the MC simulations. However, we see that for increasing zenith angles the mean composition tends to be heavier, which could be attributed to a mismatch between the predicted and the measured zenith angle evolutions of the shower muon content of EAS. This point was investigated in the analysis of the next subsection.

4.3 The muon attenuation length

In order to study the behaviour of the corrected N_μ with θ for air showers, we calculated the attenuation length of muons in the Earth's atmosphere, Λ_μ using the method of [11]. This parameter is especially useful for cross-checking hadronic interaction models. Recent analyses performed in [11] have shown that the Λ_μ predictions of the QGSJET-II-2 [12], QGSJET-II-04 [13], EPOS-LHC [14] and SIBYLL 2.1 [15] hadronic interaction models deviate from the measured value at KASCADE-Grande for data in the primary energy interval from $E \sim 10^{16.3}$ eV to about 10^{17} eV and zenitn angles < 40°. Besides, it was pointed out in [11] that statistical and systematic errors (associated to instrumental effects, reconstruction/analysis methods, EAS fluctuations, etc.) can not explain the observed anomaly. From the results of the last subsection we suspect that also the prediction of SIBYLL 2.3 could deviate from the experimental value. To test this hypothesis we performed an analysis like that described in [11].

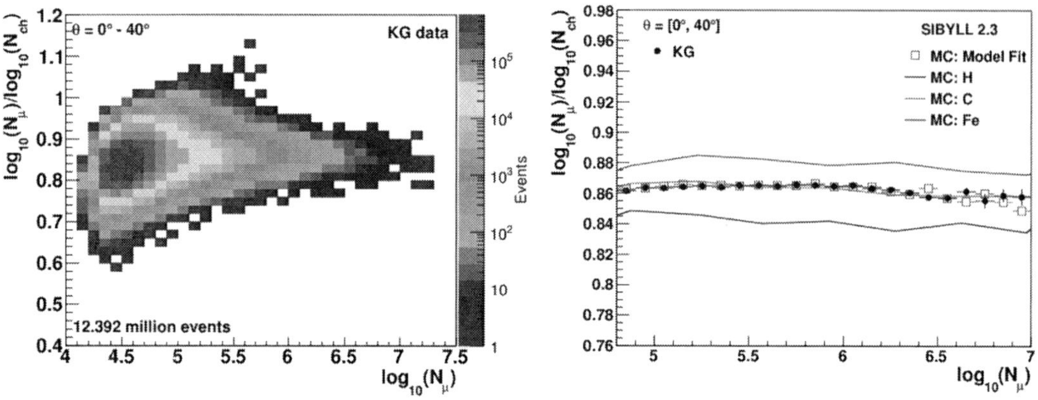

Figure 4. Left: The measured distribution for the ratio $Y = \log_{10}(N_\mu)/\log_{10}(N_{ch})$ as a function of the $\log_{10}(N_\mu)$ for selected EAS with $\theta < 40°$ [10]. The effective observation time of the data sample is $\Delta t = 1.6 \times 10^8$ s. **Right**: The mean of the experimental Y parameter for the data shown on the left against $\log_{10}(N_\mu)$ (black circles). The respective mean is also shown for predictions of SIBYLL 2.3 for protons (lower blue line), carbon (middel green line), iron nuclei (upper red line) and our fitted composition model (open squares, see text). The MC simulations include the full detector simulation and the EAS reconstruction procedures [10].

First, in order to reduce the systematic uncertainties due to the unknown cosmic ray composition in the the value of Λ_μ predicted by SIBYLL 2.3, we used a composition model derived from the data itself. The procedure is simple. By using a χ^2 fit, we find the weight parameters for the MC energy spectra of four mass groups (H, He, C, and Si+Fe in a 50 % mixture) that minimize the difference between the measured $Y = \log_{10}(N_\mu)/\log_{10}(N_{ch})$ vs $\log_{10}(N_\mu)$ event distribution (c.f. fig. 4, left) and the predicted one, the latter for the linear combination of the four mass groups. Before weighting, the MC energy spectra resemble an E^{-3} power-law function. The fit was performed for $\log_{10}(N_\mu) \geq 4.8$ with the following Chi-square function:

$$\chi^2 = \sum_{i,j} \frac{[n_{ij}^{exp} - \sum_A n_{ijA}^{MC}(\mathbf{p}_A)]^2}{(\sigma_{ij}^{MC})^2}, \tag{1}$$

where n_{ij}^{exp} is the number of experimental events for the bin $(Y_i, N_{\mu,j})$ of the respective Y distribution, while n_{ijA}^{MC} is the corresponding number of MC events for the above bin and the mass group A. The widths of the Y and $\log_{10}(N_\mu)$ bins are 0.02 and 0.1, respectively. On the other hand, σ_{ij}^{MC} is the statistical error of n_{ijA}^{MC} and $\mathbf{p}_A$ is a vector with the parameters that define the weights, $\mathbf{w}_A(E)$, of the energy spectrum for A. The weights are applied event-by-event on the MC data according to its true energy, E. They are parameterized using a double power-law formula

$$\mathbf{w}_A(E, \mathbf{p}_A) = p_{0,A}\, E^{p_{1,A}} \left[1 + \left(\frac{E}{p_{4,A}}\right)^{p_{3,A}}\right]^{(p_{2,A}-p_{1,A})/p_{3,A}}. \tag{2}$$

Here the $p_{k,A}$ are free parameters (i.e. the components of $\mathbf{p}_A$). The reason of using these expressions is due to the presence of breaks around $E = 10^{17}$ eV in the energy spectra of the light [16] and the heavy [17] components of cosmic rays. With the results of the fit and the SIBYLL 2.3 data, we built our composition model. This is shown in fig. 5. The mean values of $Y(N_\mu)$ for our fitted model and the respective measurements are compared in fig. 4, right. In fig. 5, we observe the heavy knee at

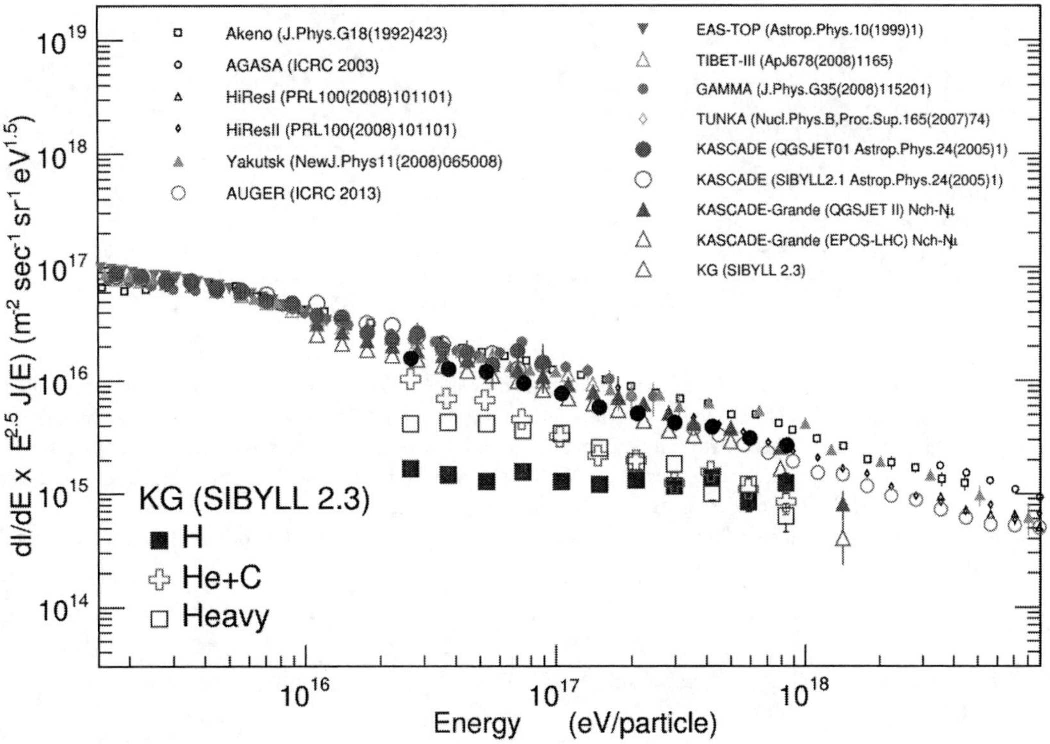

Figure 5. Results of the fit to the measured $Y(N_\mu)$ distributions using SIBYLL 2.3 data and a composition model with four mass groups (H, He, C, and Si+Fe in a 50 % mixture) each of them following a double power-law energy spectrum [10]. The derived all-particle energy spectrum is shown with black dots. The fitted spectra for protons, He+C, and the heavy component of cosmic rays are shown in the lower part of the figure with blue full squares, green open crosses and red open squares, respectively. The aforementioned curves are compared with results of other observatories and previous estimations of KASCADE-Grande for the all-particle energy spectrum.

$\log_{10}(E/\text{eV}) = 16.80 \pm 0.01$ ($\Delta\gamma = p_2 - p_1 = -0.67 \pm 0.03$). We also see the presence of a proton and a medium (He+C) component in the data.

Now, following the procedure of [11], we selected a subsample of MC data with radial distances, R, between 270 m and 440 m and within an area of 8×10^4 m^2 located at the center of the KASCADE-Grande array, as they further reduce the systematic bias on N_μ. Then, to obtain Λ_μ, we applied the constant intensity cut (CIC) method that assumes that the intensity of cosmic rays is isotropic in zenith angle [18]. The method consists of selecting N_μ data from different zenith angles by means of cuts applied at fixed frequencies on the integral intensity of N_μ for different zenith angle ranges (see fig. 6, left). Actually, data selected along these curves should correspond to events with the same primary energy but different shower sizes due to the increasing atmospheric depth at high zenith angles. This dependence is expressed in terms of a typical exponential absorption formula, which is parameterized

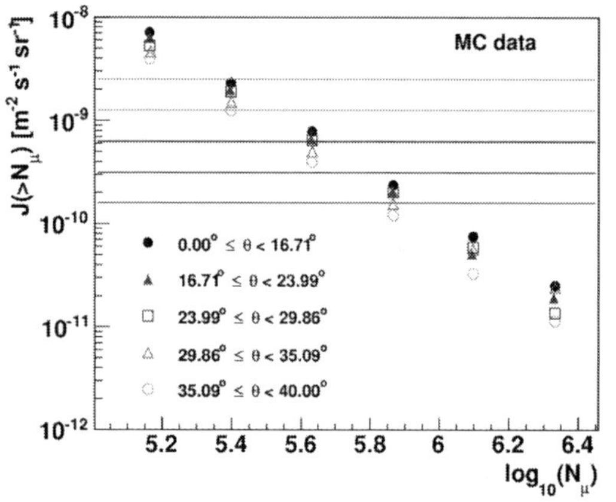

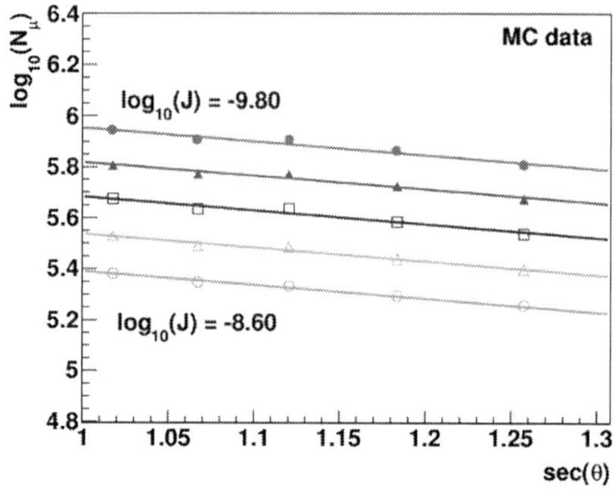

Figure 6. Left: Integral muon intensities derived from a subsample of SIBYLL 2.3 data for five zenith angle intervals using the composition model described in section 4.3. The CIC cuts employed in this work are shown as horizontal lines. Statistical error bars are smaller than the size of the data points. **Right**: Muon attenuation curves extracted with the CIC method from the MC data. The results of the global fit with equation (3) are shown with solid lines. The cuts are decreasing from bottom to top in units of $\Delta \log_{10}[J/(\mathrm{m}^{-2} \cdot \mathrm{s}^{-1} \cdot \mathrm{sr}^{-1})] = -0.30$. The vertical error bars of the data points are smaller than the size of the symbols and consider statistical uncertainties and errors from interpolation.

as a function of a single Λ_μ [11] (see fig. 6, right):

$$N_\mu(\theta) = N_\mu^\circ e^{-X_0 \sec(\theta)/\Lambda_\mu}, \tag{3}$$

where $X_0 = 1022\,\mathrm{g/cm^2}$ is the vertical column depth at KASCADE-Grande and N_μ°, a normalization factor, which depends on the attenuation curve. In our analysis, we divided the MC data into five zenith angles with approximately the same acceptance as in [11] and section 4.1 (see also fig. 6, left) and applied the same CIC cuts described in the abovementioned paper. Using our composition model and the method described, we have calculated Λ_μ for SIBYLL 2.3. The result is presented in fig. 7. The systematic error only includes the uncertainty from the CIC method.

In the same plot we show the experimental value obtained in [11] for a subsample of experimental data (collected during the period December 2003 - October 2011) along with its corresponding statistical and total uncertainties. The latter were estimated through detailed studies. The systematic error includes the uncertainty due to the calculation procedure, the uncertainty owing to the muon correction function, the systematic error of the corrected muon number and the corresponding composition and model dependence, the uncertainty owing to the size of the θ intervals and the uncertainties associated with the shower core position, R. MC predictions from QGSJET-II-2, QGSJET-II-04, EPOS-LHC and SIBYLL 2.1 for an E^{-3} model and a mixed composition assumption (all elements on equal abundances) [11] are also shown in our plot. The MC estimations involve, in addition to the ones above, uncertainties due to the spectral index of the primary spectrum, γ, and due to the primary composition. This last systematic error is estimated by comparing the nominal value, obtained using a mixed composition, with the results for the five primary nuclei employed in simulations. From figure 7, we observe a hint that the predictions for Λ_μ using SIBYLL 2.3 tend to be below the measured data.

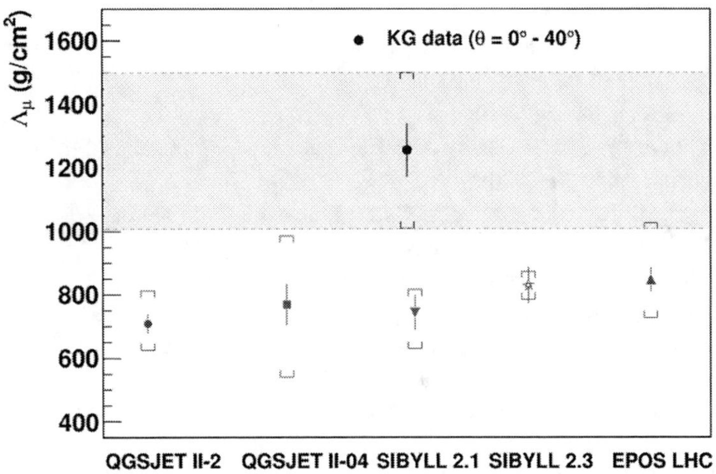

Figure 7. Muon attenuation length measured with the KASCADE-Grande experiment (upper point) compared with predictions of several pre-LHC and post-LHC high-energy hadronic interaction models (lower points) for EAS with core distances within 270 m and 440 m from the KASCADE center and with primary energies between $E \sim 10^{16}$ eV and $\sim 10^{17}$ eV. The total uncertainty (statistical and systematic errors added in quadrature) are shown with squared brackets. Error bars represent statistical uncertainties. The hollowed star is the prediction from SIBYLL 2.3 for our fitted composition model [10] (see text). The experimental result and the predictions from QGSJET-II-02 and 04, EPOS-LHC and SIBYLL 2.1 (for a mixed composition assumption and spectrum E^{-3}) are taken from [11].

5 Conclusions

We have tested muon predictions of SIBYLL 2.3 against measurements of the KASCADE-Grande observatory for $\rho_\mu(r)$ and N_μ at different θ and N_{ch} ranges, as well as for Λ_μ. In general, it seems that $\rho_\mu(r)$ measurements for $\theta < 40°$ and $6.0 \leq \log_{10}(N_{ch}) \leq 7.74$ are within the predictions of the model. For N_μ the agreement is observed in the full shower size range ($6.0 \leq \log_{10}(N_{ch}) \leq 8.0$) and zenith angle intervals ($\theta < 40°$). On the other hand, we see what appears to be a mismatch between the predicted and measured Λ_μ values.

6 Acknowledgments

The authors would like to thank the members of the engineering and technical staff of the KASCADE-Grande Collaboration, who contributed to the success of the experiment. The KASCADE-Grande experiment was supported in Germany by the BMBF and by the 'Helmholtz Alliance for Astroparticle Physics - HAP' funded by the Initiative and Networking Fund of the Helmholtz Association, by the MIUR and INAF of Italy, the Polish Ministry of Science and Higher Education, the Romanian Authority for Scientific Research UEFISCDI (PNII-IDEI grants 271/2011 and 17/2011), and the German-Mexican bilateral collaboration grants (DAAD-CONACYT 2009-2012, 2015-2016). J.C.A.V. acknowledges the partial support of CONACyT (grant CB-2008/106717) and the Coordinación de la Investigación Científica de la Universidad Michoacana.

References

[1] S. Ostapchenko, Proc. of the *XXV ECRS*, eConf **C16-09-04.3** (2016), [astro-ph/1612.09461v1].
[2] F. Riehn et al., Proc. of the *34th ICRC*, PoS (ICRC2015) **558**.
[3] W.D. Apel et al., *NIM A* **620** 202 (2010) .
[4] L. Cazon et al., *Astrop. Phys.* **21** 72 (2004); L. Cazon et al., *Astrop. Phys.* **23** 393 (2005) .
[5] W. D. Apel et al., *NIM A* **620** 20 (2010).
[6] A.Fasso et al., Report CERN-2005-10,INFN/TC-05/11, SLAC-R-773 (2005).
[7] D. Heck et al., Report FZKA 6019, Forschungszentrum Karlsruhe (1998).
[8] R. Brun, F. Carminati, CERN Program Library Long Writeup W5013 (1993).
[9] W.D. Apel et al., *Astrop. Phys.* **36** 183 (2012).
[10] J.C. Arteaga-Velazquez, D. Rivera-Rangel et al., Proc. of the *35th ICRC*, PoS (ICRC2017) **316**.
[11] W.D. Apel et al., *Astrop. Phys.* **95** 25 (2017).
[12] S.S. Ostapchenko, *Nucl. Phys. B (Proc. Suppl.)* **151** 143 (2006); S. Ostapchenko, *Phys. Rev. D* **74** 014026 (2006).
[13] S.S. Ostapchenko, *Phys. Rev. D* **83** 014018 (2011).
[14] T. Pierog et al., *Phys. Rev. C* **92** 034906 (2015).
[15] E.J. Ahn et al., *Phys. Rev D* **80** 094003 (2009).
[16] W.D. Apel et al., *PRD* **87** 081101 (2013).
[17] W.D. Apel et al., *PRL* **107** 171104 (2011).
[18] J. Hersil et al., *Phys. Rev. Lett.* **6** 22 (1961); D. M. Edge et al., *J. Phys. A* **6** 1612 (1973).

On muon bundles from the Universe

P. Kankiewicz[1], M. Rybczyński[1,*], Z. Włodarczyk[1], and G. Wilk[2]

[1] *Institute of Physics, Jan Kochanowski University, 25-406 Kielce, Poland*
[2] *National Centre for Nuclear Research, Department of Fundamental Research, 00-681 Warsaw, Poland*

Abstract. Recently the CERN ALICE experiment, in its dedicated cosmic ray run, observed muon bundles of very high multiplicities, thereby confirming similar findings from the LEP era at CERN (in the CosmoLEP project). We found significant evidence for anisotropy of arrival directions of the observed high multiplicity muonic bundles. The distribution on celestial sphere and the estimated directionality suggests their possible extragalactic source. We argue that muonic bundles of highest multiplicity are produced by strangelets, hypothetical stable lumps of strange quark matter infiltrating our Universe.

1 Introduction

This talk is based on [1], where the relevant details can be found. Cosmic ray physics is our unique source of information on events in the energy range which will never be accessible in Earth-bound experiments [2, 3]. This is why one of the most important aspects of their investigation is the understanding of the primary cosmic ray (CR) flux and its composition. In this respect the recent measurement performed by the ALICE experiment at CERN LHC in its dedicated cosmic ray run [4, 5] is of great importance. A number of events with muon bundles of high multiplicity was registered in the so called Extensive Air Showers (EAS) produced by cosmic ray interactions in the upper atmosphere [4]. A special emphasis has been given to the study of high multiplicity events containing more than 100 reconstructed muons (in particular, we were interested in their directionality). Similar events have already been studied in the previous LEP experiments at CERN (in the so-called CosmoLEP program) such as ALEPH [6], DELPHI [7] and L3 [8], and in the underground Baksan experiment [9].

In this talk we shall concentrate mainly on the first, high multiplicity events. To describe them we propose seriously to consider, for a moment, the possibility of the existence in the flux of incoming CR a component with very high atomic number, of the order of $A \sim 10^3$. In fact we propose to return to our old idea that muon bundles of extremely high multiplicities could be produced by *strangelets*, hypothetical stable lumps of strange quark matter (SQM) infiltrating our Universe [10]. Strangelets with such masses, much larger than the masses of ordinary nuclei, could easily produce extremely large groups of muons in collisions with the atmosphere.

2 Searches for Strange Quark Matter in cosmic rays

The idea of SQM originated some time ago in Refs. [11–14], but it remains alive, cf. [15–21]. In short, the basic, commonly accepted view is that that SQM (understood as a combination of roughly an equal

*e-mail: maciej.rybczynski@ujk.edu.pl

© The Authors, published by EDP Sciences. This is an open access article distributed under the terms of the Creative Commons Attribution License 4.0 (http://creativecommons.org/licenses/by/4.0/).

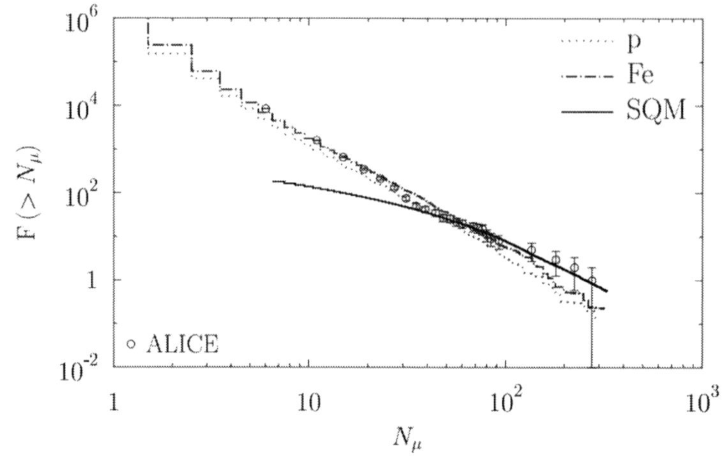

Figure 1. Integral multiplicity distribution of muons for the ALICE data (circles) published in [4]. Monte Carlo simulations for primary protons (dotted line); iron nuclei (dashed dot line) and primary strangelets with mass A taken from the $A^{-7.5}$ distribution (full line) with abundance of the order of $2 \cdot 10^{-5}$ of the total primary flux.

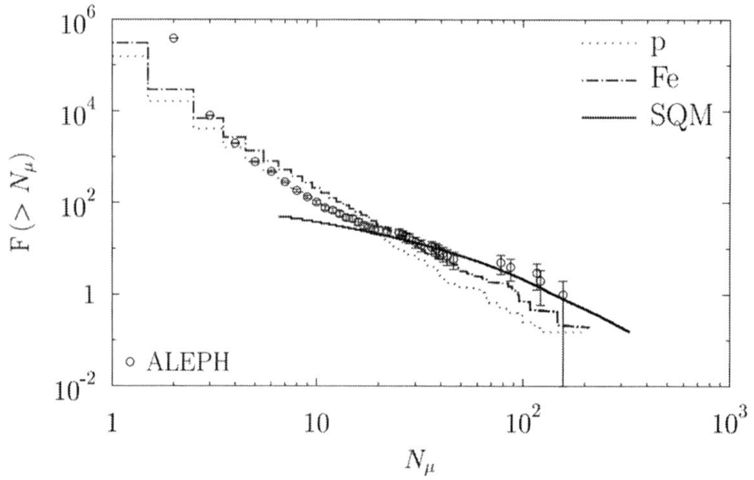

Figure 2. Integral multiplicity distribution of muons the CosmoLEP data (circles) published in [6]. Monte Carlo simulations for primary protons (dotted line); iron nuclei (dashed dot line) and primary strangelets with mass A taken from the $A^{-7.5}$ distribution (full line) with abundance of the order of $2 \cdot 10^{-5}$ of the total primary flux.

number of up, down and strange quarks) might be the true ground state of quantum chromodynamics (QCD). Therefore, it is reasonable to expect that it exists in some form in the Universe and can be detected. This supposition resulted in a number of searches for strange stars and quark stars, in which such a form of matter would be dominant and which could therefore be a possible source of strangelets penetrating outer space [22–29]. The experimental data mentioned above lead to a flux of strangelets which follows the $A^{-7.5}$ behaviour, which in turn coincides with the behaviour of the abundance of

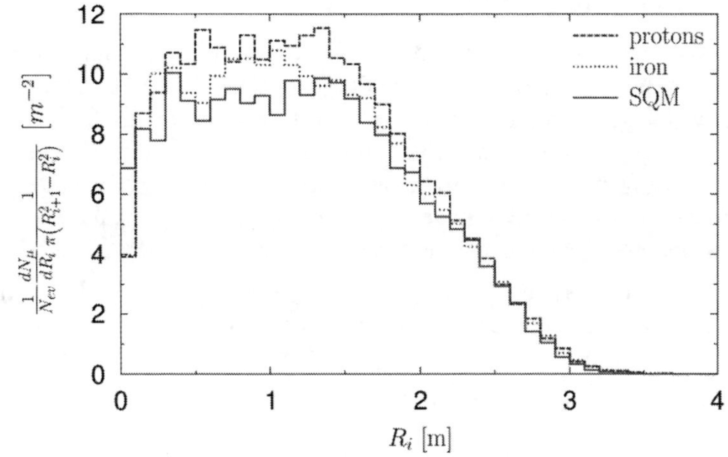

Figure 3. Lateral distribution of muons produces in bundles with $N_\mu > 100$. Monte Carlo simulations using ALICE experimental setup [4] for primary protons (dashed line); iron nuclei (dotted line) and primary strangelets with mass A taken from the $A^{-7.5}$ distribution (full line).

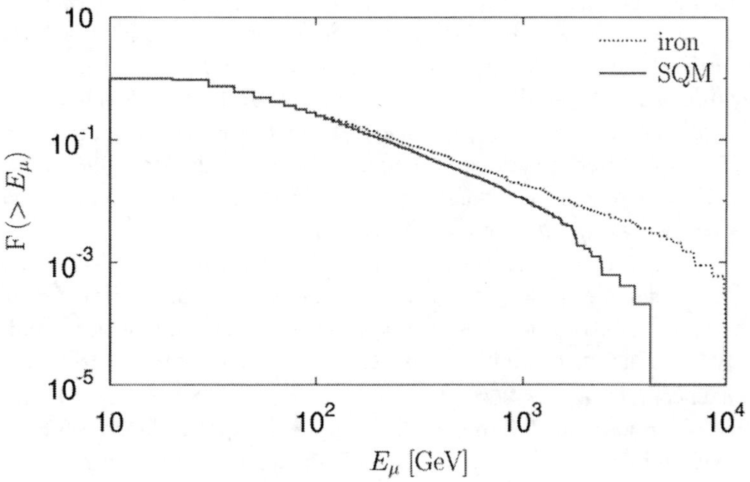

Figure 4. Energy distribution of muons produces in bundles with $N_\mu > 100$. Monte Carlo simulations using ALICE experimental setup [4] for primary iron nuclei (dotted line) and primary strangelets with mass A taken from the $A^{-7.5}$ distribution (full line).

normal nuclei in the Universe. Strange quark matter fills a vast gap in the distribution of all known forms of stable matter, cf. chart of nuclides presented in Ref. [30]. It places itself exactly between the heaviest atomic elements and neutron stars. The estimated flux of strangelets is also consistent with the astrophysical limits and with the upper limits given experimentally [31].

3 High multiplicity muon bundles from SQM

We start with a short reminder of our proposed possible scenario of propagation of strangelets through the atmosphere [32, 33]. The apparent contradiction between the large initial size of the incoming strangelets and their required strong penetrability in the atmosphere can be resolved by assuming that strangelets reaching deeply into the atmosphere are formed from the original large strangelet which loses mass in many successive interactions with air nuclei when penetrating the atmosphere. To provide numerical estimate we limit ourselves to the two most extreme pictures of the collision of a strangelet of mass number A with an air nucleus target of mass number A_t :

(i) All quarks of A_t which are located in the geometrical intersection of the two colliding projectiles are involved and one assumes that each quark from the target interacts with only one quark from the strangelet.

(ii) All quarks from both nuclei which are in their geometrical intersection region participate in the collision.

In the first case, during the interaction up to ($3A_t$) quarks from the strangelet could be used up and its mass could drop to a value of $A - A_t$, at most. The total penetration depth of the strangelet can be estimated to be in this case equal to

$$\Lambda \simeq \frac{1}{3}\lambda_{NA_t}\left(\frac{A_0}{A_t}\right)^{\frac{1}{3}}\left(1-\frac{A_{crit}}{A_0}\right)^2\left(4-\frac{A_{crit}}{A_0}\right) \simeq \frac{4}{3}\lambda_{NA_t}\left(\frac{A_0}{A_t}\right)^{\frac{1}{3}} \qquad (1)$$

(here λ_{NA_t} denotes the mean free path for $N - A_t$ collisions). One accommodates here both the most probable "normal" mean free paths for successive interactions and the final large penetration depth. This scenario is fully consistent with all present and proposed experiments and could be additionally checked only by measuring the products of the intermediate collisions, which so far is impossible.

The second case is an analogue of the so called tube model used occasionally in nuclear collisions (cf., for example, [34]). After the collision the atomic mass of the strangelet diminishes to a value equal to $A - A^{1/3} \cdot A_t^{2/3}$. While this is a rather extreme variant, it is still useful in providing an estimate of the maximum possible destruction of the quarks in the strangelet.

In this study we have used suitably modified SHOWERSIM [35] modular software. We performed Monte Carlo simulations for primary nuclei composed of protons, iron nuclei, and for primary strangelets with mass A taken from the $A^{-7.5}$ distribution. The results of our simulations are shown in Fig. 1. Integral distributions of muons originating from ordinary nuclei (protons or irons) exhibit power-law dependence, $F\left(> N_\mu\right) \sim N_\mu^{-2.8}$. Note that whereas the lower and medium multiplicities can be reproduced by the ordinary nuclei, the extremely large groups of muons can be described only by allowing (a relatively minute, of the order of 10^{-5} of the total primary flux) admixture of SQM of the same total energy. Fig. 2 shows, for comparison, similar results obtained for the CosmoLEP data [6].

It seems that there are experimental possibilities which allow to distinguish between different types of primaries. Namely lateral, cf. Fig. 3 or energy (Fig. 4) distributions of muons registered in highest multiplicity bundles (here $N_\mu > 100$) may be used for such purpose. Although the lateral distributions of high muon multiplicity groups generated by different types of primary cosmic rays are rather similar (Fig. 3), it is relatively easy to disentangle different primaries by looking at the energy distribution of these muons.

4 Anisotropy of arrival directions of strangelets

The ALICE data also turns out to be very valuable for another reason. Namely, they show the angular distribution of the muon events in the spherical reference frame with zenith angle (θ) and azimuth angle (Φ). This reference system was used in the experiment and is related to the geographical location of the detector.

Taking into account the orientation of the axes, these coordinates can be transformed into a horizontal reference frame: $(\Phi, \theta) \rightarrow (Az, h)$, where Az is the Azimuth used in navigation/astronomy (in this particular case: $Az = \Phi + \alpha_0$, where α_0 is the angle between the x axis of detector reference frame and the direction to the north), and h is the elevation above the horizon ($h = 90° - \theta$). Finally, we can transform the coordinates from the horizontal to the equatorial reference frame: $(Az, h) \rightarrow (\alpha, \delta)$. The equatorial system of coordinates, expressed by right ascension (α) and declination (δ), related to particular epoch (e.g. 2000.0) is commonly used in astronomy.

Using the usual spherical triangle formulas or rotation matrices one can convert between the horizontal and equatorial coordinates. For the five high-multiplicity muon events we obtained the celestial equatorial coordinates $(\alpha_{2000}, \delta_{2000})$ and the estimated limits of their positional errors are $\sim 10°$ [1]. The centroid (average location) of the five considered events is marked by a star in Fig. 5. The coordinates of this centroid are: $\alpha_{centr} = 11\ h\ 15\ m$, $\delta_{centr} = 39°\ 29'$. Although these five events seem dispersed near this centroid at a distance of few degrees, they are all located far from the plane of our galaxy. It is worth to notice that the position of the above mentioned centroid is located close to the center of the region registered by the Telescope Array Collaboration data [38]: $\alpha_{TA} = 9\ h\ 54\ m$, $\delta_{TA} = 44°\ 36'$. The angular separation between the centroid and center of the area determined from the TAC data is only 15°48′. Considering the dispersion of five events, possible errors and a very small statistical sample, it is possible to suggest agreement of these results.

We can conclude that the distribution of the directions of these five events suggests their possible extragalactic provenance. Most of them are located in the vicinity of the galactic north pole, far from the galactic plane. If we temporarily assume that all these events have the same origin, the most probable source would be the blazar Mrk 421 which is located close to the centroid of the five events. It is one of the brightest blazars known, with major outbursts, active at all wavebands [36]. However, there are also many other known extragalactic high energy sources located close to the coordinates of our events [36], [37] and, at the moment, they cannot be excluded as potential candidates for the cause of the investigated phenomena.

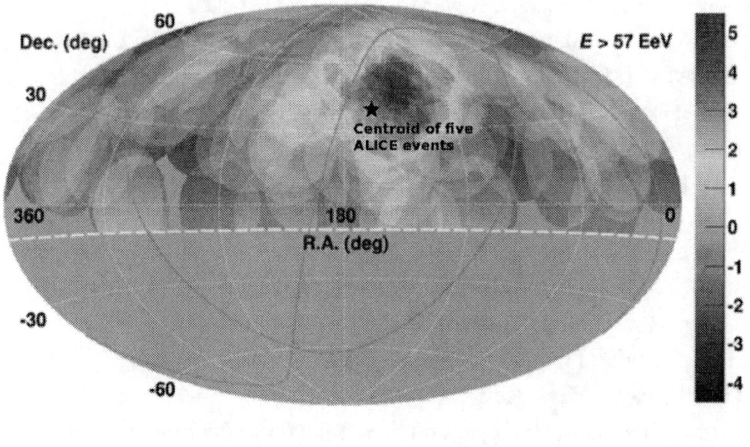

Figure 5. Aitoff projection of the UHECR map in equatorial coordinates taken from Telescope Array Collaboration data [38]. Location of the centroid of five ALICE events is additionally marked with a star.

5 Conclusions

There is an ongoing discussion concerning the possibility of the existence in the Universe of some stable forms of SQM and the feasibility of its detection. Whereas until now we do not have fully convincing observations of SQM, nevertheless there are numerous indications of the possible existence of strange stars, or even of the existence of quark stars. Strangelets could originate, for example, from violent collisions between such objects. The recent results of ALICE reporting the existence of bundles of high-multiplicity muons could serve, notwithstanding the weakness of this signal, as a direct (almost) signal of stable lumps of SQM called strangelets arriving at the Earth from outer space. As a result, we have found that whereas the low multiplicities of muon groups measured by the CERN ALICE experiment favour light nuclei as primaries and the medium multiplicities show behaviour specific for heavier primaries, the muon groups (bundles) of really high multiplicities (of the order of ~ 100) apparently cannot be described by the common interaction models. We have shown here that the situation can be rescued by allowing for a relatively small (of the order of 10^{-5} of the total primary flux) admixture of strangelets of the same total energy. Our estimation of their flux does not contradict the results obtained recently by the SLIM Collaboration [31]. The ALICE data allowed us to estimate the directionality of the five events of the highest muon multiplicities, and to identify their most probable extragalactic source(s). The compatibility with other results, especially with the direction of events published by the Telescope Array Collaboration [38] remains very interesting.

Acknowledgments

This research was supported by the Polish National Science Centre, grant 2016/23/B/ST2/00692 (MR) and contract 2016/22/M/ST2/00176 (GW).

References

[1] P. Kankiewicz, M. Rybczynski, Z. Włodarczyk and G. Wilk, Astrophys. J. **839**, no. 1, 31 (2017).
[2] A. Dar and A. De R ujula, Phys. Rep. **466** 179 (2008).
[3] A. Letessier-Selvon and T. Stanev, Rev. Mod. Phys. **83** 907 (2011).
[4] J. Adam *et al.* [ALICE Collaboration], J. Cosm. Astrop. Phys. **01** 032 (2016).
[5] J. Adam *et al.* [ALICE Collaboration], ALICE-PUBLIC-2016-003.
[6] V. Avati *et al.*, Astropart. Phys. **19** 513 (2003).
[7] J. Abdallah *et al.* [DELPHI Collaboration], Astropart. Phys. **28** 273 (2007).
[8] P. Achard *et al.* [L3 Collaboration], Phys. Lett. B **598** 15 (2004); [hep-ex/0408114].
[9] V. N. Bakatanov, Yu. F. Novoseltsev and R. V. Novoseltseva, Astrop. Phys. **8** 59 (1997).
[10] M. Rybczyński, Z. Włodarczyk, G. Wilk, Nucl. Phys. B (Proc. Suppl.) **97**, 85 (2001).
[11] H. Terazawa, INS-Report - **336** (INS, University of Tokyo, Tokyo) May, 1979.
[12] E. Witten, Phys. Rev. D **30** 272 (1984).
[13] E. Farhi and R. L. Jaffe, Phys. Rev. D **30** 2379 (1984).
[14] C. Alcock and E. Farhi, Phys. Rev. D **32** 1273 (1985).
[15] J. Madsen, in: J. Cleymans, H. Geyer and F. Scholtz (eds) *Lecture notes in physics* **516**, Springer, Berlin, p. 162.
[16] M. G. Alford, Nucl. Phys. A **830** 385c (2009).
[17] F. Weber, Prog. Part. Nucl. Phys., **54** 193 (2005).
[18] R. Xu, Chin. J. Astron. Astrophys. Suppl. **5** 353 (2005).

[19] R. Klingenberg, J. Phys. G **27** 475 (2001).
[20] H. Terezawa, Nonl. Phen. Compl. Systems **18** 25 (2015).
[21] H. Terezawa, Nonl. Phen. Compl. Systems **19** 147 (2016).
[22] C. Alcock, E. Fahri and A. Olinto, Astrop. J. **310** 261 (1986).
[23] K. S. Cheng, Z. G. Dai and T. Lu, Int. J. Mod. Phys. D **7** 139 (1998).
[24] K. S. Cheng, Z. G. Dai, D. M. Wei and T. Lu, Science **280** 407 (1998).
[25] J. J Drake *et al.*, Astroph. J. **572** 996 (2002).
[26] A. Bauswein, H.-T. Janka, R. Oechslin, G. Pagliara, I. Sagert, J. Schaffner-Bielich, M. M. Hohle and R. Neuhaüser, Phys. Rev. Lett. **103** 011101 (2009).
[27] E.Finch, J. Phys. G **32** S251 (2006).
[28] W. Kluźniak and W. H. Lee, Mon. Not. R. Astron. Soc. **335** L29 (2002).
[29] Z. Kovács, K. S. Cheng and T. Harko, Astr. Astroph. **500** 621 (2009).
[30] H. J. Crawford and C. H. Greiner, Sci. Am. **270** 58 (1994).
[31] Z. Sahnoun *et al.*, Radiat. Meas. **44** 894(2009).
[32] G. Wilk and Z. Wlodarczyk, Acta Phys. Hung. **4** 395 (1996) [hep-ph/9606401].
[33] G. Wilk and Z. Wlodarczyk, J. Phys. G **23** 2057 (1997).
[34] F. Takagi, Prog. Theor. Phys. **65** 948 (1981).
[35] A. Wrotniak, SHOWERSIM/84, University of Maryland Preprint, 85 (1984).
[36] D. Horan and T. C. Weekes, New Astr. Rev. **48** 527 (2004).
[37] C. F. Turley *et al.* [Astrophysical Multimessenger Observatory Network) Collaboration], Astrophys. J. **833**, no. 1, 117 (2016).
[38] R. U. Abbasi *et al.* [Telescope Array Collaboration], Astrophys. J. **790**, L21 (2014).

Four Channel Mini Wire Chamber to Study Cosmic Rays

J. Felix[1,*] , *G. J. Rodriguez*[1,**]

[1]*Laboratorio de Partículas Elementales. División de Ciencias e Ingenierías, Campus León, Universidad de Guanajuato*

Abstract. Multiwire proportional chamber is a conventional technique to study radiation in general, and cosmic rays in particular. To study cosmic rays, it was planned, designed, constructed, characterized, and tested a four channel mini wire chamber, based on two 3 cm × 3 cm × 0.6 cm Aluminum frames, two 3 cm × 3 cm × 0.6 cm plastic frames, two 3 cm × 3 cm × 0.3 cm Aluminum frames, two electronic planes each with two Tungsten Gold plated 1 mil diameter wires, parallel and 1 cm apart each other at 25 g stretched -each plane was 90° rotated each other in the final assemble- and two Aluminum foil window to define the gas volume; it was operated with Argon 90%-CH4 10% gas mixture at 1 atmosphere and ambient temperature (20^oC in the average). It is presented technical details, results on characterization, and preliminary results on cosmic rays detection.

1 Introduction

To detect cosmic rays, and radiation in general, there are multiwire proportional chambers, parallel plate chambers, resistive plate chambers, etc. used in high energy physics, nuclear physic, etc. to detect particles or radiation in general [1] with applications in medicine, archeology, biology, planetary science, integrated-circuit quality control, geology, environmental science, volcanology, climatology, security, and many other applications. All of them are based on especial gas systems and applied high voltages [2-18].

Here in Mexico, Universidad de Guanajuato, Guanajuato, there is the Laboratorio de Partículas Elementales, División de Ciencias e Ingenierías (DCeI), Campus León (CL), Universidad de Guanajuato (UGTO) a small facility to design, construct, test, and run small ionizing particle detectors, specially, cosmic ray detectors. See Figure 1. http://laboratoriodeparticulaselementales.blogspot.mx/.

There are many projects. One of them is related with design construction and operation of a four channel mini wire chamber to study very fast cosmic ray signals and apply cosmic rays to a broad technological areas –oil and mineral exploration, etc.- and basic science research –biology, particle physics, etc.-. But the applications are, and overall, for research, for teaching and for learning physics.

These series of experiences could lead to propose a big cosmic ray detector of about 40 000 detection channels, with particle trajectory reconstruction and particle identification capabilities to study very fast cosmic ray signals.

*e-mail: felix@fisica.ugto.mx
**e-mail: rodriguezbg2012@licifug.ugto.mx

© The Authors, published by EDP Sciences. This is an open access article distributed under the terms of the Creative Commons Attribution License 4.0 (http://creativecommons.org/licenses/by/4.0/).

Figure 1. Different views of Laboratorio de Partículas Elementales, DCeI. CL. Universidad de Guanajuato. Left, construction area; right, electronic testing area.

2 Design and construction

A layout of the 4 channel mini wire chamber is in Figure 2, left, performed with Sketchup (*https : //www.sketchup.com/*). This detector is based on two homemade electronic boards, Figure 2, right.

It contains the basic electric and electronic circuitries to operate, readout, and transmit the digital information. It is intended to detect ionizing radiation in general and cosmic rays in particular. It is our own invention, design and construction.

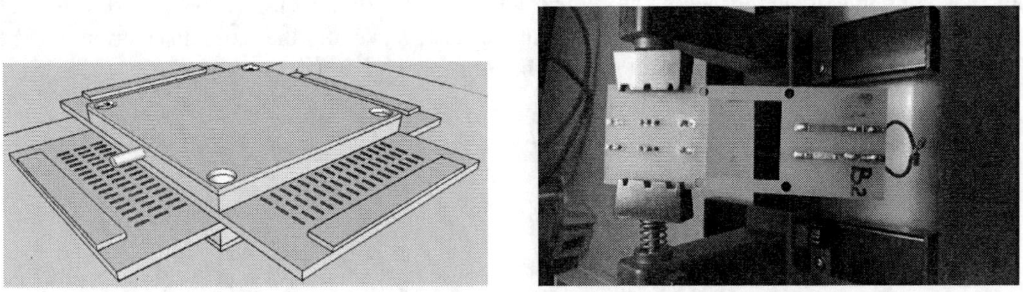

Figure 2. Left, layout of the planned mini four wire chamber. Main components –electronic boards and Aluminum frames- are shown. Right, electronic boards; the basic electronic circuit is shown.

The detector electronic board is 4 cm × 11.5 cm × 1 mm, with a 3 cm × 3 cm square hollow, as is shown in Figure 2; with two channels each, based on Tungsten 2.54 × 10^{-3} cm in diameter, gold-plated, from California Fine Wires (*http : //www.calfinewire.com/index.html*), separated each other by one cm, stretched by 25 g, and soldered to the electronic board. Each board has the HV connector and the RC circuit, see Figure 3, left.The detection regions are defined by the crossing wires, R_{ij}, $i, j = 1, 2, 3$, see Figure 3, right.

Each electronic board was placed 90° each other, with one above and other below rubber 4 cm × 4 cm × 0.6 cm square, with a symmetrical 3 cm × 3 cm square hollow. Figure 4, left. This system was supported by two 4 cm × 4 cm × 0.6 cm Aluminum frames with a symmetrical 3 cm × 3 cm square hollow, one for below and one for above, and closed with two tops, one above and one below. Figure 4, right.

The completed system was assembled by 4 screws $\frac{1}{8}$ inch diameter. A gas chamber was formed with this stack, with one $\frac{1}{8}$ in input plastic elbow pipe and $\frac{1}{8}$ in output plastic elbow pipe, to circulate,

Figure 3. Left, the basic electronic circuit to readout the electric signal leaved by the passage of the ionizing particle. Right in this draw, A1, A2, B1, B2 represent the four channels. With A1 and A2 are represented one electronic board; with B1, B2 are represented the other electronic board. The 9 detection regions (R_{ij}; $i, j = 1, 2, 3$) are shown.

take in and take out Ar90%+CH410% commercial gas. See Figure 5, left. The complete four channel mini wire chamber is in Figure 5, right.

The flow of gas was regulated by a glass bubbler constructed for this purpose and operated with mineral cosmetic oil. See Figure 6, left. The analogical signal was digitalized by an additional electronic board designed and constructed for this purpose, see Figure 6, right.

The whole system was assembled and supported by an Aluminum base to connect electrically the four channels and readout the analogical signals, and attach the digital circuit. See Figure 7, left. Final assembling is in Figure 7, right. Many details can be seen, especially the cross disposition or the two electronic boards, 90° one with respect to the other, the gas pipes.

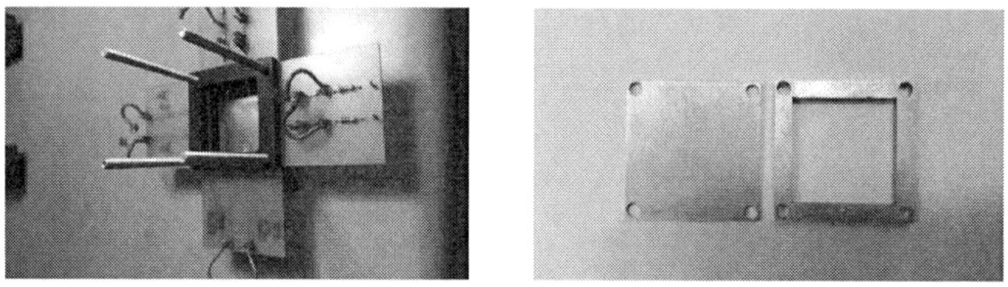

Figure 4. Left, rubber square seals. Right, Aluminum detector top and Aluminum frame.

3 Experimental Setup

The four wire channel detector, the digitalizing electronic board, the high voltage power supplier, the gas system –gas bubbler and gas container-, and one oscilloscope, or the data acquisition system were assembled to operate together. See Figures 8-10.

The four wire channel detector, the digitalizing electronic board, and the gas bubbler were home made. High voltage power supplier is a KEITHEY 2290-2 5 kV or a Home made or a POWER DESIGNS INC. Palo Alto California Model 1570; the oscilloscope is a Tektronics TDS 1001 C-EDU, or a GW INSTEK GDS-3354; and the data acquisition system is a 32 channel Compact Rio from National Instruments ($http : //www.ni.com/compactrio/$). Figure 11.

The applied high voltage was between 1900 V cd-2900 V dc, for both positive and negative voltages. The gas pressure was about 3 cm of oil above atmospheric pressure. The temperature is about

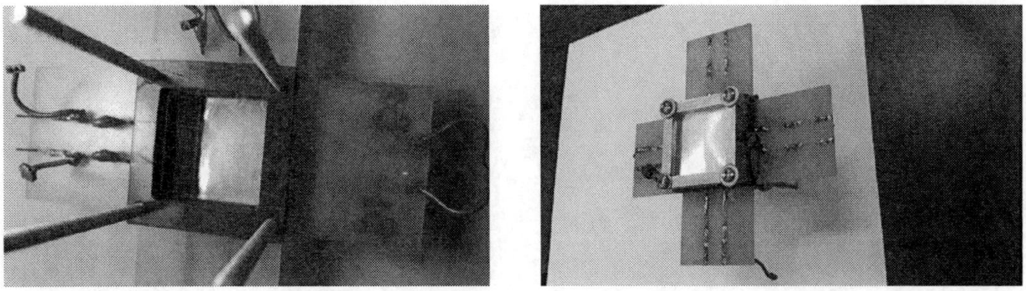

Figure 5. Left, mini four wire chamber, basic elements are shown, specially the plastic sealers. Right, completed assembled.

Figure 6. Left, home made gas bubbler operated with cosmetic oil. Right four channel digitizer electronic circuit.

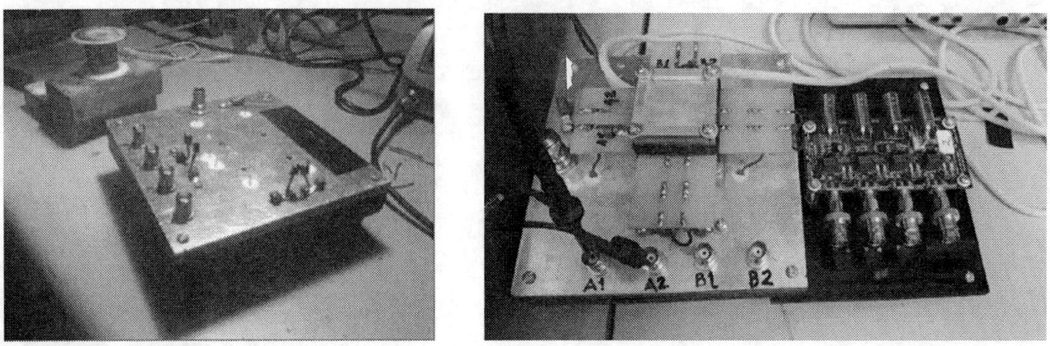

Figure 7. Left,this is the base for the mini four channel chamber and its electronic boards. Right, home made mini four wire chamber is shown; many details can be seen.

20°C. The conditions were not ideal. Some contaminants are present like the following: dust, water vapor, temperature fluctuations and electric noise.

Figure 8. One experimental setup. Power suppliers, multimeter, and oscilloscope. Displays on the oscilloscopes are seen.

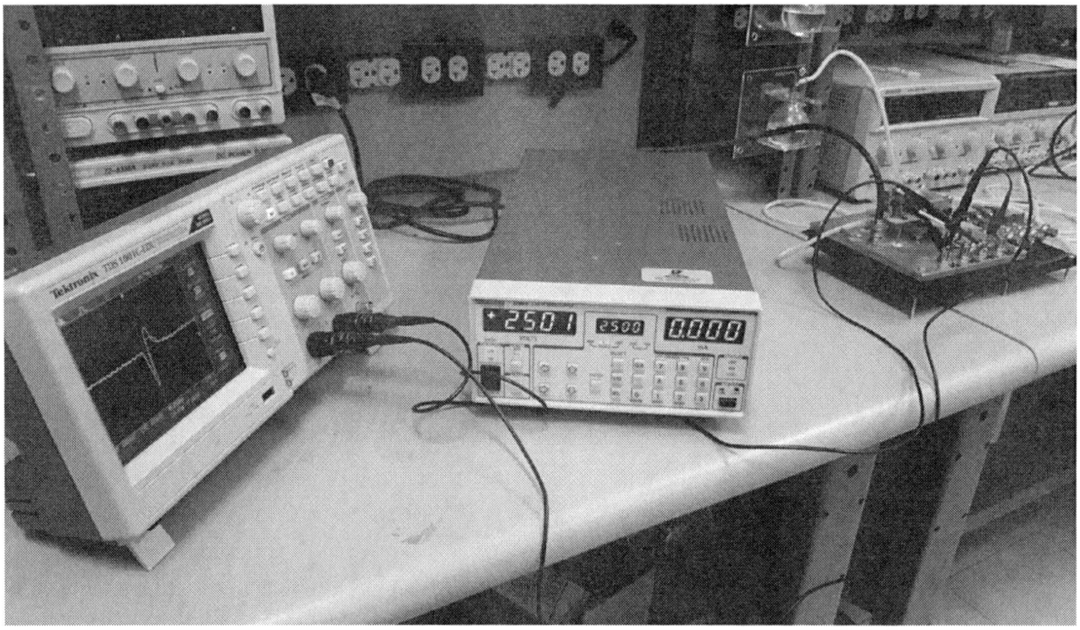

Figure 9. Another experimental setup.

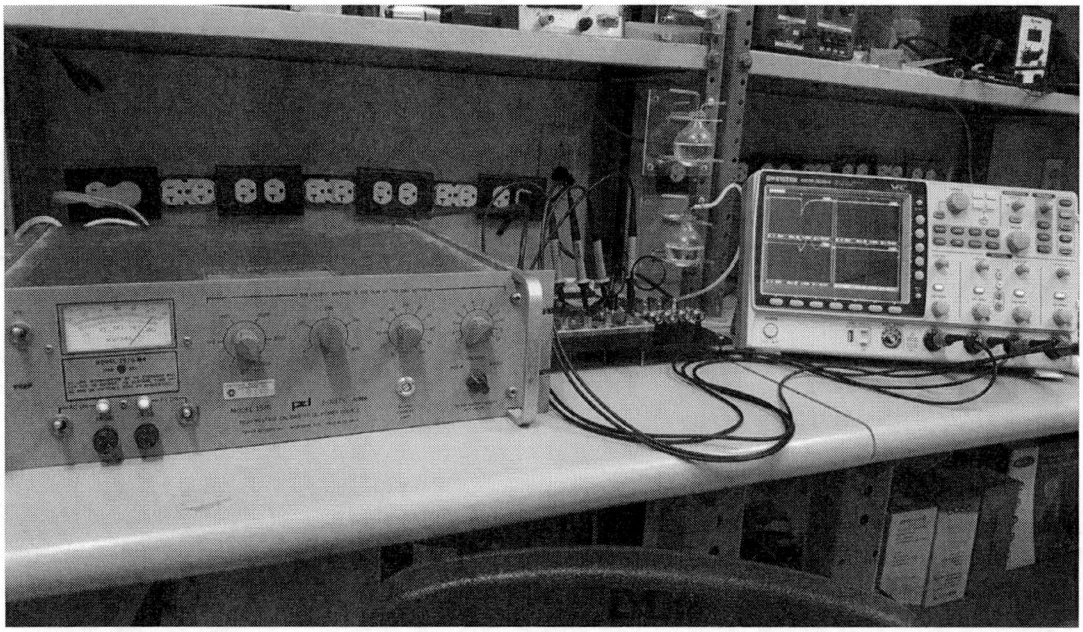

Figure 10. Another experimental setup. Many details are shown.

Figure 11. Data Acquisition system. Compact RIO from National Instruments surrounded by other instruments.

4 Characterization

The four wire chamber was purged and cleaned by running Ar90%-CH410% gas, for ten hours, at 3 cm of oil above the normal atmospheric pressure. After that, the electrical tests and detections test were performed.

The four wire chamber detector was operated between 0.00 Vcd-2900 Vcd, for both, negative and positive voltages. No signals were detected between 0.0 V cd and ±1800 V cd. Signals, probably produced by cosmic rays, are detected on the four channels, starting at close ±1800 V cd. Each channel operates at different voltages and at different paces. See Figure 12 and Figure 13, left.

Digital signals, with analogical signals, are at Figure 13, right. These signals are used to count the cosmic particles that hit the detector.

Also the input signal was simulated with an oscillator, to study the efficiency of the detector, electronic damping, the possible generation of cross talk signals, and the efficiency of the discriminator. Exponential decay, and exponential crescent, signals were injected to the four wire channel detector using a Tektronic oscillator, with frequencies between 0 Hz and 1 MHz, 1 Vpp. The detector was operated at 0.0 Vcd, during these tests. The four wire channel detector works properly with negative and positive signals; both, digital and analogical signal electronic circuits are 100 % efficient, at least inside the frequencies where it was operated, and work properly. The electronic noise is of the order or 20 mVpp. It is from the ambient, and from the power suppliers. The discriminator trigger was fixed at 3 times the noise level, about 60 mV cd. Attenuation factor is about 50 %, the output signal is about one half of the input signal, from 1 Hz up to 1 MHz. Cross talk signals were observed; they were eliminated with a 100 Ohm electrical resistor at the input, in series with the metallic fiber detector.

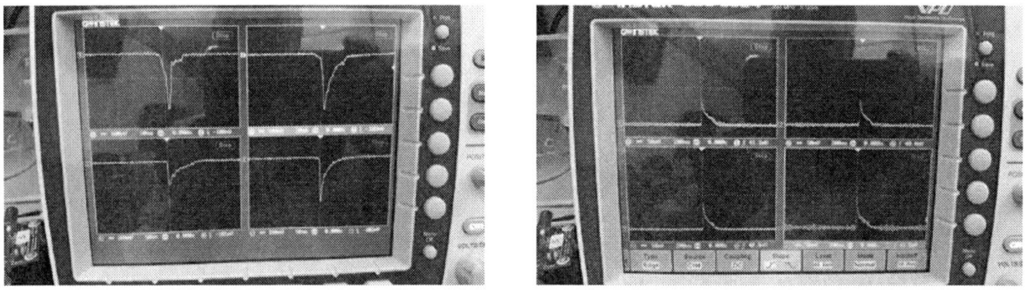

Figure 12. Left, signals obtained with positive high voltage. Right, signals with negative high voltage. They are slower compared with those from positive high voltage.

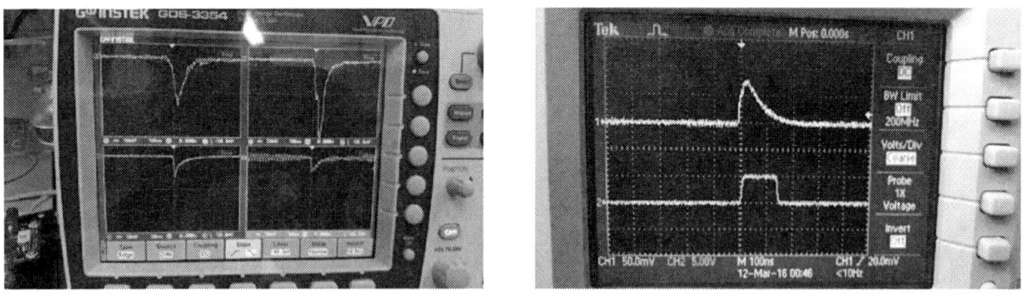

Figure 13. Left, other display of signals from cosmic rays. Right, analogical and digital signal display.

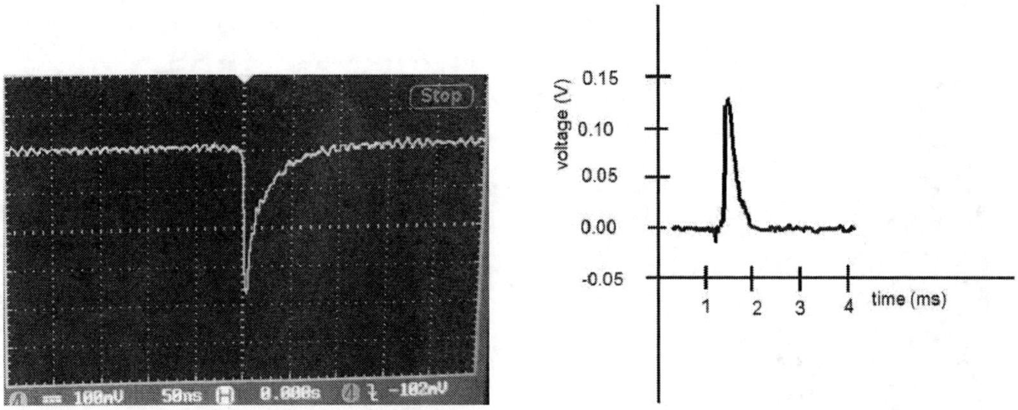

Figure 14. Left, pulse from the mini four channel chamber compared with right pulse from Z. Ahmed et al, IOP Publishing Ltd and Sissa Medialab srl. Doi: 10.1088/1748-0221/9/01/P01009.

5 Results

The amplitudes of signals are of the order of 300 mV, but are variables; they are related with the energy cosmic particles deposit on the gas chamber. The average amplitude depends on the applied high voltage close to linearly, for both negative and positive voltages. The observed rise time is about 5 ns; the decay time, about 20 ns. See Figure 14. There is a comparison with signal from reference [19]. It is seen that this four wire chamber is faster than others, which makes it very convenient to study fast signals.

The frequency of the number of counts versus the number of counts appears in Figure 15 for channels 1, 2 3, 4, respectively. In the semilog plot these distributions are roughly linear, with negative slope. The average characteristics are statistically equivalent.

6 Conclusions

The four wire chamber detector works properly, in spite of the no ideal conditions were it was constructed and operated. Amplitude signals of the order of 300 mV were acquired. No electronic amplifiers were required. It is a fast detector: rising time 5 ns; falling time 20 ns. It is good for Time of Flight studies. The digitalizing electronic board works fine. It is possible to count events produced by the detector from cosmic particles. Much more studies are needed, like efficiency, analogical signal amplitude dependence on detector high voltage, time resolution, space resolution, calibration, electronic attenuation, detector alignment, data acquisition system, triggering. Big detectors are made assembling and synchronizing hundreds, even millions, of small detectors, to collect and combine information from each one.

7 Acknowledgements

CONACyT grant CONACYT, Fondo I0017 Fondo SEP - CONACYT Solicitud 000000000223179. The mole of students which participates in these projects. Luis J. Arceo, Miguel A. Hernández, Diego A. Andrade, María L. Rodríguez, Everardo Granados, Oscar E. Moreno, Karla N. Herrera, Raúl A. Gutiérrez, Francisco J. Rosas, Alfredo Martínez, Carlos A. Cervantes.

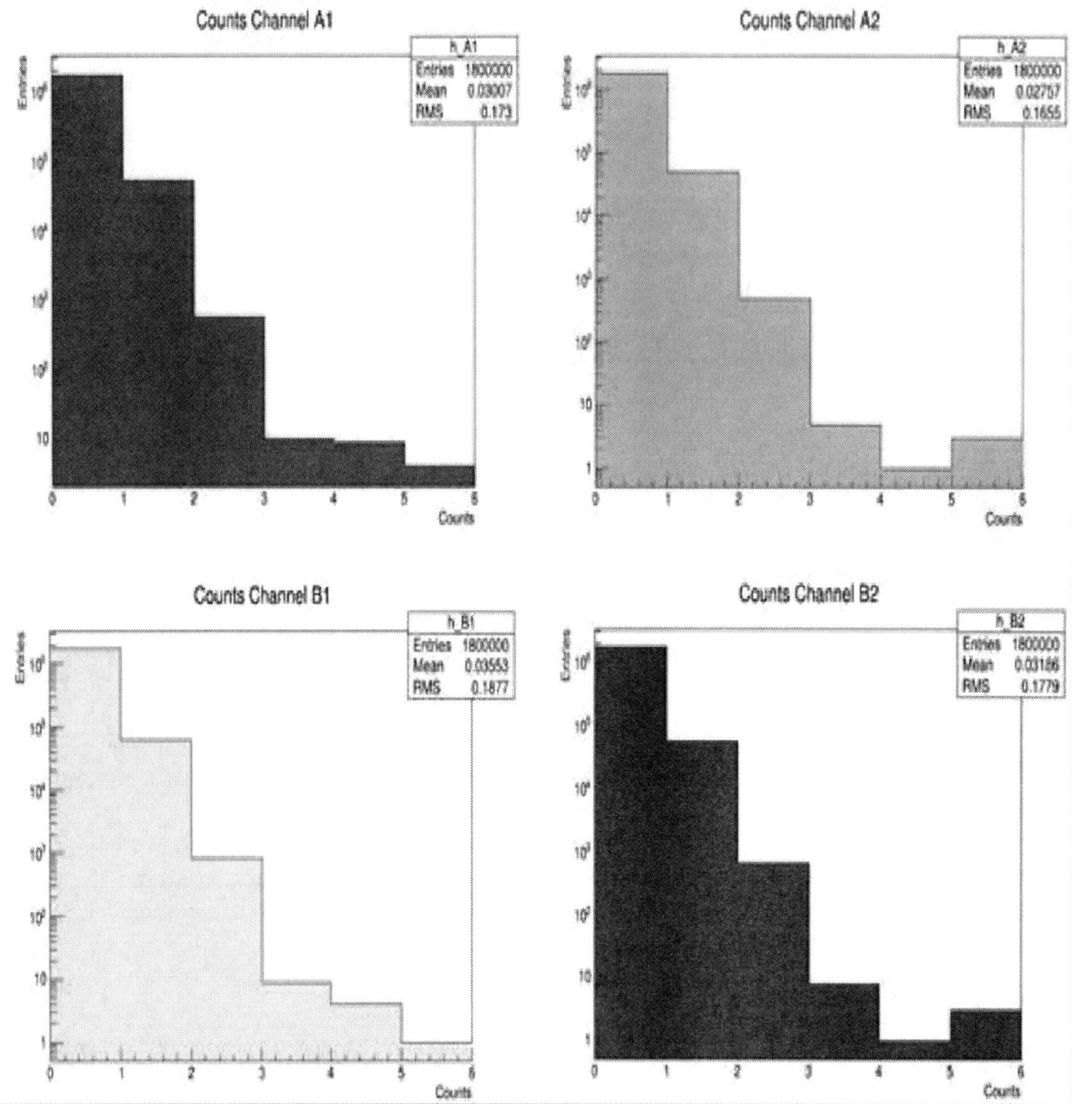

Figure 15. Frequency of counts Vs number of counts for each four mini wirechamber channel.

References

[1] H. Stelzer, Nucl. Inst. And Meth. **133**,499-413 (1976).
[2] D. V. Harrach and H. J. Specht, Nuclear Instruments and Methods 164 (1979) 477-490.
[3] R. Santonico and R. Cardarelli, Nuclear Instruments and Methods 187 (1981) 377-380.
[4] I. Crotty, et al, Nucl. Inst. and Methods in Physics Research A329 (1993) 133-139.
[5] R . Cardarelli, A. Di Ciaccio and R. Santonico, Nucl. Inst. and Meth. in Phy. Res. A 333 (1993) 399-403.
[6] C. Bacci et al, Nuclear Instruments and Methods in Physics Research A 352 (1995) 552-556.
[7] R. Cardarelli, V. Makeev, R. Santonico, Nucl. Inst. and Meth. in Phys. Res. A 382 (1996) 470-474.

[8] Frank Hartmann, Archana Sharma, Nucl. Instr. and Meth. in Phys. Res. A666 (2012) 1–9.
[9] M. Abbrescia et al, Nuclear Physics B (Proc. Suppl.) 44 (1995) 218-221.
[10] YE Yanlin et al, SCIENCE IN CHINA (Series A) Vol. 45 No. 11. November 2002.
[11] G. Charpak, F. Sauli, CERN-EP/84-35 (1984).
[12] D.F. Anderson, S. Kwan and V. Peskov, FERMILAB-Conf-93215/215.
[13] E. Bossini et al, IL NUOVO CIMENTO 39 C (2016) 294. DOI 10.1393/ncc/i2016-16294-9, Communications: SIF Congress 2015.
[14] P. Paolucci et al, Published By IOP Publishing For Sissa Medialab, Special Issue On Resistive Plate Chambers And Related Detectors Rpc2012 (2013).
[15] M. Morales et al, Published By IOP Publishing For Sissa Medialab, Special Issue On Resistive Plate Chambers And Related Detectors Rpc2012 (2013).
[16] P. Fonte, Published By IOP Publishing For Sissa Medialab, Special Issue On Resistive Plate Chambers And Related Detectors Rpc2012 (2013).
[17] R. Iuppa, Published By IOP Publishing For Sissa Medialab, Special Issue On Resistive Plate Chambers And Related Detectors Rpc2012 (2013).
[18] M. Widgoff et al, GEM TN-92-206 Rev. A. (1992).
[19] Z. Ahmed et al, IOP Publishing Ltd and Sissa Medialab srl. Doi: 10.1088/1748-0221/9/01/P01009.

Chiral asymmetry during the EWPT from CP-violating scattering off bubble walls

Alejandro Ayala[1,2,⋆], *L. A.* Hernández[1,⋆⋆], and *Jordi* Salinas[1,⋆⋆⋆]

[1] *Instituto de Ciencias Nucleares, Universidad Nacional Autónoma de México, Apartado Postal 70-543, Ciudad de México 04510, México*
[2] *Centre for Theoretical and Mathematical Physics, and Department of Physics, University of Cape Town, Rondebosch 7700, South Africa*

Abstract. We compute a net electric current during a first order EWPT arising from the asymmetric propagation of fermion chiral modes due to a CP-violating interaction with the Higgs. The interaction is quantified in terms of a CP-violating phase in the bubble wall that separate both phases. We comment on the possibility of this current to generate a seed magnetic field and its implications for primordial magnetogenesis in the early Universe.

1 Introduction

The physics of the Early Universe remains unknown on many of its aspects. One of the long standing problems of this era is the generation of the matter-antimatter asymmetry, the so called, baryogenesis. Observations from γ-ray telescopes indicate that there are no sources of this annihilation processes happening in the visible Universe. This situation leads to think that an asymmetry was created soon after the Big Bang, leaving little room for antimatter to spread.

Another outstanding problem in the realm of cosmology is the puzzling presence of large-scale magnetic fields, coherent up to galactic and intergalactic distances. Recent evidence has allowed to observe magnetic fields coherent up to galactic distances with a strength of a few μG, one millionth part of Earth's magnetic field. Despite their ample presence in the Universe, the origin of the magnetic fields of such astronomical scales remains a mystery [1].

The observational data suggest that the fields that are seen are a product of the amplification of seed magnetic fields through some dynamo process [2]. Furthermore, the experimental limits obtained from high redshift galaxies seem to indicate that this dynamo would have little time to act and would require field strengths of 10^{-6}G to 10^{-30}G, depending on the model [3].

Several scenarios have been studied to explain the possible mechanisms that may have given rise to these seed fields. The two common categories that are used to classify the possible methods of generating the seed fields are astrophysical sources which tend to amplify magnetic fields from stars and other astrophysical objects up to galactic lengths, and cosmological or primordial sources that

⋆e-mail: ayala@nucleares.unam.mx
⋆⋆e-mail: luis.hernandezr@correo.nucleares.unam.mx
⋆⋆⋆e-mail: jordissm@ciencias.unam.mx

© The Authors, published by EDP Sciences. This is an open access article distributed under the terms of the Creative Commons Attribution License 4.0 (http://creativecommons.org/licenses/by/4.0/).

propose distinct forms of generating large-scale magnetic fields before the galaxies are themselves formed, thus explaining at the same time the intracluster and void magnetic fields. One of the plausible cases of primordial generation of magnetic fields takes place during the electroweak phase transition (EWPT) when the Universe had temperatures of the order of 100 GeV. It is known that the EWPT is weakly first order as dictated by the Higgs mass, and can become stronger in the presence of magnetic fields [4, 5].

2 Dirac equation with a complex mass

The bubble nucleation during the EWPT separates the vacuum phases and the membrane separating the inside and outside of the bubble is called the wall. The properties of the wall depend on those of the electroweak effective potential. Taking into account the need for CP violation and the departure from thermal equilibrium from Sakharov's conditions, we set a complex mass that introduces explicit CP symmetry violation during the EWPT [8]. We cast the problem of fermion scattering off the bubble wall in terms of solving a Dirac equation with a position dependent fermion mass $m(z)$, proportional to the Higgs field. Explicitly,

$$m(z) = m_0 e^{i\phi}\Theta(z), \tag{1}$$

where ϕ is a phase, $\Theta(z)$ is the step function and z is the axis perpendicular to the bubble wall. As CP-violation implies that left- and right-handed chirality modes of a fermion spinor ψ couple to the Higgs field with $m(z)$ and $m^*(z)$, respectively, the Dirac equation is written as

$$\left\{ i\partial\!\!\!/ - m^* \frac{1}{2}(1-\gamma^5) - m\frac{1}{2}(1+\gamma^5) \right\} \Psi = 0. \tag{2}$$

3 Solution of the Dirac equation

Each of the fermion spinor components in Eq. (2) have to satisfy the Klein-Gordon equation. Therefore, a general solution is of the form

$$\left\{ i\partial\!\!\!/ + m^* \frac{1}{2}(1-\gamma^5) + m\frac{1}{2}(1+\gamma^5) \right\} \mathit{\Phi} = 0, \tag{3}$$

and when Eq. (3) is substituted into Eq. (2), we obtain the corresponding KG-like equations

$$\left\{ -\partial\!\!\!/^2 - i\gamma^3 m_0 e^{-i\phi}\delta(z)\frac{1}{2}(1-\gamma^5) - i\gamma^3 m_0 e^{i\phi}\delta(z)\frac{1}{2}(1+\gamma^5) - m_0^2\Theta(z) \right\} \mathit{\Phi} = 0. \tag{4}$$

By writing $\mathit{\Phi}(\vec{x},t)$ as $\mathit{\Phi}(\vec{x},t) = \xi(x,y)e^{-iEt}\Phi(z)$ and focusing on the solution for the z-direction we get

$$\left\{ E^2 - m_0^2\Theta(z) + d_z^2 - im_0\delta(z)\gamma^3 \left[e^{-i\phi}\frac{1}{2}(1-\gamma^5) + e^{i\phi}\frac{1}{2}(1+\gamma^5) \right] \right\} \Phi(z) = 0. \tag{5}$$

Expanding the function $\Phi(z)$ into a linear combination of the eigenspinors of γ^3 which are

$$u_\pm^1 = \begin{pmatrix} 1 \\ 0 \\ \pm i \\ 0 \end{pmatrix}, \qquad u_\pm^2 = \begin{pmatrix} 0 \\ 1 \\ 0 \\ \mp i \end{pmatrix}$$

where $\{1,2\}$ correspond to the spin projection and $\{+,-\}$ to the positive and negative energies, respectively. Using these spinors, $\Phi(z)$ can be written as

$$\Phi(z) = \Phi_+^1(z)u_+^1 + \Phi_-^1(z)u_-^1 + \Phi_+^2(z)u_+^2 + \Phi_-^2(z)u_-^2. \tag{6}$$

By substituting this into Eq. (5), we obtain the system of coupled equations

$$[E^2 - m_0^2\Theta(z) + d_z^2 + m_0\delta(z)\cos\phi]\Phi_+^1(z) = -im_0\delta\sin\phi\Phi_-^1(z) \tag{7}$$
$$[E^2 - m_0^2\Theta(z) + d_z^2 - m_0\delta(z)\cos\phi]\Phi_-^1(z) = im_0\delta\sin\phi\Phi_+^1(z) \tag{8}$$

with a similar set for Φ_+^2 and Φ_-^2. We note that when the fermion is in each of the phases, the equations decouple. The solution of the system of decoupled equations is given in terms of plane waves. We physically foresee that coefficients $G_\pm$ that represent an incoming wave from the RHS to the delta potential must be null. Solving the system of equations for the coefficients and implementing the continuity conditions, we get the following solutions.
For $z < 0$,

$$\begin{aligned}
\Phi_+^{1,2} &= A_+e^{iEz} + \left[\left(\frac{im_0\cos\phi}{E+\sqrt{E^2-m_0^2}}\right)A_+ - \left(\frac{m_0\sin\phi}{E+\sqrt{E^2-m_0^2}}\right)A_-\right]e^{-iEz},\\
\Phi_-^{1,2} &= A_-e^{iEz} + \left[\left(\frac{m_0\sin\phi}{E+\sqrt{E^2-m_0^2}}\right)A_+ - \left(\frac{im_0\cos\phi}{E+\sqrt{E^2-m_0^2}}\right)A_-\right]e^{-iEz},
\end{aligned} \tag{9}$$

For $z > 0$,

$$\begin{aligned}
\Phi_+^{1,2} &= \left[\left(1+\frac{im_0\cos\phi}{E+\sqrt{E^2-m_0^2}}\right)A_+ - \left(\frac{m_0\sin\phi}{E+\sqrt{E^2-m_0^2}}\right)A_-\right]e^{i\sqrt{E^2-m_0^2}z},\\
\Phi_-^{1,2} &= \left[\left(\frac{m_0\sin\phi}{E+\sqrt{E^2-m_0^2}}\right)A_+ + \left(1-\frac{im_0\cos\phi}{E+\sqrt{E^2-m_0^2}}\right)A_-\right]e^{i\sqrt{E^2-m_0^2}z}.
\end{aligned} \tag{10}$$

4 Transmission coefficients

The fermion spinor $\psi(z)$ can be decomposed into its incident, reflected and transmitted components from which we can extract the corresponding currents. With these currents, it is also possible to calculate the transmission and reflection coefficients defined as per $R = J_{\text{ref}}/J_{\text{inc}}$ and $T = J_{\text{tra}}/J_{\text{inc}}$. Moreover, by applying the chiral projectors to the spinor, we can specifically obtain the distinct chirality transmission coefficients T^L and T^R which are found to be

$$\begin{aligned}
T^L &= \frac{1}{E\left(E+\sqrt{E^2-m_0^2}\right)^2}\cdot\left[+m_0^3\sin\phi\cos\phi + \sqrt{E^2-m_0^2}\left(E+\sqrt{E^2-m_0^2}\right)(E+m_0\sin\phi)\right]\\
T^R &= \frac{1}{E\left(E+\sqrt{E^2-m_0^2}\right)^2}\cdot\left[-m_0^3\sin\phi\cos\phi + \sqrt{E^2-m_0^2}\left(E+\sqrt{E^2-m_0^2}\right)(E-m_0\sin\phi)\right].
\end{aligned} \tag{11}$$

Equation (11) contains the information on the ratio of the number of fermions for each chiral mode being transmitted through the wall to the incident ones. The asymmetry between the modes is quantified in terms of the phase ϕ. Figure 1 displays the chiral transmission coefficients for the trivial case $\phi = 0$ as a function of energy, where both probabilities coincide, as expected. Figure 2 shows the left- and right- handed transmission coefficient difference for two distinct values of ϕ, demonstrating the asymmetry between both quantities.

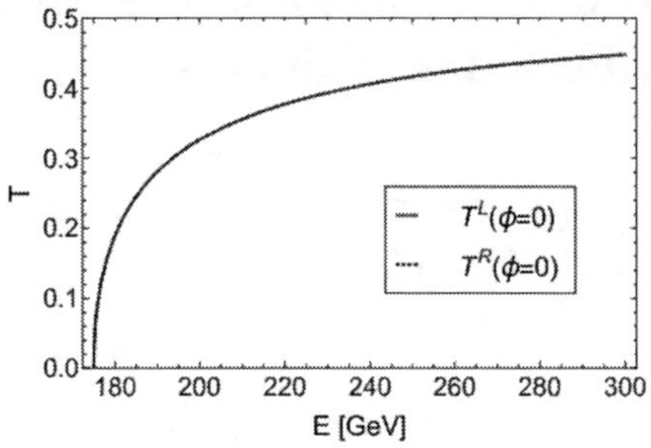

Figure 1. Chiral transmission coefficients for the case where $\phi = 0$. As expected, there is no asymmetry present between both quantities.

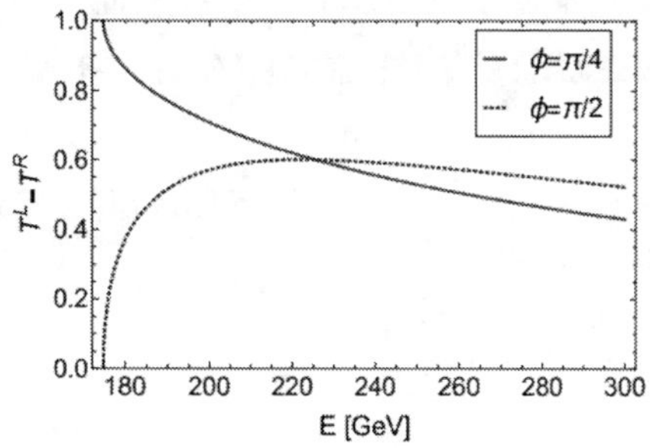

Figure 2. Chiral transmission coefficients difference for $\phi = \pi/4$ and $\phi = \pi/2$. The asymmetry is energy- and phase-dependent.

5 Summary and conclusions

In the thin-wall approximation, where the fermion mean free path is far greater than the bubble wall thickness, fermions behave as particles interacting only with the walls of a bubble nucleated during the EWPT. We have showed that the introduction of a CP-violating interaction with the bubble through a complex phase in the fermion mass can generate a net injection of chiral flux that can be quantified in terms of the difference between chiral transmission coefficients. The above implies that for fermions with an electric charge q, a local electric current transverse to the wall is generated.

In the case considered in this work, electric currents are created normal to the bubble walls with varying intensity, dependent to the particle's energy. These inhomogeneities can induce a helical magnetic field, a property that is crucial for possible future amplification processes [6]. The asymmetry is energy dependent and maximal for $\phi \simeq \pi/2$ and can be thought of to come from the complex phase in the CKM matrix. Although the magnitude of such a phase is consistent with current observations [7], one can think that the asymmetry might not be maximal.

References

[1] R. Durrer, A. Neronov, Astron. Astrophys. Rev. **21**, 62 (2013)
[2] A. Zucca, Y. Li, . Pogosian, Phys. Rev. D **95**, 063506 (2017)
[3] T. Kahniashvili, A. Brandenburg,A. G. Tevzadze, Phys. Scr. **91**, 104008 (2016)
[4] K. Subramanian, Rep. Prog. Phys. **79**, 076901 (2016)
[5] M. E. Carrington, Phys. Rev. D **45**, 2933 (1992)
[6] A. Brandemburg, K. Subramanian, Phys. Rept. **417**, 1-209 (2005)
[7] C. Patrignani, *et. al.* (Particle Data Group), Chin. Phys. C **40**, 100001 (2016)
[8] A. Ayala, L. A. Hernández, J. Salinas, Phys. Rev. D **95**, 123004 (2017)

Using the Linear Sigma Model with quarks to describe the QCD phase diagram and to locate the critical end point

Alejandro Ayala[1], *Jorge David* Castaño-Yepes[1], *José Antonio* Flores[1,⋆], *Saúl* Hernández[1] and *Luis* Hernández.[1]

[1] *Instituto de Ciencias Nucleares, Universidad Nacional Autónoma de México, Apartado Postal 70-543, Ciudad de México 04510, México.*

Abstract. We study the QCD phase diagram using the linear sigma model coupled to quarks. We compute the effective potential at finite temperature and quark chemical potential up to ring diagrams contribution. We show that, provided the values for the pseudo-critical temperature T_c = 155 MeV and critical baryon chemical potential $\mu_{Bc} \simeq 1$ GeV, together with the vacuum sigma and pion masses. The model couplings can be fixed and that these in turn help to locate the region where the crossover transition line becomes first order.

1 Introduction

The description of the QCD phase diagram on the T and μ plane reveals profound information for the different phases of strongly interacting matter under extreme conditions such that high temperatures and densities. Most of our knowledge of this phase diagram is restricted to the region for low values of μ. Lattice QCD has found values for a crossover transition with a critical temperature $T_c \sim 155$ MeV considering $2 + 1$ quark flavours Ref. [1]. On the other hand, effective models find that for $T \sim 0$ there is a first order phase transition Ref. [2]. This means that there must be a point in the diagram where both transitions converge and such point is generally refered to as the critical end point (CEP). In this work we used the Linear Sigma Model coupled to quarks (LSMq) to locate the CEP. We organize the content as follows: In Sec. 2, we give an overview of the main properties of LSMq. In Sec. 3, we show the effective potential at high and low-temperature. In Sec. 4, we use the effective potential to determine the coupling constants and to locate the CEP. Finally we summarize and conclude in Sec 5.

2 Linear Sigma Model coupled to quarks

We study the restoration of the chiral symmetry using an effective model that accounts for the physics of the spontaneous symmetry breaking at finite temperature and density, the linear sigma model. In order to account to the fermion degrees of freedom around the phase transition, we also include quarks in this model. The Lagrangian for this model is given by

⋆e-mail: jose.flores@correo.nucleares.unam.mx

© The Authors, published by EDP Sciences. This is an open access article distributed under the terms of the Creative Commons Attribution License 4.0 (http://creativecommons.org/licenses/by/4.0/).

$$\mathcal{L} = \frac{1}{2}(\partial_\mu \sigma)^2 + \frac{1}{2}(\partial_\mu \vec{\pi})^2 + \frac{a^2}{2}(\sigma^2 + \vec{\pi}^2) - \frac{\lambda}{4}(\sigma^2 + \vec{\pi}^2)^2 + i\bar{\psi}\gamma^\mu \partial_\mu \psi - g\bar{\psi}(\sigma + i\gamma_5 \vec{\tau}\cdot\vec{\pi})\psi \quad (1)$$

where ψ is an SU(2) isospin doublet, σ is an isospin singlet and $\vec{\pi}$ is an isospin triplet. λ is the boson's self-coupling and g is the fermion-boson coupling. $a^2 > 0$ is the mass parameter. The Lagrangian admits a broken symmetry vaccum solution given by the minimum of the classical potential when λ is positive, this means that the sigma field σ develops a vacuum expectation value v that becomes in the order parameter of the theory. Shifting the field as $\sigma + v_0$, notice the three pions, sigma and the constituent quarks develop the vacuum masses

$$m_\sigma^2 = 3\lambda v^2 - a^2, \quad m_\pi^2 = \lambda v^2 - a^2,$$
$$m_f = gv, \quad (2)$$

respectively. We take their conservative vacuum values as $m_\sigma = 450$ MeV, $m_\pi = 140$ MeV, $m_f = 300$ MeV Ref. [3]. Finally, we can fix the value of a using these vacuum values togheter with the first of Eqs. (2)

$$a = \sqrt{\frac{m_\sigma^2 - 3m_\pi^2}{2}}. \quad (3)$$

3 Effective potential

We compute the effective potential and the self-energies in the limit where the masses and quark chemical potential are small compared to the temperature, and we include contributions up to ring diagrams. Taking the renormalization scale as $\tilde{\mu} = ae^{-1/2}$, see Ref. [4]. For low temperatures, we compute up to the 1-loop correction which is given by the expressions in Ref. [5]. To compute 1-loop contribution within both approximations, we start from the following expressions

$$\begin{aligned} V_{\rm boson} &= T\sum_{i=\sigma,\vec{\pi}}\sum_{n=-\inf}^{\inf}\int\frac{d^3k}{(2\pi)^3}\ln\left(\frac{1}{k^2+m_i^2+\omega_n^2}\right), \\ V_{\rm fermion} &= T\sum_{i=\sigma,\vec{\pi}}\sum_{n=-\inf}^{\inf}\int\frac{d^3k}{(2\pi)^3}\ln\left(\frac{1}{k^2+m_i^2+(\tilde{\omega}_n-i\mu)^2}\right), \end{aligned} \quad (4)$$

where $\omega_n = 2n\pi T$ and $\tilde{\omega}_n = (2n+1)\pi T$.

3.1 Effective potential at high temperature

The effective potential in the high temperature approximation is computed in Ref. [3] and is given explicitly by

$$\begin{aligned} V_{eff}^h &= -\frac{1}{2}a^2v^2 + \frac{1}{4}\lambda v^4 - \sum_{i=\sigma,\pi}\left\{\frac{m_i^4}{64\pi^2}\left[\ln\left(\frac{a^2}{4\pi T^2}\right)+\frac{1}{2}-\gamma_E\right]\right\} + N_cN_f\frac{(gv)^4}{16\pi^2}\left[\ln\left(\frac{4\pi^2a^2}{(gv)^2}\right)+\frac{1}{2}-\gamma_E\right] \\ &+ \sum_{i=\sigma,\pi}\left\{\frac{T^2m_i^2}{24}-\frac{\pi^2T^2}{90}-\frac{T}{12\pi}[(m_i)^2+\Pi(T,\mu)]^{3/2}\right\} \\ &- N_cN_f\frac{T}{\pi^2}\int_0^\infty k^2\left[\ln\left(1+\exp\left(-\frac{\sqrt{k^2+(gv)^2}-\mu}{T}\right)\right)+\ln\left(1+\exp\left(-\frac{\sqrt{k^2+(gv)^2}+\mu}{T}\right)\right)\right] \end{aligned} \quad (5)$$

where the self-energy has the expression

$$\Pi[T,\mu] = \underbrace{\frac{\lambda T^2}{2}}_{bosons} \underbrace{-\frac{N_c N_f g^2 T^2}{\pi^2}\left[Li_2(-e^{\frac{\mu}{T}}) + Li_2(-e^{\frac{-\mu}{T}})\right]}_{fermions}. \tag{6}$$

3.2 Effective potential at low temperature

The effective potential in the low temperature approximation is calculated in the same fashion as Ref. [4] and the result is

$$\begin{aligned} V^l_{eff} &= -\frac{1}{2}a^2 v^2 + \frac{1}{4}\lambda v^4 - \sum_{i=\sigma,\pi}\left\{\frac{m_i^4}{64}\left[\ln\left(\frac{4\pi a^2}{(b\mu)+\sqrt{(b\mu)^2-m_i^2}}\right) - \gamma_E + \frac{1}{2}\right]\right. \\ &- \left.\frac{1}{96\pi^2}(b\mu)\sqrt{(b\mu)^2-m_i^2} - \frac{7\pi^2 T^4}{180}\frac{\mu(2(b\mu)^2-5m_i)}{((b\mu)^2-m_i^2)^{3/2}}\right\} \\ &+ N_f N_c\left[\ln\left(\frac{4\pi^2 a^2}{(\sqrt{\mu^2-(gv)^2}+\mu)^2}\right) + \frac{1}{2} - \gamma_E\right] \\ &+ N_f N_c\left\{\frac{\mu\sqrt{\mu^2-(gv)^2}(2\mu^2-5(gv)^2)}{24\pi^2} + \frac{1}{6}T^2\mu\sqrt{\mu^2-(gv)^2} + \frac{7\pi T^2}{360}\frac{\mu(2\mu^2)-5(gv)^2}{[\mu^2-(gv)^2]^{3/2}}\right\}, \end{aligned} \tag{7}$$

where the boson chemical potential μ_b is taken as a fraction b of the quark chemical potential, namelly, $b = \mu_b/\mu_q$ and the baryon chemical potential is given by $\mu_B = 3\mu_q$. The boson chemical potential μ_b provides the information on the average number of boson particles interacting between both phases at high densities.

4 Coupling Constants and location of CEP

To determine a unique transition curve of the QCD phase diagram, we compute the coupling constants and the critical values (μ_{Bc}, T_c), in order to determine the parameters λ and g, we solve a system of equations obtained from the conditions

$$\left.\frac{dV^l_{eff}}{dv}\right|_{v=0,T=0,\mu=\mu_c} = \left.\frac{dV^l_{eff}}{dv}\right|_{v=v_1,T=0,\mu=\mu_c} = 0 \tag{8}$$

$$V^l_{eff}(0,0,\mu_c) = V^l_{eff}(v_1,0,\mu_c) \tag{9}$$

$$\left.[m_\pi^2(v)+\Pi(T,\mu)]\right|_{v=0,T=T_c,\mu=0} = \left.[m_\pi^2(v)+\Pi(T,\mu)]\right|_{v=(m_q/3),T=0,\mu=\mu_c} = 0 \tag{10}$$

where the critical temperature is taken as T_c = 170 Mev for 2 light flavors, see Ref. [7], the critical quark chemical potential $\mu_c \sim$ 340 MeV Ref. [8], the dynamical quark mass m_q = 300 MeV and the v_0 is the expectation value of sigma. Notice that on the boundary of the first order phase transitions the effective potential shows two minima, one at $v = 0$ and the other at $v = v_1$. By solving the Eqs. (8), (9) and (10) the coupling constants are $\lambda = 0.897$ and $g = 1.57$. For details see Ref. [9]. Finally, we consider an interpolation between both approximations of the effective potential finding the lines that correspond to the phase order transitions. The region where they converge locates the position of the CEP which is approximately given by $(\mu^{CEP}/T_c, T^{CEP}/T_c) \sim (0.993, 0.113)$.

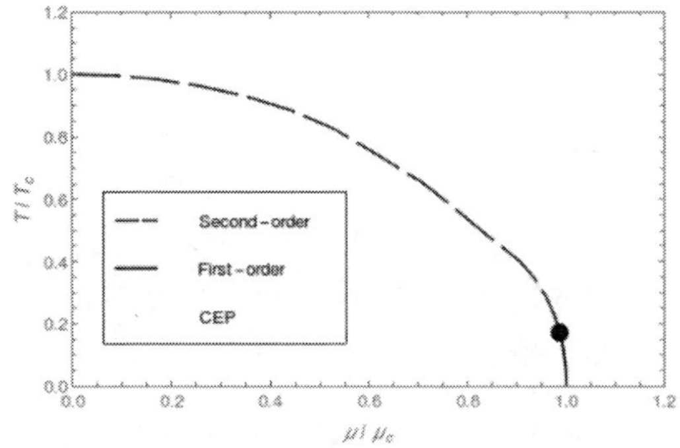

Figure 1: Effective QCD phase diagram obtained for $\lambda = 0.897$ and $g = 1.57$, taking the critical temperature for two light flavors as $T_c = 170$ MeV at $\mu = 0$ MeV and critial quark chemical potential $\mu_{qc} = 340$ MeV at $T = 0$ MeV. The dashed line represents the second order transition and the solid line represents the first order transition. The CEP is located at $(\mu^{CEP}/T_c, T^{CEP}/T_c) \sim (0.993, 0.113)$.

5 Conclusion

In this work we have explored the QCD phase diagram using the LSMq considering the approximation of the effective potential in the high temperature up to ring diagrams order and in the low temperature working up to 1-loop correction. The phase diagram derived within these approximation gives us information on the location of the CEP and we conclude that the LSMq is a suitable effective model to describe the phase transition in the temperature and density plane to understand much better the chiral restoration symmetry.

References

[1] T. Bhattacharya, *et al.*, Phys. Rev. Lett. **113**, 082001 (2014).

[2] M. Asakawa and K. Yazaki, Nucl. Phys. A **504**, 668 (1989); A. Barducci, R. Casalbuoni, S. De curtis, R. Gato and G. Pettini, Phys. Lett. B **231**, 463 (1989); Phys. Rev. D **41**, 1610 (1990).

[3] C. Pratignani, *et al.* (Particle Data Group), Chin. Phys. **C** 40, 100001 (2016).

[4] C. O. Dib, O. R. Espinosa, Nucl. Phys. B **612**, 492-518 (2001).

[5] M. Le Bellac, *Thermal Field Theory* (Cambridge University Press, Cambridge, 1996).

[6] A. Ayala, C. A. Bashir, J. J. Cobos-Martinez, S. Hernandez-Ortiz, A. Raya, Nucl. Phys. B **897**, 77-86 (2015).

[7] Y. Maezawa, S. Aoki, S. Ejiri, T. Hatsuda, N. Ishii, K. Kan aya, N. Ukita, J. Phys. G **34**, S651 (2007).

[8] R. Hagedorn, Nuovo Cim. Suppl. **3** (1965) 147; Nuovo Cim. A **56** (1968) 1027.

[9] A. Ayala, L. Hernández, S. Hernández Ortíz, e-print arXiv: 1710.09007 [hep-ph].

QGP hydrodynamical study using energy-momentum in-medium deposition by an extended source.

Julio Cesar Maldonado-Gonzalez[1,⋆], *Alejandro* Ayala[2], *Isabel* Dominguez[3], and *Maria Elena* Tejeda-Yeomans[1,4]

[1] *Departamento de Investigación en Física, Universidad de Sonora, Boulevard Luis Encinas J. y Rosales, Colonia Centro, Hermosillo, Sonora 83000, México.*
[2] *Instituto de Ciencias Nucleares, Universidad Nacional Autónoma de México, Apartado Postal 70-543, México Distrito Federal 04510, México.*
[3] *Facultad de Ciencias Físico-Matemáticas, Universidad Autónoma de Sinaloa, Avenida de las Américas y Boulevard Universitarios, Ciudad Universitaria, C.P. 80000, Culiacán, Sinaloa, México.*
[4] *Departamento de Física, Universidad de Sonora, Boulevard Luis Encinas J. y Rosales, Colonia Centro, Hermosillo, Sonora 83000, México.*

Abstract. The quark-gluon plasma (QGP) is created under extreme conditions, such as the ones prevailing in heavy ion collisions. The characterization of the QGP can be done using high-p_T probes such as the partons that are created through hard scatterings in the fireball. These fast-moving partons lose energy and momentum along their traveled path through the medium. The parton deposition of energy-momentum creates an in-medium disturbance that can be described using approximations within relativistic hydrodynamics in a defined regime of the QGP evolution. Based on earlier research in this field, we study the use of extended sources that depend on the location of the parton-jet in the initial stages of the QGP evolution. We explore this approach as a way to complement the current numerical landscape of hydrodynamical QGP studies and to eventually generate initial conditions that can be used as input of hydrodynamical numerical simulations.

1 Introduction

Current experimental evidence points to the creation of the quark-gluon plasma in heavy ion experiments such as the ones taking place at the Large Hadron Collider (LHC) at CERN and the Relativistic Heavy Ion Collider (RHIC) at BNL. These experiments have allowed a more detailed characterization of the QGP, mainly in the transition phase from quarks to hadrons: they have discovered the QGP to be a strongly-coupled fluid, with low viscosity [1]. In these heavy-ion collisions, hadrons are created and they reach the detectors with large collective velocity. In order to characterize the QGP properties, we can use high p_T probes such as hard-scattering partons, that are created in the bulk of the plasma and that travel across it, loosing energy and momentum. In this work we report on the energy-momentum deposition of partons modeled as extended sources and compare with previous results.

⋆e-mail: julio.maldonadog@gmail.com

© The Authors, published by EDP Sciences. This is an open access article distributed under the terms of the Creative Commons Attribution License 4.0 (http://creativecommons.org/licenses/by/4.0/).

2 Perturbing the medium in local equilibrium

In ultra-relativistic heavy-ion collisions the medium never reaches the equilibrium state, but it is possible to consider it in local equilibrium. Local conservation laws guarantee that $\partial_\mu \Theta_0^{\mu\nu} = 0$ and $\Theta_0^{\mu\nu} = -Pg^{\mu\nu} + (\epsilon + P)u^\mu u^\nu$, where the medium 4-velocity is $u^\mu = (1, \mathbf{u})$ and where the momentum density $\mathbf{g}$ connects with the 3-velocity as $\mathbf{u} = \mathbf{g}/\epsilon_0(1 + c_s^2)$. ϵ_0 is the static background energy density and c_s is the sound velocity. Assuming that the disturbance introduced by the parton is small, the medium energy-momentum tensor can be written as $\Theta^{\mu\nu} = \Theta_0^{\mu\nu} + \delta\Theta^{\mu\nu}$, where $\Theta_0^{\mu\nu}$ is the equilibrium energy-momentum tensor and $\delta\Theta^{\mu\nu}$ is the perturbation made by the parton [2]. The energy-momentum tensor satisfies $\partial_\mu \delta\Theta^{\mu\nu} = J^\mu$ and $\partial_\mu \Theta_0^{\mu\nu} = 0$ where J^μ is the source of the disturbance. If we consider a first order approximation in shear (η) and ignoring bulk viscosity effects, the perturbation to the energy-momentum tensor will have the following components,

$$\delta\Theta^{00} = \delta\epsilon, \quad \delta\Theta^{0i} = \mathbf{g}, \quad \delta\Theta^{ij} = \delta_{ij} c_s^2 \delta\epsilon - \tfrac{3}{4}\Gamma_s(\partial^i \mathbf{g}^j + \partial^j \mathbf{g}^i - \tfrac{2}{3}\delta_{ij}\nabla \cdot \mathbf{g}) \tag{1}$$

where the *sound attenuation length* is $\Gamma_s \equiv 4\eta/3\epsilon_0(1 + c_s^2)$ and $\epsilon(t, \mathbf{x}) = \epsilon_0 + \delta\epsilon(t, \mathbf{x})$ with ϵ_0 is the energy density of the background fluid, and $\delta\epsilon$ the disturbance energy density. Applying a Fourier Transform to Eq. (1), the longitudinal and transverse components of the momentum density as well as the energy density are given by

$$\mathbf{g}_L = \frac{i\left[\frac{\omega}{k^2}\mathbf{k} \cdot \mathbf{J}(\mathbf{k}, \omega) + c_s^2 J^0(\mathbf{k}, \omega)\right]}{\omega^2 - c_s^2 k^2 + i\Gamma_s \omega k^2}, \quad \mathbf{g}_T = \frac{i\mathbf{J}_T}{\omega + i\frac{3}{4}\Gamma_s k^2}, \quad \delta\epsilon = \frac{i\mathbf{k} \cdot \mathbf{J}(\mathbf{k}, \omega) + J^0(\mathbf{k}, \omega)(i\omega - \Gamma_s k^2)}{\omega^2 - c_s^2 + i\Gamma_s \omega k^2}, \tag{2}$$

where $\mathbf{g} = \mathbf{g}_L + \mathbf{g}_T$ and $\mathbf{g}_L = (\mathbf{g} \cdot \mathbf{k})\mathbf{k}/k^2$.

3 Localized versus extended source

The in-medium perturbation that the traveling parton generates can be modeled by a localized source $J^\nu(\mathbf{x}, t) = (dE/dx)v^\nu \delta(\mathbf{x} - \mathbf{v}t)$, as it was done in Ref. [2] where dE/dx is the parton energy-loss per unit length and $v^\nu = (1, \mathbf{v})$ is the parton velocity. In order to model a jet (instead of a parton) passing through the medium, an extended source was proposed in Ref. [3]. Inspired by this proposal, we study a simplified version of an extended source and we make a comparison with the results of Ref. [2] for a localized source. We now consider an extended source

$$J^\nu(\mathbf{x}, t) = \left(\frac{dE}{dx}\right) v^\nu \frac{1}{\left(\sqrt{2\pi}\sigma\right)^3} \exp\left\{-\frac{(\mathbf{x} - \mathbf{v}t)^2}{2\sigma^2}\right\}, \tag{3}$$

where the parameter σ allows us to control the extension of the energy-momentum deposit. For this study, we will consider $\mathbf{x} \cdot \mathbf{v} = |\mathbf{x}||\mathbf{v}| \cos\theta_0$, with fixed θ_0. The extended source of Eq. (3) in Fourier space is now,

$$J^\nu(\mathbf{k}, \omega) = \frac{1}{(2\pi)^4}\left(\frac{dE}{dx}\right)\frac{\sqrt{2\pi}\sigma}{v \sin^3\theta_0} v^\nu \exp\left\{-\frac{2\sigma^2}{v^2 \sin^2\theta_0}\left[\left(1 + \frac{1}{4}\sin^2\theta_0\right)\omega^2 - 2\mathbf{v} \cdot \mathbf{k}\,\omega + v^2 k^2\right]\right\}. \tag{4}$$

Using Eqs.(2) and (4), we can calculate, for example $(\mathbf{g}_T)_z$ as

$$(\mathbf{g}_T)_z = \frac{i}{(2\pi)^4}\left(\frac{dE}{dx}\right)\frac{\sqrt{2\pi}\sigma}{v \sin^3\theta_0} \int \frac{d^3k}{(2\pi)^3} \int \frac{d\omega}{2\pi} \frac{k_T^2/(k_T^2 + k_z^2)}{\omega + i\frac{3}{4}\Gamma_s(k_T^2 + k_z^2)} e^{-\frac{2\sigma^2}{v^2 \sin^2\theta_0}\left[\left(1 + \frac{1}{4}\sin^2\theta_0\right)\omega^2 - 2\mathbf{v}\cdot\mathbf{k}\,\omega + v^2 k^2\right]}.$$

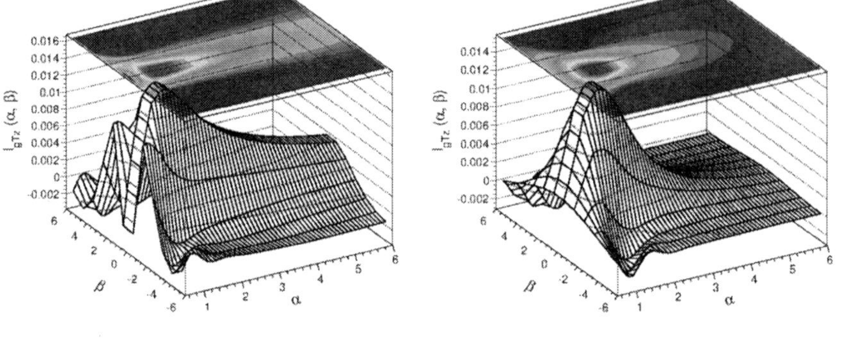

Figure 1: The integrand of $(\mathbf{g}_T)_z$ in Eq. (5) is shown on the left hand side for a localized source as was reported in Ref. [2] and, on the right hand side the resulting $(\mathbf{g}_T)_z$ with the present analysis of an extended source. We have performed the ξ integral in the $\alpha_{\min} \leq \alpha \leq +6$ and $-6 \leq \beta \leq +6$ range for $\alpha_{\min} = 0.5$.

The integrals over ω and $\mathbf{k}$ can be performed using similar methods as those reported in Ref. [2] where using cylindrical coordinates $\mathbf{k} = (k_T \cos\phi_k, k_T \sin\phi_k, k_z)$ and $\mathbf{x} = (x_T \cos\phi_x, x_T \sin\phi_k, z)$, we can describe the momenta both along and transverse to the traveled path, with $\mathbf{v} = v\hat{z}$. Using this and contour integration to calculate $(\mathbf{g}_T)_z$, $(\mathbf{g}_T)_y$ and $\delta\epsilon$ we arrive at

$$
\begin{aligned}
(\mathbf{g}_T)_z &= \frac{1}{(2\pi)^{\frac{11}{2}}}\left(\frac{dE}{dx}\right)\frac{\sigma}{2\sin^2\theta_0}\left(\frac{2v}{3\Gamma_s}\right)^3 \int_0^\infty d\xi\, \xi^2\, J_0(\beta\xi)\, \mathrm{e}^{-\alpha\xi}\,, \\
(\mathbf{g}_T)_y &= \frac{1}{(2\pi)^{\frac{11}{2}}}\left(\frac{dE}{dx}\right)\frac{\sigma}{2\sin^3\theta_0}\left(\frac{2v}{3\Gamma_s}\right)^3 \int_0^\infty d\xi\, \xi^2\, J_1(\beta\xi)\, \mathrm{e}^{-\alpha\xi}\,, \\
\delta\epsilon &= \frac{1}{(2\pi)^{\frac{11}{2}}}\left(\frac{dE}{dx}\right)\frac{9\pi\sigma}{4v\sin^3\theta_0}\left(\frac{2v}{3\Gamma_s}\right)^3 \int_0^\infty d\xi\, \left(A\xi^2 + B\xi + C\right) J_0(\beta\xi)\, \mathrm{e}^{-\alpha\xi}
\end{aligned}
\tag{5}
$$

where $A = -4\frac{\sigma^2}{\sin^2\theta_0}\left(\frac{2v}{3\Gamma_s}\right)\left(\frac{c_s}{v}\right)^2$, $B = -1$ and $C = \left(\frac{4v}{\Gamma_s}\frac{\sigma^2}{\sin^2\theta_0} + vt - \frac{3\Gamma_s}{2v} - \frac{v\alpha}{\Gamma_s}\right)\left(\frac{c_s}{v}\right)^2$ and where we have used the dimensionless variables $\xi \equiv \left(\frac{3\Gamma_s}{2v}\right)k_T$, $\alpha \equiv \left(\frac{3\Gamma_s}{2v}\right)^{-1}|z|$ and $\beta \equiv \left(\frac{3\Gamma_s}{2v}\right)^{-1}x_T$. In order to visualize the energy-momentum deposition, we construct 3D plots of the integrands $\mathrm{I}(\alpha,\beta)$ of Eq. (5), by performing a numerical integration on the dimensionless variable ξ. In Figure 1, on the left hand side we show the integrands of $(\mathbf{g}_T)_z$ for a localized source as was reported in Ref. [2] and, on the right hand side the resulting integrand of $(\mathbf{g}_T)_z$ with the present analysis of an extended source where we have integrated in the $\alpha_{\min} \leq \alpha \leq +6$ and $-6 \leq \beta \leq +6$ range for $\alpha_{\min} = 0.5$. Notice how the extended source gives a smother profile near the source, which allows for an easier numerical treatment of the collection of the signal. Figure 2, shows the integrands of $(\mathbf{g}_T)_y$ and $\delta\epsilon$ where we have performed the ξ integration in the same range as in the previous figure. On the left column we show the results for a localized source and on the right column the corresponding plots with the present analysis. For $\delta\epsilon$ we use $\Gamma_s = 0.06, \theta_0 = \pi/2$ and $\sigma = 0.001$. Again, the signal is smoother and for $\delta\epsilon$ the energy deposit is wider, which is reminiscent of a broad jet.

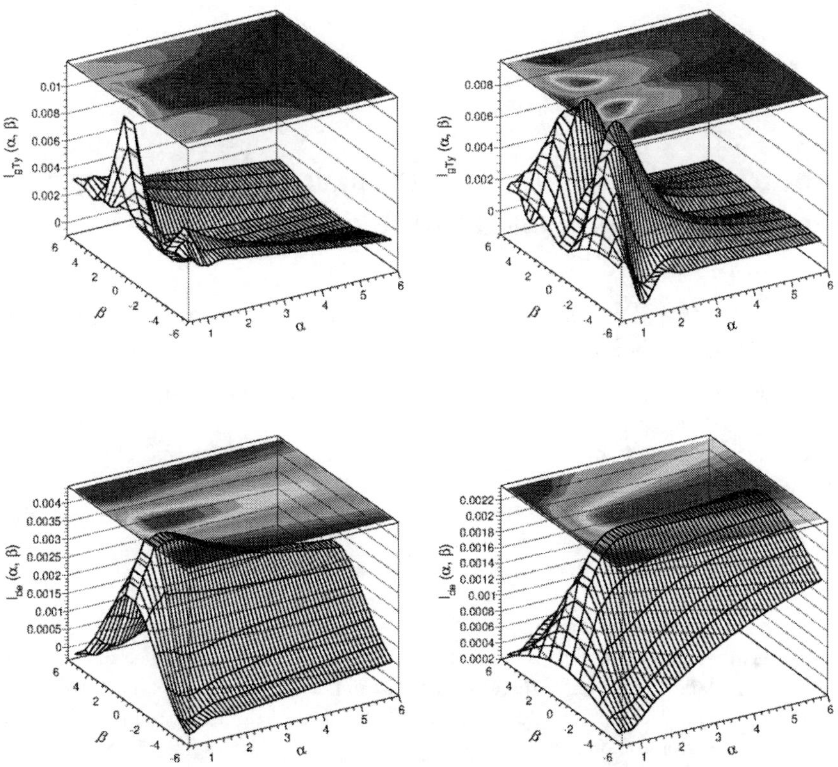

Figure 2: We plot the integrands of $(\mathbf{g}_T)_y$ and $\delta\epsilon$ in Eq. (5) where we have performed the ξ integration in the same range as in the previous figure. On the left column we show the results for a localized source as was reported in Ref. [2] and, on the right column the corresponding plots with the present analysis of an extended source. $(\mathbf{g}_T)_y$ and $\delta\epsilon$ for a localized source (left) and an extended source (right) with $\alpha_{\min} = 0.5$. For $\delta\epsilon$ we use $\Gamma_s = 0.06, \theta_0 = \pi/2$ and $\sigma = 0.001$.

In this work we report on the energy-momentum deposition of partons modeled as extended sources and compare with previous results, as a tool to study initial conditions for hydrodynamical numerical simulations and other relevant QGP observables, which will be reported elsewhere [4].

References

[1] Heavy ions and quark-gluon plasma, *home.cern/about/physics/heavy-ions-and-quark-gluon-plasma*, CERN (2017); L. Cifarelli, L.P. Csernai, and H. Stöcker, *The Quark-Gluon Plasma, a nearly perfect fluid*, Euro Physics News, DOI:10.1051/epn/2012206.

[2] A. Ayala, I. Dominguez, and M. E. Tejeda-Yeomans, Phys. Rev. C 88, 025203 (2013).

[3] B. Betz, J. Noronha, G. Torrieri, M. Gyulassy, I. Mishustin, and D. H. Rischke, Phys. Rev. C 79, 034902 (2009).

[4] A. Ayala, I. Dominguez, J. C. Maldonado-Gonzalez, M. E. Tejeda-Yeomans, in progress.

Promp photon yield and v_2 coefficent from gluon fusion induced by magnetic field in heavy-ion collisions

Jorge David Castaño-Yepes[1,★], *Alejandro* Ayala[1,2], *C. A.* Dominguez[2], *L. A.* Hernández[1], *Saúl* Hernández-Ortíz[1], and *María Elena* Tejeda-Yeomans[3]

[1] *Instituto de Ciencias Nucleares, Universidad Nacional Autónoma de México, Apartado Postal 70-543, México Distrito Federal 04510 México.*
[2] *Centre for Theoretical and Mathematical Physics, and Department of Physics, University of Cape Town, Rondebosch 7700, South Africa.*
[3] *Departamento de Física, Universidad de Sonora, Boulevard Luis Encinas J. y Rosales, Colonia Centro, Hermosillo, Sonora 83000, México*

Abstract. We compute the production of prompt photons and the v_2 harmonic coefficient in relativistic heavy-ion collisions induced by gluon fusion in the presence of an intense magnetic field, during the early stages of the reaction. The calculations take into account several parameters which are relevant to the description of the experimental transverse momentum distribution, and elliptic flow for RHIC and LHC energies. The main imput is the strenght of the magnetic field which varies in magnitude from 1 to 3 times the pion mass squared, and allows the gluon fusion that otherwise is forbidden in the absence of the field. The high gluon occupation number and the value of the saturation scale also play an important role in our calculation, as well as a flow velocity and geometrical factors. Our results support the idea that the origin of at least some of the photon excess observed in heavy-ion experiments may arise from magnetic field induced processes, and gives a good description of the experimental data.

1 Introduction

It is well established that in heavy-ion experiments carried out at the CERN Large Hadron Collider and at the BNL Relativistic Heavy-Ion Collider, magnetic fields with intensities of several times the pion mass squared are produced, both in central and peripheral collisions [1, 2]. Studies on the centrality-dependence of these short-lived magnetic fields, show that their intensity along the reaction plane is small compared with the intensity along the normal to the reaction plane.

The presence of a magnetic fields in a medium with high gluon occupation number [3] allows processes in which the photon production by gluon fusion can be achieved [4, 5]. This photon production mechanism together with the common sources calculated for syncroton radiation, bremsstrahlung, pair annihilation [6, 7] and modeled by hydrodynamical and transport calculations [8–10] could explain the experimentally measured photon excess at low momentum in the invariant

★e-mail: jorge.castano@correo.nucleares.unam.mx

© The Authors, published by EDP Sciences. This is an open access article distributed under the terms of the Creative Commons Attribution License 4.0 (http://creativecommons.org/licenses/by/4.0/).

momentum distribution [11–13].

On the other hand, the precense of a magnetic field breaks the spatial isotropy in the photon emission which must have consequences for the v_2 harmonic coefficient or elliptic flow. This coefficient has been also calculated by hydrodynamical models and compared with ALICE and PHENIX measurements but its agreement is yet incomplete [14].

In this work we compute the photon production and v_2 harmonic coefficient from the fusion of low momentum gluons in the presence of a magnetic field. Our perturbative scheme is valid during times when the magnetic field reaches its maximum values and the shattered glasma is highly occupied by gluons that can be described as quasiparticles [15]. These times are of order of $\tau_s \approx 1/\Lambda_s$ or $\Delta\tau_s \simeq 1.5$ fm, with Λ_s the saturation scale [16, 17]. We explore a region of magnetic fields between 1 to 3 times the pion mass squared and we include a phenomenological expansion factor.

2 Photon Production by gluon fusion

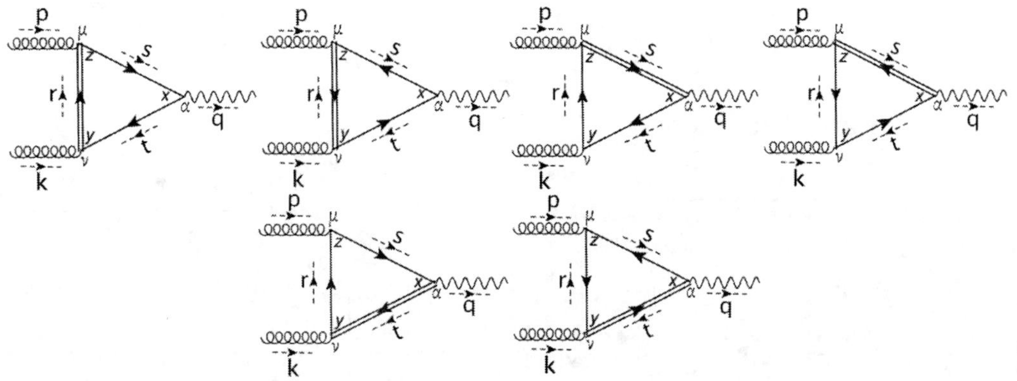

Figure 1. Dominant contribution for photon production by gluon fusion in presence of a magnetic field. The double lines represent that the corresponding propagator is in the first Landau Level $S^{(1)}$. The single lines represents the propagator in the lowest Landau Level $S^{(0)}$. The arrows in the propagators represent the direction of the flow of charge. The arrows at the sides of the propagator lines represent the momentum direction.

The sum of amplitudes of Fig. 1 is given by

$$
\begin{aligned}
\tilde{\mathcal{M}} &= -i(2\pi)^4\delta^{(4)}(q-k-p)\frac{|q_f|g^2\delta^{cd}e^{f(p_\perp,k_\perp)}}{32\pi(2\pi)^8}\left\{\left(g_\parallel^{\mu\alpha}-\frac{p_\parallel^\mu p_\parallel^\alpha}{p_\parallel^2}\right)h^\nu(a)-\left(g_\parallel^{\mu\nu}-\frac{p_\parallel^\mu p_\parallel^\nu}{p_\parallel^2}\right)h^\alpha(a)\right. \\
&+ \left.\left(g_\parallel^{\mu\nu}-\frac{k_\parallel^\mu k_\parallel^\nu}{k_\parallel^2}\right)h^\alpha(b)-\left(g_\parallel^{\alpha\nu}-\frac{k_\parallel^\alpha k_\parallel^\nu}{k_\parallel^2}\right)h^\mu(b)+\left(g_\parallel^{\alpha\nu}-\frac{q_\parallel^\alpha q_\parallel^\nu}{q_\parallel^2}\right)h^\mu(c)-\left(g_\parallel^{\mu\alpha}-\frac{q_\parallel^\mu q_\parallel^\alpha}{q_\parallel^2}\right)h^\nu(c)\right\} \\
&\times \epsilon_\mu(\lambda_p)\epsilon_\nu(\lambda_k)\epsilon_\alpha(\lambda_q). \qquad (1)
\end{aligned}
$$

with $h^\mu(a)=(i/\pi)\epsilon_{ij}a^i g_\perp^{j\mu}$, $a_i=p_i+2k_i+i\epsilon_{im}p_m$; $b_i=2p_i+k_i-i\epsilon_{im}k_m$, $c_i=k_i-p_i+i\epsilon_{im}(p_m+k_m)$ and

$$
f(p_\perp,k_\perp)=\frac{1}{8|q_fB|}\left(p_m-k_m+i\epsilon_{mj}(p_j+k_j)\right)^2-\frac{1}{2|q_fB|}\left(p_m^2+k_m^2+2i\epsilon_{jm}p_mk_j\right). \qquad (2)
$$

In order to compute Eq. (2) we considered one quark in the Lowest Landau Level ($S^{(0)}$) and two in the first excited Landau Level ($S^{(1)}$). Also we have been working in the massless quark aproximation and used the fact that the magnetic field is the dominan energy scale, i.e., $2|q_f B| \gg t_\parallel^2,\ s_\parallel^2,\ r_\parallel^2$.

Finally, when ignoring the magnetized medium dispersive properties, as in the present work, from energy-momentum conservation, the 4-vectors p and k need to be parallel to q: $p^\mu = \left(\omega_p/\omega_q\right) q^\mu$, $k^\mu = \left(\omega_k/\omega_q\right) q^\mu$, therefore the invariant photon momentum distribution is thus given by

$$\omega_q \frac{dN^{\rm mag}}{d^3q} = \frac{\chi V \Delta\tau_s}{2(2\pi)^3} \int \frac{d^3p}{(2\pi)^3 2\omega_p} \int \frac{d^3k}{(2\pi)^3 2\omega_k} n(\omega_p) n(\omega_k) (2\pi)^4 \delta^{(4)}(q-k-p) \frac{1}{4} \sum_{{\rm pol.}f} |\mathcal{M}|^2, \tag{3}$$

where the high distribution number is given as in Refs. [16, 19] and we sum over the three light flavors. The final result is shifted by the expansion factor $\omega_{p,k} \to (p,k)\cdot u$. For simplicity we allow for a constant flow velocity $u^\mu = \gamma(1,\beta)$, with $\gamma = 1/\sqrt{1-\beta^2}$. The coefficent v_2 results fom the Fourier decomposition of Eq. (3) and from the weighed average

$$v_2(\omega_q) = \frac{\frac{dN^{\rm mag}}{d\omega_q}(\omega_q)\, v_2^{\rm mag}(\omega_q) + \frac{dN^{\rm direct}}{d\omega_q}(\omega_q)\, v_2^{\rm direct}(\omega_q)}{\frac{dN^{\rm mag}}{d\omega_q}(\omega_q) + \frac{dN^{\rm direct}}{d\omega_q}(\omega_q)}. \tag{4}$$

3 Results and Discussion

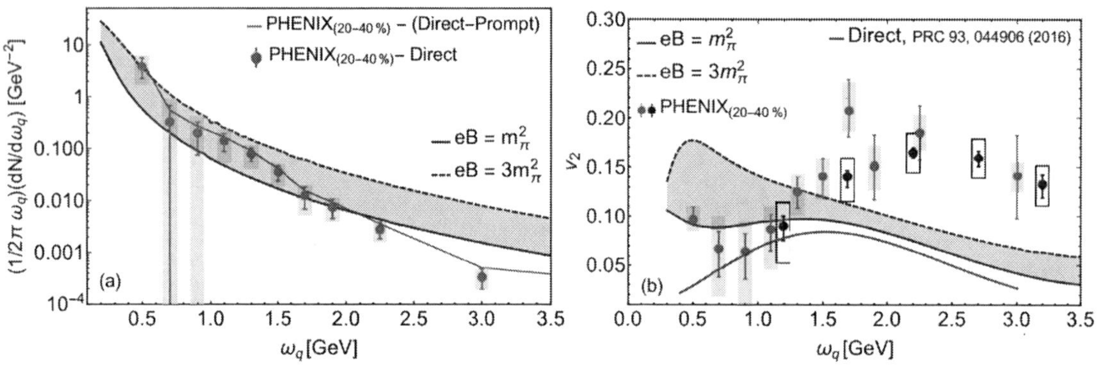

Figure 2. (Left) Difference between PHENIX photon invariant momentum distribution [11] and direct (points) or direct minus prompt (zigzag) photons from Ref. [8] compared to the yield from the present calculation. (Right) Harmonic coefficient v_2 combining the calculation of Ref. [8] and the present calculation compared to PHENIX data [12]. Curves are shown as functions of the photon energy for central rapidity and the centrality range 20-40%. Only the experimental error bars are shown. The bands show variations of the parameter eB within the indicated ranges and computed with $\alpha_s = 0.3$, $\Lambda_s = 2$ GeV, $\eta = 3$, $\Delta\tau_s = 1.5$ fm, $R = 7$ fm, $\beta = 0.25$ and $\chi = 0.8$.

Figure 2-(left) shows the results of Eq. (3) compared with the diference between PHENIX [11] data and the hydrodynamical calculations of Ref. [8], wich are represented by the points. If the present calculation is considered as a yield, it provides a very good description of the excess of photons. Figure 2-(right) shows coefficient v_2 from Eq. (4) and compared with the direct photon result of Ref. [8] together with our calculation, also compared to PHENIX data [12]. The curves are shown as functions of the photon energy for central rapidity and the centrality range 20% - 40%. We notice that the photon excess by gluon fusion helps to better describe the experimental data and highlights the importance of including the effects of magnetic fields in the early stages of the collision and its impact on the final state observables.

References

[1] D. E. Kharzeev, L. D. McLerran and H. J. Warringa, Nucl. Phys. A **803**, 227 (2008); V. Skokov, A.Y. Illarionov, V. Toneev, Int. J. Mod. Phys. A **24**, 5925 (2009); V. Voronyuk, V. D. Toneev, W. Cassing, E. L. Bratkovskaya, V. P. Konchakovski, S. A. Voloshin, Phys. Rev. C **83**, 054911 (2011); L. McLerran, V. Skokov, Nucl. Phys. A **929**, 184-190 (2014).

[2] A. Bzdak, V. Skokov, Phys. Lett. B **710**, 171-174 (2012).

[3] L. McLerran and B. Schenke, Nucl. Phys. A **946**, 158 (2016).

[4] A. Ayala, J. D. Castaño-Yepes, C. A. Dominguez, and L. A. Hernandez, EPJ Web Conf. **141**, 02007 (2017).

[5] A. Ayala, J. D. Castaño-Yepes, C. A. Dominguez, L. A. Hernandez, S. Hernández-Ortíz, and M. E. Tejeda-Yeomans, Phys. Rev. D **96**, 014023 (2017).

[6] B. G. Zakharov, Eur. Phys. J. C **76**, 609 (2016).

[7] K. Tuchin, Phys. Rev. C **91**, 014902 (2015).

[8] J.-F. Paquet, C. Shen, G. S. Denicol, M. Luzum, B. Schenke, S. Jeon, C. Gale, Phys. Rev. C **93**, 044906 (2016).

[9] H. van Hees, M. He, R. Rapp, Nucl. Phys. A **933**, 256Ð271 (2015).

[10] O. Linnyk, V. Konchakovski, T. Steinert, W. Cassing, E. L. Bratkovskaya, Phys. Rev. C **92**, 054914 (2015).

[11] A. Adare *et al.* [PHENIX Collaboration], Phys. Rev. C **91**, 064904 (2015).

[12] A. Adare *et al.* [PHENIX Collaboration], Phys. Rev. C **94**, 064901 (2016).

[13] J. Adam *et al.* [ALICE Collaboration], Phys. Lett. B **754**, 235 (2016).

[14] C. Shen, Nucl. Phys. A **956**, 184-191 (2016).

[15] J. Berges, K. Reygers, N. Tanji, and R. Venugopalan, Phys. Rev. C **95**, 054904 (2017).

[16] L. McLerran and B. Schenke, Nucl. Phys. A **929**, 71-82 (2014).

[17] T. Lappi, Proceedings, 10th Workshop on Non-Perturbative Quantum Chromodynamics: Paris, France, June 8-12, 2009, ECONF C0906083, 27 (2009); T. Lappi and L. McLerran, Nucl. Phys. A**772**, 200 (2006); H. Kowalski, T. Lappi and R. Venugopalan, Phys. Rev. Lett. **100**, 022303 (2008).

[18] J. Schwinger, Phys. Rev. **82**, 664 (1951).

[19] A. Krasnitz, Y. Nara, R. Venugopalan, Phys. Rev. Lett. **87**, 192302 (2001).

EPJ Web of Conferences **172**, 08005 (2018) https://doi.org/10.1051/epjconf/201817208005
ISMD 2017

Performance of BEBE-prototype: A BEam-BEam counter prototype for the MPD-NICA experiment at JINR

Cristian Heber Zepeda Fernández[1,⋆]

[1] *Centro de Investigación y de Estudios Avanzados del I.P.N.*

Abstract. In this work we show the arrival time resolution for the Beam Monitoring Detector (BMD). We made the study for Au+Au collision at $\sqrt{s} = 8$ Gev and a smearing of $\sigma = 300$ cm. The arrival time resolution we found is $\Delta\sigma = 57.982 \pm 0.509$ ps. We show preliminary results of the time resolution for a cell of the BMD.

1 Introduction

The Nuclotron-based Ion Collider fAcility (NICA) is a new accelerator at the Joint Institute for Nuclear Research (JINR) in Dubna, Russia. Its field of study will be the properties of dense baryonic matter. NICA will have three experiments (or detectors):

1. The Baryonic Matter at Nuclotron (BM@N).
2. The Spin Physics Detector (SPD).
3. The MultiPurpose Detector (MPD).

Each one has its own physics of study. In Figure 1 is shown a representation of NICA.
In this report we show the work we did for MPD, which consisting of the implementation of a new detector (see sections below). The MPD (as its name says) has several physics studies. One of the principal field is to measurements of the production of strange particles (particles conformed by the *strange* quark). This study is due to that they can give information about the medium and particle production mechanism. To know more about the detectors of MPD see [1].
To make MPD measurements more accurate, will be added a new detector: **The Beam Monitoring Detector (BMD)**, which will be two scintillator detectors. The main goal of BMD will be increase the pseudorapidity acceptance of MPD. Other measurements that BMD can do, are listed below.

- Optimization of events: Plane resolution.
- Centrality.
  - Interaction point location.
- Multiplicity reference estimator.
- Trigger system.

⋆e-mail: hzepeda@fis.cinvestav.mx

© The Authors, published by EDP Sciences. This is an open access article distributed under the terms of the Creative Commons Attribution License 4.0 (http://creativecommons.org/licenses/by/4.0/).

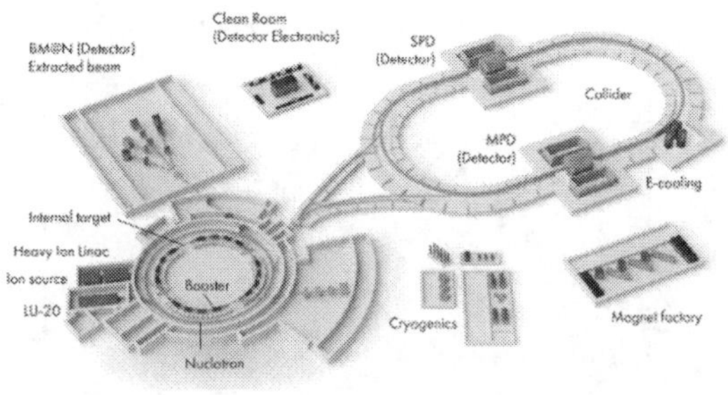

Figure 1. A picture of NICA. It is shown the three experiments and the rest of the components.

- Beam monitoring.
- Discriminate centrality events from background and beam-gas interaction.
- Determinate the absolute cross section of reaction process.

The name we will use to refer to the two detectors are "BMD-A" and "BMD-C". In the next section the properties of BMD will be descibed.

2 Characteristics of BMD

BMD-A and BMD-C will be located at -200 cm and 200 cm, respectively, with respect to the geometric center of MPD. For this study we used a circular geometry for each detector, whose minimum radius is 5.1 cm, maximum radius is 73.63 cm and the width is 1.25 cm, for each one. Each detector is divided in 5 rings and at the same time each ring is divided into 16 slices, giving a total of 160 scintillator pieces for all BMD. The Ring 1 is located in a radius range of (5.1,8.30) cm, the Raidus 2 in (8.5,14.5) cm, the Radius 3 in (14.7,23.4) cm, the Radius 4 in (23.6,42.00) cm and Radius 5 in (42.2,73.63) cm. With this geometry, the pseudorapidity range for BMD is $1.69 < |\eta| < 4.36$. There are 4 important studies that we have to do for BMD:

- Time resolution and arrival time resolution.
- Centrality and multiplicity determination.
- Trigger efficiencies.
- Background reaction.

The results of these studies will be able to give us a good "way" to construct the real BMD. For the present work, we made the study for the arrival time resolution.

3 BMD simulation and results

We simulated Au-Au collision at $\sqrt{s} = 8$ GeV. We do not restrict that the collision occurs at the origin, i.e., we let the collision occurs with a Gaussian distribution whose center is the geometrical origin and has a sigma of 300 cm, this is called "smearing". We are interested in the analysis for the arrival particles to BMD-A and BMD-C. In this section we will describe the technique we used to calculate the *arrival time resolution* for BMD. We used the *mpdroot* framework for the MPD simulation.
The first result is to locate the position of interaction point. This can be possible using the addition and difference time of arrival of the particles, when the collision occurs inside or outside of BMD. The

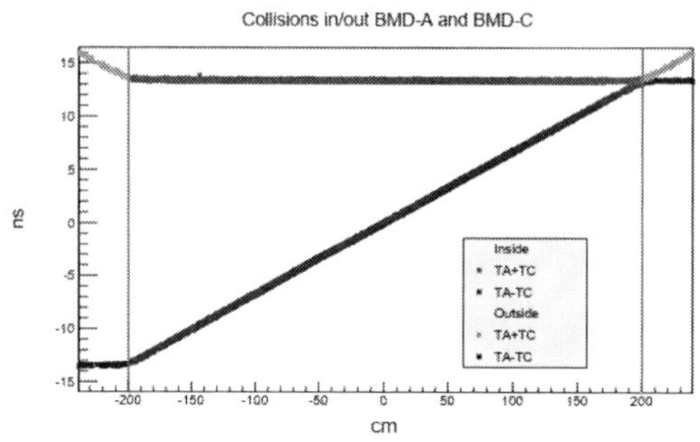

Figure 2. Comparation between TA+TC and TA-TC when the collision occurs inside and outside BMD. The vertical lines indicate the region of BMD-A and BMD-C.

position of BMD can define four zones: 1) Z_1 <-200 cm, 2) -200 cm< Z_2 <0 cm, 0 cm< Z_3 <200 cm and 200 cm< Z_4. As an example let's take Z_3. The two arrival times for BMD-A and BMD-C are: $T_A = \frac{2m + Z_3}{nc}$ and $T_C = \frac{2m - Z_3}{nc}$, respectively, The quantity $n(< 1)$ represents the percentage of the velocity that the particles travel with respect to the speed of light. Z_3 are the points in this Zone. Then we have:

$$T_A - T_C = \frac{2}{nc} Z_3. \tag{1}$$

It means that the differences is propotional to de distance. If we take the Z_2 we will have the negative of Eq. 1. We see that the difference, in middle Zone, is linear with respect to the position of the interaction point. This simple calculation can be made for the other zones and for the sum. This theorical prediction are agree with the results obtained by MC, as is shown in Figure 2. Note that each operation has a different shape. Then, it is possible to distinguish if the collision occurs inside or outside BMD. In Figure2 are shown all arrival particles (primary and secondary).

3.1 Arrival time resolution

We worked inside BMD and selected the fastest particle for each event for BMD-A and BMD-C and we made a fit for their respective TA and TC. The fit was made with a Gaussian function, then, the difference in their σ is the arrival time resolution, i.e.

$$\Delta\sigma = |\sigma_A - \sigma_C| = 57.982 \pm 0.509 ps \tag{2}$$

The resolution time of all detector can be divided in three resolution times: 1) The arrival resolution times of particles from the collision (Eq. 2), 2) The arrival resolution time of the scintillator-photons created by the scintillator material of BMD and 3) The resolution time of the electronic. Reading this information must be done before the other event occurs. We expected that BMD will have a resolution time at most 50 ps.

3.2 Geant4 simulation

As is mentioned above, the circular geometry is not the final for the BMD, it was used as a preliminary results. The final geometry we will use is a hexagonal. We simulated Bc404 scintillator material in a hexagonal geometry (which represent a cell of BMD) using Geant4. In Figure 3 is shown the

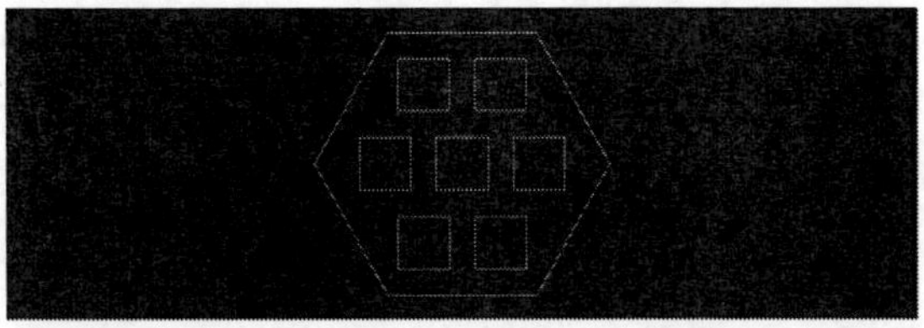

Figure 3. Ilustration for one scintillator cell of BMD-A and BMD-C. The yellow squares represent the APD to collect the scintillator photons. They are located behind the scintillator. Each one covers 2 cm × 2 cm of area.

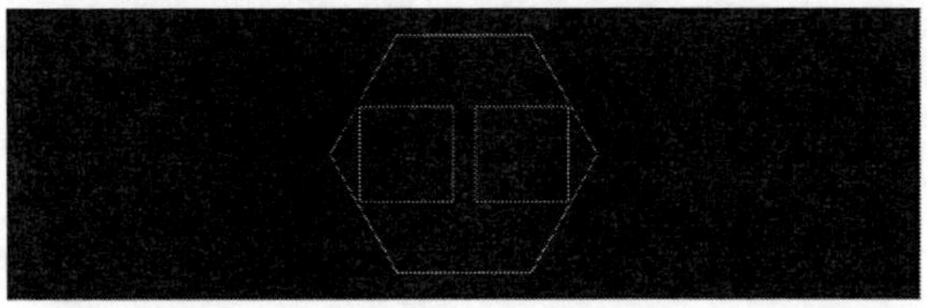

Figure 4. Same description as in Figure 3.

cell. The yellow squares are the representation of APDs. The dimentions of the scintillator material are 10 cm×10 cm and 2 cm of thick. The material we used is Bc404. We considered a π^+ as a primary particle. The energy deposited is calculated by mpdroot and is arround 5 MeV, this results was obtained using the MPDroot simulation. We condired the fastest photon collectet by each APD and we obtained that 133.579 ± 21.803 ps $\leq \Delta\sigma \leq 226.409 \pm 37.821$ ps. This resolution time is too bigger for the ideal resolution time.
It is natural to think that if the cell is more smaller, the time resolution will be lower. We made the simulation considering the dimentions 5 cm×5 cm and 2 cm. Due to the dimentions, it is only possible to use two scorers, as is shown in Figure 4. The rsolution time for this arrangement is $\Delta\sigma = 12.908 \pm 4.762$ ps, which is obviously very good. However, we need to cover 1m×1m of area, so, using this dimentions, will be expensive. We will continue to this study, finding the geometry to obtain the ideal resolution time.

4 Conclusions

The implementation of BMD code in the mpdroot framework was successfully. With this new framework, we could calculate in first aproximation the time resolution for BMD. The circular geometry will not be the final geometry for BMD, however this results will not depend on it. The arrival resolution time analysis can be done for each cell to and choose the better sections or cells to detemine the better arrival resolution time and improve the resolution time of BMD. Also, with this "puntual" study can be used to determine the centrality, interaction point position, etc. And obtain a better measurement.

References

[1] L. Yordanova, J. Phys.: Conf. Ser. **503**, 012041, (2014)

AUTHOR INDEX

AUTHOR INDEX